Robotics

Robotics

Science and Systems III

edited by Wolfram Burgard, Oliver Brock, and Cyrill Stachniss

The MIT Press
Cambridge, Massachusetts
London, England

For information on quantity discounts, email special_sales@mitpress.mit.edu.

Printed and bound in the United States of America.

Library of Congress Cataloging-in-Publication Data

Robotics : science and systems III / edited by Wolfram Burgard, Oliver Brock, and Cyrill Stachniss.
p. cm.
Includes bibliographical references.
ISBN 978-0-262-52484-1 (pbk. : alk. paper)
1. Robotics. I. Burgard, Wolfram. II. Brock, Oliver. III. Stachniss, Cyrill.
TJ211.R57485 2008
629.8'92—dc22

2007041215

10 9 8 7 6 5 4 3 2 1

Contents

Preface

This volume contains the papers presented at the third *Robotics Science and Systems (RSS)* Conference, held in Atlanta, Georgia, from June 27 to June 30, 2007. Set in the intricately ornamented Egyptian Ballroom of the Fox Theatre, the conference brought together over 230 researchers from around the world. *RSS* strives to showcase the leading scientific research and the most advanced systems from all areas of robotics. This goal is reflected in the diversity of the topics covered in this volume. The papers collected here cover work in mechanism design, motion planning, control, computer vision, mobile robotics, mapping, navigation, machine learning, manufacturing, distributed systems, sensor networks, self-reconfigurable systems, dynamic simulation, robot safety, and field applications.

The *RSS* area chairs and program committee carefully reviewed 137 submissions and selected 41 papers for publication. The review process was double-blind, that is, program committee members were not told authors' identities in the reviewing process. Authors were given the opportunity to respond to the reviewers' comments in a rebuttal process. The final program was selected during an in-person meeting at MIT. This meeting was attended by the general chair, the program chair, and the area chairs. Decisions were based on the recommendations of three to seven program committee members (most papers received four reviews), discussions among the reviewers, the rebuttals of the authors, and intense discussions among the area chairs. Twenty-three of the papers were presented orally in a single conference track; eighteen additional papers were presented at a vibrant and stimulating poster session. Each poster was advertised in a plenary spotlight presentation session.

In addition to the contributed papers, the technical program also featured five invited talks. Shree Nayar from Columbia University gave an inspirational presentation on *Computational Cameras: Redefining the Image*; Arthur Horwich from Yale University took his audience on a breathtaking journey into the miraculous world of proteins, reporting on *GroEL: A Protein Folding Machine*; Atsushi Iriki from the Riken Brain Science Institute gave an insightful and fascinating account of *Latent Precursors of Human Intelligence in Monkey Tool Use Actions*; Daniel Wolpert from the University of Cambridge gave an exciting and thought-provoking presentation on *Probabilistic Models of Human Sensorimotor Control*; and Mathieu Desbrun from California Institute of Technology shared his deep insights in a superb presentation on *Calculus Ex Geometrica: Structure-Preserving Computational Foundations for Graphics and Simulation*.

This year's *RSS* introduced a new feature: the Early Career Spotlight Presentations. These presentations allow promising early-career researchers to present their research vision beyond the scope of a single conference or journal paper. Noah Cowan from Johns Hopkins University gave a fascinating account of his work on *Sensorimotor Integration in Robots and Animals: Signals, Geometry and Mechanics*; and Hod Lipson from Cornell University presented surprising and intriguing work on *Morphological and Behavioral Adaptation in Robotics*.

Ken Goldberg from the University of California at Berkeley gave a truly amazing banquet presentation about *Earth Art with Robots and Networks*. The aquariums surrounding the banquet room at the Georgia Aquarium provided a perfect setting for the symbiosis of art and science.

For the second year in a row, *RSS* featured a best student paper award, sponsored by Springer on behalf of the journal Autonomous Robots. At the area chair meeting, the area chairs had selected the following three finalists among the many nominees: *An Implicit Time-Stepping Method for Multibody Systems with Intermittent Contact* by Nilanjan Chakraborty, Stephen Berard, Srinivas Akella, and Jeff Trinkle; *Dimensionality Reduction Using Automatic Supervision for Vision-based Terrain Learning* by Anelia Angelova, Larry Matthies, Daniel Helmick, and Pietro Perona; and *A Fundamental Tradeoff Between Performance and Sensitivity Within Haptic Rendering* by Paul Griffiths, Brent Gillespie, and Jim Freudenberg. Congratulations to these authors for their outstanding contribution! The best student paper award went to Nilanjan Chakraborty, Stephen Berard, Srinivas Akella, and Jeff Trinkle from the Rensselaer Polytechnic Institute.

Three days of technical sessions were followed by four one-day workshops: *Robot Manipulation: Sensing and Adapting to the Real World*, organized by Charles Kemp, Aaron Edsinger, Robert Platt, and Neo Ee Sian; *Robotic Sensor Networks: Principles and Practice*, organized by Gaurav Sukhatme and Wolfram Burgard; *Algorithmic Equivalences Between Biological and Robotic Swarms*, organized by James McLurkin and Paulina Varshavskaya; and *Research in Robots for Education*, organized by Doug Blank, Maria Hybinette, Keith O'Hara, and Daniela Rus. These workshops were complemented by the tutorial *Microsoft Robotics Studio (MSRS): A Technical Introduction*, organized by Stewart Tansley and Joseph Fernando. We would like to thank the organizers of these events.

The success of *RSS* is fueled by the dedication, enthusiasm, and effort of members of the robotics community. We express our gratitude to the area chairs, who spent endless hours reading papers, managing reviews, and debating acceptance decisions; some of them traveled between continents to attend the in-person area chair meeting: Nancy Amato (Texas A&M University), Darius Burschka (Technical University Munich), Jaydev P. Desai (University of Maryland, College Park), Dieter Fox (University of Washington), Hiroshi Ishiguro (University of Osaka), Yokoi Kazuhito (AIST), Yoky Matsuoka (University of Washington), Brad Nelson (ETH Zurich), Paul Newman (University of Oxford), Allison Okamura (Johns Hopkins University), Nicholas Roy (MIT), Roland Siegwart (ETH Zurich), Jeff Trinkle (Rensselaer Polytechnic Institute), Jing Xiao (University of North Carolina Charlotte), and Katsu Yamane (University of Tokyo). We would also like to thank the approximately 150 program committee members for providing extensive, detailed, and constructive reviews.

We thank the workshop chair, Udo Frese from the University of Bremen, for orchestrating the workshop proposal and selection process.

Frank Dellaert and Magnus Egerstedt from Georgia Tech were the local arrangement chairs this year. They combined Southern hospitality with excellent organization and fabulous venues. Their attention to detail ensured that the meeting and workshops went smoothly. We would like to especially thank Teri Russell, Jacque Berry, Vivian Chandler, Katrien Hemelsoet, and Jennifer Beattie for helping with the organization of poster sessions, the banquet, registration, and many other aspects of the conference, assisted by a wonderful group of student volunteers.

We thank the researchers from the following Georgia Tech robotics labs for live demonstrations: the Sting Racing Team (Dave Wooden, Matt Powers), the GRITS Lab (Jean-Pierre de la Croix, Magnus Egerstedt), the BORG Lab (Roozbeh Mottaghi, Frank Dellaert), and the HumAnSlab (Ayanna Howard). We also thank Henrik Christensen for the KUKA demo.

We are extremely grateful for the generosity of our sponsors. We would like to thank Stewart Tansley from Microsoft Corporation, Jeff Walz from Google, Rainer Bischoff from Kuka, Paolo Pirjanian from Evolution Robotics, and Melissa Fearon from Springer for their continued support of *RSS*. We would also like to thank the exhibitors from Microsoft Research (Stewart Tansley) and MIT Press (John Costello).

Finally, we would like to thank the members of the robotics community for their contributions. By submitting their outstanding research and by attending the conference, they provided the most essential ingredients for a successful scientific meeting. Through their active participation, *RSS* has become a forum for intellectual exchange as well as a high-quality robotics conference. The high ratio of attendees to accepted papers attests to this. We are looking forward to the next event in Zurich, Switzerland, in June 2008.

Wolfram Burgard, Albert-Ludwigs Universität Freiburg
Oliver Brock, University of Massachusetts Amherst
Cyrill Stachniss, Albert-Ludwigs Universität Freiburg
September 2007

Conference Organizers

General Chair	Wolfram Burgard, University of Freiburg
Program Chair	Oliver Brock, University of Massachusetts Amherst
Local Arrangement Chairs	Frank Dellaert, Georgia Tech Magnus Egerstedt, Georgia Tech
Publicity Chair	Cyrill Stachniss, University of Freiburg
Publication Chair	Cyrill Stachniss, University of Freiburg
Workshop Chair	Udo Frese, University of Bremen
Web Masters	Cyrill Stachniss, University of Freiburg Christian Plagemann, University of Freiburg
Conference Management System	Oliver Brock, University of Massachusetts Amherst
Childcare	Ayanna Howards, Georgia Tech Maren Bennewitz, University of Freiburg
Area Chairs	Nancy Amato, Texas A&M University Darius Burschka, TU Munich Jaydev P. Desai, University of Maryland, College Park Dieter Fox, University of Washington Hiroshi Ishiguro, University of Osaka Yokoi Kazuhito, AIST Yoky Matsuoka, University of Washington Brad Nelson, ETH Zurich Paul Newman, University of Oxford Allison Okamura, Johns Hopkins University Nicholas Roy, MIT Roland Siegwart, ETH Zurich Jeff Trinkle, Rensselaer Polytechnic Institute Jing Xiao, University of North Carolina Charlotte Katsu Yamane, University of Tokyo
Conference Board	Oliver Brock, University of Massachusetts Amherst Wolfram Burgard, University of Freiburg Greg Dudek, McGill University Dieter Fox, University of Washington Lydia Kavraki, Rice University Eric Klavins, University of Washington Nicholas Roy, Massachusetts Institute of Technology Daniela Rus, Massachusetts Institute of Technology Sven Koenig, University of Southern California John Leonard, Massachusetts Institute of Technology Stefan Schaal, University of Southern California Gaurav Sukhatme, University of Southern California Sebastian Thrun, Stanford University

Advisory Board

George Bekey, University of Southern California
Raja Chatila, Centre National de la Recherche Scientifique
Henrik Christensen, Royal Institute of Technology
Hugh Durrant-Whyte, Australian Centre for Field Robotics
Ken Goldberg, UC Berkeley
John Hollerbach, University of Utah
Hiroshi Ishiguro, University of Osaka
Vijay Kumar, University of Pennsylvania
Matt Mason, Carnegie Mellon University
Yoshihiko Nakamura, University of Tokyo
Brad Nelson, ETH Zurich
Bruno Siciliano, Universit degli Studi di Napoli Federico II
Roland Siegwart, ETH Zurich

Program Committee

Farhad Aghili
Peter Allen
Sean Andersson
Kai Arras
James Bagnell
Tim Bailey
Ravi Balasubramanian
Devin Balkcom
Stacy Bamberg
Christoph Bartneck
Burchan Bayazit
Patrick Beeson
Michael Beetz
Mike Benjamin
Peter Berkelman
Antonio Bicchi
Philippe Bidaud
Michael Bosse
Michael Bowling
Jean-Marc Breguet
Bambi Brewer
Herman Bruyninckx
Karl Bhringer
Cenk Cavusoglu
Raja Chatila
Francois Chaumette
Gordon Cheng
Woojin Chung
Jacob Crandall
Andrew Davison
Matt Deans
Rdiger Dillmann
Zachary Dodds
Lixin Dong
Gregory Dudek
Ryan Eustice
Ronald Fearing
John Folkesson
Emilio Frazzoli
Antonio Frisoli
Edwardo Fukushima
Darren Gawley
Maria Gini
Ken Goldberg
Geoff Gordon
Giorgio Grisetti
Horst-Michael Gross
Li Han
Kensuke Harada
Matthias Harders
Hideki Hashimoto
Shinichi Hirai
Yasuhisa Hirata
Koh Hosoda
David Hsu
Tetsunari Inamura
Pekka Isto
Patric Jensfelt
Yan-Bin Jia
Chris Jones
Shuuji Kajita
Takayuki Kanda
Imin Kao
Lydia Kavraki
Alonzo Kelly
Abderrahmane Kheddar
Georg Klein
Atsushi Konno
Danica Kragic
Venkat Krovi
Ben Krse
James Kuffner
Steve LaValle
Cecilia Laschi
John Leonard

Program Committee (cont.)

Zexiang Li
Jyh-Ming Lien
Ming Lin
Guanfeng Liu
Yunhui Liu
Max Lungarella
Yusuke Maeda
Ezio Malis
Dinesh Manocha
Eric Marchand
Agostino Martinelli
Stephen Mascaro
Yoshio Matsumoto
Takashi Minato
Bud Mishra
Mark Moll
Michael Montemerlo
Taketoshi Mori
Jun Nakanishi
Jos Neira
Andrew Ng
Gunter Niemeyer
Jenkins Odest
Tetsuya Ogata
Paul Oh
Edwin Olson
Shinsuk Park
William Peine
Roland Philippsen
Joelle Pineau
Ingmar Posner
Domenico Prattichizzo
Thomas Preuss
Fabio Ramos
Rajesh Rao
Christopher Rasmussen
Iead Rezek
Chris Roman
Stergios Roumeliotis
Jee-Hwan Ryu
Jesus Savage
Derik Schroeter
Luis Sentis
Tomohiro Shibata
Gabe Sibley
Rob Sim
Thierry Simeon
Metin Sitti
Dezhen Song
Guang Song
Mark Spong
Cyrill Stachniss
Raul Suarez
Thomas Sugar
Salah Sukkarieh
Yu Sun
Yasutake Takahashi
Hong Tan
Juan Tardos
Russ Tedrake
Frank Tendick
Sebastian Thrun
Rudolph Triebel
Markus Vincze
Michael Wang
Alfredo Weitzenfeld
Stefan Williams
Dirk Wollherr
Robert Wood
Jizhong Xiao
Yuichiro Yoshikawa
Nakamura Yutaka

Semantic Modeling of Places using Objects

Ananth Ranganathan and Frank Dellaert
College of Computing, Georgia Institute of Technology
{ananth, dellaert}@cc.gatech.edu

Abstract— **While robot mapping has seen massive strides recently, higher level abstractions in map representation are still not widespread. Maps containing semantic concepts such as objects and labels are essential for many tasks in manmade environments as well as for human-robot interaction and map communication. In keeping with this aim, we present a model for places using objects as the basic unit of representation. Our model is a 3D extension of the constellation object model, popular in computer vision, in which the objects are modeled by their appearance and shape. The 3D location of each object is maintained in a coordinate frame local to the place. The individual object models are learned in a supervised manner using roughly segmented and labeled training images. Stereo range data is used to compute 3D locations of the objects. We use the Swendsen-Wang algorithm, a cluster MCMC method, to solve the correspondence problem between image features and objects during inference. We provide a technique for building panoramic place models from multiple views of a location. An algorithm for place recognition by comparing models is also provided. Results are presented in the form of place models inferred in an indoor environment. We envision the use of our place model as a building block towards a complete object-based semantic mapping system.**

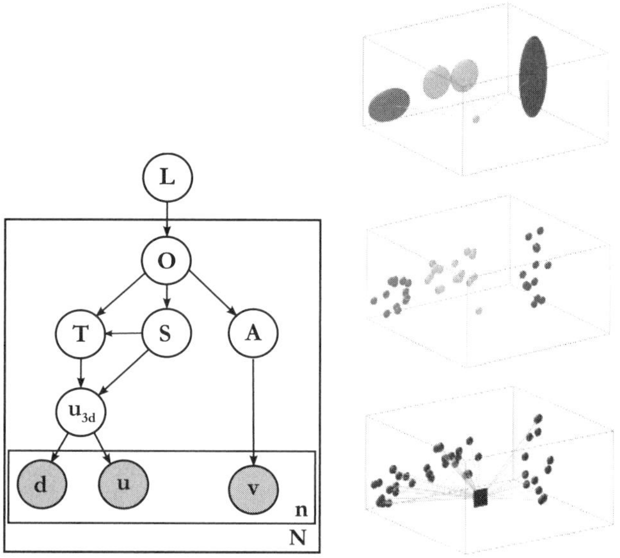

Fig. 1. A generative model for representing places in a mapped environment. The place label L generates a set of N objects O, each having a shape S, an appearance A, and a 3D location T. The objects, transformed to location T , give rise to 3D feature points u_{3d}. These features are observed in an image as n features, each with pixel location u, appearance v, and stereo depth d (shaded variables). The schematic on the right illustrates this process.

I. Introduction

Robot mapping has in recent years reached a significant level of maturity, yet the level of abstraction used in robot-constructed maps has not changed significantly. Simultaneous Localization and Mapping (SLAM) algorithms now have the capability to accurately map relatively large environments [8], [22]. However, grid-based and feature-based maps constructed using lasers or cameras remain the most common form of representation. Yet higher level abstractions and advanced spatial concepts are crucial if robots are to successfully integrate into human environments.

We have chosen objects and their location to be the semantic information in our maps based on the object-centricness of most man-made environments. People tend to associate places with their use or, especially in workspaces, by the functionality provided by objects present there. Common examples are the use of terms such as "printer room", "room with the coffee machine", and "computer lab". Even in outdoor spaces, people often remember locations by distinguishing features that most often turn out to be objects such as store signs and billboards [12]. Thus, objects form a natural unit of representative abstraction for man made spaces. While we do not claim that representing objects captures all the salient information in a place of interest, it is an important dimension that is useful in a wide variety of tasks.

A major concern in constructing maps with objects is object detection, which has been a major area of research in computer vision for a significant period of time. Due to the difficulty in general purpose object detection, many semantic mapping algorithms assume object detection as a black box [5], and so, sidestep a major component of the mapping problem. However, recent advances in stable feature detection algorithms have enabled featurebased object detection methods that have revolutionized the field. In particular, use of SIFT descriptors [11] along with affine invariant features [13] has been particularly popular since binaries for detecting these features are publicly available.

In this paper, we present one of the first instances of semantic mapping in robotics that integrates state-of-the-art object detection techniques from computer vision. We present a 3D generative model for representing places using objects and develop learning and inference algorithms for the construction of these models. The model for a place is constructed using images and depth information obtained from a stereo camera. Our model is a 3D extension of the constellation models popular in the computer vision literature [4]. In particular, as illustrated in Figure 1, a place is represented as a set of objects O with 3D locations T specified in a local coordinate frame.

In turn, an object is modeled as having a particular shape and appearance, and gives rise to features in the image. This generative model is discussed in detail in Section III below.

The models for the objects are learned in a supervised manner, as will be developed in Section IV. The shape models for the objects are learned using stereo range data, while corresponding appearance models are learned from features detected on the images. Training data is provided to the learning algorithm in the form of roughly segmented images from the robot camera in which objects have been labeled. While unsupervised learning is preferable from the standpoint of automation, it requires that the objects to be learned be prominent in the training images, a condition that is not satisfied by our training images. Supervised learning results in more reliable and robust models.

Once the object models have been learned, inference for the place model is performed at runtime using the Swendsen-Wang algorithm, a cluster Markov Chain Monte Carlo (MCMC) technique [1]. Approximate inference through MCMC is required since finding the correspondence between image features and objects is a combinatorial problem. Features in the test image are connected in a Markov Random Field (MRF) which promotes clustering based on appearance and locality in 3D space. Subsequently, we employ the Swendsen-Wang algorithm to perform sampling-based inference on the MRF. Range data and feature appearance provide strong constraints on the distributions resulting in rapid convergence. The details are given in Section V.

Finally, the place models can be used to perform place recognition, for which purpose we provide an algorithm in Section VI. We also describe a technique to build 360° panoramic models from multiple views of a place using relative robot pose.

Experiments are presented on a robot mounted with a Triclops stereo camera system. We present place models constructed using a fixed object vocabulary in an indoor environment to validate our learning and inference algorithms, in Section VII. We start with related work in the next section.

II. Related Work

The need to include semantic information in robot representations of the environment has long been recognized [10]. A number of instances of functional and semantic representations of space exist in the literature, of which we mention a few recent examples. [18] builds a map of the environment consisting of regions and gateways and augments it to include rectangular objects [17]. Supervised learning of labels on regions is performed in [16] using cascade detectors to detect objects of interest. [5] describes a hierarchical representation that includes objects and semantic labeling of places in a metric map but assumes the identities of objects to be known. [14] performs 3D scene interpretation on range data and computes a map consisting of semantically labeled surfaces. [24] lays out a detailed program for creating cognitive maps with objects as the basic unit of representation. However, their object detection technique uses simple SIFT feature matching

Fig. 2. Features detected on a typical training image using (a) Harris-affine corners (b) Canny edges (c) MSER detector.

which does not scale to larger objects. Our method is more comprehensive than the above-mentioned techniques since it incorporates inference for the 3D location, type, and number of objects in the scene.

More computer vision oriented approaches to the problem of scene modeling also exist. [7] gives a technique for 3D scene analysis and represents the analyzed 3D scene as a semantic net. [23] implements place recognition using features extracted from an image and uses place context to detect objects without modeling their location. More similar to our approach is the one in [20] that creates 3D models of scenes using the Transformed Dirichlet Process. In contrast to this, our approach is simpler and more robust as it uses supervised learning and Swendsen-Wang sampling.

III. Constellation Models for Places

We model places using a 3D extension of the popular constellation model in computer vision. More precisely, we use the "star" modification of the constellation model [4]. The model represents each place as a set of objects with 3D locations in a local coordinate frame. We assume that given the origin of this coordinate frame, hereafter called the *base location*, the objects are conditionally independent of each other. While full constellation models can also model relative locations between objects, the associated increase in complexity is substantial. More discussion on this subject can be found in [4].

A graphical representation of the generative place model is given in Figure 1. The place label L generates a set of objects O, where the number of objects N is variable. Each object gives rise to a set of 3D feature points u_{3d} according to a shape distribution S. Further, the 3D location of the object in local coordinates is represented using the translation variable T, which is unique for each object and also depends on the place label. Finally, the 3D points, transformed in space according to T, give rise to image features at locations u with appearance v distributed according to an object specific distribution A. The 3D points also produce range measurements d, obtained using stereo, corresponding to the observed features.

The shape distribution S and the appearance distribution A taken together model an object class. In our object models, we represent the shape of an object using a Gaussian distribution in 3D, while its appearance is modeled as a multinomial distribution on vector quantized appearance words.

A. Feature detection and representation

For appearance measurements, we use three complementary types of features in our work that are subsequently

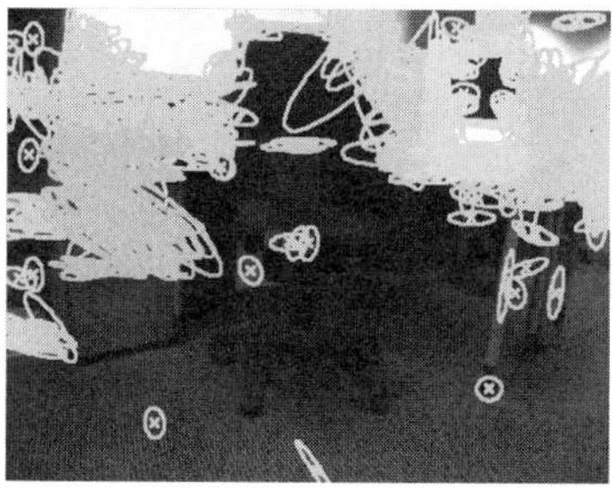

Fig. 3. An example training image with roughly segmented and labeled objects, a monitor in cyan and a drawer in brown.

discretized to facilitate learning. Following previous work [20], we use Harris-affine corners [13], maximally stable extremal regions [13], and clustered edges obtained from a Canny edge detector [2]. We used the public implementations of the first two feature detectors available at *http://www.robots.ox.ac.uk/~vgg/research/affine*. Figure 2 highlights the complementary characteristics of the detected features of each type for a sample image. As noted in [20], edge features are crucial for modeling objects with textureless surfaces such as monitors. The features are represented using SIFT descriptors in a 128 dimensional vector space. We vector quantize the SIFT descriptors using K-means clustering to produce a set of appearance "words". Each feature is subsequently described by the bin corresponding to its closest appearance word, and its 2D pixel location in the image.

IV. Supervised Learning of 3D Object Models

The individual object models in the object vocabulary of the robot are learned in a supervised manner. While unsupervised learning is the sought-after goal in most applications, learning objects in this manner poses many problems. Firstly, the objects of interest have to be displayed prominently in the images with almost no background for reliable unsupervised object discovery to be possible. Further, currently most unsupervised object recognition methods rely on some form of "topic" discovery on the set of training images to learn object models [19]. In images with varied background, the topic discovered by the algorithm may not correspond to objects of interest. Supervised learning sidesteps these issues and is accordingly more suitable for our application.

Training data is presented to the algorithm in the form of roughly segmented stereo images along with the associated stereo range data. Objects in the image are labeled while all unlabeled regions are assumed to be background. An example training image is given in Figure 3.

The shape Gaussian for an object is estimated from the range data labeled as belonging to the object, and the appearance is similarly learned from the features corresponding to the object. Learning the object models is thus straight-forward.

V. Inference for constructing place models

During the testing phase, the robot observes an image along with associated stereo range information, and has to infer the label and model for the place, i.e the types of objects and their 3D locations. We denote the set of appearance and shape models learned during the training phase as $\mathcal{A} = \{A_{1:m}\}$ and $\mathcal{S} = \{S_{1:m}\}$ respectively, where m is the number of objects in the robot's vocabulary. The pixel locations of the features observed in the image are denoted by the set $U = \{u_{1:n_f}\}$, while the corresponding appearance descriptors are written as $V = \{v_{1:n_f}\}$, n_f being the number of features. The depth from stereo corresponding to each of these features is represented as $D = \{d_{1:n_f}\}$. In the interest of brevity, in the following exposition we will compress the set of measurements to $Z = \{U, V, D\}$ whenever possible.

We infer the place model and label in a Bayesian manner by computing the joint posterior distribution on the place label, the types of objects, and their locations. This posterior can be factorized as

$$p(L,O,T|\mathcal{A},\mathcal{S},Z) \quad = \quad p(O,T|\mathcal{A},\mathcal{S},Z)p(L|O,T,\mathcal{A},\mathcal{S},Z) \qquad (1)$$

where L is the place label, O is the set of object types, and T is the corresponding 3D locations. Note that the number of objects at a place, i.e the cardinality of set O, is unknown.

The inference problem can be divided into two parts, namely place *modeling* and place *recognition*, which correspond to the two terms on the right side of (1) respectively. The modeling problem consists of inferring the objects and their locations, while the recognition problem involves finding the label of the place given the objects and their locations. In this section we focus on the modeling problem and return to the recognition problem in the next section.

Since the measurements are in the form of image features, inference cannot proceed without the correspondence between features and the object types they are generated from. We incorporate correspondence by marginalizing over it so that the model posterior of interest can be written as

$$p(O,T|\mathcal{A},\mathcal{S},Z) \quad = \quad \sum_{\Pi} p(O,T|\mathcal{A},\mathcal{S},Z,\Pi)p(\Pi|\mathcal{A},\mathcal{S},Z) \qquad (2)$$

where Π is a discrete correspondence vector that assigns each image feature to an object type. We call (2) the *place posterior*.

Since computing the place posterior analytically is intractable, we employ a sampling-based approximation. The intractability arises from the need to compute the distribution over correspondences, which is a combinatorial problem. One technique for overcoming this intractability is using Monte Carlo EM (MCEM) [21], in which a Monte Carlo estimate of the distribution over correspondences is used to maximize the posterior over the other hidden variables iteratively. In our case, this would involve a maximization over a possibly large discrete space of object types. Further, multi-modal distributions in this space cannot be discovered using MCEM, which only computes the MAP solution. These reasons motivate our use of Markov Chain Monte Carlo (MCMC) methods for computing a sampling-based approximation to the posterior.

To compute the distribution over correspondences, we note that features corresponding to a particular object type occur in clusters in the image. Hence, appearance and stereo depth provide important clues to the correspondence of a feature in

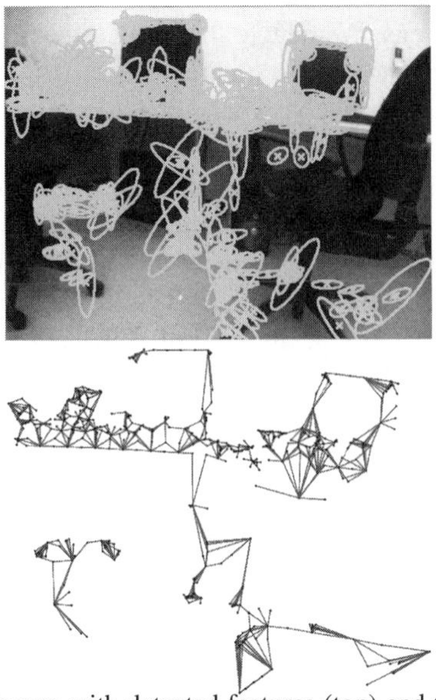

Fig. 4. A test image with detected features (top) and the corresponding MRF of features on which sampling is performed (bottom). Note that the MRF may be disconnected as shown here.

the sense that if a feature is similar to its neighboring features with respect to its appearance and 3D location, it is highly likely that it belongs to the same object type as its neighbors.

We take advantage of this spatially clustered nature of the correspondences by placing the image features in a Markov Random Field (MRF). Each feature is connected to its k closest neighbors in the image, where the neighborhood k is a parameter. Larger values of k make large scale correlations visible while increasing complexity ($k = n_f$ gives a fully connected graph). The value of k depends on the expected size of the projected objects in the images. Figure 4 shows the MRF corresponding to features in an image for $k = 10$.

We define discriminative probabilities, also called edge compatibilities, on the edges of the MRF. These are defined as functions of the depth and appearance of the features involved in the edge, where both functions are the Mahalanobis distance between the respective feature values. Denoting the functions on depth and appearance as f_d and f_a, the discriminative probability is

$$p_e \quad \propto \quad f_d(d_i, d_j) \times f_a(v_i, v_j) \qquad (3)$$

where

$$-\log f_d = \left(\frac{d_i - d_j}{\sigma_d}\right)^2 \text{ and } -\log f_a = (v_i - v_j)^T \Sigma_a^{-1} (v_i - v_j)$$

and σ_d and Σ_a are depth variance and appearance covariance respectively, that encode the size of objects and their uniformity of appearance.

The overall sampling scheme to compute the posterior can now be seen to have the following form. We sample clusters in the MRF according to the edge compatibilities and subsequently assign an object type to each cluster according to some prior distribution. The sample configuration is then evaluated using the measurement likelihoods based on the learned object models.

We employ a cluster MCMC sampling algorithm to efficiently implement the above scheme. Common MCMC sampling techniques such as Gibbs sampling and Metropolis-Hastings change the value of only a single node in a sampling step, so that mixing time for the chain, i.e the expected time to move between widely differing states, is exponential. Cluster MCMC methods change the value of multiple nodes at each sampling step, leading to fast convergence.

A. The Swendsen-Wang algorithm

We now describe the Swendsen-Wang (SW) algorithm, which we use to compute the place posterior with fast convergence. The SW algorithm has been interpreted in many ways - as a random cluster model, as an auxiliary sampling method, and as a graph clustering model using discriminative edge probabilities [1]. It is in the latter manner that we use the algorithm.

A sample is produced using the SW algorithm by independently sampling the edges of the MRF to obtain connected components. Consider the graph $G = (V, E)$ of image features, as defined above, with discriminative probabilities p_e, $e \in E$ defined in (3) on the edges. We sample each edge of the graph independently and turn "on" each edge with probability p_e. Now only considering the edges that are on, we get a second graph which consists of a number of disjoint connected components. If the discriminative edge probabilities encode the local characteristics of the objects effectively, the connected components will closely correspond to a partition Π of the graph into various objects and the background. The distribution over partitions Π is given as

$$p(\Pi|\mathcal{A}, \mathcal{X}, Z) \quad = \quad \prod_{e \in E_0} p_e \prod_{e \in E \backslash E_0} (1 - p_e) \qquad (4)$$

Samples obtained from a typical image feature graph are shown in Figure 5.

To sample over correspondence between image features and object types, we assign an object type to each component of the partition according to the posterior on object types $p(O, T|\mathcal{A}, \mathcal{S}, Z, \Pi)$. Computation of the posterior on object types involves only the appearance measurements since the other measurements also depend on the 3D location of the object T. Applying Bayes Law on the posterior, we get the distribution on object types as

$$p(O_c|\mathcal{A}, \mathcal{S}, Z, \Pi) \quad \propto \quad p(Z|O_c, \mathcal{A}, \mathcal{S}, \Pi) p(O_c|\mathcal{A}, \mathcal{S}, \Pi) \quad (5)$$

where the second term on the right is a prior on object types that can potentially incorporate information regarding the size and appearance of the component, and the frequency with which the object type has been observed in the past. We employ a uniform prior on object types for simplicity since the prior is largely overshadowed by the data in this case. The object type affects only appearance measurements and so, the measurement likelihood in (5) collapses to just the conditionally independent appearance likelihoods on the

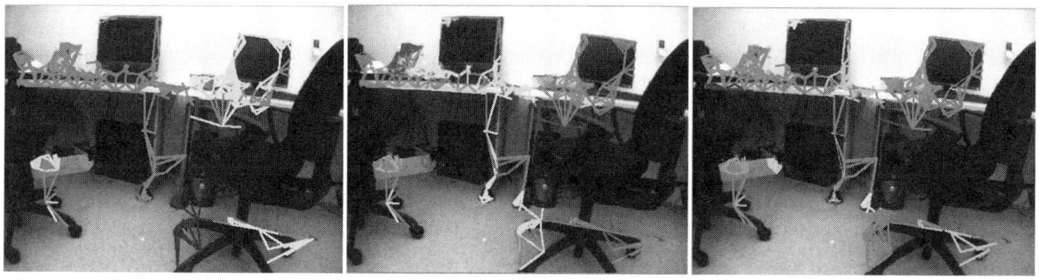

Fig. 5. Samples from the SW algorithm for the MRF corresponding to Figure 4, obtained by independently sampling the MRF edges according to edge compatibilities. Each connected component is colored differently and the edges that are turned "off" are shown in gray. The SW algorithm works by assigning an object label to each connected component and subsequently scoring the sample based on the learned object models. In practice, most components get labeled as background.

Algorithm 1 The Swendsen-Wang algorithm for sampling from the place posterior

1) Start with a valid initial configuration (Π, O) and repeat for each sample
2) Sample the graph G according to the discriminative edge probabilities (3) to obtain a new partition Π'
3) For each set in the partition $c \in \Pi'$, assign the object type by sampling from $p(O_c|Z, \Pi')$ as given by (5)
4) Accept the sample according to the acceptance ratio computed using (7)

features that are evaluated using the multinomial appearance distribution for the object type

$$p(Z|O_c, \mathcal{A}, \mathcal{S}, \Pi) = \prod_{i=1}^{n_f} p(v_i|O_c, \mathcal{A}, \mathcal{S}, \Pi) \qquad (6)$$

The sample thus generated, consisting of a graph partition and object type assignments to each component in the partition, is accepted based on the Metropolis-Hastings acceptance ratio [6], given as

$$a(\Pi' \to \Pi) = \min\left(1, \frac{\prod_{c'\in\Pi'} p(T_{c'}|O_{c'}, \mathcal{A}, \mathcal{S}, Z, \Pi')}{\prod_{c\in\Pi} p(T_c|O_c, \mathcal{A}, \mathcal{S}, Z, \Pi)}\right) (7)$$

The acceptance ratio (7) can be understood by considering the factorization of the place posterior (2) as

$$p(O_c,T_c|\mathcal{A},\mathcal{S},Z,\Pi) = p(O_c|\mathcal{A},\mathcal{S},Z,\Pi)p(T_c|\mathcal{A},\mathcal{S},Z,\Pi,O_c) \quad (8)$$

and noting that only the second term is involved in the acceptance ratio since the first term, the posterior on object types, has been used in the proposal distribution for the sample above. The acceptance ratio also makes the assumption that the partition components are independent of each other. Note that common factors in the calculation of the acceptance ratio can be omitted to improve efficiency. A summary of the SW algorithm for sampling from the place posterior is given in Algorithm 1.

B. Computing the target distribution

Computing the acceptance ratio (7), involves evaluating the posterior on object locations given a partition of the feature graph. The location posterior, which is the second term on the right in (8), can be factorized into a prior on object locations and the stereo depth likelihood using Bayes law

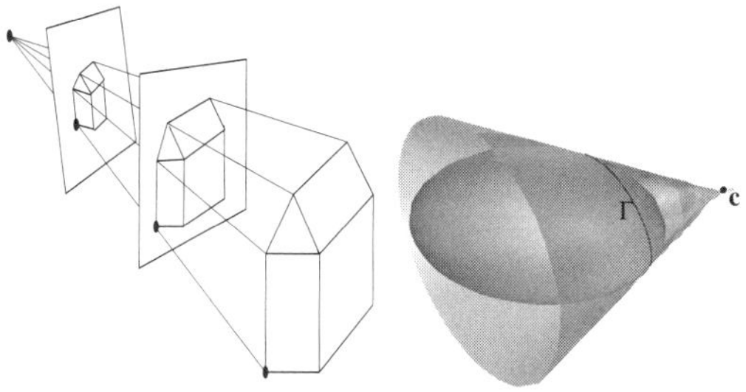

Fig. 6. An illustration of the use of scale to obtain a prior on object location. (a) The scale of the projection in the image gives a good estimate of location if the size of the object in 3D is known. (b) The projection of a 3D object model (shown as a Gaussian ellipsoid) is a 2D covariance, the size of which is used to estimate the object's location. Images taken from [9].

$$p(T_c|\mathcal{A},\mathcal{S},Z,\Pi,O_c) \propto p(D|\mathcal{S},\Pi,O_c,T_c)p(T_c|\mathcal{S},U,\Pi,O_c) \quad (9)$$

where appearance measurements have been neglected since they are assumed independent of the location of the object.

The prior on object locations $p(T_c|\mathcal{S}, U, \Pi, O_c)$ incorporates the 2D pixel locations of the features in the graph component and is evaluated using projection geometry. The graph component is a projection of the object ellipsoid in space and the scale of the projection thus gives an estimate of the 3D object location. Assuming a linear camera projection model, the mean location of the object is projected onto the mean of the feature locations. Analogously, the covariance of the feature locations is a projection of the 3D covariance of the object shape model, which is known. Hence, a proposed 3D location for the object can be evaluated through the norm-1 error between the predicted 2D feature location covariance and the actual covariance. Our use of scale to estimate location in this manner is illustrated in Figure 6. If the observed 2D covariance is Σ_f, the 3D object shape covariance is Σ_O and the Jacobian of the camera projection matrix at T_c is represented as P, the object location prior can be written as

$$\log p(T_c|\mathcal{S}, U, \Pi, O_c) \propto -\left\|\Sigma_f - P\Sigma_O P^T\right\| \qquad (10)$$

In practice, the location distribution (9) displays a peaked nature, so that its evaluation is approximated by a maximum

Algorithm 2 Inference algorithm for constructing the place model

For each image I and stereo depth map D obtained at a landmark location, do

1) Detect features using the 3 types of feature detectors on the image I.
2) Create an MRF using the edge potentials as defined in (3)
3) Sample over partitions of the graph using the Swendsen-Wang algorithm 1 and obtain the posterior (8) over object types and their locations
4) The posterior on object types and locations is the required place model

a priori (MAP) estimate. This also saves us from sampling over a large continuous space. The maximum of the posterior defined in (9) is obtained by performing a line search along the projection ray of the mean of the feature locations of the graph component. The map value for the object location, $T_c^\star$, is given as

$$T_c^\star = \underset{T_c}{\operatorname{argmax}}\ p(D|\mathcal{S}, U, \Pi, O_c, T_c)p(T_c|\mathcal{S}, U, \Pi, O_c) \quad (11)$$

We compute the individual stereo depth likelihoods by marginalizing over the "true" depth of the feature as measured along the projection ray

$$p(D|\mathcal{S},\Pi,O_c,T_c^\star) = \prod_{i=1}^{n_f} \int_{u_{3d}^i} p(d_i|u_{3d}^i)p(u_{3d}^i|\mathcal{S},U,O_c,T_c^\star) \quad (12)$$

where u_{3d}^i is the true depth measured as d_i by the stereo. The stereo measurement model $p(d_i|u_{3d}^i)$ is modeled as a Gaussian distribution that is learned from measurements obtained in a scenario where ground-truth is known.

The prior on true depth $p(u_{3d}^i|\mathcal{S}, U, O_c, T_c^\star)$ is obtained as a 1D marginal along the projection ray of the object shape model, i.e it is the Gaussian on the projection ray induced by its intersection with the object. This is necessitated by the fact that each feature in the graph component picks its own true depth based on its projection ray. Since both the depth likelihood and true depth prior are Gaussian, the integral in (12) can be computed analytically.

We now have all the pieces to compute the target distribution (8) and the acceptance ratio (7) in the Swendsen-Wang Algorithm 1. Appearance and stereo depth likelihoods are computed using equations (6) and (12) respectively, while a MAP estimate of the object locations is given by (11). A summary of the inference algorithm is given in Algorithm 2.

C. *Extension to panoramic models*

The algorithm discussed thus far computes the place model for a single image but cannot find a 360° model for a place unless a panoramic image is used. To overcome this limitation, we propose a simple extension of the algorithm.

We compute the panoramic model of a place from multiple images by "chaining" the models from the individual images using odometry information. For example, the robot is made to spin around at the place of interest to capture multiple images. We designate robot pose corresponding to the first of these images as the base location for the model and marginalize out the poses corresponding to all the other images to create a combined model from all the images.

Denoting the measurements from each of the n images of a place as $Z_1, Z_2, \ldots, Z_n$, and the corresponding detected objects and locations by $O_1, \ldots, O_n$ and $T_1, \ldots, T_n$ respectively, the panoramic model of the place is computed as

$$\begin{aligned} p(O,T|Z_{1:n},o^n) &= \int_{x_{1:n}} p(O_{1:n},T_{1:n}|Z_{1:n},x_{1:n})p(x_{1:n}|o^n) \\ &= p(O_1,T_1|Z_1) \\ &\quad \prod_{i=2}^{n} \int_{x_i} p(O_i,T_i|Z_i,x_i)p(x_i|o_{i-1}) \quad (13) \end{aligned}$$

where x_i is the pose corresponding to the ith image, $o_{i-1}$is the odometry between poses x_{i-1} and x_i, and $o^n = o_{1:n-1}$ is the set of all odometry. x_1 is assumed to be the origin as it is the base location and the pose distribution $p(x_i|o_{i-1})$ is evaluated using the odometry model. Note that (13) uses the fact that the (O_i, T_i) are conditionally independent of each other given the robot pose.

VI. PLACE RECOGNITION

Place recognition involves finding the distribution on place labels given the detected objects and their locations, i.e. finding the *recognition posterior* $p(L|O, T, \mathcal{A}, \mathcal{S}, Z)$ from (1). While robust techniques for place recognition using feature matching are well-known [15], [3], the detected objects can be effectively used to localize the robot, and can be expected to improve place recognition as they provide higher-level distinguishing information. We now give a technique to accomplish this. Applying Bayes law to the recognition posterior from (1)

$$p(L|O,T,\mathcal{A},\mathcal{S},Z) \propto p(O|L,\mathcal{A},\mathcal{S},Z)p(T|L,O,\mathcal{A},\mathcal{S},Z)p(L) \quad (14)$$

If the sequence of labels observed in the past is available, a Dirichlet label prior $p(L)$ is suitable. We assume that such history is unavailable and so use a uniform label distribution.

The object likelihood $p(O|L, \mathcal{A}, \mathcal{S}, Z)$ is evaluated as the distance between the observed discrete distribution on object types and prediction assuming the label L. Denoting the object type distribution corresponding to L as O_L, the likelihood is

$$\log p(O|L, \mathcal{A}, \mathcal{S}, Z) = -||O - O_L||_2$$

The location likelihood $p(T|L, O, \mathcal{A}, \mathcal{S}, Z)$ is computed by minimizing the distance between corresponding objects in T and the model for the place label L. Nearest neighbor (NN) correspondence is used for this purpose. However, since the robot is unlikely to be in exactly the same location even if it visits the same place, we also optimize over a 2D rigid transformation that determines the current pose of the robot in the local coordinates of the place model for the label L

$$p(T|L,O,\mathcal{A},\mathcal{S},Z) = \int_{X_r} p(T|L,O,\mathcal{A},\mathcal{S},Z,X_r)p(X_r) \quad (15)$$

where X_r, for which we use a flat Gaussian prior, is the location of the robot in the coordinates of the base location of the place L. In practice, we iterate between optimizing for X_r given object correspondence and finding NN correspondence

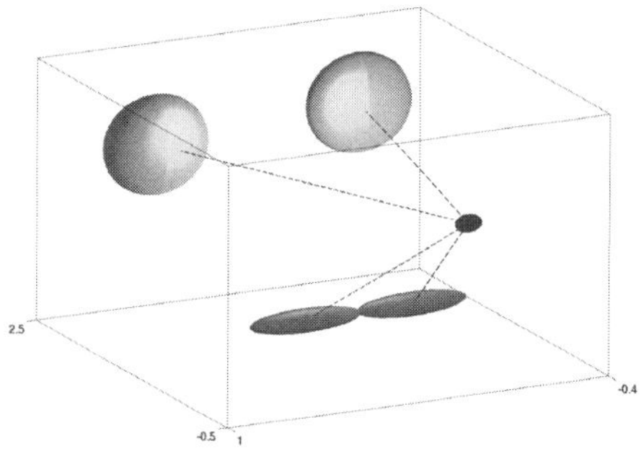

Fig. 7. The most likely model corresponding to the place shown in Figure 5. Detected computer monitors are displayed in green while chairs are shown in red. The origin, where all the dashed lines emanate from, is the location of the camera mounted on the robot. All axes are in meters.

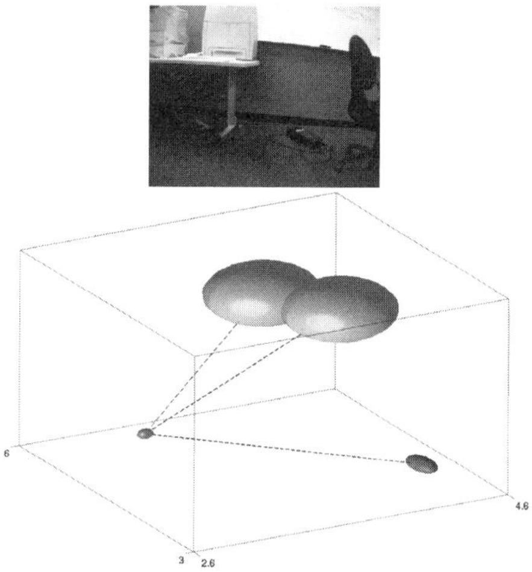

Fig. 8. Image of a place (top) and corresponding place model with two printers (cyan) and a chair (red). The camera is at the origin.

to minimize the object location errors given X_r. Subsequently, the MAP estimate for X_r is used in (15).

VII. Results

We now present results from tests conducted on a robot mounted with a Triclops stereo system. In the test setup, a total of 58 manually segmented training images, obtained from the robot cameras, were presented to the robot with the purpose of learning 5 different object models commonly visible in indoor office environments, namely computer monitors, printers, chairs, cupboards, and drawers. Subsequently, these models were used to construct place models. The problem of deciding which places to model is a non-trivial one, and is out of the scope of this paper. In these experiments, the places to be modeled are selected manually.

The features obtained from the training images were vector quantized after a preliminary classification based on shape (as described in Section III-A) into 1250 bins, which consequently was also the size of appearance histogram for the object models. We used shape and appearance variances for the discriminative edge probabilities (3) in accordance with the expected size and uniformity in appearance of the objects. For example, standard deviations of 0.25 meters for x and

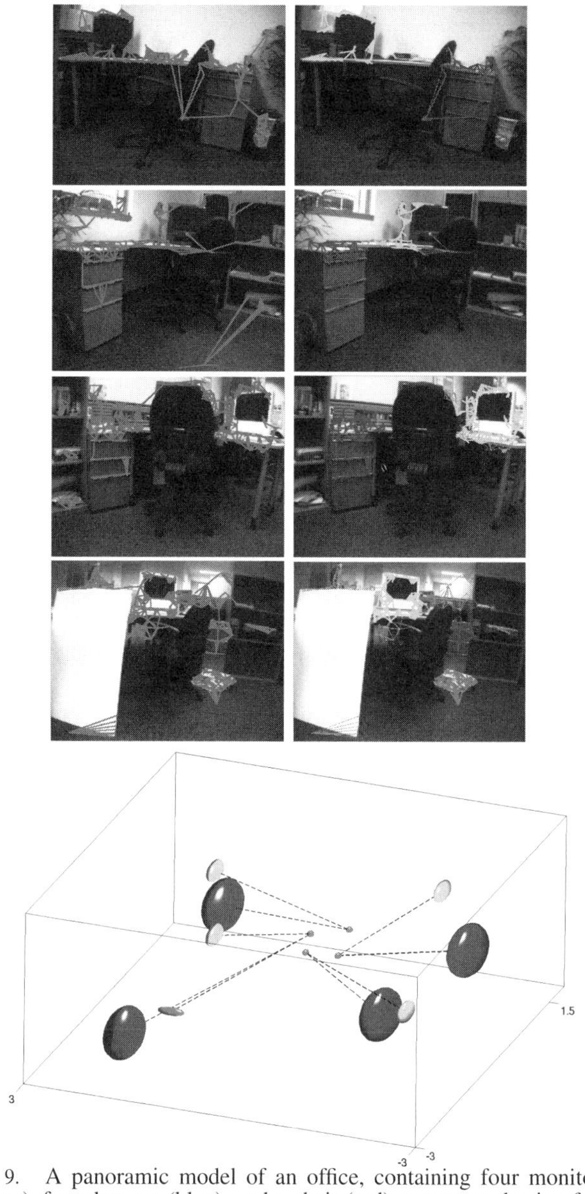

Fig. 9. A panoramic model of an office, containing four monitors (green), four drawers (blue), and a chair (red), constructed using four images. The top four rows show the four images, where the first image of each pair contains the MRF and the second shows the components associated with the objects. The robot poses corresponding to the four images are shown as small brown spheres in the model.

y directions, and 0.1 for the z-direction were used. The variability in the z-direction is less because most objects are observed from only a single direction, with the resulting lack of depth information about the object.

Typical results obtained are given in Figure 7 and 8. These results correspond to the most likely samples obtained from the Swendsen-Wang algorithm. Note that the model for chairs that was learned largely only models the base since few features are detected on the uniform and textureless upper portions. The ROC curves for the objects, shown in Figure 10, quantifies the object detection performance. A panoramic model learned from four images as described in Section V-C is given in Figure 9, which also contains the MAP robot poses.

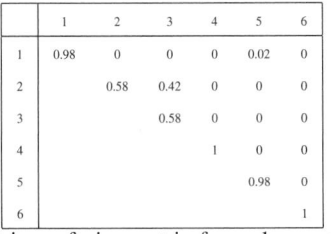

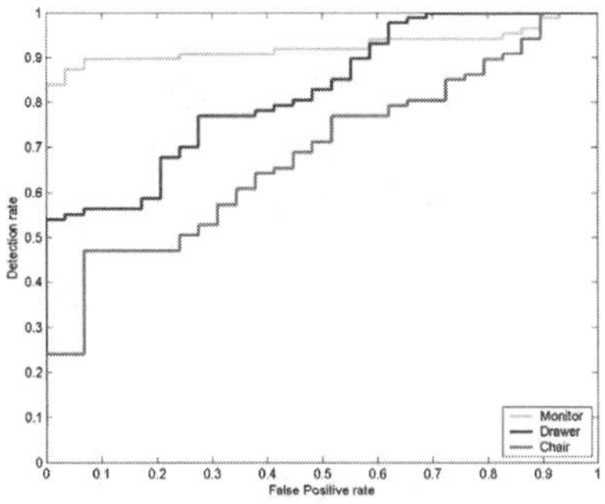

Fig. 10. ROC curves for object detection for some objects used in the experiments. An object is considered detected when it has a large posterior probability in the place model computed from a stereo image.

To test place recognition, we performed an experiment where models for six places were learned and the recognition was assessed on the recall rate of these places when the robot revisited them. A confusion matrix containing the resulting probabilites, calculated using (15), is given in Figure 11. As can be seen from Row 2 therein, a problem with the method is the eagerness of the algorithm to match any two places with the same number and type of objects regardless of their relative location. This problem arises because we do not model the variation in size of individual objects, so that objects of the same type are indistinguishable. We are currently implementing a technique to update the posterior shape of each object from measurements that will alleviate this shortcoming.

VIII. Discussion

We described a technique to model and recognize places using objects as the basic semantic concept. A place is modeled as a constellation of parts with respect to a base location, where each part is an object. Each object is modeled as a combination of 3D Gaussian shape and a multinomial appearance. The object models are learned during a supervised training phase. Our algorithm uses affine-invariant features and stereo range data as measurements to compute a posterior on objects, their locations, and the place label during runtime.

This, in many ways, is preliminary work and large number of improvements and extensions are possible. As far as the model is concerned, representing the objects themselves using a parts model will increase the representational power and robustness of the model [20]. We have not used any pose information in this work. Odometry provides a rough estimate of the robot's location and, hence, a basic idea of context, which has been shown to improve object and place recognition [23]. It is also future work to use our technique to build object-based semantic maps.

	1	2	3	4	5	6
1	0.98	0	0	0	0.02	0
2		0.58	0.42	0	0	0
3			0.58	0	0	0
4				1	0	0
5					0.98	0
6						1

Fig. 11. Symmetric confusion matrix for a place recognition experiment with 6 places. Each row gives the recognition result for a place as a distribution on all the labels.

References

[1] A. Barbu and S.-C. Zhu. Generalizing Swendsen-Wang to sampling arbitrary posterior probabilities. *IEEE Trans. Pattern Anal. Machine Intell.*, 27(8):1239–1253, August 2005.

[2] J. Canny. A computational approach to edge detection. *IEEE Trans. Pattern Anal. Machine Intell.*, 8(6):679–698, November 1986.

[3] G. Dudek and D. Jugessur. Robust place recognition using local appearance based methods. In *IEEE Intl. Conf. on Robotics and Automation (ICRA)*, pages 1030–1035, 2000.

[4] R. Fergus, P. Perona, and A. Zisserman. A sparse object category model for efficient learning and exhaustive recognition. In *Intl. Conf. on Computer Vision (ICCV)*, 2005.

[5] C. Galindo, A. Saffiotti, S. Coradeschi, P. Buschka, J.A. FernÃ¡ndez-Madrigal, and J. GonzÃ¡l ez. Multi-hierarchical semantic maps for mobile robotics. In *IEEE/RSJ Intl. Conf. on Intelligent Robots and Systems (IROS)*, pages 3492–3497, 2005.

[6] W.R. Gilks, S. Richardson, and D.J. Spiegelhalter, editors. *Markov chain Monte Carlo in practice*. Chapman and Hall, 1996.

[7] O. Grau. A scene analysis system for the generation of 3-D models. In *Proc. Intl. Conf. on Recent Advances in 3-D Digital Imaging and Modeling*, pages 221–228, 1997.

[8] D. Hähnel, W. Burgard, and S. Thrun. Learning compact 3D models of indoor and outdoor environments with a mobile robot. *Robotics and Autonomous Systems*, 44(1):15–27, 2003.

[9] R. Hartley and A. Zisserman. *Multiple View Geometry in Computer Vision*. Cambridge University Press, 2000.

[10] B. Kuipers. Modeling spatial knowledge. In S. Chen, editor, *Advances in Spatial Reasoning (Volume 2)*, pages 171–198. The University of Chicago Press, 1990.

[11] D.G. Lowe. Distinctive image features from scale-invariant keypoints. *Intl. J. of Computer Vision*, 60(2):91–110, 2004.

[12] K. Lynch. *The Image of the City*. MIT Press, 1971.

[13] K. Mikolajczyk, T. Tuytelaars, C. Schmid, A. Zisserman, J. Matas, F. Schaffalitzky, T. Kadir, and L. Van Gool. A comparison of affine region detectors. *Intl. J. of Computer Vision*, 65(1/2):43–72, 2005.

[14] A. Nüchter, H. Surmann, K. Lingemann, and J. Hertzberg. Semantic scene analysis of scanned 3D indoor environments. In *Proceedings of the 8th International Fall Workshop Vision, Modeling, and Visualization 2003*, pages 215–222, 2003.

[15] I. Posner, D. Schroeter, and P. Newman. Using scene similarity for place labeling. In *International Symposium of Experimental Robotics*, 2006.

[16] A. Rottmann, O. Martinez Mozos, C. Stachniss, and W. Burgard. Semantic place classification of indoor environments with mobile robots using boosting. In *Nat. Conf. on Artificial Intelligence (AAAI)*, 2005.

[17] D. Schröter and M. Beetz. Acquiring models of rectangular objects for robot maps. In *IEEE Intl. Conf. on Robotics and Automation (ICRA)*, 2004.

[18] D. Schröter, M. Beetz, and B. Radig. RG Mapping: Building object-oriented representations of structured human environments. In *6-th Open Russian-German Workshop on Pattern Recognition and Image Understanding (OGRW)*, 2003.

[19] Josef Sivic, Bryan Russell, Alexei A. Efros, Andrew Zisserman, and Bill Freeman. Discovering objects and their location in images. In *Intl. Conf. on Computer Vision (ICCV)*, 2005.

[20] E. Sudderth, A. Torralba, W. Freeman, and A. S. Willsky. Depth from familiar objects: A hierarchical model for 3d scenes. In *IEEE Conf. on Computer Vision and Pattern Recognition (CVPR)*, 2006.

[21] M.A. Tanner. *Tools for Statistical Inference*. Springer Verlag, New York, 1996. Third Edition.

[22] S. Thrun, M. Montemerlo, and et al. Stanley, the robot that won the DARPA grand challenge. *Journal of Field Robotics*, 2006. In press.

[23] A. Torralba, K. P. Murphy, W. T. Freeman, and M. A. Rubin. Context-based vision system for place and object recognition. In *Intl. Conf. on Computer Vision (ICCV)*, volume 1, pages 273–280, 2003.

[24] S. Vasudevan, S. Gachter, M. Berger, and R. Siegwart. Cognitive maps for mobile robots: An object based approach. In *Proceedings of the IROS Workshop From Sensors to Human Spatial Concepts (FS2HSC 2006)*, 2006.

Design of a Bio-inspired Dynamical Vertical Climbing Robot

Jonathan E. Clark[1], Daniel I. Goldman[2], Pei-Chun Lin[1], Goran Lynch[1],
Tao S. Chen[3], Haldun Komsuoglu[1], Robert J. Full[3], and Daniel Koditschek[1]
[1]GRASP Laboratory, University of Pennsylvania
Department of Electrical and Systems Engineering
200 S. 33rd Street, Philadelphia, PA, USA
[2]School of Physics, Georgia Institute of Technology
[3]PolyPedal Laboratory, Department of Integrative Biology
University of California at Berkeley, Berkeley, CA 94720, USA
email: jonclark@seas.upenn.edu

Abstract— **This paper reviews a template for dynamical climbing originating in biology, explores its stability properties in a numerical model, and presents empirical data from a physical prototype as evidence of the feasibility of adapting the dynamics of the template to robot that runs vertically upward.**

The recently proposed pendulous climbing model abstracts remarkable similarities in dynamic wall scaling behavior exhibited by radically different animal species. The present paper's first contribution summarizes a numerical study of this model to hypothesize that these animals' apparently wasteful commitments to lateral oscillations may be justified by a significant gain in the dynamical stability and, hence, the robustness of their resulting climbing capability.

The paper's second contribution documents the design and offers preliminary empirical data arising from a physical instantiation of this model. Notwithstanding the substantial differences between the proposed bio-inspired template and this physical manifestation, initial data suggest the mechanical climber may be capable of reproducing both the motions and ground reaction forces characteristic of dynamical climbing animals. Even without proper tuning, the robot's steady state trajectories manifest a substantial exchange of kinetic and potential energy, resulting in vertical speeds of 0.30 m/s (0.75 bl/s) and claiming its place as the first bio-inspired dynamical legged climbing platform.

I. Introduction

Past climbing robots have been slow and in most instances restricted to targeted surfaces where specific attachment mechanisms such as suction and electromagnetic adhesion can be brought to bear [1], [2]. Recently, robots have been built that are capable of more broadly effective attachment, for example by means of footholds [3, 4, 5] or vectored thrust [6, 7]. The last few years have also seen the revival [8], [9] of an older design [10] that used rimless wheels with sticky toes to intermittently 'roll' up smooth walls. To our best knowledge, no legged machine has climbed vertically in a dynamical manner, i.e., exploiting a controlled exchange of potential and kinetic energy in order to gain elevation.

The unremitting cost of work against gravity seems dramatically less constraining in the animal kingdom which boasts a variety of species that can dynamically speed their way up vertical environments surfaced in a broad variety of materials, textures, and geometries. Recent bio-mechanical studies of small, agile, climbing animals reveal a striking similarity in locomotion dynamics that belies stark differences in attachment mechanisms, morphology, and phylogeny [11]. These unexpectedly common patterns can be abstracted in a simple numerical model that raises the prospect of a "template" [12] for dynamical climbing analogous to the ubiquitous Spring-Loaded Inverted Pendulum (SLIP) model [13, 14, 15] in sagittal level ground runners and Lateral-Leg Spring (LLS) [16] in sprawled level ground (horizontal) runners. In this paper we explore the value and applicability of this new biological climbing template to the domain of robotics. Specifically, we desire to build a fast, agile climbing robot capable of dynamical operation across a broad variety of scansorial regimes, and we wish to test the proposition that adapting this new biological template will prove both viable and effective to that end.

We present preliminary empirical evidence that such a robot may be fast and hypothesize about why it may prove agile as well. Specifically, we review the dynamical model of interest, introduce the design of a simple physical instantiation, and describe its early implementation in an account organized as follows: Section II reviews the template model, describing its origins in animal studies and exploring the effects of altering the leg sprawl (and therefore its lateral inpulling foot forces) on the speed and stability of climbing. Section III begins with a discussion of how the template can be scaled to our target robot mass, and evaluated the consequent power requirements. We next introduce a design for a simple legged climbing platform that reconciles the limitations of off-the shelf actuators with the power demands of the scaled model [17]. Section IV describes a physical implementation of this very simple template-inspired climber and compares initial data taken from early climbing experiments to the simulation studies. We conclude by commenting on some of the broader issues associated with robot climbers, and discuss future work including limb coupling dynamics, energetics, and adaptation to a more utilitarian polypedal morphology.

II. Template and Stability

Organisms as diverse as arthropods and vertebrates use differing limb number, attachment mechanism and body morphology to achieve performance on vertical substrates that rivals level ground running. Therefore, we expected that diverse animals would necessarily use different climbing strategies. In previous work [11] we have discovered common dynamics in quite different rapidly climbing organisms, a cockroach and a gecko. Surprisingly, neither climbs straight up a vertical flat wall. Both organisms generate substantial lateral forces during climbs over 4 bodylengths per second that produce substantial changes in lateral as well as fore-aft velocity [11, 18].

Significantly, the resultant ground reaction forces generated by these animals while climbing are distinct from the forces that the organisms use on the ground. The lateral forces are of opposite sign. On the ground both cockroaches and geckos limbs push away from their midlines, while on the level they pull toward the midline [11, 19, 18].

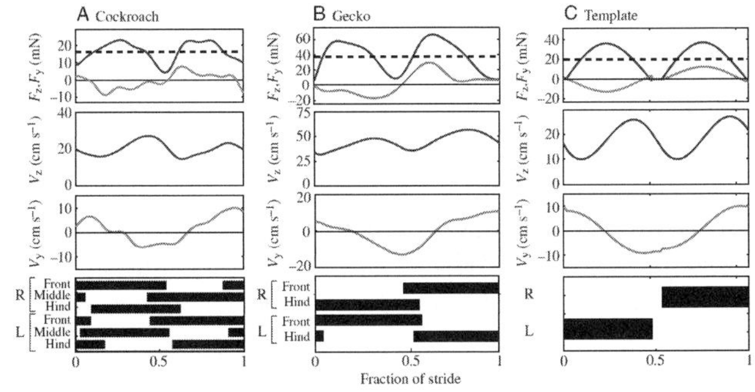

Fig. 1. Force, vertical velocity, lateral velocity, and foot fall patterns for the cockroach, gecko, and template. Broken lines indicate body weight. Data are shown for a normalized stride, with black bars representing foot contact. In each force plot F_z is the magnitude in the vertical direction and F_y is in lateral direction. Reproduced with permission from [11]

A. Template Description

A simple model which generates the template dynamics of vertical climbing is shown in Fig. 2a and a schematic of its motion in Fig. 2b. The model consists of a rigid body that is pulled upward and side-to-side through the action of a spring in series with a linear actuator.

As shown in Fig. 2, in the first step with the right leg, at touchdown ($t = 0$) the right actuator is maximally extended, and the spring is un-extended with zero rest length. Touchdown is created by establishment of a rotationally free pin joint with the wall. As the actuator length $L(t)$ decreases, the spring in the leg extends, the foot freely pivots about the point of contact and the center of mass (COM) is translated vertically and laterally. The cycle repeats for the left leg. The actuator changes length sinusoidally such that $L(t) = L_0(1 + z\sin(2\pi f t))$, where z is the fractional length change, f is the stride frequency. The solid vertical line in each panel indicates the fixed lateral position about which the center of mass laterally oscillates. The angular excursion of the body and extension of the spring are exaggerated for clarity. Actual angular excursion of the body relative to vertical is approximately ± 3 degrees. The model was coded and integrated in the Working Model 2D (Design Simulation Technologies, Inc) simulation environment. The parameters used to generate Fig. 1C were body mass=2 g, body dimensions=4 cm x 0.95 cm, $l_1 = 0.71$ cm, $l_2 = 0.84$ cm, $\beta = 10$ degrees, $L_0 = 1.54$ cm, $z = 0.6$, $k = 6Nm^{-1}$, $\gamma = 0.09N - sm^1$, $f = 9$ Hz. The attachment duty factor in the model is 0.46. The rigid body has a moment of inertia of $8 \times 10^{-7}kg - m^2$, the order of magnitude of cockroaches ($2 \times 10^{-7}kg - m^2$) [20].

The forces and resulting center of mass velocities generated by the model are shown in Fig. 1 and agree well with the pattern measured in cockroaches and the geckos.

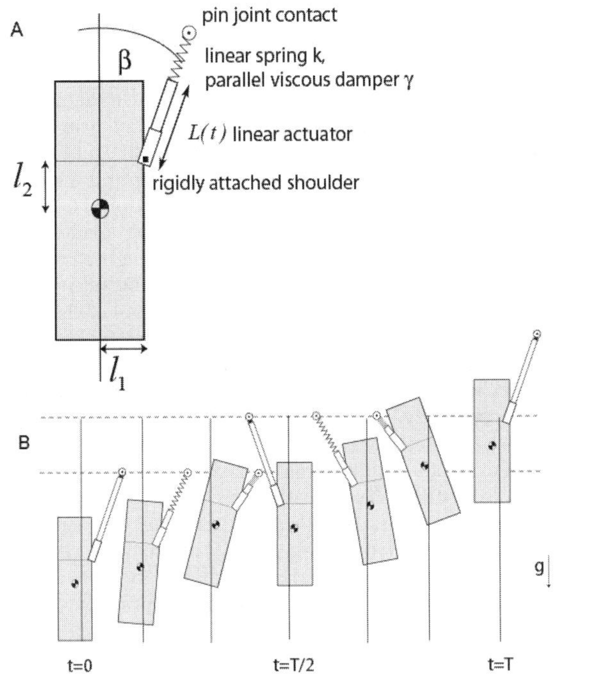

Fig. 2. A dynamic template for climbing. The two degree of freedom model that generates the template climbing dynamics shown in Fig. 1C. (A) Schematic of the model. (B) Schematic of the motion of the model during two steps. Reproduced with permission from [11].

B. Lateral Force and Sprawl Angle

One of the most intriguing findings from the study of vertically running animals is the presence of large lateral velocities and in-pulling forces. At first blush these seem to be wasteful, since substantial muscular force is developed in a direction orthogonal to the climbing direction. One possible explanation for their presence is that generation of lateral forces is necessary for proper adhesion at the feet for both the dry adhesive toe pads of geckos and the pads and claws of cockroaches. Another possible reason is that the addition of lateral dynamics aids stability. In the template model the lateral motions and forces are due to the alternating pendular nature of the climbing dynamics. To what extent does the effective splay or sprawl angle of the legs of the template affect the motion and stability of the climber? Is their a dynamic benefit to having large lateral forces? To begin to answer these questions we fix the dimensions of the template, parameters k, γ, and touchdown timing, but vary the sprawl angle β (Fig. 2).

The dynamics generated are shown in Fig. 3. As β increases, the speed reaches a maximum (Fig. 3b) near $\beta \approx 30$ degrees, then decreases as the template no longer is generating enough fore-aft force to pull against gravity, but instead produces only lateral forces (for $\beta = 90$, the climber cannot ascend).

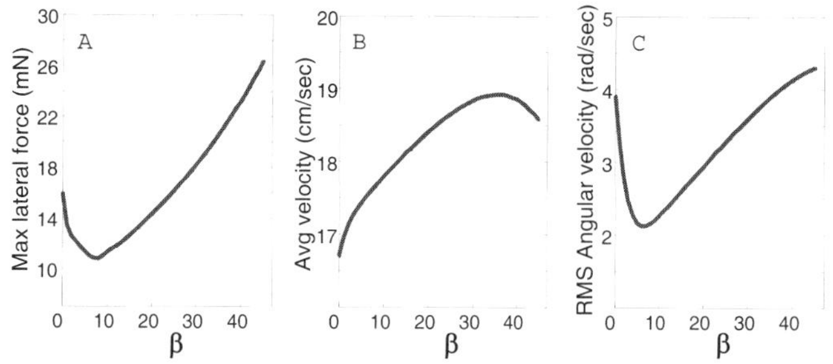

Fig. 3. Variation of template dynamics with as β, the angle of the leg, increases from 0 to 45 degrees. (a) Lateral force (b) Average climbing velocity. (c) Angular velocity.

Surprisingly, we find that as β changes, the template's angular velocity and peak lateral force do not increase monotonically, but instead display minimum around $\beta \approx 10$, see Fig. 3a,c. Furthermore, this minimum corresponds to the approximate effective angle or force ratio that the organisms use to climb, see Fig. 4.

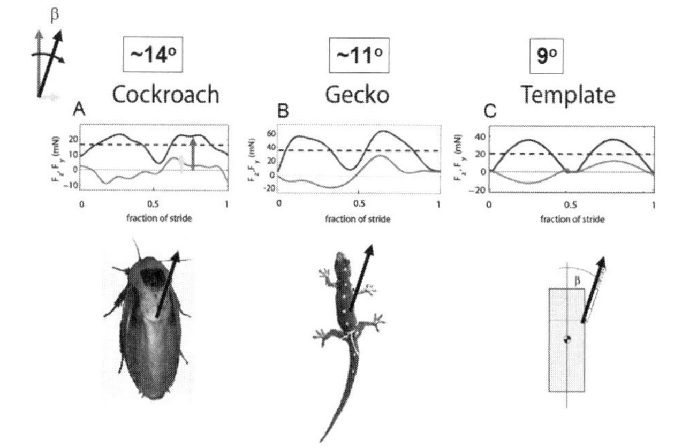

Fig. 4. The β, or effective leg angle, used by the many-legged organisms during vertical climbing is similar to that selected to match COM dynamics.

C. Perturbations

A preliminary investigation of the template response to perturbation reveals hints as to why large lateral forces may be advantageous in climbing. We note that for an organism (or template) of length l the pendular frequency associated with a single limb in contact is on the order of $1/2\pi\sqrt{g/l}$. For an organism or template of length of 5 cm, this is about 3 Hz. Since the animals climb with stride frequency on the order of 10 Hz, this slow pendular oscillation has potential to dominate the dynamics and must be either actively or passively controlled.

In Fig. 5, we demonstrate that generation of lateral forces while climbing results in passive self-stability to the low frequency pendulum oscillation. Upon a lateral impulsive perturbation of about 2 mNs, the climber that only pulls straight up the wall is still swinging wildly 20 steps (1 sec) later. The template that generates lateral forces with $\beta \approx 10$ has corrected for the perturbation within a few strides. This correction does not result from active sensing of the perturbation, but instead emerges from the dynamics.

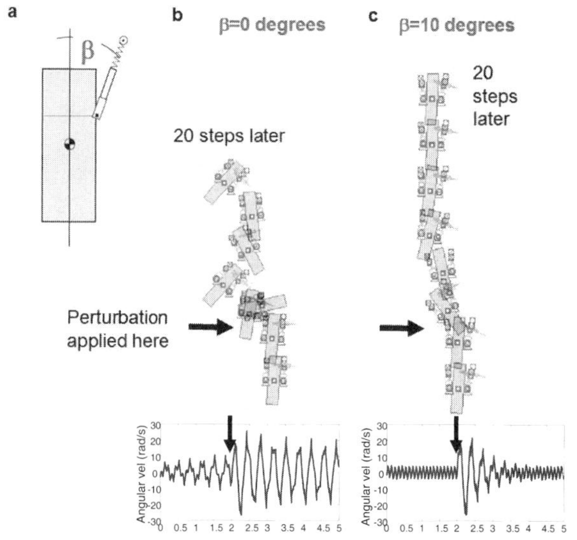

Fig. 5. (a) Template with $\beta = 10$ degrees recovers rapidly after perturbation. Template that pull directly above ($\beta = 0$) swings out of control and never recovers.

The effect of leg angle (β) on the response to perturbation is summarized in Fig. 6. Here we see that 1.5 second (20 steps) after perturbation, the dynamics for $\beta < 10$ still differ from before the perturbation, but for $\beta > 10$, the dynamics have returned to the natural gait.

Thus it appears that a $\beta = 10$ is about the smallest leg sprawl angle for the pendulous template climber that will automatically reject large perturbations quickly. Operation at this posture also results in a minimum RMS angular velocity and the lowest lateral forces (as shown in Fig. 3). We suspect that it is not accidental that this is the same effective β where the ratio lateral/vertical wall reaction forces matches those found in the dynamic animal climbers, as shown in Fig. 4.

III. Scaling The Template

With a biologically inspired template in hand, the first step in constructing a robot based on this model was to determine

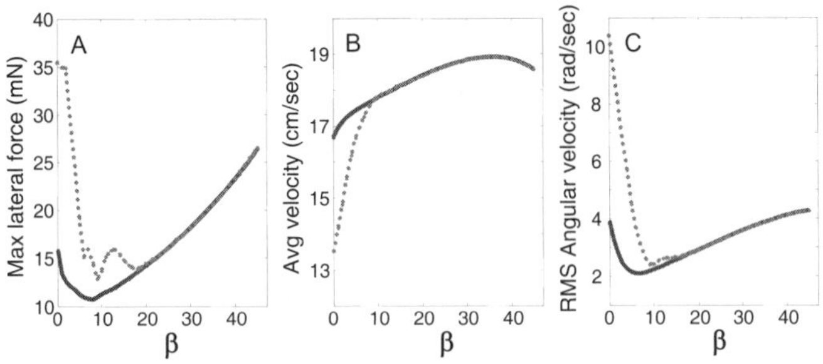

Fig. 6. The effect of β on the recovery from a lateral perturbation with impulse on the order of 2mN-sec. (a) Lateral force (b) Average climbing velocity. (c) Angular velocity. In each plot the solid line is the steady-state behavior (before impulse) and the dotted line is the state 1.5 seconds after impulse.

the appropriate scale for construction. As a target we chose the size and power-density of RiSE [5, 21] — the first general purpose legged machine to exhibit reliable climbing behavior on a variety of natural and synthetic surfaces. In order to preserve the motions and stability results for the template, the scaled model should preserve dynamic similarity.

A. Scaling Rules

Geometric scaling is preserved by changing all length by the same amount. Stiffness, mass, damping, and frequencies, however need to change by different ratios to ensure dynamic similarity. If dynamic similarity is preserved then all of the resulting displacements, times, and forces are scaled versions of the original [22], and the stability characteristics are invariant [20].A more detailed discussion of dynamic similarity and a derivation of the associated scaling laws adapted here is given in [23]. The necessary transformations are summarized in table I.

TABLE I
DYNAMIC SIMILARITY SCALING FACTORS

Quantity	α_x	Relation	$f(\alpha_L)$	$\alpha_L = 10$
Mass	α_m		α_L^3	1000
Frequency	α_ω	$\sqrt{\frac{1}{\alpha_L}}$	$\alpha_L^{-1/2}$	0.316
Stiffness	α_k	$\alpha_\omega^2 \alpha_L^3$	α_L^3	100
Velocity	α_v	$\alpha_\omega \alpha_L$	$\alpha_L^{1/2}$	3.16
Damping	α_b	$\frac{\alpha_L^3}{\alpha_v}$	$\alpha_L^{5/2}$	316
Power	α_P	$\alpha_L^3 \alpha_v$	$\alpha_L^{7/2}$	3160

Each term has a scaling factor ($\alpha_x = x_2/x_1$) where (x_2) is the scaled quantity and (x_1) is the original. The last column shows how much the various quantities need to change for a length scale (α_L) increase of 10 (i.e. the change between a 2g cockroach and a 2kg robot).

When we increase the length (α_l) by a factor of 10, we also increase the needed power to weight ratio by:

$$\frac{\Delta power}{\Delta weight} = \frac{\alpha_P}{\alpha_m} = 3.16$$

Since the power density of actuators is relatively mass independent, it becomes more difficult to provide the power necessary to achieve the template-based dynamics as our scale increases.

The historically achievable power density of climbing and running robots varies greatly, but as a point of reference both the hexapedal RiSE and RHex [24] robots have a specific power of approximately 10W/kg per tripod. The template model, at the cockroach scale, requires a peak of 6.3W/kg per leg. However, scaling to a 2kg climber increases this power demand by 3.16 to about 20W/kg per leg. Thus the model requires the availability of twice as much peak power from the motors as has been available in these previous robotic designs.

B. Design modifications

A numerical study [17] investigating how to reconcile the template's power requirements with the considerably reduced power densities to be found in high performance commercial-off-the-shelf electromechanical servo motors suggested a recourse to three independent design modifications.

The first proposed modification substitutes a DC motor model and crank slider mechanism to replace the prismatic actuators of the template. The second introduces a passive-elastic element in parallel with the leg actuator to store energy during the swing-recirculation phase of the leg motion, and then to release the energy during stance to aid with accelerating the body upwards. The third modification substitutes a force-based rather than position-based control scheme for the legs. Fig. 7 depicts these changes.

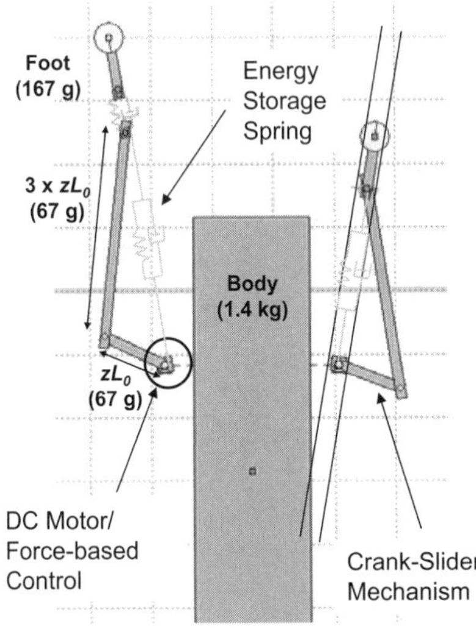

Fig. 7. Schematic of the crank-slider mechanism used to convert rotary (motor) output into linear motion. The relative lengths and masses of the links are indicated.

With the the altered kinematics and additional leg springs (k=130 N/m, b=3 N-s/m) we found [17] that the peak power required for each legs dropped from 40W to a more reasonable 25W during steady state climbing.

C. *Results with force-maximizing control*

The consequences of these combined modifications in the numerical model are detailed in [17] and summarized in Fig. 8.

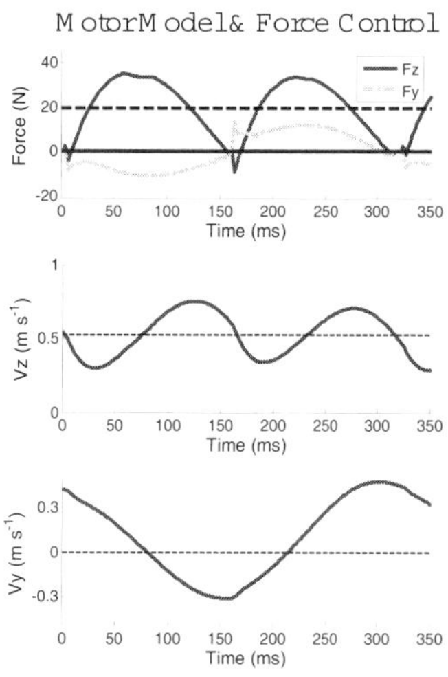

Fig. 8. Force, vertical velocity, and lateral velocity for proposed modified template climber with a motor model and force-based control scheme.

A comparison of the dynamics reveals that the modified model is, despite the scaling, essentially the same as the original template. The switch from a trajectory-tracking to a force-based control scheme releases our control of the resulting actuation frequency. While this frequency shifting during climbing can increase the performance of robot, it also complicates the dynamic coupling between the leg switching, body rotation, and the wrist-spring extension. While this could alter the motion and efficiency of the model, the simulation results suggest that for the motor model chosen the resulting steady-state trajectories work well. In any case, the transients of the dynamics are an additional factor to consider when designing the controller.

The net result, however, is a realistically sized and powered dynamic climber that is very close to the template derived from animal studies. The overall projected climbing speed of 0.55 m/s compares very favorably to that of the scaled template (0.60 m/s).

IV. Robot Design and Experiments

The general design philosophy guiding the development of this initial physical model was to investigate the efficacy of the proposed template with respect to realistically available power density. To maintain this focus, we sought to decouple the vertical and lateral climbing dynamics from other key environmental interactions required for climbing such as adhesion to the wall. Initially, we considered using electromagnets or a specially prepared surface to ensure that each foot hold was secure while climbing. While such "cheating" with respect to foot attachment mechanics would have been viable, we feared that it would unduly hinder our eventual goal of integration of this physical climbing model into a versatile climbing robot. We chose instead to work with claw-like feet on a carpet substrate, a combination that proved effective as a starting point for RiSE [5], and on which that robot's fastest climbs have been recorded [25]. This initial setting gives us confidence that the attachment developments that have enabled RiSE to move from carpet to brick, stucco, concrete, etc. [21] might be adaptable to our climbing model as well. It also provides for an equitable comparison of the robots' relative performance.

A. *Mechanical Structure and Design*

The basic mechanical design is adapted directly from the two-dimensional simulation presented in [17], which in principle is comprised of a rigid body and two linear-moving hands with springs. The resulting robot, depicted in Fig. 9, features two motors, each driving a crank-slider mechanism attached to an arm. As in simulation, each leg has a energy-storage spring in parallel to the crank-slider mechanism with a stiffness and set point designed to assist the leg while retracting during stance. Each foot also features a pair of passive-wrist springs which act in series with the drive actuation. These passively connect the hand to arm and are extended during the beginning of the stance phase, acting to mitigate the loading forces on the robot. Heavy components like motors are located below the cranks in order to shift the location of center of mass lower to match configuration in the template. The frame of robot is constructed from ABS plastic, and the transmission system is mainly comprised of a pair of bevel gears, a pulley pair, sliders (steel shafts and linear bearings), and aluminum links. The sprawl angles β of both arms are adjustable with several pre-settings, including the nominal setting of $\beta = 10$.

We implemented a bar across the rear of the robot extending laterally 20 cm on both sides to reduce the roll of the robot. This bar approximates the function of multiple legs in reducing the dynamics to the planar case considered in the template. The robot's physical parameters are summarized in table II.

TABLE II
Physical Parameters for the Robot

Body size	400 × 116 × 70 mm (excluding cables)
Body mass	2.4 kg
Effective Stride frequency	1.9 Hz
Wrist Spring Stiffness	640 N/m
Arm Spring Stiffness	140 N/m
Motor	Maxon RE 25 118752 Power: 20 watts Size: ϕ25 mm Nominal voltage: 24 V No load speed: 9660 rpm Maximum torque: 240 mNm Maximum current: 1230 mA
Gear head	Maxon Planetary Gearhead GP 32A 114473 33:1 Gear ratio
Encoder	Maxon digital encoder HEDS55 110515 500 count/turn
Bevel Gear	2:1 reduction
Leg stroke	120 mm

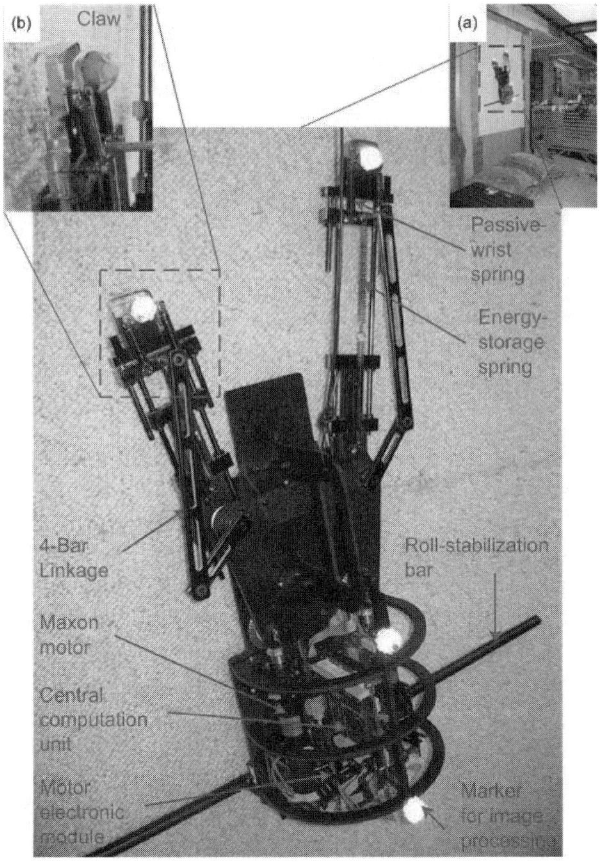

Fig. 9. Picture of the robot with annotations.

We chose to implement a passive attachment strategy where the claw is rigidly attached to the hand. The bent shape highlighted in the blow-up (b) in Fig. 9 engages the substrate when the leg is being pulled down, and releases when pushed. A slight pitch angle introduced by a block under the tail of the robot ensures that the extended foot is closer to the wall than the retracted foot and aids in attachment. Limitations in this technique, however, result in occasionally lost footholds, and a (sometimes significantly) shortened stride length.

Future foot designs include an actuator at the hand which will improve the reliability of attachment, and provide for control of the phasing of attachment and detachment.

B. Electronics and Software

Our robot employs a family of electronic and software components designed for building modular and extensible mechatronic systems governed by distributed control architectures. These components are members of a larger general purpose robot building infrastructure that consists of electronic and software building blocks and standardized interfaces facilitating easy composition.

The present implementation uses three components: 1) a Linux-based 366 MHz CPU card; 2) a carrier board; and 3) two 1-DOF motor controllers. The carrier board interfaces the CPU board with all of the peripheral components. High level control commands are executed on the CPU board, with local feedback control occurring at the motor controllers.

The high level control software is written using RHexLib — a software library which has been developed and used for other legged robots [24], [5] to implement time trigger architecture (TTA) digital control software. This code base provides a library of functions as well as a GUI.

The modular nature of this implementation will become more important in the later stages of this study as we incorporate additional actuators and sensors.

C. Controller

In contrast to the RiSE and RHex machines whose initial leg coordination obtains from a purely feedforward centralized "clock," the alternating leg oscillations arise in the present model from two self-stabilizing mechanical limit cycles that are coupled through the mechanics of the leg-body-ground interactions — a purely decentralized feedback paradigm [26, 27]. Numerical exploration of these oscillators suggests that in their natural state of mutual coupling they stabilize around an undesirable in-phase synchronized gait.

Guided by early numerical studies and general principles from past work [27, 28, 29], we use a 4-state hybrid controller to prevent synchronization of the legs while maximizing the power driving the stance leg at all times.

Imposing de-synchronizing influences on these decentralized feedback driven hybrid oscillators can be achieved in an intuitively appealing manner by recourse to the notion of a "mirror law" first introduced in [30]. A leg in stance mode is commanded the highest permissible torque (T_{max}), while the leg in flight mode is controlled to follow the stance leg with an offset of π. Specifically:

$$T_{\text{stance}} = T_{\text{max}}$$

$$T_{\text{flight}} = k_p * (\text{mod}(\theta_f - \theta_s - \frac{\pi}{2}, 2\pi) - \pi) + k_d * (\dot{\theta}_f - \dot{\theta}_s)$$

where k_p and k_d are controller gains and θ_f is the position of the leg in flight mode, and θ_s is for stance. Ideally, the robot would have both legs transition their states simultaneously, resulting in a situation in which one leg is always in flight mode and the other always in stance mode. However, frequently (and, indeed, systematically), both legs are in the same mode at the same time. In these cases, the controller continues to advance the leg which is closer to transitioning to a new state, while temporarily halting the other leg's progress. Though clearly sub-optimal, this algorithm empirically demonstrates convergence of the robot's gait to a stable limit cycle.

D. Experimental Setup and Procedure

To evaluate the robot climbing performance, a 2m x 0.8m carpet-surface vertical climbing wall was built as shown in section detail (a) of Fig. 9. A commercial 6-axis force sensor (AMTI HE6x6) is installed on the wall to collect interacting forces between the left foot and the wall. A vision system composed by a commercial HD video camera (SONY HDR-SR1) and two spotlights for robot motion traction is located

2m away facing climbing wall. In order to simplify the off-line analysis of the visual data, the robot is "blackened" and 4 spherical markers coated with reflective tape (3M) are installed: two on the body for size calibration and one on each hand for hand position.

Both force data and video data are collect while the robot climbs. Video streams are exported into sequential images for post processing in Matlab. Each color image is converted to black and white by setting threshold empirically and the "white" objects in the image are distinguished from each other by a labeling function and by their geometrical relations.

E. Climbing Results

Despite some significant limitations in this early physical instantiation, the robot climbs remarkably well. Figure 10 shows the trajectory and velocity of the center of mass of the robot while climbing.

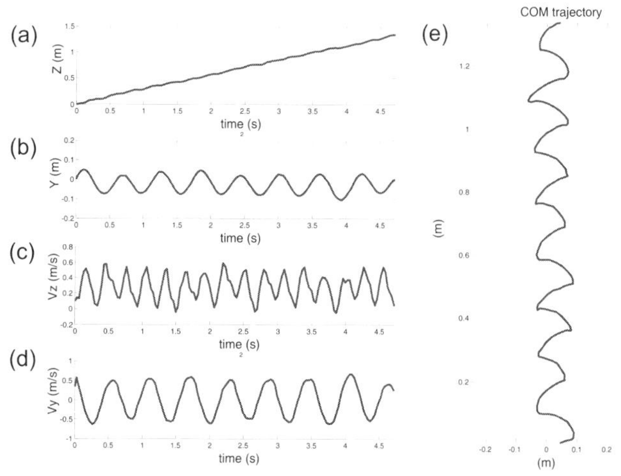

Fig. 10. Video-based marker data for six strides of the robot climbing. (a) vertical displacement, (b) lateral displacement, (c)vertical velocity, (d) lateral velocity, and (e) the path of the center of mass during climbing.

The COM position plot (e) shows that despite the limitations from attachment, convergent and repeatable gaits emerge. The magnitude of the lateral velocity is about 1.5 as large as than the simulation predicts, and the robot climbs at about 30 cm/s.

The physical climber is clearly not as effective as the simulation or the animals, but it does demonstrate that dynamic vertical climbing is in fact possible. Figure 11 compares the ground reaction forces and velocities of the robot and a scaled version of the template.

The magnitude of the vertical oscillations and velocity are considerably lower than expected. This is due to two factors: (1) we are operating at a conservative gear ratio, and (2) the effective stroke length due to attachment issues is reduced. With these issues resolved we believe that we should be able to significantly increase our effective climbing velocity.

For comparison the results of the 2D simulation and the robot are summarized in Table III.

Due to early power limitation issues, we utilized a conservativegear reduction (66 vs 50). This enabled us to climb,

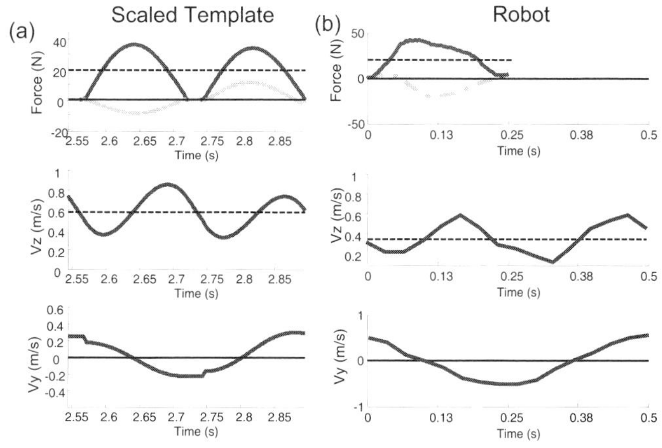

Fig. 11. Ground reaction forces and COM velocities for template and robot.

TABLE III
MODEL-ROBOT PERFORMANCE COMPARISON

	Simulation (2kg)	Robot (2.4kg)
Peak Fz	38 N	46 N
Peak Fy	11 N	22 N
Stroke Length	12 cm	8.5 cm
Stride Freq	2.5 Hz	1.9 Hz
Velocity	55 cm/s	30 cm/s

but reduced the peak operational speed by 25%. This in turn disturbs the balance between vertical extension of the leg and the rotation of the body–resulting in wider swings than predicted.

More pernicious in terms of upward velocity, however, is the effect of attachment on our effective stride length. The nominal length extension for each arm is 12 cm, yet we are currently only seeing an effective stroke length for each step of 8.5 cm. The improper passive attachment and detachment of the limbs is largely responsible for this discrepancy.

Despite these differences we still see a remarkable similarity between the robot and the model's wall reaction forces and in the phasing between the center of mass velocity and the foot-reaction forces, as shown in Fig. 11.

The result is that we are ill-tuned, but still climbing in a dynamic manner. Ongoing work to add active foot attachment, alter the gearing, and improve the control scheme should result in a much better match with the template, and faster and more reliable climbing.

V. CONCLUSION AND FUTURE WORK

A fundamental discovery of a common dynamical pattern, we term a template, for climbing in diverse animal species inspired the design of, first a simulation [17, 11], and now in he present study, a novel bipedal dynamic climber. A simulation of the template and initial experimental results from the robot testing show a remarkable correspondence. The addition of a force-assist spring in parallel with the actuator in the legs and the switch to a force maximizing control scheme allow for a dynamic climber to climb at slightly more than our target mass of 2kg. In addition, it appears that some of the characteristic force and motion patterns of the animals and the steady gaits exhibited by the template are reproducible in the physically anchored version, the robot.

Future theoretical work is required in the development of a robust leg synchronization scheme and in understanding how the structure and control scheme of the robot contribute to its lateral stability. Future experimental work will need to be carried out that includes investigating the effect of increased climbing speed on the attachment and detachment of feet, developing methods to deal with foot slippage or failed attachment, and integrating with RiSE into a dynamic climbing robot that can operate on a number of substrates and in a variety of situations.

Additional experiments will also be needed to support our hypothesis and simulation results that lateral and rotational stability for climbers is improved by generation of large steady-state lateral inpulling forces. For the robot this entails showing whether in fact utilizing a sprawl angle $(\beta) > 10°$ improves stability while climbing.

In conclusion, we have built the first, bio-inspired, dynamical climbing robot. The successful scaling and implementation of the bio-inspired template has enabled us to explore the possible advantages of this novel design that uses passive-dynamic elements. Because the robot has the same dynamics measured in diverse animal species, we can also use it as a physical model to generate the next set of testable hypotheses that will lead to new discoveries in animal climbing.

ACKNOWLEDGMENTS

This work was supported in part by DARPA/SPAWAR Contract N66001-05-C-8025. Jonathan Clark is supported by the IC Postdoctoral Fellow Program under grant number HM158204-1-2030. In addition we would like to thank Al Rizzi, Martin Buehler, and Aaron Saunders for a number of helpful discussions bearing on design and control of the physical platform.

REFERENCES

[1] C. Balaguer, A. Gimenez, J. Pastor, V. Padron, and C. Abderrahim, "A climbing autonomous robot for inspection applications in 3d complex environments." *Robotica*, vol. 18, pp. 287–297, 2000.

[2] G. La Rosa, M. Messina, G. Muscato, and R. Sinatra, "A low-cost lightweight climbing robot for the inspection of vertical surfaces." *Mechatronics*, vol. 12, no. 1, pp. 71–96, 2002.

[3] D. Bevly, S. Dubowsky, and C. Mavroidis, "A simplified cartesian-computed torque controller for highly geared systems and its application to an experimental climbing robot." *Transactions of the ASME. Journal of Dynamic Systems, Measurement and Control*, vol. 122, no. 1, pp. 27–32, 2000.

[4] T. Bretl, S. Rock, and J. C. Latombe, "Motion planning for a three-limbed climbing robot in vertical natural terrain," in *IEEE International Conference on Robotic and Automotion (ICRA 2003)*, vol. 205, Taipei, Taiwan, 2003, pp. 2946 – 295.

[5] K. Autumn, M. Buehler, M. R. Cutkosky, R. Fearing, R. Full, D. Goldman, R. Groff, W. Provancher, A. Rizzi, U. Saranli, A. Saunders, and D. Koditschek, "Robotics in scansorial environments," in *Unmanned Systems Technology VII*, D. W. G. G. R. Gerhart, C. M. Shoemaker, Ed., vol. 5804. SPIE, 2005, pp. 291–302.

[6] "http://www.vortexhc.com/vmrp.html."

[7] J. Xiao, A. Sadegh, M. Elliot, A. Calle, A. Persad, and H. M. Chiu, "Design of mobile robots with wall climbing capability," in *Proceedings of IEEE AIM, Monterey, CA, Jul. 24-28*, 2005, pp. 438–443.

[8] D. K.A., H. A.D., G. S., R. R.E., and Q. R.D., "A small wall-walking robot with compliant, adhesive feet," in *Proceedings of IROS, Edmonton, Canada*.

[9] C. Menon, M. Murphy, and M. Sitti, "Gecko inspired surface climbing robots," in *Proceedings of IEEE ROBIO, Aug. 22-26*, 2004, pp. 431–436.

[10] "http://www.irobot.com."

[11] D. I. Goldman, T. S. Chen, D. M. Dudek, and R. J. Full, "Dynamics of rapid vertical climbing in a cockroach reveals a template," *Journal of Experimental Biology*, vol. 209, pp. 2990–3000, 2006.

[12] R. J. Full and D. E. Koditschek, "Templates and anchors: Neuromechanical hypotheses of legged locomotion on land," *Journal of Experimental Biology*, vol. 202, no. 23, pp. 3325–3332, 1999.

[13] R. Blickhan and R. J. Full, "Similarity in multilegged locomotion: Bounding like a monopod," *Journal of Comparative Physiology*, vol. 173, no. 5, pp. 509–517, 1993.

[14] G. A. Cavagna, N. C. Heglund, and T. C. R., "Mechanical work in terrestrial locomotion: Two basic mechanisms for minimizing energy expenditure," *American Journal of Physiology*, vol. 233, 1977.

[15] T. A. MCMahon and G. C. Cheng, "Mechanics of running. how does stiffness couple with speed?" *Journal of Biomechanics*, vol. 23, no. 1, pp. 65–78, 1990.

[16] J. Schmitt and P. Holmes, "Mechanical models for insect locomotion: Dynamics and stability in the horizontal plane i. theory," *Biological Cybernetics*, vol. 83, no. 6, pp. 501–515, 2000.

[17] J. E. Clark, D. I. Goldman, T. S. Chen, R. J. Full, and D. Koditschek, "Toward a dynamic vertical climbing robot," in *International Conference on Climbing and Walking Robots (CLAWAR)*. Brussels, Belgium: Professional Engineering Publishing, 2006, vol. 9.

[18] K. Autumn, S. T. Hsieh, D. M. Dudek, J. Chen, C. Chitaphan, and R. J. Full, "Dynamics of geckos running vertically," *Journal of Experimental Biology*, vol. 209, pp. 260–270, 2006.

[19] L. H. Ting, R. Blickhan, and R. J. Full, "Dynamic and static stability in hexapedal runners," *Journal of Experimental Biology*, vol. 197, pp. 251–269, 1994.

[20] J. Schmitt, M. Garcia, R. Razo, P. Holmes, and R. Full, "Dynamics and stability of legged locomotion in the horizontal plane: a test case using insects." *Biological Cybernetics*, vol. 86, no. 5, pp. 343–53, 2002.

[21] M. Spenko, M. R. Cutkosky, R. Majidi, R. Fearing, R. Groff, and K. Autumn, "Foot design and integration for bioinspired climbing robots," in *In review for Unmanned Systems Technology VIII*. SPIE, 2006.

[22] R. M. Alexander and A. S. Jayes, "A dynamic similarity hypothesis for the gaits of quadrupedal mammals," *Journal of Zoology*, vol. 201, pp. 135–152, 1983.

[23] R. M. Alexander, *Principles of Animal Locomotion*. Princeton University Press, 2003.

[24] U. Saranli, M. Buehler, and D. E. Koditschek, "Rhex: A simple and highly mobile hexapod robot," *International Journal of Robotics Research*, vol. 20, no. 7, pp. 616–631, 2001.

[25] G. C. Haynes and A. Rizzi, "Gait regulation and feedback on a robotic climbing hexapod," in *Proceedings of Robotics: Science and Systems*, August 2006.

[26] E. Klavins, H. Komsuoglu, R. J. Full, and D. E. Koditschek, *The Role of Reflexes Versus Central Pattern Generators in Dynamical Legged Locomotion*. MIT Press, 2002, ch. Neurotechnology for Biomimetic Robots, pp. 351–382.

[27] E. Klavins and D. E. Koditschek, "Phase regulation of decentralized cyclic robotic systems," *International Journal of Robotics Reserach*, vol. 21, no. 3, pp. 257–275, 2002.

[28] M. Buehler, D. E. Koditschek, and P. J. Kindlmann, "Planning and control of a juggling robot," *International Journal of Robotics Research*, vol. 13, no. 2, pp. 101–118, 1994.

[29] A. A. Rizzi and D. E. Koditschek, "An active visual estimator for dexterous manipulation," *IEEE Transactions on Robotics and Automation*, vol. 12, no. 5, pp. 697–713, 1996.

[30] M. Buehler, D. E. Koditschek, and P. J. Kindlmann, "A simple juggling robot: Theory and experimentation," in *Experimental Robotics I*. Spring-Verlag, 1990, pp. 35–73.

Online Learning for Offroad Robots: Using Spatial Label Propagation to Learn Long-Range Traversability

Raia Hadsell*, Pierre Sermanet*†, Ayse Naz Erkan*, Jan Ben†,
Jefferson Han*, Beat Flepp†, Urs Muller†, and Yann LeCun*
* The Courant Institute of Mathematical Sciences, New York University, New York, NY, USA
† Net-Scale Technologies, Morganville, NJ, USA
email: {raia,naz,jhan,yann}@cs.nyu.edu, {psermanet,jben,flepp,urs}@net-scale.com

Abstract— **We present a solution to the problem of long-range obstacle/path recognition in autonomous robots. The system uses sparse traversability information from a stereo module to train a classifier online. The trained classifier can then predict the traversability of the entire scene. A distance-normalized image pyramid makes it possible to efficiently train on each frame seen by the robot, using large windows that contain contextual information as well as shape, color, and texture. Traversability labels are initially obtained for each target using a stereo module, then propagated to other views of the same target using temporal and spatial concurrences, thus training the classifier to be view-invariant. A ring buffer simulates short-term memory and ensures that the discriminative learning is balanced and consistent. This long-range obstacle detection system sees obstacles and paths at 30-40 meters, far beyond the maximum stereo range of 12 meters, and adapts very quickly to new environments. Experiments were run on the LAGR robot platform.**

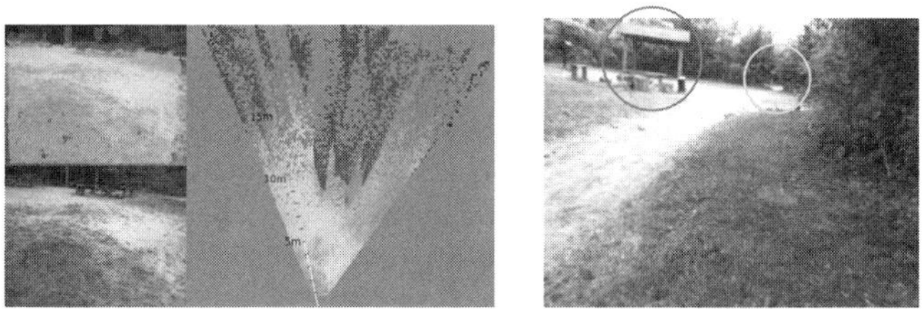

Fig. 1. *Left:* Top view of a map generated from stereo (stereo is run at 320x240 resolution). The map is "smeared out" and sparse at long range because range estimates from stereo become inaccurate above 10 to 12 meters.
Right: Examples of human ability to understand monocular images. The obstacles in the mid-range are obvious to a human, as is the distant pathway through the trees. Note that for navigation, *directions* to obstacles and paths are more important than exact distances.

I. INTRODUCTION

The method of choice for vision-based driving in offroad mobile robots is to construct a traversability map of the environment using stereo vision. In the most common approach, a stereo matching algorithm, applied to images from a pair of stereo cameras, produces a *point-cloud*, in which the most visible pixels are given an XYZ position relative to the robot. A traversability map can then be derived using various heuristics, such as counting the number of points that are above the ground plane in a given map cell. Maps from multiple frames are assembled in a global map in which path finding algorithms are run [7, 9, 3]. The performance of such stereo-based methods is limited, because stereo-based distance estimation is often unreliable above 10 or 12 meters (for typical camera configurations and resolutions). This may cause the system to drive as if in a self-imposed "fog", driving into dead-ends and taking time to discover distant pathways that are obvious to a human observer (see Figure 1 left). Human visual performance is not due to better stereo perception; in fact, humans are excellent at locating pathways and obstacles in monocular images (see Figure 1 right).

We present a learning-based solution to the problem of long-range obstacle and path detection, by designing an approach involving *near-to-far learning*. It is called near-to-far learning because it learns traversability labels from stereo-labeled image patches in the near-range, then classifies image patches in the far-range. If this training is done online, the robot can adapt to changing environments while still accurately assessing the traversability of distant areas.

In order to be effective, the long-range obstacle detection system must overcome some obstacles. A normalization scheme must be used because of the problem of relative sizes of objects in the near and far ranges. We use a distance-normalized pyramid to train on large, context-rich windows from the image. This allows for improved path and obstacle detection (compared to just learning from color or texture). Secondly, the traversability labels from the stereo module may be sparse or noisy, so we maximize their usefulness by using them to label not only the target window, but also *all other previously seen views of that target*. Thus the classifier can train on far-away views that were taken before the stereo label was available. This process of *spatial label propagation* allows the system to learn view-invariant classifications of scenes and objects. Finally, we use a ring buffer to simulate short term memory. The system allows the robot to reliably "see" 35-40m away and opens the door to the possibility of human-level navigation.

Experiments were run on the LAGR (Learning Applied to Ground Robots) robot platform. Both the robot and the reference "baseline" software were built by Carnegie Mellon University and the National Robotics Engineering Center. In this program, in which all participants are constrained to use the given hardware, the goal is to drive from a given start to a predefined (GPS) goal position through unknown, offroad terrain using only passive vision.

II. Previous Work

Considerable progress has been made over the last few years in designing autonomous offroad vehicle navigation systems. One direction of research involves mapping the environment from multiple active range sensors and stereo cameras [10, 14], and simultaneously navigating and building maps [9, 21] and classifying objects.

Estimating the traversability of an environment constitutes an important part of the navigation problem, and solutions have been proposed by many; see [1, 5, 15, 17, 18, 23]. However, the main disadvantage of these techniques is that they assume that the characteristics of obstacles and traversable regions are fixed, and therefore they cannot easily adapt to changing environments. The classification features are hand designed based on the knowledge of properties of terrain features like 3-D shape, roughness etc. Without learning, these systems are constrained to a limited range of predefined environments. By contrast, the vision system presented in this paper uses online learning and adapts quickly to new environments.

A number of systems that incorporate learning have also been proposed. These include ALVINN [16] by Pomerlau, MANIAC [6] by Jochem et al., and DAVE [11] by LeCun et al. Many other systems have been proposed that rely on supervised classification [13, 4]. These systems are trained offline using hand-labeled data, with two major disadvantages: labeling requires a lot of human effort and offline training limits the scope of the robot's expertise to environments seen during training.

More recently, *self-supervised* systems have been developed that reduce or eliminate the need for hand-labeled training data, thus gaining flexibility in unknown environments. With self-supervision, a reliable module that determines traversability can provide labels for inputs to another classifier. This is known as *near-to-far* learning. Using this paradigm, a classifier with broad scope and range can be trained online using data from the reliable sensor (such as ladar or stereo). Not only is the burden of hand-labeling data relieved, but the system can robustly adapt to changing environments. Many systems have successfully employed near-to-far learning in simple ways, primarily by identifying ground patches or pixels, building simple color histograms, and then clustering the entire input image.

The near-to-far strategy has been used successfully for autonomous vehicles that must follow a road. In this task, the road appearance has limited variability, so simple color/texture based classifiers can often identify road surface well beyond sensor range. Using this basic strategy, self-supervised learning helped win the 2005 DARPA Grand Challenge: the winning approach used a simple probabilistic model to identify road surface based on color histograms extracted immediately ahead of the vehicle as it drives [2]. In a slightly more complicated approach by Thrun et al.; previous views of the road surface are computed using reverse optical flow, then road appearance templates are learned for several target distances [12].

Several other approaches have followed the self-supervised, near-to-far learning strategy. Stavens and Thrun used self-supervision to train a terrain roughness predictor [20]. An online probabilistic model was trained on satellite imagery and ladar sensor data for the Spinner vehicle's navigation system [19]. Similarly, online self-supervised learning was used to train a ladar-based navigation system to predict the location of a load-bearing surface in the presence of vegetation [24]. A system that trains a pixel-level classifier using stereo-derived traversability labels is presented by Ulrich [22]. Recently Kim et al. [8] proposed an autonomous offroad navigation system that estimates traversability in an unstructured, unknown outdoor environment.

The work presented here uses self-supervised online learning to generalize traversability classification. Unlike other methods, our method relies solely on visual data and is efficient enough to re-train and re-classify each frame in realtime (roughly 4-5 frames/second). The system requires no human labeling or supervision.

III. The LAGR Vehicle: Overview of Path Planning and Local Navigation

This section gives an overview of the full navigation system developed for the LAGR robot. Although reference "baseline" software was provided, none was used in our system. Our LAGR system consists of 4 major components (see Figure 2).

- **Vehicle Map.** The vehicle map is a local map in polar coordinates that is fixed relative to the robot position. It is 100° wide and has a 40m radius. It stores cost and confidence data which is delivered by the different obstacle detectors.
- **Local Navigation.** The local navigation is based on the vehicle map. It determines a set of candidate waypoints based on cost, confidence, and steering requirements. The candidate waypoint is picked which lets the vehicle progress toward the goal. Driving commands are issued based on this choice.
- **Global Map.** The global map is a Cartesian grid map into which cost and confidence information from the vehicle map is copied after each processed frame. The global map is the system's "memory".
- **Global Planner.** The global planner finds a route to the goal in the global map, starting with candidate points proposed by the local navigation module. The algorithm is a modified A-Star algorithm which operates on rays rather than grid cells.

IV. Long-Range Vision from Distance-Normalized Monocular Images

A. Motivation and Overview

Humans can easily locate pathways from monocular views, e.g. trails in a forest, holes in a row of bushes. In this section, we present a vision system that uses online learning to provide the same capability to a mobile robot. Our approach, using *self-supervised* learning, is to use the short-range output of

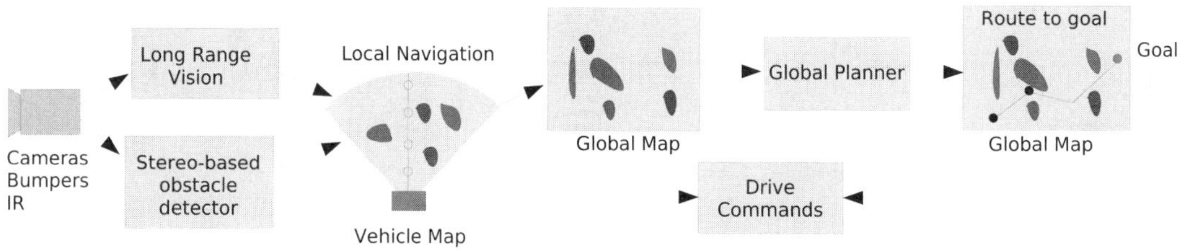

Fig. 2. A flow chart of the full navigation system. The long-range obstacle detector and the stereo obstacle detector both populate the vehicle map, where local navigation is done. The local map gets written to the global map after every frame, where route planning is done with the global planner.

a reliable module (stereo) to provide labels for a trainable module (a window-based classifier). There are three key components to the approach. First, we do horizon leveling and distance normalization in the image space, producing a multi-resolution pyramid of sub-images. This transformation is essential for generalizing the classifier to long-range views. Second, the system propagates labels temporally using spatial concurrences in a quad-tree. This allows us to directly train on previously seen views that are out of stereo range. Last, we use a ring buffer to hold a balanced set of traversable and non-traversable training samples.

The sequence of operations are summarized here and details are discussed in the following sections.

1) All points in the current frame are inserted into a quad-tree according to their XYZ coordinates. The XYZ coordinates are determined by mapping from image space to world space, and can be computed if the parameters of the ground plane are known. This is discussed in section IV-B.
2) The stereo module labels the traversability of visible points up to 12 meters.
3) Each point that was given a stereo label in the previous step is now used as a *query point* for label propagation. The quad-tree is queried to find all previously seen points that are within a radius r of the queried location. These points are labeled with the stereo label of the *query point*.
4) All stereo-labeled query points and points returned by the quad-tree are added to a ring buffer.
5) A discriminative classifier trains on the samples in the ring buffer, which are labeled with -1 (ground) or 1 (obstacle).
6) The trained module classifies all windows in the pyramid, at a range of 1 to 35 meters.

B. Horizon Leveling and Distance Normalization

Recent vision-based autonomous navigation systems have trained classifiers using small image patches or mere pixel information, thus limiting the learning to color and texture discrimination. However, it is beneficial to use larger windows from the image, thus providing a richer context for more accurate training and classification. Recognizing the feet of objects is critical for obstacle detection, and the task is easier with larger windows.

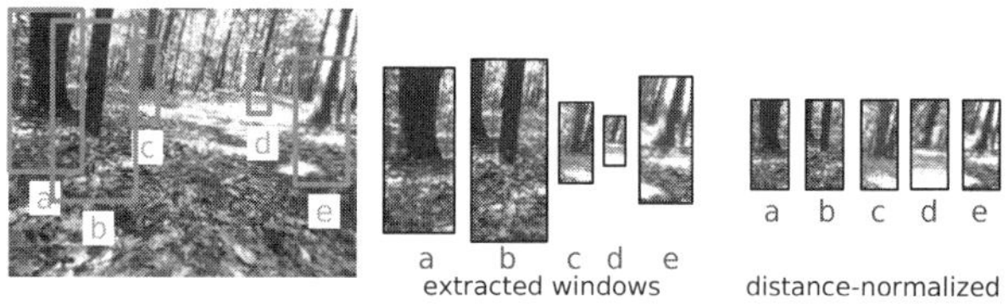

Fig. 3. This figure demonstrates the problem of distance scaling. *left*: The trees in the image vary widely in size because of different distances from the camera. This makes near-to-far learning extremely difficult. *center and right:* If windows are cropped from the image and resized such that the subsampling is proportional to their distance from the camera, then a classifier can train on more uniformly sized and positioned objects.

There is an inherent difficulty with training on large windows instead of color/texture patches. In image space, the size of obstacles, paths, etc. varies greatly with distance, making generalization from near-range to far-range unlikely. Our approach deals with this problem by building a *distance-invariant pyramid of images at multiple scales*, such that the appearance of an object sitting on the ground X meters away is identical to the appearance of the same object when sitting on the ground Y meters away (see Figure 3). This also makes the feet of objects appear at a consistent position in a given image, allowing for easier and more robust learning.

In order to build a pyramid of horizon-leveled sub-images, the ground plane in front of the robot must first be identified by performing a robust fit of the point cloud obtained through stereo. A Hough transform is used to produce an initial estimate of the plane parameters. Then, a least-squares fit refinement is performed using the points that are within a threshold of the initial plane estimate.

To build the image pyramid, differently sized sub-images are cropped from the original RGB frame such that each is centered around an imaginary *footline* on the ground. Each footline is a predetermined distance (using a geometric progression) from the robot's camera. For [*row, column, disparity, offset*] plane parameters $P = [p_0, p_1, p_2, p_3]$ and desired disparity d, the image coordinates (x_0, y_0, x_1, y_1) of the footline can be directly calculated.

After cropping a sub-image from around its footline, the sub-image is then subsampled to make it a uniform height (20 pixels), resulting in image bands in which the appearance of an object on the ground is independent of its distance from the

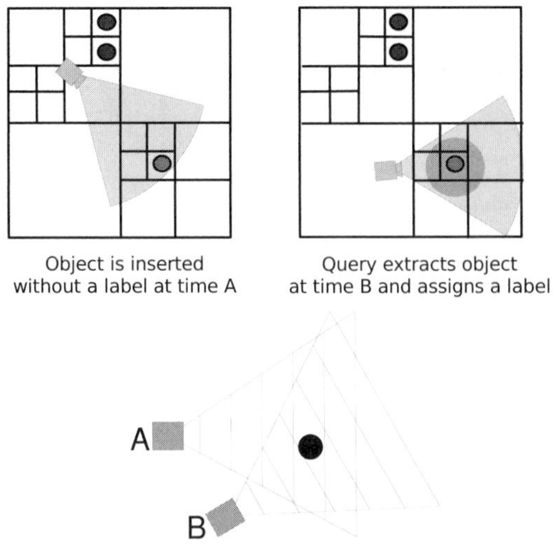

Fig. 5. Spatial label propagation. At position/time **A**, the robot can see the black location, but it is out of stereo range, so it cannot be labeled. It is inserted into the quad-tree. At position/time **B**, the black area is seen from a different, closer view and a stereo label is obtained. Now the view of the location from position **A** can be extracted from the quad-tree and trained on using the label from position **B**.

camera (see Figure 4). These uniform-height, variable-width bands form a distance-normalized image pyramid whose 36 scales are separated by a factor of $2^{\frac{1}{6}}$.

C. Spatial Label Propagation

We expand the number of labeled training samples per frame by propagating labels backward in time. This is done using a quad-tree that indexes XYZ locations in the world surrounding the robot. The quad-tree is a very efficient data structure for storing spatial information, and concurrent views can be inserted and queried in $O(lgn)$ time. Given a labeled window and its corresponding world XYZ location, we can query the quad-tree to retrieve all previously stored views of the same location (see Figure 5 and 6). Label propagation on a graph is a variant of semi-supervised learning that exploits knowledge about the robot's position and heading to expand the training set on every frame.

The stereo-labeled points and the query-extracted points are stored in 2 large ring buffers, one for traversable points, and one for non-traversable points. On each frame, the classifier trains on all the data in both ring buffers. The ring buffer acts like short-term memory, since samples from previous frames persist in the buffer until replaced. The ring buffer also balances the training, ensuring that the classifier always trains on a constant ratio of traversable/non-traversable points.

D. Training Architecture and Loss Function

The long-range obstacle detector goes through a labeling, training, and classification cycle on every frame. First, each overlapping RGB window from the right camera is assigned a traversability label (ground or obstacle) if it is within stereo range ($<$ 12 meters) and if stereo data is available. Then feature vectors are computed for the windows in the pyramid

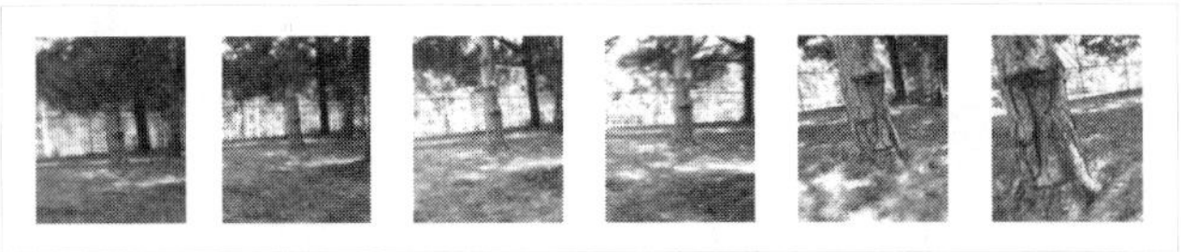

Fig. 6. Multiple views of a single object. The same label (non-traversable, in this case) is propagated to each instance.

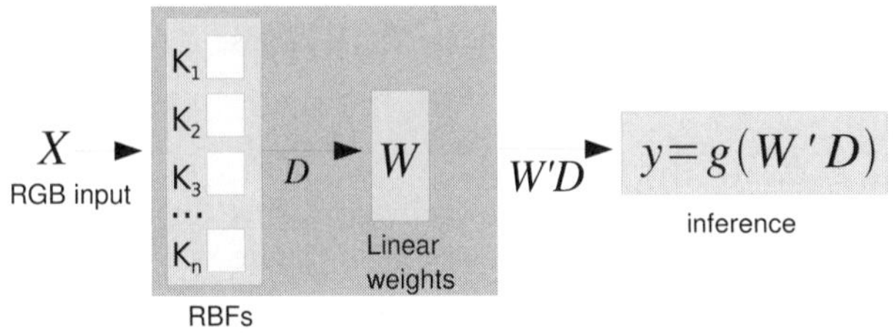

Fig. 7. The online learning architecture for the long-range vision system.X is the input window, D is the feature vector calculated from the RBF layer, and W is a weight vector. The function g is a logistic function for inference on y

using a small convolutional network that is trained offline. A set of 120 radial basis functions was also implemented as a feature extractor, but tests showed that the trained convolutional network produced more discriminative feature vectors. Details on the feature extraction process, for the convolutional network and the RBF (radial basis function) network, are provided. The classification error rates, after offline training of the different feature extractors, is given in Table I.

The RBF feature vectors are constructed using Euclidean distances between a 12x3 RGB window and the 120 fixed RBF centers. For an input window X and a set of n radial basis centers $\mathbf{K} = [K_1...K_n]$, the feature vector D is $D = [\exp(-\beta_1||X-K_1||^2) \ldots \exp(-\beta_n||X-K_n||^2)$ where β_i is the variance of RBF center K_i. The radial basis function centers **K** are trained in advance with K-means unsupervised learning.

Several convolutional networks were compared. The one chosen is a three-layer network with a 15x15 pixel field of view and an output vector of 120. The first layer (C0) has 16 6x6 convolutional kernels; the second layer (S0) is a 2x2 subsampling layer, and the third layer (C1) has 64 5x5 convolutional kernels. Although this network did not achieve the lowest offline error rate, it had the best speed/performance balance out of all the tested networks.

A logistic regression on the feature vectors is trained using the labels provided by stereo. The resulting classifier is then applied to all feature vectors in the pyramid, including those with stereo labels. The training architecture is shown in Figure 7.

The classifier is a logistic regression trained with stochastic gradient descent to produce binary labels (0 for traversable, 1 for non-traversable). Weight decay towards a previously learned set of default weights provides regularization. The loss function is cross-entropy loss. For weights W, feature vector D, label y, and logistic function $g(z) = \frac{1}{1+e^z}$, the loss function

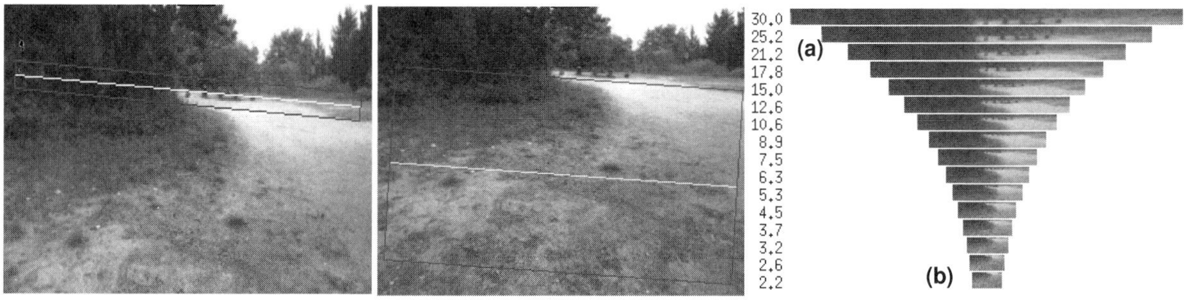

(a). sub-image extracted from far range. (21.2 m from robot).

(b). sub-image extracted at close range. (2.2 m from robot).

(c). the pyramid, with rows (a) and (b) corresponding to sub-images at left.

Fig. 4. Sub-images are extracted according to imaginary lines on the ground (computed using the estimated ground plane). *(a)* Extraction around a footline that is 21m away from the vehicle. *(b)* Extraction around a footline that is 1.1m away from the robot. The extracted area is large, because it is scaled to make it consistent with the size of the other bands. *(c)* All the sub-images are subsampled to 20 pixels high.

Test	Baseline (sec)	Our system(sec)	% Improved
11.1	327	310	5.48
11.2	275	175	57.14
11.3	300	149	101.34
12.1	325	135	140.74
12.2	334	153	118.3
12.3	470	130	261.54
12.4	318	155	105.16

TABLE II

GOVERNMENT TESTING RESULTS. A DARPA TEAM TESTED OUR SOFTWARE AGAINST THE BASELINE SYSTEM AT PERIODIC INTERVALS, AT UNKNOWN LOCATIONS WITH UNKNOWN ENVIRONMENTS AND OBSTACLES. THE PERFORMANCE OF EACH SYSTEM WAS MEASURED AS THE TIME TAKEN TO REACH THE GOAL. OUR SYSTEM TYPICALLY REACHED THE GOAL 2 TO 4 TIMES FASTER THAN THE BASELINE SYSTEM

and gradient of the loss function are

$$\mathcal{L} = -\sum_{i=1}^{n} \log(g(y \cdot W'D)) \quad \frac{\partial L}{\partial W} = -(y - g(W'D)) \cdot D$$

Learning is done by adjusting the weights with stochastic gradient descent. Although the loss function is convex and optimal weights could be found exactly, using stochastic gradient descent gives the system a natural and effective inertia that is necessary for generalizing well over successive frames and environments.

After training on the feature vectors from labeled windows, all the windows are classified. Inference on y is simple and fast: $y = sign(g(W'D))$ where $g(z) = \frac{1}{1+e^z}$ as before.

V. RESULTS

The robot drives smoothly and quickly under the full navigation system described in section 3. It typically gets to a goal 2 to 4 times faster than the baseline system (see Table II). The long-range vision module is efficient; it runs at 3-4 frames/second.

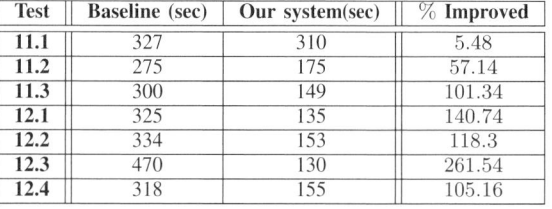

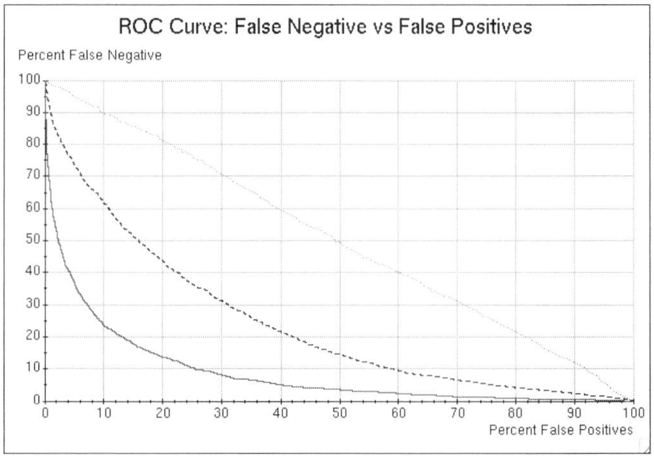

Fig. 8. ROC curves comparing classifier labels vs. stereo labels. *dotted/blue* The classifier was initialized with random weights, and the online learning was turned off. *dashed/black:* The classifier was initialized with default trained weights, but online learning was turned off. *solid/red:* The full system: trained default weights and online learning.

To specifically assess the accuracy of the long-range classifier, the error of the classifier was measured against stereo labels. If the classifier was initialized with random weights and no online learning was used, then the error rate was, predictably, 52.23%. If the classifier was initialized with a set of default parameters (average learned weights over many logfiles) but with no online learning, then the classifier had an error of 32.06%. If the full system was used, i.e., initialized with default weights and online learning on, then the average error was 15.14%. ROC curves were computed for each test error rate (see Figure 8).

Figure 9 shows examples of the maps generated by the long-range obstacle detector. It not only yields surprisingly accurate traversability information at distance up to 30 meters (far beyond stereo range), but also produces smooth, dense traversability maps for areas that are within stereo range. The

Feature extractor	Input window size	% Test error (offline)	% Error (online)
RBF	24x6	45.94	23.11
RBF	6x6	47.34	22.65
RBF	12x3	48.49	21.23
CNN (3-layer)	15x15	20.71	11.61
CNN (3-layer)	20x11	20.98	12.89
CNN (3-layer)	12x9	24.65	14.56
CNN (1-layer)	24x11	16.35	11.28

TABLE I

A COMPARISON OF ERROR RATES FOR DIFFERENT FEATURE EXTRACTORS. EACH FEATURE EXTRACTOR WAS TRAINED, OFFLINE, USING THE SAME DATA. RBF = RADIAL BASIS FUNCTION FEATURE EXTRACTOR. CNN = CONVOLUTIONAL NEURAL NETWORK FEATURE EXTRACTOR.

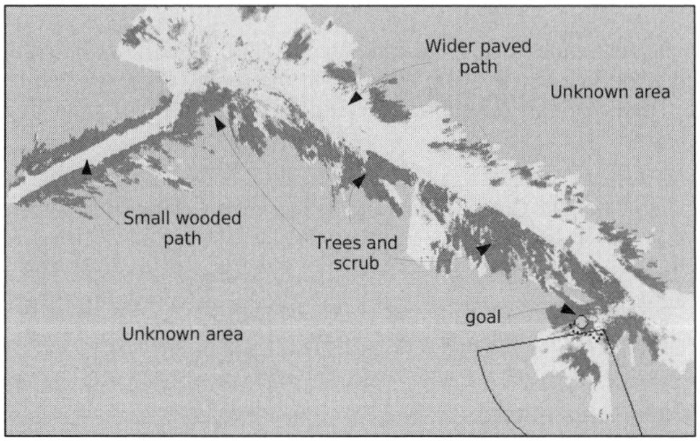

Fig. 10. This is a global map produces by the robot while navigating toward a goal. The robot sees and follows the path and resists the shorter yet non-traversable direct path through the woods.

stereo maps often have noisy spots or holes - disastrous to a path planner - but the long-range module produces maps that are smooth and accurate, without holes or noise.

The long-range module, when integrated with the whole driving system and tested in offroad environments, generally produced clear and accurate global maps (see Figure 10).

VI. Summary and Further Work

We demonstrated an autonomous, online learning method for obstacle detection beyond the range of stereo. At each frame, a pyramid of image bands is built in which the size of objects on the ground is independent of their distance from the camera. Windows on locations beyond stereo range are stored in a spatially-indexed quad-tree for future reference. Windows within stereo range are assigned labels, together with windows at the same spatial location stored in the quad-tree during previous frames. A simple online logistic regression classifier fed with features extracted using a trained convolutional network is trained on a ring buffer of such windows. The system requires no human intervention, and can produce accurate traversability maps at ranges of up to 40 meters.

The method, not including the label propagation strategy, was implemented as part of a complete vision-based driving system for an offroad robot. The system navigates through unknown complex terrains to reach specified GPS coordinates about 2 to 4 times faster than the baseline system provided with the robot platform.

Acknowledgments

The authors wish to thank Larry Jackel, Dan D. Lee, and Martial Hébert for helpful discussions. This work was supported by DARPA under the Learning Applied to Ground Robots program.

References

[1] P. Bellutta, R. Manduchi, L. Matthies, K. Owens, and A. Rankin. Terrain perception for demo iii. 2000.

[2] H. Dahlkamp, A. Kaehler, D. Stavens, S. Thrun, and G. Bradski. Self-supervised monocular road detection in desert terrain. June 2006.

[3] S. B. Goldberg, M. Maimone, and L. Matthies. Stereo vision and robot navigation software for planetary exploration. March 2002.

[4] T. Hong, T. Chang, C. Rasmussen, and M. Shneier. Road detection and tracking for autonomous mobile robots. In *Proc. of SPIE Aeroscience Conference*, 2002.

[5] A. Huertas, L. Matthies, and A. Rankin. Stereo-based tree traversability analysis for autonomous off-road navigation. 2005.

[6] T. M. Jochem, D. A. Pomerleau, and C. E. Thorpe. Vision-based neural network road and intersection detection and traversal. *Intelligent Robot and Systems (IROS)*, 03:344–349, 1995.

[7] A. Kelly and A. Stentz. Stereo vision enhancements for low-cost outdoor autonomous vehicles. *Int'l Conf. on Robotics and Automation, Workshop WS-7, Navigation of Outdoor Autonomous Vehicles*, May 1998.

[8] D. Kim, J. Sun, S. M. Oh, J. M. Rehg, and B. A. F. Traversibility classification using unsupervised on-lne visual learning for outdoor robot navigation. May 2006.

[9] D. J. Kriegman, E. Triendl, and T. O. Binford. Stereo vision and navigation in buildings for mobile robots. *IEEE Trans. Robotics and Automation*, 5(6):792–803, 1989.

[10] E. Krotkov, , and M. Hebert. Mapping and positioning for a prototype lunar rover. pages 2913–2919, May 1995.

[11] Y. LeCun, U. Muller, J. Ben, E. Cosatto, and B. Flepp. Off-road obstacle avoidance through end-to-end learning. In *Advances in Neural Information Processing Systems (NIPS 2005)*. MIT Press, 2005.

[12] D. Leib, A. Lookingbill, and S. Thrun. Adaptive road following using self-supervised learning and reverse optical flow. June 2005.

[13] R. Manduchi, A. Castano, A. Talukder, and L. Matthies. Obstacle detection and terrain classification for autonomous off-road navigation. *Autonomous Robot*, 18:81–102, 2003.

[14] L. Matthies, E. Gat, R. Harrison, V. R. Wilcox, B and, , and T. Litwin. Mars microrover navigation: Performance evaluation and enhancement. volume 1, pages 433–440, August 1995.

[15] R. Pagnot and P. Grandjea. Fast cross country navigation on fair terrains. pages 2593 –2598, 1995.

[16] D. A. Pomerlau. Knowledge based training of artificial neural networks for autonomous driving. *J. Connell and S. Mahadevan, editors, Robot Learning. Kluwer Academic Publishing*, 1993.

[17] A. Rieder, B. Southall, G. Salgian, R. Mandelbaum, H. Herman, P. Render, and T. Stentz. Stereo perception on an off road vehicle. 2002.

[18] S. Singh, R. Simmons, T. Smith, A. Stentz, V. Verma, A. Yahja, and K. Schwehr. Recent progress in local and global traversability for planetary rovers. pages 1194–1200, 2000.

[19] B. Sofman, E. Lin, J. Bagnell, N. Vandapel, and A. Stentz. Improving robot navigation through self-supervised online learning. June 2006.

(a) 2 paths (and pavilion) detected at long range

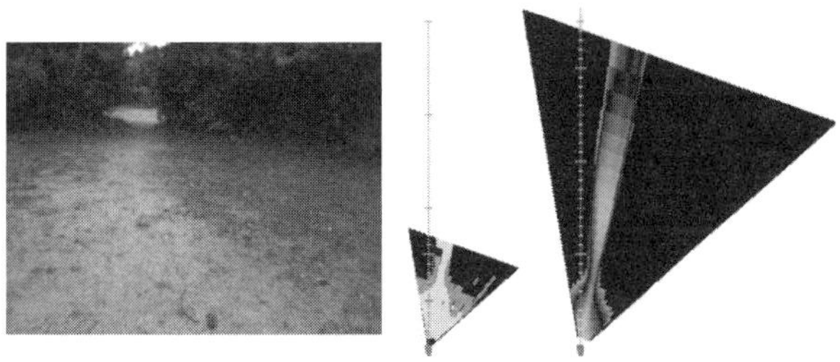

(b) Path and distant trees detected

(c) Several hay-bales detected accurately

(d) Smooth road and tree barrier is seen

(e) Distant picnic tables detected

(f) Path seen at long range

Fig. 9. Examples of performance of the long-range vision system. Each example shows the input image (left), the stereo labels (middle), and the map produced by the long-range vision (right). Green indicates traversable, and red/pink indicates obstacle. Note that stereo has a range of 12 meters, whereas the long-range vision has a range of 40 meters.

[20] D. Stavens and S. Thrun. A self-supervised terrain roughness estimator for off-road autonomous driving. 2006.

[21] S. Thrun. Learning metric-topological maps for indoor mobile robot navigation. *Artificial Intelligence*, pages 21–71, February 1998.

[22] I. Ulrich and I. R. Nourbakhsh. Appearance-based obstacle detection with monocular color vision. pages 866–871, 2000.

[23] N. Vandapel, D. F. Huber, A. Kapuria, and M. Hebert. Natural terrain classification using 3d ladar data. 2004.

[24] C. Wellington and A. Stentz. Online adaptive rough-terrain navigation in vegetation. April 2004.

Composition of Vector Fields for Multi-Robot Manipulation via Caging

Jonathan Fink, Nathan Michael, and Vijay Kumar
University of Pennsylvania
Philadelphia, Pennsylvania 19104-6228
Email: {jonfink, nmichael, kumar}@grasp.upenn.edu

Abstract— **This paper describes a novel approach for multi-robot caging and manipulation, which relies on the team of robots forming patterns that trap the object to be manipulated and dragging or pushing the object to the goal configuration. The controllers are obtained by sequential composition of vector fields or behaviors and enable decentralized computation based only on local information. Further, the control software for each robot is identical and relies on very simple behaviors. We present our experimental multi-robot system as well as simulation and experimental results that demonstrate the robustness of this approach.**

I. Introduction

We address the problem of cooperative manipulation with multiple mobile robots, a subject that has received extensive treatment in the literature. Most approaches use the notions of force and form closure to perform the manipulation of relatively large objects [1, 2, 3]. Force closure is a condition that implies that the grasp can resist any external force applied to the object while form closure can be viewed as the condition guaranteeing force closure, without requiring the contacts to be frictional [4]. In general, robots are the agents that induce contacts with the object, and are the only source of grasp forces. But, when external forces acting on the object, such as gravity and friction, are used together with contact forces to produce force closure, we get conditional force closure. It is possible to use conditional closure to transport an object by pushing it from an initial position to a goal [5, 6]. Caging or object closure is a variation on the form closure theme. It requires the less stringent condition that the object be trapped or caged by the robots and confined to a compact set in the configuration space. Motion planning for circular robots manipulating a polygonal object is considered in [7]. Decentralized control policies for a group of robots to move toward a goal position while maintaining a condition of object closure were developed in [8].

We are interested in distributed approaches to multirobot manipulation that have the following attributes. (1) The coordination between robots must be completely decentralized allowing scaling up to large numbers of robots and large objects. (2) There must be no labeling or identification of robots. In other words, the algorithm should not explicitly encode the number of robots in the team, and the identities of the robots. Indeed the instruction set for each robot must be identical. This allows robustness to failures, ease of programming and modularity enabling addition and/or deletion of robots from the team. (3) We are interested in an approach that requires minimal communication and sensing and controllers that are based only on local information. It is often impractical for large numbers of robots to share information and to have every robot access global information. Indeed these three attributes are seen frequently in nature. As seen in Figure 1, relatively small agents are able to manipulate objects that are significantly larger in terms of size and payload by cooperating with fairly simple individual behaviors. We believe these three attributes are key to the so-called *swarm paradigm* for multi-robot coordination.

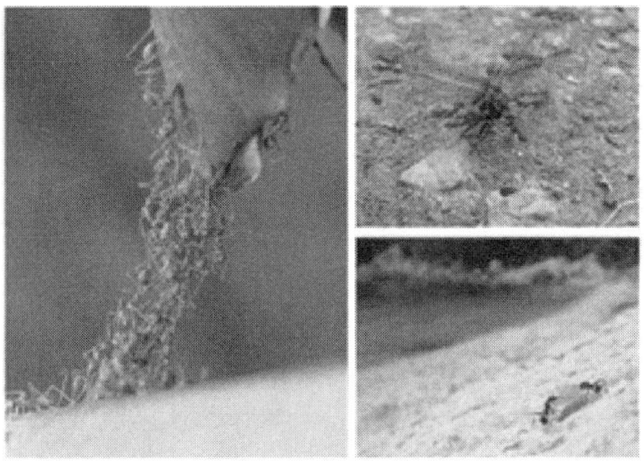

Fig. 1. Ants are able to cooperatively manipulate and transport objects often in large groups, without identified or labeled neighbors, and without centralized coordination.

In the work above, none of the papers discuss these three attributes, with the exception of [8] which does incorporate the first attribute. We note that [9] incorporates the three swarm attributes for multirobot manipulation but it makes the unrealistic assumption of point robots without considering inter-robot collisions. In this paper, we show how simple vector fields can be designed and composed to enable a team of robots to approach an object, surround it to cage it, and transport it to a destination, while avoiding inter-robot collisions. We provide the theoretical justification for the construction of the vector fields and their composition, and demonstrate the application of this approach with dynamic simulation which incorporates robot-object interactions and through experimentation.

II. Architecture and Framework

Our primary interest is in performing complex tasks with large teams of distributed robotic agents. Independent of the

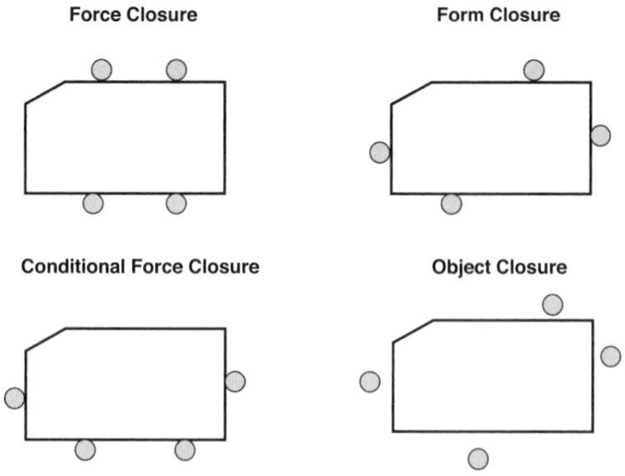

Fig. 2. The caging paradigm only requires that the object be confined to some compact set in the plane. The requirements for object closure are less stringent than form or force closure.

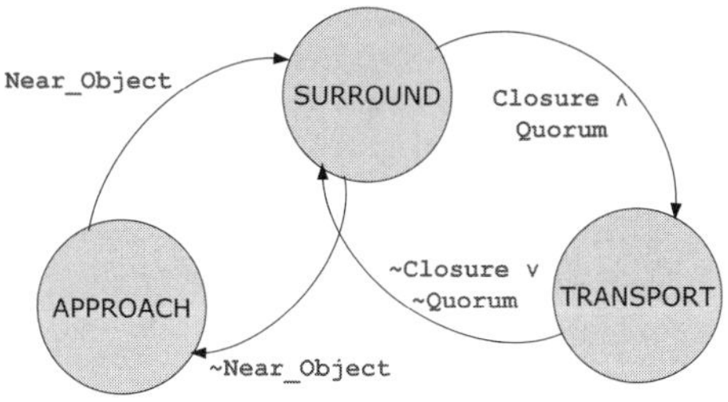

Fig. 3. Behavior Architecture. The software for each robot is identical and consists of several simple modes and the sequential composition of these modes.

specific task, scalability is of primary concern. This means: (a) Control computations must be decentralized; (b) Each robot must only rely on local information; (c) Robot controllers must be simple (the performance of a complex model-based controller does not always degrade gracefully with respect to variations in the environment). Requirement (c) essentially means that formation control based methods (for example, [3]) that rely on accurate dynamic modeling of the robots and the objects being manipulated cannot be used. In this paper, we rely on the formations that maintain closure around the object and then transport the object simply by moving along a desired trajectory while maintaining closure. See Figure 2.

A. Behaviors

Our approach to caging and manipulation of objects is summarized in the behavior architecture in Figure 3. The architecture relies on three behaviors.

1) **Approach**: The robot approaches the object while avoiding collisions with obstacles and other robots in the environment;
2) **Surround**: The robot stabilizes to a trajectory that orbits the object while avoiding collisions with other robots; and
3) **Transport**: The robot moves toward the goal configuration and/or tracks a reference trajectory that is derived from the object's reference trajectory.

As shown in the figure, the transitions between these behaviors are based on very simple conditions derived from

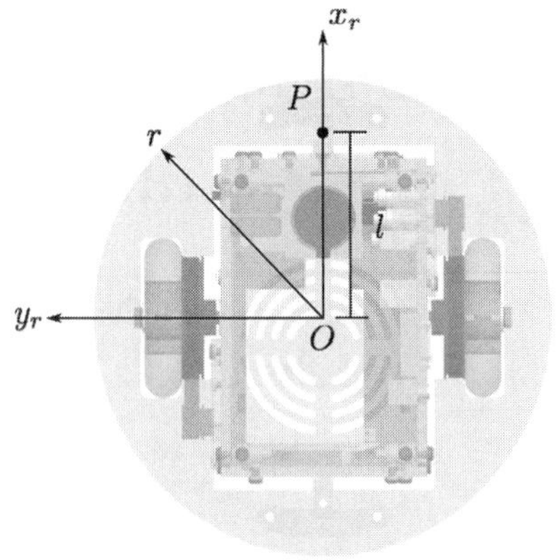

Fig. 4. A top view of the robot showing the body-fixed coordinate system. P is a reference point on the robot whose position is regulated by the vector fields.

simple sensor abstractions. If a robot is near an object, a sensor sets its `Near_Object` flag causing the robot to switch to the SURROUND mode. A `Quorum` flag is set based upon the number of neighbors within their field of view. The `Closure` flag is set when the robot network surround the object. When `Closure` and `Quorum` are both set, the robot enters the TRANSPORT mode and starts transporting the object. As the figure shows, reseting the flags can cause the robot to regress into a different mode.

B. Robot Model

We consider a simple model of a point robot with coordinates (x, y) in the world coordinate system. In the differential-drive robot in Figure 4, these are the coordinates of a reference point P on the robot which is offset from the axle by a distance l. We consider a simple kinematic model for this point robot:

$$\begin{aligned} \dot{x} &= u_1, \\ \dot{y} &= u_2, \end{aligned} \tag{1}$$

with the understanding that velocities of the reference point can be translated to commanded linear and angular velocities for the robot through the equations:

$$\begin{bmatrix} \dot{x} \\ \dot{y} \end{bmatrix} = \begin{bmatrix} \cos\theta & -l\sin\theta \\ \sin\theta & l\cos\theta \end{bmatrix} \begin{bmatrix} v \\ \omega \end{bmatrix}. \tag{2}$$

It is well-known if a robot's reference point is at point P, and if r is the radius of a circle circumscribing the robot, all points on the robot lie within a circle of radius is $l + r$ centered at P. In other words, if the reference point tracks a trajectory $(x_d(t), y_d(t))$, the physical confines of the robot are within a circle of radius $l + r$ of this trajectory. In what follows, we will use the simple kinematic model of Equation (1) to design vector fields for our controllers relying on our ability to invert the model in Equation (2) for $l \neq 0$ to implement controllers on the real robot.

We will assume that each robot can sense the current relative position of the manipulated object and the state of its neighbors through local communication if necessary.

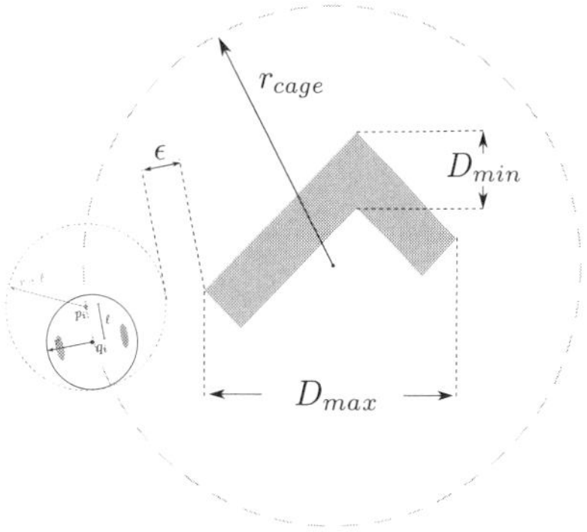

Fig. 5. An L-shaped object with circular caging shape.

C. Manipulated Object and Task

In this paper, we will consider both convex and concave objects for manipulation. But to keep the controllers simple, we will only use robot formations that are circular in shape. We will require robots to converge to circular trajectories surrounding the object. The shape of this *caging* circle is defined by three parameters. First, two parameters are defined by the manipulated object. We define the minimum diameter of the object $D_{min}(obj)$ to be the smallest gap through which the object will fit and the maximum diameter $D_{max}(obj)$ to be the maximum distance between any two points in the object. The third parameter, ϵ, is a tolerance that in turn specifies the radius of the caging circle:

$$r_{cage} = \frac{1}{2}D_{max}(obj) + (r + l) + \epsilon. \tag{3}$$

Setting ϵ too small will lead to robots bumping into the object and potentially excessive frictional forces leading to jamming. Setting ϵ too large leads to errors in following the desired trajectory. These parameters are depicted in Figure 5. For this work, we will assume all three parameters are known to all robots.

D. Command and control

While our system architecture relies heavily on autonomous agents making control decisions in a decentralized way based only on local sensing, it also depends upon a supervisory agent (which could be a human operator) that can provide task descriptions and feedback on task completion via broadcast communication. We will require that this supervisory agent can talk to all agents through broadcast commands. The agent knows the task and is able to plan trajectories for the manipulated object. However, it does not know specifics in terms of the number of robots or their initial configurations. Indeed, it does not rely on identifying individual robots in the team. This agent specifies the desired trajectory for the manipulated object and provides information on the three parameters characterizing the task ($D_{min}(obj)$, $D_{max}(obj)$, ϵ) using broadcast commands.

Our framework is inherently decentralized — individual robots make decisions on which behaviors to employ and when to take state transitions based only on local sensing. However, with a broadcast architecture it is possible for the supervisory agent to observe the global state of the system (from say an over head camera network) and provide some global information allowing the team to coordinate the robots. For example, flags could be concurrently set or reset for all the robots through broadcast commands ensuring synchronization. Note that this can be done without identifying or even enumerating the number of robots. And the amount of broadcast information is minimal and independent of the size of the team.

III. Robot Control

As shown in Figure 3, all robots are identical and the control software for each robot consists of several simple modes (behaviors) and a rule set that enables the sequential composition of these modes. In this section we will describe the distinct behaviors necessary to achieve the task of caging and manipulation and this rule set for transitions between modes.

A. Shape Control

The objective of our shape-controller and the behaviors derived from it is to surround and cage the object so that it cannot escape as the shape is moved through the workspace. Generally, the caging shape could be any enlarged approximation of the object to be manipulated but as mentioned earlier, we use a circular caging shape with radius r_{cage}.

Distributed controllers for general implicitly defined shapes are given in [10, 11]. We base our behaviors on the controller presented in [11] since it provides an orbiting component that synthesizes the Surround behavior. The basic idea is as follows.

For a desired shape given by $s(x, y) = 0$ with $s(x, y) < 0$ for (x, y) inside $\partial\mathcal{S}$ and $s(x, y) > 0$ for (x, y) outside $\partial\mathcal{S}$, we consider $\gamma = s(x, y)$ and define the navigation function

$$\varphi(q) = \frac{\gamma^2}{\gamma^2 + \beta_0}, \tag{4}$$

where $\beta_0 = R_0 - \|q\|^2$ is a representation of the world boundary. The function φ is positive-semidefinite, zero only when $s(x, y) = 0$, uniformly maximal ($\varphi = 1$ on boundary of workspace), and real analytic.

If $\psi = [0 \quad 0 \quad \gamma]^T$ such that $\nabla \times \psi$ is a vector tangent to the level set curves of φ, a decentralized control law is given by:

$$u_i = -\nabla_i \varphi_i \cdot f(N_i) - \nabla_i \times \psi_i \cdot g(T_i), \tag{5}$$

where $f(N_i)$ and $g(T_i)$ are functions used to modulate the agent's velocity towards $\partial\mathcal{S}$ and along the level sets of φ to avoid collisions with other agents via local sensing.

To construct the inter-agent terms, consider the scalar functions

$$N_{ij}(k) = \frac{(q_i - q_j)^T(-\nabla_j \varphi_j)}{\|q_i - q_j\|^k - (r_i + r_j)^k}, \tag{6}$$

$$T_{ij}(k) = \frac{(q_i - q_j)^T(-\nabla_j \times \psi_j)}{\|q_i - q_j\|^k - (r_i + r_j)^k}, \tag{7}$$

where k is a positive even number and r_i is the radius of the ith element. Incorporating two switching functions

$$\sigma_+(w) = \frac{1}{1+e^{1-w}}, \quad (8)$$

$$\sigma_-(w) = \frac{1}{1+e^{w-1}}, \quad (9)$$

we can define the functions $f(N_i)$ and $g(T_i)$ to be:

$$f(N_i) = \sigma_+(N_i), \quad (10)$$

$$g(T_i) = 1 - \sigma_-(T_i), \quad (11)$$

where N_i and T_i are given by

$$N_i = \sum_{j \in \mathcal{N}_i} \left(\frac{\sigma_+(N_{ij}(2))}{\|q_i - q_j\|^2 - (r_i + r_j)^2} - \frac{\sigma_-(N_{ij}(4))}{\|q_i - q_j\|^4 - (r_i + r_j)^4} \right) \quad (12)$$

$$T_i = \sum_{j \in \mathcal{N}_i} \left(\frac{\sigma_+(T_{ij}(2))}{\|q_i - q_j\|^2 - (r_i + r_j)^2} - \frac{\sigma_-(T_{ij}(4))}{\|q_i - q_j\|^4 - (r_i + r_j)^4} \right) \quad (13)$$

Note that $f(N_i)$ and $g(T_i)$ have been constructed so that as q_i approaches q_j, $f(N_i) \to 0$ and $g(T_i) \to 0$.

Several compelling results have been shown for this controller. First, it is scalable - each controller has a computational complexity that is linear in the number of neighbors $|\mathcal{N}_i|$. Second, the system is safe ensuring that no collisions can occur between robots. Note that we do allow contact between the robot and the object — indeed it is essential to do so for manipulation. Finally, the stability and convergence properties of the controller guarantee the robots converge to the desired shape, provided certain conditions relating the maximum curvature of the boundary to the robot geometry and the number of robots are met.

In summary, the basic shape controller for agent i is given by

$$u_i = -K_N \nabla_i \varphi_i \cdot f(N_i) - \nabla_i \times \psi_i \cdot g(T_i). \quad (14)$$

The gain K_N controls the rate of descent to the specified surface relative the orbiting velocity and is used to help generate different behaviors.

B. Approach

The APPROACH behavior is characterized by a larger gain K_N on the gradient descent component of the controller to yield agent trajectories that efficiently approach the object from a distance while avoiding collisions with neighboring agents.

C. Surround

In the SURROUND mode, agents are near the desired shape and K_N is decreased so that the agents are distributed around the object. Given enough robots, this behavior will lead to object closure. For a given r_{cage} and $D_{min}(obj)$, the minimum necessary number of robots to acheive object closure is

$$n_{min} = \frac{2\pi r_{cage}}{2r + D_{min}(obj)}. \quad (15)$$

We make the assumption that there will always be at least n_{min} robots available. Additionally, for the shape controller convergence guarantees given in [11] to hold, we must maintain that no more than

$$n_{max} = \frac{\pi r_{cage}}{r} \quad (16)$$

agents attempt to surround the shape. In practice however, this is not a stringent requirement. As we will see, if the number of robots is greater than this number, because the state transitions are robust to the number of robots, the additional robots do not disrupt the caging property.

D. Transport

The controller for the TRANSPORT mode relies on the parameterization of the smooth shape $s(x, y)$ with the reference trajectory. Given a trajectory for the object $(x^d_{obj}(t), y^d_{obj}(t))$, the reference trajectory for the shape is written as

$$s(x - x^d_{obj}(t), y - y^d_{obj}(t)) = 0.$$

The vector field for the robots is otherwise unchanged. The reference trajectory adds a time-varying component to the vector field that is computed independently by each robot. The reference trajectory is computed by the supervisory agent and can be modified on the fly if needed because the broadcast architecture allows this modified trajectory to be communicated to all the robots.

E. Composition of Behaviors

As described above, our manipulation system for multiple robots is based on a decentralized shape controller with proved stability and convergence results [10, 11]. By varying certain properties of this controller, each mobile agent in the system can operate in one of several modes including APPROACH, SURROUND, and TRANSPORT. Composition of these controller modes results in a global (non smooth) vector field that, when combined with local interactions, achieves the desired task (see Figure 6). Transitions between these modes as well as exceptions in the case of failure can be defined that allow the system to be robust while keeping individual agent decisions a function of local sensing.

In general, each agent's transition between modes will result from local observations of neighbors as well as the current distance to the manipulated object. By utilizing mode transitions that rely on locally sensed data, we can be confident this system is deployable with larger numbers of agents without any modifications.

An agent will initialize to the APPROACH mode if $d_{\mathrm{obj}}(q_i) > d_{\mathtt{near\_object}}$ (i.e. it is far from the object). As the agent approaches the desired caging shape, $d_{\mathrm{obj}}(q_i) \leq d_{\mathtt{near\_object}}$ will result in a transition to the SURROUND mode.

In the SURROUND mode, the orbiting term of the shape controller will be favored so that the robots are distributed around the object to be manipulated. Given at least n_{min} agents, this mode will converge on an equilibrium where object closure is attained. While closure is a global property of the system, it can be determined in a distributed manner by computing homology groups for the network [12, 13]. Alternately, we define an algorithm whereby closure can be locally estimated.

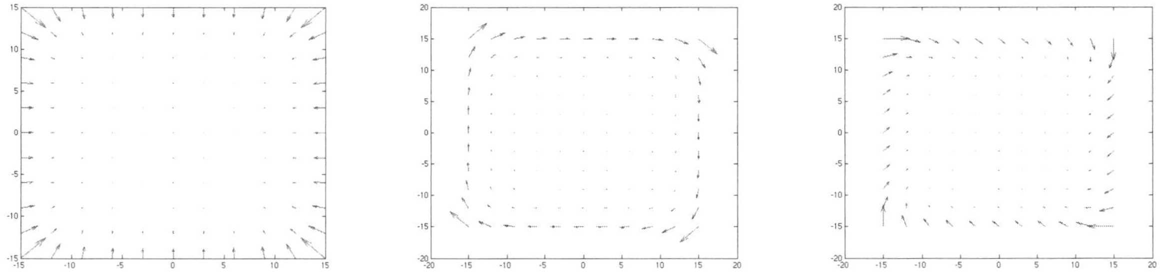

(a) (b) (c)

Fig. 6. The two components of the vector field, the gradient descent and rotational vector fields (Figure 6(a) and Figure 6(b), respectively). The resulting composition is shown in Figure 6(c).

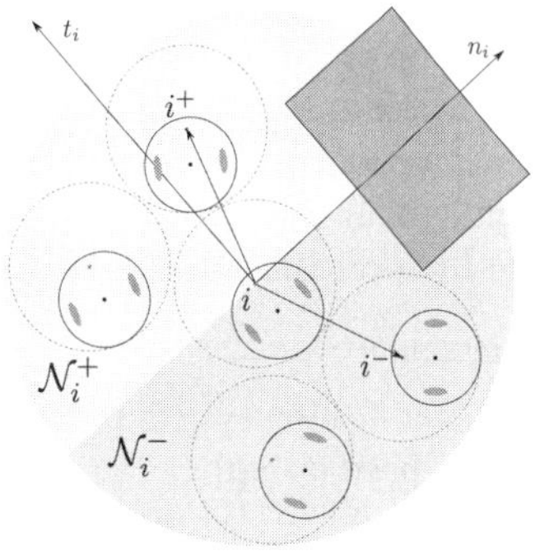

Fig. 7. Agent i's neighborhoods $\mathcal{N}_i^+$ and $\mathcal{N}_i^-$ with i^+ and i^-

1) Quorum: In order to locally define quorum, we will introduce the concept of a *forward* neighborhood $\mathcal{N}^+$ and a *backward* neighborhood $\mathcal{N}^-$ with respect to the manipulated object and the SURROUND mode. For agent i, define the normal and tangential unit vectors, $\mathbf{n}_i$ and $\mathbf{t}_i$, based on the vectors $\nabla_i \varphi_i$ and $\nabla_i \times \psi_i$ respectively. Recall that the SURROUND mode includes an approach component (in the direction $\mathbf{n}_i$) and rotation component (in the direction $\mathbf{t}_i$). Thus we can define some set of robots to be *in front of* agent i and another set *behind*. If a neighborhood N_i represents the agents within a distance $D_{min}(obj)$, then

$$\mathcal{N}_i^+ = \{j \in \mathcal{N}_i \mid 0 < (q_j - q_i)^T t_i\}, \quad (17)$$

and

$$\mathcal{N}_i^- = \{j \in \mathcal{N}_i \mid 0 > (q_j - q_i)^T t_i\}, \quad (18)$$

Furthermore, we can define agents from $\mathcal{N}_i^+$ and $\mathcal{N}_i^-$,

$$i^+ = \text{argmax}_{k \in \mathcal{N}_i^+} \frac{(q_k - q_i)^T n_i}{\|q_k - q_i\|}, \quad (19)$$

$$i^- = \text{argmax}_{k \in \mathcal{N}_i^-} \frac{-(q_k - q_i)^T n_i}{\|q_k - q_i\|}, \quad (20)$$

that will be the adjacent agents in the potential cage around the object as depicted in Figure 7.

Remembering that our agents only have the ability to observe/communicate their neighbor's state (including the value of the quorum variable), we propose the following update equation for $\texttt{quorum}_i$,

$$\texttt{quorum}_i = \begin{cases} 0 & \text{if } (\mathcal{N}_i^+ = \emptyset) \vee (\mathcal{N}_i^- = \emptyset), \\ n_{min} & \text{if } f(i^+, i^-) > n_{min}, \\ f(i^+, i^-) & \text{otherwise,} \end{cases} \quad (21)$$

with $f(j,k) = \min(\texttt{quorum}_j, \texttt{quorum}_k) + 1$. $\texttt{quorum}_i$ is a measure of how many robots are near agent i and in position to perform a manipulation task. Note that n_{min} is a heuristic bound on the quorum variable and there may be better choices. We shall use $\texttt{quorum}_i$, $\texttt{quorum}_{i^+}$, and $\texttt{quorum}_{i^-}$ to determine when there is object closure.

2) Closure: If there is no closed loop around the object, $\texttt{quorum}_i$ will converge to the minimum of n_{min} and the shorter of the forward/backward chain of agents. On the other hand, if there is a complete loop around the object, $\texttt{quorum}_i$ will grow as large as the imposed bound n_{min}.

We will define *local* closure to be

$$\begin{aligned} \texttt{closure}_i = &(\texttt{quorum}_i \geq n_{min}) \wedge \\ &(\texttt{quorum}_i = \texttt{quorum}_{i^+}) \wedge \\ &(\texttt{quorum}_i = \texttt{quorum}_{i^-}). \end{aligned} \quad (22)$$

Our condition for *local* closure will coincide with *global* closure for any situation where up to $2n_{min}$ agents are used to form a cage around the object (which should not happen if agents are correctly controlling onto the specified caging shape).

When an agent estimates that *local* closure has been attained, it will switch in the TRANSPORT mode and begin attempting manipulation of the object. The quorum and closure events are defined such that they represent a kind of distributed consensus and as a result a set of manipulating agents will switch into the TRANSPORT mode in a nearly simultaneous fashion.

During manipulation an exception will occur if closure is lost and each agent in the system will return to the SURROUND mode to reaquire the object.

IV. EXPERIMENTAL TESTBED

The methodology for manipulation discussed in the preceeding sections was implemented in both simulation and on

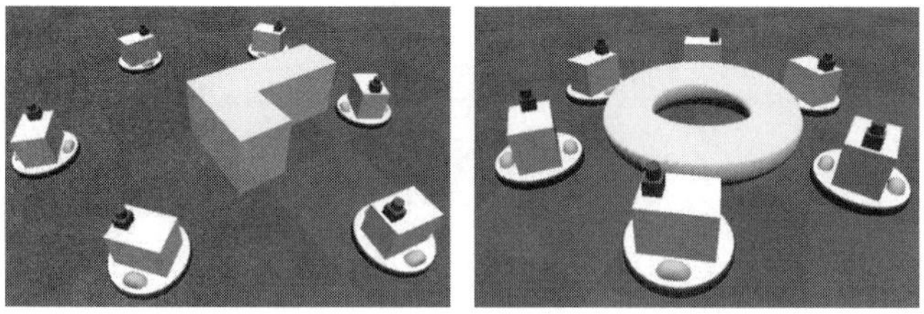

Fig. 8. Two images captured during GAZEBO simulations. Figure 8(a) shows six SCARAB models manipulating an "L-shaped" object in a physically correct environment. Figure 8(b) depicts the simulated robots moving a round object of the same shape as the inner-tube object shown in Figure 13(h).

hardware. We describe here the experimental testbed developed for testing scalable distributed algorithms that was used to test the proposed methodology.

PLAYER, an open source networking middleware and part of the PLAYER/STAGE/GAZEBO project [14] interfaces the distributed system and provides communication between the robots of the system. Additionally, PLAYER provides a layer of hardware abstraction that permits algorithms to be tested in simulated 2D and 3D environments (STAGE and GAZEBO, respectively) and on hardware. For this reason, all algorithm implementations discussed in Section V, both in simulation and on hardware were the same.

A. Simulation Environment

The open source 3D simulator GAZEBO was used to verify the algorithm. GAZEBO incorporates dynamic interactions between models via the *Open Dynamics Engine* [15]. Models of the environment of the local laboratory and hardware (discussed in Section IV-B) were reproduced in a simulated world. The robot models accurately reflect the geometric, kinematic, and dynamic descriptions of the local robots used in the hardware implementation. Frictional coefficients between the agents and the manipulated object were set to realistic values (as feasible via the *Open Dynamics Engine*).

B. Robot Platform

Our small form-factor robot is called the SCARAB. The SCARAB is a $20 \times 13.5 \times 22.2$ cm^3 indoor ground platform with a fender for manipulation with a diameter of 30 cm. Each SCARAB is equipped with a differential drive axle placed at the center of the length of the robot with a 21 cm wheel base. Each of the two stepper motors drive 10 cm (standard rubber scooter) wheels with a gear reduction (using timing belts) of 4.4:1 resulting in a nominal holding torque of 28.2 kg-cm at the axle. The weight of the SCARAB as shown in Figure 9(a) is 8 kg.

Each robot is equipped with a Nano ITX motherboard with a 1 GHz. processor and 1 GB of ram. A power management board and smart battery provide approximately two hours of experimentation time (under normal operating conditions). A compact flash drive provides a low-energy data storage solution. The on-board embedded computer supports IEEE 1394 firewire and 802.11a/b/g wireless communication.

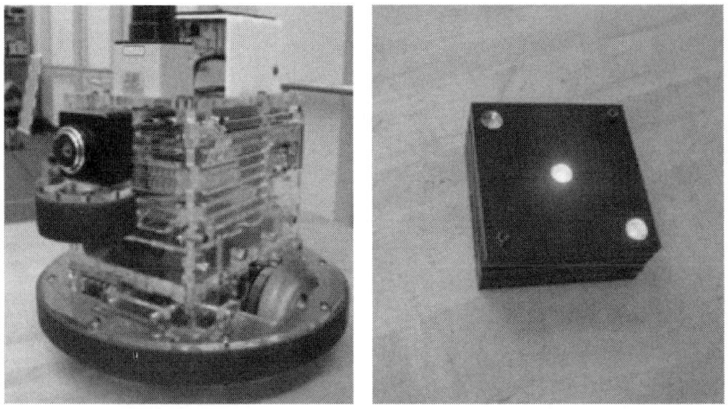

Fig. 9. The $20 \times 13.5 \times 22.2$ cm^3 SCARAB platform is shown in Figure 9(a). Figure 9(b) depicts a LED target used for localization.

The sensor models on the SCARAB are a Hokuyo URG laser range finder, a motor controller that provides odometry information from the stepper motors, and power status (through the power management board). Additionally, each robot is able to support up to two firewire cameras.

C. Other Instrumentation

The ground-truth verification system permits the tracking of LED markers with a position error of approximately 2.5 cm and an orientation error of $5°$. The tracking system consists of LED markers and two overhead IEEE 1394 Point Grey Color Dragonfly cameras.

The LED marker contains three LEDs of the colors red, green, and blue. The red and blue LEDs maintain continuous illumination which are detected and tracked by the overhead cameras. The green LED flashes an eight bit pattern at a fixed interval with error checking. The pattern defines an identification number associated with each robot.

D. Algorithm Implementation

Every robot is running identical modularized software with well defined abstract interfaces connecting modules via the PLAYER robot architecture system. We process global overhead tracking information but hide the global state of the system from each robot. In this way, we use the tracking system in lieu of an inter-robot sensor implementation. Each robot receives only its state and local observations of its neighbors. An overview of the system implementation is shown in Figure 10.

V. RESULTS

The algorithms and framework presented in Sections II and III were tested through simulation and hardware implementation.

A. Simulation

We used the simulation environment to quantifiy the effectiveness of the algorithms for different object shapes and numbers of agents. The task specified was simply to manipulate the object within a distance $\epsilon + \ell$ of a waypoint. Failure was identifed by simulations that did not complete the task

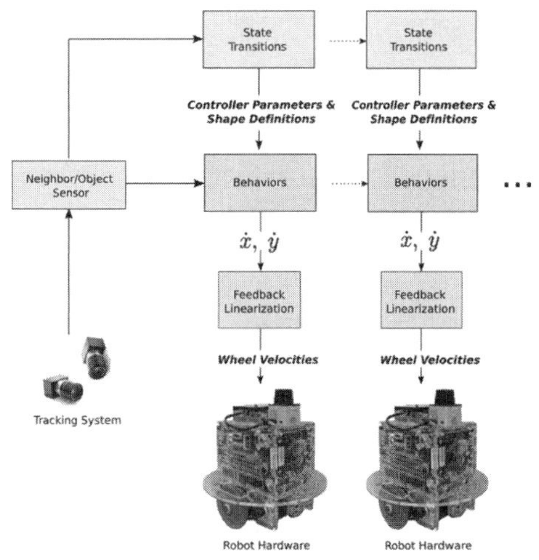

Fig. 10. Overview of the system implementation.

	Torus shape	L shape
D_{min}	1.0	0.5
D_{max}	1.0	1.0
n_{min}	4	6
n_{max}	12	12

TABLE I

OBJECT PARAMETERS (DIMENSIONS IN METERS).

within some conservative timeout period. Repeated trials in simulation were conducted for N robots manipulating both an L-shaped and torus shaped object with N ranging from n_{min} to 20. Parameters for each object are listed in Table I.

Success results from many trials in simulation are enumerated in Table II. First, it should be noted that failures only occur when $n > n_{max}$. When $n > n_{max}$ our shape controller cannot provide convergence guarantees to the specified shape which can lead to the agents orbiting a circle with radius larger than r_{cage}. With a larger caging radius, object placement within the $\epsilon + \ell$ bounds is not guaranteed and the task will not always succeed.

Representative images from a simulation trial with the L-shaped object are shown in Figures 13(a)–13(e). Note that since a trial is considered to be successful when the object is within $\epsilon + \ell$ of the goal, there is no meaningful error metric that can be reported.

B. Experimental

The methodology for manipulation was tested on the experimental testbed presented in Section IV. Four to eight SCARAB robots were instructed to manipulate a rubber inner-tube with diameter of 0.60 m and an L-shaped object with

		Success	Trials
Torus	$n_{min} \leq n \leq n_{max}$	100%	63
	$n_{max} < n \leq 20$	90%	40
L	$n_{min} \leq n \leq n_{max}$	100%	70
	$n_{max} < n \leq 20$	95%	80

TABLE II

SIMULATION RESULTS

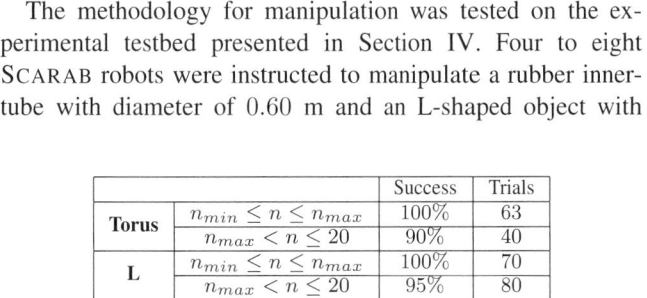

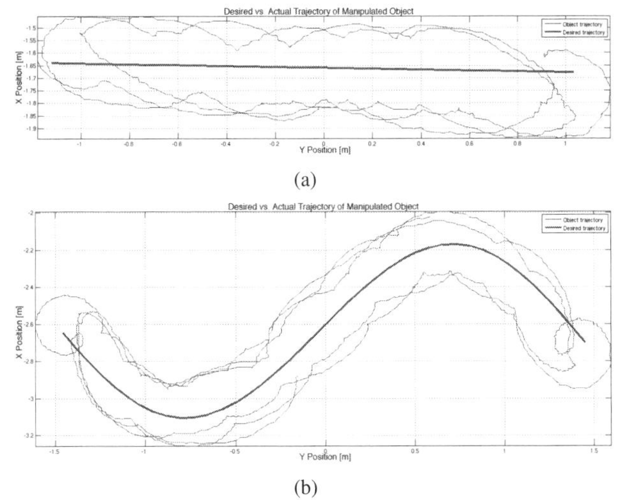

Fig. 11. Experimental Results. The linear (Figure 11(a)) and sinusoidal (Figure 11(b)) reference trajectories for the inner-tube (shown in red) and the actual trajectories traced by the geometric center of the inner-tube (shown in blue).

$D_{min} = 0.6$ m, $D_{max} = 1.2$ m. Figures 13(f)–13(j) show the robots manipulating the inner-tube object during a trial run.

Two different types of trajectories, linear and sinusoidal, were tested over ten trials. A subset of these trajectories as well as the resulting inner-tube trajectories (as defined by the center of the inner-tube) are shown in Figure 11. During experimentation no failures occurred where the object escaped form-closure. These results as well as the mean squared error (MSE) are provided in Table III.

Trajectory	Success Rate (%)	Average MSE (m)
Linear	100	0.33
Sinusoidal	100	0.38

TABLE III

EXPERIMENTATION RESULTS USING FOUR SCARABS OVER TEN TRIALS.

The robots were instructed to enclose the object in a circular shape with a radius $r_{cage} = 0.58$ m for both the linear and sinusoidal trajectories. These values (given the fender dimensions and error via the tracking system) correspond to the computed MSE.

The individual trajectories of each of the robots during the approach, surround, and transport modes for a linear trajectory are shown in Figure 12.

VI. DISCUSSION

We present a framework, the architecture, and algorithms for multi-robot caging and manipulation. The team of robots forms patterns that result in the manipulated object being trapped and dragged or pushed to the goal configuration. The controllers are obtained by sequential composition of vector fields or behaviors and enable decentralized computation based only on local information. Further, the control software for each robot is identical and relies on very simple behaviors. We present our experimental multi-robot system and simulation and experimental results that demonstrate the robustness of this approach.

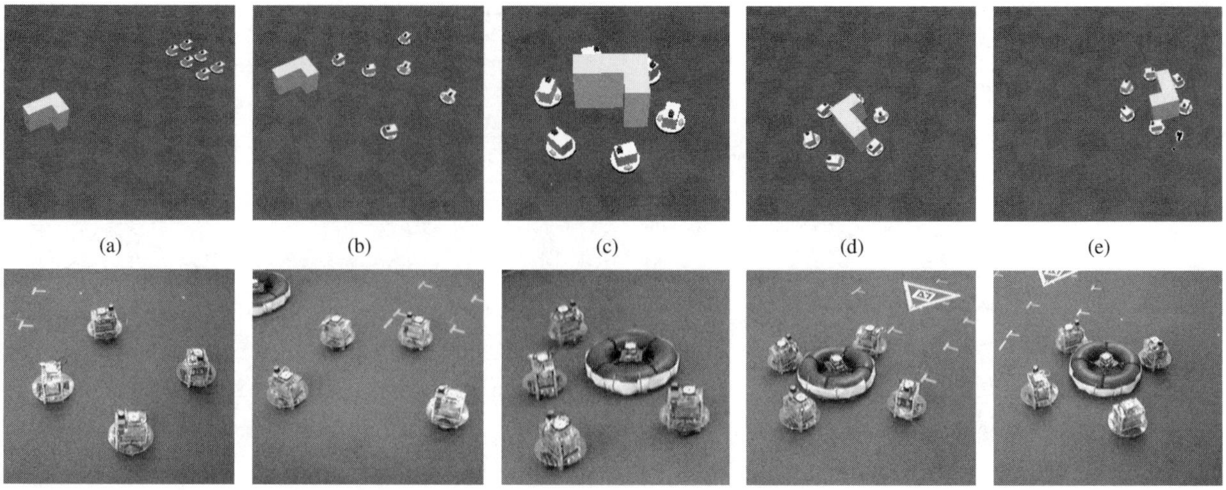

(a) (b) (c) (d) (e)

(f) (g) (h) (i) (j)

Fig. 13. A representative trial run is simulated in GAZEBO including (a) the starting formation, (b) approach, (c) surround, and (d-e) transport behaviors. (f-j) depict a similar scenario but with four SCARAB robots. The object being manipulated is a 0.60 m diameter rubber inner-tube.

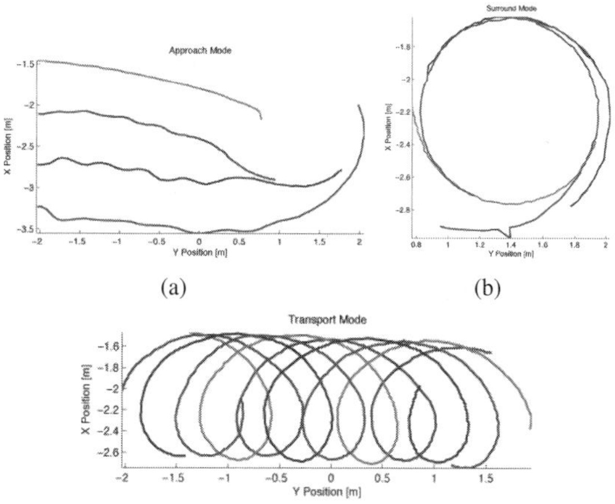

(a) (b)

(c)

Fig. 12. Experimental trajectories of individual SCARAB robots during (a) approach, (b) surround, and (c) transport modes.

While the individual controllers and behaviors APPROACH, SURROUND and TRANSPORT have stability and convergence properties, there are no formal guarantees for the switched system, especially since each robot may switch behaviors at different times. However, our experimental results with circular objects and dynamic simulation results with objects based on hundreds of trials with more complex shapes show that this approach is robust. Though the experimental results presented in this paper were generated by transitions derived from global information, the simulated results relied on autonomous transitions and local estimates of closure. In the future we are interested in exploring exact algorithms to determine closure with without any metric information based only on local proximity sensing as in [12].

VII. ACKNOWLEDGMENT

This research was supported by: NSF grants CCR02-05336, NSF IIS-0413138, and IIS-0427313, and ARO Grants W911NF-04-1-0148 and W911NF-05-1-0219.

REFERENCES

[1] K. Kosuge, Y. Hirata, H. Asama, H. Kaetsu, and K. Kawabata, "Motion control of multiple autonomous mobile robot handling a large object in coordination," in *Proc. of the IEEE Int. Conf. on Robotics and Automation*, Detroit, Michigan, May 1999, pp. 2666–2673.

[2] D. Rus, "Coordinated manipulation of objects in a plane," *Algorithmica*, vol. 19, no. 1, pp. 129–147, 1997.

[3] T. Sugar and V. Kumar, "Multiple cooperating mobile manipulators." in *Proc. of the IEEE Int. Conf. on Robotics and Automation*, Detroit, Michigan, May 1999, pp. 1538–1543.

[4] J. K. Salisbury and B. Roth, "Kinematic and force analysis of articulated hands," *ASME Journal of Mechanisms, Transmissions, and Automation in Design*, vol. 104, no. 1, pp. 33–41, 1982.

[5] M. J. Mataric, M. Nilsson, and K. Simsarian, "Cooperative multi-robot box-pushing," in *Proc. of the IEEE/RSJ Int. Conf. on Intelligent Robots and Systems*, Pittsburgh, Pennsylvania, August 1995, pp. 556–561.

[6] K. M. Lynch and M. T. Mason, "Stable pushing: Mechanics, controllability, and planning," *The Int. Journal of Robotics Research*, vol. 16, no. 6, pp. 533–556, 1996.

[7] A. Sudsang and J. Ponce, "Grasping and in-hand manipulation: Experiments with a reconfigurable gripper," *Advanced Robotics*, vol. 12, no. 5, pp. 509–533, 1998.

[8] G. A. S. Pereira, V. Kumar, and M. F. Campos, "Decentralized algorithms for multi-robot manipulation via caging," *The Int. Journal of Robotics Research*, vol. 23, no. 7/8, pp. 783–795, 2004.

[9] P. Song and V. Kumar, "A potential field based approach to multi-robot manipulation," in *Proc. of the IEEE Int. Conf. on Robotics and Automation*, Washington, DC, May 2002, pp. 1217–1222.

[10] L. Chaimowicz, N. Michael, and V. Kumar, "Controlling swarms of robots using interpolated implicit functions," in *Proc. of the IEEE Int. Conf. on Robotics and Automation*, Barcelona, Spain, April 2005, 2487-2492.

[11] M. A. Hsieh, S. G. Loizou, and V. Kumar, "Stabilization of multiple robots on stable orbits via local sensing," in *Proc. of the IEEE Int. Conf. on Robotics and Automation*, Rome, Italy, April 2007, To Appear.

[12] V. de Silva and R. Ghrist, "Homological sensor networks," *Notices of the American Mathematical Society*, vol. 54, no. 1, pp. 10–17, 2007.

[13] A. Muhammad and A. Jadbabaie, "Decentralized computation of homology groups in networks by gossip," in *Proc. of the American Control Conf.*, New York, July 2007, To Appear.

[14] B. Gerkey, R. T. Vaughan, and A. Howard, "The Player/Stage project: Tools for multi-robot and distributed sensor systems," in *Proc. of the Int. Conf. on Advanced Robotics*, Coimbra, Portugal, June 2003, pp. 317–323.

[15] "Open Dynamics Engine," http://www.ode.org.

Closed Loop Control of a Gravity-assisted Underactuated Snake Robot with Application to Aircraft Wing-Box Assembly

Binayak Roy and H. Harry Asada
d'Arbeloff Laboratory for Information Systems and Technology
Department of Mechanical Engineering
Massachusetts Institute of Technology
Cambridge, MA 02139, USA
{binayak, asada}@mit.edu

Abstract— **Stable, nonlinear closed-loop control of a gravity-assisted underactuated robot arm with 2nd order non-holonomic constraints is presented in this paper. The joints of the hyper articulated arm have no dedicated actuators, but are activated with gravity. By tilting the base link appropriately, the gravitational torque drives the unactuated links to a desired angular position. With simple locking mechanisms, the hyper articulated arm can change its configuration using only one actuator at the base. This underactuated arm design was motivated by the need for a compact snake-like robot that can go into aircraft wings and perform assembly operations using heavy end-effecters. The dynamics of the unactuated links are essentially 2nd order non-holonomic constraints, for which there are no general methods for designing closed loop control. We propose an algorithm for positioning the links of an n-link robot arm inside an aircraft wing-box. This is accomplished by sequentially applying a closed loop point-to-point control scheme to the unactuated links. We synthesize a Lyapunov function to prove the convergence of this control scheme. The Lyapunov function also provides us with lower bounds on the domain of convergence of the control law. The control algorithm is implemented on a prototype 3-link system. Finally, we provide some experimental results to demonstrate the efficacy of the control scheme.**

I. Introduction

Most assembly operations in aircraft manufacturing are currently done manually. Although aircraft are small in lot size, numerous repetitive assembly operations have to be performed on a single aircraft. The conditions are often ergonomically challenging and these result in low productivity as well as frequent injuries. Thus, there is a need to shift from manual assembly to automated robotic assembly. The following wing-box assembly illustrates this.

Fig. 1 shows a mock-up of the cross-section of an aircraft wing-box. Several assembly operations, such as burr-less drilling and fastener installations, have to be carried out inside the wing-box after the upper and lower skin panels are in place. The interior of the wing-box is accessible *only* through small portholes along its length. The portholes are roughly rectangular with dimensions of 45 cm by 23 cm. The wing-box also has a substantial span, which varies from 1 m to 3 m depending upon the size of the aircraft. The height of the wing-box varies from about 20 cm to 90 cm, depending upon the size of the aircraft. Presently, the assembly operations are carried out manually. A worker enters the wing-box through the small portholes and lies flat on the base, while carrying out the assembly operations. Evidently, the working conditions are ergonomically challenging.

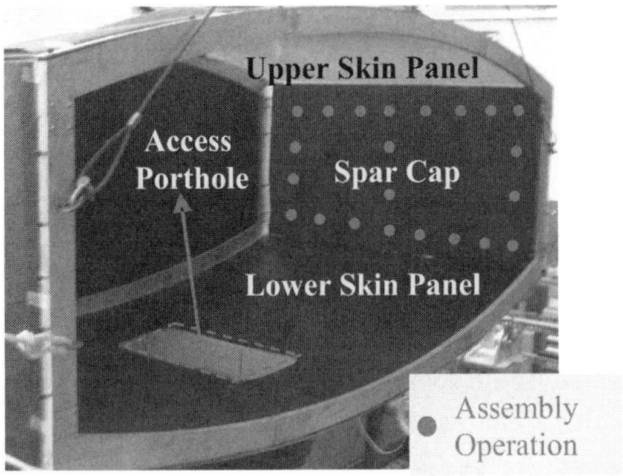

Fig. 1. Cross section of aircraft wing-box

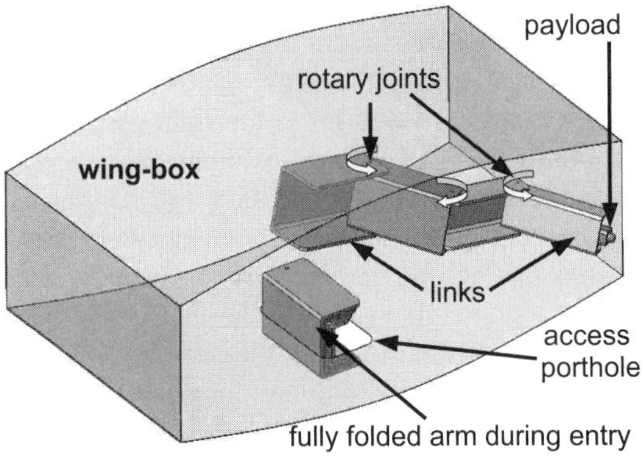

Fig. 2. Structure of robot arm

We have proposed a "Nested-Channel" serial linkage mechanism capable of operating inside an aircraft wing box [20].

The links are essentially C-channels with successively smaller base and leg lengths, as shown in Fig. 2. They are connected by 1 d.o.f rotary joints, the axes of which are parallel. The use of channel structures is advantageous for a number of reasons. The channels can fold into each other resulting in an extremely compact structure during entry through the porthole, as shown in Fig. 2. Once inside the wing-box, the links may be deployed to access distal points in the assembly space. The open channel structure also facilitates the attachment of a payload to the last link without increasing the overall dimensions of the arm.

The lack of a compact, powerful and high stroke actuation mechanism is the primary bottleneck in the development of the hyper articulated arm. In our previous work, we have proposed an underactuated design concept, which obviates the use of dedicated actuators for each joint. Instead, we utilize gravity for driving individual joints. This drastically reduces the size and weight of the manipulator arm. The methodology requires a single actuator for tilting the arm at the base. This single actuator can be placed *outside* the wing-box and can be used in conjunction with simple locking mechanisms to reconfigure the serial linkage structure.

The reconfiguration scheme is illustrated in Fig. 3, which shows a schematic of an n-link robot arm. The base link (link 1) is the only servoed link. It may be rotated about a fixed axis Z_0, which is orthogonal to the direction of gravity. All other joint axes (Z_j, $j \neq 0$) are orthogonal to Z_0. They are equipped with simple *on-off* locking mechanisms only. The goal is to rotate link i about Z_{i-1} by actuating link 1 appropriately. All unactuated links except link i are locked. Link 1 starts in the vertical upright position. Then it is rotated, first clockwise and then counter-clockwise, before being brought back to its vertical position. This tends to accelerate and then decelerate link i due to gravity and dynamic coupling with link 1. By controlling the tilting angle of link 1, link i can be brought to a desired position with zero velocity. Link i may be locked thereafter. This procedure can be repeated sequentially for the other unactuated links. Contraction of the arm can be performed by reversing the above deployment procedure.

A considerable amount of work has been done in the area of underactuated systems [3]-[10]. Most of the work in this area deals with the planar (vertical or horizontal) case where the actuated and unactuated joint axes are parallel. In our approach, the actuated and unactuated joints are orthogonal and we can modulate the effects of gravity by controlling the actuated joint. The presence of gravity renders the nonlinear system locally controllable, as can be seen from local linearization. This ensures that we can go from any initial point to any final point in the configuration space of the unactuated coordinate. However, it is inefficient to patch together local linear control laws to traverse the entire configuration space. Moreover, any control design must ensure that the range of motion of the actuated coordinate is small, because the arm operates inside an aircraft wing-box. Earlier approaches [8]-[10] to the control of underactuated systems generate constructive global control laws applied to specific systems. Such constructive control laws cannot be directly applied to our system.

In our earlier work [21], we have proposed several motion planning algorithms suitable for the gravity-assisted under-actuated robot arm. They include parameterized trajectory planning for the actuated joint and feed-forward optimal control. These are *open-loop* techniques and work well in the absence of disturbances. Also, an exact knowledge of the system dynamics is needed. In particular, a good estimate of Coulomb friction is necessary for accurate position control. However, it is unrealistic to assume prior knowledge of such state dependent unknown parameters. This necessitates the development of a *closed-loop* control strategy for our system.

In this paper, we first explore the system dynamics to develop an understanding of the relationship between the actuated and unactuated degrees of freedom. We make important approximations to capture the dominant effects in the system dynamics so as to facilitate control design. Next, we propose a closed loop control strategy for point to point control of the unactuated coordinate. We synthesize a Lyapunov function to prove the convergence of the control law. The Lyapunov function also provides us with lower bounds on the domain of convergence of the control law. Finally, we present some experimental results which demonstrate the efficacy of the control law.

II. System Dynamics

Fig. 3 shows a schematic of an n-link robot arm with one actuated (link 1) and $n-1$ unactuated links. $X_0 Y_0 Z_0$ denotes the World Coordinate Frame. The coordinate frames are attached according to the Denavit-Hartenberg convention with the ith coordinate frame fixed to the ith link. We seek rotation of link i ($i \geq 2$) about the axis Z_{i-1} by rotating link 1 about the horizontal axis Z_0. The angle θ_1 denotes the tilt of link 1 relative to the fixed vertical axis X_0 and the angle θ_i denotes the angular position of link i relative to link $i-1$.

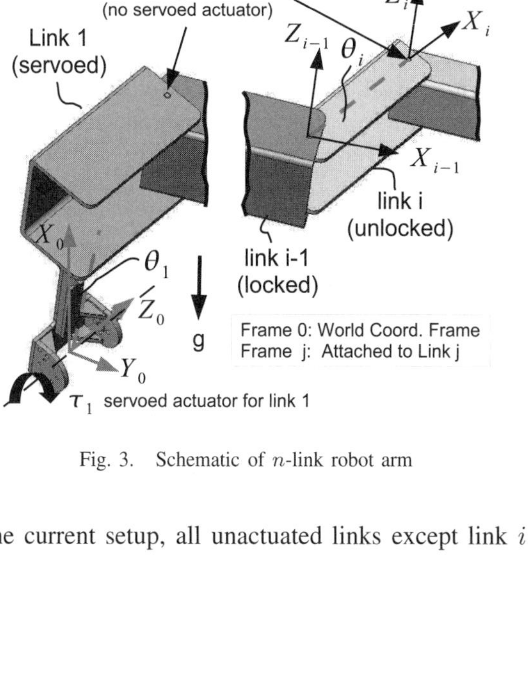

Fig. 3. Schematic of n-link robot arm

In the current setup, all unactuated links except link i are locked. The system dynamics may be written as:

$$\begin{bmatrix} H_{11} & H_{i1} \\ H_{i1} & H_{ii} \end{bmatrix} \begin{bmatrix} \ddot{\theta}_1 \\ \ddot{\theta}_i \end{bmatrix} + \begin{bmatrix} F_1 \\ F_i \end{bmatrix} + \begin{bmatrix} G_1 \\ G_i \end{bmatrix} = \begin{bmatrix} \tau_1 \\ 0 \end{bmatrix} \quad (1)$$

$$\theta_j = \theta_{j0} \quad j \neq 1, i \quad (2)$$

Here $q = [\theta_2, \ldots, \theta_n]^T$, $[H_{kl}(q)]$ is the $n \times n$ symmetric positive-definite inertia matrix, $[F_1(q, \dot{q}, \dot{\theta}_1), F_i(q, \dot{q}, \dot{\theta}_1)]^T$ represents the 2×1 vector of centrifugal and coriolis effects and $[G_1(q, \theta_1), G_i(q, \theta_1)]^T$ represents the 2×1 vector of gravitational effects. The torque on the actuated joint axis Z_0 is represented by τ_1. We note that θ_{j0} is a constant because the jth link ($j \neq 1, i$) is locked. Using $F_i(q, \dot{q}, \dot{\theta}_1) = f_i(q)\dot{\theta}_1^2$ and $G_i(q, \theta_1) = g_i(q) g \sin\theta_1$, the second row of (1) may be written as:

$$\ddot{\theta}_i = -\frac{H_{i1}(q)}{H_{ii}(q)}\ddot{\theta}_1 - \frac{f_i(q)}{H_{ii}(q)}\dot{\theta}_1^2 - \frac{g_i(q)}{H_{ii}(q)} g \sin\theta_1. \quad (3)$$

As shown in [3], (3) is a 2nd order non-holonomic constraint and thus cannot be integrated to express θ_i as a function of θ_1. Also, at any given time only one unactuated link (link i) is in motion. Thus, the n-link problem can be treated as a 2-link problem without loss of generality. Hereafter, to simplify the algebra, we deal exclusively with the 2-link problem. For the 2-link case, we may write (3) as:

$$\ddot{\theta}_2 = -\frac{H_{21}(\theta_2)}{H_{22}(\theta_2)}\ddot{\theta}_1 - \frac{f_2(\theta_2)}{H_{22}(\theta_2)}\dot{\theta}_1^2 - \frac{g_2(\theta_2)}{H_{22}(\theta_2)} g \sin\theta_1, \quad (4)$$

where:

$$\begin{aligned} H_{12} =& M_2(z_{c2} + d_2)(y_{c2}\cos\theta_2 + (x_{c2} + a_2)\sin\theta_2) \\ &+ I_{yz2}\cos\theta_2 + I_{xz2}\sin\theta_2, \quad (5) \\ H_{22} =& I_{zz2} + M_2((x_{c2} + a_2)^2 + y_{c2}^2), \quad (6) \\ f_2 =& I_{xy2}\cos 2\theta_2 + 0.5(I_{yy2} - I_{xx2})\sin 2\theta_2 \\ &+ M_2(a_1 + (x_{c2} + a_2)\cos\theta_2 - y_{c2}\sin\theta_2) \\ &((x_{c2} + a_2)\sin\theta_2 + y_{c2}\cos\theta_2), \quad (7) \\ g_2 =& - M_2((x_{c2} + a_2)\sin\theta_2 + y_{c2}\cos\theta_2). \quad (8) \end{aligned}$$

M_2 denotes the mass of link 2. I_{xy2} etc. denote the moments of inertia of link 2 about a centroidal coordinate frame. The parameters x_{c2}, y_{c2}, z_{c2} are the coordinates of the C.O.M of link 2 in the link-attached frame. Also, a_2, d_2 refer to the corresponding Denavit-Hartenberg parameters.

As seen in the next section, we may choose the control torque τ_1 in (1) so as to converge exponentially to any bounded trajectory for the actuated coordinate θ_1. We refer to θ_1 and its derivatives $(\dot{\theta}_1, \ddot{\theta}_1)$ in (4) as the *pseudo input*. The terms involving θ_2 (H_{12}/H_{22}, f_2/H_{22} and g_2/H_{22}) in (4) are referred to as the *modulating coefficients*. These *modulating coefficients* scale the various components of the *pseudo input* $(\theta_1, \dot{\theta}_1, \ddot{\theta}_1)$ depending on the position of the unactuated link 2.

Fig. 4 shows the variation of the *modulating coefficients* in the configuration space of the unactuated coordinate. The simulation is based on parameter values taken from a 2-link version of our prototype system shown in Fig. 7. The dominant term is the modulating coefficient due to gravity (g_2/H_{22}), followed by the contribution of the inertial coupling (H_{12}/H_{22}) and finally the contribution of the centrifugal coupling (f_2/H_{22}). In view of these observations, we make the following assumptions:

1. Inertial coupling is neglected.
2. Centrifugal coupling is neglected.

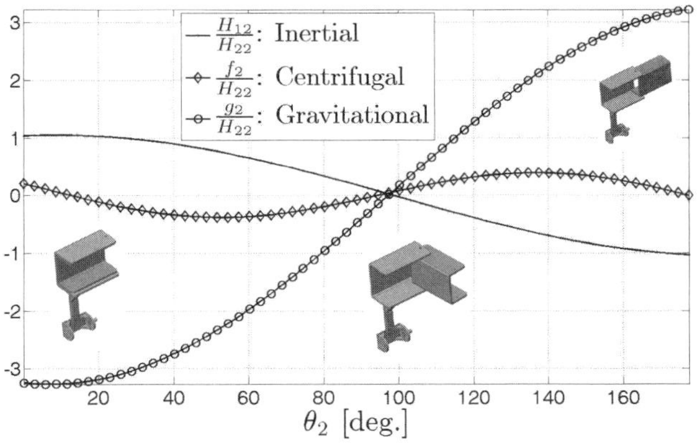

Fig. 4. Comparison of modulating coefficients over configuration space

These assumptions are valid as long as the gravitational component of acceleration $|g\sin\theta_1|$ is of the same (or higher) order of magnitude as compared to $|\ddot{\theta}_1|$ and $|\dot{\theta}_1^2|$. We validate these approximations *a posteriori* in the section on experimental results. Under these assumptions, the dynamics (4) may be simplified as:

$$\ddot{\theta}_2 = -\frac{g_2(\theta_2)}{H_{22}} g \sin\theta_1 \quad (9)$$

Using (6) and (8), we may write (9) as:

$$\ddot{\theta} = A \sin\theta \sin\theta_1, \quad (10)$$

where:

$$\begin{aligned} \theta &= \theta_2 + \alpha, \\ A &= \frac{M_2 g\sqrt{y_{c2}^2 + (x_{c2} + a_2)^2}}{I_{zz2} + M_2(y_{c2}^2 + (x_{c2} + a_2)^2)}, \\ \alpha &= \mathrm{atan2}(y_{c2}, x_{c2} + a_2). \end{aligned}$$

It is worthwhile to examine the physical significance of the dynamics (10). It represents a pendulum in a modulated "gravity" field. The strength of this field can be modulated as $g\sin\theta_1$ by controlling the angle θ_1. The pendulum behaves as a regular or inverted pendulum depending on the sign of $\sin\theta\sin\theta_1$. Also, the "gravity" field may be switched off by setting $\theta_1 = 0$. This gives rise to a continuum of equilibria given by $[\theta = \bar{\theta}, \dot{\theta} = 0, \theta_1 = 0]$, where $\bar{\theta}$ is arbitrary.

III. Closed Loop Control

A. Control Law

In this section, we propose a closed loop control law for point-to-point control of the unactuated link. The goal is to transfer the unactuated link from an initial angular position θ_0

($= \theta_{20}+\alpha$) with zero initial velocity to a final angular position θ_f ($= \theta_{2f}+\alpha$) with zero final velocity. We treat the actuated coordinate θ_1 as a *pseudo input* and prescribe a feedback law in terms of the *pseudo input*. The formal justification of this treatment is deferred to Appendix A.

From (10), we see that the input θ_1 has a bounded effect on the acceleration because $|\sin\theta_1| \leq 1$. We propose a feedback control law of the form:

$$\sin\theta_1 = \frac{\sin(k_1(\theta_f-\theta) - k_2\dot{\theta})\sin\theta}{k}, \qquad (11)$$

where $k \geq 1$ and $k_1, k_2 > 0$ are constants. Also θ_f is the desired final angular position of the unactuated link. We note that θ_1 exists because $|\sin(k_1(\theta_f-\theta) - k_2\dot{\theta})\sin\theta/k| \leq 1$. Using (11) in (10) we get:

$$\ddot{\theta} = \frac{A}{k}\sin(k_1(\theta_f-\theta) - k_2\dot{\theta})\sin^2\theta. \qquad (12)$$

The intuition behind the control law (11) is to introduce a virtual non-linear spring and damper into the system. These virtual elements introduce a stable equilibrium point $[\theta,\dot{\theta}] = [\theta_f, 0]$ in the system dynamics. In the vicinity of the equilibrium point $[\theta_f, 0]$, the dynamics (12) may be approximated as:

$$\ddot{\theta} \approx \frac{A\sin^2\theta_f}{k}(k_1(\theta_f-\theta) - k_2\dot{\theta}). \qquad (13)$$

The ratios k_1/k and k_2/k are measures of stiffness and damping respectively. Further, the multiplicative term $\sin\theta$ in (11) ensures that the *sign* of the acceleration $\ddot{\theta}$ in (12) is not affected by the regime of motion ($\sin\theta > 0$ or $\sin\theta < 0$). It is only affected by the deviation from the desired final state $[\theta,\dot{\theta}] = [\theta_f, 0]$. These intuitive notions are formalized in the proof below.

B. Proof of Convergence

Let us consider a domain $\Omega = \{[\theta,\dot{\theta}] : |k_1(\theta_f-\theta) - k_2\dot{\theta}| \leq \pi/2 \text{ and } |\theta| \leq \pi/2\}$, and a Lyapunov function candidate (defined on Ω):

$$V(\theta,\dot{\theta}) = \frac{B}{k_1}\int_0^{\psi} \sin x \sin^2(\frac{x+k_2\dot{\theta}}{k_1} - \theta_f)dx + \frac{1}{2}\dot{\theta}^2, \quad (14)$$

where $\psi = k_1(\theta_f-\theta) - k_2\dot{\theta}$, $B = A/k$.

Proposition:
The control law (11) guarantees *local* asymptotic convergence of the state $[\theta,\dot{\theta}]$ in (12) to $[\theta_f, 0]$ ($\theta_f \neq 0$). Further, $\exists\ l > 0$ for which a domain of attraction of the control law is the *largest* connected region $\Omega_l = \{[\theta,\dot{\theta}] : V(\theta,\dot{\theta}) < l\} \subset \Omega$.

Proof:
The scalar function $V(\theta,\dot{\theta})$ defined in (14) is positive definite in Ω because it satisfies the following conditions:

1. $V(\theta_f, 0) = 0$.
2. $V(\theta,\dot{\theta}) > 0$ in $\Omega\ \forall\ [\theta,\dot{\theta}] \neq [\theta_f, 0]$.

The 1st condition follows from direct substitution in (14) and noting that $[\theta,\dot{\theta}] = [\theta_f, 0]$ implies $\psi = 0$. The 2nd condition follows by noting that $\sin x > 0$ for $\pi/2 \geq x > 0$ and $\sin x < 0$ for $-\pi/2 \leq x < 0$. Thus, for $0 < |\psi| \leq \pi/2$:

$$\int_0^{\psi} \sin x \sin^2(\frac{x+k_2\dot{\theta}}{k_1} - \theta_f)dx > 0.$$

Henceforth, we abbreviate $V(\theta,\dot{\theta})$ as V. It is convenient to rewrite (14) as:

$$\begin{aligned} V =& \frac{B}{2}[\frac{k_1(\cos\psi\cos 2\theta - \cos 2(\frac{\psi}{k_1}+\theta)) - 2\sin\psi\sin 2\theta}{k_1^2-4}] \\ &+ \frac{B}{2k_1}(1-\cos\psi) + \frac{1}{2}\dot{\theta}^2, \qquad k_1 \neq 2. \end{aligned} \qquad (15)$$

The time derivative of (15) is given by:

$$\begin{aligned} \dot{V} =& \frac{\partial V}{\partial\theta}\dot{\theta} + \frac{\partial V}{\partial\dot{\theta}}\ddot{\theta} \\ =& -\frac{B^2k_2\sin^2\theta}{k_1(4-k_1^2)}[(2-k_1^2\sin^2\theta)\sin^2\psi \\ &+ k_1\sin\psi(\sin 2\theta\cos\psi - \sin 2(\frac{\psi}{k_1}+\theta))] \end{aligned} \qquad (16)$$

It may be shown that $\dot{V} \leq 0$ in Ω for all $k_1, k_2 > 0$. In the interest of brevity, we just prove this assertion for $k_1 = 1$ and $k_2 > 0$. We further show that $\exists\ l_0 > 0$, such that $\Omega_{l_0} \subset \Omega$. Substituting $k_1 = 1$ in (16) and after some rearrangement we get:

$$\begin{aligned} \dot{V} =& -\frac{B^2k_2}{3}\sin^2\theta(1-\cos\psi)[3\sin^2\theta\cos\psi(1-\cos\psi) \\ &+ (2\sin\theta\cos\psi + \sin\psi\cos\theta)^2 + (\sin\psi\cos\theta + \sin\theta)^2 \\ &+ \sin^2\theta\cos^2\psi] \end{aligned} \qquad (17)$$

We note the $0 \leq \cos\psi \leq 1$ in Ω. Thus, the expression in square brackets in (17) is always non-negative. Hence, $\dot{V} \leq 0$ in Ω. Also, from (17):

$$\begin{aligned} &\dot{V} = 0 \\ \Rightarrow\ &\theta = 0 \text{ or } \psi = 0 \end{aligned} \qquad (18)$$

Using (18) in (12) we get:

$$\dot{V} = 0 \Rightarrow \ddot{\theta} = 0 \qquad (19)$$

From (18) and (19), the largest invariant set where $\dot{V} = 0$ is given by $\{[\theta,\dot{\theta}] = [0,0] \cup [\theta_f, 0]\}$. Using *La Salle's invariant set theorem*, we conclude that the state $[\theta,\dot{\theta}]$ converges to $[\theta = 0, \dot{\theta} = 0]$ or $[\theta = \theta_f, \dot{\theta} = 0]$.

The choice of l_0 is illustrated graphically in Fig. 5. We used (14) for the simulation with the parameters $k_1 = 1$, $k_2 = 1$, $B = 32$ and $\theta_f = 30°$. For these parameters, we obtain $l_0 = 0.54$ and Ω_{l_0} is the largest connected region within Ω such that $V(\theta,\dot{\theta}) < l_0$. Once again, it follows from *La Salle's invariant set theorem* that Ω_{l_0} is a domain of attraction for the largest invariant set.

It remains to establish the stability of the equilibrium points. We show that $[\theta = 0, \dot{\theta} = 0]$ is unstable and $[\theta = \theta_f, \dot{\theta} = 0]$ is a stable equilibrium point for $k_1 = k_2 = 1$. We note that there are other choices of k_1, k_2 for which these conclusions

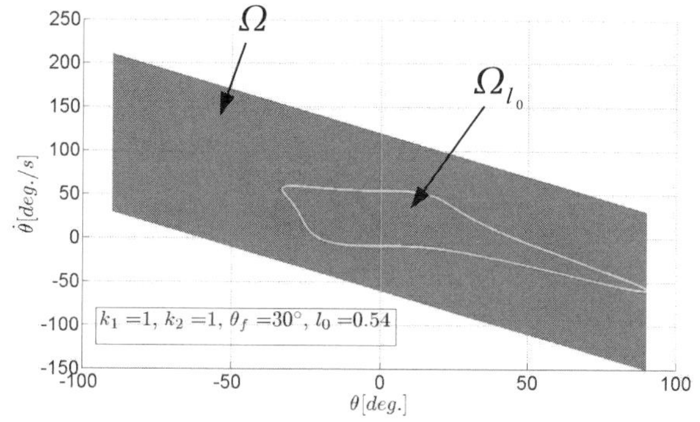

Fig. 5. Domain of Convergence

hold and the current choice only serves to simplify the algebra. From (15):

$$\frac{\partial^2 V}{\partial \theta^2} = 0 \text{ and } \frac{\partial^3 V}{\partial \theta^3} \neq 0 \text{ at } [\theta = 0, \dot{\theta} = 0].$$

This implies that $[\theta = 0, \dot{\theta} = 0]$ is not a local minimum

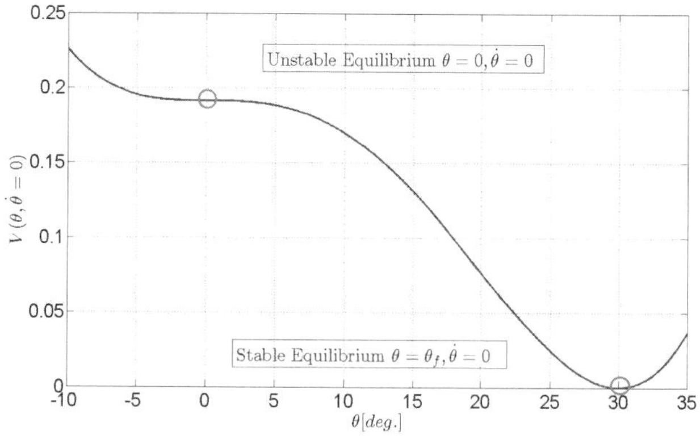

Fig. 6. Stable and unstable equilibria of system dynamics

for V and thus an *unstable* equilibrium point. We note that this conclusion does not follow from linearization because the linearized system has zero eigenvalues at $[\theta = 0, \dot{\theta} = 0]$. Once again, from (15):

$$\nabla^2 V = \begin{bmatrix} B\sin^2\theta_f & B\sin^2\theta_f \\ B\sin^2\theta_f & B\sin^2\theta_f + 1 \end{bmatrix} \text{ at } [\theta = \theta_f, \dot{\theta} = 0].$$

This implies that $\nabla^2 V$ is positive definite and $[\theta = \theta_f, \dot{\theta} = 0]$ is a local minimum for V and thus a *stable* equilibrium point. These ideas are illustrated in Fig. 6 for the case $k_1 = 1$, $k_2 = 1$, $B = 32$ and $\theta_f = 30°$. Thus the state $[\theta, \dot{\theta}]$ in (12) converges to $[\theta_f, 0]$ as long as it does not start from $[0, 0]$.

IV. Implementation and Experiments

We conducted position control experiments on a prototype system with 3 links which is shown in Fig. 7. The link mechanism, which operates inside the wing-box, is shown in Fig. 7a. The links are essentially C-channels which are serially connected by 1 d.o.f rotary joints. Link 1 is the *only* servoed link. Links 2 and 3 are equipped with *on-off* pneumatic brakes. The relative angular position of the links are measured using optical encoders placed at the rotary joints. They have a resolution of 1000 ppr.

The actuation mechanisms for link 1 operate completely outside the wing-box and are shown in Fig. 7b. They comprise a servoed *tilting mechanism* and a servoed *azimuthal positioning mechanism*. The *tilting mechanism* is used to tilt link 1 relative to a vertical axis. Depending on the state (*on* or *off*) of the pneumatic brakes, the unactuated links (2 and 3) may be deployed by exploiting gravity and dynamic coupling with link 1. The *azimuthal positioning mechanism* is used for angular positioning of the entire link mechanism inside the wing-box and serves to expand the workspace of the robot arm. This mechanism is used after the links have been deployed using the *tilting mechanism*. The pneumatic brakes are in the *on* state when the *azimuthal positioning mechanism* is in use. Both mechanisms have harmonic drive gearing (100:1) coupled to AC servomotors (0.64 Nm, 3000rpm). In the experiments that follow, the *azimuthal positioning mechanism* is not used. We only use the tilting mechanism to deploy the links and verify the proposed control law.

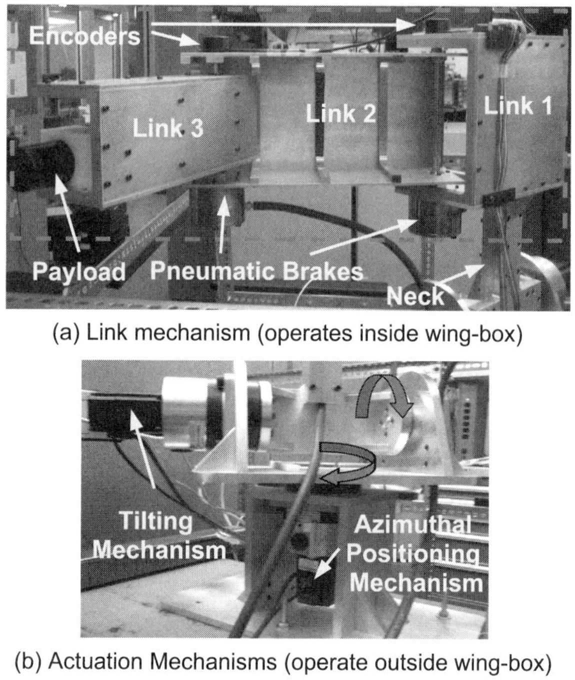

Fig. 7. 3-link prototype arm

The dynamical system (10) corresponding to our experimental setup has the parameters $A = 32.8s^{-2}$ and $\alpha = -3.2°$. The experimental results are illustrated in Fig. 8. The goal was to move link 2 from an initial position $\theta_{20} = 35°$ to a desired final position of $\theta_{2f} = 50°$. Link 3 was kept fixed at 30° relative to link 2. The controller parameter values in (11) were set at $k = 12$, $k_1 = 1.2$ and $k_2 = 0.2s$. It may be verified that these parameters ensure that the initial position lies within the domain of convergence. The scaling factor of

$k = 12$ was used to restrict the amplitude of θ_1 to less than $1.5°$. A small amplitude of θ_1 is very important in practice because the arm operates inside an aircraft wing. There are other choices of k_1 and k_2 which ensures convergence of the control law. For example, a higher value of k_2 would imply less overshoot and slower convergence.

The actual final position of the arm was $\theta_2 = 50.5°$ as shown in Fig. 8a. The tilt trajectory of link 1 is shown in Fig. 8b. The maximum tilt is $1.3°$ which is small enough for operation inside the wing-box. Fig. 8c shows a comparison of the gravitational, inertial and centrifugal contributions on the angular acceleration of link 2. The gravitational contribution clearly dominates the other effects. This demonstrates, *a posteriori*, the validity of the approximations made in our dynamic modeling.

The control law (11) demonstrates reasonable positioning accuracy of the unactuated links. The performance is achieved without any knowledge of Coulomb friction or the dynamics introduced by the flexible hose supplying air to the pneumatic brakes. This is a significant improvement compared to the open loop motion planning schemes explored in our earlier work. Such schemes required frequent tuning of friction coefficients and other parameters related to the dynamics of the hose.

A primary drawback of the proposed control law arises from the conflicting requirements of small amplitude of tilt of link 1 and small steady state error for link 2. This is readily seen from (11). If link 2 starts at θ_0 with zero initial velocity, the initial tilt of link 1 is given by:

$$\sin\theta_{10} = \frac{\sin(k_1(\theta_f - \theta_0))\sin\theta}{k} \tag{20}$$

θ_{10} may be large if the amplitude of motion $|\theta_f - \theta_0|$ is large. To achieve smaller values of θ_{10}, the scaling factor k may be increased or the gain k_1 may be reduced. As noted before, the ratio k_1/k is a measure of the stiffness of the virtual non-liner spring introduced by the controller. Increasing k and reducing k_1 would result in lower stiffness. This would lower the speed of convergence and also increase the steady state error induced by Coulomb friction.

We address this issue by replacing the fixed reference θ_f in (11) by a time varying reference $\theta_{ref}(t)$ starting at θ_0 and changing smoothly to θ_f. In particular, the reference may be a sigmoidal trajectory given by:

$$\theta_{ref}(t) = \begin{cases} \theta_0 + (10\mu^3 - 15\mu^4 + 6\mu^5)(\theta_f - \theta_0) \\ \quad \mu = \frac{t}{t_{f1}}, \quad 0 \le t \le t_{f1} \\ \theta_f \quad\quad t \ge t_{f1} \end{cases} \tag{21}$$

We may choose t_{f1} to set a desired average speed of motion $|\theta_f - \theta_0|/t_{f1}$. Substituting (21) in (11), we obtain the modified control law:

$$\sin\theta_1 = \frac{\sin(k_1(\theta_{ref}(t) - \theta) - k_2\dot{\theta})\sin\theta}{k}. \tag{22}$$

We applied the control law (22) to our prototype system. The goal was to move link 2 from an initial position $\theta_{20} = 10°$ to a desired final position of $\theta_{2f} = 70°$. Link 3 was kept fixed at $0°$ relative to link 2. The controller parameter values in (22) were set at $k = 5$, $k_1 = 1$ and $k_2 = 0.2s$, $t_{f1} = 12s$. The experimental results are shown in Fig. 9. The actual final position was $69.7°$ at the end of $12s$, as shown in Fig. 9a. The tilt trajectory of link 1 is shown in Fig. 9b. The maximum amplitude of tilt of link 1 was $1.1°$ which is within the acceptable limits.

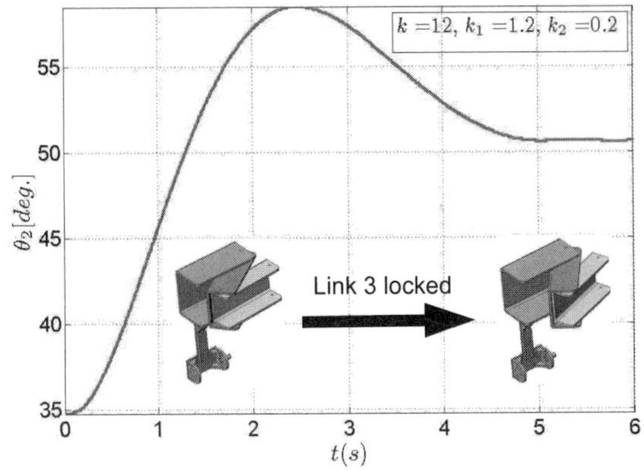

(a) Link 2 trajectory (unactuated joint)

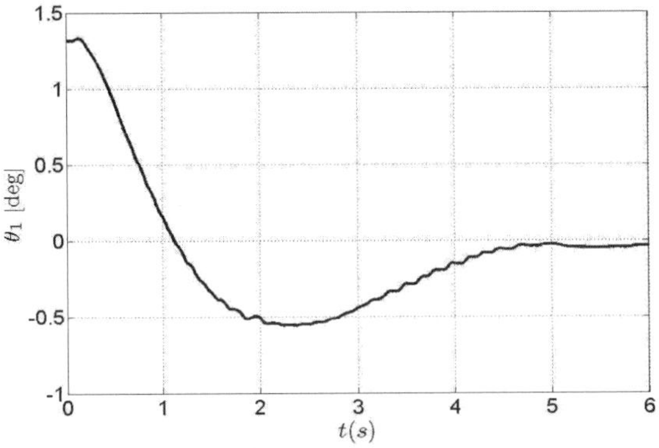

(b) Link 1 trajectory (servoed joint)

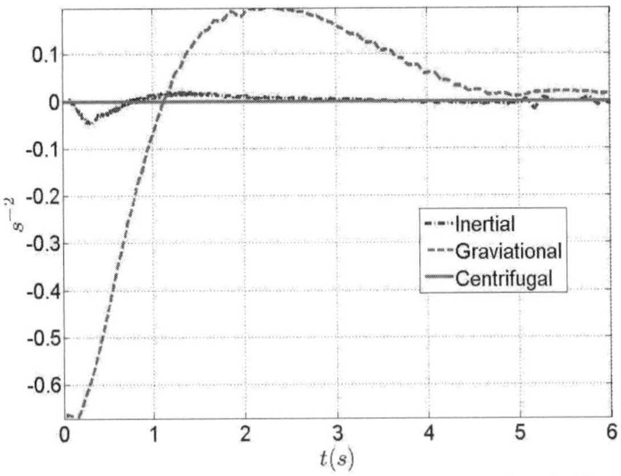

(c) Comparison of gravitational, inertial and centrifugal effects

Fig. 8. Position control experiment on 3-link prototype

V. Conclusion

We have addressed the problem of closed loop point-to-point control of a gravity assisted underactuated robot arm. The arm is particularly well suited to high payload assembly

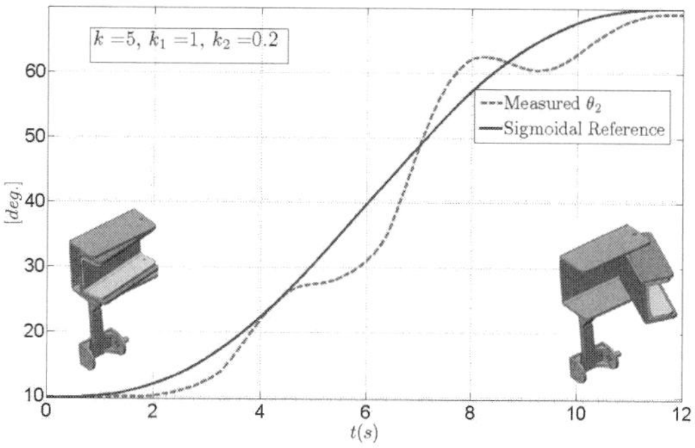

(a) Link 2 trajectory (unactuated joint)

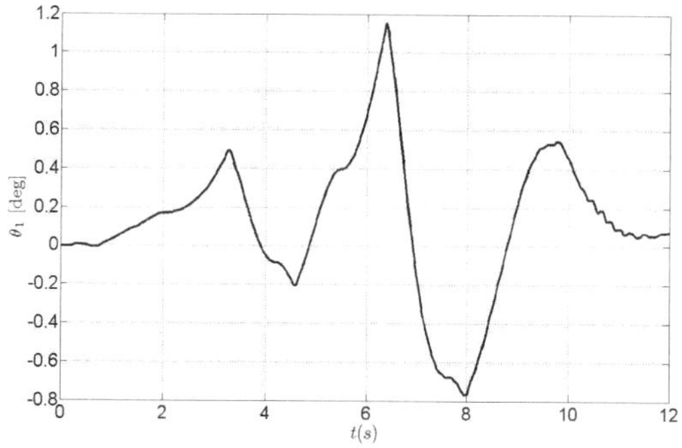

(b) Link 1 trajectory (servoed joint)

Fig. 9. Experimental results for modified control law using sigmoidal reference trajectory

operations inside an aircraft wing-box. We proposed a closed loop control algorithm for point-to-point control of the unactuated links. A Lyapunov function was synthesized to prove the convergence of the control law. The Lyapunov function also provides us with lower bounds on the domain of convergence of the control law.

The control algorithm was applied to a prototype 3-link robot arm. The experimental results showed reasonable performance of the control law in the absence of prior knowledge of friction and other unmodelled dynamical effects. We further proposed a modified control law to handle the conflicting requirements of small tilt of the actuated link and low steady state error of the unactuated links. The efficacy of the modified control law was demonstrated on the prototype system.

The modified control law results in a non-autonomous dynamical system. Our current proof has to be suitably modified to prove asymptotic convergence of the output using the modified control scheme. Also, a lower steady state error in the position of the unactuated links is desirable. These issues will be addressed in future work.

ACKNOWLEDGMENTS

The authors gratefully acknowledge the support provided by the Boeing Company. The first author would like to thank Dr. Jun Ueda for many helpful discussions.

APPENDIX A

We justify the treatment of the actuated coordinate θ_1 as a *pseudo input*. We denote the desired trajectory of θ_1 by θ_{1d}. From (11), $\theta_{1d} = \sin^{-1}(\sin(k_1(\theta_f - \theta) - k_2\dot{\theta})\sin^2\theta / k)$. The dynamics of the actuated coordinate θ_1 may always be feedback linearized by choosing the control torque as:

$$\tau_1 = \frac{\ddot{\theta}_{1d} - 2\lambda\dot{\tilde{\theta}}_1 - \lambda^2\tilde{\theta}_1}{N_{11}} + F_1 + G_1 + \frac{N_{12}}{N_{11}}(F_2 + G_2), \tag{23}$$

where:

$$\begin{bmatrix} N_{11} & N_{12} \\ N_{12} & N_{22} \end{bmatrix} = \begin{bmatrix} H_{11} & H_{i1} \\ H_{i1} & H_{ii} \end{bmatrix}^{-1},$$

$$\tilde{\theta}_1 = \theta_1 - \theta_{1d} \text{ and } \lambda > 0.$$

Using (23) in (1), the error dynamics of the actuated coordinate is given by:

$$\ddot{\tilde{\theta}}_1 + 2\lambda\dot{\tilde{\theta}}_1 + \lambda^2\tilde{\theta}_1 = 0. \tag{24}$$

Let us define $x = [\theta, \dot{\theta}]^T$ and $y = [\tilde{\theta}_1, \dot{\tilde{\theta}}_1]^T$. The dynamics of the unactuated coordinate (x) and the error dynamics of the actuated coordinate (y) may be written in cascade form as $\dot{x} = f(x, y)$ and $\dot{y} = g(y)$. Here, $f(x, y) = [\dot{\theta}, A\sin\theta\sin(\theta_{1d} + \tilde{\theta}_1)]^T$ and $g(y) = [\dot{\tilde{\theta}}_1, -2\lambda\dot{\tilde{\theta}}_1 - \lambda^2\tilde{\theta}_1]^T$. We note that $f(x, y)$ is globally Lipschitz and the linear subsystem $\dot{y} = g(y)$ is globally exponentially stable. Also, we have proved that the non-linear subsystem $\dot{x} = f(x, 0)$ is asymptotically stable using La Salle's Theorem. It follows from Sontag's Theorem [22], [24] that the cascade system is locally asymptotically stable for an appropriate choice of λ.

REFERENCES

[1] S. Hirose and M. Mori, "Biologically Inspired Snake-like Robots Robotics and Biomimetics," in *Proc. of ROBIO*, Aug. 2004, pp 1–7.

[2] A. Wolf, H. B. Brown, R. Casciola, A. Costa, M. Schwerin, E. Shamas and H. Choset, "A mobile hyper redundant mechanism for search and rescue tasks," *Proc. of IROS*, Oct. 2003, vol. 3, pp 2889–2895.

[3] G. Oriolo and Y. Nakamura, "Free-joint manipulators: motion control under second-order nonholonomic constraints," in *Proceedings of IROS*, Nov. 1991, vol. 3 pp 1248–1253.

[4] J. Hauser and R. M. Murray, "Nonlinear controller for nonintegrable systems: the acrobot example," in *Proc. American Contr. Conf.*, 1990, pp 669–671.

[5] M. W. Spong, "Partial feedback linearization of underactuated mechanical systems," in *Proc. of IROS*, Sep. 1994 v 1, pp 314–321.

[6] M. W. Spong, "Swing up control of the acrobot," in *Proc. of ICRA*, Mar. 1994, v 3, pp 2356–2361.

[7] K. M. Lynch and M. T. Mason, "Dynamic underactuated nonprehensile manipulation," in *Proc of IROS*, Nov. 1996 v 2, pp 889–896.

[8] H. Arai, K. Tanie and N. Shiroma, "Nonholonomic control of a three-dof planar manipulator," in *IEEE Trans. Robot. Automat.*, vol. 14, no. 5, pp 681–695, Oct. 1998.

[9] H. Arai, "Position control of a 3-dof manipulator with a passive joint under a Nonholonomic constraint," in *Proc. of IROS*, Nov. 1996 v 1, pp 74–80.

[10] K. M. Lynch, N. Shiroma, H. Arai and K. Tanie, "Collision-free trajectory planning for a 3-dof robot with a passive joint," in *Int. J. Robot. Res.* vol. 19, no. 12, Dec. 2000, pp. 1171–1184.

[11] R. W. Brockett, "Asymptotic stability and feedback stabilization," in Differential Geometric Control Theory, R. W. Brockett, R. S. Millman, and H. J. Sussmann, Eds. Boston, MA: Birkhäuser, 1983, pp. 181–191.

[12] T. Suzuki, M. Koinuma, and Y. Nakamura, "Chaos and nonlinear control of a nonholonomic free-joint manipulator," in *Proc. of IROS*, 1996, pp. 2668–2675.

[13] T. Suzuki and Y. Nakamura, "Nonlinear control of a nonholonomic free joint manipulator with the averaging method," in *Proc. 35th IEEE Int. Conf. Decision Contr.*, 1996, pp. 1694–1699.

[14] A. De Luca and G. Oriolo, "Motion planning under gravity for underactuated three-link robots, in Proc. of IROS, 2000, pp.139–144.

[15] A. Jain and G. Rodriguez, "An analysis of the kinematics and dynamics of underactuated manipulators," in *IEEE Trans. Robot. Automat.*, vol. 9, pp. 411–422, Aug. 1993

[16] R. Mukherjee and D. Chen, "Control of free-flying underactuated space manipulators to equilibrium manifolds," in *IEEE Trans. Robot. Automat.*, vol. 9, pp. 561–570, Oct. 1993.

[17] Y. Nakamura and R. Mukherjee, "Nonholonomic path planning of space robots via a bidirectional approach," in *IEEE Trans. Robot. Automat.*, vol. 7, pp. 500–514, Aug. 1991.

[18] E. Papadopoulos and S. Dubowsky, "Failure recovery control for space robotic systems," in *Proc. American Contr. Conf.*, 1991, pp. 1485–1490.

[19] E. Papadopoulous, Path planning for space manipulators exhibiting nonholonomic behavior, in *Proc. IEEE/RSJ Int.Workshop Intell. Robots Syst. (IROS92)*, 1992, pp. 669–675.

[20] B. Roy and H. Asada, "An underactuated robot with a hyper-articulated deployable arm working inside an aircraft wing-box," in *Proc. of IROS*, August 2005, pp 4046–4050.

[21] B. Roy and H. Asada, "Dynamics and control of a gravity-assisted underactuated robot arm for assembly operations inside an aircraft wing-box," in *Proc. of ICRA*, May 2006, pp 701–706.

[22] E. D. Sontag, "Remarks on stabilization and input-to-state stability," in *Proc. 28th IEEE Conf. Decision Contr.*, vol. 2, pp. 1376–1378, 13-15 Dec. 1989.

[23] H. J. Sussmann, "A general theorem on local controllability," in *SIAM J. Contr. Optimization*, vol. 25, no. 1, pp. 158–194, 1987.

[24] H. J. Sussmann and P.V. Kokotovic, "The peaking phenomenon and the global stabilization of nonlinear systems," in *IEEE Trans. Automat. Contr.*, vol. 36, no. 4, pp. 424–440, April 1991.

[25] I. Kolmanovsky and N. H. McClamroch, "Developments in nonholonomic control problems," in *IEEE Contr. Syst.*, vol. 15, pp. 20–36, 1995.

[26] R. M. Murray, Z. Li and S. S. Sastry, *A Mathematical Introduction to Robotic Manipulation*, CRC Press.

Predicting Partial Paths from Planning Problem Parameters

Sarah Finney, Leslie Kaelbling, Tomás Lozano-Pérez
CSAIL, MIT
{sjf,lpk,tlp}@csail.mit.edu

***Abstract*— Many robot motion planning problems can be described as a combination of motion through relatively sparsely filled regions of configuration space and motion through tighter passages. Sample-based planners perform very effectively everywhere but in the tight passages. In this paper, we provide a method for parametrically describing workspace arrangements that are difficult for planners, and then learning a function that proposes partial paths through them as a function of the parameters. These suggested partial paths are then used to significantly speed up planning for new problems.**

I. Introduction

Modern sample-based single-query robot-motion planners are highly effective in a wide variety of planning tasks. When they do encounter difficulty, it is usually because the path must go through a part of the configuration space that is constricted, or otherwise hard to sample.

One set of strategies for solving this problem is to approach it generically, and to develop general methods for seeking and sampling in constricted parts of the space. Because these planning problems are ultimately intractable in the worst case, the completely general strategy cannot be made efficient. However, it may be that a rich set of classes of more specific cases can be solved efficiently by learning from experience.

The basic question addressed in this paper is whether it is possible to learn a function that maps directly from parameters describing an arrangement of obstacles in the workspace, to some information about the configuration space that can be used effectively by a sample-based planner. We answer this question in the affirmative and describe an approach based on task templates, which are parametric descriptions (in the workspace) of the robot's planning problem. The goal is to learn a function that, given a new planning problem, described as an instance of the template, generates a small set of suggested partial paths in the constricted part of the configuration space. These suggested partial paths can then be used to "seed" a sample-based planner, giving it guidance about paths that are likely to be successful.

Task templates can be used to describe broad classes of situations. Examples include: moving (all or part of) the robot through an opening and grasping something; moving through a door carrying a large object; arriving at a pre-grasp configuration with fingers "straddling" an object. Note that the task template parameters are in the workspace rather than the configuration space of the robot, so knowing the parameters is still a long way from knowing how to solve the task.

One scenario for obtaining and planning with task templates would be a kind of teleoperation via task-specification. A human user, rather than attempting to control the robot's end-effector or joints directly, would instead choose an appropriate task template, and "fit" it to the environment by, for instance, using a graphical user interface to specify the location of a car-window to be reached through and a package on the car seat to be grasped. Given this task-template instance the planner can generate a solution and begin to execute it. During the process of execution the operator might get additional sensory information (from a camera on the arm, for example), and modify the template dynamically, requiring a very quick replanning cycle. Eventually, it might be possible to train a recognition system, based on 2D images or 3D range data to select and instantiate templates automatically.

Of course, we cannot get something for nothing: the power of learning comes from exploiting the underlying similarity in task instances. For this reason, template-based learning cannot be a general-purpose solution to motion planning. It is intended to apply to commonly occurring situations where a very fast solution is required. Although one might be tempted to encode a general workspace as a single task template, with a finite (but very large) set of parameters that indicate which voxels are occupied, there is no reason to believe that the resulting learning problem for such a template is tractable.

II. Related work

A good deal of work has been done on improving multiple-query roadmaps by biasing their sampling strategy to generate samples in difficult areas of configuration space and avoid over-sampling in regions that can be fairly sparsely covered. Several techniques involve rejection sampling intended to concentrate the samples in particular areas, such as near obstacles [14], or near configuration-space bottlenecks [4]. Pushing samples toward the medial axis of the configuration freespace is another method for generating samples that are concentrated in the difficult narrow passages [7]. Visibility-based approaches also strive to increase sampling near bottlenecks [13].

Each of these techniques involves computationally expensive operations, as they are intended to be used in a multiple-query setting, in which the cost is amortized over a large number of queries within a static environment. We are focused instead on a single-query setting, in which these techniques are prohibitively time-consuming. However, the collective body of work argues persuasively that a relatively small number of carefully selected samples can make the planning job much easier. Our approach is similarly intended to address these difficult regions of the planning space.

Other similarly motivated techniques temporarily expand the freespace by contracting the robot to identify the narrow passages [11, 6], and these methods are fast enough to apply to single-query problems. Nonetheless, they are required to discover the structure of the environment at planning time. We would like to use learning to move much of this work offline.

A related class of techniques looks at aspects of the workspace, rather than configuration space, to bias sampling toward finding difficult narrow passages [8, 16, 15, 5]. This work builds on the intuition that constrained areas in configuration space are also often constrained in the workspace, and are easier to identify in the lower dimensional space. Our work shares this intuition.

There are also a number of strategies for guiding the search in a single-query setting. Voronoi-based [17] and entropy-guided [2, 10] exploration are among these techniques. Since we are proposing a learning method that predicts partial paths in order to help a single-query planner, our approach is compatible with any single-query probabilistic roadmap exploration strategy. In particular, in this paper we integrate our approach with Stanford's single-query, bidirectional, lazy (SBL) planner [12].

Other approaches have also taken advantage of learning. Features of configuration space can be used to classify parts of the space and thereby choose the sampling strategies that are most appropriate for building a good multiple-query roadmap in that region [9]. Burns and Brock learn a predictive model of the freespace to better allocate a roadmap's resources and avoid collision checking [1]. This use of learning is related to ours, but is again in a multiple-query setting, and so does not address generalizing to different environments.

III. Task-template planning

The task-template planner is made up of two phases: first, the task is described parametrically as an instance of the template, and used as input to a learned *partial-path suggestion function* that generates a set of candidate partial paths for solutions; second, these partial paths are used to initialize a sample-based planner, suitably modified to take advantage of these suggestions. If the suggested partial paths are indeed in parts of the configuration space that are both useful and constricted, then they will dramatically speed up the sample-based planner.

To make the discussion concrete, we will use as a running example a mobile two-degree-of-freedom arm, operating in essentially a planar environment. One common task for such a robot will be to go through doors. A door in a planar wall can be described using 3 parameters $(x_{door}, y_{door}, \theta_{door})$ as shown in figure 1 (top left). Note that the environment need not be exactly conformant to the template, as shown in figure 1 (top right). As long as the suggestions to the planner help it with the

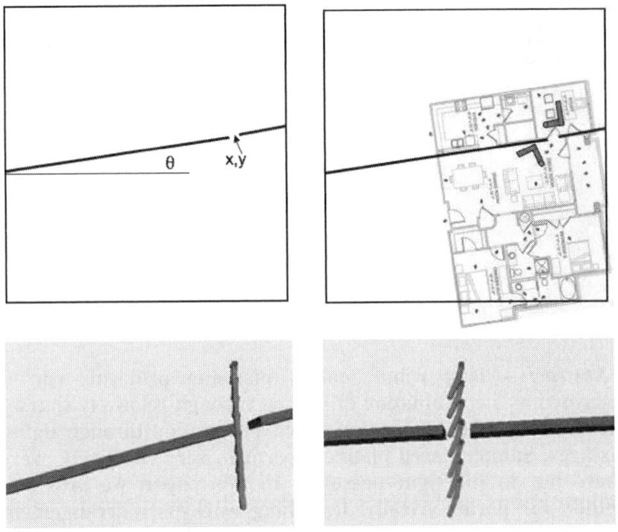

Fig. 1. A task template. Top left: Parameterization of the template. Top right: Application of task template to a complex environment. Bottom: Suggested partial paths generated by the same strategy for different environments.

difficult parts of the environment, the planner can handle other sparsely arranged obstacles. The remaining panels in figure 1 show some useful suggested partial paths.

The overall learning approach generates training examples by calling a single-query planner on a number of planning problems, each of which is an instance of the same task template. Each plan is then examined to extract the most constrained portion of the path, and the partial path through this "tight spot" is stored along with the task-template parameters describing the world. We then train several partial-path generators on this data. The idea is that the partial-path suggestion function will learn the parametric dependence of the useful partial paths on the task-template instance parameters, and therefore apply to previously unseen examples.

Additionally, we must adapt the single-query planner so that it can take in, along with the start and goal configuration for a particular query, a set of suggested partial paths.

A. Generating training data

For each task template, we assume a source of training task instances of that template. These instances might be synthetically generated at random from some plausible distribution, or be drawn from some source of problems encountered in the world in which the robot will be operating.

We begin with a set of task instances, $t^1, \ldots, t^n$, where each task instance is a specification of the detailed problem, including start and goal configurations and a description of the workspace obstacles. These descriptions will generally be non-parametric in the sense of not having a fixed size. We will convert each of these tasks and their solution paths into a training example, $\langle x^i, y^i \rangle$, where x^i is a parametric representation of t^i as an instance of the task template, and y^i is a path segment. Intuitively, we instantiate the task template for the particular task at hand. In the case of the doorway example this means identifying the door, and finding its position and orientation, x_{door}, y_{door} and θ_{door}.

Currently, we assume the each t^i in the training set is "labeled" with a parametric description x^i; in the future, we expect that the labels might be computed automatically. To generate the y^i values, we begin by calling the single-query planner on task instance t^i. If the planner succeeds, it returns a path $p = \langle c_1, ..., c_r \rangle$ where the c's are configurations of the robot, connected by straight-line paths.

We do not, however, want to use the entire path to train our learning algorithm. Instead we want to focus in on the parts of the path that were most difficult for the planner to find, so we will extract the most constrained portion of the path, and use just this segment as our training data. Before we analyze the path, we would like to reduce the number of unnecessary configurations generated by the probabilistic planner, so we begin by smoothing the path. This process involves looking for non-colliding straight-line replacements for randomly selected parts of the path, and using them in place of the original segment. We made use of the SBL planner's built-in smoother to do this.

We also want to sample the paths at uniform distance intervals, so that paths that follow similar trajectories through the workspace are similar when represented as sample sequences. Thus, we first resample the smoothed path at a higher resolution, generating samples by interpolating along p at some fixed distance between samples, resulting in a more finely sampled path.

Given this new, finely sampled path p', we now want to extract the segment that was most constrained, and therefore difficult for the planner to find. In fact, if there is more than one such segment, we would like to find them all. For each configuration in the path, we draw some number of samples from a normal distribution centered on that configuration (sampling each degree of freedom independently). Each sample is checked for collision, and the ratio of colliding to total samples is used as a measure of how constrained the configuration is.

We then set a threshold to determine which configurations are tight, and find the segments of contiguous tight configurations. Our outputs need to have a fixed length l, so for all segments less than l, we pad the end with the final configuration. For segments with length greater than l, we skip configurations in cases where skipping does not cause a collision.

In our more complicated test domains, we found that it was useful to add a step that pushes each configuration away from collisions, by sampling randomly around each configuration in the path, and replacing it if another is found that has greater clearance. This makes it less likely that we will identify spurious tight spots, and it also makes the segment we extract more likely to generalize, since it has a larger margin of error. This step also requires that we are careful not to decrease the path's clearance when we smooth it. In this way, we extract each of the constrained path segments from the solution path. Each of these, paired with the parametric description of the task, becomes a separate data point.

B. Learning from successful plans

Each of our n training instances $\langle x^i, y^i \rangle$ consists of a task-template description of the world, x^i, described in terms of m parameters, and the constrained segment from a successful path, y^i, as described above. The configurations have consistent length d (the number of degrees of freedom that the robot has), and each constrained segment has been constructed to have consistent length l, therefore y^i is a vector of length ld.

At first glance, this seems like a relatively straightforward non-linear regression problem, learning some function f^* from an m-dimensional vector x to an ld-dimensional vector y, so that the average distance between the actual output and the predicted output is minimized. However, although it would be useful to learn a single-valued function that generates one predicted y vector given an input x, our source of data does not necessarily have that form. In general, for any given task instance, there may be a large set of valid plans, some of which are qualitatively different from one another. For instance, members of different homotopy classes should never be averaged together.

Consider a mobile robot going around an obstacle. It could choose to go around either to the left of the obstacle, or to the right. In calling a planner to solve such problems, there is no general way to bias it toward one solution or the other; and in some problems there may be many such solutions. If we simply took the partial paths from all of these plans and tried to solve for a single regression function f from x to y, it would have the effect of "averaging" the outputs, and potentially end up suggesting partial paths that go through the obstacle.

1) Mixture model: To handle this problem, we have to construct a more sophisticated regression model, in which we assume that data are actually drawn from a mixture of regression functions, representing qualitatively different "strategies" for negotiating the environment. We learn a generative model for the selection of strategies and for the regression function given the strategy, in the form of a probabilistic mixture of h regression models, so that the conditional distribution of an output y given an input x is

$$\Pr(y|x) = \sum_{k=1}^{h} \Pr(y|x, s = k) \Pr(s = k|x) \ .$$

where s is the mixture component responsible for this example.

For each strategy, we assume that the components of the output vector are generated independently, conditioned on x; that is, that

$$\Pr(y|x, s = k) = \prod_{j=1}^{ld} \Pr(y_j|x, s = k) \ .$$

Note that this applies to each configuration on the path as well as each coordinate of each configuration. Thus, the whole partial path is being treated as a point in an ld-dimensional space and not a sequence of points in the configuration space. The parameter-estimation model will cluster paths such

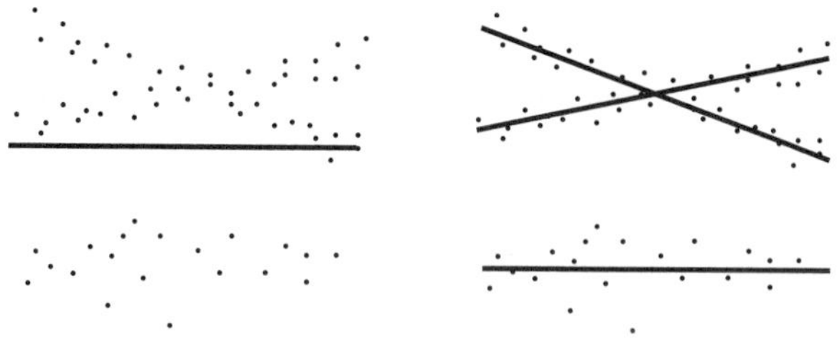

Fig. 2. Data with linear regression (left) and mixture regression (right) fits.

that this independence assumption is satisfied, to the degree possible, in the model.

Then, we assume that each y_j has a linear dependence on x with Gaussian noise, so that

$$\Pr(y_j|x, s=k) = \frac{1}{\sigma_{jk}\sqrt{2\pi}} \exp\left(-\frac{(y_j - w_{jk}\cdot x)^2}{2\sigma_{jk}^2}\right) ,$$

where σ_{jk} is the standard deviation of the data in output dimension j of strategy k from the nominal line, and w_{jk} is an m-dimensional vector of weights specifying the dependence of the mean of the distribution of y_j in strategy k given x as the dot product of w_{jk} and x.

It would be possible to extend this model to contain non-linear regression models for each mixture components, and it might be useful to do so in the future. However, the current model can approximate a non-linear strategy function by using multiple linear components.

In our current model, we assume that $\Pr(s = k|x)$ is actually independent of x (we may wish to relax this in future), and define $\pi_k = \Pr(s = k)$. So, finally, we can write the log likelihood of the entire training set $LL(\sigma, w, \pi)$, as a function of the parameters σ, w, and π, (note that each of these is a vector or matrix of values), as

$$\sum_{i=1}^{n} \log \sum_{k=1}^{h} \pi_k \prod_{j=1}^{ld} \frac{1}{\sigma_{jk}\sqrt{2\pi}} \exp\left(-\frac{(y_j^i - w_{jk}\cdot x^i)^2}{2\sigma_{jk}^2}\right) .$$

We will take a simple maximum likelihood approach and attempt to find values of σ, w, and π that maximize this likelihood, and use that parameterization of the model to predict outputs for previously unseen values of x.

Figure 2 illustrates a data set in which the x and y are one dimensional. In the first frame, we show the best single linear regression line, and in the second frame, a mixture of linear regression models. It is clear that a single linear (or non-linear, for that matter) regression model is inappropriate for this data.

The model described in this section is essentially identical to one used for clustering trajectories in video streams [3], though their objective is primarily clustering, where ours is primarily regression.

2) Parameter estimation: If we knew which training points to assign to which mixture component, then the maximum likelihood parameter estimation problem would be a simple matter of counting to estimate the π_k and linear regression to estimate w and σ. Because we don't know those assignments, will will have to treat them as hidden variables. Let $\gamma_k^i = \Pr(s^i = k|x^i, y^i)$ be the probability that training example i belongs to mixture component k. With this model we can use the expectation-maximization (EM) algorithm to estimate the maximum likelihood parameters.

We start with an initial random assignment of training examples to mixture components, ensuring that each component has enough points to make the linear regression required in the M step described below be well-conditioned.

In the expectation (E) step, we temporarily assume our current model parameters are correct, and use them and the data to compute the responsibilities:

$$\gamma_k^i := \frac{\pi_k \Pr(y^i|x^i, s^i = j)}{\sum_{a=1}^{h} \pi_a \Pr(y^i|x^i, s^i = a)} .$$

In the maximization (M) step, we temporarily assume the responsibilities are correct, and use them and the data to compute a new estimate of the model parameters. We do this by solving, for each component k and output dimension j, a weighted linear regression, with the responsibilities γ_k^i as weights. Weighted regression finds the weight vector, w_{jk}, minimizing

$$\sum_i \gamma_k^i (w_{jk}\cdot x^i - y_j^i)^2 .$$

When the regression is numerically ill-conditioned, we use a ridge parameter to regularize it.

In addition, for each mixture component k, we re-estimate the standard deviation:

$$\sigma_{jk} := \frac{1}{\sum_i \gamma_j^i} \sum_i \gamma_j^i (w_{jk}\cdot x^i - y_j^i)^2 .$$

Finally, we reestimate the mixture probabilities:

$$\pi_j := \frac{1}{n}\sum_{i=1}^{n} \gamma_j^i .$$

This algorithm is guaranteed to find models for which the log likelihood of the data is monotonically increasing, but has the potential to be trapped in local optima. To ameliorate this effect, our implementation does random re-starts of EM and selects the solutions with the best log likelihood.

The problem of selecting an appropriate number of components can be difficult. One standard strategy is to try different values and select one based on held-out data or cross-validation. In this work, since we are ultimately interested in regression outputs and not the underlying cluster structure, we simply use more clusters than are likely to be necessary and ignore any that ultimately have little data assigned to them.

C. Generating and using suggestions

Given the model parameters estimated from the training data, we can generate suggested partial paths from each strategy, and use those to bias the search of a sample-based planner. In our experiments, we have modified the SBL planner to accept these suggestions, but we expect that most sample-based planners could be similarly modified. We describe the way in which we modified the SBL planner below.

For a new planning problem with task-template description x, each strategy k generates a vector $y^k = w_k \cdot x$ that represents a path segment. However, in the new planning environment described by x, the path segment may have collisions. We want to avoid distracting the planner with bad suggestions, so we collision-check each configuration and each path between consecutive configurations, and split the path into valid subpaths. For example, if y is a suggested path, consisting of configurations $\langle c_1, \ldots, c_{15} \rangle$, imagine that configuration c_5 collides, as do the paths between c_9 and c_{10} and between c_{10} and c_{11}. We would remove the offending configurations, and split the path into three non-colliding segments: $\langle c_1 \ldots c_4 \rangle$, $\langle c_6 \ldots c_9 \rangle$, and $\langle c_{11} \ldots c_{15} \rangle$. All of the non-colliding path suggestions from each strategy are then given to the modified SBL planner, along with the initial start and goal query configurations.

The original SBL planner searches for a valid plan between two configurations by building two trees: T_s, rooted at the start configuration, and T_g, rooted at the goal configuration. Each node in the trees corresponds to a robot configuration, and the edges to a linear path between them. At each iteration a node n, from one of the trees is chosen to be expanded.[1] A node is expanded by sampling a new configuration n_{new} that is near n (in configuration space) and collision-free. In the function CONNECTTREES, the planner tries to connect T_s and T_g via n_{new}. To do this, it finds the node n_{close} in the other tree that is closest to n_{new}. If n_{close} and n_{new} are sufficiently close to one another, a candidate path from the start to the goal through the edge between n_{new} and n_{close} is proposed. At this point the edges along the path are checked for collision. If they are all collision-free, the path is returned. If an invalid edge is found, the path connecting the two trees is broken at the colliding edge, possibly moving some nodes from one tree to the other.

We extended this algorithm slightly, so that the query may include not only start and goal configurations, c_s and c_g, but a set of h path segments, $\langle p_1, ..., p_h \rangle$, where each path segment p_i is a list of configurations $\langle c_{i1}, ..., c_{ir} \rangle$. Note that r may be less than l, since nodes in collision with obstacles were removed. We now root trees at the first configuration c_{i1} in each of the path segments p_i, adding h trees to the planner's collection of trees. The rest of the configurations in each path segment are added as linear descendants, so each suggestion tree starts as a trunk with no branches.

Importantly, the order of the suggestions (barring those that were thrown out due to collisions) is preserved. This means that the suggested path segments are potentially more useful than a simple collection of suggested collision-free configurations, since we have reason to believe that the edges between them are also collision-free.

The original SBL algorithm chose with equal probability to expand either the start or goal tree. If we knew that the suggestions were perfect, we would simply require them to be connected to the start and goal locations by the planner. However, it is important to be robust to the case in which some or all of the suggestions are unhelpful. So, our planner chooses between the three *types* of trees uniformly: we choose the start tree and goal tree each with probability $1/3$, and one of the k previous path trees with probability $1/(3k)$. When we have generated a new node, we must now consider which, if any, of our trees to connect together. Algorithms 1 and 2 show pseudo-code for this process. The overall goal is to consider, for each newly sampled node n_{new}, any tree-to-tree connection that n_{new} might allow us to make, but still to leave all collision checking until the very end, as in the SBL planner.

T_{expand} is the tree that was just expanded with the addition of the newly sampled node n_{new}. If T_{expand} is either the start or goal tree (T_s and T_g respectively), we first try to make a connection to the other endpoint tree, through the CONNECTTREES function in the SBL planner. If this function succeeds, we have found a candidate path and determined that it is actually collision free, so we have solved the query and need only return the path.

If we have not successfully connected the start and goal trees, we consider making connections between the newly expanded tree and each of the previous path trees, T_i. The MERGETREES function tries to make these connections. The paths between these trees are not checked for collisions at this point. As mentioned above, the original algorithm has a distance requirement on two nodes in different trees before the nodes can be considered a link in a candidate path. We use the same criterion here. If the new node is found to be close enough to a node in another tree, we reroot T_i at node n_{close}, and then graft the newly structured tree onto T_{expand} at node n_{new}. If one connection is successfully made, we consider no more connections until we have sampled a new node.

Algorithm 1 TRYCONNECTIONS(n_{new}, T_{expand}) : Boolean

```
success ← false
if T_expand == T_g then
    success ← CONNECTTREES(n_new, T_expand, T_s)
else if T_expand == T_s then
    success ← CONNECTTREES(n_new, T_expand, T_g)
else
    for previous path trees T_i ≠ T_expand do
        merged ← MERGETREES(n_new, T_expand, T_i)
        if merged then
            break
        end if
    end for
end if
return success
```

IV. EXPERIMENTS

We have experimented with task-templates in several different types of environment, to show that learning generalized paths in this way can work, and that the proposed partial paths

[1]The SBL planning algorithm has many sophisticated details that make it a high-performance planner, but which we will not describe here, such as how to decide which configuration to expand. While these details are crucial for the effectiveness of the planner, we did not alter them, and so omit them for the sake of brevity.

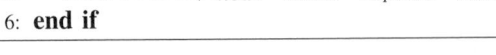

Algorithm 2 MERGETREES(n_{new}, T_{expand}, T_{elim}) : Boolean

```
1: n_close ← T_elim.CLOSESTNODE(n_new)
2: success ← CLOSEENOUGH(n_close, n_new)
3: if success then
4:     REROOTTREE(n_close, T_elim)
5:     GRAFTTREE(n_new, n_close, T_expand, T_elim)
6: end if
```

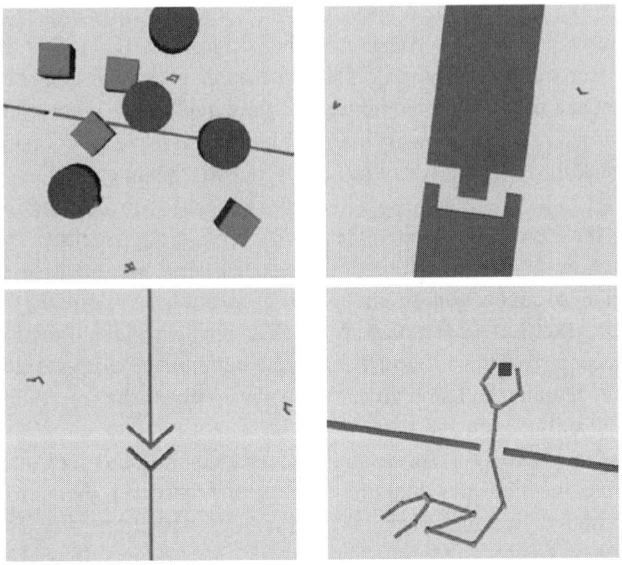

Fig. 3. Four types of environment for experimental testing. The robot shown in the first two illustrations is a planar mobile arm with varying number of degrees of freedom; in the last, it is a 8-DOF planar arm, with a simple hinged gripper.

can be used to speed up single-query planning. The simplest environment is depicted in figure 1, and the rest are shown in figure 3. Each illustration shows the obstacle placement and the start and goal locations for a particular task instance.

The first environment (depicted in figure 1, bottom) is simply a wall with a doorway in it. The robot is a mobile arm with a variable number of links in the arm, so that we can carefully explore how well the algorithms scale. Experiments done in this very simple domain were intended to be a proof of concept for both the generalized path learning from probabilistically generated paths, and the integration of suggested paths with the SBL planner.

The second domain extends the doorway domain slightly by introducing random obstacles during testing. The task template for this domain is the same as for the previous one, and in both cases training data was gathered in the absence of obstacles. Since randomly placing obstacles may completely obscure the door, we report planning times only for problems for which either the planner alone, or the modified planner, was able to find a solution.

The third domain was designed to show that the learning algorithm can learn more complicated path segments. This domain is tricky because the constrained portion is longer than in the previous examples. This means the learning is more difficult, as the training data is less well-aligned than in the

domain	average planner time	average time with suggestions
4 DOF door	1.5 s	0.7 s
5 DOF door	6.6 s	0.7 s
6 DOF door	38.1 s	1.0 s
7 DOF door	189.9 s	1.6 s
6 DOF cluttered door	82.3 s	5.3 s
4 DOF zig-zag corr.	12.9 s	2.3 s
5 DOF angle corr.	25.8 s	1.7 s
8 DOF simple gripper	78.6	52.0 s
8 DOF simple gripper with 2000 data points	74.8 s	13.7 s

Fig. 4. Planning performance results.

previous cases, where the "tight spot" is generally easy to identify.

The first three domains all share the attribute that there exists a global rigid transform which could be applied to a good path segment for one task instance to generate a good path segment for a new task instance. This is not true of motion planning tasks in general, and our approach does not rely on the existence of such a transform; nonetheless, the fourth domain was designed to show that the technique applies in cases for which such a transform does not exist.

The final domain involves a robot with different kinematics, that has to have a substantially different configuration in going through the tight spot from that at the goal.

A summary of the results in each domain is shown in figure 4. We compare the time spent by the basic SBL planner to the time required by the partial-path-suggestion method, applying each to the same planning problem. Each planning time measurement was an average of 10 training trials, in which the model was trained on 500 data points (except the last experiment, which used 2000 data points), and then tested, along with the unmodified planner, on 100 planning problems. The running times reported throughout the paper for the version of the planner with suggestions includes the time required to generate suggestions for the new problem instance.

Figure 5 shows the time required to extract the constrained portion of each path in order to generate training data, and the time required to train all of the randomly initialized EM models. We used 500 samples with a sample variance of 0.01 for estimating the tightness of each configuration. The zig-zag corridor, angle corridor and simple gripper domains use the additional step for increasing collision clearance during the tight spot identification, which makes the process substantially slower. We used 20 random restarts of the EM algorithm. We discuss the details of each domain below.

A. The door task

The doorway in this domain, and the following one, is parameterized by the x, y position of the door, and the angle of

domain	time to find tight spots	time to train models
4 DOF door	1086 s	204 s
5 DOF door	1410 s	221 s
6 DOF door	2043 s	283 s
7 DOF door	3796 s	321 s
4 DOF zig-zag corr.	14588 s	1003 s
5 DOF angle corr.	25439 s	928 s
8 DOF simple gripper	42142 s	3396 s

Fig. 5. Offline learning time.

the wall relative to horizontal. We chose a tight spot segment length of 7 for this domain.

These simple experiments show that the suggestions can dramatically decrease the running time of the planner, with speed-ups of approximately 2, 9, 38 and 118 for the different number of degrees of freedom. For the robots with 4, 5, and 6 degrees of freedom, both methods returned an answer for every query. In the 7-DOF case, the unmodified planner found a solution to 88% of the queries, while the suggestions allowed it to answer 100% successfully within the allotted number (100,000) of iterations.

B. The door task with obstacles

We also did limited experiments by adding obstacles to the test environments after training on environments with just the door described above. The goal of these experiments was to show that the suggestions generated by the learner are often helpful even in the presence of additional obstacles.

For a mobile arm robot with 6 degrees of freedom, the suggestions allow the planner to find a solution approximately 15 times faster than the planner alone, on problems where either method finds an answer. The unmodified planner fails to find a solution over 4% of the time, whereas the suggestion-enhanced planner fails less just 0.3% of the time.

C. The zig-zag corridor

The zig-zag corridor domain was designed to require more interesting paths in order to test the robustness of the parameterized path learning. The robot for this domain has 4 degrees of freedom. The domain has two parameters, the vertical displacement of the corridor, and its rotation from vertical. The rotation parameter is encoded as three separate entries in the input vector: the angle, its sine and its cosine, since each of these may be necessary for the path to be expressed as a linear function of the input. The shape of the corridor is constant. The tight spot segment length is 15.

In this domain, we found that our results were initially unimpressive, due to difficulty in identifying the tight spot in our training paths. The step of pushing paths away from collisions before extracting the tight spot allowed us to improve our results to the same level as those achieved by hand-picking the tight spot. With the improved data generated in this way, we find that the speedups are substantial, but less dramatic than the doorway cases. Figure 6 shows an example

Fig. 6. A suggested path segment for the zig zag corridor.

suggestion for this domain. It is clear that the suggestions are providing useful guidance to the planner

If the corridor is no longer allowed to rotate, leading to a one-dimensional input parameter vector, we find that planning time with suggestions is reduced to 1.2 seconds. This suggests that we have increased the difficulty of the planning problem by adding the additional parameters.

D. The angle corridor

The angle corridor domain was designed to be a simple domain that has the property that there is not a single rigid transformation for transforming paths for one task instance into paths for another task instance. In this domain, the robot has 5 degrees of freedom. The task parameters are the vertical displacement of the corridor, and the angle of the bend in the middle of the corridor, which varied from horizontal ($0°$) by up to $\pm 50°$. We again see a speedup of roughly 15 times. This demonstrates that our technique does not require that a rigid path transformation exist in order to make useful suggestions in new environments based on experience.

E. Simple gripper

The simple gripper domain was designed to test another kind of added path complexity. This domain is the simple hole in the wall (parameterized by x, y, and θ), but the robot has different kinematics and must ultimately finish by "grasping" a block. We again chose a tight spot segment length of 15.

In this domain, we find that more training data is required to achieve substantial speedup over the unmodified planner; figure 8 shows the performance as the amount of training data is increased. Even with 500 training points, however, the suggestions allow the planner to find a solution in 97.5% of the problem for which either method succeeds, compared to about 92% for the unmodified planner. As the training data increases, the success rate climbs above 99%.

Figure 7 shows example path segments from different strategies in different environments. The path segments in figure 7 are in roughly the right part of the configuration space, but they tend to involve collisions. The colliding configurations are removed, and so the suggested path segments require patching by the planner, they are less beneficial to the planner than a completely connected path segment would have been.

V. Conclusions

The simple experiments described in this paper show that it is possible to considerably speed up planning time, given experience in related problems, by describing the problems in terms of an appropriate parametric representation and then

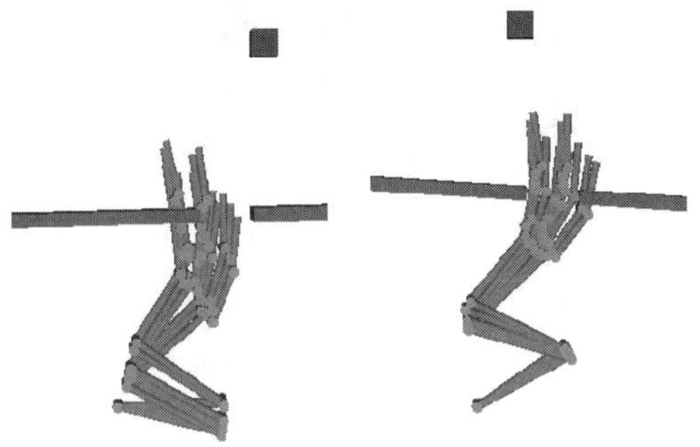

Fig. 7. Suggested path segments for the simple gripper in a single environment.

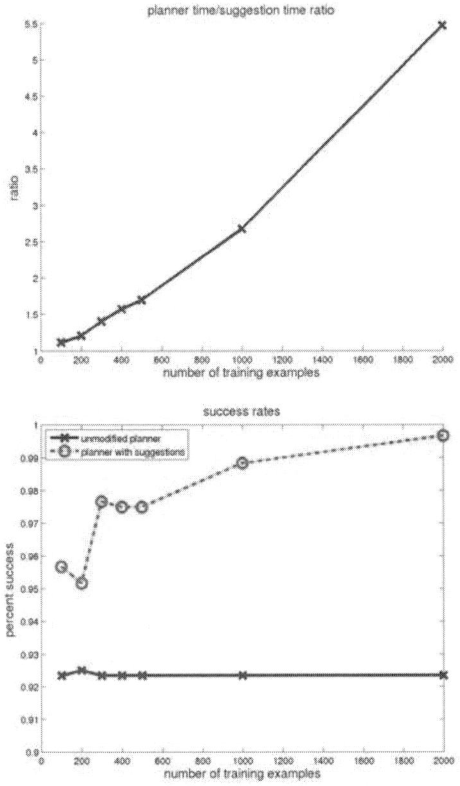

Fig. 8. Results for the simple gripper as the size of the training set is increased. The top plot shows the ratio of the unmodified average planning time to the average planning time with suggestions. The bottom plot shows the percentage of the time that each method finds a plan successfully.

learning to suggest path sub-sequences that are in the most constrained parts of the space.

To improve the performance of this method, we will need to develop a more efficient way to extract and align the sub-sequences of paths from the training examples. In addition, it would broaden the applicability of this method if we could automatically derive task-template parameters from environment descriptions and, ultimately, learn a small set of generally useful task templates from a body of experience in a variety of environments. Lastly, the issue of how to reduce the data requirements in the more complicated domains must be addressed.

Ultimately, this suggests a different direction for path-planning: systems that can routinely and quickly identify the hard spots in any new planning instance and then fill in the rest of the plan, dramatically improving their performance from experience.

ACKNOWLEDGMENT

This research was supported in part by DARPA IPTO Contract FA8750-05-2-0249, "Effective Bayesian Transfer Learning"

REFERENCES

[1] B. Burns and O. Brock. Sampling-based motion planning using predictive models. *Proceedings of the 2005 IEEE International Conference on Robotics and Automation (ICRA 2005)*, pages 3120–3125, 2005.

[2] B. Burns and O. Brock. Single-query entropy-guided path planning. *Proceedings of the 2005 IEEE International Conference on Robotics and Automation (ICRA 2005)*, 2005.

[3] S. Gaffney and P. Smyth. Trajectory clustering with mixtures of regression models. In *Conference in Knowledge Discovery and Data Mining*, pages 63–72, 1999.

[4] D. Hsu, T. Jiang, J. Reif, and Z.Sun. The bridge test for sampling narrow passages with probabilistic roadmap planners. *EIII/RSJ International Conference on Intelligent Robots and Systems*, pages 4420–4426, 2003.

[5] D. Hsu and H. Kurniawati. Workspace-based connectivity oracle: An adaptive sampling strategy for prm planning. *Proceedings of the International Workshop on the Algorithmic Foundations of Robotics (WAFR)*, 2006.

[6] D. Hsu, G. Sánchez-Ante, H l. Cheng, and J-C. Latombe. Multi-level free-space dilation for sampling narrow passages in prm planning. *Proceedings of the 2006 IEEE International Conference on Robotics and Automation (ICRA 2006)*, pages 1255–1260, 2006.

[7] N. M. Amato J-M Lien, S. L. Thomas. A general framework for sampling on the medial axis of the free space. *Proceedings of the 2003 IEEE International Conference on Robotics and Automation (ICRA 2003)*, pages 4439–4444, 2003.

[8] H. Kurniawati and D. Hsu. Workspace importance sampling for probabilistic roadmap planning. *EIII/RSJ International Conference on Intelligent Robots and Systems*, 2004.

[9] M. Morales, L. Tapia, R. Pearce, S. Rodriguez, and N. Amato. A machine learning approach for feature-sensitive motion planning. *Proceedings of the International Workshop on the Algorithmic Foundations of Robotics (WAFR)*, pages 361–376, 2004.

[10] S. Rodriguez, S. Thomas, R. Pearce, and N. Amato. Resampl: A region-sensitive adaptive motion planner. *Proceedings of the International Workshop on the Algorithmic Foundations of Robotics (WAFR)*, 2006.

[11] M. Saha and J-C. Latombe. Finding narrow passages with probabilistic roadmaps: The small step retraction method. *EIII/RSJ International Conference on Intelligent Robots and Systems*, 2005.

[12] G. Sanchez and J-C. Latombe. A single-query bi-directional probabilistic roadmap planner with lazy collision checking. In *International Symposium on Robotics Research*, 2001.

[13] T. Siméon, J. Laumond, and C. Nissoux. Visibility-based probabilistic roadmaps. *Proceedings of the IEEE International Conference on Intelligent Robots and Systems*, pages 1316–1321, 1999.

[14] A. F. van der Stappen V. Boor, M. H. Overmars. The gaussian sampling strategy for probabilistic roadmap planners. *Proceedings of the 1999 IEEE International Conference on Robotics and Automation (ICRA 1999)*, pages 1018–1023, 1999.

[15] J. P. van den Berg and M. H. Overmars. Using workspace information as a guide to non-uniform sampling in probabilistic roadmap planners. *The International Journal of Robotics Research*, 24(12):1055–1071, 2005.

[16] Y. Yang and O. Brock. Efficient motion planning based on disassembly. *Proceedings of Robotics: Science and Systems*, 2005.

[17] A. Yershova, L. Jaillet, T. Siméon, and S. LaValle. Dynamic-domain rrts: Efficient exploration by controlling the sampling domain. *Proceedings of the 2005 IEEE International Conference on Robotics and Automation (ICRA 2005)*, 2005.

Emergent Task Allocation for Mobile Robots

Nuzhet Atay
Department of Computer Science and Engineering
Washington University in St. Louis
Email: atay@cse.wustl.edu

Burchan Bayazit
Department of Computer Science and Engineering
Washington University in St. Louis
Email: bayazit@cse.wustl.edu

Abstract—**Multi-robot systems require efficient and accurate planning in order to perform mission-critical tasks. However, algorithms that find the optimal solution are usually computationally expensive and may require a large number of messages between the robots as the robots need to be aware of the global spatiotemporal information. In this paper, we introduce an emergent task allocation approach for mobile robots. Each robot uses only the information obtained from its immediate neighbors in its decision. Our technique is general enough to be applicable to any task allocation scheme as long as a utilization criteria is given. We demonstrate that our approach performs similar to the integer linear programming technique which finds the global optimal solution at the fraction of its cost. The tasks we are interested in are detecting and controlling multiple regions of interest in an unknown environment in the presence of obstacles and intrinsic constraints. The objective function contains four basic requirements of a multi-robot system serving this purpose:** *control regions of interest*, *provide communication between robots*, *control maximum area* **and** *detect regions of interest*. **Our solution determines optimal locations of the robots to maximize the objective function for small problem instances while efficiently satisfying some constraints such as avoiding obstacles and staying within the speed capabilities of the robots, and finds an approximation to global optimal solution by correlating solutions of small problems.**

I. Introduction

Several real life scenarios, such as fire fighting, search and rescue, surveillance, etc., need multiple mobile robot coordination and task allocation. Such scenarios generally include distinct regions of interest that require the attention of some robots. If the locations of these regions are not known, the mobile robots need to explore the environment to find them. In this paper, we propose a solution to the problem of detecting and controlling multiple regions of interest in an unknown environment using multiple mobile robots. In our system, we assume a bounded environment that is to be controlled by a group of heterogeneous robots. In this environment, there are regions of interest which need to be tracked. These regions are dynamic, i.e. they can appear at any point, anytime and can move, spread or disappear. Each region may require more than one robot to track and control. Robots do not have initial information about the environment, and the environment is only partially-observable by the robots. Each robot has wireless communication capability, but its range is not uniform. Two robots can communicate between each other only if both of them are in the communication range of each other. They can have different speed limits and are equipped with the sensors to identify the obstacles and the regions of interest if they are within robots' sensing range. Sensor ranges on these robots are not necessarily uniform. The environment can have static or dynamic obstacles, and the robots need to avoid them in order to perform their tasks.

We propose an emergent solution to the task allocation problem for heterogeneous robots. The tasks we are interested in are: (i) *covering all regions of interest*, (ii) *providing communication between as many robots as possible*, (iii) *controlling maximum total surface by all the robots*, (iv) *exploring new regions*. Our objective is to maximize these items while satisfying the constraints such as avoiding the obstacles or moving within the speed capabilities of individual robots. Additional constraints we are considering are the communication between two robots (which exists only if either two robots are in the communication range of each other or there is a route between them through other robots satisfying the communication constraints), and, the sensing of the obstacles and regions of interest when they are within the robots' sensor range. Our approach is general enough to be easily adapted to additional constraints and objectives, making it customizable for various mobile robot problems.

Several linear programming based solutions have been proposed for mobile robot task allocation problem. Although these proposals are generally successful in finding the optimal solution, they usually require collecting information about all robots and regions of interest, and processing this information at a central location. As a result, these approaches can be infeasible in terms of the computation time and communication cost for large groups. In order to provide scalability and efficiency, we are proposing an emergent approach. In this approach, each robot solves a partial problem based on its observations, then exchanges information (such as intentions and directives) with the robots in the communication range to maintain coordination. The system is fully distributed which allows this technique to be applied to any number of robots with computation and communication cost limited by constant parameters which can be defined according to the application requirements. We experimentally show that this approach gives results comparable to global optimal solution, and performs hundreds of times faster with little communication cost.

Since we use mixed integer linear programming for the solution of the partial problems, our contributions also include a customizable multi-robot task allocation solver which can be used to find global optimal solution under the given constraints. In contrast to other linear programming solutions, we also present an efficient way to check obstacle collisions.

While we are concentrated on the mobile robots, our solution is applicable to other distributed task allocation problem as long as a function to evaluate the goodness of the solution is defined. The technical report version of this paper that includes the details of the mixed integer linear programming solution with the description of constraints and variables, as well as some proofs including the convergence of our approach to the global solution, extensions that show the flexibility of the approach, and a larger set of experiments on different environments can be found at [1].

The rest of the paper is organized as follows. The next section gives a summary of the related research and brief comparison to our approach when it is applicable. Section III gives the problem definition. Section IV describes our mixed integer linear programming solution, and Section V explains the emergent behavior task allocation approach. We present simulation results in Section VI and Section VII concludes our paper.

II. Related Work

Multi-robot task allocation has been studied extensively because of the importance of application areas. One quite popular approach to this problem is utilizing negotiation or auction based mechanisms. In this approach, each distributed agent computes a cost for completing a task, and broadcasts the bid for that task. Auctioneer agent decides the best available bid, and winning bidder attempts to perform this task. Following the contract-net protocol [2], several variations of this method has been proposed [3]–[7]. Another important approach is using behavior based architecture. ALLIANCE [8] is a behavior-based architecture where robots use motivational behaviors such as robot impatience and robot acquiescence. These behaviors motivate robots to perform tasks that cannot be done by other robots, and give up the tasks they cannot perform efficiently. BLE [9] is another behavior-based architecture which uses continuous monitoring of tasks among robots and best fit robot is assigned to each task. A detailed analysis and comparison of these methods can be found at [10], [11]. These methods propose distributed algorithms where resource allocation is an approximation to the global optimum. The main difference between these methods and our approach is that we are using a formulation that can provide global optimum solution when information propagation is not limited. However, instead of finding the global optimal solution using all the information which has high computation and communication cost, we distribute computation and information processing among robots and reach an approximation to the global optimal solution through iteration.

Task allocation problem is also studied in the context of cooperation of Unmanned Aerial Vehicles (UAVs). Several methods are proposed for search and attack missions of UAVs [12]–[20]. Our method is similar to the methods proposed in [13], [14], [17], [20], since these methods are also using mixed-integer linear programming task allocation. However, in these papers, the problem is defined as minimizing mission completion time while UAVs visiting predetermined waypoints and avoiding no-fly zones. The solution to this problem is formulated as finding all possible combinations of task allocations, and choosing the best combination. This definition of task allocation is actually quite different than our problem definition. Our aim is to explore environment, find regions of interest, and assign tasks optimally obeying the constraints imposed at that moment. In other words, we are finding a solution in real-time, instead of finding an initial plan and executing it.

III. Problem Definition

In our problem definition, there are regions of interest we want robots to explore and cover. In the rest of the paper, we will call these regions “targets”. Since larger areas can be represented with multiple points, without loss of generality, we assume targets are represented as points in planar space. A target is assumed to be covered if there are enough robots that have the target in their sensing range. The number of robots required to cover a target varies for each target. We assume the future locations of known targets after a time period can be predicted. Our primary purpose is to find locations of robots in order to cover as many targets as possible using the estimated locations of targets. While covering all the targets, it is also desirable to provide communication between as many robots as possible because this will allow robots to exchange the information about the environment and the targets. In a centralized approach, this also leads to a better solution since the solver will be aware of more information. It is also preferable that robots need to cover as much area as possible in addition to covering targets to increase the chances of detecting other undiscovered targets. Similarly, in order to discover new targets and avoid waiting at the same location when no targets are being tracked, the robots are expected to explore new regions.

We define the state of the system as current locations of targets, number of robots needed to cover a target, current positions of the robots, positions of the obstacles, previously explored regions, and each robot’s speed, communication range and sensing range. The output of our algorithm is the optimal locations of the robots for the next state of the system after a brief period of time. Please note that, we assume we can predict the location of the targets at the next step. There are approaches for motion prediction that can be used for this purpose [21]. We also assume that there are no sensor or odometry errors, however, implementation of our method on real robots can introduce these errors. The method we are planning to utilize for handling noisy measurements, sensor errors and mechanical errors like slippage or odometry errors takes advantage of communication among nearby robots. We believe our approach promotes robots to stay in the contact as much as possible and make it possible to share as much sensor information as possible.

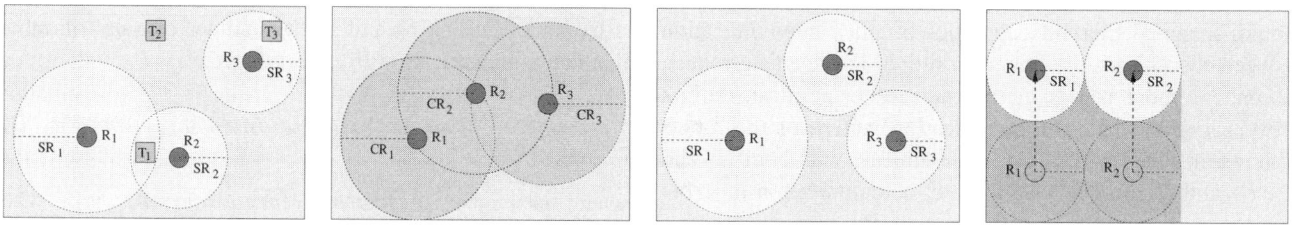

Fig. 1. SR stands for sensing range, and CR stands for communication range (a) A target is covered when it is in sensing range of some robots, where number of robots is determined according to the requirements of the target. Robots R_1 and R_2 cover T_1, while R_3 covers T_3. T_2 is not covered. (b) Two robots can communicate if both robots are in communication range of each other. R_2 can communicate with R_1 and R_3, and works as a hub between R_1 and R_3 which cannot communicate directly. (c) Maximum area coverage is obtained if sensing range of robots do not overlap. In the figure, sensing regions of robots barely touch each other (d) Robots mark regions they explored before, and move towards unexplored regions. R_1 and R_2 move upward toward unexplored region after marking dark (blue) region as explored

IV. Mixed Integer Linear Programming for Task Allocation

Although our main contribution is the emergent task allocation, we first would like to show how a centralized approach can be utilized to find the optimal placement of robots after a defined time period. In the next section, we will show how individual robots can use the same approach to solve their partial problems to achieve emergent task allocation.

Our centralized approach utilizes a mixed integer linear program. Either a designated robot runs the solver or each robot in a group executes the same solver with the same data to find its placement. A group consists of the robots that are in the communication range of each other, hence states and observations of all the robots are known to the solver(s). If there are multiple groups of robots that cannot communicate with each other, each group will have its own task allocation based on its world view. If two groups merge, they can share their knowledge. The program runs periodically to find the best placements for each robot. It also runs if a new event happens, such as the discovery of an obstacle or a target. The linear program should satisfy some constraints: (i) an evaluated location is not acceptable if the robot cannot reach there either because of its speed limits or because of an obstacle, (ii) two robots cannot communicate if one of them is outside the communication range of the other, (iii) an obstacle or target is detectable only if it is within the sensing range of the robot. Our goal is then to maximize the number of targets tracked, the number of robots that can communicate with each other, the area of the environment covered by the robot sensors, and the area of the environment that was explored. In the next subsections, we will first discuss different objective functions and constraints, then we will show our overall optimization criterion and we will discuss the complexity. We give only the overview of the linear program because of space limitations, but detailed formulations and explanations can be found in the technical report version [1].

A. Obstacle Avoidance

In our system, we assume there are only rectangular shaped obstacles for the sake of simplicity of defining linear equations. However, more general shaped obstacles can be represented as rectangular meshes. When considering obstacles, we are not finding a path to avoid them, but we are finding whether or not it is possible to avoid them with the robot speed and timestep as the constraints. As it is mentioned before, output of the linear program is the final positions of the robots. When computing these positions, we utilize Manhattan paths to identify if there is a way for a robot to avoid an obstacle. As long as there is a Manhattan path that bypasses the obstacle and has a length that is possible for the robot to traverse under the given speed constraints, we consider the final position of the robot as a feasible configuration. Otherwise, that configuration is eliminated. Once a position is selected, more advanced navigation algorithms can be utilized to find more efficient paths. The alternative approach, i.e., finding exact path, requires finding intermediate states of the system at a fine resolution which increases complexity drastically. Please note that we are not aware of any other linear programming approach that addresses navigation problem.

B. Target Coverage

A target can be considered covered only if the number of robots following it is greater than or equal to its coverage requirement. [1] A robot can sense and control a target only if its sensing range is greater than or equal to the distance between itself and the target. A sample organization of the robots and targets is shown in Fig. 1(a). R_1 and R_2 are covering target T_1 and R_3 is covering T_3 while T_2 is not covered by any of the robots.

C. Communication

Each robot has a communication range. A robot can have a duplex communication link to another robot only if each robot is in the communication range of the other one. However, robots can communicate between each other with the help of other robots. So, if two robots cannot directly communicate with each other, but they share a common robot both of which can communicate, we assume that they can communicate. In other words, transitive links are allowed in the system. It

[1] Please see the technical report [1] for the proof that our optimization criterion results in continuous target coverage of all targets, if this optimization has highest priority.

should be noted that this condition implies communication between robots with the help of multiple intermediate robots, i.e. one or more robots can participate in a transitive link between two robots. A communication pattern of the robots is shown in Fig. 1(b). R_2 can communicate with both R_1 and R_3. R_1 and R_3 do not have a direct communication link, but they can communicate with the help of R_2.

D. Area Coverage

Robots have limited and constant sensing range, so the only way to maximize area coverage is by preventing the overlap of sensing ranges of robots. An ideal area coverage for the robots is represented in Fig. 1(c), where robots have no overlapping sensing range.

E. Exploration

In order to explore the environment, robots need to know places they have visited recently. We store this information as rectangular regions defining explored areas. Then the linear program tries to move robots into unexplored regions by checking the final position of the robots. So, the program gives a final position not located in an explored region. [2] A sample exploration scenario is shown in Fig. 1(d). Dark (blue) region is explored in the first step, so robots try to locate themselves outside of the explored area.

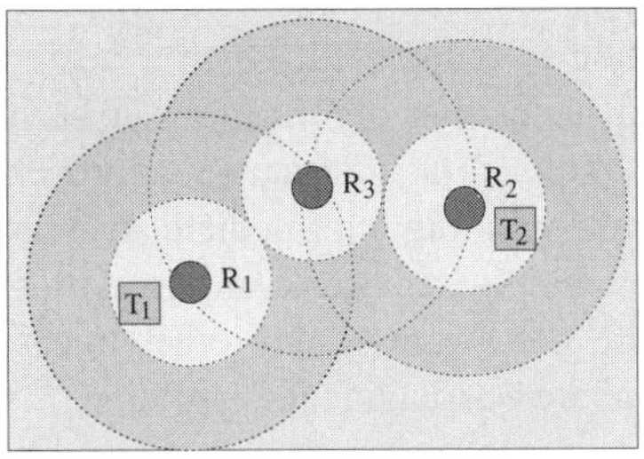

Fig. 2. An example distribution of robots providing optimum target coverage, communication and area coverage. Robot R_1 covers target T_1 and R_2 covers target T_2. R_3 is located to provide communication between them, and its sensing range does not overlap with others. Dark colored circles represent communication range, light colored circles represent sensing range.

F. Optimization Criterion

Optimization criterion consists of four components, *target coverage*, *communication between robots*, *area covered by the robots* and *the number of robots located in unexplored regions.*

Target Coverage: We utilize the number of targets that are covered, i.e.,

$$T = \sum_{j=1}^{n} coverage_j \tag{1}$$

where n=number of targets, $coverage_j$ is 1 when the number of robots that are covering $target_j$ is greater than or equal to the minimum requirement for that target, 0 otherwise.

Communication: We utilize the number of pairs of robots that can communicate with each other, i.e.,

$$C = \sum_{i=1}^{n}\sum_{j=1}^{n} communication_{ij} \tag{2}$$

where n=number of robots, $communication_{ij}$ is 1 when robots i and j are within their communication range or they can communicate with the help of other robots, 0 otherwise.

Area Coverage: We utilize the number of pairs of robots whose sensor ranges do not intersect, i.e.,

$$A = \sum_{i=1}^{n}\sum_{j=1}^{n} area_{ij} \tag{3}$$

where n=number of robots, $area_{ij}$ is 1 when robots i and j cover non-overlapping regions, 0 otherwise.

Exploration: We utilize the number of robots in unexplored regions, i.e.,

$$E = \sum_{i=1}^{n}\sum_{j=1}^{m} exploration_{ij} \tag{4}$$

where n=number of robots, m=number of explored regions, $exploration_{ij}$ is 1 if the robot i is not in the explored region j, 0 otherwise.

Optimization Criterion: Our objective function is weighted sum of the above components.

$$maximize\ \alpha T + \beta C + \gamma A + \delta E \tag{5}$$

where α, β, γ, and δ are constants defining priorities.

Figure 2 represents an optimal distribution of robots according to this optimization criterion. Robots arrange themselves so that they cover all targets, provide communication between each other, and cover as much area as possible.

G. Complexity

Our formulation results in a mixed-integer linear program, which is NP-Hard in the number of binary variables, so complexity of our program is dominated by the number of binary variables. Definitions and properties of binary variables can be found at the technical report [1]. For a problem with n targets, m robots, p obstacles and q explored regions, there are $n + nm + 2nn + 5mq + 4mp$ binary variables. So, the complexity can be stated as $O(n + nm + n^2 + mq + mp)$.

V. Emergent Task Allocation

As we have mentioned in the previous section, finding the optimal solution is an NP-Hard problem. While it may be possible to solve simple problems with on-board processors, finding solution for larger networks is very expensive even for a more powerful central server (because of both the cost of computation and the number of messages). In order to overcome this problem, we propose a distributed approach where each robot in the network finds a local solution based on the information from the vicinity of the robot. This approach utilizes the mixed integer linear program we described in Section IV. The local vicinity of the robot contains the region covered by

[2]Please see the technical report [1] for the proof that given sufficient number of robots for communication and target tracking, our algorithm will result in the exploration of the all environment.

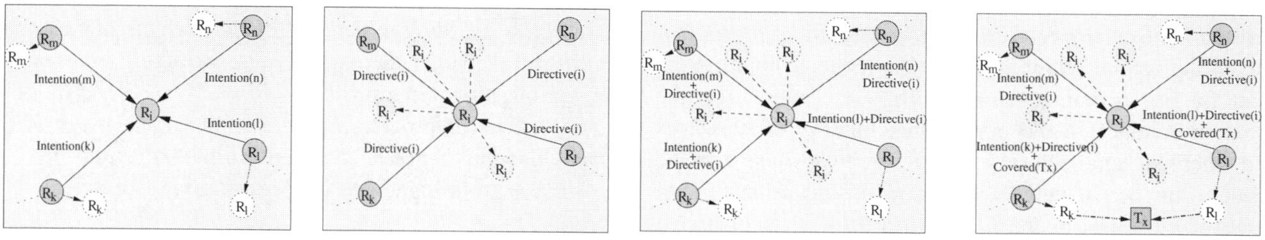

(a) (b) (c) (d)

Fig. 3. Information exchange for emergent task allocation: (a) Intentions, (b) Directives, (c) Intentions and Directives, (d) Intentions, Directives and Target Assignment. The dashed-circles connected to the neighboring robots $R_{k,l,m,n}$ represent their intentions, the dashed-circles connected to the R_i represent the directives to that robot by its neighbors.

the robot and its first-degree neighbors (1-hop away). Each robot uses the information about targets and obstacles that can be sensed by the robot itself and 1-hop neighbor robots in its computation. In order to increase efficiency, we further restrict the vicinity to k-closest robots if the number of 1-hop neighbors is large. While this segmentation of the problem makes the individual problems solvable by the mobile robots, each robot is concentrated on its own problem which is usually different than those of neighboring robots. As a result, its solution may be different from another robot's solution. In order to provide coordination between the neighboring robots, the robots exchange information among the neighbors (mainly contains intentions and/or directives) and update their local solutions based on this information. This exchange makes the solution of emergent task allocation comparable to that of centralized approach. Algorithm 1 summarizes this approach.

Algorithm 1 Coordination (robot i)

```
1: Find a solution with local information
2: for all k-closest 1-hop neighbor j do
3:    Send solution information to j
4:    Receive solution information from j
5: end for
6: Update solution according to new information
7: return position(i)
```

Although it is possible to iterate through lines $2-6$ several times, i.e., continuously updating the solution until it converges, we are interested in only a single exchange for efficiency purposes. In the technical report [1], we show that as the number of iterations increases, the solution converges to the global optimum. Similarly, if there is sufficient computational power on individual robots, the size of neighborhood can be increased to include the robots that are more hops away for obtaining better solution.

The information exchange between the robots could range from single position information which may require a single message between the robots to all the state information which may require multiple messages. We have selected the following methods for the information exchange:

A. Intentions

In the most simple approach, after finding a position that maximizes its utility (based on the current sensor information and neighbor information), each robot sends this location to its neighbors as its intended location. When a robot gets intentions from all neighbors, it assumes that these locations are final, and computes its own location that would maximize the utility. Note that, we still use the algorithm of Section IV, however, other robots' positions now become constraints of the system. Figure 3(a) represents this approach for robot i.

B. Directives

In the second approach, each robot computes a location for its neighbor, and sends this location to the neighbor as a directive. When a robot gets location information from all neighbors, it uses the list of locations as the potential locations, and finds the one that gives the highest value of the objective function using the linear program. The information transferred for robot i is shown in Figure 3(b).

C. Intentions and Directives

In the third approach, each robot computes optimal locations of itself and its neighbors, and sends these locations to the neighbors. When a robot gets these locations, for each potential location given by the neighbors, it evaluates the utility of that directive based on the intended locations of all neighbors. The directive that gives the highest value of the objective function is selected as the next location for that robot. This is represented in Figure 3(c) for robot i.

D. Intentions, Directives and Target Assignment Information

The last approach is similar to the third approach, but in addition to the information about locations, target assignment information is also sent to the neighbors. Target assignment states whether or not a robot is assigned to cover a target. This information can be used in different ways, but we use this so that no two robots try to cover the same target, unless that target needs to be covered by more than one robot. This approach provides better exploration and better area coverage, as robots can ignore a target and spread out when the target is covered by another robot. Figure 3(d) represents this approach for robot i.

E. Comparison to Centralized Global Optimization

Global optimization through centralized computation requires all information about the environment to be collected at one location. Assuming the central server is physically located in the center of the network and average hop count from other robots to the central server is p, average message count in the system for one planning phase is $O(p * n)$, where n is the number of robots. On the other hand, number of messages at the emergent approach is k for each robot, where k is the maximum number of neighbors that a robot can have. Total number of messages in the system is $O(k * n)$ at emergent approach. It should be noted that p is dependent on the network size, whereas k is a constant and for practical applications $p >> k$. Average delay for transmitting messages at the global approach is $O(p)$, whereas average delay is constant and 1 at emergent approach when each robot communicates to only 1-hop neighbors.

Once all the information is collected at a central location, the linear program can find the global optimal solution if the problem instance is not too big for the processing capability and the memory available. On the other hand, the solution with emergent approach is found using limited information, so the solution may not be optimal. However, as the information is shared among neighbors, the quality of the solution improves and optimal solution can be obtained if information sharing is continued until the system reaches a stable state, which is when all robots find the same solution. The proof showing that these iterations finally converge can be found at [1].

VI. Simulation Results

In our simulations, we want to evaluate how well emergent task allocation (ETA) behaves with respect to centralized global optimization approach (CGO) using mixed integer linear programming. For this purpose we have designed an experimental scenario and run ETA with different information exchange methods and CGO. Next, we will discuss the environment, present the behaviors of individual techniques and compare them. Since our main application is mobile sensors, we are interested in finding how well either technique can cover targets. For this purpose we compared the number of targets covered by each technique as well as the solution times. We also experimented with larger networks of robots and targets on bigger environments to show the scalability of ETA. Simulation results with 20 robots - 10 targets and 30 robots - 15 targets can be found at [1].

A. Environment

The environment is bounded and has size 12×12. There are three rectangular obstacles, which are located at $\{(0,4),(5,6)\}$, $\{(4,8),(8,10)\}$ and $\{(8,2),(10,6)\}$ (darkest (dark blue) regions in Figs. 4 and 5). In the environment there are 8 robots which are located at point $(0,0)$, and 6 targets whose locations are unknown initially. The targets follow predefined paths and we assume we can predict their locations for the next timestep, if their locations are known at the current step. Robots are heterogeneous with sensing range and speed differing between 1-2, and communication range 4. Detailed parameters can be found at [1]. All targets except the target t_3 require a single robot for coverage, whereas t_3 requires two robots. Timestep is selected to be 4, so robots arrange themselves according to the environment which they estimate to be in 4 steps. In the experiments, we chose constants at the optimization criterion as $\alpha > \beta > \gamma > \delta$. In other words, the linear program optimizes (1) *target coverage*, (2) *communication between robots*, (3) *area coverage* and (4) *exploration* from highest to lowest priority, respectively.

B. Centralized Global Optimization (CGO)

We show a sample execution of our program to highlight the properties of the solution. Robots start exploring the environment by moving out of the region they explored when they were all at $(0,0)$. The initial explored region is the rectangle $\{(0,0),(1,1)\}$ because the robot with highest sensing range can sense a region of radius 2.

Since there are no targets detected yet, and the communication constraints are satisfied, the robots try to cover as much area as possible while obeying the movement constraints. The new environment is shown in Fig. 4(a) where blue (darker) areas indicate explored regions. Exploration reveals targets t_1 and t_2, and predicts their positions to be $(0,4)$ and $(2,2)$, respectively. Optimal allocation is shown in Fig. 4(b). Robots r_6 and r_8 cover targets, and other robots continue exploration while staying within the communication range. Next, target t_3 is found, which requires two robots to be covered. Robots r_2, r_3 and r_7 continue exploration and r_6 works as the communication bridge while remaining robots are assigned to the targets. Distribution of robots is shown in Fig. 4(c). Two other targets, t_4 and t_5 are discovered at the next step. Moreover, targets t_1 and t_2 move faster than their controller robots, r_1 and r_4, which cannot catch them. However, global optimization finds a solution to this problem by assigning the covering task to other robots that can reach the targets (Fig. 4(d)). Target t_6 is discovered at the next step. At this time, it is not possible to cover all the targets while keeping the communication between all robots. Since target coverage is given more importance, robots are distributed into two independent groups. Robots r_3 and r_5 form one team, while others form the other team. Each team has communication in itself, but cannot reach to the other team. An optimal solution is found and applied for each team. Fig. 4(e) represents result of two optimal solutions. Targets t_1 and t_5 leave the environment at the next step. Team of robots r_3 and r_5 has one target to follow, so while one robot follows target, the other robot, in this case r_3, which is the faster robot, continues exploration. The other team covers all targets, and provides communication in itself. Fig. 4(f) shows the final state of the environment which is totally explored.

Our experiment shows that we can successfully assign tasks to the robots. We can successfully cover individual targets, keep communication distance as long as possible, provide maximum area coverage and explore the environment.

C. Emergent Task Allocation

In this section, we present the performance of the distributed emergent approach under the same scenario. We have run emergent approach for each information exchange method described in Section V with k-closest neighbors where $k = 4$.

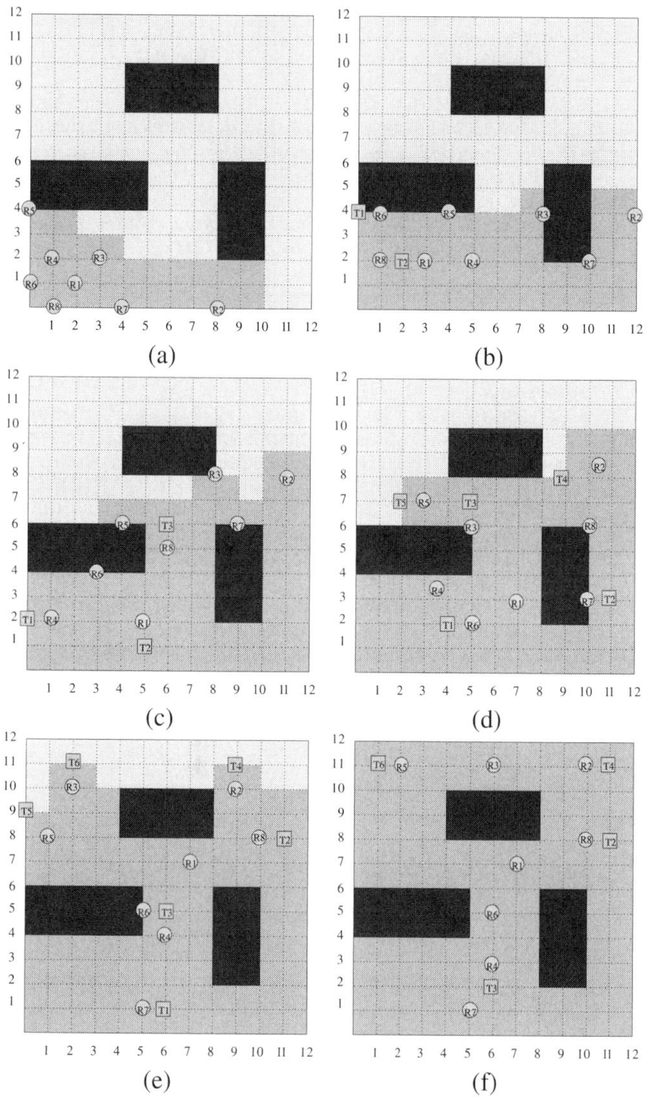

Fig. 4. Sample execution of the Centralized Global Optimization. Robots are represented as circles, and targets are represented as squares. Dark blue (darkest) regions are obstacles, blue (darker) regions are explored regions, and gray (light gray) regions are unexplored regions.

Table I presents running times for each method. It can be seen that there is no significant difference in computation times among ETA methods. On the other hand, as the amount of shared information increase, the performance of ETA increases (see Table II which shows the number of targets covered at each time step). We obtain the worst performance if we just utilize "Intentions", i.e., the least number of targets is covered. The performance of the "Directives" and "Intentions and Directives" are similar and both are better than "Intentions" which suggests that "Directives" are more important. However, both fail to capture all targets. This is because no target information is shared among neighbors, so multiple robots can assign themselves to the same target independently. Finally when the target information is distributed, we obtain the best performance with "Intentions, Directives and Target Assignment" where ETA can cover all the targets. Figures 5 (a) to (f) shows the behavior of ETA in this case. We also run ETA on larger environments and networks to measure the scalability of this approach [1]. These experiments show that the quality of the solution is satisfactory also in large networks, and execution time per robot stays constant irrespective of the network size.

TABLE I
AVERAGE, MAXIMUM AND MINIMUM EXECUTION TIMES PER ROBOT FOR EACH METHOD

	avg. time	*max. time*	*min. time*
ETA w/ Int.	4 s	11 s	<1 s
ETA w/ Dir.	7 s	15 s	<1 s
ETA w/ Int.Dir.	7 s	16 s	<1 s
ETA w/ Int.Dir.Tgt	5 s	16 s	<1 s
CGO	36 min	120 min	9 min

TABLE II
RATIO OF TARGETS COVERED BY ROBOTS FOR EACH METHOD

steps	1	2	3	4	5
ETA w/Int.	2/2	2/3	2/5	2/6	2/4
ETA w/Dir.	2/2	3/3	3/5	4/6	2/4
ETA w/Int.Dir.	2/2	3/3	3/5	4/6	2/4
ETA w/Int.Dir.Tgt	2/2	3/3	5/5	6/6	4/4
CGO	2/2	3/3	5/5	6/6	4/4

Please remember that we chose to exchange information among neighbors only once for each planning phase because of the time limitations of real world applications. However, each update increases the performance and if updates are continued until the system reaches a stable state, the final state will be closer to the global optimal solution.

D. Comparison of CGO and ETA

As it is seen at Table II, the performance of ETA with "Intentions, Directives and Target Assignment" is similar to CGO. On the other hand, ETA is 400 times faster than CGO (Table I). This shows the main drawback of CGO which is the infeasible computation time as the number of robots and targets increase (e.g., when the number of robots is 8 and number of targets is 6, the execution time can reach 2 hours).

VII. CONCLUSIONS

We have presented an emergent task allocation method to solve the task allocation problem of multiple heterogeneous robots for detecting and controlling multiple regions of interest in an unknown environment under defined constraints. We compared our results to a mixed integer linear programming approach which finds the global optimal solution for the given state of the robots, targets and environment. Emergent approach guarantees that each robot in the system computes a limited sized problem, no matter what the number of robots

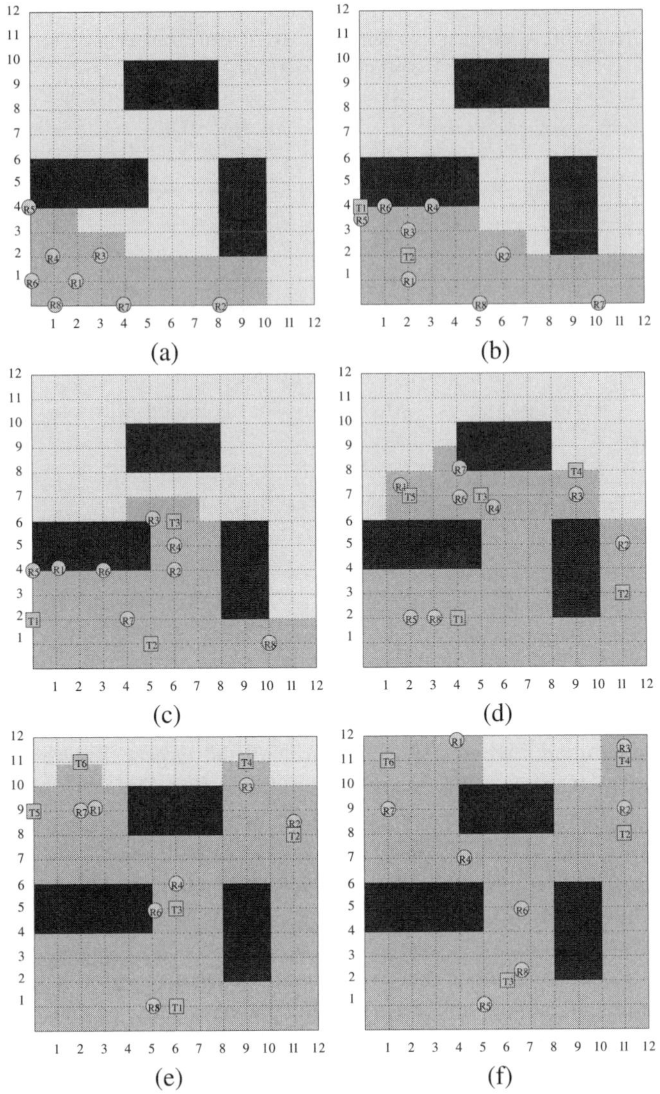

Fig. 5. Sample execution of the Emergent Task Allocation. Robots are represented as circles, and targets are represented as squares. Dark blue (darkest) regions are obstacles, blue (darker) regions are explored regions, and gray (light gray) regions are unexplored regions.

or targets in the environment is. Our simulation results and analysis show that our approach performs similar to global optimal solution at the fraction of its cost (hundreds of times faster). We are planning to implement this approach to in-network task allocation for sensor networks.

References

[1] N. Atay and B. Bayazit, "Emergent task allocation for mobile robots through intentions and directives," Dept. of Computer Science and Engineering, Washington University in St. Louis, Tech. Rep. WUCSE-2007-2, Jan 2007.

[2] R. Davis and R. G. Smith, "Negotiation as a metaphor for distributed problem solving," *Artificial Intelligence*, vol. 20, pp. 63–109, 1983.

[3] B. P. Gerkey and M. J. Matarić, "Sold!: Auction methods for multirobot coordination," *IEEE Transactions on Robotics and Automation*, vol. 18, no. 5, pp. 758–786, October 2002.

[4] S. Botelho and R. Alami, "M+: a scheme for multi-robot cooperation through negotiated task allocation and achievement," in *Proc. IEEE Int. Conf. Robot. Autom. (ICRA)*, Detroit, Michigan, May 1999, pp. 1234–1239.

[5] R. Zlot and A. Stentz, "Complex task allocation for multiple robots," in *Proc. IEEE Int. Conf. Robot. Autom. (ICRA)*, Barcelona, Spain, April 2005, pp. 1515–1522.

[6] G. Thomas, A. M. Howard, A. B. Williams, and A. Moore-Alston, "Multi-robot task allocation in lunar mission construction scenarios," in *IEEE International Conference on Systems, Man and Cybernetics*, vol. 1, Hawaii, October 2005, pp. 518–523.

[7] T. Lemaire, R. Alami, and S. Lacroix, "A distributed tasks allocation scheme in multi-uav context," in *Proc. IEEE Int. Conf. Robot. Autom. (ICRA)*, New Orleans, LA, April 2004, pp. 3822–3827.

[8] L. E. Parker, "Alliance: An architecture for fault tolerant multirobot cooperation," *IEEE Transactions on Robotics and Automation*, vol. 14, no. 2, pp. 220–240, April 1998.

[9] B. B. Werger and M. J. Matarić, "Broadcast of local eligibility for multi-target observation," in *5th International Symposium on Distributed Autonomous Robotic Systems (DARS)*, Knoxville, TN, October 4-6 2000, pp. 347–356.

[10] B. P. Gerkey and M. J. Matarić, "Multi-robot task allocation: Analyzing the complexity and optimality of key architectures," in *Proc. IEEE Int. Conf. Robot. Autom. (ICRA)*, Taipei, Taiwan, September 17-22 2003, pp. 3862–3867.

[11] ——, "A formal analysis and taxonomy of task allocation in multi-robot systems," *Intl. Journal of Robotics Research*, vol. 23, no. 9, pp. 939–954, September 2004.

[12] K. Nygard, P. Chandler, and M. Pachter, "Dynamic network flow optimization models for air vehicle resource allocation," in *The American Control Conference*, Arlington, Texas, June 25-27 2001, pp. 1853–1858.

[13] J. Bellingham, M. Tillerson, A. Richards, and J. How, "Multi-task allocation and path planning for cooperating uavs," in *Conference on Coordination, Control and Optimization*, November 2001, pp. 1–19.

[14] C. Schumacher, P. Chandler, M. Pachter, and L. Pachter, "Uav task assignment with timing constraints," in *AIAA Guidance, Navigation, and Conference and Exhibit*, Arlington, Texas, 2003.

[15] Y. Jin, A. Minai, and M. Polycarpou, "Cooperative real-time search and task allocation in uav teams," in *42nd IEEE Conference on Decision and Control*, Maui, Hawaii USA, December 2003, pp. 7–12.

[16] C. Schumacher, P. Chandler, S. Rasmussen, and D. Walker, "Task allocation for wide area search munitions with variable path length," in *The American Control Conference*, Denver, Colorado, June 2003, pp. 3472–3477.

[17] M. Alighanbari, Y. Kuwata, and J. How, "Coordination and control of multiple uavs with timing constraints and loitering," in *The American Control Conference*, vol. 6, Denver, Colorado, June 4-6 2003, pp. 5311–5316.

[18] D. Turra, L. Pollini, and M. Innocenti, "Fast unmanned vehicles task allocation with moving targets," in *43rd IEEE Conference on Decision and Control*, Atlantis, Paradise Island, Bahamas, December 14-17 2004, pp. 4280–4285.

[19] P. B. Sujit, A. Sinha, and D. Ghose, "Multi-uav task allocation using team theory," in *44th IEEE International Conference on Decision and Control, and the European Control Conference*, Seville, Spain, December 12-15 2005, pp. 1497–1502.

[20] M. A. Darrah, W. Niland, and B.M.Stolarik, "Multiple uav dynamic task allocation using mixed integer linear programming in a sead mission," in *Infotech@Aerospace*, Arlington, Virginia, September 26-29 2005.

[21] A. Elganar and K. Gupta, "Motion prediction of moving objects based on autoregressive model," *IEEE Transactions on Systems, Man and Cybernetics-Part A:Systems and Humans*, vol. 28, no. 6, pp. 803–810, November 1998.

Passivity-Based Switching Control for Stabilization of Wheeled Mobile Robots

Dongjun Lee
Department of Mechanical, Aerospace and Biomedical Engineering
University of Tennessee - Knoxville
502 Dougherty Hall, 1512 Middle Dr.
Knoxville, TN 37996 USA
E-mail: djlee@utk.edu

Abstract— **We propose a novel switching control law for the posture stabilization of a wheeled mobile robot, that utilizes the (energetic) passivity of the system's open-loop dynamics with non-negligible inertial effects. The proposed passivity-based switching control law ensures that the robot's (x, y)-position enters into an arbitrarily small (as specified by user-designed error-bound) level set of a certain navigation potential function defined on the (x, y)-plane, and that its orientation converges to a target angle. Under this passivity-based switching control, the robot moves back and forth between two submanifolds in such a way that the navigation potential is strictly decreasing during this inter-switching move. Once the system's (x, y)-trajectory enters such a desired level set, at most one more switching occurs to correct orientation. Simulation is performed to validate/highlight properties of the presented control law.**

I. Introduction

Wheeled mobile robots define one of the most important classes of robotic systems in practice. Let alone the ubiquitous automobiles, we can find them in such a variety of practical applications as material handling (e.g. Cobot [1]), space exploration (e.g. NASA Rover [2]), smart wheelchairs (e.g. NavChair [3]), and, recently, mobile sensor networks (e.g. [4]).

In addition to its practical importance, this wheeled mobile robot also constitutes a theoretically rich dynamical system due to the presence of nonholonomic constraints (i.e. no-slip condition of wheels). These constraints only restrict the admissible velocity space but not that of the configuration [5]. Because of these nonholonomic constraints, control design and analysis become substantially more involved. For instance, as shown in the celebrated work [6], any continuous time-invariant state feedback, which would work just fine if there is no nonholonomic constraints, now can not stabilize the position and orientation of the wheeled mobile robot at the same time (i.e. posture stabilization problem).

On the other hand, (energetic) passivity of open-loop robotic systems (i.e. passive with kinetic energy and mechanical power as storage function and supply rate [7]) has been a very powerful concept in many control problems in robotics: general motion control problem including adaptive and robust controls [8], teleoperation and haptic interface [9], [10], biped walking robot [11], and multirobot cooperative control [12], [13], to name a few. This is one of the most fundamental properties of robotic systems, holding for any choice of coordinate systems (e.g. euler-angle or quaternion for $SO(3)$). However, so far, it has been largely overlooked for the control problem of wheeled mobile robots and nonholonomic mechanical systems in general.

In this paper, we aim to bring this fundamental passivity property[1] into the posture stabilization problem of a wheeled mobile robot with second-order Lagrangian dynamics. The outcome is a novel switching control law, which, by utilizing the passivity property and energetic structure of the wheeled mobile robot, can ensure that the robot's orientation converges to a target value, while driving the robot's (x, y)-trajectory to a desired position on the plane within a user-specified error-bound.

The main idea of our control design can be roughly summarized as follows. Consider a wheeled mobile robot in Fig, 1. We first decouple its (x, y)-dynamics and orientation dynamics from each other by a certain feedback. Then, on the (x, y)-plane, we design a navigation potential [14], which may also incorporate other control objectives on the top of the posture stabilization (e.g. obstacle avoidance). With this potential, we switch the orientation angle between θ_1 and θ_2 (with some damping control) in such a way that the wheeled mobile robot will move between two one-dimensional submanifolds (each specified by θ_1 and θ_2) while guaranteeing that the navigation potential is strictly decreasing between the switchings. Thus, with these switchings, the system's (x, y)-position moves toward the minimum of the navigation potential. Once the system's (x, y)-position is close enough (as specified by a user-designed error-bound) to this minimum, at most one more switching occurs for correction of orientation while keeping the system's position still close to the minimum as specified by the error-bound.

Numerous feedback control methods have been proposed for the posture stabilization of wheeled mobile robots (e.g. [15], [16], [17], [18], [19], [20], [21]). However, to our best knowledge, none of them (other than one exception of

[1]Here, we utilize this passivity property mainly as an intrinsic structural-property of the (open-loop) system relating mechanical power and energy. This passivity may also be related to the energy efficiency of the controller system (e.g. biped walking), although this is not the direction taken in this paper.

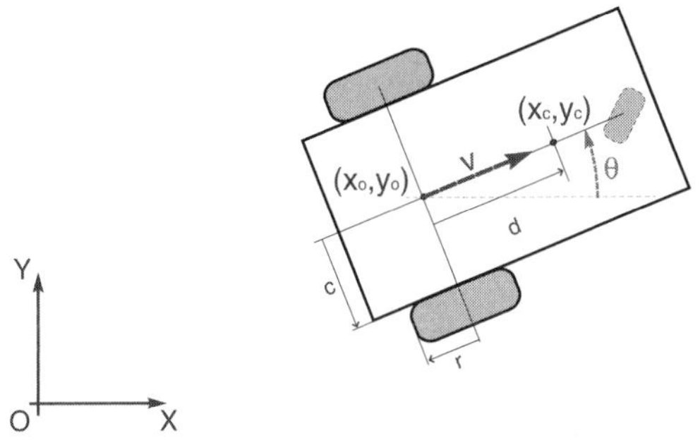

Fig. 1. Wheeled mobile robot with the center-of-mass at (x_c, y_c) and the geometric center at (x_o, y_o), both w.r.t. the reference frame (O, X, Y).

[22]) exploits this intrinsic passivity property of (open-loop) wheeled mobile robots. In [22], the authors proposed a control law, which stands upon the port-controlled Hamiltonian structure of the nonholonomic system and a time-varying canonical transformation. This transformation preserves the open-loop passivity, which is central to enure their stabilization. However, a nontrivial partial differential equation needs to be solved there to find this transformation. Moreover, convergence of their obtained control law is very slow, as often found to be so with time-varying controls (e.g. [16]). Some switching control laws for the posture stabilization of wheeled mobile robots have also been proposed. However, to our best knowledge, all of them (e.g. [18], [19], [20]) are derived for the first-order kinematic model of wheeled mobile robots, thus, it is not clear how (of if) we can use them in many practical applications where the actual control inputs are torques and/or the inertial effect of the robot is not negligible.

Our passivity-based switching control relies on intrinsic energetic/geometric concepts and entities (e.g. passivity, submanifolds, dissipation, decoupling), which are not necessarily limited to the wheeled mobile robot and its posture stabilization problem. Rather, we think our passivity-based switching control idea must be applicable to more general control problems of possibly more general mechanical systems with nonholonomic constraints as well. As an example of such extensions, in this paper, we briefly present a preliminary result of an application of this passivity-based switching idea to the coordination problem of multiple wheeled mobile robots (with no motion planning). More detailed exposition of this and other extensions will be reported in future publications.

The rest of this paper is organized as follows. Dynamics of the wheeled mobile robot and its passivity property will be derived in Sec. II, and the passivity-based switching control law will be presented and analyzed in Sec. III. Simulation results to validate/highlight properties of the proposed control law will be given in Sec. IV. A preliminary result of extension to the multiple wheeled mobile robots will be presented in Sec. V, and summary and some concluding remarks on future research will be made in Sec. VI.

II. Wheeled Mobile Robot

In this section, we derive the dynamics of the wheeled mobile robot projected on the admissible distribution (i.e. along the direction of the velocity, which does not violate the nonholonomic constraints). For more details, refer to [5].

Consider a 3-degree of freedom (DOF) wheeled mobile robot with the configuration $q = [x, y, \theta] \in SE(2)$, where $(x, y) := (x_o, y_o)$ is the position of the geometric center and $\theta \in S$ is the orientation of the robot w.r.t. the inertial frame (O, X, Y). See Fig. 1. Here, we assume that the wheeled mobile robot has two independently-controlled rear wheels, and one front passive castor to avoid tipping over of the robot. We also assume that their inertias are negligible.

Then, the nonholonomic constraints (i.e. no slip condition) can be written by $A(\theta)\dot{q} = 0$ with

$$A(\theta) := \begin{bmatrix} \sin\theta & -\cos\theta & 0 \end{bmatrix} \in \Re^{1\times 3}$$

and the admissible velocity (i.e. not violating the nonholonomic constraints) of the wheeled mobile robot is given by

$$\dot{q} = S(q)\begin{pmatrix} v \\ w \end{pmatrix}, \quad S(q) := \begin{bmatrix} \cos\theta & 0 \\ \sin\theta & 0 \\ 0 & 1 \end{bmatrix} \quad (1)$$

where $v \in \Re$ is the forward-velocity of the robot's geometric center and $w = \dot{\theta}$ (see Fig. 1). Here, the column vectors of $S(q) \in \Re^{3\times 2}$ constitute the basis of the admissible velocity space at q, i.e., $\Delta_q := \{\dot{q} \in \Re^3 \mid A(q)\dot{q} = 0\}$. This Δ_q is a 2-dimensional linear vector space. By collecting this Δ_q over all q, we can get the (regular) admissible distribution Δ.

Then, the dynamics of the wheeled mobile robot projected on Δ can be written by

$$D(\theta)\dot{\nu} + Q(\theta, \dot{\theta})\nu = u \quad (2)$$

with $\nu := [v, w]^T$, $u = [u_v, u_w]^T$,

$$D := \begin{bmatrix} m & 0 \\ 0 & I \end{bmatrix}, \quad Q := \begin{bmatrix} 0 & -md\dot{\theta} \\ md\dot{\theta} & 0 \end{bmatrix}$$

where $d \geq 0$ the distance between (x_c, y_c) and (x_o, y_o), m is the mass of the robot, and $I := I_c + md^2$ is the moment of inertia of the robot w.r.t. (x_o, y_o) with I_c being that of the robot w.r.t. (x_c, y_c). Also, $u_v = \frac{1}{r}(\tau_r + \tau_l)$ and $u_w = \frac{c}{r}(\tau_r - \tau_l)$, where $\tau_r, \tau_l \in \Re$ are the torques of the right and left rear wheels, $r > 0$ is the radius of the wheels, and $c > 0$ is the half of the cart width. See Fig. 1. Since this is the dynamics projected on the admissible distribution Δ, no constraint force (i.e. terms with Lagrange multiplier [5]) shows up here.

Using the skew-symmetricity of $\dot{D} - 2Q$, we can show that this wheeled mobile robot possesses the (energetic) passivity property [10]:

$$\frac{d}{dt}\kappa = u^T\nu \quad (3)$$

where $\kappa := \frac{1}{2}m(\dot{x}^2 + \dot{y}^2) + \frac{1}{2}Iw^2 = \frac{1}{2}mv^2 + \frac{1}{2}Iw^2$ is the total kinetic energy. Here, κ and $u^T\nu$ serve as the storage function and the supply rate of the standard passivity definition [7].

Note that the dynamics of v and w in (2) are coupled with each other via the Coriolis terms $Q\nu$. Since the Coriolis matrix Q is skew-symmetric, this coupling is energetically passive, i.e. does not generate nor dissipate energy. This can be shown s.t.: with $\kappa_v := \frac{1}{2}mv^2$ and $\kappa_w := \frac{1}{2}Iw^2$,

$$\frac{d}{dt}\kappa_v = u_v v + md\dot{\theta}wv \quad (4)$$

$$\frac{d}{dt}\kappa_w = u_w w - md\dot{\theta}vw \quad (5)$$

where the last terms in (4) and (5), which represent the coupling effects, are summed to be zero (i.e. define energetically conservative internal energy shuffling). These coupling terms disappear if $d = 0$ (i.e. (x_c, y_c) and (x_o, y_o) coincide with each other) or if the Coriolis terms $Q\nu$ in (2) are canceled out as done in the next Sec. III. Then, we will get (4)-(5) without the last terms. This implies that the (decoupled) dynamics of v and w will then individually possess the passivity property similar to (3).

III. Passivity-Based Switching Control Design for Posture Stabilization

In this section, we design the switching control u in (2) s.t. $\lim_{t\to\infty}(x(t), y(t))$ and the desired position (x_d, y_d) is close enough in the sense that a certain distance measure between these two point is less than or equal to a user-specific performance specification $\delta_o > 0$ and $\theta(t) \to \theta_d$. Without losing generality, here, we assume that $(x_d, y_d, \theta_d) = (0, 0, 0)$.

Now, we design the control law (u_v, u_w) s.t.

$$\begin{pmatrix} u_v \\ u_w \end{pmatrix} = \begin{bmatrix} 0 & -md\dot{\theta} \\ md\dot{\theta} & 0 \end{bmatrix} \begin{pmatrix} v \\ w \end{pmatrix} - \begin{pmatrix} bv + \frac{\partial \varphi_v}{\partial x} \mathrm{c}\,\theta + \frac{\partial \varphi_v}{\partial y} \mathrm{s}\,\theta \\ b_w\dot{\theta} + k_w(\theta - \theta_{\sigma(t)}) \end{pmatrix} \quad (6)$$

where $\mathrm{s}\,\theta = \sin\theta$, $\mathrm{c}\,\theta = \cos\theta$, $b, b_w > 0$ are the damping gains, $k_w > 0$ is the spring gain, and $\sigma(t) \in \{1, 2\}$ is the switching signal s.t. $\theta_1 = 0$ and $\theta_2 = \theta_o$ with $\theta_o \neq \theta_1 + n\pi$ being a constant ($n = 0, \pm1, \pm2, ...$). Switching law for $\sigma(t)$ will be designed below (see Theorem 1). Here, $\varphi_v(x, y) \geq 0$ is a smooth navigation potential function [14] defined on the (x, y)-plane s.t. 1) $\varphi_v(x, y) = 0$ if and only if $(x, y) = 0$; 2) $(\frac{\partial \varphi_v}{\partial x}, \frac{\partial \varphi_v}{\partial y}) = 0$ if and only if $(x, y) = 0$; and 3) for any finite constants $l \geq 0$, the level set

$$\mathcal{L}_l := \{(x, y) \in \Re^2 \mid \varphi_v(x, y) \leq l\} \quad (7)$$

is a compact set containing $(0, 0)$ and $\mathcal{L}_{l_1} \subseteq \mathcal{L}_{l_2}$ if $l_2 \geq l_1 \geq 0$. In addition to the stabilization $(x, y) \to (0, 0)$, this navigation potential $\varphi_v(x, y)$ can also incorporate other control objectives such as obstacle avoidance [23], although how to design such a potential field without unwanted local minima is beyond the scope of this paper. Here, we want to emphasize that this potential function $\varphi_v(x, y)$ can be designed without considering the nonholonomic constraints as if the given system is just a (unconstrained) point mass on (x, y)-plane. Thus, we can use many of already available results for the generation of this $\varphi_v(x, y)$ (e.g. [14]).

Under this control (6), the closed-loop dynamics of (2) becomes

$$m\dot{v} + bv + \frac{\partial \varphi_v}{\partial x} \mathrm{c}\,\theta + \frac{\partial \varphi_v}{\partial y} \mathrm{s}\,\theta = 0 \quad (8)$$

$$I\ddot{\theta} + b_w\dot{\theta} + k_w(\theta - \theta_{\sigma(t)}) = 0 \quad (9)$$

where, due to the decoupling control in (6) (first term) and the fact that φ_v is a function of (x, y), these two dynamics (8)-(9) are energetically decoupled from each other: 1) for the v-dynamics (8), the total energy $V_v := \kappa_v + \varphi_v(x, y)$ is a function of only (v, x, y) and, from (4),

$$\frac{d}{dt}V_v = -bv^2 - \frac{\partial \varphi_v}{\partial x} v\,\mathrm{c}\,\theta - \frac{\partial \varphi_v}{\partial y} v\,\mathrm{s}\,\theta + \frac{d\varphi_v}{dt} = -bv^2 \quad (10)$$

with $v\,\mathrm{c}\,\theta = \dot{x}$ and $v\,\mathrm{s}\,\theta = \dot{y}$ (from (1)); and 2) for the w-dynamics (9), the total energy $V_w := \kappa_w + \varphi_w$ with $\varphi_w = k_w(\theta - \theta_{\sigma(t)})^2/2$ is a function of only $(\theta, \theta_{\sigma(t)}, w)$ and, from (5), $dV_w/dt = -b_w w^2$ between two consecutive switchings. Due to this energetic decoupling between (8)-(9), switchings in $\theta_{\sigma(t)}$, which induce jumps in V_w, neither change the value of V_v at the switching nor affect the dissipation relation of (10). Also, note that, by the decoupling control in (6), the w-dynamics (9) is completely decoupled from the v-dynamics (8).

Proposition 1 *Suppose that $\sigma(t)$ is fixed with $\sigma(t) = \sigma$ $\forall t \geq 0$. Then, under the control (6), $(v, w, \theta - \theta_\sigma) \to 0$ and the the robot's (x, y)-position converges to the following submanifold*

$$\mathcal{M}_\sigma := \{(x, y) \in \Re^2 \mid \frac{\partial \varphi_v}{\partial x} \mathrm{c}\,\theta_\sigma + \frac{\partial \varphi_v}{\partial y} \mathrm{s}\,\theta_\sigma = 0\}.$$

Proof: From (10), we have $V_v(t) \leq V_v(0)$. Therefore, for all $t \geq 0$, v (also, $\dot{x}, \dot{y}$ from (1)) is bounded and $(x, y) \in \mathcal{L}_{V_v(0)}$, where $\mathcal{L}_{V_v(0)}$ is a compact set containing $(0, 0)$. Also, for (9), if $\sigma(t) = \sigma$ $\forall t \geq 0$, we have $dV_w/dt = -b_w w^2$ $\forall t \geq 0$, thus, $(w, \theta - \theta_\sigma)$ is bounded $\forall t \geq 0$, too. Therefore, applying LaSalle's Theorem [24] to (8)-(9) with $V := V_v + V_w$ and $dV/dt = -bv^2 - b_w w^2$ for all $t \geq 0$, we have $(v, w, \theta - \theta_\sigma) \to 0$ and $\frac{\partial \varphi_v}{\partial x} \mathrm{c}\,\theta_\sigma + \frac{\partial \varphi_v}{\partial y} \mathrm{s}\,\theta_\sigma \to 0$. ■

The main idea of our switching control is to make the system to move back and forth between these two submanifolds $\mathcal{M}_1$ and $\mathcal{M}_2$ (by switching θ_σ between θ_1 and θ_2), while guaranteeing that the navigation potential φ_v is strictly decreasing during these inter-switching moves. Here, note that $\mathcal{M}_1$ and $\mathcal{M}_2$ intersect with each other only at $(x, y) = 0$.

Now, suppose that two consecutive switchings occur at t_i and t_{i+1} and $\sigma(t) = \sigma_i$ for $I_i := [t_i, t_{i+1})$. Then, for this time-interval I_i, by integrating (10), we have

$$\varphi_v(t_i) - \varphi_v(t_{i+1}) = \kappa_v(t_{i+1}) - \kappa_v(t_i) + \int_{t_i}^{t_{i+1}} bv^2 dt$$

$$\geq -\kappa_v(t_i) + \int_{t_i}^{t_{i+1}} bv^2 dt \quad (11)$$

since $\kappa_v(t) \geq 0$. Thus, if we can ensure that, during the system's inter-switching move between the two submanifolds, the energy dissipation via the damping b is strictly larger than the initial kinetic energy $\kappa_v(t_i)$, $\varphi_v(t)$ will be strictly decreasing between the switchings, thus, we can achieve $\varphi_v(t) \to 0$. This would be trivially achieved if $\kappa_v(t_i) = 0$. However, detecting this requires perfect velocity sensing and, even with that, it requires the time-interval between two consecutive switchings be infinite.

The next Lemma shows that, if the wheeled mobile robot's moving distance is large enough and the initial velocity is small enough, the damping can always dissipate all the initial kinetic energy.

Lemma 1 *Suppose that, on the (x, y)-plane, the wheeled mobile robot under the control* (6) *moves from (x_0, y_0) to (x_1, y_1) with the distance between them being $D > 0$. Suppose further that the robot's (x, y)-trajectory is twice-differentiable. Then, if $\kappa_v(t_0) < 2b^2D^2/m$,*

$$\int_{t_0}^{t_1} bv^2 dt > \kappa_v(t_0) \tag{12}$$

where t_0 and t_1 are the initial and (unspecified) final times.

Proof: This is trivially ensured if $\kappa_v(t_0) = 0$, since $D > 0$. Now, suppose that $\kappa_v(t_0) > 0$. Then, from the standard result of the calculus of variation [25], among the twice-differentiable trajectories $a(t) = (x_a(t), y_a(t))$ on $\Re^2$ connecting (x_0, y_0) and (x_1, y_1), the one that extremizes $\int bv_a^2 dt = \int b(\dot{x}_a^2 + \dot{y}_a^2)dt$ is given by the Lagrangian equation, that is, in this case, simply given by $\ddot{x}_a = \ddot{y}_a = 0$. This implies that this extremizing trajectory (i.e. geodesics) is the straight line connecting the two points with constant velocity (and kinetic energy) along the line. This extremizer is also the minimizer, since we can find another curves with higher damping dissipation (e.g. fast oscillatory curve).

Thus, if the wheeled mobile robot moves from (x_0, y_0) to (x_1, y_1) along a twice-differentiable trajectory with $\kappa_v(t_0) > 0$, we have: with $\bar{v}_a := \sqrt{2\kappa_v(t_0)/m}$,

$$\int_{t_0}^{t_1} bv^2 dt \geq \min_{a(t)} \int bv_a^2 dt = b\bar{v}_a^2 \frac{D}{\bar{v}_a} = bD\sqrt{\frac{2\kappa_v(t_0)}{m}}$$

which is strictly larger than $\kappa_v(t_0)$ if $0 < \kappa_v(t_0) < 2b^2D^2/m$. ■

This Lemma 1, thus, enables us to switch even when $\kappa_v \neq 0$ while enforcing strict decrease of φ_v between the switchings, providing that the switching occurs when the velocity is small and the moving distance of the robot between two consecutive switchings is not small. Here, the former condition can be ensured simply by waiting for the system to slow down enough into the switching submanifold (see Proposition 1), while the latter by designing the two submanifolds $\mathcal{M}_1$ and $\mathcal{M}_2$ far enough from each other. In many cases, this separation would be possible (at least locally) except very near the origin, where $\mathcal{M}_1$ and $\mathcal{M}_2$ intersect with each other.

The Lemma 1 is directly applicable for the wheeled mobile robot between two consecutive switchings, since, with $\theta_{\sigma(t)}$ being a constant, both the dynamics (2) and the control (6) are smooth, therefore, the system's (x, y)-trajectory is also smooth. This is all we need for our switching control design. However, it is also interesting to see that this Lemma 1 is still valid even in the presence of switchings. To check this, recall from (1) that $\dot{x} = v\,c\,\theta$. Thus, $\ddot{x} = \dot{v}\,c\,\theta - v\,s\,\theta\dot{\theta}$, where, from (8)-(9), $\dot{v}, \dot{\theta}$ are both bounded and continuous. Therefore, $\ddot{x}(t)$ is also bounded and continuous $\forall t \geq 0$. Similar fact can be shown for $\ddot{y}(t)$, too. This implies that, although the non-smooth switching $\theta_{\sigma(t)}$ occurs in (9), the robot's (x, y)-trajectory is always twice-differentiable, therefore, Lemma 1, which only requires that candidate trajectory $a(t)$ is twice-differentiable, still holds regardless of the switching.

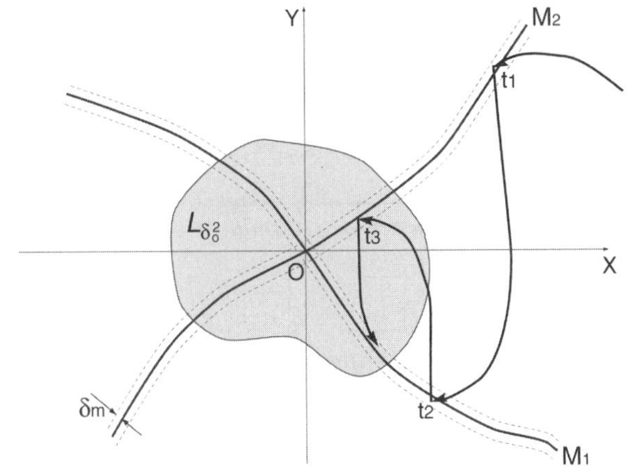

Fig. 2. Illustration of level set $\mathcal{L}_{\delta_o^2}$, submanifolds $\mathcal{M}_1$ and $\mathcal{M}_2$, and their strips with thickness of $\delta_m > 0$ on the (x, y)-plane. Also, shown is an exemplary system trajectory with switching times.

We now present our main result with passivity (or energy) based switching logic.

Theorem 1 *Suppose that the navigation potential φ_v is designed to satisfy the following condition: $\exists c_m > 0$ s.t., for any $(x, y) \notin \mathcal{L}_{\delta_o^2}$, if $(x, y) \in \bar{\mathcal{M}}_i$,*

$$\text{dist}((x, y), \bar{\mathcal{M}}_j) \geq c_m\sqrt{\varphi_v(x, y)} \tag{13}$$

where $\mathcal{L}_{\delta_o^2}$ is the level set of φ_v in (7) with $\delta_o > 0$ being a user-specified performance measure, $\bar{\mathcal{M}}_k$ is the "strip" of $\mathcal{M}_k$ with a thickness $\delta_m > 0$ s.t. $c_m\delta_o \gg \delta_m$ ($k = 1, 2$, see Fig. 2), and $(i, j) \in \{(1, 2), (2, 1)\}$. Trigger the switching at a time $t > 0$, if

1) $\kappa_v(t) \leq 2b^2c_m^2\delta_o^2/m$; *and*
2) $\text{dist}((x(t), y(t)), \mathcal{M}_{\sigma(t^-)}) \leq \delta_m$; *and*
3) a) $(x(t), y(t)) \notin \mathcal{L}_{\delta_o^2}$; *or*
   b) $(x(t), y(t)) \in \mathcal{L}_{\delta_o^2}$ *and* $\sigma(t^-) \notin 1$.

Then, $\lim_{t\to\infty}(x(t), y(t)) \in \mathcal{L}_{\delta_o^2}$ and $\theta(t) \to \theta_d$. Also, once $(x(t), y(t)) \in \mathcal{L}_{\delta_o^2}$, at most one more switching occurs.

Proof: The system's initial condition is given by $(\dot{q}(0), q(0), \sigma(0))$, where $\sigma(0) \in \{1, 2\}$. Then, from Prop. 1, $\theta \to \theta_{\sigma(0)}$ and $(x, y) \to \mathcal{M}_{\sigma(0)}$. Let us denote the time when the switching conditions 1)-2) are satisfied by $t_1 > 0$, i.e. $\kappa_v(t_1) \leq 2b^2c_m^2\delta_o^2/m$ and $\text{dist}((x(t_1), y(t_1)), \bar{\mathcal{M}}_{\sigma(0)}) \leq \delta_m$. Consider the following two cases separately, 1) when $(x(t_1), y(t_1)) \notin \mathcal{L}_{\delta_o^2}$; and 2) otherwise.

1) **Outside of** $\mathcal{L}_{\delta_o^2}$: If $(x(t_1), y(t_1)) \notin \mathcal{L}_{\delta_o^2}$, according to the switching logic, there will be a switching at this t_1 s.t. $\sigma(t_1) = 1$ if $\sigma(0) = 2$ or $\sigma(t_1) = 2$ if $\sigma(0) = 1$. Then, following Prop. 1, $(x, y) \to \bar{\mathcal{M}}_{\sigma(t_1)}$ and $\theta \to \theta_{\sigma(t_1)}$. Denote the time when the system again satisfies the above switching conditions 1)-2) by $t_2 > t_1$, i.e. the system converges into $\bar{\mathcal{M}}_{\sigma(t_1)}$ and slows down enough. During this time-interval $I_1 := [t_1, t_2)$, the potential function $\varphi_v(t)$ is strictly decreasing, since, following (11), we have

$$\varphi_v(t_1) - \varphi_v(t_2) \geq -\kappa_v(t_1) + \int_{t_1}^{t_2} bv^2 dt > 0 \qquad (14)$$

where the last inequality is a direct consequence of Lemma 1, since, from (13) with $(x(t_1), y(t_1)) \notin \mathcal{L}_{\delta_o^2}$, the moving distance $D_{I_1} > 0$ of the wheeled mobile robot from $(x(t_1), y(t_1)) \in \bar{\mathcal{M}}_{\sigma(0)}$ to $(x(t_2), y(t_2)) \in \bar{\mathcal{M}}_{\sigma(t_1)}$ satisfies $D_{I_1} \geq c_m\sqrt{\varphi_v(x(t_1), y(t_1))} > c_m\delta_o$, but, $\kappa_v(t_1) \leq 2b^2c_m^2\delta_o^2/m < 2b^2D_{I_1}^2/m$. If $(x(t_2), y(t_2)) \notin \mathcal{L}_{\delta_o^2}$, another switching will be triggered.

By continuing this process, since the navigation potential $\varphi_v(t_i)$ is strictly decreasing, a sequence of the times can be generated $(t_1, t_2, ..., t_{n-1}, t_n)$, where $(x(t_k), y(t_k)) \notin \mathcal{L}_{\delta_o^2}$ $\forall k = 1, 2, ..., n-1$ (i.e. $t_1, t_2, ..., t_{n-1}$ are the switching times) and, at time t_n (switching time candidate: $n \geq 2$), the switching conditions 1)-2) are satisfied with $(x(t_n), y(t_n)) \in \mathcal{L}_{\delta_o^2}$. Then, it becomes the case 2) discussed below.

2) **Inside of** $\mathcal{L}_{\delta_o^2}$: Now, suppose that, at some t_n with $n \geq 1$, $(x(t_n), y(t_n)) \in \mathcal{L}_{\delta_o^2}$ and the two conditions 1)-2) are satisfied. Then, if $\sigma(t_n^-) = 1$, we are done and no more switching will be triggered. If $\sigma(t^-) = 2$, to correct the orientation, another switching will be activated at t_n with $\sigma(t_n) = 1$ and the system will again converge into $\bar{\mathcal{M}}_1$. Denote by t_{n+1} the time when the system again satisfies the two switching conditions 1)-2) in $\bar{\mathcal{M}}_1$. In this case, $(x(t_{n+1}), y(t_{n+1}))$ must be still contained in $\mathcal{L}_{\delta_o^2}$. This can be shown by the following contradiction. Suppose that $(x(t_{n+1}), y(t_{n+1})) \notin \mathcal{L}_{\delta_o^2}$. Then, from (13), the wheeled mobile robot should move the distance strictly larger than $c_m\delta_o$, since $\text{dist}((x(t_n), y(t_n)), (x(t_{n+1}), y(t_{n+1}))) \geq c_m\sqrt{\varphi_v(x(t_{n+1}), y(t_{n+1}))} > c_m\delta_o$. Thus, with $\kappa_v(t_n) \leq 2b^2c_m^2\delta_o^2/m$, the inequality (14) still holds and $\varphi_v(t_{n+1}) < \varphi_v(t_n)$. This implies that $(x(t_{n+1}), y(t_{n+1})) \in \mathcal{L}_{\varphi_v(t_{n+1})} \subset \mathcal{L}_{\varphi_v(t_n)} \subset \mathcal{L}_{\delta_o^2}$, which is contradictory to the above supposition. Therefore, we have $\lim_{t\to\infty}(x(t), y(t)) \in \mathcal{L}_{\delta_o^2}$ and $\theta(t) \to \theta_d$. This completes the proof. ■

If the objective is only posture stabilization (without obstacle avoidance), the frequently-used quadratic function $k_v(x^2 + y^2)$ with $k_v > 0$ can be directly used as the navigation

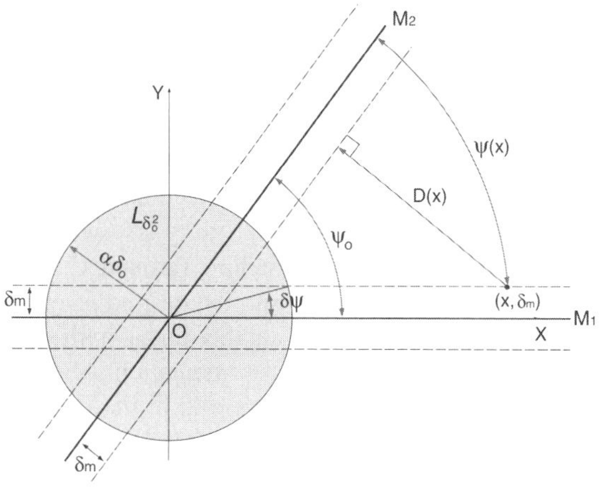

Fig. 3. Illustration of some geometric entities of $\varphi_v(x, y) = k_v(x^2 + y^2)/2$: with $\alpha = \sqrt{2/k_v}$, 1) $\mathcal{L}_{\delta_o^2}$ is circle with radius $\alpha\delta_o > 0$; 2) if $(x, y) \in \mathcal{M}_1 - \mathcal{L}_{\delta_o^2}$, $|y| \leq \delta_m$ and $|x| > \alpha\delta_o$; and 3) $\sin\psi(x) = (D(x) + \delta_m)/\sqrt{x^2 + \delta_m^2}$ and $0 < \psi_o - \delta\psi \leq \psi(x) \leq \psi_o \leq \pi/2$. Here, we conveniently choose an arbitrarily x-axis as aligned to $\mathcal{M}_1$.

potential φ_v. This is because 1) this quadratic function satisfies all the conditions for φ_v given in the beginning of this Sec. III; and 2) if we set $\alpha\delta_o \gg \delta_m$ and choose θ_1, θ_2 s.t. the two switching submanifolds $\mathcal{M}_i = \{(x, y) \mid k_v x\,\text{c}\,\theta_i + k_v y\,\text{s}\,\theta_i = 0\}$ (i.e. straight line obtained by rotating y-axis by θ_i counterclockwise w.r.t. origin) are far enough from each other, it also ensures the condition (13) with $c_m \leq \alpha[\sin(\psi_o - \delta\psi) - \sin\delta\psi]$, where $\alpha = \sqrt{2/k_v}$, $0 < \psi_o \leq \pi/2$ is the angle between $\mathcal{M}_1, \mathcal{M}_2$, and $\delta\psi = \sin^{-1}(\delta_m/(\alpha\delta_o))$. This is because, from Fig. 3, if $(x, y) \in \bar{\mathcal{M}}_1 - \mathcal{L}_{\delta_o^2}$,

$$\begin{aligned} \frac{\text{dist}((x, y), \bar{\mathcal{M}}_2)}{\sqrt{\varphi_v(x, y)}} &\geq \alpha\frac{D(x)}{\sqrt{x^2 + \delta_m^2}} \\ &= \alpha\left(\sin\psi(x) - \frac{\delta_m}{\sqrt{\alpha^2\delta_o^2 + \delta_m^2}}\right) \\ &> \alpha[\sin(\psi_o - \delta\psi) - \sin\delta\psi]. \end{aligned} \qquad (15)$$

On the other hand, in practice, the switching conditions 1)-2) in Theorem 1 can be easily ensured by separating two consecutive switchings by a large enough dwell-time $\tau_D > 0$ [26].

The decoupling control in (6) is necessary here, since, if we omit it, energy can be transferred between (8) and (9) via the coupling terms in (4)-(5). Then, some portion of the θ-spring energy $\varphi_w = \frac{1}{2}k_w(\theta - \theta_{\sigma(t)})^2$, which jumps at every switching, may flow back to the navigation potential φ_v and recharge it. If this amount is more than the damping dissipation via b, we may lose the strict decrease of $\varphi_v(t)$ between switchings. More detailed analysis on the effect of this (uncompensated or partially-compensated) coupling is a topic for future research.

IV. Simulation

For this simulation, we use the quadratic navigation potential $\varphi_v(x, y) = \frac{1}{2}k_v(x^2 + y^2)$ with $k_v > 0$. We also use a long enough dwell-time $\tau_D > 0$ to ensure the switching conditions 1)-2) of Theorem 1. For the first simulation, we choose $\theta_1 = 0$ and $\theta_2 = \pi/2$ so that the two submanifolds $\mathcal{M}_1$ and $\mathcal{M}_2$ are respectively given by the y-axis and x-axis. As stated in the paragraph after Theorem 1, these chosen φ_v and θ_1, θ_2 are legitimate for use in our switching control. Simulation results are presented in Fig. 4. Total thirty-six switchings are occurred. After the thirty-fifth switching (around 17.5sec), the system's (x, y)-trajectory enters into the desired level set, but, its orientation is not correct. So, another thirty-sixth switching occurs (around 18sec.) to correct the orientation. After this last switching, the system's (x, y)-position is in the desired level set, the orientation converges to the target value, and no more switching occurs.

For the second simulation, we use $\theta_1 = -\pi/2$ and $\theta_2 = \pi/2$, thus, $\mathcal{M}_1$ and $\mathcal{M}_2$ are the same: both of them are given by $k_v y = 0$ (i.e. x-axis). However, by tuning the gains, we make the v-dynamics (8) much faster than that of w (9), so that the robot's (x, y)-trajectory converges on the straight line $\mathcal{M}_{\theta(t)} = \{(x, y) \mid k_v x \,\mathrm{c}\,\theta(t) + k_v x \,\mathrm{s}\,\theta(t) = 0\}$ fast enough, as this line rotates slowly back and force between the positive x-axis and the negative x-axis. By doing so, although $\mathcal{M}_1 = \mathcal{M}_2$, we can ensure that, between the switchings, the robot moves between the two strips on the positive x-axis and the negative x-axis separated by $\mathcal{L}_{\delta_o^2}$. This ensures the condition (13), since, if $(x, y) \in \mathcal{M}_i - \mathcal{L}_{\delta_o^2}$ and $\sqrt{2/k_v}\delta_o \gg \delta_m$, $\mathrm{dist}((x, y), \bar{\mathcal{M}}_j)/\sqrt{\varphi_v(x, y)} \geq \sqrt{2/k_v}(|x| + \delta_m)/\sqrt{x^2 + \delta_m^2} > \sqrt{2/k_v}$, thus, any $0 < c_m \leq \sqrt{2/k_v}$ would work to enforce (13). Results for this simulation are shown in Fig. 5. Compared to the results in Fig. 4, much less number of switchings (only ten switchings) is required to move into the desired level set, since the moving distance between two switchings is larger than that of Fig. 4.

Similar to the second simulation, we may slowly rotate the submanifold more than 2π. For this, again, by tuning the dynamics of v much faster than that of w, we could get an even faster convergence and less number of switchings. A detailed analysis and exposition for this rotating submanifold result will be reported in a future publication.

V. Application to Multiple Wheeled Mobile Robots Coordination

In this section, as one example of extensions to more general problems/systems, we apply the passivity-based switching control for the coordination problem of multiple wheeled mobile robots. No motion planning is necessarily here. More detailed exposition will be reported in future publications.

Consider n (strongly-connected) wheeled mobile robots, and define $p_e := (x_e, y_e) \in \Re^{2(n-1)}$ and $\theta_e \in \Re^{n-1}$, with $\star_e := (\star_1 - \star_2, ..., \star_{n-1} - \star_n) \in \Re^{n-1}$. Then, for simplicity (while without losing generality), by the coordination, we mean $(p_e, \theta_e) \to 0$. Then, for each k-th agent, we design its control to be (6) with its second term replaced by

$$-\begin{pmatrix} b v_k + \frac{\partial \bar{\varphi}_v}{\partial x_k} \,\mathrm{c}\,\theta_k + \frac{\partial \bar{\varphi}_v}{\partial y_k} \,\mathrm{s}\,\theta_k \\ b_w \dot{\theta}_k + \frac{\partial \bar{\varphi}_w}{\partial \theta_k} + k_w(\theta_k - \theta_{\sigma(t)}) \end{pmatrix} \quad (16)$$

where $\bar{\varphi}_v(p_e)$ and $\bar{\varphi}_w(\theta_e)$ are (smooth) navigation potentials defined on the p_e and θ_e spaces respectively.

Then, similar to Prop. 1, using Barbalat's lemma with smoothness of suitable terms, we can show that, with fixed $\sigma(t) = \sigma$, the system converges in the p_e-space to the submanifold $\mathcal{T}_\sigma := \{p_e \mid \frac{\partial \bar{\varphi}_v}{\partial x} \,\mathrm{c}\,\theta_\sigma + \frac{\partial \bar{\varphi}_v}{\partial y} \,\mathrm{s}\,\theta_\sigma = 0\}$, where $\frac{\partial \bar{\varphi}_v}{\partial \star} := (\frac{\partial \bar{\varphi}_v}{\partial \star_1}, ..., \frac{\partial \bar{\varphi}_v}{\partial \star_n}) \in \Re^n$. Also, being unconstrained, $\theta_k \to \theta_\sigma$. Moreover, between switchings, similar to (11), we have

$$\begin{aligned} \bar{\varphi}_v(t_i) - \bar{\varphi}_v(t_{i+1}) &\geq -\bar{\kappa}_v(t_i) + \sum_{k=1}^{n} \int_{t_i}^{t_{i+1}} b v_k^2 dt \\ &\geq -\bar{\kappa}_v(t_i) + \int_{t_i}^{t_{i+1}} \bar{b}||v_e||^2 dt \end{aligned}$$

where $\bar{b} > 0$, $v_e = dp_e/dt$ (i.e. system velocity on p_e-space), and $\bar{\kappa}_v = \sum \frac{1}{2} m_k v_k^2$. Here, we can obtain the last inequality by using the passive decomposition [13].

Therefore, we can achieve similar results as those in Sec. III on the p_e-space. In other words, if we design $\bar{\varphi}_v$ on the p_e-space s.t. the switching submanifolds $\mathcal{T}_i$ are far enough from each other and trigger the switchings when $\bar{\kappa}_v$ is small enough, we can ensure that $\theta_e \to 0$ and p_e approaches to 0 within some user-specific performance bound. See Fig. 6 for simulation results of this as applied to four wheeled mobile robots. Here, our coordination control is centralized, although it can be partially decentralized by defining $\bar{\varphi}_v$ as the sum of the potential between two robots. Its complete decentralization is beyond the scope of this paper and will be published elsewhere.

VI. Summary and Future Works

In this paper, we propose a novel passivity-based switching control law for the posture stabilization of a wheeled mobile robot. The proposed control law is derived using the fundamental (open-loop) passivity property, which has been extensively used in other control problems in robotics, but not been so at all for systems with nonholonomic constraints.

Since it is based on fairly intrinsic concepts and entities (e.g. passivity, dissipation, decoupling, submanifolds), we believe that our proposed framework could be extended for more general control problems (e.g. coordination problem of multiple wheeled mobile robots as presented in Sec. V), or even further, control of general mechanical systems with nonholonomic constraints on a differential manifold. The latter may require that those systems have dynamics/energetic structure similar to, but probably more generalized than, that of the wheeled mobile robot. Real implementation of this passivity-based switching control and its experimental comparison with other schemes may also further shed lights on its strength/weakness and robustness/practicality as well.

We also wish that this work serves as an initiating step toward fully utilizing the passivity property in the control of wheeled mobile robots and more general robotic systems with nonholonomic constraints.

REFERENCES

[1] M. Peshkin and J. E. Colgate. Cobots. *Industrial Robot*, 26(5):335–341, 1999.

[2] C.R. Weisbin and D. Lavery. Nasa rover and telerobotics technology program. *IEEE Robotics & Automation Magazine*, pages 14–21, December 1994.

[3] S. P. Levine, D. A. Bell, L. A. Jaros, R. C. Simpson, Y. Koren, and J. Borenstein. The navchair assistive wheelchair navigation system. *IEEE Transactions on Rehabilitation Engineering*, 7(4):443–451, 1999.

[4] J. Cortes, S. Martinez, T. Karatas, and F. Bullo. Coverage control for mobile sensing networks. *IEEE Transactions on Robotics and Automation*, 20(2):243–255, 2004.

[5] R. M. Murray, Z. Li, and S. S. Sastry. *A mathematical introduction to robotic manipulation*. CRC, Boca Ranton, FL, 1993.

[6] R. W. Brockett. Asymptotic stability and feedback stabilization. In R. W. Brockett, R. S. Milman, and H. J. Sussmann, editors, *Differential Geometric Control Theory*, pages 181–191, Boston, MA, 1983. Birkhauser.

[7] J. C. Willems. Dissipative dynamical systems part1: general theory. *Arch. Rational Mech. Anal.*, 45(22):321-351, 1972.

[8] M. W. Spong, S. Hutchinson, and M. Vidyasaga. *Robot modeling and control*. John Wiley & Sons, Hoboken, NJ, 2006.

[9] J. E. Colgate and G. Schenkel. Passivity of a class of sampled-data systems: application to haptic interfaces. *Journal of Robotic Systems*, 14(1):37–47, 1997.

[10] D. J. Lee and M. W. Spong. Passive bilateral teleoperation with constant time delay. *IEEE Transactions on Robotics*, 22(2):269–281, 2006.

[11] M. W. Spong, J. K. Holm, and D. J. Lee. Passivity-based control of biped locomotion. *IEEE Robotics & Automation Magazine*. To appear.

[12] M. Arcak. Passivity as a design tool for group coordination. In *Proceeding of the American Control Conference*, pages 29–23, 2006.

[13] D. J. Lee and P. Y. Li. Passive decomposition approach to formation and maneuver control of multiple agents with inertias. *ASME Journal of Dynamic Systems, Measurement & Control*. To appear.

[14] E. Rimon and D. E. Koditschek. Exact robot navigation using artificial potential functions. *IEEE Transactions on Robotics and Automation*, 8(5):501–518, 1992.

[15] C. Canudas de Wit and O. J. Sordalen. Exponential stabilization of mobile robots with nonholonomic constraints. *IEEE Transactions on Automatic Control*, 13(11):1791–1797, 1992.

[16] C. Samson. Time-varying feedback stabilization of car-like wheeled mobile robots. *International Journal of Robotics Research*, 12(1):55–64, 1993.

[17] R. Fierro and F. L. Lewis. Control of a nonholonomic mobile robot: backstepping kinematics into dynamics. *Journal of Robotic Systems*, 14(3):149–163, 1997.

[18] I. Kolmanovsky, M. Reyhanoglu, and N. H. McClamroch. Switched mode feedback control laws for nonholonomic systems in extended power form. *Systems & Control Letters*, 27:29–36, 1996.

[19] J. P. Hespanha and A. S. Morse. Stabilization of nonholonomic integrators via logic-based switching. *Automatica*, 35:385–393, 1999.

[20] J. P. Hespanha, D. Liberzon, and A. S. Morse. Logic-based switching control of a nonholonomic system with parametric modeling uncertainty. *Systems & Control Letters*, 38:167–177, 1999.

[21] A. Astolfi. Exponential stabilization of a wheeled mobile robots via discontinuous control. *ASME Journal of Dynamic Systems, Measurement & Control*, 121(1):121–126, 1999.

[22] K. Fujimoto and T. Sugie. Stabilization of hamiltonian systems with nonholonomic constraints based on time-varying generalized canonical transformation. *Systems & Control Letters*, 44:309–319, 2001.

[23] H. G. Tanner, A. Jadbabaic, and G. J. Pappas. Flocking in teams of nonholonomic agents. In S. Morse, N. Leonard, and V. Kumar, editors, *Cooperative Control*, pages 229–239, New York, NY, 2005. Springer Lecture Notes in Control & Information Sciences.

[24] H. K. Khalil. *Nonlinear systems*. Prentice-Hall, Upper Saddle River, NJ, second edition, 1995.

[25] C. Lanczos. *The variational principles of mechanics*. Dover, Mineola, NY, fourth edition, 1986.

[26] J. P. Hespanha and A. S. Morse. Stability of switched systems with average dwell-time. In *Proceedings of the 38th IEEE Conference on Decision & Control*, pages 2655–2660, 1999.

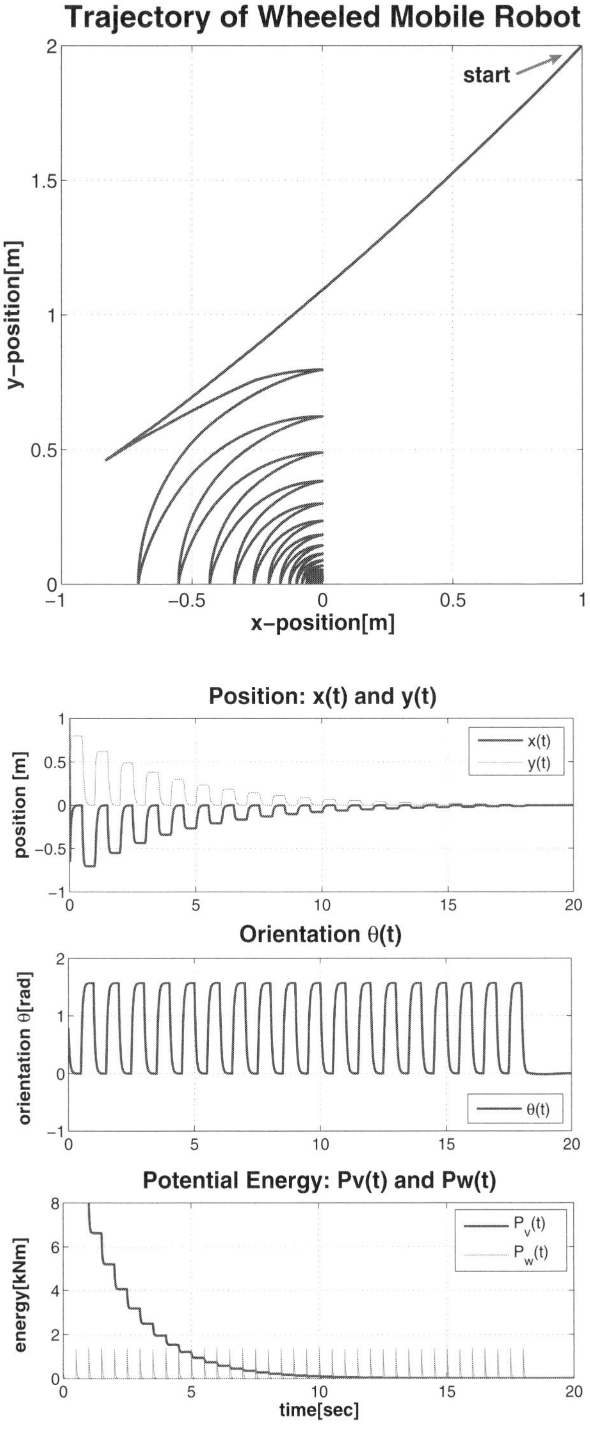

Fig. 4. First simulation results with $\theta_1 = 0$ and $\theta_2 = \pi/2$.

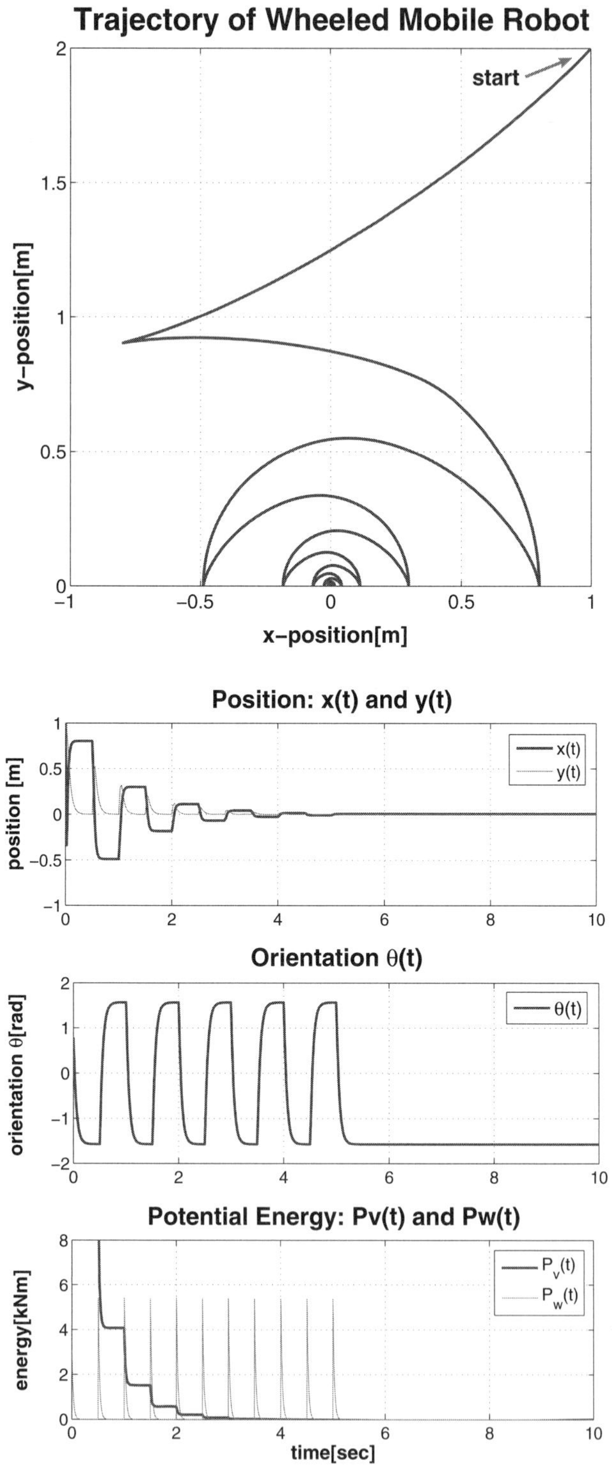

Fig. 5. Second simulation results with $\theta_1 = -\pi/2$ and $\theta_2 = \pi/2$.

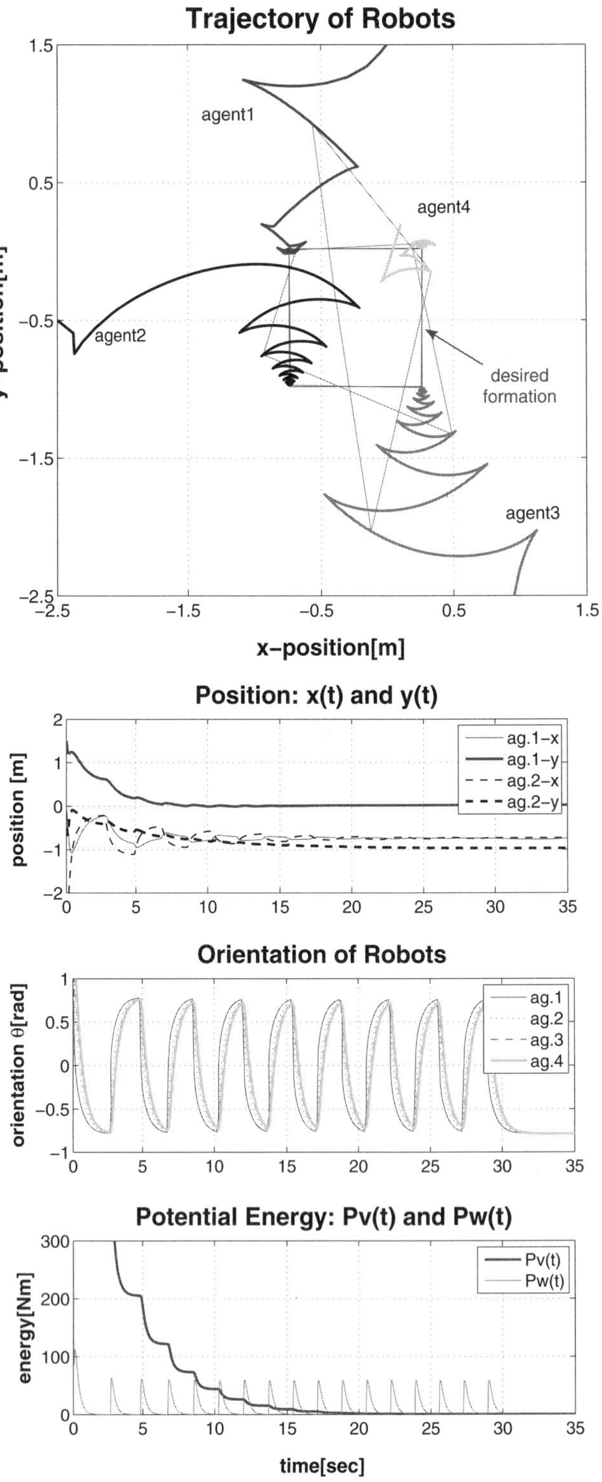

Fig. 6. Coordination of multiple wheeled mobile robots.

A Tree Parameterization for Efficiently Computing Maximum Likelihood Maps using Gradient Descent

Giorgio Grisetti Cyrill Stachniss Slawomir Grzonka Wolfram Burgard
University of Freiburg, Department of Computer Science, 79110 Freiburg, Germany

***Abstract*— In 2006, Olson *et al.* presented a novel approach to address the graph-based simultaneous localization and mapping problem by applying stochastic gradient descent to minimize the error introduced by constraints. Together with multi-level relaxation, this is one of the most robust and efficient maximum likelihood techniques published so far. In this paper, we present an extension of Olson's algorithm. It applies a novel parameterization of the nodes in the graph that significantly improves the performance and enables us to cope with arbitrary network topologies. The latter allows us to bound the complexity of the algorithm to the size of the mapped area and not to the length of the trajectory as it is the case with both previous approaches. We implemented our technique and compared it to multi-level relaxation and Olson's algorithm. As we demonstrate in simulated and in real world experiments, our approach converges faster than the other approaches and yields accurate maps of the environment.**

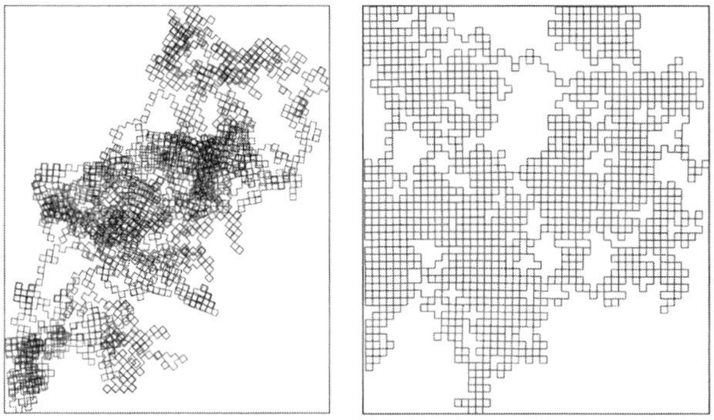

Fig. 1. The left image shows an uncorrected network with around 100k poses and 450k constraints. The right image depicts the network after applying our error minimization approach (100 iterations, 17s on a P4 CPU with 1.8GHz).

I. Introduction

Models of the environment are needed for a wide range of robotic applications, including search and rescue, automated vacuum cleaning, and many others. Learning maps has therefore been a major research focus in the robotics community over the last decades. Learning maps under uncertainty is often referred to as the simultaneous localization and mapping (SLAM) problem. In the literature, a large variety of solutions to this problem can be found. The approaches mainly differ due to the underlying estimation technique such as extended Kalman filters, information filters, particle filters, or least-square error minimization techniques.

In this paper, we consider the so-called "graph-based" or "network-based" formulation of the SLAM problem in which the poses of the robot are modeled by nodes in a graph [2, 5, 6, 7, 11, 13]. Constraints between poses resulting from observations or from odometry are encoded in the edges between the nodes.

The goal of an algorithm designed to solve this problem is to find the configuration of the nodes that maximizes the observation likelihood encoded in the constraints. Often, one refers to the negative observation likelihood as the error or the energy in the network. An alternative view to the problem is given by the spring-mass model in physics. In this view, the nodes are regarded as masses and the constraints as springs connected to the masses. The minimal energy configuration of the springs and masses describes a solution to the mapping problem. Figure 1 depicts such a constraint network as a motivating example.

A popular solution to this class of problems are iterative approaches. They can be used to either correct all poses simultaneously [6, 9, 11] or to locally update parts of the network [2, 5, 7, 13]. Depending on the used technique, different parts of the network are updated in each iteration. The strategy for defining and performing these local updates has a significant impact on the convergence speed.

Our approach uses a tree structure to define and efficiently update local regions in each iteration. The poses of the individual nodes are represented in an incremental fashion which allows the algorithm to automatically update successor nodes. Our approach extends Olson's algorithm [13] and converges significantly faster to a network configuration with a low error. Additionally, we are able to bound the complexity to the size of the environment and not to the length of the trajectory.

The remainder of this paper is organized as follows. After discussing the related work, Section III explains the graph-based formulation of the mapping problem. Subsequently, we explain the usage of stochastic gradient descent to find network configurations with small errors. Section V introduces our tree parameterization and in Section VI we explain how to obtain such a parameterization tree from robot data. We finally present our experimental results in Section VII.

II. Related Work

Mapping techniques for mobile robots can be classified according to the underlying estimation technique. The most popular approaches are extended Kalman filters (EKFs), sparse extended information filters, particle filters, and least square error minimization approaches. The effectiveness of the EKF approaches comes from the fact that they estimate a fully correlated posterior about landmark maps and robot poses [10, 14]. Their weakness lies in the strong assumptions that have to be made on both, the robot motion model and the sensor noise. Moreover, the landmarks are assumed to be uniquely identifiable. There exist techniques [12] to deal with unknown data association in the SLAM context, however, if certain assumptions are violated the filter is likely to diverge [8].

Frese's TreeMap algorithm [4] can be applied to compute nonlinear map estimates. It relies on a strong topological assumption on the map to perform sparsification of the information matrix. This approximation ignores small entries in the information matrix. In this way, Frese is able to perform an update in $\mathcal{O}(\log n)$ where n is the number of features.

An alternative approach is to find maximum likelihood maps by least square error minimization. The idea is to compute a network of relations given the sequence of sensor readings. These relations represent the spatial constraints between the poses of the robot. In this paper, we also follow this way of formulating the SLAM problem. Lu and Milios [11] first applied this approach in robotics to address the SLAM problem using a kind of brute force method. Their approach seeks to optimize the whole network at once. Gutmann and Konolige [6] proposed an effective way for constructing such a network and for detecting loop closures while running an incremental estimation algorithm. Howard *et al.* [7] apply relaxation to localize the robot and build a map. Duckett *et al.* [2] propose the usage of Gauss-Seidel relaxation to minimize the error in the network of constraints. In order to make the problem linear, they assume knowledge about the orientation of the robot. Frese *et al.* [5] propose a variant of Gauss-Seidel relaxation called multi-level relaxation (MLR). It applies relaxation at different resolutions. MLR is reported to provide very good results and is probably the best relaxation technique in the SLAM context at the moment.

Note that such maximum likelihood techniques as well as our method focus on computing the best map and assume that the data association is given. The ATLAS framework [1] or hierarchical SLAM [3], for example, can be used to obtain such data associations (constraints). They also apply a global optimization procedure to compute a consistent map. One can replace such optimization procedures by our algorithm and in this way make ATLAS or hierarchical SLAM more efficient.

The approach closest to the work presented here is the work of Olson *et al.* [13]. They apply stochastic gradient descent to reduce the error in the network. They also propose a representation of the nodes which enables the algorithm to perform efficient updates. The approach of Olson *et al.* is one of the current state-of-the-art approaches for optimizing networks of constraints. In contrast to their technique, our approach uses a different parameterization of the nodes in the network that better takes into account the topology of the environment. This results in a faster convergence of our algorithm.

Highly sophisticated optimization techniques such as MLR or Olson's algorithm are restricted to networks that are built in an incremental way. They require as input a sequence of robot poses according to the traveled path. First, this makes it difficult to use these techniques in the context of multi-robot SLAM. Second, the complexity of the algorithm depends on the length of the trajectory traveled by the robot and not on the size of the environment. This dependency prevents to use these approaches in the context of lifelong map learning.

One motivation of our approach is to build a system that depends on the size of the environment and not explicitely on the length of the trajectory. We designed our approach in a way that it can be applied to arbitrary networks. As we will show in the remainder of this paper, the ability to use arbitrary networks allows us to prune the trajectory so that the complexity of our approach depends only on the size of the environment. Furthermore, our approach proposes a more efficient parameterization of the network when applying gradient descent.

III. On Graph-based SLAM

Most approaches to graph-based SLAM focus on estimating the most-likely configuration of the nodes and are therefore referred to as maximum-likelihood (ML) techniques [2, 5, 6, 11, 13]. They do not consider to compute the full posterior about the map and the poses of the robot. The approach presented in this paper also belongs to this class of methods.

The goal of graph-based ML mapping algorithms is to find the configuration of the nodes that maximizes the likelihood of the observations. For a more precise formulation consider the following definitions:

- $\mathbf{x} = (x_1 \cdots x_n)^T$ is a vector of parameters which describes a configuration of the nodes. Note that the parameters x_i do not need to be the absolute poses of the nodes. They are arbitrary variables which can be mapped to the poses of the nodes in real world coordinates.
- δ_{ji} describes a constraint between the nodes j and i. It refers to an observation of node j seen from node i. These constraints are the edges in the graph structure.
- Ω_{ji} is the information matrix modeling the uncertainty of δ_{ji}.
- $f_{ji}(\mathbf{x})$ is a function that computes a zero noise observation according to the current configuration of the nodes j and i. It returns an observation of node j seen from node i.

Given a constraint between node j and node i, we can define the *error* e_{ji} introduced by the constraint as

$$e_{ji}(\mathbf{x}) \quad = \quad f_{ji}(\mathbf{x}) - \delta_{ji} \tag{1}$$

as well as the *residual* r_{ji}

$$r_{ji}(\mathbf{x}) \quad = \quad -e_{ji}(\mathbf{x}). \tag{2}$$

Note that at the equilibrium point, e_{ji} is equal to 0 since $f_{ji}(\mathbf{x}) = \delta_{ji}$. In this case, an observation perfectly matches the current configuration of the nodes. Assuming a Gaussian observation error, the negative log likelihood of an observation f_{ji} is

$$\begin{aligned} F_{ji}(\mathbf{x}) &\propto (f_{ji}(\mathbf{x}) - \delta_{ji})^T \, \Omega_{ji} \, (f_{ji}(\mathbf{x}) - \delta_{ji}) && (3) \\ &= e_{ji}(\mathbf{x})^T \Omega_{ji} e_{ji}(\mathbf{x}) && (4) \\ &= r_{ji}(\mathbf{x})^T \Omega_{ji} r_{ji}(\mathbf{x}). && (5) \end{aligned}$$

Under the assumption that the observations are independent, the overall negative log likelihood of a configuration $\mathbf{x}$ is

$$\begin{aligned} F(\mathbf{x}) &= \sum_{\langle j,i \rangle \in \mathcal{C}} F_{ji}(\mathbf{x}) && (6) \\ &= \sum_{\langle j,i \rangle \in \mathcal{C}} r_{ji}(\mathbf{x})^T \Omega_{ji} r_{ji}(\mathbf{x}). && (7) \end{aligned}$$

Here $\mathcal{C} = \{\langle j_1, i_1\rangle, \dots, \langle j_M, i_M\rangle\}$ is set of pairs of indices for which a constraint $\delta_{j_m i_m}$ exists.

The goal of a ML approach is to find the configuration $\mathbf{x}^*$ of the nodes that maximizes the likelihood of the observations. This can be written as

$$\mathbf{x}^* = \underset{\mathbf{x}}{\operatorname{argmin}}\, F(\mathbf{x}). \tag{8}$$

IV. Stochastic Gradient Descent for Maximum Likelihood Mapping

Olson *et al.* [13] propose to use a variant of the preconditioned stochastic gradient descent (SGD) to address the SLAM problem. The approach minimizes Eq. (8) by iteratively selecting a constraint $\langle j, i\rangle$ and by moving the nodes of the network in order to decrease the error introduced by the selected constraint. Compared to the standard formulation of gradient descent, the constraints are not optimized as a whole but individually. The nodes are updated according to the following equation:

$$\mathbf{x}^{t+1} \quad = \quad \mathbf{x}^t + \underbrace{\lambda \cdot \mathbf{H}^{-1} J_{ji}^T \Omega_{ji} r_{ji}}_{\Delta \mathbf{x}_{ji}} \tag{9}$$

Here $\mathbf{x}$ is the set of variables describing the locations of the poses in the network and $\mathbf{H}^{-1}$ is a preconditioning matrix. J_{ji} is the Jacobian of f_{ji}, Ω_{ji} is the information matrix capturing the uncertainty of the observation, and r_{ji} is the residual.

Reading the term $\Delta \mathbf{x}_{ji}$ of Eq. (9) from right to left gives an intuition about the sequential procedure used in SGD:

- r_{ji} is the residual which is the opposite of the error vector. Changing the network configuration in the direction of the residual r_{ji} will decrease the error e_{ji}.
- Ω_{ji} represents the information matrix of a constraint. Multiplying it with r_{ji} scales the residual components according to the information encoded in the constraint.
- J_{ji}^T: The role of the Jacobian is to map the residual term into a set of variations in the parameter space.
- $\mathbf{H}$ is the Hessian of the system and it represents the curvature of the error function. This allows us to scale the variations resulting from the Jacobian depending on the curvature of the error surface. We actually use an approximation of $\mathbf{H}$ which is computed as

$$\mathbf{H} \quad \simeq \quad \sum_{\langle j,i\rangle} J_{ji} \Omega_{ji} J_{ji}^T. \tag{10}$$

  Rather than inverting the full Hessian which is computationally expensive, we approximate it by

$$\mathbf{H}^{-1} \quad \simeq \quad [\operatorname{diag}(\mathbf{H})]^{-1}. \tag{11}$$

- λ is a learning rate which decreases with the iteration of SGD and which makes the system to converge to an equilibrium point.

In practice, the algorithm decomposes the overall problem into many smaller problems by optimizing the constraints individually. Each time a solution for one of these subproblems is found, the network is updated accordingly. Obviously, updating the different constraints one after each other can have opposite effects on a subset of variables. To avoid infinitive oscillations, one uses the learning rate to reduce the fraction of the residual which is used for updating the variables. This makes the solutions of the different sub-problems to asymptotically converge towards an equilibrium point that is the solution reported by the algorithm.

This framework allows us to iteratively reduce the error given the network of constraints. The optimization approach, however, leaves open how the nodes are represented (parameterized). Since the parameterization defines also the structure of the Jacobians, it has a strong influence on the performance of the algorithm.

The next section addresses the problem of how to parameterize a graph in order to efficiently carry out the optimization approach.

V. Network Parameterizations

The poses $\mathbf{p} = \{p_1, \dots, p_n\}$ of the nodes define the configuration of the network. The poses can be described by a vector of parameters $\mathbf{x}$ such that a bijective mapping g between $\mathbf{p}$ and $\mathbf{x}$ exists

$$\mathbf{x} = g(\mathbf{p}) \qquad\qquad \mathbf{p} = g^{-1}(\mathbf{x}). \tag{12}$$

As previously explained, in each iteration SGD decomposes the problem into a set of subproblems and solves them successively. In this work, a subproblem is defined as the optimization of a single constraint. Different solutions to the individual subproblems can have antagonistic effects when combining them.

The parameterization g defines also the subset of variables that are modified by a single constraint update. A good parameterization defines the subproblems in a way that the combination step leads only to small changes of the individual solutions.

A. Incremental Pose Parameterization

Olson *et al.* propose the so-called incremental pose parameterization. Given a set of node locations p_i and given a fixed order on the nodes, the incremental parameters x_i can be computed as follows

$$x_i \quad = \quad p_i - p_{i-1}. \tag{13}$$

Note that x_i is computed as the difference between two subsequent nodes and not by motion composition. Under this parameterization, the error in the global reference frame (indicated by primed variables) has the following form

$$e'_{ji} \quad = \quad p_j - (p_i \oplus \delta_{ji}) \tag{14}$$

$$= \quad \left(\sum_{k=i+1}^{j} x_k\right) + \underbrace{\left(\prod_{k=1}^{i} \tilde{R}_k\right)}_{R_i} \delta_{ji}, \tag{15}$$

where $\oplus$ is the motion composition operator according to Lu and Milios [11] and $\tilde{R}_k$ the homogenous rotation matrix of the *parameter* x_k. The term R_k is defined as the rotation matrix of the *pose* p_k. The information matrix in the global reference frame can be computed as

$$\Omega'_{ji} \quad = \quad R_i \Omega_{ji} R_i^T. \tag{16}$$

According to Olson *et al.* [13], neglecting the contribution of the angular terms of $x_0, \ldots, x_i$ to the Jacobian results in the following simplified form

$$J'_{ji} = \sum_{k=i+1}^{j} \mathcal{I}_k \quad \text{with} \quad \mathcal{I}_k = (0 \cdots 0 \underbrace{I}_{k} 0 \cdots 0). \quad (17)$$

Here 0 is the 3 by 3 zero matrix and I is the 3 by 3 identity.

Updating the network based on the constraint $\langle j, i \rangle$ with such an Jacobian results in keeping the node i fixed and in distributing the residual along all nodes between j and i.

Olson *et al.* weight the residual proportional to $j-i$ which is the number of nodes involved in the constraint. The parameter x_k of the node k with $k = i+1, \ldots, j$ is updated as follows

$$\Delta x_k = \lambda w_k \Omega'_{ji} r'_{ji}, \quad (18)$$

where the weight w_k is computed as

$$w_k = (j-i) \left[\sum_{m=i+1}^{j} D_m^{-1} \right]^{-1} D_k^{-1}. \quad (19)$$

In Eq. (19), D_k are the matrices containing the diagonal elements of the kth block of the Hessian $\mathbf{H}$. Intuitively, each variable is updated proportional to the uncertainty about that variable. Note that the simple form of the Jacobians allows us to update the parameter vector for each node individually as expressed by Eq. (18).

The approach presented in this section is currently one of the best solutions to ML mapping. However, it has the following drawbacks:

- In practice, the incremental parameterization cannot deal with arbitrarily connected networks. This results from the approximation made in Eq. (17), in which the angular components are ignored when computing the Jacobian. This approximation is only valid if the subsequent nodes in Eq. (13) are spatially close. Furthermore, the way the error is distributed over the network assumes that the nodes are ordered according to poses along the trajectory. This results in adding a large number of nodes to the network whenever the robot travels for a long time in the same region. This requirement prevents an approach from merging multiple nodes into a single one. Merging or pruning nodes, however, is a necessary precondition to allow the robot lifelong map learning.
- When updating a constraint between the nodes j and i, the parameterization requires to change the j-i nodes. As a result, each node is likely to be updated by several constraints. This leads to a high interaction between constraints and will typically reduce the convergence speed of SGD. For example, the node k will be updated by all constraints $\langle j', i' \rangle$ with $i' < k \leq j'$. Note that using an intelligent lookup structure, this operation can be carried out in $\mathcal{O}(\log n)$ time where n is the number of nodes in the network [13]. Therefore, this is a problem of convergence speed of SGD and not a computational problem.

B. Tree Parameterization

Investigating a different parameterization which preserves the advantages of the incremental one but overcomes its drawbacks is the main motivation for our approach. First, our method should be able to deal with arbitrary network topologies. This would enable us to compress the graph whenever robot revisits a place. As a result, the size of the network would be proportional to the visited area and not to the length of the trajectory. Second, the number of nodes in the graph updated by each constraint should mainly depend on the topology of the environment. For example, in case of a loop-closure a large number of nodes need to be updated but in all other situations the update is limited to a small number of nodes in order to keep the interactions between constraints small.

Our idea is to first construct a spanning tree from the (arbitrary) graph. Given such a tree, we define the parameterization for a node as

$$x_i = p_i - p_{\text{parent}(i)}, \quad (20)$$

where $p_{\text{parent}(i)}$ refers to the parent of node i in the spanning tree. As defined in Eq. (20), the tree stores the differences between poses. As a consequence, one needs to process the tree up to the root to compute the actual pose of a node in the global reference frame.

However, to obtain only the difference between two arbitrary nodes, one needs to traverse the tree from the first node upwards to the first common ancestor of both nodes and then downwards to the second node. The same holds for computing the error of a constraint. We refer to the nodes one needs to traverse on the tree as the path of a constraint. For example, $\mathcal{P}_{ji}$ is the path from node i to node j for the constraint $\langle j, i \rangle$. The path can be divided into an ascending part $\mathcal{P}_{ji}^{[-]}$ of the path starting from node i and a descending part $\mathcal{P}_{ji}^{[+]}$ to node j. We can then compute the error in the global frame by

$$\begin{aligned} e'_{ji} &= p_j - (p_i \oplus \delta_{ji}) & (21) \\ &= p_j - (p_i + R_i \delta_{ji}) & (22) \\ &= \sum_{k^{[+]} \in \mathcal{P}_{ji}^{[+]}} x_{k^{[+]}} - \sum_{k^{[-]} \in \mathcal{P}_{ji}^{[-]}} x_{k^{[-]}} - R_i \delta_{ji}. & (23) \end{aligned}$$

Here R_i is the rotation matrix of the pose p_i. It can be computed according to the structure of the tree as the product of the individual rotation matrices along the path to the root.

Note that this tree does not replace the graph as an internal representation. The tree only defines the parameterization of the nodes. It can furthermore be used to define an order in which the optimization algorithm can efficiently process the constraints as we will explain in the remainder of this section. For illustration, Figure 2 (a) and (b) depict two graphs and possible parameterization trees.

Similar to Eq. (16), we can express the information matrix associated to a constraint in the global frame by

$$\Omega'_{ji} = R_i \Omega_{ji} R_i^T. \quad (24)$$

As proposed in [13], we neglect the contribution of the rotation matrix R_i in the computation of the Jacobian. This approximation speeds up the computation significantly. Without

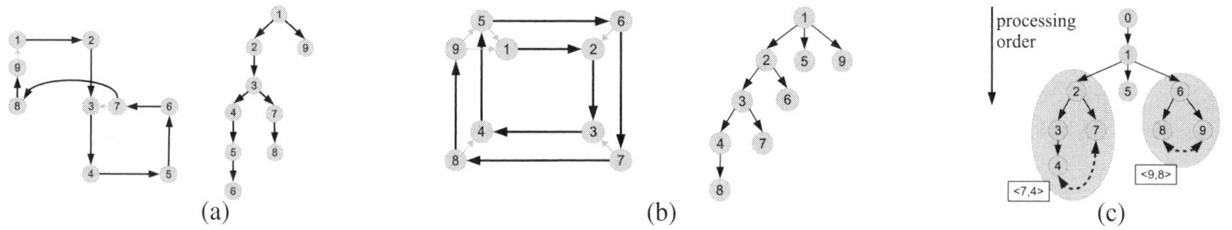

(a) (b) (c)

Fig. 2. (a) and (b): Two small example graphs and the trees used to determine the parameterizations. The small grey connections are constraints introduced by observations where black ones result from odometry. (c) Processing the constraints ordered according to the node with the smallest level in the path avoids the recomputation of rotational component of all parents. The same holds for subtrees with different root nodes on the same level.

this approximation the update of a single constraint influences the poses of all nodes up to the root.

The approximation leads to the following Jacobian:

$$J'_{ji} \quad = \quad \sum_{k^{[+]} \in \mathcal{P}_{ji}^{[+]}} \mathcal{I}_{k^{[+]}} - \sum_{k^{[-]} \in \mathcal{P}_{ji}^{[-]}} \mathcal{I}_{k^{[-]}} \qquad (25)$$

Compared to the approach described in the previous section, the number of updated variables per constraint is in practice smaller when using the tree. Our approach updates $|\mathcal{P}_{ji}|$ variables rather than $j - i$. The weights w_k are computed as

$$w_k = |\mathcal{P}_{ji}| \left[\sum_{m \in \mathcal{P}_{ji}}^{j} D_m^{-1} \right]^{-1} D_k^{-1}, \qquad (26)$$

where D_k is the k-th diagonal block element of $\mathbf{H}$. This results in the following update rule for the variable x_k

$$\Delta x_k \quad = \quad \lambda w_k \cdot \mathrm{s}(x_k, i, j) \cdot \Omega'_{ji} r'_{ji}, \qquad (27)$$

where the value of $\mathrm{s}(x_k, j, i)$ is $+1$ or -1 depending on where the parameter x_k is located on the path $\mathcal{P}_{ji}$:

$$\mathrm{s}(x_k, j, i) \quad = \quad \begin{cases} +1 & \text{if } x_k \in \mathcal{P}_{ji}^{[+]} \\ -1 & \text{if } x_k \in \mathcal{P}_{ji}^{[-]} \end{cases} \qquad (28)$$

Our parameterization maintains the simple form of the Jacobians which enables us to perform the update of each parameter variable individually (as can be seen in Eq. (27)). Note that in case one uses a tree that is degenerated to a list, this parameterization is equal to the one proposed by Olson *et al.* [13]. In case of a non-degenerated tree, our approach offers several advantages as we will show in the experimental section of this paper.

The optimization algorithm specifies how to update the nodes but does not specify the order in which to process the constraints. We can use our tree parameterization to sort the constraints which allows us to reduce the computational complexity of our approach.

To compute the residual of a constraint $\langle j, i \rangle$, we need to know the rotational component of the node i. This requires to traverse the tree up to the first node for which the rotational component is known. In the worst case, this is the root of the tree.

Let the *level* of a node be the distance in the tree between the node itself and the root. Let z_{ji} be the node in the path of the constraint $\langle j, i \rangle$ with the smallest level. The level of the constraint is then defined as the level of z_{ji}.

Our parameterization implies that updating a constraint will never change the configuration of a node with a level smaller than the level of the constraint. Based on this knowledge, we can sort the constraints according to their level and process them in that order. As a result, it is sufficient to access the parent of z_{ji} to compute the rotational component of the node i since all nodes with a smaller level than z_{ji} have already been corrected.

Figure 2 (c) illustrates such a situation. The constraint $\langle 7, 4 \rangle$ with the path $4, 3, 2, 7$ does not change any node with a smaller level than the one of node 2. It also does not influence other subtrees on the same level such as the nodes involved in the constraint $\langle 9, 8 \rangle$.

In the following section, we describe how we actually build the tree given the trajectory of a robot or an arbitrary network as input.

VI. Construction of the Spanning Tree

When constructing the parameterization tree, we distinguish two different situations. First, we assume that the input is a sequence of positions belonging to a trajectory traveled by the robot. Second, we explain how to build the tree given an arbitrary graph of relations.

In the first case, the subsequent poses are located closely together and there exist constraints between subsequent poses resulting from odometry or scan-matching. Further constraints between arbitrary nodes result from observations when revisiting a place in the environment. In this setting, we build our parameterization tree as follows:

1) We assign a unique id to each node based on the timestamps and process the nodes accordingly.
2) The first node is the root of the tree (and therefore has no parent).
3) As the parent of a node, we choose the node with the smallest id for which a constraint to the current node exists.

This tree can be easily constructed on the fly. The Figures 2 (a) and (b) illustrates graphs and the corresponding trees. This tree has a series of nice properties when applying our optimization algorithm to find a minimal error configuration of the nodes. These properties are:

- The tree can be constructed incrementally: when adding a new node it is not required to change the existing tree.
- In case the robot moves through nested loops, the interaction between the updates of the nodes belonging to the individual loops depends on the number of nodes the loops have in common.

- When retraversing an already mapped area and adding constraints between new and previously added nodes, the length of the path in the tree between these nodes is small. This means that only a small number of nodes need to be updated.

The second property is illustrated in Figure 2 (a). The two loops in that image are only connected via the constraint between the nodes 3 and 7. They are the only nodes that are updated by constraints of both loops.

The third property is illustrated in Figure 2 (b). Here, the robot revisits a loop. The nodes 1 to 4 are chosen as the parents for all further nodes. This results in short paths in the tree when updating the positions of the nodes while retraversing known areas.

The complexity of the approach presented so far depends on the length of the trajectory and not on the size of the environment. These two quantities are different in case the robot revisits already known areas. This becomes important whenever the robot is deployed in a bounded environment for a long time and has to update its map over time. This is also known as lifelong map learning. Since our parameterization is not restricted to a trajectory of sequential poses, we have the possibility of a further optimization. Whenever the robot revisits a known place, we do not need to add new nodes to the graph. We can assign the current pose of the robot to an already existing node in the graph.

Note that this can be seen as an approximation similar to adding a rigid constraint neglecting the uncertainty of the corresponding observation. However, in case local maps (e.g., grid maps) are used as nodes in the network, it makes sense to use such an approximation since one can localize a robot in an existing map quite accurately.

To also avoid adding new constraints to the network, we can refine an existing constraint between two nodes in case of a new observation. Given a constraint $\delta_{ji}^{(1)}$ between the nodes j and i in the graph and a new constraint $\delta_{ji}^{(2)}$ based on the current observation. Both constraints can be combined to a single constraint which has the following information matrix and mean:

$$\Omega_{ji} = \Omega_{ji}^{(1)} + \Omega_{ji}^{(2)} \quad (29)$$

$$\delta_{ji} = \Omega_{ji}^{-1}(\Omega_{ji}^{(1)} \cdot \delta_{ji}^{(1)} + \Omega_{ji}^{(2)} \cdot \delta_{ji}^{(2)}) \quad (30)$$

As a result, the size of the problem does not increase when revisiting known locations. As the experiments illustrate, this node reduction technique leads to an increased convergence speed.

In case the input to our algorithm is an arbitrary graph and no natural order of the nodes is provided, we compute a minimal spanning tree to define the parameterization. Since no additional information (like consecutive poses according to a trajectory) is available, we cannot directly infer which parts of the graph are well suited to form a subtree in the parameterization tree. The minimal spanning tree appears to yield comparable results with respect to the number of iterations needed for convergence in all our experiments.

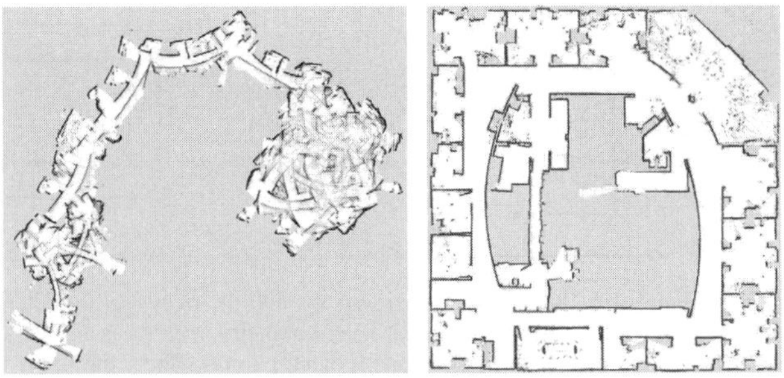

Fig. 3. The map of the Intel Research Lab before (left) and after (right) execution of our algorithm (1000 nodes, runtime <1s).

VII. Experiments

This section is designed to evaluate the properties of our tree parameterization for learning maximum likelihood maps. We first show that such a technique is well suited to generate accurate occupancy grid maps given laser range data and odometry from a real robot. Second, we provide simulation experiments on large-scale datasets. We furthermore provide a comparison between our approach, Olson's algorithm [13], and multi-level relaxation by Frese *et al.* [5]. Finally, we analyze our approach and investigate properties of the tree parameterization in order to explain why we obtain better results then the other methods.

A. Real World Experiments

The first experiment is designed to illustrate that our approach can be used to build maps from real robot data. The goal was to build an accurate occupancy grid map given the laser range data obtained by the robot. The nodes of our graph correspond to the individual poses of the robot during data acquisition. The constraints result from odometry and from the pair-wise matching of laser range scans. Figure 3 depicts two maps of the Intel Research Lab in Seattle. The left one is constructed from raw odometry and the right one is the result obtained by our algorithm. As can be seen, the corrected map shows no inconsistencies such as double corridors. Note that this dataset is freely available on the Internet.

B. Simulated Experiments

The second set of experiments is designed to measure the performance of our approach quantitatively. Furthermore, we compare our technique to two current state-of-the-art SLAM approaches that work on constraint networks, namely multi-level relaxation by Frese *et al.* [5] and Olson's algorithm [13]. In the experiments, we used the two variants of our method: the one that uses the node reduction technique described in Section VI and the one that maintains all the nodes in the graph.

In our simulation experiments, we moved a virtual robot on a grid world. An observation is generated each time the current position of the robot was close to a previously visited location. We corrupted the observations with a variable amount of noise for testing the robustness of the algorithms. We simulated different datasets resulting in graphs with a number of constraints between around 4,000 and 2 million.

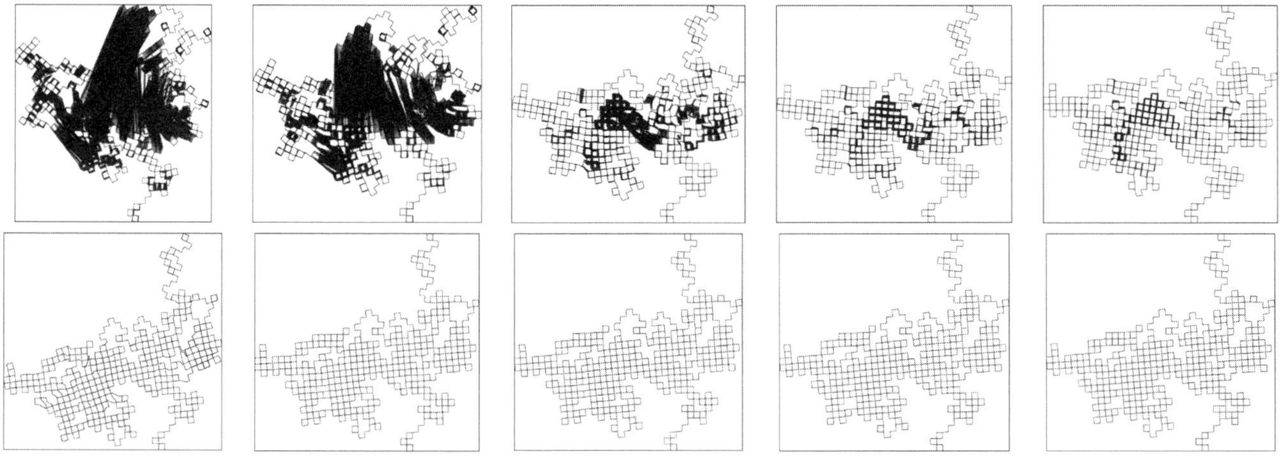

Fig. 4. Results of Olson's algorithm (first row) and our approach (second row) after 1, 10, 50, 100, 300 iterations for a network with 64k constraints. The black areas in the images result from constraints between nodes which are not perfectly corrected after the corresponding iteration (for timings see Figure 6).

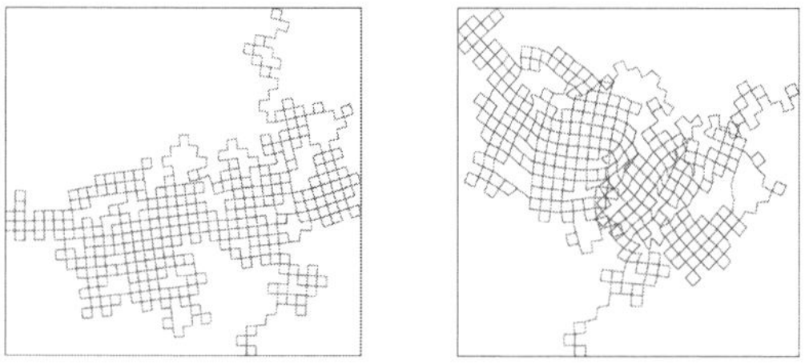

Fig. 5. The result of MLR strongly depends on the initial configuration of the network. Left: small initial pose error, right: large initial pose error.

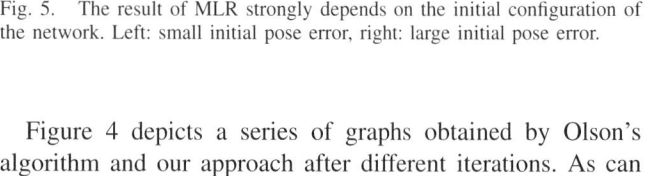

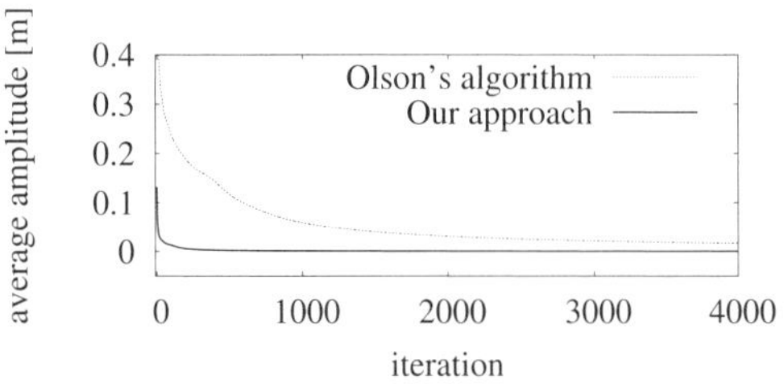

Fig. 7. The average amplitude of the oscillations of the nodes due to the antagonistic effects of different constraints.

Figure 4 depicts a series of graphs obtained by Olson's algorithm and our approach after different iterations. As can be seen, our approach converges faster. Asymptotically, both approaches converge to a similar solution.

In all our experiments, the results of MLR strongly depended on the initial positions of the nodes. In case of a good starting configuration, MLR converges to an accurate solution similar to our approach as shown in Figure 5 (left). Otherwise, it is likely to diverge (right). Olson's approach as well as our technique are more or less independent of the initial poses of the nodes.

To evaluate our technique quantitatively, we first measured the error in the network after each iteration. The left image of Figure 6 depicts a statistical experiments over 10 networks with the same topology but different noise realizations. As can be seen, our approach converges significantly faster than the approach of Olson *et al.* For medium size networks, both approaches converge asymptotically to approximatively the same error value (see middle image). For large networks, the high number of iterations needed for Olson's approach prevented us from showing this convergence experimentally. Due to the sake of brevity, we omitted comparisons to EKF and Gauss Seidel relaxation because Olson *et al.* already showed that their approach outperforms such techniques.

Additionally, we evaluated in Figure 6 (right) the average computation time per iteration of the different approaches. As a result of personal communication with Edwin Olson, we furthermore analyzed a variant of his approach which is restricted to spherical covariances. It yields similar execution times *per iteration* than our approach. However, this restricted variant has still the same converge speed with respect to the number of iterations than Olson's unrestricted technique. As can be seen from that picture, our node reduction technique speeds up the computations up to a factor of 20.

C. Analysis of the Algorithm

The experiments presented above illustrated that our algorithm offers significant improvements compared to both other techniques. The goal of this section is to experimentally point out the reasons for these improvements.

The presented tree parameterization allows us to decompose the optimization of the whole graph into a set of weakly interacting problems. A good measure for evaluating the interaction between the constraints is the average number l of updated nodes per constraint. For example, a network with a large value of l has typically a higher number of interacting constraints compared to networks with low values of l. In all experiments, our approach had a value between 3 and 7. In contrast to that, this values varies between 60 and 17,000 in Olson's approach on the same networks. Note that such a high average path length reduces the convergence speed of Olson's algorithm but does not introduce a higher complexity.

The optimization approach used in this paper as well as in Olson's algorithm updates for each constraint the involved nodes to minimize the error in the network. As a result, different constraints can update poses in an antagonistic way during one iteration. This leads to oscillations in the position

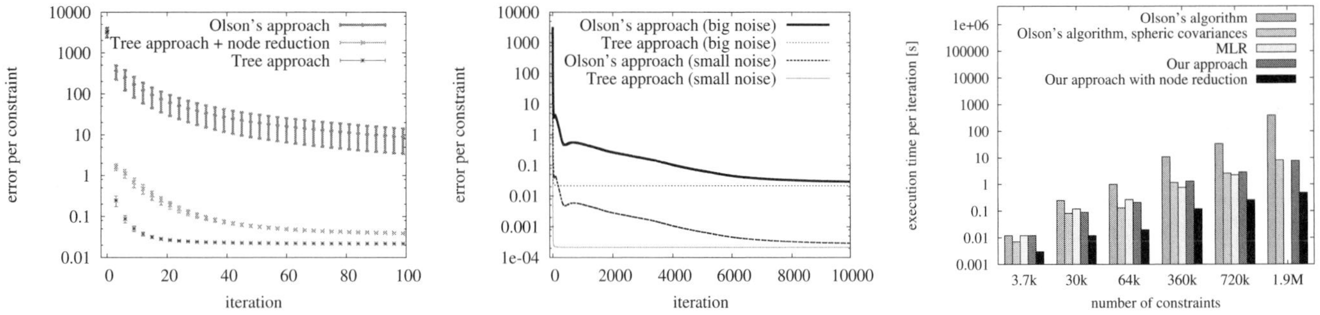

Fig. 6. The left image illustrates shows the error of our and Olson's approach in a statistical experiment ($\sigma = 0.05$ confidence). The image in the middle shows that both techniques converge asymptotically to the same error. The right image shows the average execution time *per iteration* for different networks. For the 1.9M constraints network, the executing of MLR required memory swapping and the result is therefore omitted.

of a node before convergence. Figure 7 illustrates the average amplitude of such an oscillations for Olson's algorithm as well as for our approach. As can be seen, our techniques converges faster to an equilibrium point. This a further reason for the higher convergence speed of our approach.

D. Complexity

Due to the nature of gradient descent, the complexity of our approach per iteration depends linearly on the number of constraints. For each constraint $\langle j, i \rangle$, our approach modifies exactly those nodes which belong to the path $\mathcal{P}_{ji}$ in the tree. Since each constraint has an individual path length, we consider the average path length l. This results in an complexity per iteration of $\mathcal{O}(M \cdot l)$, where M is the number of constraints. In all our experiments, l was approximatively $\log N$, where N is the number of nodes. Note that given our node reduction technique, M as well as N are bounded by the size of the environment and not by the length of the trajectory.

A further advantage of our technique compared to MLR is that it is easy to implement. The function that performs a single iteration requires less than 100 lines of C++ code. An open source implementation, image and video material, and the datasets are available at the authors' web-pages.

VIII. Conclusion

In this paper, we presented a highly efficient solution to the problem of learning maximum likelihood maps for mobile robots. Our technique is based on the graph-formulation of the simultaneous localization and mapping problem and applies a gradient descent based optimization scheme. Our approach extends Olson's algorithm by introducing a tree-based parameterization for the nodes in the graph. This has a significant influence on the convergence speed and execution time of the method. Furthermore, it enables us to correct arbitrary graphs and not only a list of sequential poses. In this way, the complexity of our method depends on the size of the environment and not directly on the length of the input trajectory. This is an important precondition to allow a robot lifelong map learning in its environment.

Our method has been implemented and exhaustively tested on simulation experiments as well as on real robot data. We furthermore compared our method to two existing, state-of-the-art solutions which are multi-level relaxation and Olson's algorithm. Our approach converges significantly faster than both approaches and yields accurate maps with low errors.

Acknowledgment

The authors would like to gratefully thank Udo Frese for his insightful comments and for providing us his MLR implementation for comparisons. Further thanks to Edwin Olson for his helpful comments on this paper. This work has partly been supported by the DFG under contract number SFB/TR-8 (A3) and by the EC under contract number FP6-2005-IST-5-muFly and FP6-2005-IST-6-RAWSEEDS.

References

[1] M. Bosse, P.M. Newman, J.J. Leonard, and S. Teller. An ALTAS framework for scalable mapping. In *Proc. of the IEEE Int. Conf. on Robotics & Automation*, pages 1899–1906, Taipei, Taiwan, 2003.
[2] T. Duckett, S. Marsland, and J. Shapiro. Fast, on-line learning of globally consistent maps. *Autonomous Robots*, 12(3):287 – 300, 2002.
[3] C. Estrada, J. Neira, and J.D. Tardós. Hierachical slam: Real-time accurate mapping of large environments. *IEEE Transactions on Robotics*, 21(4):588–596, 2005.
[4] U. Frese. Treemap: An $o(logn)$ algorithm for indoor simultaneous localization and mapping. *Autonomous Robots*, 21(2):103–122, 2006.
[5] U. Frese, P. Larsson, and T. Duckett. A multilevel relaxation algorithm for simultaneous localisation and mapping. *IEEE Transactions on Robotics*, 21(2):1–12, 2005.
[6] J.-S. Gutmann and K. Konolige. Incremental mapping of large cyclic environments. In *Proc. of the IEEE Int. Symp. on Comp. Intelligence in Robotics and Automation*, pages 318–325, Monterey, CA, USA, 1999.
[7] A. Howard, M.J. Matarić, and G. Sukhatme. Relaxation on a mesh: a formalism for generalized localization. In *Proc. of the IEEE/RSJ Int. Conf. on Intelligent Robots and Systems*, pages 1055–1060, 2001.
[8] S. Julier, J. Uhlmann, and H. Durrant-Whyte. A new approach for filtering nonlinear systems. In *Proc. of the American Control Conference*, pages 1628–1632, Seattle, WA, USA, 1995.
[9] J. Ko, B. Stewart, D. Fox, K. Konolige, and B. Limketkai. A practical, decision-theoretic approach to multi-robot mapping and exploration. In *Proc. of the IEEE/RSJ Int. Conf. on Intelligent Robots and Systems*, pages 3232–3238, Las Vegas, NV, USA, 2003.
[10] J.J. Leonard and H.F. Durrant-Whyte. Mobile robot localization by tracking geometric beacons. *IEEE Transactions on Robotics and Automation*, 7(4):376–382, 1991.
[11] F. Lu and E. Milios. Globally consistent range scan alignment for environment mapping. *Autonomous Robots*, 4:333–349, 1997.
[12] J. Neira and J.D. Tardós. Data association in stochastic mapping using the joint compatibility test. *IEEE Transactions on Robotics and Automation*, 17(6):890–897, 2001.
[13] E. Olson, J.J. Leonard, and S. Teller. Fast iterative optimization of pose graphs with poor initial estimates. In *Proc. of the IEEE Int. Conf. on Robotics & Automation*, pages 2262–2269, 2006.
[14] R. Smith, M. Self, and P. Cheeseman. Estimating uncertain spatial realtionships in robotics. In I. Cox and G. Wilfong, editors, *Autonomous Robot Vehicles*, pages 167–193. Springer Verlag, 1990.

Spatially-Adaptive Learning Rates for Online Incremental SLAM

Edwin Olson, John Leonard, and Seth Teller
MIT Computer Science and Artificial Intelligence Laboratory
Cambridge, MA 02139
Email: eolson@mit.edu, jleonard@mit.edu, teller@csail.mit.edu
http://rvsn.csail.mit.edu

***Abstract*—Several recent algorithms have formulated the SLAM problem in terms of non-linear pose graph optimization. These algorithms are attractive because they offer lower computational and memory costs than the traditional Extended Kalman Filter (EKF), while simultaneously avoiding the linearization error problems that affect EKFs.**

In this paper, we present a new non-linear SLAM algorithm that allows incremental optimization of pose graphs, i.e., allows new poses and constraints to be added without requiring the solution to be recomputed from scratch. Our approach builds upon an existing batch algorithm that combines stochastic gradient descent and an incremental state representation. We develop an incremental algorithm by adding a spatially-adaptive learning rate, and a technique for reducing computational requirements by restricting optimization to only the most volatile portions of the graph. We demonstrate our algorithms on real datasets, and compare against other online algorithms.

I. Introduction

Simultaneous Localization and Mapping (SLAM) algorithms compute a map of an environment using feature observations and estimates of robot motion. SLAM can be viewed as an optimization problem: find a configuration of features and a robot trajectory that is maximally probable given the *constraints* (the sensor observations and robot motion estimates).

The Kalman filter (and its dual, the information filter) are classical approaches to the SLAM problem that assume that the map estimation problem is *linear*, i.e., that uncertainties can be modeled as Gaussians and that the constraint equations are linear in the state variables. Neither is true in practice, but the resulting approximations permit closed-form optimization of the posterior using straight-forward linear algebra. While these algorithms are simple in *form*, their run-time and memory requirements increase quadratically in the number of poses. Many authors have attempted to address these costs [1, 2]. The Iterated [3] and Unscented filters [4] improve the performance of these classical filters in non-linear domains.

Particle filter approaches like FastSLAM [5] explicitly sample the posterior probability distribution, allowing any distribution to be approximated. Unfortunately, large numbers of particles are required to ensure that an acceptable posterior estimate is produced. Supporting large particle populations leads to computational and memory consumption issues.

Non-linear constraints can also be handled by iteratively updating an estimate, each time linearizing around the current state estimate. Lu and Milios suggested a brute-force method [6] that is impractical for all but small problems. Sparse factorizations of the information matrix permit faster updates; TreeMap [7] and Thin Junction Tree Filters [8] truncate small values to enable efficient factorizations, while Square-root SAM searches for a variable reordering that produces sparse

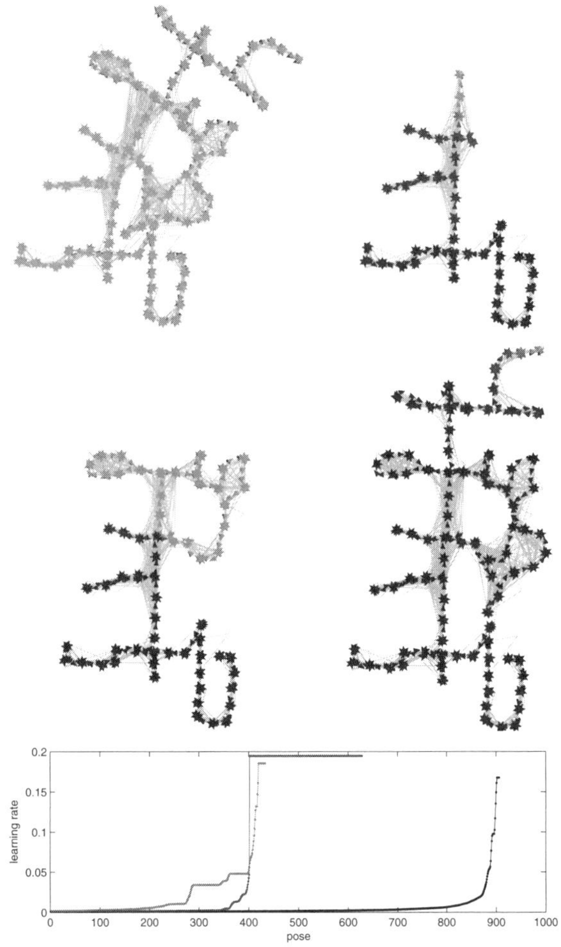

Fig. 1. Incremental Processing of Freiburg dataset. The open-loop graph (top-left) is incrementally optimized; the state of the graph is shown at two intermediate configurations and the final configuration. The colors used in the map indicate the learning rates Λ_i, which are also plotted on the bottom. When closing large loops (middle-left figure), the learning rate is increased over a larger portion of the graph.

but still exact factorizations [9].

Maps with non-linear constraints can also be iteratively improved without computing a factorization of the information matrix. Dellaert proposed a simple relaxation based scheme [10], which was improved by Frese [11]; both of these methods iteratively improve the state variables (the poses), considering a subset of them at a time. More recently, we proposed an alternative method [12] similar to stochastic gradient descent [13]; this method approaches optimization by considering the constraints, rather than the poses. However, these algorithms are *batch* algorithms, and thus are not well-suited to online use.

In this paper, we develop an incremental non-linear optimization algorithm extending our previous batch algorithm [12]; namely, the method is based upon stochastic gradient descent operating on the incremental state representation. The central contributions of this paper are:

- An on-line (incremental) variation of an algorithm that could previously only be used in batches;
- A spatially-varying learning rate that allows different parts of the graph to converge at different rates in correspondence with the impact of new observations;
- A method for accelerating convergence by iterating only on the most volatile parts of the graph, reducing the number of constraints that need to be considered during an iteration.

Iterative methods, like the one described in this paper, are well suited for on-line use: they can incorporate new observations very quickly, and can produce successively better posterior estimates using as much or as little CPU time as the robot can afford. Because the CPU requirements can be throttled, and because the memory requirements are linear in the number of poses and constraints, our approach is well-suited to computationally constrained robots. Data association algorithms also benefit from online algorithms, as the partial maps they produce can be used to help make new associations.

II. Previous Method

This section briefly reviews the batch optimization algorithm described in [12]. The algorithm takes modified gradient steps by considering a single constraint at time. An alternative state space representation is also employed.

Consider a set of robot poses x and a set of constraints that relate pairs of poses. Let J_i be the Jacobian of the ith constraint, and J be the Jacobian of all constraints. Similarly, let Σ_i^{-1} be the information matrix for the ith constraint, and r_i the ith residual. In this paper, we assume that constraints are rigid-body transformations (though generalizations are possible): this means that if there are C constraints and N poses, J will be $3C \times 3N$ and Σ^{-1} will be $3C \times 3C$. The factors of three reflect the degrees-of-freedom inherent in a 2D rigid-body transformation (translation in $\hat{x}$, $\hat{y}$, and rotation).

Given some small step d from the current state estimate, we can write the χ^2 error for all the constraints as:

$$\chi^2 = (Jd - r)^T \Sigma^{-1} (Jd - r) \tag{1}$$

Differentiating with respect to d results in the normal equations for the system:

$$J^T \Sigma^{-1} J d = 2 J^T \Sigma^{-1} r \tag{2}$$

Note that solving this expression for d would yield a least-squares iteration. Now, considering the effects of a single constraint i (i.e., setting $r_j = 0$ for all $j \neq i$), we obtain the step:

$$d = (J^T \Sigma^{-1} J)^{-1} J_i^T \Sigma_i^{-1} r_i \tag{3}$$

This expression cannot be easily evaluated, as it requires the inversion of the information matrix. The quantity $J_i^T \Sigma_i^{-1} r_i$ corresponds to the pure gradient step: the inverted information matrix can be interpreted as a weighting term that accelerates convergence by incorporating knowledge of the relative importance of other constraints on each state variable.

We can accelerate convergence versus a pure gradient step by approximating the information matrix with a matrix M. As in [12], we use the diagonal elements of the information matrix (which are easily computed). This approximation is coarse; in partial compensation we scale all matrix-vector products such that the magnitude of the resulting vector is the same as the original vector. In other words, we use the *shape* of M, but use the magnitude of a gradient-descent step.

The approach in [12] also employed a novel state representation that leads to Jacobians with a particularly simple form that permits fast updates. For each pose, the three unknowns are rewritten as the sum of global-relative increments. Each variable (x, y, and θ) is handled independently; for example:

$$x_i = \sum_{j=0}^{i-1} \Delta x_j \tag{4}$$

This change of variables is motivated by the fact that robot motion is cumulative: the position of a given pose is a function of the motions that preceded it. This could also be accomplished by using rigid-body transformations as the state variables, but the incremental representation leads to a particularly simple Jacobian whereas rigid-body motions lead to complex Jacobians.

Consider a constraint connecting poses a and b, which is a function of the motions between a and b. The Jacobian is well-approximated by zero for the poses between $[0, a]$, block-wise constant for the poses $[a+1, b]$, and zero for the poses after b. This special structure allows a step to be taken in $O(\log N)$ time, as described in [12].

As with stochastic gradient descent, a learning rate λ is employed with each step. Without a learning rate, antagonistic constraints would cause the state estimate to forever oscillate; the learning rate allows these constraints to find an equilibrium by forcing them to compromise. Over time, the learning rate is decreased according to a harmonic progression, the standard rate schedule for stochastic gradient descent [13].

Gradient steps are scaled by the magnitude of the covariance matrix, but the maximum likelihood solution is affected only by their *relative* magnitudes: this results in different convergence behavior for problems differing only by a scale factor. In a least-squares iteration, the correct scaling is determined via inversion of the information matrix, but in our case, this is too costly to compute (and our estimate M is far too coarse). We can, however, rescale the problem such that the magnitudes of the covariance matrices are approximately 1; we write this scale factor as Ω. The parameter Ω is not critical; the average value of Σ_i is generally a reasonable choice.

Combining all of these elements, the step size used in [12] can be written:

$$d_i = \lambda \Omega M^{-1} J_i^T \Sigma_i^{-1} r_i \tag{5}$$

Recall that the scaling by M^{-1} is really a more complicated operation that preserves the amount by which the residual will be reduced. In [12], Eqn. 5 is implemented by constructing a binomial tree from the scaling weights M, then distributing the total *residual reduction* over its leaves (the poses). Consequently, we actually need to calculate the total reduction in residual, Δr_i that results from adding d_i to the state estimate.

Because M^{-1} preserves the residual reduction, Δr_i is independent of M^{-1}. Recall that the Jacobian J_i is well-approximated as zero, except for a block matrix that is repeated $(b-a)$ times. The repeated matrix is in fact a rotation matrix, which we will call R. Multiplying out Eqn. 5 and summing the incremental motion between each pose, we can compute Δr_i:

$$\Delta r_i = \lambda (b-a) \Omega R \Sigma_i^{-1} r_i \tag{6}$$

If necessary, we clamp Δr_i to r_i, to avoid stepping past the solution. Stepping past *might* result in faster convergence (as in the case of successive over-relaxation), but increases the risk of divergence.

This method can rapidly optimize graphs, even when the initial state estimate is poor. This robustness arises from considering only one constraint at a time: the large noisy steps taken early in the optimization allow the state estimate to escape local minima. However, once the solution lands in the basin of the global minimum, the constraints tend to be well-satisfied and smaller steps result.

However, as described above and in [12], the algorithm operates in *batch* mode: new poses and constraints cannot be added to the graph once optimization begins. This makes the algorithm poorly suited for online applications.

III. Incremental Extension

A. Overview

This paper presents a generalization of the batch algorithm that allows new poses and new constraints to be added without restarting the optimization.

When adding a new constraint to a graph, it is desirable to allow the state estimate to reflect the new information fairly quickly. This, in general, requires an increase in the learning rate (which can otherwise be arbitrarily small, depending on how many optimization iterations have been performed so far).

However, large increases in the learning rate cause large steps, obliterating the fine-tuning done by previous iterations. The challenge is to determine a learning rate increase that allows a new constraint to be rapidly incorporated into the state estimate, but that also preserves as much of the previous optimization effort as possible.

Intuitively, a good approach would have the property that a constraint that contained little new information would result in small learning rate increases. Conversely, a new constraint that radically alters the solution (i.e., the closure of a large loop) would result in a large learning rate increase.

When new constraints are added to a graph, their effects are often limited to only a portion of the graph. A good approach should insulate stable parts of the graph from those parts that are being reconfigured due to the addition of new constraints.

To be worthwhile, any candidate approach must be faster than the batch algorithm. It would also be compelling if the incremental algorithm was equivalent to the batch algorithm when the set of constraints is fixed.

This section describes our approach, which has the desirable properties outlined above. Note that we do not discuss how graph constraints are computed (or where they come from): we assume that they are produced by some external sensor system, such as a laser scan-matching algorithm [14] or vision system [15].

B. Spatially-varying learning rates

It is desirable to be able to insulate one area of the graph from the effects of another area of the graph. Suppose that a robot is traveling in a world with two buildings: first it explores building A, then building B. Suppose that the robot discovers that two rooms are in fact the same room in building B: we intuitively expect that the map of building B might require a substantial modification, but the map of building A should be virtually unchanged.

If a significant reconfiguration of building B needlessly causes a violent reconfiguration of building A, the optimization effort previously expended to optimize the map of building A would be wasted. This is to be avoided.

We can isolate one part of the graph from other parts of the graph by spatially varying the learning rate. Instead of a global learning rate λ, we give each pose a different learning rate Λ_i. This allows the learning rate to be varied in different parts of the graph. Managing these learning rates is the subject of this section.

C. Adding a new constraint

When adding a new constraint, we must estimate how large a step should be taken. Once determined, we can compute the learning rate that will permit a step of that size by using Eqn. 6. This learning rate will be used to update the Λ_i's that are affected by the constraint.

The graph's current state estimate already reflects the effects of a number of other constraints. The step resulting from the addition of a new constraint should reflect the certainty of the new constraint and the certainty of the constraints already incorporated into the graph. Let gain β be the fraction of a full-step that would optimally fuse the previous estimate and the new constraint. β can be derived by differentiating the χ^2 cost of two Gaussian observations of the same quantity, or manipulated from the Kalman gain equation:

$$\beta = \Sigma_i^{-1}(\Sigma_i^{-1} + \Sigma_{graph}^{-1})^{-1} \tag{7}$$

We can estimate Σ_{graph}^{-1} from the diagonals of the information matrix: the graph's uncertainty about the transformation from pose a to b is the sum of the uncertainty of the motions between them. We have already approximated these uncertainties in our diagonal approximation to the information matrix M. In truth, the motions are correlated, but we arrive at a serviceable approximation of Σ_{graph} by summing the inverse of the diagonal elements of M between a and b. Because the Jacobians change very slowly in comparison to the state estimate, both M and these sums can be cached (rather than recomputing them every iteration). In our implementation, M (and the quantities derived from it) are updated on iterations that are powers of two.

Using Eqn. 6, we can solve for the learning rate that would result in a step of size $\sum d_i = \beta r_i$. Because there are three degrees-of-freedom per pose, we obtain three simultaneous equations for λ; we could maintain separate learning rates for each, but we use the maximum value for all three. With $\oslash$ representing row-by-row division, we write:

$$\lambda = \text{maxrow}\left(\frac{1}{b-a}\left(\beta r_i \oslash \Omega R \Sigma_i^{-1} r_i\right)\right) \tag{8}$$

This value of λ is then propagated to all of the poses after pose a:

$$\Lambda_i' = \max(\Lambda_i, \lambda) \quad \text{for } i > a \tag{9}$$

D. Processing an old constraint

When processing an old constraint, we must determine what *effective learning rate* should be used when calculating its step size. If no new constraints have ever been added to the graph, then all of the poses have identical learning rates Λ_i: the effective learning rate is just Λ_i. But if new constraints have been added, then the poses affected by the constraint might have different learning rates.

A learning rate increase caused by a new constraint can cause a large change in the state estimate, upsetting the equilibrium of other constraints in the graph. Increasing the effective learning rate of these constraints will decrease the amount of time it takes for the graph to reach a new equilibrium. If the learning rate of these older constraints was not increased, the graph would still converge to an equilibrium; however, because the learning rate could be arbitrarily small (depending on how long the optimization has been running), it could take arbitrarily long for it to do so.

A constraint between poses a and b is sensitive to changes to any of the poses between a and b: the more the poses have been perturbed (i.e., the larger the Λ_i's), the larger the effective learning rate should be. We can interpret each of the poses belonging to a constraint "voting" for the learning rate that should be used. Consequently, the effective learning rate for a constraint can be reasonably set to the average value of the learning rates between a and b. Notably, this rule has the property that it reduces to the batch case when no new constraints are added (and thus all the Λ_i's are equal).

Once the effective learning rate is computed, it can be used to compute a step according to Eqn. 5.

The effective learning rate may be greater than some of the Λ_i's of the affected poses; this must be accounted for by increasing the Λ_i's to be at least as large as the effective learning rate, as was the case when adding a new constraint.

Note that in order to avoid erroneously increasing the learning rate of poses more than necessary, any changes to the learning rate should not take effect until all constraints have been processed. For example, if there are two constraints between poses a and b, the learning rates should not be doubly increased: both constraints are responding to the same perturbation caused by a new edge.

Consider an example with three constraints: a newly-added constraint X between poses 100 and 200, and existing constraints Y (between poses 50 and 150) and Z (between poses 25 and 75). The learning rate increase caused by constraint X will cause an increase in the effective learning rate for constraint Y. On the next iteration, constraint Z will also see an increase in its effective learning rate, because constraint Y perturbed it on the previous iteration. In other words, constraint Z will be affected by constraint X, even though they have no poses in common. Their interaction is mediated by constraint Y.

This "percolating" effect is important in order to accommodate new information, even though the effect is generally small. It is, in essence, an iterative way of dealing with the correlations between constraints.

Returning to the example of buildings A and B, learning rate increases due to loop closures will propagate back toward building A in direct relationship to how tightly coupled buildings A and B are (in terms of constraints interconnecting the two). If they are coupled only by an open-loop path with no loop closures, then building A will not be affected by volatility in building B. This is because learning rates are propagated backward only via constraints involving overlapping sets of poses.

E. Algorithm Summary

In summary, adding a new constraint (or constraints) to the graph requires the following initialization:

1) Compute the constraint's effective learning rate λ using Eqn. 8 and perform a step according to Eqn. 5.
2) Increase the learning rates, as necessary:

$$\Lambda_i' = \max(\Lambda_i, \lambda) \quad \text{for } i > a \tag{10}$$

Updating an existing constraint between poses a and b involves three steps:

1) Compute the constraint's effective learning rate λ by computing the mean value of the learning rates for each

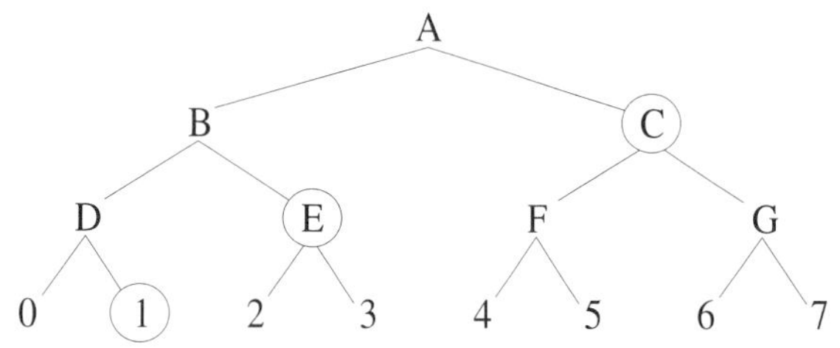

Fig. 2. Learning Rate Tree. Learning rates for each pose are stored in the leaves. Contiguous ranges of nodes can be set by modifying at most $O(\log N)$ nodes. For example, the learning rate for nodes 1-7 can be modified by adjusting three nodes: 1, E, and C. Nodes D, B, and A then must be updated as well. Similarly, cumulative sum can be implemented in $O(\log N)$ time; for example, the sum of Λ_i for $i \in [0, 5]$ can be determined by adding the sums of nodes B and F.

pose spanned by the constraint:

$$\lambda = \frac{1}{b-a} \sum_{a+1}^{b} \Lambda_i \tag{11}$$

2) Compute and apply the constraint step, using learning rate λ;
3) Update the learning rates of the affected poses, as in Eqn. 10.

After processing all of the constraints, the learning rates Λ_i are decreased according to a generalized harmonic progression, e.g.:

$$\Lambda_i' = \frac{\Lambda_i}{1 + \Lambda_i} \tag{12}$$

Note that these rules guarantee the increasing monotonicity of the Λ_i at any given time step. In other words, $\Lambda_i \leq \Lambda_j$ if $i < j$. While any particular Λ_i tends to decrease over time, it does not necessarily decrease monotonically due to the learning rate increases caused by new constraints.

F. Learning Rate Data Structure

An obvious data structure for maintaining the values Λ_i is an array. The operations required by Eqn. 10 and Eqn. 11 would run in $O(N)$ time for a graph with N poses. This would be worse than the batch algorithm, which has $O(\log N)$ complexity per constraint.

Fortunately, an augmented balanced binary tree can be used to implement both of these operations in $O(\log N)$ time. Each pose is represented by a leaf node. Other nodes maintain the minimum, maximum, and sum of their children, with the special caveat that in the case when the minimum and maximum values are the same, the child nodes are overridden. It is this "overriding" behavior which allows the write operations to be performed in $O(\log N)$ time.

For example, implementing Eqn. 10 will affect only a contiguous set of indices starting at some index $j \geq i$ (see Fig. 2). Every member of this set can be overridden by modifying no more than $O(\log N)$ ancestor nodes in the tree. The ancestors' parents also need to be visited so that their min/max/sum fields can be updated, but there are at most $O(\log N)$ parents that require updating.

Eqn. 11 is most easily implemented using a primitive operation that computes the cumulative sum $\sum_{j=0}^{i} \Lambda_j$; this is done by adding together the $O(\log N)$ sums that contribute to the cumulative sum (taking care to handle those nodes that override their children). The mean over an interval is then the difference of two cumulative sums divided by the number of indices in the interval.

The implementation of this data structure is relatively straightforward, if tedious. We refer you to our source code for the implementation details, at http://rvsn.csail.mit.edu.

G. Analysis

On a graph with N poses and M constraints, memory usage of the algorithm is $O(N + M)$, the same as the batch algorithm. Runtime per constraint also remains $O(\log N)$, though constraints are usually processed in "full iterations", in which all the constraints are processed in a rapid succession.

The actual CPU requirements in order to obtain a "good" result are difficult to specify. Our approach does not guarantee how much the error will decrease at each step; the error can even increase. Consequently, we cannot provide a bound on how many iterations are required for a particular level of performance.

That said, the convergence of the algorithm is typically quite rapid, especially in correcting gross errors. Quality requirements naturally vary by application, and iterative approaches, like the one presented here, offer flexibility in trading quality versus CPU time.

The classical stochastic gradient descent algorithm picks the constraints at random, however, we typically process the constraints in a fixed order. The additional randomization caused by processing the constraints in different orders may have a small positive effect on convergence rate. Processing the constraints in a fixed order, however, causes the graph to vary more smoothly after each full iteration.

IV. Constraint Scheduling

In the case of the batch algorithm, each *full iteration* includes an update step for *every* constraint in the graph. The learning rate is controlled globally, and the graph converges more-or-less uniformly throughout the graph.

In the incremental algorithm that we have described, the learning rate is not global, and different parts of the graph can be in dramatically different states of convergence. In fact, it is often the case that older parts of the graph have much lower learning rates (and are closer to the minimum-error configuration) than newer parts, which are farther from a minimum.

This section describes how the least-converged parts of the graph can be optimized, without the need to further optimize distant and already well-converged parts.

The least-converged part of the graph is typically the most important because they generally contain the robot itself. The area around the robot is usually the least-converged part of the graph because new constraints are added to the graph based on the observations of the robot. Consequently, it is critically important for the robot to be able to improve its local map, and it is relatively unimportant to "fine tune" some distant (and often not immediately relevant) area.

Our method does not require a map to be explicitly segmented (into buildings, for example): rather, we automatically identify the subgraph that is the most volatile (i.e., has the largest learning rates), then determine the set of constraints that must be considered in order to reduce the learning rates of that subgraph. This subset of constraints is typically much smaller than the total set of constraints in the graph, resulting in significant CPU savings.

Here's the basic idea: suppose we want to reduce the maximum Λ_i to Λ'_{max} during an iteration. All of the constraints in the graph have an effective learning rate either larger or smaller than Λ'_{max}. Those constraints with larger effective learning rates must be processed before the Λ_i's are reduced, because those constraints still need to take larger steps.

In contrast, those constraints that have smaller effective learning rates are taking comparatively small steps: their small steps tend to be ineffective because other constraints are taking larger steps. Processing constraints with small effective learning rates will generally not achieve any χ^2 reduction when there are other constraints taking large steps; we can save CPU time by skipping them.

The following algorithm implements this heuristic to optimize only a recent subgraph of the pose graph:

1) Look up the maximum learning rate in the graph, e.g., $\Lambda_{max} = \Lambda_{nposes-1}$. If we performed a full iteration, the maximum learning rate after the iteration would be $\Lambda'_{max} = \Lambda_{max}/(1 + \Lambda_{max})$. We use this as our target value.
2) Perform update steps only on those constraints whose effective learning rate is greater than Λ_{max}.
3) Set $\Lambda'_i = \Lambda'_{max}$ for all $i \geq p$.

In other words, this algorithm reduces the *maximum* learning rate in the graph by a full harmonic progression by computing and operating on only a subset of the graph's constraints. The procedure conservatively identifies the set of constraints that should be considered. Note that the operation in step 3 can also be implemented in $O(\log N)$ time using the previously-described learning rate data structure.

In many cases, large reductions in learning rate can be achieved by considering only a handful of constraints. In the Freiburg data set, the technique is very effective, with under 10% of the constraints updated at each iteration (see Fig. 3). Since computational cost is linear in the number of constraints processed, this yields a speed-up of almost 10x. Despite the fact that a only a small fraction of the constraints are considered, the χ^2 error is essentially the same as the much slower implementation that considers all constraints (see Fig. 4). This is because the bulk of error in the graph is concentrated in the more recent portions of the graph. The "all constraints" method spends large amounts of CPU time tweaking distant and already well-converged portions of the graph, which generally does not yield very large χ^2 reductions.

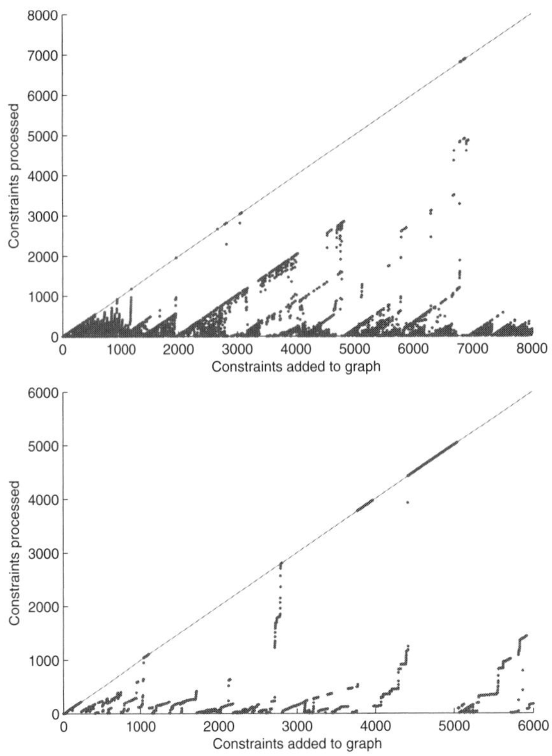

Fig. 3. Constraint Scheduling. Top: Freiburg, Bottom: Intel Research Center. Selective processing of constraints leads to large speed-ups. In the figure, the line indicates the total number of constraints in the graph; the points show the actual number of constraints that were processed. In the Intel dataset, more constraints tend to be processed since the robot repeatedly revisits the same areas, creating new constraints that span many poses, thus disturbing larger portions of the graph.

In contrast, the "selected constraints" method focuses solely on the parts of the graph that will lead to the largest χ^2 reductions.

The same approach on the Intel Research Center data set processes 27% of the constraints on average. The lower performance is due to the fact that the robot is orbiting the entire facility, frequently creating new constraints between poses that are temporally distant. This causes learning rate increases to propagate to more poses. Despite this, a significant speedup is achieved.

The approach outlined here is somewhat greedy: it attempts to reduce the worst-case learning rate to Λ'_{max}. It is Λ'_{max} that determines the number of constraints that are necessary to perform the partial update. It is possible that a slightly larger value of Λ_{max} would result in significantly fewer constraints to process, resulting in larger χ^2 reduction per CPU time. This is an area of future work.

V. Results

We compared the runtime performance characteristics of our approach to that of LU Decomposition (non-linear least squares), the Extended Kalman Filter (EKF) and Gauss-Seidel Relaxation (GS). Our testing was performed on a modern desktop system with a 2.4GHz CPU, and our code was written in Java.

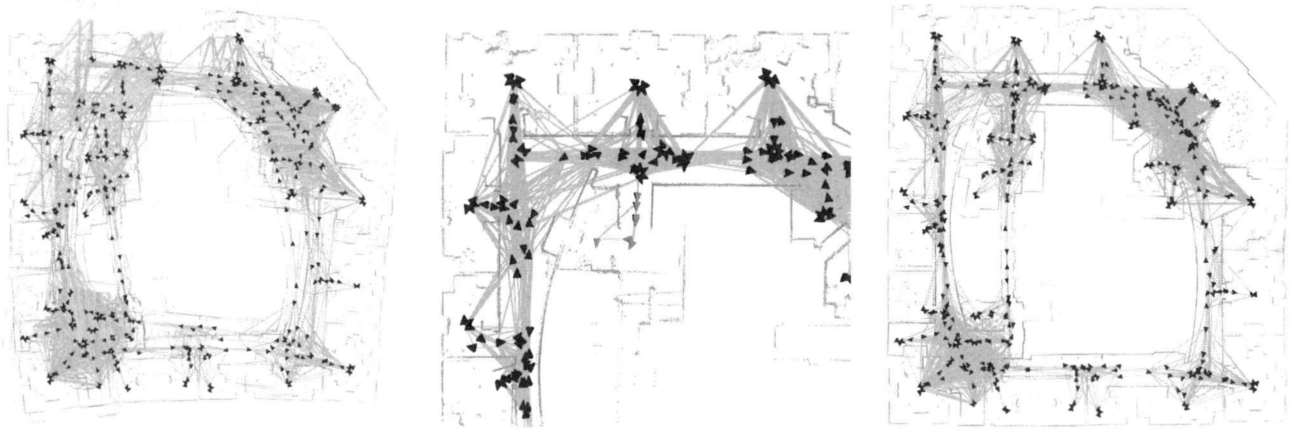

Fig. 6. Intel Research Center. Left: the open-loop trajectory. Middle: After orbiting the facility three times, the robot is entering a new area; the well-explored area has a low learning rate while the newly entered area has a high learning rate. Right: the posterior map.

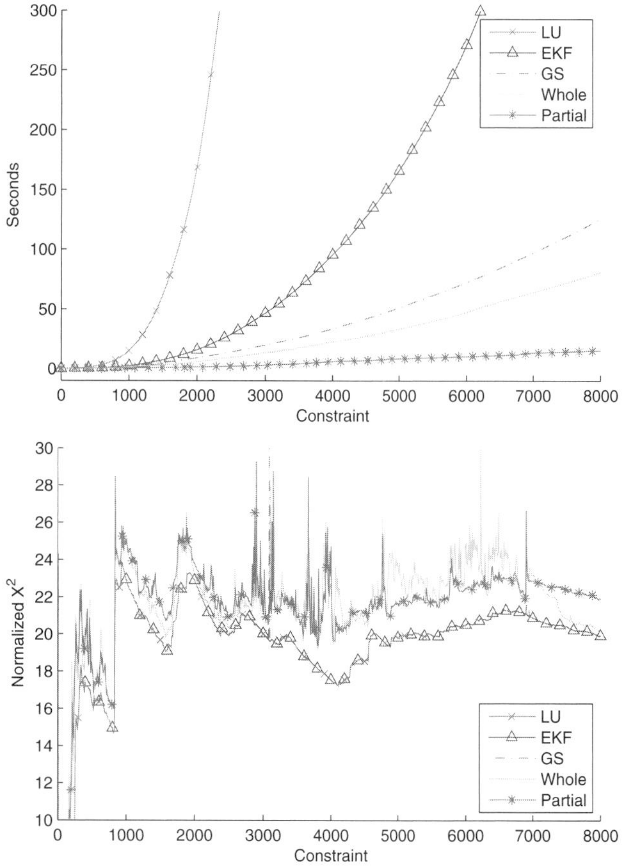

Fig. 4. Cumulative Runtime and Error Comparisons, Freiburg dataset. Each constraint was added one at a time. EKF and LU computational time dwarfs the others. Our proposed method (with constraint selection) is by far the fastest at 21 seconds; our method (without constraint selection) beats out Gauss-Seidel relaxation. In terms of quality, LU, EKF, and Gauss-Seidel all produce nearly optimal results; our methods have marginally higher χ^2 error, as expected, but the maps are subjectively difficult to distinguish. (Partial: 15.5s, Whole: 82.5s, Gauss Seidel: 128.3s, EKF: 650s.)

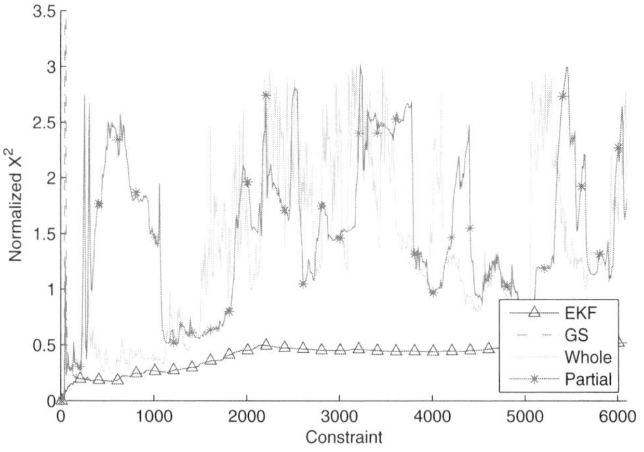

Fig. 5. Error Comparisons, Intel Research Center. Our methods are able to stay relatively close to the nearly-optimal χ^2 error produced by the EKF and Gauss-Seidel, however, did so at a fraction of the run time. (Partial: 14.5s, Whole: 45.4s, Gauss-Seidel: 65.2s, EKF: 132.3s)

Since this algorithm is targeted at on-line applications, we assume that the robot requires a full state estimate after every observation; this makes the performance of the EKF no worse than that of an information-form filter which would require frequent inversions of the information matrix. The CPU time and χ^2 results on the Freiburg data set are shown in Fig. 4. Similar behavior occurred on the Intel Research Center dataset (see Fig. 5).

To mitigate any potential advantage of the iterative algorithms (they could, after all, be extremely fast by simply doing no work), they were forced to continue to iterate until the χ^2 was reduced below a threshold (Freiburg 25, Intel 3.0).

Our approach, especially with constraint selection enabled, is significantly faster than any of the other methods. In terms of quality (as measured by χ^2 error), our approaches produced somewhat worse maps. However, the difference is very subtle. When using the constraint selection algorithm (figures labeled "partial"), our algorithm is significantly faster than the other approaches.

Even if all constraints must be processed, the algorithm is very fast. After adding the last constraint, processing all 8000 constraints on the Freiburg graph with 906 poses required 16.8ms. When using the constraint selection algorithm, only a small fraction of these constraints need to be processed: the algorithm took an average of 1.1ms to add each of the final 10 constraints on the Freiburg graph: for each, it considered an average of 73 constraints. Several Freiburg maps are illustrated in Fig. 1, including the learning rates (as a function of pose).

Putting these numbers in perspective, the Intel Research Center data set represents 45 minutes of data; incorporating observations one at a time (and outputting a posterior map after every observation) required a total cumulative time of 14.5s with constraint selection enabled, and 45.4s without. This would consume about 0.6% of the robot's CPU over the lifetime of the mission, making the CPU available for other purposes. These maps were of fairly high quality, with χ^2 errors only marginally larger than that of the EKF.

Several maps from the Intel Research Center are shown in Fig. 6. The open-loop trajectory is shown, as well as several intermediate maps. In the first map, a loop closure is just about to occur; prior to this, the learning rate is low everywhere. After the loop closure, the learning rate is high everywhere. The final map exhibits sharp walls and virtually no feature doubling; the incremental algorithm matches the quality of the batch algorithm.

As with the batch algorithm, the optimization rapidly finds a solution *near* the global minimum, but once near the minimum, other algorithms can "fine tune" more efficiently. On a large synthetic data set, our method requires 24.18s for incremental processing; after a post-processing using 2.0s of Gauss-Seidel relaxation, the normalized χ^2 is 1.18. This is far better than Gauss-Seidel can do on its own: it requires 168s to achieve the same χ^2 on its own.

VI. Conclusion

We have presented an incremental non-linear SLAM algorithm, generalizing an existing batch algorithm. Our introduction of a spatially-dependent learning rate improves CPU efficiency by limiting learning rate increases to only those areas of the map that require them. We also showed how to optimize only the subgraph that has the largest learning rate, which leads to significant performance improvements.

Iterative non-linear methods, like the one presented here, offer many advantages over conventional SLAM algorithms including faster operation, lower memory consumption, and the ability to dynamically trade CPU utilization for map quality.

References

[1] S. Thrun, Y. Liu, D. Koller, A. Ng, Z. Ghahramani, and H. Durrant-Whyte, "Simultaneous localization and mapping with sparse extended information filters," April 2003.

[2] M. Bosse, P. Newman, J. Leonard, and S. Teller, "An Atlas framework for scalable mapping," *IJRR*, vol. 23, no. 12, pp. 1113–1139, December 2004.

[3] A. Gelb, *Applied Optimal Estimation.* Cambridge, MA: MIT Press, 1974.

[4] S. Julier and J. Uhlmann, "A new extension of the Kalman filter to nonlinear systems," in *Int. Symp. Aerospace/Defense Sensing, Simul. and Controls, Orlando, FL*, 1997, pp. 182–193. [Online]. Available: citeseer.ist.psu.edu/julier97new.html

[5] M. Montemerlo, "FastSLAM: A factored solution to the simultaneous localization and mapping problem with unknown data association," Ph.D. dissertation, Robotics Institute, Carnegie Mellon University, Pittsburgh, PA, July 2003.

[6] F. Lu and E. Milios, "Globally consistent range scan alignment for environment mapping," *Autonomous Robots*, vol. 4, no. 4, pp. 333–349, 1997.

[7] U. Frese, "Treemap: An $O(log(n))$ algorithm for simultaneous localization and mapping," in *Spatial Cognition IV*, 2004.

[8] M. Paskin, "Thin junction tree filters for simultaneous localization and mapping," Ph.D. dissertation, Berkeley, 2002.

[9] F. Dellaert, "Square root SAM," in *Proceedings of Robotics: Science and Systems*, Cambridge, USA, June 2005.

[10] T. Duckett, S. Marsland, and J. Shapiro, "Learning globally consistent maps by relaxation," in *Proceedings of the IEEE International Conference on Robotics and Automation (ICRA'2000)*, San Francisco, CA, 2000.

[11] U. Frese, P. Larsson, and T. Duckett, "A multilevel relaxation algorithm for simultaneous localisation and mapping," *IEEE Transactions on Robotics*, 2005.

[12] E. Olson, J. Leonard, and S. Teller, "Fast iterative optimization of pose graphs with poor initial estimates," in *Proceedings of ICRA 2006*, 2006, pp. 2262–2269.

[13] H. Robbins and S. Monro, "A stochastic approximation method," *Annals of Mathematical Statistics*, vol. 22, pp. 400–407, 1951.

[14] F. Lu and E. Milios, "Robot pose estimation in unknown environments by matching 2d range scans," in *CVPR94*, 1994, pp. 935–938. [Online]. Available: citeseer.ist.psu.edu/lu94robot.html

[15] S. Se, D. Lowe, and J. Little, "Vision-based mobile robot localization and mapping using scale-invariant features," in *Proceedings of the IEEE International Conference on Robotics and Automation (ICRA)*, Seoul, Korea, May 2001, pp. 2051–2058. [Online]. Available: citeseer.ist.psu.edu/se01visionbased.html

Adaptive Non-Stationary Kernel Regression for Terrain Modeling

Tobias Lang Christian Plagemann Wolfram Burgard

Albert-Ludwigs-University of Freiburg, Department for Computer Science, 79110 Freiburg, Germany
{langt,plagem,burgard}@informatik.uni-freiburg.de

***Abstract*— Three-dimensional digital terrain models are of fundamental importance in many areas such as the geo-sciences and outdoor robotics. Accurate modeling requires the ability to deal with a varying data density and to balance smoothing against the preservation of discontinuities. The latter is particularly important for robotics applications, as discontinuities that arise, for example, at steps, stairs, or building walls are important features for path planning or terrain segmentation tasks. In this paper, we present an extension of the well-established Gaussian process regression approach that utilizes non-stationary covariance functions to locally adapt to the structure of the terrain data. In this way, we achieve strong smoothing in flat areas and along edges and at the same time preserve edges and corners. The derived model yields predictive distributions for terrain elevations at arbitrary locations and thus allows to fill gaps in the data and to perform conservative predictions in occluded areas.**

I. Introduction

The modeling of three-dimensional terrain has been widely studied across different research areas like the geo-sciences or robotics. Important applications in the latter case include mobile robotics for agriculture, search and rescue, or surveillance. In these domains, accurate and dense models of the three-dimensional structure of the environment enable the robot to estimate the traversability of locations, to plan its path to a goal location, or to localize itself using range sensor measurements. Building a digital terrain model means to transform a set of sensory inputs, typically a 3D point cloud or the raw range sensor readings, to a function mapping 2-dimensional pose coordinates to elevation values. While geological applications often operate on a larger spatial scale, in which local terrain features can be neglected, autonomous robots greatly rely on distinct structural features like edges or corners to guide navigation, localization, or terrain segmentation. We therefore have two, at the first glance contradicting requirements for terrain models: First, raw sensory data needs to be smoothed in order to remove noise and to be able to perform elevation predictions at all locations and, second, discontinuities need to be preserved as they are important features for path planning, localization, and object recognition.

In this paper, we present a novel terrain modeling approach based on an extended Gaussian process formulation. Our model uses non-stationary covariance functions as proposed by Paciorek *et al.* [9] to allow for local adaptation of the regression kernels to the underlying structure. This adaptation is achieved by iteratively fitting the local kernels to the structure of the underlying function using local gradient features and the local marginal data likelihood (see Figure 1 for an illustration). Indeed, this idea is akin to adaptive image smoothing studied in computer vision, where the task is to achieve de-noising of an image without reducing the contrast of edges and corners [17, 8]. Although these approaches from the computer vision literature are not specifically designed for dealing with a varying density of data points or with potential gaps to fill, they nevertheless served as an inspiration for our kernel adaptation approach.

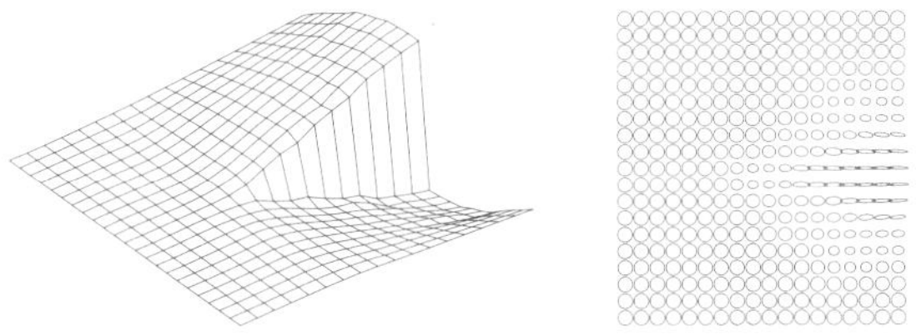

Fig. 1. A hard, synthetic regression problem (left). The continuous regions should be smoothed without removing the strong edge feature. Our approach achieves this by adapting local kernels to the terrain data (right).

The paper is structured as follows. We first discuss related work in the next section. In Section III, we formalize the terrain modeling problem using Gaussian processes and introduce our approach to non-stationary adaptive regression. Section IV presents our experimental results on real and simulated terrain data sets.

II. Related Work

A broad overview over methods used for modeling terrain data is given by Hugentorp [7]. Elevation maps have been used as an efficient data structure for representing dense terrain data [1, 10] and have later been extended to multi-level probabilistic surface maps [19]. Früh *et al.* [5] present an approach to filling local gaps in 3D models based on local linear interpolation. As their approach has yielded promising results in city mapping applications, we compare its modeling accuracy to our approach in Section IV.

Gaussian processes (GPs) have a long tradition in the geo-sciences and statistics literature [14]. Classical approaches for dealing with non-stationarity include input-space warping [15, 16] and hierarchical modeling using local kernels [9]. The latter approach provides the general framework for this work.

Recently, GPs have become popular in robotics, e.g., for learning measurement models [4, 2, 12], action models [6], or failure models [11]. To deal with varying target function properties in the context of perception problems, Williams [20] uses mixtures of GPs for segmenting foreground and background in images in order to extract disparity information from binocular stereo images. Rasmussen and Ghahramani [13] extend ideas of Tresp [18] and present an infinite mixture of experts model where the individual experts are made up from different GP models. A gating network assigns probabilities to the different expert models based completely on the input. Discontinuities in wind fields have been dealt with by Cornford *et al.* [3]. They place auxiliary GPs along the edge on both sides of the discontinuity. These are then used to learn GPs representing the process on either side of the discontinuity. In contrast to our work, they assume a parameterized segmentation of the input space, which appears to be disadvantageous in situations such as depicted in Figure 1 and on real-world terrain data sets.

The problem of adapting to local structure has also been studied in the computer vision community. Taketa *et al.* [17] perform non-parametric kernel regression on images. They adapt kernels according to observed image intensities. Their adaptation rule is thus based on a nonlinear combination of both spatial and intensity distance of all data points in the local neighborhood. Based on singular value decompositions of intensity gradient matrices, they determine kernel modifications. Middendorf and Nagel [8] propose an alternative kernel adaptation algorithm. They use estimates of gray value structure tensors to adapt smoothing kernels to gray value images.

III. Digital Terrain Modeling

Data for building 3-dimensional models of an environment can be acquired from various sources. In robotics, laser range finders are popular sensors as they provide precise, high-frequency measurements at a high spatial resolution. Other sensors include on-board cameras, which are chosen because of their low weight and costs, or satellite imagery, which covers larger areas, e.g., for guiding unmanned areal vehicles (UAVs) or autonomous cars. After various preprocessing steps, the raw measurements are typically represented as 3D point clouds or are transformed into a 3D occupancy grid or elevation map [1]. In this work, we introduce a technique for constructing continuous, probabilistic elevation map models from data points, that yield predictive distributions for terrain elevations at arbitrary input locations.

The terrain modeling problem can be formalized as follows. Given a set $\mathcal{D} = \{(\mathbf{x}_i, y_i)\}_{i=1}^n$ of n location samples $\mathbf{x}_i \in \mathbb{R}^2$ and the corresponding terrain elevations $y_i \in \mathbb{R}$, the task is to build a model for $p(y^*|\mathbf{x}^*, \mathcal{D})$, i.e., the predictive distribution of elevations y^* at new input locations $\mathbf{x}^*$. This modeling task is a hard one for several reasons. First, sensor measurements are inherently affected by noise, which an intelligent model should be able to reduce. Second, the distribution of available data points is typically far from uniform. For example, the proximity of the sensor location is usually more densely sampled than areas farther away. Third, small gaps in the data should be filled with high confidence while more sparsely sampled locations should result in higher predictive uncertainties. To illustrate the last point, consider an autonomous vehicle navigating in off road terrain. Without filling small gaps, even single missing measurements may lead to the perception of an un-traversable obstacle and consequently the planned path might differ significantly from the optimal one. On the other hand, the system should be aware of the increased uncertainty when filling larger gaps to avoid overconfidence at these locations. As a last non-trivial requirement, the model should preserve structural elements like edges and corners as they are important features for various applications including path planning or object recognition.

In this paper, we propose a model to accommodate for all of the above-mentioned requirements. We build on the well-established framework of Gaussian processes, which is a non-parametric Bayesian approach to the regression problem. To deal with the preservation of structural features like edges and corners, we employ non-stationary covariance functions as introduced by Paciorek and Schervish [9] and present a novel approach to local kernel adaptation based on gradient features and the local marginal data likelihood.

In the following, we restate the standard Gaussian process approach to non-parametric regression before we introduce our extensions to local kernel adaptation.

A. Gaussian Process Regression

As stated in the previous section, the terrain modeling task is to derive a model for $p(y^*|\mathbf{x}^*, \mathcal{D})$, which is the predictive distribution of terrain elevations y^*, called *targets*, at input locations $\mathbf{x}^*$, given a training set $\mathcal{D} = \{(\mathbf{x}_i, y_i)\}_{i=1}^n$ of elevation samples. The idea of Gaussian processes (GPs) is to view any finite set of samples y_i from the sought after distribution as being jointly normally distributed,

$$p(y_1, \ldots, y_n \mid \mathbf{x}_1, \ldots, \mathbf{x}_n) \sim \mathcal{N}(\boldsymbol{\mu}, K) , \quad (1)$$

with mean $\boldsymbol{\mu} \in \mathbb{R}^n$ and covariance matrix K. $\boldsymbol{\mu}$ is typically assumed $\mathbf{0}$ and K is specified in terms of a parametric covariance function k and a global noise variance parameter σ_n, $K_{ij} := k(\mathbf{x_i}, \mathbf{x_j}) + \sigma_n^2 \delta_{ij}$. The covariance function k represents the prior knowledge about the target distribution and does not depend on the target values y_i of $\mathcal{D}$. A common choice is the squared exponential covariance function

$$k(\mathbf{x}_i, \mathbf{x}_j) = \sigma_f^2 \exp\left(-\frac{1}{2}\sum_{k=1}^{2}\frac{(\mathbf{x}_{i,k} - \mathbf{x}_{j,k})^2}{\ell_k}\right) , \quad (2)$$

where σ_f denotes the amplitude (or signal variance) and ℓ_k are the characteristic length-scales of the individual dimensions (see [14]). These parameters plus the global noise variance σ_n are called hyperparameters of the process. They are typically denoted as $\boldsymbol{\Theta} = (\sigma_f, \boldsymbol{\ell}, \sigma_n)$. Since any set of samples from the process is jointly Gaussian distributed, the prediction of a new target value y^* at a given location $\mathbf{x}^*$ can be performed by conditioning the $n+1$-dimensional joint Gaussian on the known target values of the training set $\mathcal{D}$. This yields a predictive normal distribution $y^* \sim \mathcal{N}(\mu^*, v^*)$ defined by

$$\mu^* = E(y^*) = \mathbf{k}^T \left(K + \sigma_n^2 I\right)^{-1} \mathbf{y} \,, \quad (3)$$

$$v^* = V(y^*) = k^* + \sigma_n^2 - \mathbf{k}^T \left(K + \sigma_n^2 I\right)^{-1} \mathbf{k} \,, \quad (4)$$

with $K \in \mathbb{R}^{n \times n}$, $K_{ij} = k(\mathbf{x}_i, \mathbf{x}_j)$, $\mathbf{k} \in \mathbb{R}^n$, $k_j = k(\mathbf{x}^*, \mathbf{x}_j)$, $k^* = k(\mathbf{x}^*, \mathbf{x}^*) \in \mathbb{R}$, and the training targets $\mathbf{y} \in \mathbb{R}^n$. Learning in the Gaussian process framework means finding the parameters $\mathbf{\Theta}$ of the covariance function k. Throughout this work we use a conjugate gradient based algorithm [14] that fixes the parameters by optimizing the marginal data likelihood of the given training data set. Alternatively, the parameters could be integrated over using parameter-specific prior distributions, which results in a fully Bayesian model but which is also computationally more demanding as one has to employ Markov-Chain Monte Carlo sampling for approximating the intractable integral.

The standard model introduced so far already accounts for three of the requirements discussed in the previous section, namely de-noising, dealing with non-uniform data densities, and providing predictive uncertainties. As a major drawback, however, by using the stationary covariance function of Equation (2), which depends only on the *differences* between input locations, one basically assumes the same covariance structure on the whole input space. In practice, this significantly weakens important features like edges or corners. The left diagram of Figure 1 depicts a synthetic data-set which contains homogenous regions which should be smoothed, but also a sharp edge that has to be preserved. Our model, which is detailed in the next section, addresses this problem by adapting a non-stationary covariance function to the local terrain properties.

B. Non-Stationary Covariance Functions

Most Gaussian process based approaches found in the literature use stationary covariance functions that depend on the difference between input locations $\mathbf{x} - \mathbf{x}'$ rather than on the absolute values $\mathbf{x}$ and $\mathbf{x}'$. A powerful model for building non-stationary covariance functions from arbitrary stationary ones has been proposed by Paciorek and Schervish [9]. For the Gaussian kernel, their non-stationary covariance function takes the simple form

$$k(\mathbf{x}_i, \mathbf{x}_i) = |\Sigma_i|^{\frac{1}{4}} |\Sigma_j|^{\frac{1}{4}} \left| \frac{\Sigma_i + \Sigma_j}{2} \right|^{-\frac{1}{2}} \cdot \exp\left[-(\mathbf{x}_i - \mathbf{x}_j)^T \left(\frac{\Sigma_i + \Sigma_j}{2} \right)^{-1} (\mathbf{x}_i - \mathbf{x}_j) \right] , \quad (5)$$

where each input location $\mathbf{x}'$ is assigned an individual Gaussian kernel matrix Σ' and the covariance between two targets y_i and y_j is calculated by averaging between the two individual kernels at the input locations $\mathbf{x}_i$ and $\mathbf{x}_j$. In this way, the local characteristics at both locations influence the modeled covariance of the corresponding target values. In this model, each kernel matrix Σ_i is internally represented by its eigenvectors and eigenvalues. Paciorek and Schervish build a hierarchical model by placing additional Gaussian process priors on these kernel parameters and solve the integration using Markov-Chain Monte Carlo sampling. While the model presented in [9] provides a flexible and general framework, it is, as also noted by the authors, computationally demanding and clearly not feasible for the real world terrain data sets that we are aiming for in this work. As a consequence, we propose to model the kernel matrices in Equation (5) as independent random variables that are initialized with the learned kernel of the corresponding stationary model and then iteratively adapted to the local structure of the given terrain data. Concretely, we assign to every input location $\mathbf{x}_i$ from the training set $\mathcal{D}$ a local kernel matrix Σ_i, which in turn is represented by one orientation parameter and two scale parameters for the length of the axes. Given these parameters, the evaluation of Equation (5) is straightforward. In the following section, we will discuss in detail, how the kernel matrices Σ_i can be adapted to the local structure of the terrain.

C. Local Kernel Adaptation

The problem of adapting smoothing kernels to local structure has been well studied in the computer vision community. It is therefore not surprising that, although image processing algorithms are typically restricted to dense and uniformly distributed data, we can use findings from that field as an inspiration for our terrain adaptation task. Indeed, Middendorf and Nagel [8] present a technique for iterative kernel adaptation in the context of optical flow estimation in image sequences. Their approach builds on the concept of the so called grey-value structure tensor (GST), which captures the local structure of an image or image sequence by building the locally weighted outer product of grey-value gradients in the neighborhood of the given image location. Analogously to their work, we define the elevation structure tensor (EST) for a given location $\mathbf{x}_i$ as

$$EST(\mathbf{x}_i) := \overline{\nabla y (\nabla y)^T}(\mathbf{x}_i) \,, \quad (6)$$

where $y(\mathbf{x})$ denotes the terrain elevation at a location $\mathbf{x}$ and $\overline{\cdot}$ stands for the operator that builds a locally weighted average of its argument according to the kernel Σ_i. For two-dimensional $\mathbf{x}_i$, Equation (6) calculates the locally weighted average of the outer product of $\nabla y = (\frac{\partial y}{\partial x_1}, \frac{\partial y}{\partial x_2})^T$. This local elevation derivative can be estimated directly from the raw elevation samples in the neighborhood of the given input location $\mathbf{x}_i$. We cope with the noise stemming from the raw data by averaging over the terrain gradients in the local neighborhood.

Equation (6) yields a tensor, representable as a 2×2 real-valued matrix, which describes how the terrain elevation changes in the local neighborhood of location $\mathbf{x}_i$. To get an intuition, what $EST(\mathbf{x}_i)$ encodes and how this can guide the adaptation of the local kernel Σ_i, consider the following situations. Let λ_1 and λ_2 denote the eigenvalues of $EST(\mathbf{x}_i)$ and β be the orientation angle of the first eigenvector. If $\mathbf{x}_i$

is located in a flat part of the terrain, the elevation gradients ∇y are small in the neighborhood of $\mathbf{x}_i$. This results in two equally small eigenvalues of $EST(\mathbf{x}_i)$. In contrast, if $\mathbf{x}_i$ was located in an ascending part of the terrain, the first eigenvalue of $EST(\mathbf{x}_i)$ would be clearly greater than the second one and the orientation β would point towards the strongest ascent.

Intuitively and as discussed in more detail by Middendorf and Nagel [8], the kernel Σ_i describing the extent of the local environment of $\mathbf{x}_i$ should be set to the inverse of $EST(\mathbf{x}_i)$. In this way, flat areas are populated by large, isotropic kernels, while sharp edges have long, thin kernels oriented along the edge directions. Corner structures, having strong elevation gradients in all dimensions, result in relatively small local kernels. To prevent unrealistically large kernels, Middendorf and Nagel describe how this inversion can be bounded to yield kernels, whose standard deviations lie between given values σ_{min} and σ_{max}. Based on their findings, we give three concrete local adaptation rules that have been compared in our experimental evaluation. To simplify notation, we introduce $\overline{\lambda_k} = \lambda_k/(\lambda_1 + \lambda_2)$, $k = 1, 2$ and the re-parameterization

$$\Sigma_i = R^{-T} \begin{pmatrix} \alpha_1 & 0 \\ 0 & \alpha_2 \end{pmatrix} R^{-1} \qquad (7)$$

where α_1 and α_2 scale in orthogonal directions and R is a rotation matrix specified by the orientation angle θ.

1) *Direct Inverse Adaptation*: $\Sigma_i = EST(\mathbf{x}_i)^{-1}$
2) *Bounded Linear Adaptation*:

$$\alpha_k = \overline{\lambda_k}\,\sigma^2_{min} + (1 - \overline{\lambda_k})\,\sigma^2_{max} \quad , k = 1, 2$$

3) *Bounded Inverse Adaptation*:

$$\alpha_k = \frac{\sigma^2_{max}\sigma^2_{min}}{\overline{\lambda_k}\,\sigma^2_{max} + (1 - \overline{\lambda_k})\,\sigma^2_{min}} \quad , k = 1, 2$$

The two *bounded* adaptation procedures prevent unrealistically small and large kernels. The *Bounded Inverse* strongly favors the larger eigenvalue dimension and produces more pronounced kernels (larger difference between semiaxes) while the *Bounded Linear* Linear tends to produce more balanced and larger kernels. This is why *Bounded Linear* performs better in the presence of sparse data as it is less vulnerable to overfitting. In this work, the bounds σ_{min} and σ_{max} are estimated empirically. We are currently working on determining optimal values with respect to the marginal data likelihood.

So far, we have described how to perform one local adaptation step for an arbitrary kernel Σ_i. As the complete learning and adaptation procedure, which is summarized in Algorithm 1, we propose to assign to each input location $\mathbf{x}_i$ of the training set $\mathcal{D}$ a kernel matrix Σ_i, which is initialized with a global parameter vector $\boldsymbol{\Theta}$, that in turn has been learned using standard GP learning with the corresponding stationary covariance function. The local kernels are then iteratively adapted to the elevation structure of the given terrain data set until their parameters have converged. To quickly adapt the kernels at locations where the regression error is high (relative to the given training data set), we propose to make the adaptation speed for each Σ_i dependent on the local data fit df$(\mathbf{x}_i)$, which is the normalized observation likelihood of the corresponding y_i from the training set relative to the current predictive distribution (see Equation (III-A)), and the kernel complexity approximated as $c_i = 1/|\Sigma_i|$. Both quantities are used to form a learning rate parameter calculated by means of a modified sigmoid function, $\eta_i = \text{sigmoid}(-\text{df}(\mathbf{x}_i) \cdot c_i; \boldsymbol{\delta})$, where the additional parameters $\boldsymbol{\delta}$ are determined empirically. Intuitively, we get a high adaptation speed when the data-fit relative to the kernel size is small. Algorithm 1 summarizes the adaptation procedure.

Algorithm 1 Local Kernel Adaptation

```
Learn global parameters Θ for the stationary squared exponential covariance function.
Initialize all local kernels Σ_i with Θ.
while not converged do
  for all Σ_i do
    Estimate the local learning rate η_i
    Estimate EST(x_i) according to Σ_i
    Σ_i^* ← ADAPT(EST(x_i))
    Σ_i ← η_i Σ_i^* + (1 − η_i) Σ_i
  end for
end while
```

IV. Experimental Evaluation

The goals of the experimental evaluation presented in this section are (a) to show that our terrain modeling approach is indeed applicable to real data sets, (b) that our model is able to remove noise while at the same time preserving important structural features, and (c) that our model yields more accurate and robust elevation predictions at sparsely sampled input locations than an alternative approach to this problem.

As an evaluation metric, we use the mean squared error $\text{MSE}(\mathcal{X}) = \frac{1}{m}\sum_{i=1}^{m}(y_i - y_i^*)^2$ of predicted elevations y_i^* relative to ground truth elevations y_i on a set of input locations $\mathcal{X} = \{\mathbf{x}_i\}_{i=1}^m$.

A. Evaluation on Artificial Terrain Data

The first set of experiments was designed to quantify the benefits of local kernel adaptation and to compare the three different adaptation rules. As a test scenario, we took the artificial terrain data set depicted in Figure 2 consisting of 441 data points, which contains uniform regions as well as sharp edges and corners, which are hard to adapt to locally. Note, for example, that the edge between the lowest and the second lowest plateau has a curvature and that three different height levels can be found in the local neighborhood of the corner in the middle of the diagram. We set $\sigma_{min} = 0.001$ and $\sigma_{max} = 5.0$ for the bounded adaptation rules.

To generate training data sets for the different experiments reported on here, we added white noise of a varying standard deviation σ to the true terrain elevations and randomly removed a portion of the samples to be able to assess the model's predictive abilities.

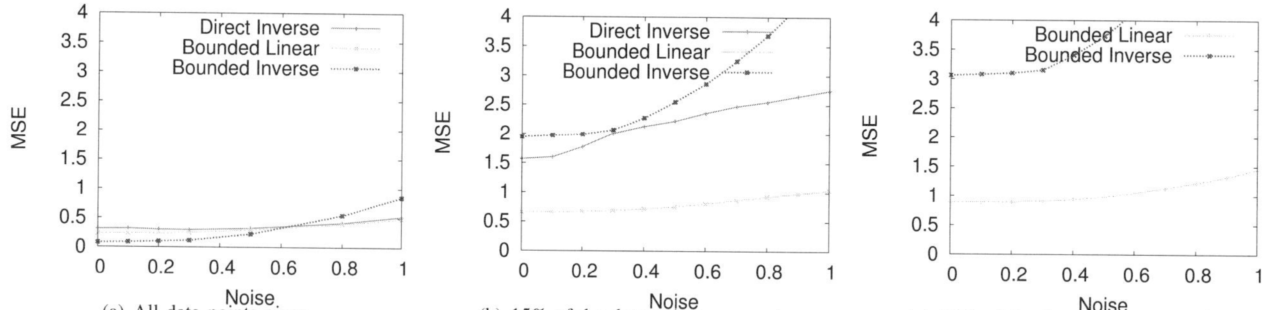

(a) All data points given. (b) 15% of the data points removed. (c) 30% of the data points removed.

Fig. 3. Prediction accuracy for the scenario depicted in Figure 4 with (a) all data points available, (b) 15% of the data-points randomly removed and (c) 30% randomly removed. Each figure plots the mean squared error of elevation predictions for a varying level of added white noise. The values are averaged over 10 independent runs per configuration. (In the case of (c), the error of *Direct Inverse* was always greater than 4.0).

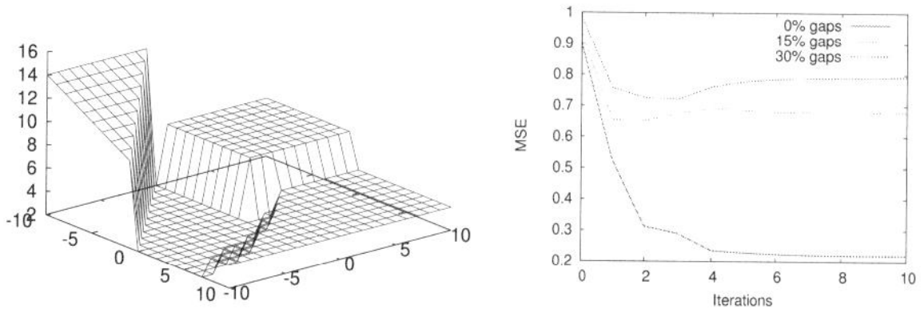

(a) Terrain (b) Convergence

Fig. 2. An artificial terrain data set used in the experimental evaluation, that exhibits several local features that are hard to adapt to (a). Test data sets are generated by adding white noise and randomly removing a portion of the data points. The mean squared error (MSE) of predicted elevations converges with an increasing number of adaptation steps (b). Iteration 0 gives the MSE for the learned standard GP. Values are averaged over ten independent runs.

Figure 4 visualizes a complete adaptation process for the case of a data set generated using a noise rate of $\sigma = 0.3$. On average, a single iteration per run took 44 seconds on this data-set using a PC with a 2.8 GHz CPU and 2 GB of RAM. Figures 4(c)-4(f) show the results of standard GP regression which places the same kernels at all input locations. While this leads to good smoothing performance in homogeneous regions, the discontinuities within the map are also smoothed as can be seen from the absolute errors in the third column. Consequently, those locations get assigned a high learning rate, see right column, used for local kernel adaption.

The first adaptation step leads to the results depicted in Figures 4(g)-4(j). It is clearly visible, that the steps and corners are now better represented by the regression model. This has been achieved by adapting the kernels to the local structure, see the first column of this row. Note, how the kernel sizes and orientations reflect the corresponding terrain properties. Kernels are oriented along discontinuities and are small in areas of strongly varying elevation. In contrast, they have been kept relatively large in homogeneous regions. After three iterations, the regression model has adapted to the discontinuities accurately while still de-noising the homogeneous regions (Figures 4(k)-4(n)). Note, that after this iteration, the local learning rates have all settled at low values.

Figure 2 gives the convergence behavior of our approach using the *Bounded Linear* adaptation rule in terms of the mean squared prediction error for different amounts of points removed from the noisy data set. After at most 6 iterations, the errors have settled close to their final value.

In a different set of experiments, we investigated the prediction performance of our approach for all three adaptation rules presented in Section III-C. For this experiment, we added white noise of a varying noise level to the artificial terrain given in Figure 2. The diagrams in Figure 3 give the results for different amounts of points removed from the noisy data set. When no points are removed from the test set, the *Bounded Inverse* adaptation rule performs best for small noise values. For large noise values, *Bounded Linear* and *Direct Inverse* achieve better results. In the case of 15% and 30% data points removed, *Direct Inverse* and *Bounded Inverse* are not competitive. In contrast, *Bounded Linear* still achieves very good results for all noise levels.

Thus, *Bounded Linear* produces reliable predictions for all tested noise rates and data densities. This finding was supported by experiments on other real data sets not presented here.

B. Evaluation on Real Terrain Data

In order to demonstrate the usefulness of our approach on real data sets, we acquired a set of 3D scans of a scene using a mobile robot equipped with a laser range finder, see Figure 5(a). We compared our prediction results to an approach from the robotics literature [5] that has been applied successfully to the problem of 3-dimensionally mapping urban areas. We employed the *Bounded Linear* adaptation procedure for our learning algorithm where we set $\sigma_{min} = 0.25$ and $\sigma_{max} = 4.0$. Figure 5 gives the results of this experiment. An obstacle, in this case a person, is placed in front of the robot and thus occludes the sloped terrain behind.

We evaluated our approach for the situation depicted in the figure as well as for three similar ones and compared its prediction accuracy to the approach of Früh *et al.* [5], who perform horizontal linear interpolation orthogonally to the robot's view. These scenarios used are actually rather easy ones for [5], as the large gaps can all be filled orthogonally to

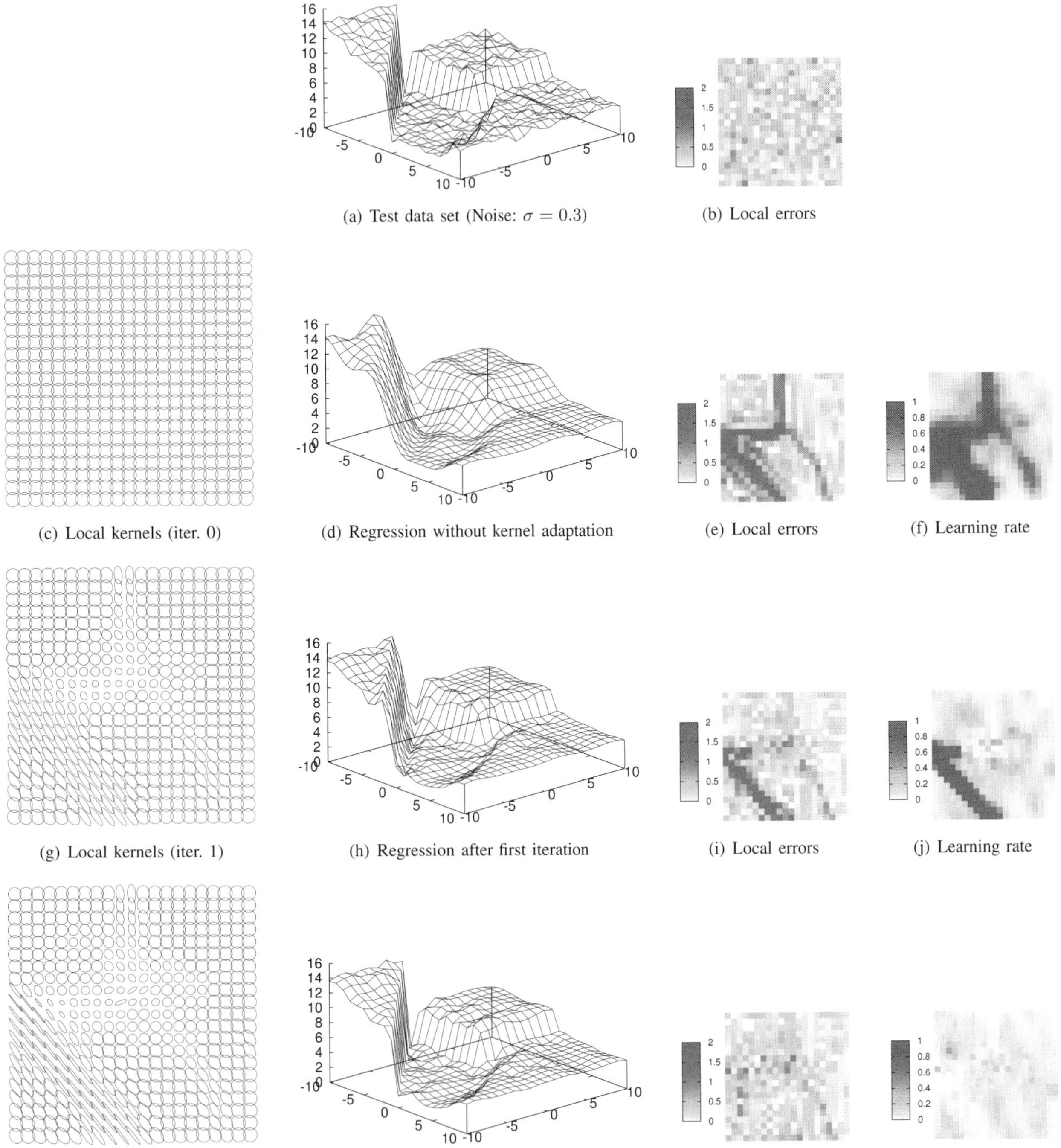

(a) Test data set (Noise: $\sigma = 0.3$)

(b) Local errors

(c) Local kernels (iter. 0)

(d) Regression without kernel adaptation

(e) Local errors

(f) Learning rate

(g) Local kernels (iter. 1)

(h) Regression after first iteration

(i) Local errors

(j) Learning rate

(k) Local kernels (iter. 3)

(l) Regression after third iteration

(m) Local errors

(n) Learning rate

Fig. 4. The local kernel adaptation process on an artificial terrain data set: the original data set, depicted in Figure 2, exhibits several local features that are hard to adapt to. The test data set (a) was generated by adding white noise, resulting in the errors shown in (b). The second row of diagrams gives information about the initialization state of our adaptation process, i.e. the results of standard GP learning and regression. The following two rows depict the results of our approach after the first and after the third adaptation iteration respectively. In the first column of this figure, we visualize the kernel dimensions and orientations after the corresponding iteration. The second column depicts the predicted means of the regression. The third column gives the absolute errors to the known ground truth elevations and the right-most column gives the resulting learning rates η_i for the next adaptation step resulting from the estimated data likelihoods.

the robot's view, which is not the case in general. To estimate the kernels at unseen locations, we built a weighted average over the local neighborhood with an isotropic two-dimensional Gaussian with a standard deviation of 3 which we had found to produce the best results. Table I gives the results. In all four cases, our approach achieved higher prediction accuracies, reducing the errors by 30% to 70%. Figure 5(b) depicts the predictions of our approach in one of the situations. In contrast to Früh *et al.*, our model is able to also give the predictive uncertainties. These variances are largest in the center of the occluded area as can be seen in Figure 5(c).

In a second real-world experiment illustrated in Figure 6, we investigated the ability of our terrain model approach to preserve and predict sharp discontinuities in real terrain data. We positioned the robot in front of a rectangular stone block such that the straight edges of the block run diagonally to the robot's line of view. A person stood in between the robot and the block, thereby occluding parts of the block and of the area in front of it. This scenario is depicted in 6(a). The task is to recover the linear structure of the discontinuity and fill the occluded area consistent with the surrounding terrain elevation levels. The adaptation procedure converged already after two iterations. The learned kernel structure, illustrated in Figure 6(c), enables the model to correctly represent the stone blocks as can be seen from the predicted elevations visualized in 6(d). This figure also illustrates the uncertainties of these predictions, corresponding to the variances of the predictive distributions, by means of two contour lines. This indicates that a mobile robot would be relatively certain about the block structure within the gap although not having observed it directly. In contrast, it would be aware that it cannot rely upon its terrain model in the occluded areas beyond the blocks: there are no observations within a reasonable distance and thus, the predictive variances are large.

To show that our approach is applicable to large, real-world problems, we have tested it on a large data-set recorded at the University of Freiburg campus[1]. The raw terrain data was preprocessed, corrected, and then represented in a multi-level surface map with a cell size of 10cm $\times$ 10cm. The scanned area spans approximately 299 by 147 meters. For simplicity, we only considered the lowest data-points per location, i.e., we removed overhanging structures like tree tops or ceilings. The resulting test set consists of 531,920 data-points. To speed up computations, we split this map into 542 overlapping sub-maps. This is possible without loss of accuracy as we can assume compact support for the local kernels involved in our calculations (as the kernel sizes in our model are bounded). We randomly removed 20% of the data-points per sub-map. A full run over the complete data-set took about 50 hours. Note that the computational complexity can be reduced substantially by exploiting the sparsity of our model (due to the bounded kernels) and by introducing additional sparsity using approximative methods, e.g., sparse GPs. Table II gives the results of this experiment for the different adaptation rules. The *Bounded Linear* and the *Bounded Inverse* adaptation procedures outperform the *Standard GP* model where kernels are not adapted, while *Direct Inverse* is not competitive. Together with the results of the other experiments, this leads to the conclusion that *Bounded Linear* is an adequate choice as an adaptation rule in synthetic and real-world scenarios.

[1]Additional material for the campus experiment can be found at http://www.informatik.uni-freiburg.de/~plagem/rss07terReg

Scenario	Linear Interp. [5]	Adapted GP	Improvement
1 (Fig. 5)	0.116	0.060	48.3%
2	0.058	0.040	31.0%
3	0.074	0.023	69.9%
4	0.079	0.038	51.9%

TABLE I

PREDICTION PERFORMANCE IN TERMS OF MSE RELATIVE TO A SECOND, NOT OCCLUDED SCAN.

Adaptation procedure	MSE
Standard GP	0.071
Direct Inverse	0.103
Bounded Linear	0.062
Bounded Inverse	0.059

TABLE II

PREDICTION PERFORMANCE ON A LARGE CAMPUS ENVIRONMENT.

V. CONCLUSIONS

In this paper, we propose an adaptive terrain modeling approach that balances smoothing against the preservation of structural features. Our method uses Gaussian processes with non-stationary covariance functions to locally adapt to the structure of the terrain data. In experiments on synthetic and real data, we demonstrated that our adaptation procedure produces reliable predictions in the presence of noise and is able to fill gaps of different sizes. Compared to a state-of-the-art approach from the robotics literature we achieve a prediction error reduced by approximately 30%-70%.

In the future, we intend to evaluate our approach in online path planning applications for mobile robots. Since our approach retrieves terrain properties in terms of kernels, its application to terrain segmentation is promising. Another direction of further research are SLAM techniques where the trajectory of the robot is also unknown and the model has to be updated sequentially. We also intend to evaluate our approach on typical test cases in computer vision and to compare it with the algorithms of this community. Finally, we work on an analytical derivation for optimal kernels based solely on data likelihoods and model complexity.

VI. ACKNOWLEDGMENTS

The authors would like to thank Kristian Kersting for the stimulating discussion as well as Rudolph Triebel and Patrick Pfaff for providing the campus data-set and their source code for multi-level surface maps. This work has been supported by the EC under contract number FP6-004250-CoSy and by the German Federal Ministry of Education and Research (BMBF) under contract number 01IMEO1F (project DESIRE).

(a) The first test scenario

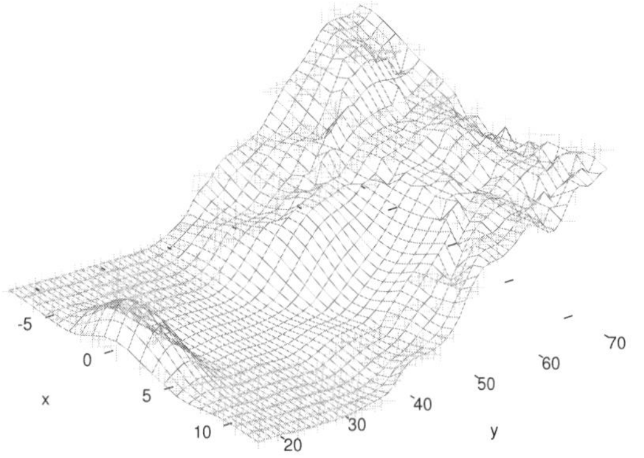

(b) Observations (points) and predicted means (lines)

(c) Predictive uncertainties (white: zero)

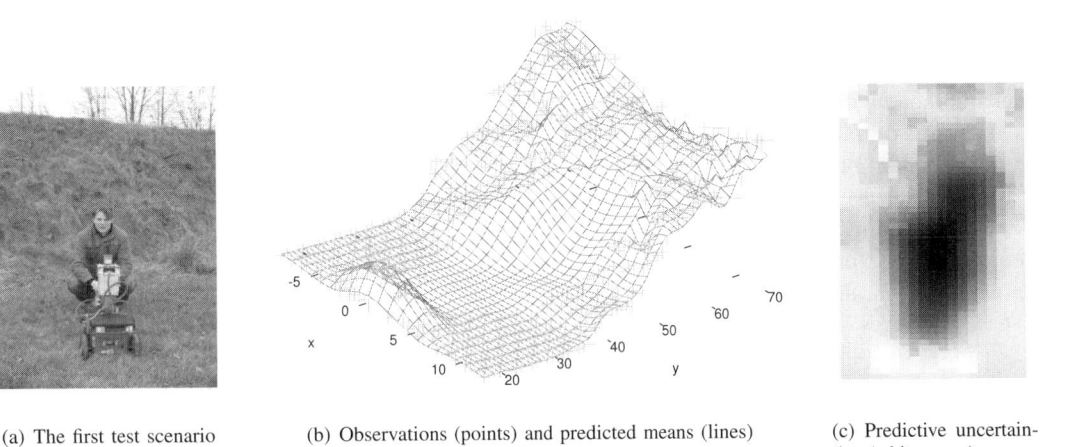

Fig. 5. A real-world scenario, where a person blocks the robot's view on an inhomogeneous and sloped terrain (a). Figure (b) gives the raw data points as well as the predicted means of our adapted non-stationary regression model. Importantly, our model also yields the predictive uncertainties for the predicted elevations as depicted in Figure (c).

(a) The second test scenario

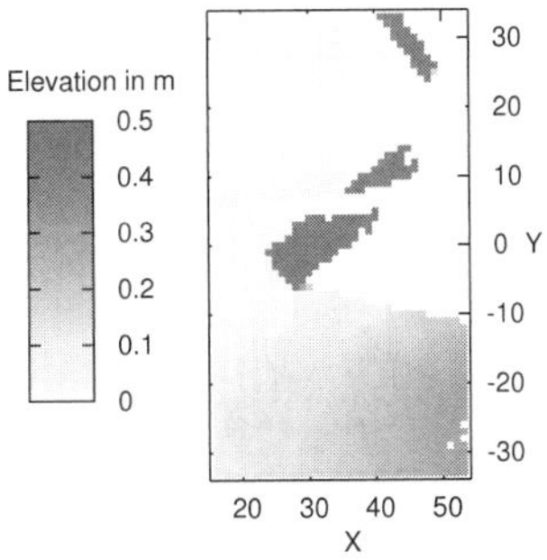

(b) Observed elevations

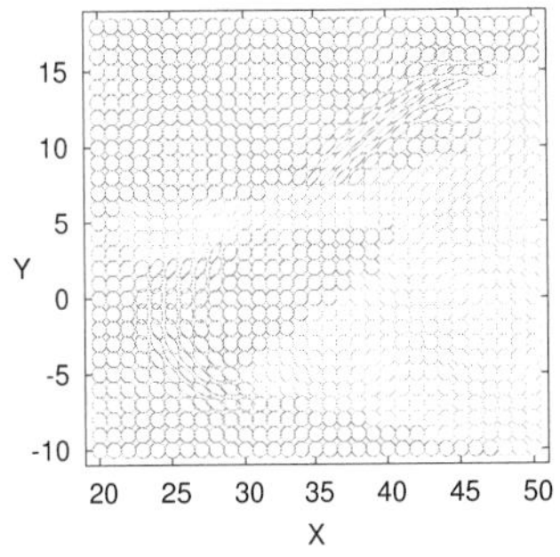

(c) Adapted kernels at observed (red) and occluded locations (green)

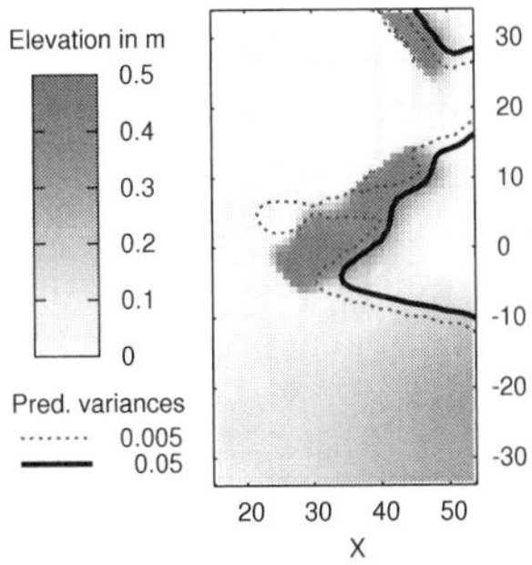

(d) Predicted elevations

Fig. 6. A real-world scenario where a person blocks the robot's view on a stone block, i.e., a sharp linear discontinuity (a). Figure (b) visualizes the kernels that have adapted to the observed block edges illustrated in (c). Figure (d) illustrates the predicted terrain elevations and two contour lines for two different predictive uncertainty thresholds.

REFERENCES

[1] J. Bares, M. Hebert, T. Kanade, E. Krotkov, T. Mitchell, R. Simmons, and W. Whittaker. Ambler: An autonomous rover for planetary exploration. *IEEE Computer*, 22(6):18–26, June 1989.

[2] A. Brooks, A. Makarenko, and B. Upcroft. Gaussian process models for sensor-centric robot localisation. In *Proc. of the IEEE Int. Conf. on Robotics & Automation (ICRA)*, 2006.

[3] D. Cornford, I. Nabney, and C. Williams. Adding constrained discontinuities to gaussian process models of wind fields. In *Proc. of the Conf. on Neural Information Processing Systems (NIPS)*, 1999.

[4] B. Ferris, D. Haehnel, and D. Fox. Gaussian processes for signal strength-based location estimation. In *Proceedings of Robotics: Science and Systems*, Philadelphia, USA, August 2006.

[5] C. Früh, S. Jain, and A. Zakhor. Data processing algorithms for generating textured 3d building facade meshes from laser scans and camera images. *Int. Journal of Computer Vision*, 61(2):159–184, 2005.

[6] D. Grimes, R. Chalodhorn, and R. Rao. Dynamic imitation in a humanoid robot through nonparametric probabilistic inference. In *Proc. of Robotics: Science and Systems*, Philadelphia, USA, August 2006.

[7] Marco Hugentobler. *Terrain Modelling with Triangle Based Free-Form Surfaces*. PhD thesis, University of Zurich, 2004.

[8] M. Middendorf and H. Nagel. Empirically convergent adaptive estimation of grayvalue structure tensors. In *Proc. of the 24th DAGM Symposium on Pattern Recognition*, pages 66–74, London, UK, 2002.

[9] C. Paciorek and M. Schervish. Nonstationary covariance functions for Gaussian process regression. In *Proc. of NIPS 2004*.

[10] P. Pfaff and W. Burgard. An efficient extension of elevation maps for outdoor terrain mapping. In *Proc. of the Int. Conf. on Field and Service Robotics (FSR)*, pages 165–176, Port Douglas, QLD, Australia, 2005.

[11] C. Plagemann, D. Fox, and W. Burgard. Efficient failure detection on mobile robots using particle filters with gaussian process proposals. In *Proc. of the Twentieth International Joint Conference on Artificial Intelligence (IJCAI)*, Hyderabad, India, 2007.

[12] C. Plagemann, K. Kersting, P. Pfaff, and W. Burgard. Gaussian beam processes: A nonparametric bayesian measurement model for range finders. In *Robotics: Science and Systems (RSS)*, June 2007.

[13] C. Rasmussen and Z. Ghahramani. Infinite mixtures of Gaussian process experts. In *Proc. of NIPS 2002*.

[14] C. E. Rasmussen and C. K.I. Williams. *Gaussian Processes for Machine Learning*. The MIT Press, Cambridge, Massachusetts, 2006.

[15] P. D. Sampson and P. Guttorp. Nonparametric estimation of nonstationary spatial covariance structure. *Journal of the American Statistical Association*, 87(417):108–119, 1992.

[16] A. M. Schmidt and A. O'Hagan. Bayesian inference for nonstationary spatial covariance structure via spatial deformations. *Journal of the Royal Statistical Society, Series B*, 65:745–758, 2003.

[17] H. Takeda, S. Farsiu, and P. Milanfar. Kernel regression for image processing and reconstruction. *IEEE Trans. on Image Processing*, 2006.

[18] V. Tresp. Mixtures of Gaussian processes. In *Proc. of the Conf. on Neural Information Processing Systems (NIPS)*, 2000.

[19] R. Triebel, P. Pfaff, and W.Burgard. Multi-level surface maps for outdoor terrain mapping and loop closing. In *Proc. of the International Conference on Intelligent Robots and Systems (IROS)*, 2006.

[20] O. Williams. A switched Gaussian process for estimating disparity and segmentation in binocular stereo. In *Proc. of the Conf. on Neural Information Processing Systems (NIPS)*, 2006.

Fishbone Model for Belt Object Deformation

Hidefumi Wakamatsu and Eiji Arai
Dept. of Materials and Manufacturing Science,
Graduate School of Eng.,
Osaka University
2-1 Yamadaoka, Suita, Osaka 565-0871, Japan
Email: {wakamatu, arai}@mapse.eng.osaka-u.ac.jp

Shinichi Hirai
Dept. of Robotics,
Ritsumeikan University
1-1-1 Noji Higashi, Kusatsu, Shiga 525-8577, Japan
Email: hirai@se.ritsumei.ac.jp

***Abstract*— A modeling method for representing belt object deformation is proposed. Deformation of belt objects such as film circuit boards or flexible circuit boards must be estimated for automatic manipulation and assembly. In this paper, we assume that deformation of an inextensible belt object can be described by the shape of its central axis in a longitudinal direction called "the spine line" and lines with zero curvature called "rib lines". This model is referred to as a "fishbone model" in this paper. First, we describe deformation of a rectangular belt object using differential geometry. Next, we propose the fishbone model considering characteristics of a developable surface, i.e., a surface without expansion or contraction. Then, we formulate potential energy of the object and constraints imposed on it. Finally, we explain a procedure to compute the deformed shape of the object and verify the validity of our proposed method by comparing some computational results with experimental results.**

I. Introduction

According to downsizing of various electronic devices such as note PCs, mobile phones, digital cameras, and so on, more film circuit boards or flexible circuit boards illustrated in Fig.1 are used instead of conventional hard circuit boards. It is difficult to assemble such flexible boards by a robot because they can be easily deformed during their manipulation process and they must be deformed appropriately in the final state. For example, the flexible circuit board shown in Fig.1-(a) must deform to the objective shape illustrated in Fig.1-(b) to install into the hinge part of a mobile phone. Therefore, analysis and estimation of deformation of film/flexible circuit boards is required.

In solid mechanics, Kirchhoff theory for thin plates and Reissner-Mindlin theory for thick plates have been used[1]. For very thin plates, the inextensional theory was proposed[2]. In this theory, it is assumed that the middle surface of a plate is inextensional, that is, the surface of the plate is developable. Displacement of plates can be calculated using FEM based on these theories. However, the high aspect ratio of thin objects often causes instability in computation of deformed shapes. In computer graphics, a deformable object is represented by a set of particles connected by mechanical elements[3]. Recently, fast algorithms have been introduced to describe linear object deformation using the Cosserat formulation[4]. Cosserat elements possess six degrees of freedom; three for translational displacement and three for rotational displacement. Flexure, torsion, and extension of a linear object can be described by use of Cosserat elements. In robotics, insertion of a wire into a hole in 2D space has been analyzed using a beam model of the wire to derive a strategy to perform the insertion successfully[5][6]. Kosuge et al. have proposed a control algorithm of dual manipulators handling a flexible sheet metal[7]. Lamiraux et al. have proposed a method of path planning for elastic object manipulation with its deformation to avoid contact with obstacles in a static environment[8]. Dynamic modeling of a flexible object with an arbitrary shape has been proposed to manipulate it without vibration[9]. In differential geometry, curved lines in 2D or 3D space have been studied to describe their shapes mathematically[10]. Moll et al. have proposed a method to compute the stable shape of a linear object under some geometrical constraints quickly based on differential geometry[11]. It can be applied to path planning for flexible wires. We have proposed a modeling method for linear object deformation based on differential geometry and its applications to manipulative operations[12]. In this method, linear object deformation with flexure, torsion, and extension can be described by only four functions. We can simulate various deformation such as torsional buckling, knotted shape, and so on. This method can be applied to a sheet object if the shape of the object is regarded as rectangle, namely, the object has belt-like shape. However, in [12], it is assumed that the shape of cross-section of a linear object is fixed. This assumption is not appropriate to represent 3D shape of a belt object because the shape of its cross-section can change due to deformation.

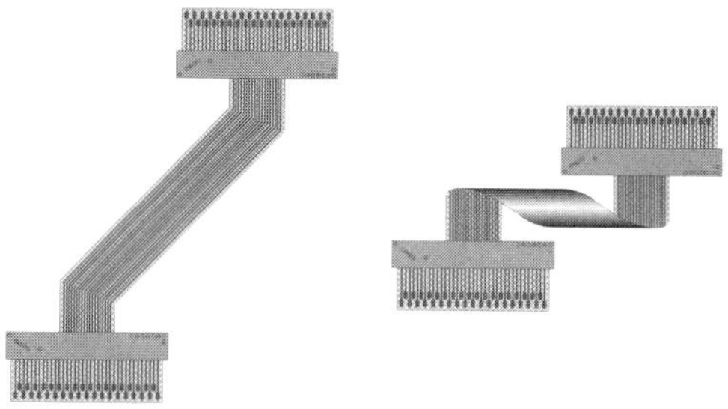

(a) natural shape (b) objective shape

Fig. 1. Example of flexible circuit board

In this paper, we propose *a fishbone model* based on differential geometry to represent belt object deformation. In

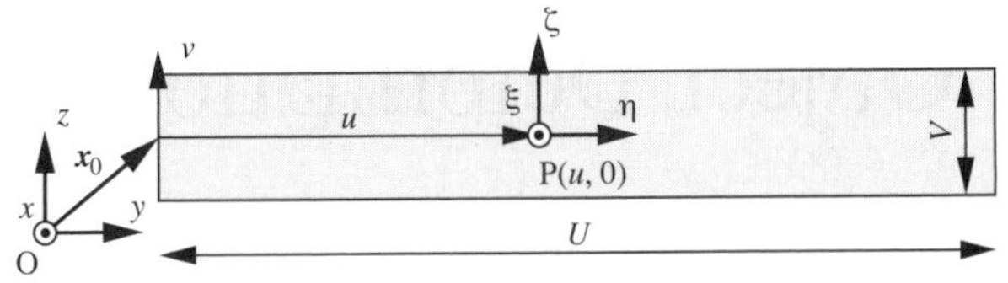

Fig. 2. Coordinates of belt object

this model, deformation of a belt object is represented using its central axis in a longitudinal direction referred to as *the spine line* and lines with zero curvature referred to as *rib lines*. The objective of manipulation of a flexible circuit board is to connect its ends to other devices. So, it is important to estimate position and orientation of ends of the board. This implies that we have to estimate more accurately its deformation in a longitudinal direction than that in a transverse direction. The fishbone model is suitable for representation of deformation in a longitudinal direction, that is, deformed shape of the spine line. Moreover, we can estimate belt object deformation if only the flexural rigidity of the object along the spine line is given. It indicates that we can easily identify the actual parameter of the object from experiment. First, we describe deformation of a rectangular belt object using differential geometry. Next, we propose the fishbone model considering characteristics of a developable surface, *i.e.*, a surface without expansion or contraction. After that, we formulate potential energy of the object and constraints imposed on it. Finally, a procedure to compute the deformed shape of the object was explained and some computational results are compared with experimental results.

II. Modeling of Belt Object

A. Differential Geometry Coordinates

In this section, we formulate the deformation of a belt object in 3D space. Assumptions in this paper are as follows:

- A belt object has rectangular shape.
- The width of the belt object is sufficiently small compared to its length.
- The object is inextensible. Namely, it can be bent and twisted but cannot be expanded or contracted.
- Both ends of the object cannot be deformed because connectors are attached to the ends.

In this paper, we focus on deformation of the central axis in a longitudinal direction of a belt object and attempt to represent the whole shape of the object using it.

Let U and V be the length and the width of the object, respectively. Let u be the distance from one end of the object along the central axis in its longitudinal direction and let v be the distance from the central axis in a transverse direction of the object. Let $\mathrm{P}(u,v)$ be a point on the object. In order to describe deformation of the central axis of a belt object, the global space coordinate system and the local object coordinate systems at individual points on the object are introduced as shown in Fig.2. Let O-xyz be the coordinate system fixed in space and P-$\xi\eta\zeta$ be the coordinate system fixed at an arbitrary point $\mathrm{P}(u,0)$ on the central axis of the object. Assume that the central axis in a longitudinal direction of the object is parallel to the y-axis and the normal vector of any point on the object is parallel to the x-axis in its natural state whereby the object has no deformation. Select the direction of coordinates so that the ξ-, η-, and ζ-axes are parallel to the x-, y-, and z-axes, respectively, at the natural state. Deformation of the object is then represented by the relationship between the local coordinate system P-$\xi\eta\zeta$ at each point on the object and the global coordinate system O-xyz. This is referred to as differential geometry coordinate representation. Let us describe the orientation of the local coordinate system with respect to the space coordinate system by use of Eulerian angles, $\phi(u,0)$, $\theta(u,0)$, and $\psi(u,0)$. The rotational transformation from the coordinate system P-$\xi\eta\zeta$ to the coordinate system O-xyz is expressed by the following rotational matrix:

$$A(\phi,\theta,\psi) = \left[\begin{array}{c|c|c} C_\theta C_\phi C_\psi - S_\phi S_\psi & -C_\theta C_\phi S_\psi - S_\phi C_\psi & S_\theta C_\phi \\ C_\theta S_\phi C_\psi + C_\phi S_\psi & -C_\theta S_\phi S_\psi + C_\phi C_\psi & S_\theta S_\phi \\ -S_\theta C_\psi & S_\theta S_\psi & C_\theta \end{array}\right]. \quad (1)$$

For the sake of simplicity, $\cos\theta$ and $\sin\theta$ are abbreviated as C_θ and S_θ, respectively. Note that Eulerian angles depend on distance u. Let $\boldsymbol{\xi}$, $\boldsymbol{\eta}$, and $\boldsymbol{\zeta}$ be unit vectors along the ξ-, η-, and ζ-axes, respectively, at point $\mathrm{P}(u,0)$. These unit vectors are given by the first, second, and third columns of the rotation matrix, respectively. Namely,

$$A(\phi,\theta,\psi) = [\ \boldsymbol{\xi} \mid \boldsymbol{\eta} \mid \boldsymbol{\zeta}\ ]. \quad (2)$$

Let $\boldsymbol{x}(u,0) = [\ x(u,0),\ y(u,0),\ z(u,0)\ ]^{\mathrm{T}}$ be the position vector of point $\mathrm{P}(u,0)$. The position vector can be computed by integrating vector $\boldsymbol{\eta}(u,0)$. Namely,

$$\boldsymbol{x}(u,0) = \boldsymbol{x}_0 + \int_0^u \boldsymbol{\eta}(u,0)\,\mathrm{d}u, \quad (3)$$

where $\boldsymbol{x}_0 = [\ x_0,\ y_0,\ z_0\ ]^{\mathrm{T}}$ is the position vector at the end point $\mathrm{P}(0,0)$.

Let $\omega_{u\xi}$, $\omega_{u\eta}$, and $\omega_{u\zeta}$ be infinitesimal ratios of rotational angles around the ξ-, η-, and ζ-axes, respectively, at point $\mathrm{P}(u,0)$. They correspond to differentiation of rotational angles around these three axes with respect to distance u and they are described as follows:

$$\begin{bmatrix} \omega_{u\xi} \\ \omega_{u\eta} \\ \omega_{u\zeta} \end{bmatrix} = \begin{bmatrix} -S_\theta C_\psi \\ S_\theta S_\psi \\ C_\theta \end{bmatrix} \frac{\mathrm{d}\phi}{\mathrm{d}u} + \begin{bmatrix} S_\psi \\ C_\psi \\ 0 \end{bmatrix} \frac{\mathrm{d}\theta}{\mathrm{d}u} + \begin{bmatrix} 0 \\ 0 \\ 1 \end{bmatrix} \frac{\mathrm{d}\psi}{\mathrm{d}u}. \quad (4)$$

Note that $\omega_{u\zeta}$ corresponds to bend of the object, $\omega_{u\eta}$ represents torsion of the object, and $\omega_{u\xi}$ indicates curvature of the central line in a longitudinal direction on the object.

B. Description of Surface Bending

Next, we consider general description of 3D surface. Let $\boldsymbol{x}(u,v)$ be the position vector of point $\mathrm{P}(u,v)$ on a surface. Let $\boldsymbol{x}_u(u,v)$ and $\boldsymbol{x}_v(u,v)$ be tangent vectors at point $\mathrm{P}(u,v)$ along u- and v-axes, respectively, and let $\boldsymbol{e}(u,v)$ be the normal vector at point $\mathrm{P}(u,v)$. According to differential geometry, the normal curvature κ in direction $\boldsymbol{d} = a\boldsymbol{x}_u + b\boldsymbol{x}_v$ is represented as follows:

$$\kappa = \frac{La^2 + 2Mab + Nb^2}{Ea^2 + 2Fab + Gb^2}, \tag{5}$$

where E, F, and G are coefficients of the first fundamental form and L, M, and N are those of the second fundamental form of the surface. These coefficients are defined as follows:

$$E = \boldsymbol{x}_u \cdot \boldsymbol{x}_u, \tag{6}$$
$$F = \boldsymbol{x}_u \cdot \boldsymbol{x}_v, \tag{7}$$
$$G = \boldsymbol{x}_v \cdot \boldsymbol{x}_v, \tag{8}$$
$$L = \frac{\partial \boldsymbol{x}_u}{\partial u} \cdot \boldsymbol{e}, \tag{9}$$
$$M = \frac{\partial \boldsymbol{x}_u}{\partial v} \cdot \boldsymbol{e}, \tag{10}$$
$$N = \frac{\partial \boldsymbol{x}_v}{\partial v} \cdot \boldsymbol{e}. \tag{11}$$

The normal curvature κ depends on the direction $\boldsymbol{d}$ and its maximum value κ_1 and its minimum value κ_2 are called the principal curvatures. Direction $\boldsymbol{d}_1$ of the maximum curvature κ_1 and direction $\boldsymbol{d}_2$ of the minimum curvature κ_2 are referred to as principal directions. The principal curvatures and the principal directions specify bend of a surface. A surface is also characterized by Gaussian curvature $K(u, v)$ and the mean curvature $H(u, v)$. They are related to the principal curvatures κ_1 and κ_2 by

$$K = \kappa_1 \kappa_2 = \frac{LN - M^2}{EG - F^2}, \tag{12}$$
$$H = \frac{\kappa_1 + \kappa_2}{2} = \frac{EN - 2FM + GL}{2(EG - F^2)}. \tag{13}$$

Vectors $\boldsymbol{x}_u$, $\boldsymbol{x}_v$, and $\boldsymbol{e}$ correspond to $\boldsymbol{\eta}$, $\boldsymbol{\zeta}$, and $\boldsymbol{\xi}$ in this paper, respectively. Then, coefficients of the first fundamental form are $E = 1$, $F = 0$, and $G = 1$, respectively. Moreover, the derivation of unit vectors $\boldsymbol{\eta}$ and $\boldsymbol{\zeta}$ can be described using infinitesimal ratios of rotational angles as follows:

$$\frac{\partial \boldsymbol{\eta}}{\partial u} = -\omega_{u\zeta}\boldsymbol{\xi} + \omega_{u\xi}\boldsymbol{\zeta}, \tag{14}$$
$$\frac{\partial \boldsymbol{\zeta}}{\partial u} = \omega_{u\eta}\boldsymbol{\xi} - \omega_{u\xi}\boldsymbol{\eta} = \frac{\partial \boldsymbol{\xi}}{\partial v}. \tag{15}$$

Substituting eqs.(14) and (15) into eqs.(9) and (10), L and M can be represented as a function of infinitesimal angle ratios as follows:

$$L = (-\omega_{u\zeta}\boldsymbol{\xi} + \omega_{u\xi}\boldsymbol{\zeta}) \cdot \boldsymbol{\xi} = -\omega_{u\zeta}, \tag{16}$$
$$M = (\omega_{u\eta}\boldsymbol{\xi} - \omega_{u\xi}\boldsymbol{\eta}) \cdot \boldsymbol{\xi} = \omega_{u\eta}. \tag{17}$$

In contrast, N cannot be described by Eulerian angles. So, we introduce the fourth parameter $\delta(u, 0)$: $N = \delta(u, 0)$. It corresponds to the curvature in a transverse direction. Consequently, Gaussian curvature K and the mean curvature H is described by

$$K = -\omega_{u\zeta}\delta - \omega_{u\eta}^2, \tag{18}$$
$$H = \frac{-\omega_{u\zeta} + \delta}{2}. \tag{19}$$

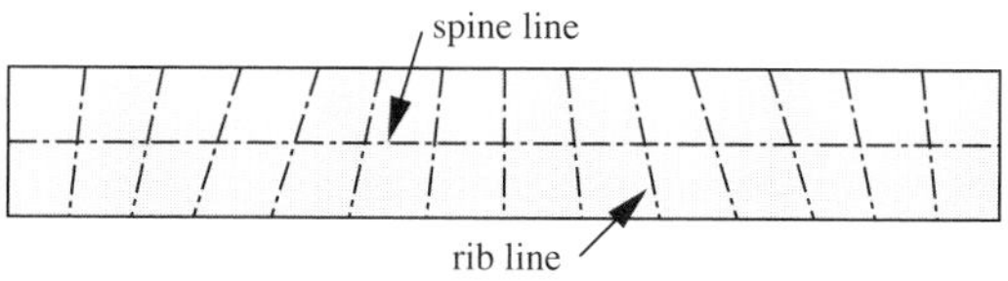

Fig. 3. Fishbone model

Thus, bending of a surface is characterized by Eulerian angles $\phi(u, 0)$, $\theta(u, 0)$, and $\psi(u, 0)$ and the curvature in a transverse direction $\delta(u, 0)$. Note that K and H depends on not only coordinate u but also coordinate v. In this paper, we assume that the whole shape of a belt object can be described by the shape of the central axis in a longitudinal direction because the width of a belt object is sufficiently small compared to its length.

If a principal curvature κ_2, *i.e.*, the minimum value of the normal curvature is equal to zero, the surface is developable. Namely, it can be flattened without its expansion or contraction. Such surface is referred to as a *developable surface*. In this paper, we assume that a belt object is inextensible. Then, the deformed shape of the object corresponds to a developable surface. It means that the object bends in direction $\boldsymbol{d}_1$ and it is not deformed in direction $\boldsymbol{d}_2$. Namely, a line the direction of which coincides with direction $\boldsymbol{d}_2$ is kept straight after deformation. In this paper, the central axis in a longitudinal direction of the object is referred to as the *spine line* and a line with zero curvature at a point on the object is referred to as a *rib line* as shown in Fig.3. We assume that bend and torsion of the spine line and direction of the rib line of each point specifies deformation of a belt object. This model is referred to as a *fishbone model* in this paper. Let $\alpha(u, 0)$ be rib angle, which is the angle between the spine line and direction $\boldsymbol{d}_1$ as shown in Fig.4-(a). Let $\boldsymbol{r}(u)$ be a unit vector along a rib line at point $\mathrm{P}(u, 0)$ on the spine line. It is described by

$$\boldsymbol{r} = -\boldsymbol{\eta}\sin\alpha + \boldsymbol{\zeta}\cos\alpha. \tag{20}$$

Then, coordinates of a point on a rib line and on either longitudinal edge $\boldsymbol{x}(u', \pm V/2)$ is represented as follows:

$$\boldsymbol{x}(u', \pm V/2) = \boldsymbol{x}(u, 0) \pm \frac{V}{2\cos\alpha(u, 0)}\boldsymbol{r}(u, 0), \tag{21}$$

where u' satisfies

$$u' = u + \frac{V}{2}\tan\alpha(u, 0). \tag{22}$$

Consequently, the whole shape of a belt object can be represented using five variables $\phi(u)$, $\theta(u)$, $\psi(u)$, $\delta(u)$, and $\alpha(u)$. Note that they depend on only the distance u from one end of the object along the spine line.

C. Constraints on Belt Object Variables

Let us consider conditions which five variables must satisfy so that the surface of a belt object is developable. Gaussian curvature K of a developable surface must be zero at any point. So, the following constraint is imposed on the object.

$$K = -\omega_{u\zeta}\delta - \omega_{u\eta}^2 = 0, \quad \forall u \in [\,0,\; U\,]. \tag{23}$$

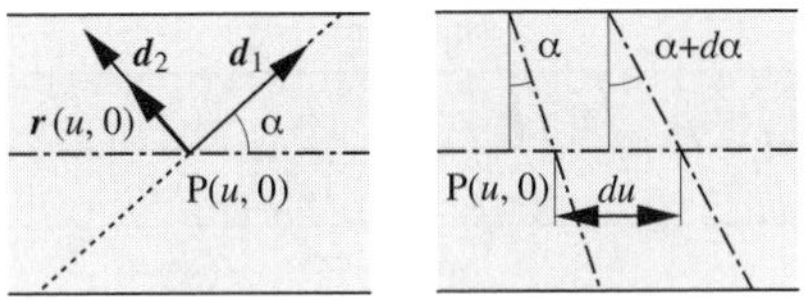

(a) (b)

Fig. 4. Rib angle and rib lines

From eq.(23), δ is described by

$$\delta = -\frac{\omega_{u\eta}^2}{\omega_{u\zeta}}. \tag{24}$$

Recall that infinitesimal ratio of rotational angle around ξ-axis $\omega_{u\xi}$ indicates curvature of the spine line on the object. In the initial state, the spine line is straight, that is, its curvature is constantly equal to zero. So, $\omega_{u\xi}$ must be satisfied the following equation after any deformation because of the inextensibility of a belt object:

$$\omega_{u\xi} = 0, \quad \forall u \in [\,0,\ U\,]. \tag{25}$$

Moreover, as shown in Fig.4-(b), to prevent rib lines from intersecting with themselves on a belt object, the following inequalities must be satisfied:

$$\frac{V}{2}\tan\alpha + du \geq \frac{V}{2}\tan(\alpha + \mathrm{d}\alpha), \tag{26}$$

$$\frac{V}{2}\tan(\alpha + \mathrm{d}\alpha) + du \geq \frac{V}{2}\tan\alpha. \tag{27}$$

Then, rib angle α at any point on the spine line must be satisfied

$$-\frac{2\cos^2\alpha}{V} \leq \frac{\mathrm{d}\alpha}{\mathrm{d}u} \leq \frac{2\cos^2\alpha}{V}, \quad \forall u \in [\,0,\ U\,]. \tag{28}$$

Substituting eq.(24) into eqs.(5) and (19), The normal curvature in direction $\boldsymbol{d}_1 = \boldsymbol{\xi}\cos\alpha + \boldsymbol{\eta}\sin\alpha$, *i.e.*, a principal curvature κ_1 is as follows:

$$\kappa_1 = -\omega_{u\zeta}\cos^2\alpha + 2\omega_{u\eta}\cos\alpha\sin\alpha - \frac{\omega_{u\eta}^2}{\omega_{u\zeta}}\sin^2\alpha$$
$$= -\omega_{u\zeta} - \frac{\omega_{u\eta}^2}{\omega_{u\zeta}} \tag{29}$$

Then, α can be described as follows:

$$\alpha = -\tan^{-1}\frac{\omega_{u\eta}}{\omega_{u\zeta}}. \tag{30}$$

Now, let us introduce parameter $\beta(u)$:

$$\beta = \tan\alpha. \tag{31}$$

Then, β must satisfy the following equation from eq.(30):

$$\omega_{u\eta} + \omega_{u\zeta}\beta = 0, \quad \forall u \in [\,0,\ U\,]. \tag{32}$$

Moreover, eq.(28) is described as follows by substituting eq.(31):

$$-\frac{2}{V} \leq \frac{\mathrm{d}\beta}{\mathrm{d}u} \leq \frac{2}{V}, \quad \forall u \in [\,0,\ U\,]. \tag{33}$$

Consequently, the shape of a belt object can be represented by four functions $\phi(u)$, $\theta(u)$, $\psi(u)$, and $\beta(u)$. And, they must satisfy eqs.(25), (32), and (33) in any state to maintain developability.

D. Potential Energy and Geometric Constraints

Let us formulate the potential energy of a deformed belt object. We can assume that a belt object bends along direction $\boldsymbol{d}_1$ without torsional deformation. This implies that the shape of cross-section along rib line is fixed while that along a transverse direction can change. Then, the potential energy I can be described as follows assuming that the flexural energy is proportional to the bending moment at each point $\mathrm{P}(u)$:

$$I = \int_0^U \frac{R_f}{2\cos\alpha}\kappa_1^2\,\mathrm{d}u = \int_0^U \frac{R_f}{2\cos\alpha}\frac{(\omega_{u\zeta}^2 + \omega_{u\eta}^2)^2}{\omega_{u\zeta}^2}\mathrm{d}u, \tag{34}$$

where R_f represents the flexural rigidity of a belt object along the spine line at point $\mathrm{P}(u)$. If rib angle α is equal to zero, the width of an infinitesimal region for integration coincides with object width V. Then, $R_f/\cos\alpha$ corresponds to the flexural rigidity along the spine line. If α is not equal to zero, the width of a infinitesimal region becomes longer than the object width. This means that the latter region has larger potential energy than the former region even if they have the same principal curvature κ_1.

Next, let us formulate geometric constraints imposed on a belt object. The relative position between two points on the spine line of the object is often controlled during a manipulative operation of the object. Consider a constraint that specifies the positional relationship between two points on the object. Let $\boldsymbol{l} = [\,l_x,\ l_y,\ l_z\,]^T$ be a predetermined vector describing the relative position between two operational points on the spine line, $\mathrm{P}(u_a)$ and $\mathrm{P}(u_b)$. Recall that the spatial coordinates corresponding to distance u are given by eq.(3). Thus, the following equation must be satisfied:

$$\boldsymbol{x}(u_b) - \boldsymbol{x}(u_a) = \boldsymbol{l}. \tag{35}$$

The orientation at one point on the spine line of the object is often controlled during an operation as well. This constraint is simply described as follows:

$$A(\phi(u_c), \theta(u_c), \psi(u_c)) = A(\phi_c, \theta_c, \psi_c), \tag{36}$$

where ϕ_c, θ_c, and ψ_c are predefined Eulerian angles at one operational point $\mathrm{P}(u_c)$.

Therefore, the shape of a belt object is determined by minimizing the potential energy described by eq.(34) under necessary constraints for developability described by eqs.(25), (32), and (33) and geometric constraints imposed on the object described by eqs.(35) and (36). Namely, computation of the deformed shape of the object results in a variational problem under equational and inequality constraints.

III. Computation of Belt Object Deformation

A. Computation Algorithm

Computation of the deformed shape of a belt object results in a variational problem as mentioned in the previous section.

In [12], we developed an algorithm based on Ritz's method[13] and a nonlinear programming technique to compute linear object deformation. In this paper, we apply such algorithm to the computation of belt object deformation.

Let us express functions $\phi(u)$, $\theta(u)$, $\psi(u)$, and $\beta(u)$ by linear combinations of basic functions $e_1(u)$ through $e_n(u)$:

$$\phi(u) = \sum_{i=1}^{n} a_i^{\phi} e_i(u) \triangleq \boldsymbol{a}^{\phi} \cdot \boldsymbol{e}(u), \qquad (37)$$

$$\theta(u) = \sum_{i=1}^{n} a_i^{\theta} e_i(u) \triangleq \boldsymbol{a}^{\theta} \cdot \boldsymbol{e}(u), \qquad (38)$$

$$\psi(u) = \sum_{i=1}^{n} a_i^{\psi} e_i(u) \triangleq \boldsymbol{a}^{\psi} \cdot \boldsymbol{e}(u), \qquad (39)$$

$$\beta(u) = \sum_{i=1}^{n} a_i^{\beta} e_i(u) \triangleq \boldsymbol{a}^{\beta} \cdot \boldsymbol{e}(u), \qquad (40)$$

where $\boldsymbol{a}^{\phi}$, $\boldsymbol{a}^{\theta}$, $\boldsymbol{a}^{\psi}$, and $\boldsymbol{a}^{\beta}$ are vectors consisting of coefficients corresponding to functions $\phi(u)$, $\theta(u)$, $\psi(u)$, and $\beta(u)$ respectively, and vector $\boldsymbol{e}(u)$ is composed of basic functions $e_1(u)$ through $e_n(u)$. Substituting the above equations into eq.(34), potential energy I is described by a function of coefficient vectors $\boldsymbol{a}^{\phi}$, $\boldsymbol{a}^{\theta}$, $\boldsymbol{a}^{\psi}$, and $\boldsymbol{a}^{\beta}$. Constraints are also described by conditions involving the coefficient vectors. Especially, discretizing eqs.(25), (32), and (33) by dividing interval $[0, U]$ into n small intervals yields a finite number of conditions. As a result, a set of the constraints is expressed by equations and inequalities in terms of the coefficient vectors.

Consequently, the deformed shape of a belt object can be derived by computing a set of coefficient vectors $\boldsymbol{a}^{\phi}$, $\boldsymbol{a}^{\theta}$, $\boldsymbol{a}^{\psi}$, and $\boldsymbol{a}^{\beta}$ that minimizes the potential energy under the constraints. This minimization problem can be solved by the use of a nonlinear programming technique such as the multiplier method[14]. In this method, Lagrange multipliers are introduced as variables for optimization to satisfy given constraints.

B. Examples of Computation

In this section, numerical examples demonstrate how the proposed method computes the deformed shape of a belt object. The following set of basic functions are used in the computation of these examples:

$$e_1 = 1, \quad e_2 = u, \qquad (41)$$

$$e_{2i+1} = \sin\frac{\pi i u}{U}, \qquad (42)$$

$$e_{2i+2} = \cos\frac{\pi i u}{U}, \quad (i = 1, 2, 3, 4). \qquad (43)$$

Assume that the length of the object U is equal to 1, its width V is equal to 0.1, and its flexural rigidity along the spine line R_f is constantly equal to 1. Necessary constraints for developability described by eqs.(25), (32), and (33) are divided into 16 conditions at point $P(iU/15)$ $(i = 0, \cdots, 15)$ respectively in the following examples. All computations were performed on a 750MHz Alpha 21264 CPU with 512MB memory operated by Tru64UNIX. Programs were compiled by a Compaq C Compiler V6.1 with optimization option -O4.

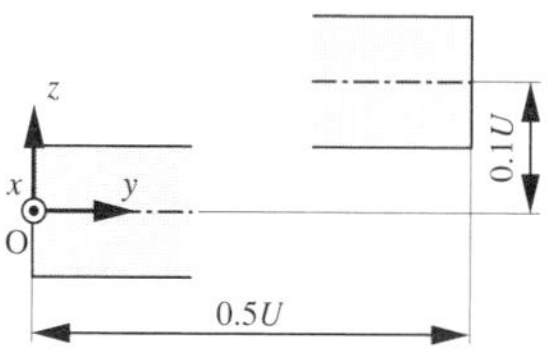

Fig. 5. Example 1

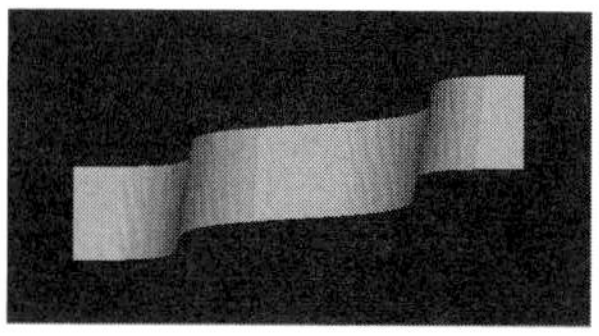

(a) Top view

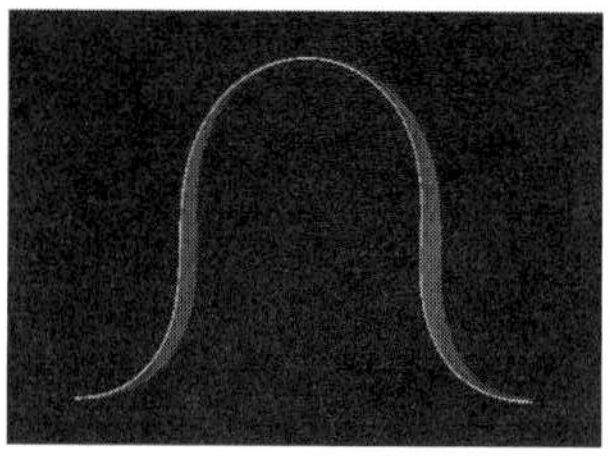

(b) Front view

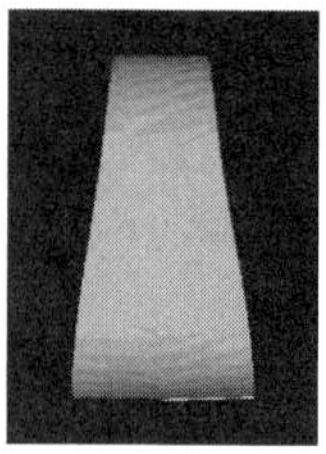

(c) Side view

Fig. 6. Computational result of example 1

Fig.5 shows the first example of belt object deformation. In this example, positional constraints imposed on a belt object are described by

$$\boldsymbol{x}(U) = \int_0^U \boldsymbol{\eta}(u)\,\mathrm{d}u = \begin{bmatrix} 0 \\ 0.5 \\ 0.1 \end{bmatrix} U. \qquad (44)$$

Orientational constraints are represented as follows:

$$\phi(0) = \theta(0) = \psi(0) = \beta(0) = 0, \qquad (45)$$

$$\phi(U) = \theta(U) = \psi(U) = \beta(U) = 0. \qquad (46)$$

This means that directions of the spine line at both ends are parallel but they are not collinear. Then, this optimization problem has 40 variables for Eulerian angles and the rib angle, 11 for geometrical constraints, and 64 for necessary constraints for developability. Fig.6 shows computational results. Fig.6-(a), -(b), and -(c) illustrate the top, front, and side view of the object, respectively. As shown in this figure, the object is bent and twisted to satisfy given geometric constraints. This implies that rib angle α varies with distance u. Fig.7 shows the relationship between α and u. Considering eq.(34), it is found that α becomes smaller, that is, $\cos\alpha$ becomes larger at a point with a large curvature such as the midpoint of the object to reduce its potential energy. The maximum height of the object is $0.35U$. The computation time was about 1200 seconds.

Fig.8 shows the second example. Positional and orienta-

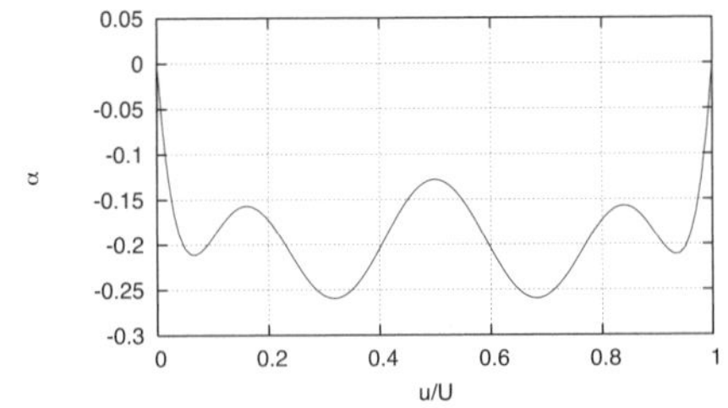

Fig. 7. Rib angle in example 1

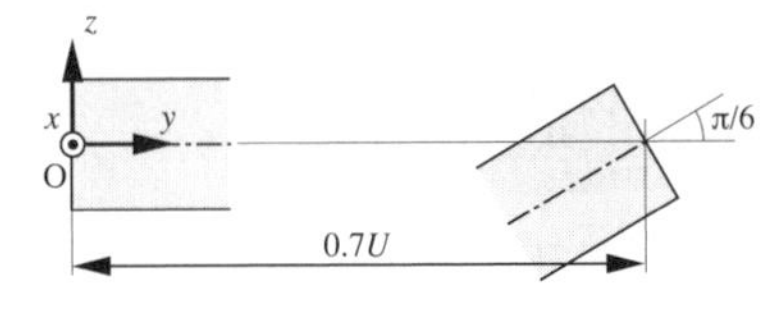

Fig. 8. Example 2

tional constraints are described by

$$\boldsymbol{x}(U) = \int_0^U \boldsymbol{\eta}(u)\,\mathrm{d}u = \begin{bmatrix} 0 \\ 0.7 \\ 0 \end{bmatrix} U, \tag{47}$$

$$\phi(0) = \theta(0) = \psi(0) = \beta(0) = 0, \tag{48}$$

$$\boldsymbol{\eta}(U) = \begin{bmatrix} 0 \\ \cos(\pi/6) \\ \sin(\pi/6) \end{bmatrix},\ \boldsymbol{\zeta}(U) = \begin{bmatrix} 0 \\ -\sin(\pi/6) \\ \cos(\pi/6) \end{bmatrix}, \tag{49}$$

$$\beta(U) = 0. \tag{50}$$

Namely, both end of the spine line are on the same line but directions of the spine line at these points are different. Fig.9 shows computational results and Fig.10 shows the relationship between α and u. As shown in these figures, at parts close to both ends of the object, where the object kinks, α has a large value. Coordinates of the object peak are $[\,0.3U,\ 0.4U,\ -0.01U\,]^T$. This computation took about 1500 seconds.

IV. Experiments of Belt Object Deformation

In this section, the computation results are experimentally verified by measuring the deformed shape of a belt object. We

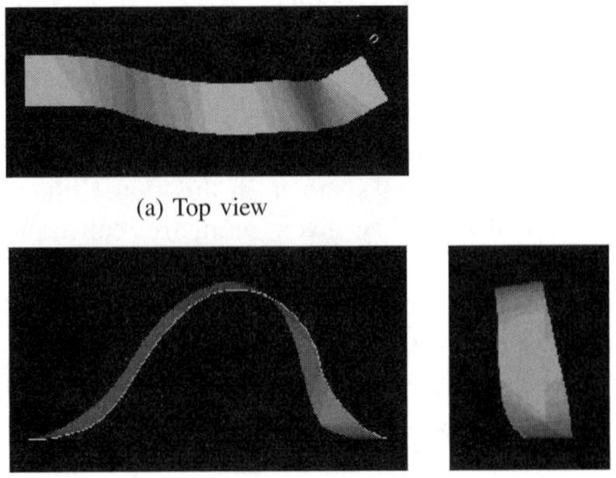

(a) Top view

(b) Front view (c) Side view

Fig. 9. Computational result of example 2

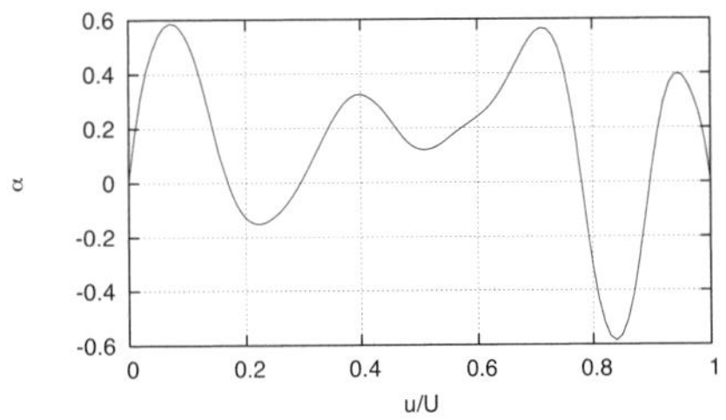

Fig. 10. Rib angle in example 2

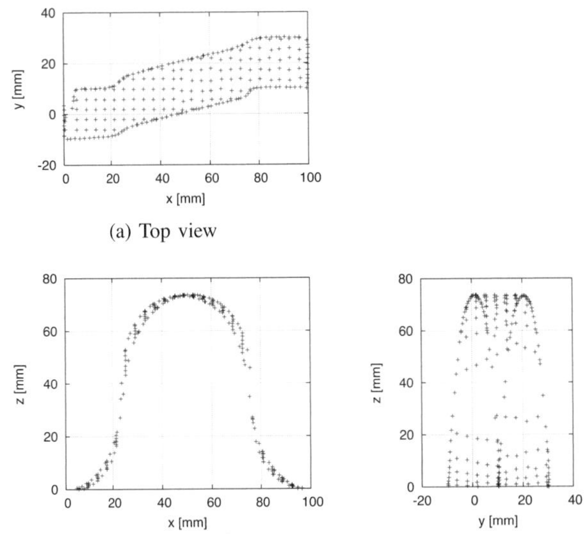

(a) Top view

(b) Front view (c) Side view

Fig. 11. Experimental result of example 1

measured the shape of a rectangular polystyrol sheet which is 200mm long, 20mm wide, and 140μm thick with a 3D scanner. Their flexural rigidity is unknown but from eq.(34), it is found that the deformed shape is independent of it when it is constant along the spine line. Fig.11 shows the experimental result of deformation illustrated in Fig.5. The computational result shown in Fig.6 almost coincide with this experimental result. Fig.12 shows the experimental result of deformation illustrated in Fig.8. As shown in this figure, the computed shape on xy- and xz-planes is qualitatively similar to the actual shape and x- and y-coordinates of the object peak almost coincide. Thus, our method can estimate bending and torsional deformation of a rectangular belt object using only flexural rigidity of the object along its spine line if the object is isotropic.

V. Discussion of Fishbone Model

In this section, we discuss our proposed model. Recall that the surface of an inextensible belt object corresponds to a developable surface, which is a kind of ruled surfaces. A ruled surface is a surface that can be swept out by moving a straight line, which is called a ruling, in 3D space and it can be formulated as follows:

$$\boldsymbol{x}(u, v) = \boldsymbol{p}(u) + v\boldsymbol{q}(u), \tag{51}$$

where $\boldsymbol{p}(u)$ and $\boldsymbol{q}(u)$ are referred to as the base curve and the director curve, respectively. Rib lines in the fishbone model

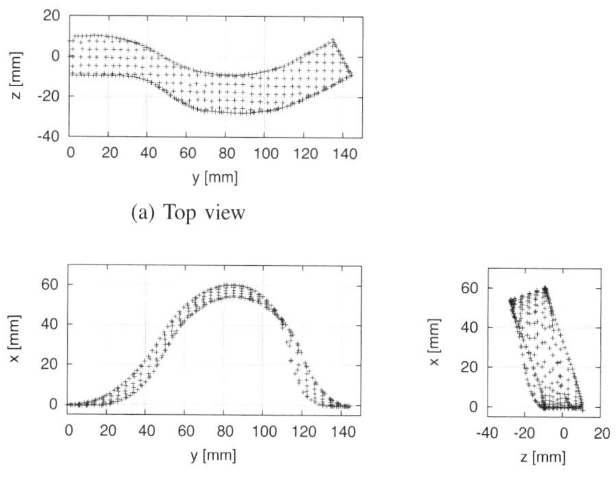

(a) Top view

(b) Front view (c) Side view

Fig. 12. Experimental result of example 2

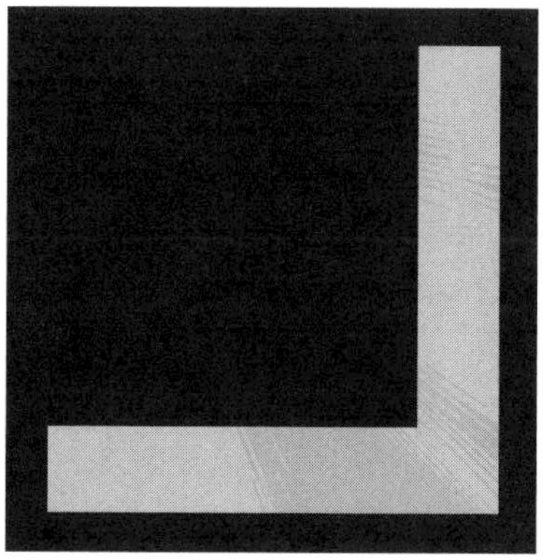

Fig. 13. Bent Belt Object

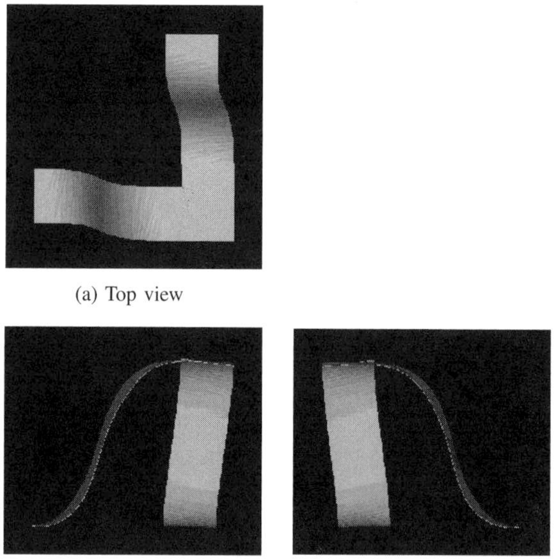

(a) Top view

(b) Front view (c) Side view

Fig. 14. Deformation of Bent Belt Object

correspond to rulings. Moreover, x and r in eq.(21) are similar to the base curve and the director curve, respectively. The formulation described by eq.(51) is sufficient to represent the object surface after deformation. However, it is not suitable for representation of energy increment from the initial shape. To estimate potential energy of the object and to derive its stable shape, we have to specify dependent parameters on deformation and independent parameters of deformation. As a belt object is assumed to be inextensible, its shape in uv-space is not changed by any deformation. This means that the length, width, and angle between u- and v-axes are constant. So, $E = 1$, $G = 0$, $F = 1$. Furthermore, the constraint described by eq.(25) is added for straightness of the spine line in uv-space. Then, the object only can be bent around ζ-axis and twisted around η-axis, and the rib angle is determined from these bend and torsion. As mentioned before, the object shape is represented by four variables $\phi(u)$, $\theta(u)$, $\psi(u)$, and $\beta(u)$. Note that they must satisfy constraints described by eqs.(25) and (32). Therefore, we can conclude that deformation of an inextensible belt object is described by two independent variables.

Some flexible circuit boards bend like a polygonal line as shown in Fig.1 or curve like a circular arc. Let us discuss application of our model to such bent/curved boards. First, to represent a belt object with multiple bends, Eulerian angles and rib angles of straight parts between bends should be defined separately. The deformed shape of the object is then derived by minimizing total potential energy of each part. But, continuity of the rib line at each bend should be discussed. Fig.14 shows a computational result of deformation of a belt object with one bend illustrated in Fig.13. Next, let us consider a curved belt object. As we assume that the spine line is straight in this paper, $\omega_{u\xi}$ is constantly equal to zero. If an object is curved with a certain curvature, $\omega_{u\xi}$ must be equal to that curvature even if the object deforms. We can impose this constraint on the object instead of eq.(25). This implies that our proposed method can be applied to a curved belt object. Fig.16 shows a computational result of deformation of a curved belt object illustrated in Fig.15. Thus, our proposed method can represent deformation of various belt objects.

VI. Conclusions

A fishbone model based on differential geometry to represent belt object deformation was proposed toward manipulation/assembly of film/flexible circuit boards. First, deformation of a rectangular belt object was described using differential geometry. Next, the fishbone model was proposed by considering characteristics of a developable surface. In this model, deformation of a belt object is represented using the shape of the spine line and the direction of straight rib lines. Then, we can estimate belt object deformation if only the flexural rigidity of the object along the spine line is given. After that, we formulate potential energy of the object and constraints imposed on it. Finally, a procedure to compute the deformed shape of the object was explained and some computational results were compared with experimental results. They demonstrated that the fishbone model can represent deformation of a belt object qualitatively well.

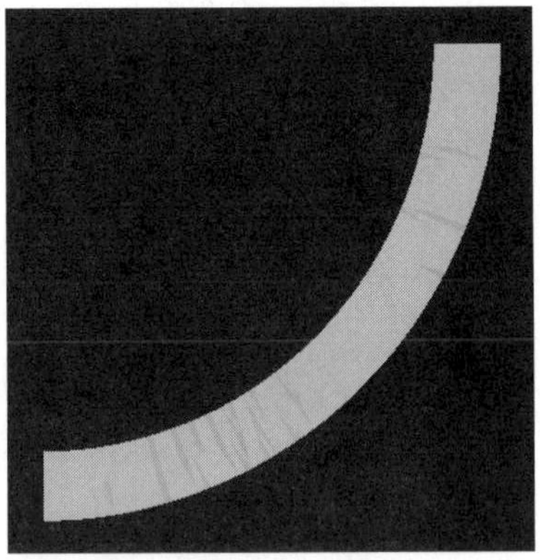

Fig. 15. Curved Belt Object

(a) Top view

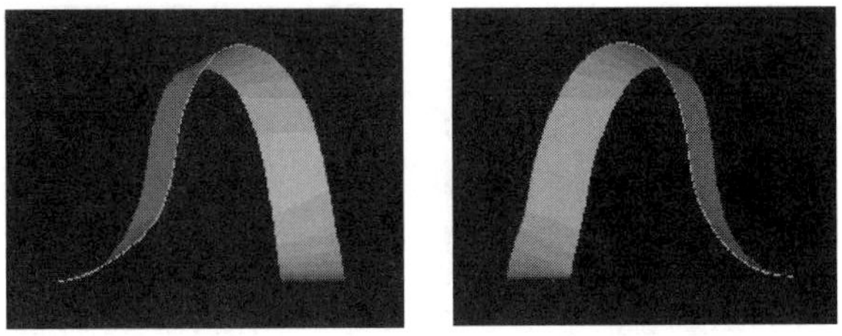

(b) Front view (c) Side view

Fig. 16. Deformation of Curved Belt Object

References

[1] S. Timoshenko, "Theory of Plates and Shells", McGraw-Hill Book Company, Inc., 1940.

[2] E. H. Mansfield, "The Inextensional Theory for Thin Flat Plates", The Quarterly J. Mechanics and Applied Mathematics, Vol.8, pp.338-352, 1955.

[3] A. Witkin and W. Welch, "Fast Animation and Control of Nonrigid Structures", Computer Graphics, Vol.24, pp.243-252, 1990.

[4] D. K. Pai, "STRANDS: Interactive Simulation of Thin Solids using Cosserat Models", Computer Graphics Forum, Vol.21, No.3, pp.347-352, 2002.

[5] Y. E. Zheng, R. Pei, and C. Chen, "Strategies for Automatic Assembly of Deformable Objects", Proc. of IEEE Int. Conf. Robotics and Automation, pp.2598-2603, 1991.

[6] H. Nakagaki, K. Kitagaki, and H. Tsukune, "Study of Insertion Task of a Flexible Beam into a Hole", Proc. of IEEE Int. Conf. Robotics and Automation, pp.330-335, 1995.

[7] K. Kosuge, M. Sasaki, K. Kanitani, H. Yoshida, and T. Fukuda, "Manipulation of a Flexible Object by Dual Manipulators", Proc. of Int. Conf. Robotics and Automation, pp.318-323, 1995.

[8] F. Lamiraux and L. E. Kavraki, "Planning Paths for Elastic Objects under Manipulation Constraints", Int. J. Robotics Research, Vol.20, No.3, March, pp.188-208, 2001.

[9] D. Sun, J. K. Mills, and Y. Liu, "Position Control of Robot Manipulators Manipulating a Flexible Payload", Int. J. Robotics Research, Vol.18, No.3, March, pp.319-332, 1999.

[10] A. Gray, "Modern Differential Geometry of Curves and Surfaces", CRC Press, 1993.

[11] M. Moll and L. E. Kavraki, "Path Planning for Variable Resolution Minimal-Energy Curves of Constant Length", Proc. IEEE Int. Conf. Robotics and Automation, pp.2142-2147, 2005.

[12] H. Wakamatsu and S. Hirai, "Static Modeling of Linear Object Deformation Based on Differential Geometry", Int. J. Robotics Research, Vol.23, No.3, pp.293-311, 2004.

[13] L. E. Elsgolc, "Calculus of Variations", Pergamon Press, 1961.

[14] M. Avriel, "Nonlinear Programming: Analysis and Methods", Prentice-Hall, 1976.

Context and Feature Sensitive Re-sampling from Discrete Surface Measurements

David M. Cole and Paul M. Newman
Oxford University Mobile Robotics Group, Department of Engineering Science, Oxford, OX1 3PJ, UK
Email: dmc,pnewman@robots.ox.ac.uk

Abstract— **This paper concerns context and feature-sensitive re-sampling of workspace surfaces represented by 3D point clouds. We interpret a point cloud as the outcome of repetitive and non-uniform sampling of the surfaces in the workspace. The nature of this sampling may not be ideal for all applications, representations and downstream processing. For example it might be preferable to have a high point density around sharp edges or near marked changes in texture. Additionally such preferences might be dependent on the semantic classification of the surface in question. This paper addresses this issue and provides a framework which given a raw point cloud as input, produces a new point cloud by re-sampling from the underlying workspace surfaces. Moreover it does this in a manner which can be biased by local low-level geometric or appearance properties and higher level (semantic) classification of the surface. We are in no way prescriptive about what justifies a biasing in the re-sampling scheme — this is left up to the user who may encapsulate what constitutes "interesting" into one or more "policies" which are used to modulate the default re-sampling behavior.**

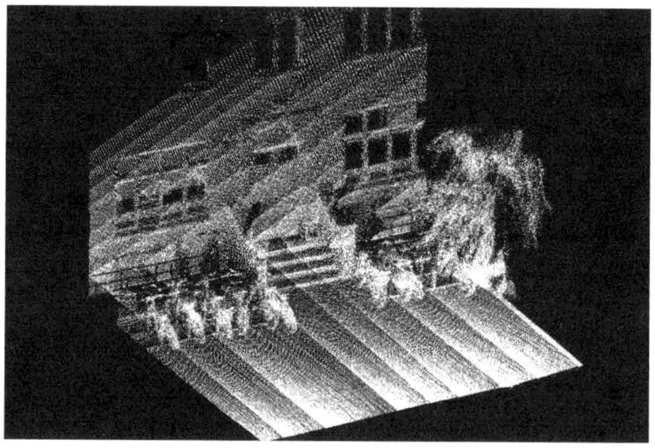

Fig. 1. An example of a large 3D point cloud generated with an outdoor mobile robot. In the foreground we can observe a rack of bicycles, to the right a large tree, and in the background a building with protruding staircase. Notice the large variation in local point density.

I. Introduction and Motivation

This paper is about scene representation using point clouds with application to mobile robot navigation. Sensors like laser range scanners and stereo pairs are becoming ubiquitous and finding substantial application to navigation and mapping tasks. They return sets of 3D points which are discrete samples of continuous surfaces in the workspace, which we shall refer to as 'point clouds'. There are two prominent and distinct schools of thought regarding how point clouds might be used to infer vehicle location and/or workspace structure. On one hand, segmentation and consensus techniques can be used to extract subsets of the data which support the existence of geometric primitives - planes, edges, splines or quadrics for example. Vehicle state inference is performed using the parameters of these primitives and the world is modelled as the agglomeration of geometric primitives. All raw data is subsequently disregarded. More recently the 'view based' approach has become very popular. Here the raw data is left untouched. Location estimation is achieved by perturbing discrete instances of the vehicle trajectory until the overlapping 'views' match in a maximum likelihood sense. The workspace estimate is then the union of all views rendered from the vehicle trajectory.

There is a stark contrast between these two approaches - one strives to explain measurements in terms of parameterizations resulting in terse, perhaps prescriptive representations, but which we can then reason with at higher levels. The other makes no attempt to explain the measurements, no data is thrown away and every data point is treated with equal importance. The ensuing work-space representations look visually rich but the maps are semantically bland.

This paper examines the ground between these two camps and contributes a framework by which raw point clouds can be used to generate new samples from the surfaces they implicitly represent in a selective way. We will refer to this as 'feature sensitive re-sampling'. The approach we describe here allows us to avoid fitting data to *a-priori* prescriptive models while still embracing the concept of a feature. Importantly it retains the ability of view based approaches to capture irregular aspects of the workspace as anonymous sets of points. The framework consists of a generalized re-sampling mechanism and a set of hooks to which any number of user-defined sampling 'policies' may be attached. The central idea is that the point cloud is a discrete sampling of continuous surfaces in the workspace. Depending on application (mapping, matching, localizing, indexing etc.), some parts of the surfaces in the workspace are more useful, interesting, salient, or descriptive than others. Such inclinations are entirely encapsulated in the individual sampling policies which have access to the local properties of the implicit workspace surfaces. For example they may preferentially select regions of a particular color, curvature or normal direction. In the light of this information each policy modulates the generalized re-sampling mechanism — biasing samples to originate from regions that, to it, are 'interesting'.

To make this abstraction more tangible we proceed with a few motivating examples. Imagine we are in possession of a substantial point cloud obtained from a 3D laser scanner like the one shown in Figure 1. Consider first the task of matching such a scan to another. There is an immediate concern that the workspace's surfaces have not been sampled uniformly - the fan-out of the laser and the oscillating nature of the device mean that close surfaces have more samples per unit area. Furthermore, surface patches close to the intersection of the workspace and the nodding axis also receive disproportionate representation. It would be advantageous to re-sample the data so that surfaces are sampled uniformly and that an ICP-like scan-matching process is not biased towards aligning regions of greater point density. Secondly, consider the case in which we are presented with both a 3D laser point cloud of a scene and color images of it from a camera - allowing the point cloud to be colored as shown in Figure 6a. A workspace region that may have appeared to be bland from a geometric perspective might now be marked with transitions in color space. We might wish to pay special attention to regions of a particular color, gradient, or neighborhood texture - the resulting data points might yield better registration results, provide discriminative indexes into appearance based databases or simply decimate the overall scene representation.

It is important to emphasize that the behavior of the framework can also be influenced by semantic information as well as low-level geometry and appearance. *A-priori* classification and segmentation of point cloud regions (glass, shrubbery or floor for example) can be used to inhibit existing policies as needs dictate. For example, a policy for maintaining regions of high curvature may be inhibited, if it is found to lie on a shrub, rather than a building - corners on a building being deemed more interesting than corners on a plant. Before proceeding, it is worth noting that the method we propose is different from that which simply sweeps through a point cloud and retains points in interesting regions. To see why, we must clarify what we mean by re-sampling. We do not draw new points from the measurements themselves but from an implicit representation of the workspace surfaces. If we constrained ourselves to re-sampling from the original point cloud we could not so easily address issues of under (or over) sampling. We would also be ignoring the fact that a single range point is just a sample from a surface - it is not an observation of an actual point in the workspace. In light of this it appears overly restrictive to limit ourselves to only working with the observed 3D points. Instead we work with the surfaces that they represent implicitly.

The rest of the paper is structured as follows: Section II begins by providing a brief summary of previous work. Section III follows, split into sub-section III-A, which provides a brief summary of the approach we have adopted and its relationship to existing literature, sub-sections III-B and III-C, which describe some of the key techniques in detail and sub-section III-D, which examines the overall framework. Section IV then shows some of the initial results obtained, and we conclude in Section V with some conclusions and final thoughts.

II. Previous Work

Over the last decade, there has been an enormous amount of high quality research in mobile robot navigation. Many papers have focused on building complex 3D environments using vision or laser scanned point clouds. However, relatively few have examined efficient representations that improve performance, or facilitate task accomplishment. One approach which has had considerable success extracts planar surfaces from the data - for example [2], [3], [9] and [10]. This has been taken a step further in [6], where a semantic net selects particular plane configurations, and forces them to conform to pre-conceived notions of how the world should look. This can correct planes corrupted by noise. However, many typical facades, surfaces and objects in realistic outdoor environments cannot be simplified using planes alone. In contrast, there has been a large amount of research in the computer graphics community on point cloud simplification, for example [7]. In the next section we take just one of these and adapt it into a context and feature sensitive re-sampling framework for mobile robotics.

III. Sampling Implicit Surfaces in 3D Point Clouds

A. Approach Summary

Let us assume we have a set of 3D points, lying on an unknown 2D manifold. The fundamental idea, a modified and extended version of Moenning and Dodgson's work in [5], is to approximate the manifold as the set of 3D grid cells which lie inside a union of ellipsoids, where an ellipsoid is fitted to each data point and a fixed number of its neighbors. This could be considered to represent a 'crust' over the data points, and is similar to the approach of Memoli and Sapiro in [4]. After randomly selecting a small number of 'seed' points, to initiate the process, we propagate fronts from each of their corresponding grid cells. This is performed using the Fast Marching Level Set Method [8] (described later in Section III-C), and continues until fronts collide or reach the edge of the problem domain. By storing the time at which a front arrived in each cell, we effectively generate a distance function over the cells considered (ie. the manifold). Furthermore, by looking for local maxima over this function, we are able to extract the vertices of a Voronoi diagram painted over the manifold. This enables us to select a location for a new re-sample point as prescribed by the Farthest Point Strategy in [1] (which is described further in Section III-B). After selection, this process can be repeated to find as many new points as required, or until the surface density reaches a predefined threshold. Note however, on subsequent iterations, it is not necessary to recompute the distance function (Voronoi diagram) over the entire problem domain. By maintaining front arrival times over all cells, only the front belonging to the most recently generated point needs propagating.

Another attractive and significant feature of this approach is that fronts can move at different speeds through different regions of the crust. These speeds can be found using the set of re-sampling 'policies' or user-defined hooks (described

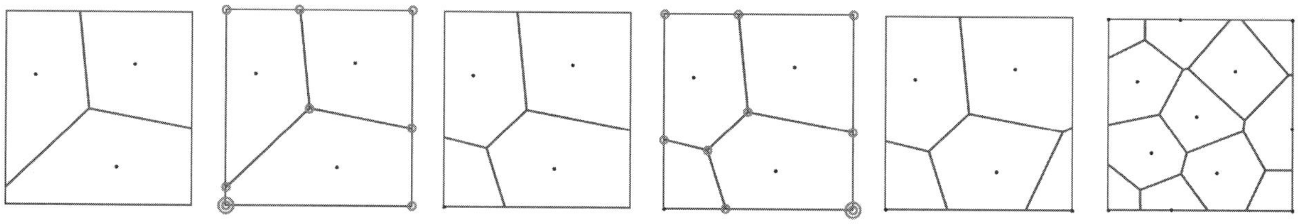

Fig. 2. Given a square image with three initial sample locations, one can construct a corresponding BVD, as shown in (a). Sub-figure (b) shows the results of a BVD vertex extraction (shown as circles), which are efficiently searched to yield the vertex with greatest distance to its closest point (circled twice). This forms the next iteration's sample point, which can be observed in the BVD in (c). Sub-figures (d) and (e) progress similarly. Sub-figure (f) shows the sample points generated after 9 iterations - note the distribution is becoming increasingly uniform.

earlier) which incorporate local surface attributes together with local class information. Consequently, when the generalized re-sampling mechanism (described above) comes to choose a new site, it extracts a vertex of a *weighted* Voronoi diagram, which is biased to be in the vicinity of a user-defined feature and/or class.

B. The Farthest Point Sampling Strategy

Farthest Point Sampling [1] was originally developed as a progressive method for approximating images. After initially selecting a set of 'seed' pixels, the idea is to select each subsequent 'sample' pixel to lie in the least known region of the image. This progressively generates high quality image approximations, *given* the current number of 'sample' pixels available. Furthermore, it can be shown that the next 'sample' pixel to select lies on a vertex of the previous iteration's Bounded Voronoi Diagram (BVD) - which provides an efficient means for future sample selection, as shown in [1]. This is illustrated in Figure 2. Given a square image with three initial sample locations, one can construct a corresponding BVD, as shown in (a). Sub-figure (b) shows the results of a BVD vertex extraction (shown as circles), which are efficiently searched to yield the vertex with greatest distance to its closest point (circled twice). This forms the next iteration's sample point, which can be observed in the BVD in (c). Sub-figures (d) and (e) progress similarly. Sub-figure (f) shows the sample points generated after 9 iterations - note the distribution is becoming increasingly uniform.

C. The Fast Marching Level Set Method

Level Set Methods [8] are efficient techniques for calculating the evolution of propagating fronts. This is achieved by considering a front Γ, moving normal to itself in $\mathbf{x}$, at time t, with speed function F, as the zeroth level set of a function $\Phi(\mathbf{x}, t)$ (ie. $\Phi(\mathbf{x}, t) = 0$). Given that the front's initial position is known, for the purpose of initialization, we can write the function as:

$$\Phi(\mathbf{x}, t = 0) = \pm d \quad (1)$$

where d is the distance from $\mathbf{x}$ to Γ and the plus or minus sign indicates whether the point is outside or inside the front

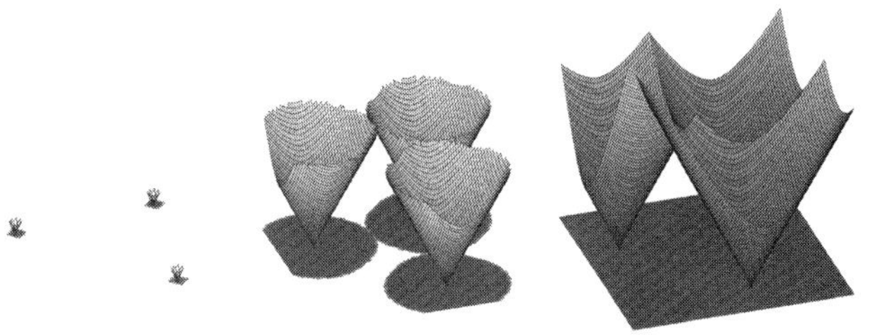

Fig. 3. These examples illustrate how the Fast Marching Level Set Method can be used to generate a 2D distance function. Sub-figure (a) shows 3 initial points with fronts propagating from each. The surface being generated above the plane represents each cell's arrival time, whilst the colors on the plane show the status of each cell. Red signifies 'dead', whilst green signifies 'active'. Sub-figure (b) shows how the front propagates over time, until the full distance function is generated in (c).

respectively. Assuming we observe a 'particle' at position $\mathbf{x}(t)$, which always lies on the propagating front, we can write:

$$\Phi(\mathbf{x}(t), t) = 0 \quad (2)$$

After differentiation, this becomes:

$$\frac{\partial \Phi(\mathbf{x}(t), t)}{\partial t} + \nabla \Phi(\mathbf{x}(t), t) . \frac{\partial \mathbf{x}(t)}{\partial t} = 0 \quad (3)$$

Given that the function F gives the speed normal to the front, $F = \frac{\partial \mathbf{x}(t)}{\partial t} . \mathbf{n}$, and that $\mathbf{n} = \nabla \Phi / |\nabla \Phi|$, this can be simplified to:

$$\frac{\partial \Phi(\mathbf{x}(t), t)}{\partial t} + F |\nabla \Phi(\mathbf{x}(t), t)| = 0 \quad (4)$$

$$\text{given } \Phi(\mathbf{x}, t = 0) \quad (5)$$

Generalized Level Set Methods proceed by iterating over a grid in $(\mathbf{x}, \Phi)$ space to solve this initial value partial differential equation at each time step (ensuring that a careful approximation is made for the spatial gradient to generate a realistic weak solution). This implicitly yields the zeroth level set or propagating front for all times considered. As well as being straightforward to numerically solve, one of this technique's major advantages is that it can easily cope with sharp discontinuities or changes in front topology - as 'slices' of a smooth, higher dimensional volume.

As an example, consider a two dimensional, initially circular front. The initial three dimensional surface in $(\mathbf{x}, \Phi)$ space would form a cone. At each time step, after sufficient iterations, the solution would represent the cone moving and/or 'morphing' within this space. Simultaneously, the zeroth level set (or intersection with the plane, $\Phi(\mathbf{x}, t) = 0$) would show how the true front propagates, correctly modelling discontinuities or changes in front topology.

Unfortunately, given the dimension of a typical $(\mathbf{x}, \Phi)$ space and a reasonable temporal and spatial resolution, this can be computationally expensive. However, under the assumption that the speed function F is always positive, we can guarantee that the front passes each point in $\mathbf{x}$ once and once only (ie. Φ is single valued). This means that the Level Set Equation (Equation 4) can be expressed in terms of a front arrival time function $T(\mathbf{x})$, as an example of a stationary Eikonal equation:

$$|\nabla T(\mathbf{x})|F = 1 \tag{6}$$

This time, $T(\mathbf{x})$ need only be solved over a discrete grid in $\mathbf{x}$ (rather than $\mathbf{x}$ *and* Φ). Additionally, with careful choice of spatial gradient operator, updates can become 'one way' (rather than depending on *all* neighboring cells) - whilst still enforcing the realistic weak solution. This facilitates the efficient Fast Marching algorithm, which is only required to 'process' *each* grid cell once. In effect, it is able to calculate the front's arrival time, by working outwards over the discrete grid from its initial position.

If we now assume the front propagates in 3D (ie. $\mathbf{x} \in \mathbb{R}^3$) and substitute an expression for the spatial gradient operator, $\nabla T(\mathbf{x})$, into Equation 6, we can write the 3D Fast Marching Level Set Method's update equation as follows:

$$\begin{aligned}&[\max(\max(D^{-x}_{ijk}T, 0), -\min(D^{+x}_{ijk}T, 0))^2\\ &+\max(\max(D^{-y}_{ijk}T, 0), -\min(D^{+y}_{ijk}T, 0))^2\\ &\quad+\max(\max(D^{-z}_{ijk}T, 0), -\min(D^{+z}_{ijk}T, 0))^2] = \frac{1}{F^2_{ijk}}\end{aligned} \tag{7}$$

Where i,j and k are grid cell indices along the x,y and z axes respectively, $T_{i,j,k}$ refers to the value of T at grid cell i, j, k and $D^{+x}_{ijk}T$ refers to the finite difference $+x$ gradient of T at grid cell i, j, k, which can be expressed as $D^{+x}_{ijk}T = (T_{i+1,j,k} - T_{i,j,k})/\Delta x$, where Δx is the unit grid cell length in the x direction. F_{ijk} is then the speed of wave propagation through grid cell i, j, k.

Before describing how to put this into practice, let us assume our domain is discretized into grid cells. We initialize the process by marking all cells which coincide with the front's initial position as 'active' and giving them a T value of zero (whilst all others are marked 'far', with T undefined). The front can then start marching forward, progressively calculating $T(\mathbf{x})$ for each 'active' cell's neighbors (except the 'dead' ones), using the update shown in Equation 7. If one of the neighboring cells is 'active', it will already possess a value for $T(\mathbf{x})$. In this instance, the smallest value of the existing and newly calculated $T(\mathbf{x})$ should be accepted, as it can be assumed this front reached that cell first. Conversely, if the neighboring cell is 'far', it should be made 'active' and assigned a value for $T(\mathbf{x})$. After all neighbors are dealt with, the processed 'active' cell should itself be considered 'dead'. It is important to note that when moving to process the next 'active' cell (ie. one on the current front) it is important to always process the one with a minimum value of $T(\mathbf{x})$ first. This ensures that any of the other 'active' cells could not possibly have had any influence on the processed cell's $T(\mathbf{x})$ (as they have greater arrival time values) - allowing (after its neighbors $T(\mathbf{x})$ values are calculated) it to be declared 'dead' (ie. fixed). This process continues until all cells are declared 'dead' and consequently all have $T(\mathbf{x})$ values. As an example, Figure 3 shows snapshots of this process propagating fronts on a plane. The full 3D technique is then formally described in Algorithm 1 where the fronts propagate over a grid $\mathbf{G}$ in which each cell $g(i, j, k)$ has the following properties: a front speed

Inputs : Grid $\mathbf{G}$, with cells $g(i, j, k)$, $i \in [1 : I]$, $j \in [1 : J]$ and $k \in [1 : K]$. Set of cells in initial front Γ.

Output: Arrival times: $g(i, j, k).T \quad \forall g(i, j, k) \in \mathbf{G}$

Initialization:
```
for g(i,j,k) ∈ Γ do
    g(i,j,k).active ← 1;
    g(i,j,k).T ← 0;
    H.PUSH(g(i,j,k));
end
while ¬H.empty() do
    g(i_min, j_min, k_min) ← H.POP();
    N ← NEIGHBOURHOOD(i_min, j_min, k_min);
    for g(i,j,k) ∈ N do
        if ∃g(i,j,k) ∧ ¬g(i,j,k).dead then
            F_ijk ← g(i,j,k).F;
            if g(i,j,k).active then
                τ ← SOLVEPDE(F_ijk, N)
                g(i,j,k).T ← min(g(i,j,k).T, τ);
                H ← HEAPIFY(H);
            else
                g(i,j,k).Active ← 1;
                g(i,j,k).T ← SOLVEPDE(F_ijk, N);
                H.PUSH(g(i,j,k))
            end
        end
    end
    g(i_min, j_min, k_min).dead ← 1;
end
```

Algorithm 1: Pseudo-code for an efficient implementation of the 3D fast marching algorithm for propagating fronts. The algorithm propagates a front Γ through a set of cells recording the time at which the front first passes through each of them. These times can be used to calculate the vertices of a bounded Voronoi diagram over the grid $\mathbf{G}$ which need not have homogenous propagation speeds.

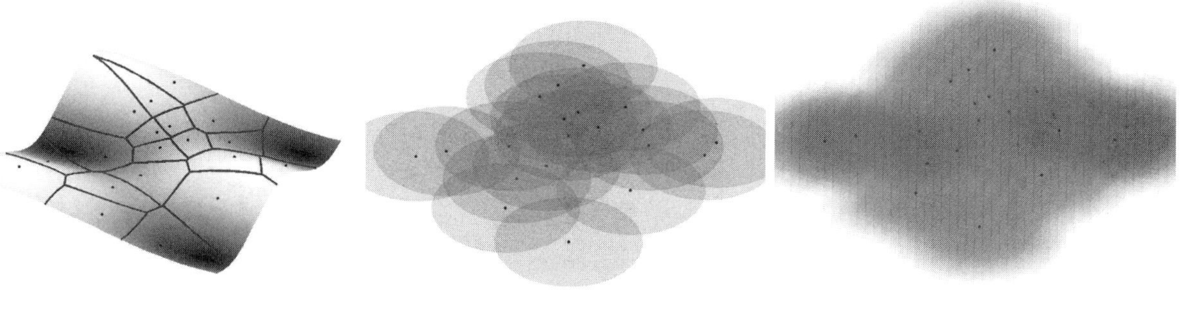

Fig. 4. Sub-figure (a) shows a 2D manifold, with a set of points on its surface. The blue lines represent surface Voronoi edges. This is what we aim to reproduce with our manifold approximation, to find where to place the next sample. Sub-figure (b) shows a series of spheres, with one centered on each original data point. If we take all 3D grid cells that lie inside this union, as in (c), we find an approximation to the original manifold.

field F, an arrival time T (initialized to ∞) and three status flags DEAD ACTIVE and FAR. The term $\mathcal{H}$ is a minheap of cells sorted by arrival time T. The function SOLVEPDE refers to solving $T_{i,j,k}$ using the discretized P.D.E. in Equation 7.

Whilst this technique has many uses, its primary function in this paper is to efficiently generate a Voronoi diagram over the original point cloud manifold, as illustrated in the example in Figure 4(a). This is implemented by approximating the manifold with a thin 'crust' of 3D grid cells. These can be found by locating spheres at each original point, as illustrated in Figure 4(b), and taking all those cells which lie within this union (as shown in Figure 4(c)). However, this approach tends to introduce large amounts of lateral noise — especially if regions of the true surface are truly planar. Consequently, we choose to fit ellipsoids to each point and its local neighborhood, and take the union of these instead. This approach seems to work particularly well, though some care does need taking in very sparse regions, when a fixed number of nearest neighbors span a large distance - and therefore produce excessively large ellipsoids. We can then take this bounded 3D space, initialize fronts in those cells which correspond to the set of points concerned, and propagate fronts outwards. This continues until all cells have been visited, when the resulting distance function can be processed to yield the Bounded Voronoi Diagram.

It is worth noting that this technique also allows efficient incremental construction of distance functions, when adding points to existing clouds. This can be achieved by propagating a front from each new point, until it reaches a region of the distance function which has values less than the front's predicted arrival time. Furthermore, by changing the speed at which fronts propagate through the discretized space, we are able to bias the positions of the Voronoi vertices. This will be examined in more detail in the next sub-section.

D. Imposing a Sampling Bias

Assume we have set of N_p data points, $P_{N_p} = \{p_1, p_2 .... p_{N_p}\}$, which are non-uniformly distributed samples of a 2D manifold M embedded in $\mathbb{R}^3$, each belonging to a single class of a total set of N_c classes, $C_{N_c} = \{c_1, c_2 .... c_{N_c}\}$, and each with a single associated vector of attributes, forming the total set of N_p attribute vectors, $A_{N_p} = \{a_1, a_2 .... a_{N_p}\}$. Our aim is to generate a new set of N_T points, $\tilde{P}_{N_T} = \{\tilde{p}_1, \tilde{p}_2 .... \tilde{p}_{N_T}\}$ with a local surface density at each original point proportional to a weight given by:

$$w_m = \mathcal{F}(a_m, \mathcal{C}(p_m)) \tag{8}$$

Where $\mathcal{F}$ expresses how interesting the neighborhood of p_m appears to be in the context of its geometry and appearance, a_m, and its semantic classification, c_m. This aim can be achieved by modifying Algorithm 1 in a straightforward way. The step $F_{ijk} \leftarrow g(i,j,k).F$ can be replaced with $F_{ijk} \leftarrow \mathcal{F}(a(g(i,j,k)), \mathcal{C}(g(i,j,k))$. Here we have slightly abused notation and used $a(g(i,j,k))$ to mean the attributes of the region surrounding the raw data point closest to cell $g(i,j,k)$ (and similarly for $\mathcal{C}(g(i,j,k))$). The upshot of this is that now the distance function calculated by the solution of the wave front P.D.E. will be biased (propagate slower) around interesting places. Because the wavefront determines the Voronoi vertices which in turn determine the new surface samples generated, the inclusion of Equation 8 in Algorithm 1 constitutes context and feature-sensitive sampling.

IV. RESULTS

Initial experiments have been performed using the data shown earlier in Figure 1, collected with a mobile robot. One of the key things to notice is the large variation in local point density. This is particularly evident on the floor, close to the vehicle, and on object faces sensed perpendicular to the scanning laser. The scene actually contains part of a building, a protruding staircase (which is sandwiched between two bicycle racks), and on the right, a large tree. Due to the nature of laser ranging - points are particularly dense around the tree's branches, as many mixed (multiple return) measurements were recorded. Figure 5a shows the same data, but includes semi-transparent colored and lettered control volumes for analysis. The number of points present is written adjacent to each (in a corresponding color), and is included in a bar chart (top right) for comparison. The total number of points in the entire cloud is shown bottom left.

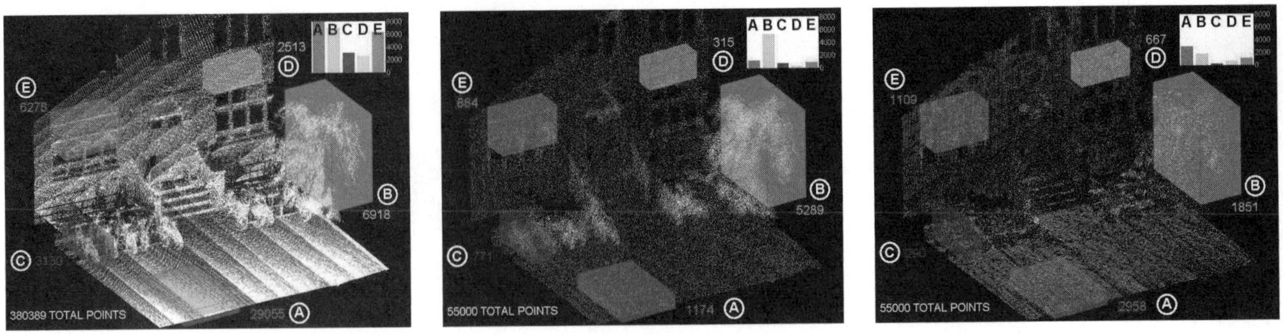

Fig. 5. Sub-figure (a) shows the original point cloud (as shown earlier in Figure 1), although this time includes semi-transparent colored and lettered control volumes for analysis. Sub-figure (b) shows the result of uniform re-sampling (maintaining a uniform weight over all cells), terminated when 55000 points had been produced. Sub-figure (c) shows the result of feature sensitive re-sampling based on modified 'normal variation', and was terminated at the same point.

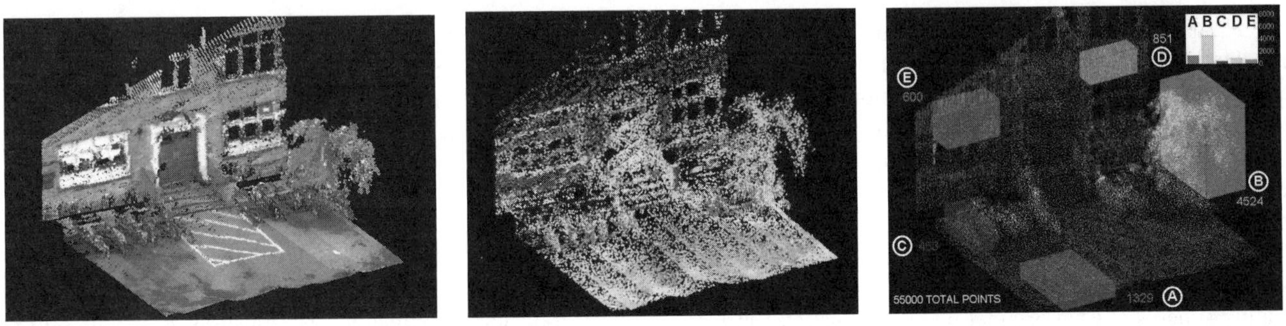

Fig. 6. Sub-figure (a) shows the result of projecting each laser point into an image of the scene and assigning it with color information. Sub-figure (b) shows a distribution of weights over the same cloud, proportional to a distance from each point (in RGB space) to a manually selected 'building' color. Sub-figure (c) then shows the result of feature sensitive re-sampling using these weights, preferentially choosing points of a particular shade.

The first experimental result is shown in Figure 5b, and later summarized in Table 1. This was produced using the re-sampling algorithm described earlier (maintaining a uniform weight over all cells), and was terminated when 55000 points had been produced. Notice that as a consequence of such weighting, the front propagated through all cells at the same speed, and results in an output which is a uniform representation of the implicit manifold. We also observe a 96% reduction in the number of points used to represent the red (A) control volume's section of floor. Similar re-sampling behavior throughout this cloud resulted in an overall representational overhead cut of 86% (from 380389 down to 55000), whilst preserving significant geometry.

The second experiment began with the same initial point cloud, and the resulting re-sampled point cloud is shown in Figure 5c. It is also summarized in Table 1. This shows that by weighting front propagation speeds through each grid cell, we can bias re-sample points to be in certain regions of the manifold. Whilst such user-defined 'policies' are generally functions of any number of local point attributes, here we make use of modified 'normal variation' alone. After generating a surface normal for each point (based on a close set of nearest neighbors), 'normal variation' can be defined as the variance of the angle between the point concerned and a fixed set of nearest neighbor surface normals. These are then transformed to ensure the distribution maps to a wide range of grid cell weights. If this is not performed, a few, very dominant points with very high 'normal variation' would tend to spread over a large range of the weights considered, leaving the majority of the grid cells with near identical propagation speeds. In this particular point cloud, the tree's foliage, the bicycles, the building's steps and its window frames all score highly. This leads to faster local front propagation and a tendency to inhibit nearby Voronoi vertices and the subsequent re-sample points. As in the last experiment, we stopped re-sampling when 55000 points were selected. Notice in particular the red (A) control volume located on the floor, and the green (B) control volume incorporating foliage, and contrast to the uniformly re-sampled cloud. The number of points in the predominantly planar red (A) control volume increased by 152% from 1174 to 2958, whilst those in the high curvature green (B) control volume decreased by 65% from 5289 to 1851.

The next series of experiments aims to highlight the versatility of our approach, as re-sampling policies can be based on any individual or set of local point cloud attributes. In this example, we choose to project each laser point into an image of the scene and assign it with color information as shown in Figure 6a. Figure 6b then shows a distribution of weights over the same cloud, proportional to a distance from each point (in RGB space) to a manually selected 'building' color. These

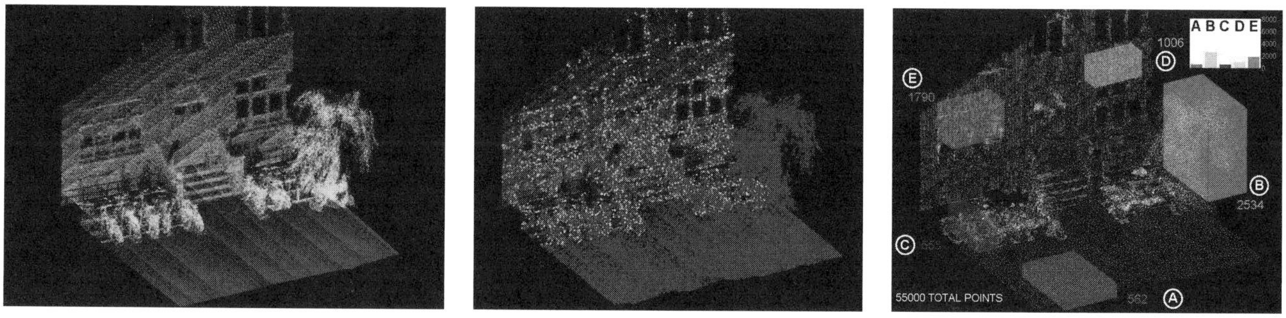

Fig. 7. Sub-figure (a) shows the point cloud from earlier, segmented into three distinct classes: 'floor', 'foliage' and 'other'. This was performed manually for the sake of demonstration. Sub-figure (b) then shows a weight distribution generated using local 'normal variation', along with the new class information. The intention here is to preferentially re-sample points on the building by artificially boosting the 'normal variation' weights of points classified as 'floor' or 'foliage'. Note that other 'non-building' points (on various objects) already have high 'normal variation' weights. Sub-figure (c) shows the result of context and feature sensitive re-sampling using these weights.

were then used to bias the generalized re-sampling mechanism, which preferentially chose points of this particular shade. Once again, re-sampling was stopped after 55000 points had been selected. The results of this process can be seen in 6c, and are summarized in Table 1. A careful examination of these results reveals some success in extracting points on the building - the number of points in the cyan (D) control volume, located on the building's wall, has increased 170% from 315 to 851 (compared to uniform re-sampling). Furthermore, the dramatic increase in the density of building points (to the left of the cyan (D) control volume, all the way across to the magenta (E) control volume) is clearly visibly, with a reduction in points on bicycles and foliage. However, a 13% increase in red (A) control volume points indicates that the color of the floor is too close to that of the building for this approach to disambiguate between the two. Additionally, we see a 32% decrease in blue (C) control volume points, as colors around the window vary dramatically compared to the main brickwork.

The third series of results demonstrates how the re-sampling process can be influenced by semantic information. Figure 7a shows the point cloud from earlier, segmented into three distinct classes: 'floor', 'foliage' and 'other'. This was performed manually for the sake of demonstration. Figure 7b shows a weight distribution generated using local 'normal variation' (described earlier), along with the new class information. The intention here is to preferentially re-sample points on the building by artificially boosting the 'normal variation' weights of points classified as 'floor' or 'foliage'. Note that other 'non-building' points (on various objects) already have high 'normal variation' weights. These were subsequently used to bias the re-sampling mechanism, which continued until the standard 55000 point target had been reached. The results generated can be seen in Figure 7c, and are also summarized in Table 1. Notice the massive increase in the number of points in the cyan (D) and magenta (E) control volumes (located on the building), when compared to the equivalent uniformly re-sampled cloud. The number of points in the cyan (D) control volume increased by 219% from 315 to 1006, whilst the number of points in the magenta (E) control volume increased by 102% from 884 to 1790. There was also a reduction in the number of points in all other control volumes (ie. those on the objects, foliage and the floor).

The final experiment highlights one use for this technique: re-sampling point clouds attached to vehicle poses in delayed state SLAM formulations. This could offer superior registration convergence properties, reduce overall processing time (if the point clouds are large or used frequently), reduce overall storage requirements, and allow efficient viewing of the maps constructed. Whilst this area does need significant further study, we include a few results to demonstrate this technique's potential. Figure 8a shows two point clouds (belonging to separate vehicle poses), each containing approximately 190,000 points. Figure 8b then shows the same point clouds uniformly re-sampled, each reduced to 50,000 points. Note that whilst we choose uniform weighting in this example (for the sake of clarity), there could be greater justification for another re-sampling policy. The pair of graphs in Figure 8c then compare the original pair of point clouds, and the re-sampled pair of point clouds when each was aligned with a typical registration algorithm. The top plot shows the normalized residual error at registration termination versus independent x, y and theta perturbations around the initial transformation estimate. Note that the re-sampled registration (shown in green) not only matches the performance of the original registration (shown in black) - the bounds of the convergence basin are actually increased. The lower plot shows corresponding processing times for the registrations, and it is apparent that with less points - the re-sampled registrations are significantly faster. Given that in this example the re-sampling process took 50 minutes (2998 seconds) for pose 1, and 23 minutes (1380 seconds) for pose 2, re-sampling would be beneficial if performing at least 6-8 registrations per point cloud (in terms of overall processing time alone). Note that these times are likely to reduce significantly when we engineer away the large time constant in our initial exploratory implementation (which scales $O(N \log N)$, where N is the number of grid cells, which is proportional to the workspace surface area - not the number of data points).

	Original	Uniform	Normals	Color	Color and Class
Entire Point Cloud	380389	55000	55000	55000	55000
Red Volume (A) (Floor)	29055	1174	2958	1329	562
Green Volume (B) (Foliage)	6918	5289	1851	4524	2534
Blue Volume (C) (Bicycle)	3130	771	290	453	555
Cyan Volume (D) (Building Wall)	2513	315	667	851	1006
Magenta Volume (E) (Building Window)	6278	884	1109	600	1790

TABLE I

EACH ENTRY IN THIS TABLE CORRESPONDS TO THE NUMBER OF POINTS IN AN ENTIRE POINT CLOUD, OR ONE OF THE CONTROL VOLUMES. THE COLUMN HEADINGS SIGNIFY WHICH EXPERIMENT PRODUCED EACH SET OF RESULTS. THE RED AND BLUE NUMBERS INDICATE WHETHER THE NUMBER OF POINTS INCREASED OR DECREASED RESPECTIVELY, COMPARED TO THE EQUIVALENT UNIFORMLY RE-SAMPLED POINT CLOUD.

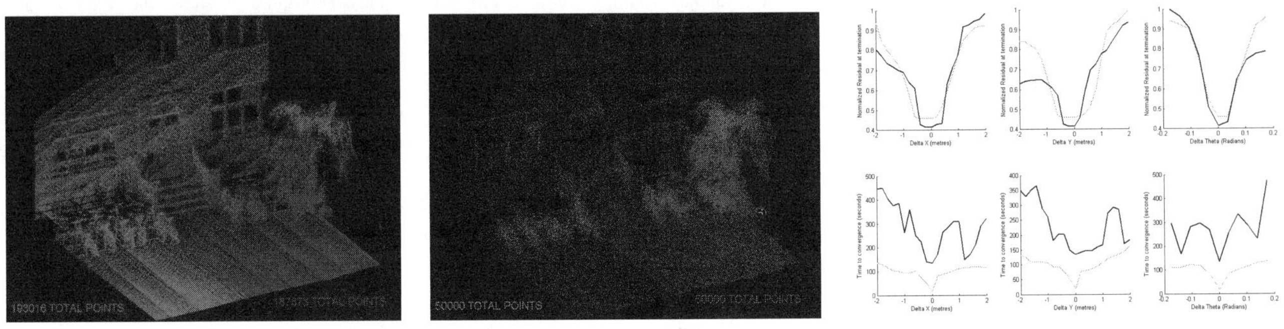

Fig. 8. Sub-figure (a) shows two point clouds (belonging to separate vehicle poses), each containing approximately 190,000 points. Sub-figure (b) shows the same point clouds uniformly re-sampled, each reduced to 50,000 points. Sub-figure (c) finishes with a pair of graphs that compare registration performance - first using the original pair of point clouds, and then using the re-sampled pair. The top plot shows the normalized residual error at registration termination versus independent x, y and theta perturbations around the initial transformation estimate, whilst the bottom one shows corresponding processing times.

V. CONCLUSIONS

In this paper we have described how one may use large 3D point clouds to generate samples from the underlying workspace surfaces. We have adopted the technique of solving a discretized P.D.E. to find a Voronoi diagram over a surface approximated by a thin crust of cells. An iterative scheme uses the Voronoi-vertices to generate new samples and update the Voronoi diagram. We have suggested in this paper that this approach has substantial value to the robotics community in terms of using 3D laser range data. By using the qualities of the original data and prior information regarding the type of surfaces being sampled to bias the solution of the governing P.D.E. we can mitigate sensor aliasing issues, select regions of interest and apply feature extraction algorithms all within a single framework. We have not yet fully explored the spectrum of uses for this technique in outdoor navigation (which is the domain which motivates this work) especially when using both laser and camera images. Nevertheless we feel it is a promising and elegant approach which comfortably occupies the middle ground between the feature-based (prescriptive) and view-based (indifferent) approaches to interpreting and using point clouds for navigation.

ACKNOWLEDGMENT

This work was supported by the Engineering and Physical Sciences Research Council Grant #GR/S62215/01.

REFERENCES

[1] Y. Eldar, M. Lindenbaum, M. Porat, and Y. Zeevi. The farthest point strategy for progressive image sampling. *IEEE Trans. on Image Processing*, 6(9):1305 – 1315, 1997.

[2] D. Hähnel, W. Burgard, and S. Thrun. Learning compact 3D models of indoor and outdoor environments with a mobile robot. *Robotics and Autonomous Systems*, 44(1):15–27, 2003.

[3] Y. Liu, R. Emery, D. Chakrabarti, W. Burgard, and S. Thrun. Using EM to learn 3D models of indoor environments with mobile robots. In *ICML '01: Proceedings of the Eighteenth International Conference on Machine Learning*, pages 329–336, San Francisco, CA, USA, 2001.

[4] F. Memoli and G. Sapiro. Fast computation of weighted distance functions and geodesics on implicit hyper-surfaces: 730. *J. Comput. Phys.*, 173(2):764, 2001.

[5] C. Moenning and N. Dodgson. A new point cloud simplification algorithm. *The 3rd IASTED International Conference on Visualization, Imaging and Image Processing (VIIP)*, 2003.

[6] A. Nuchter, H. Surmann, and J. Hertzberg. Automatic model refinement for 3D reconstruction with mobile robots. *3DIM*, 00:394, 2003.

[7] M. Pauly, M. Gross, and L. P. Kobbelt. Efficient simplification of point-sampled surfaces. In *IEEE Visualization (VIS)*, Washington, DC, USA, 2002.

[8] J. Sethian. A fast marching level set method for monotonically advancing fronts. *Proc. Nat. Acad. Sci.*, 93(4):1591–1595, 1996.

[9] J. Weingarten, G. Gruener, and R. Siegwart. Probabilistic plane fitting in 3D and an application to robotic mapping. In *IEEE International Conference on Robotics and Automation (ICRA)*, volume 1, pages 927–932, New Orleans, USA, 2004.

[10] D. F. Wolf, A. Howard, and G. S. Sukhatme. Towards geometric 3D mapping of outdoor environments using mobile robots. In *IEEE/RSJ International Conference on Intelligent Robots and Systems (IROS)*, pages 1258–1263, Edmonton Canada, 2005.

Simultaneous Localisation and Mapping in Dynamic Environments (SLAMIDE) with Reversible Data Association

Charles Bibby
Department of Engineering Science
Oxford University
Email: cbibby@robots.ox.ac.uk

Ian Reid
Department of Engineering Science
Oxford University
Email: ian@robots.ox.ac.uk

Abstract— **The conventional technique for dealing with dynamic objects in SLAM is to detect them and then either treat them as outliers [20][1] or track them separately using traditional multi-target tracking [18]. We propose a technique that combines the least-squares formulation of SLAM and sliding window optimisation together with generalised expectation maximisation, to incorporate both dynamic and stationary objects directly into SLAM estimation. The sliding window allows us to postpone the commitment of model selection and data association decisions by delaying when they are marginalised permanently into the estimate. The two main contributions of this paper are thus: (i) using reversible model selection to include dynamic objects into SLAM and (ii) incorporating reversible data association. We show empirically that (i) if dynamic objects are present our method can include them in a single framework and hence maintain a consistent estimate and (ii) our estimator remains consistent when data association is difficult, for instance in the presence of clutter. We summarise the results of detailed and extensive tests of our method against various benchmark algorithms, showing its effectiveness.**

I. Introduction

SLAM in dynamic environments is essentially a model selection problem. The estimator constantly needs to answer the question: is a landmark moving or is it stationary? Although there are methods for doing model selection in recursive filtering frameworks such as interacting multiple model estimation or generalised pseudo-Bayesian estimation [9], these methods always have some lag before the model selection parameter(s) converge to the correct steady state. This means that for a period of time the filter could classify a target as dynamic when it is stationary or vice versa. This is potentially catastrophic for SLAM because incorrectly modeling a dynamic or stationary landmark will lead to biased measurements and hence map corruption and inconsistency.

We propose a framework that combines least-squares with sliding window optimisation [17] and generalised expectation maximisation [13]. This allows us to include reversible model selection and data association parameters in the estimation and hence include dynamic objects in the SLAM map robustly. At the heart of our method is the use of sliding window optimisation, which delays the point when information is marginalised out, allowing the filter a period of time to get the model selection and data association parameters correct before marginalisation. Although something similar could be achieved with a more traditional Extended Kalman Filter (EKF) using delayed decision making [11], the difference is that our method uses reversible as opposed to delayed decision making i.e. decisions can change many times in light of new information before being committed to the estimate.

The adverse effects of poor data association in SLAM, namely inconsistent estimates and divergence, are normally unacceptable and hence a suitable method must be selected. A common approach is the chi-squared nearest neighbour (NN) test, which assumes independence between landmarks and then probabilistically chooses the best measurement which falls within the gate of an individual landmark. This method can work well for sparsely distributed environments with good sensors; however, once the proximity between landmarks approaches the sensor noise or clutter is present, ambiguous situations arise and a more sophisticated method is required. One such method is joint compatibility branch and bound (JCBB) [14] which takes into account the correlations between landmarks by searching an interpretation tree [7] for the maximum number of jointly compatible associations. This method produces very good results when ambiguities are present, but still suffers from problems in the presence of clutter and is slow for large numbers of measurements. More recently the data association problem has been treated as a discrete optimisation over multiple time steps [15]. We also treat data association as a discrete optimisation, but include model selection and propose an alternative method that uses sliding window optimisation and generalised expectation maximisation [13](more specifically an approximate method called classification expectation maximisation [3]).

We begin by introducing our notation and the background on least-squares SLAM in Section II; Section III describes the background on sliding window optimisation; Section IV describes our method for doing SLAM with reversible data association; Section V extends this to SLAM in dynamic environments; Section VI compares our methods to an Iterated Extended Kalman Filter (IEKF) with either NN or JCBB for data association and finally in Section VII we conclude and discuss our ideas for future work.

II. NOTATION AND LEAST-SQUARES SLAM

Although this section and the next cover background work [6][17], they have been included because they form the basis for our techniques.

A. Notation

- $\mathbf{x}_t$: A state vector describing the vehicle's pose (location and orientation $[x, y, \theta]$) at time t.
- $\mathbf{u}_t$: A control vector (odometry $[\dot{x_v}, \dot{y_v}, \dot{\theta_v}]$ in vehicle coordinates where $\dot{x_v}$ is in the direction the vehicle is pointing) that was applied to vehicle at time $t-1$ to take it to time t.
- $\mathbf{z}_t$: A measurement made by the vehicle at time t of a landmark in the world.
- $\mathbf{m}_k$: A state vector describing the location of landmark k.
- $\mathbf{X} = \{\mathbf{x}_0, \ldots, \mathbf{x}_t\}$: A set of vehicle poses.
- $\mathbf{U} = \{\mathbf{u}_1, \ldots, \mathbf{u}_t\}$: A set of odometry.
- $\mathbf{Z} = \{\mathbf{z}_1, \ldots, \mathbf{z}_t\}$: A set of measurements.
- $\mathbf{M} = \{\mathbf{m}_0, \ldots, \mathbf{m}_k\}$: A set of all landmarks.

B. Least-Squares SLAM

Consider the Bayesian network in Figure 1 where each node represents a stochastic variable in the system. The grey nodes represent observed variables, the white nodes represent hidden variables and the arrows in the graph represent the dependence relationships between variables, for instance $\mathbf{z}_{t-1}$ depends upon $\mathbf{x}_{t-1}$ and $\mathbf{M}$. For the benefit of the reader let us reduce notation by making two assumptions: (i) only one observation per time step; (ii) we assume known data association i.e. which landmark generated a given measurement (we will relax this assumption from Section IV onwards).

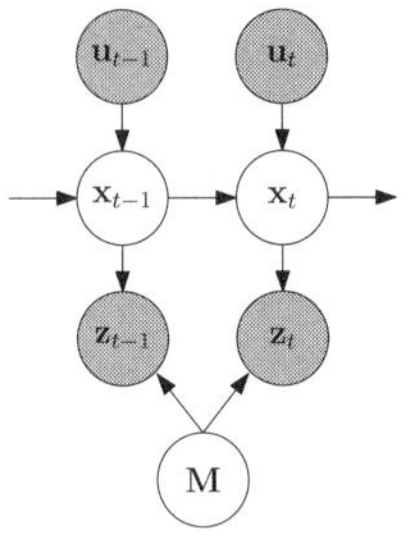

Fig. 1. A Bayesian network representing the SLAM problem.

The joint probability of $\mathbf{X}$, $\mathbf{M}$, $\mathbf{U}$ and $\mathbf{Z}$ can be factorised, using the independence structure depicted by the Bayesian network in Figure 1, as follows:

$$P(\mathbf{X}, \mathbf{M}, \mathbf{U}, \mathbf{Z}) = P(\mathbf{x}_0)P(\mathbf{M}) \times \prod_{t=1}^{T} P(\mathbf{z}_t|\mathbf{x}_t, \mathbf{M})P(\mathbf{x}_t|\mathbf{x}_{t-1}, \mathbf{u}_t). \quad (1)$$

where:

- T is the number of time steps.
- $P(\mathbf{x}_0)$ is the prior on vehicle state, which has a mean $\tilde{\mathbf{x}}_0$ and covariance $\mathbf{P}_0$.
- $P(\mathbf{M})$ is the prior on the map, which is normally taken to be the uninformative uniform distribution.
- $P(\mathbf{z}_t|\mathbf{x}_t, \mathbf{M})$ is the measurement model i.e. the probability of the measurement $\mathbf{z}_t$ given the vehicle pose $\mathbf{x}_t$, the map $\mathbf{M}$ and the correct data association.
- $P(\mathbf{x}_t|\mathbf{x}_{t-1}, \mathbf{u}_t)$ is the motion model i.e. the probability of the new pose $\mathbf{x}_t$ given the last vehicle pose $\mathbf{x}_{t-1}$ and the odometry $\mathbf{u}_t$.

If we take $P(\mathbf{M})$ to be the uninformative uniform distribution then (1) reduces to:

$$P(\mathbf{X}, \mathbf{M}, \mathbf{U}, \mathbf{Z}) = P(\mathbf{x}_0) \prod_{t=1}^{T} P(\mathbf{z}_t|\mathbf{x}_t, \mathbf{M})P(\mathbf{x}_t|\mathbf{x}_{t-1}, \mathbf{u}_t). \quad (2)$$

Let us now also make Gaussian assumptions and define the prior term, motion model and measurement model respectively as:

$$\mathbf{x}_0 = \tilde{\mathbf{x}}_0 + \mathbf{p}_0 \Leftrightarrow$$
$$P(\mathbf{x}_0) \propto exp(-\frac{1}{2}\,\|\tilde{\mathbf{x}}_0 - \mathbf{x}_0\|^2_{\mathbf{P}_0}) \quad (3)$$
$$\mathbf{x}_t = f(\mathbf{x}_{t-1}, \mathbf{u}_t) + \mathbf{q}_t \Leftrightarrow$$
$$P(\mathbf{x}_t|\mathbf{x}_{t-1}, \mathbf{u}_t) \propto exp(-\frac{1}{2}\,\|f(\mathbf{x}_{t-1}, \mathbf{u}_t) - \mathbf{x}_t\|^2_{\mathbf{Q}_t}) \quad (4)$$
$$\mathbf{z}_t = h(\mathbf{x}_t, \mathbf{M}) + \mathbf{r}_t \Leftrightarrow$$
$$P(\mathbf{z}_t|\mathbf{x}_t, \mathbf{M}) \propto exp(-\frac{1}{2}\,\|h(\mathbf{x}_t, \mathbf{M}) - \mathbf{z}_t\|^2_{\mathbf{R}_t}) \quad (5)$$

where $\mathbf{p}_0$, $\mathbf{q}_t$ and $\mathbf{r}_t$ are normally distributed, zero mean, noise vectors with covariances $\mathbf{P}_0$, $\mathbf{Q}_t$ and $\mathbf{R}_t$ respectively. The $\|\mathbf{e}\|^2_{\mathbf{\Sigma}}$ notation represents the squared Mahalanobis distance $\mathbf{e}^T\mathbf{\Sigma}^{-1}\mathbf{e}$, where $\mathbf{\Sigma}$ is a covariance. We can now perform inference on the Bayesian network in Figure 1 to find the maximum a posteriori (MAP) estimate $\{\hat{\mathbf{X}}, \hat{\mathbf{M}}\} = \arg\max_{\{\mathbf{X},\mathbf{M}\}} P(\mathbf{X}, \mathbf{M}|\mathbf{U}, \mathbf{Z})$. This can be done by minimising the negative log of the joint distribution (2):

$$\{\hat{\mathbf{X}}, \hat{\mathbf{M}}\} \triangleq \underset{\{\mathbf{X},\mathbf{M}\}}{\arg\min}\,(-\log(P(\mathbf{X}, \mathbf{M}, \mathbf{U}, \mathbf{Z}))). \quad (6)$$

By substituting Equations (3), (4) and (5) into (6) we get a non-linear least-squares problem of the form:

$$\{\hat{\mathbf{X}}, \hat{\mathbf{M}}\} \triangleq \underset{\{\mathbf{X},\mathbf{M}\}}{\arg\min}\Big\{\|\tilde{\mathbf{x}}_0 - \mathbf{x}_0\|^2_{\mathbf{P}_0} + \sum_{t=1}^{T}\big(\|f(\mathbf{x}_{t-1}, \mathbf{u}_t) - \mathbf{x}_t\|^2_{\mathbf{Q}_t} + \|h(\mathbf{x}_t, \mathbf{M}) - \mathbf{z}_t\|^2_{\mathbf{R}_t}\big)\Big\}. \quad (7)$$

Let us now linearise the non-linear terms and re-write as a matrix equation:

$$\{\hat{\mathbf{X}}, \hat{\mathbf{M}}\} \triangleq \underset{\{\mathbf{X},\mathbf{M}\}}{\arg\min}\Big\{\|\mathbf{B}\boldsymbol{\delta}\mathbf{x}_0 - \{\mathbf{x}_0 - \tilde{\mathbf{x}}_0\}\|^2_{\mathbf{P}_0} + \sum_{t=1}^{T}\big(\|\{\mathbf{F}_{t-1}\boldsymbol{\delta}\mathbf{x}_{t-1} + \mathbf{B}\boldsymbol{\delta}\mathbf{x}_t\} - \{\mathbf{x}_t - f(\mathbf{x}_{t-1}, \mathbf{u}_t)\}\|^2_{\mathbf{Q}_t} + \|\{\mathbf{H}_t\boldsymbol{\delta}\mathbf{x}_t + \mathbf{J}_t\boldsymbol{\delta}\mathbf{M}\} - \{\mathbf{z}_t - h(\mathbf{x}_t, \mathbf{M})\}\|^2_{\mathbf{R}_t}\big)\Big\}, \quad (8)$$

where $\mathbf{F}_{t-1}$ is the Jacobian of $f(.)$ w.r.t. $\mathbf{x}_{t-1}$, $\mathbf{H}_t$ is the Jacobian $h(.)$ w.r.t. $\mathbf{x}_t$ and $\mathbf{J}_t$ is the Jacobian of $h(.)$ w.r.t. $\mathbf{M}$. We have also introduced $\mathbf{B} = -\mathbf{I}_{d\times d}$ so that $\boldsymbol{\delta}\mathbf{x}_t$ which is a small change in the states corresponding to the pose at time t is treated in the same way as the other terms; where d is the dimension of a single vehicle pose.

We can now factorise and write a standard least-squares matrix equation:

$$\mathbf{A}^T\boldsymbol{\Sigma}^{-1}\mathbf{A}\delta = \mathbf{A}^T\boldsymbol{\Sigma}^{-1}\mathbf{b}, \quad (9)$$

where $\mathbf{A}$ is a matrix of Jacobians, $\boldsymbol{\Sigma}$ is a covariance matrix and $\mathbf{b}$ is an error vector, for a detailed look at the structure of these matrices refer to [6]. We solve for $\boldsymbol{\delta}$ in (9) using direct sparse methods [4][6].

C. *The Hessian, Information Matrix and Inverse Covariance*

If the Cramer Rao lower bound [16] is reached, then given that we have normally distributed zero mean variables the following condition is satisfied:

$$\mathbf{P}^{-1} = \mathbf{Y} = \mathbf{A}^T\boldsymbol{\Sigma}^{-1}\mathbf{A}$$

where $\mathbf{A}^T\boldsymbol{\Sigma}^{-1}\mathbf{A}$ is the approximation to the Hessian calculated when solving the least-squares problem (9). This is why the information matrix $\mathbf{Y}$ in the information filter version of SLAM [19] and the Hessian in the least-squares formulation of SLAM [17] are equivalent to the inverse of the covariance matrix $\mathbf{P}$ in the more traditional Kalman filter based SLAM systems. Interestingly, as explained fully in [6], the non-zero elements of the information matrix $\mathbf{Y}$ correspond to links in the Markov Random Field (MRF) that is equivalent to the Bayesian network in Figure 1. Each of these links represents a constraint or relationship between two nodes in the MRF, e.g. a measurement equation linking a vehicle pose to a landmark or an odometry equation linking one vehicle pose to the next. The first row of Figure 2 shows the structure of the information matrix and MRF for a simple 2D example. Let us now consider the structure of the information matrix in the first row of Figure 2: Y_v is block tridiagonal and represents the information from odometry between vehicle poses; Y_m is block diagonal and represents the information about landmarks in the map and Y_{vm} and Y_{vm}^T represent the information associated with measuring a landmark from a given pose. We will revisit this simple example in Section III-A when discussing how marginalisation affects the structure of both $\mathbf{Y}$ and the MRF.

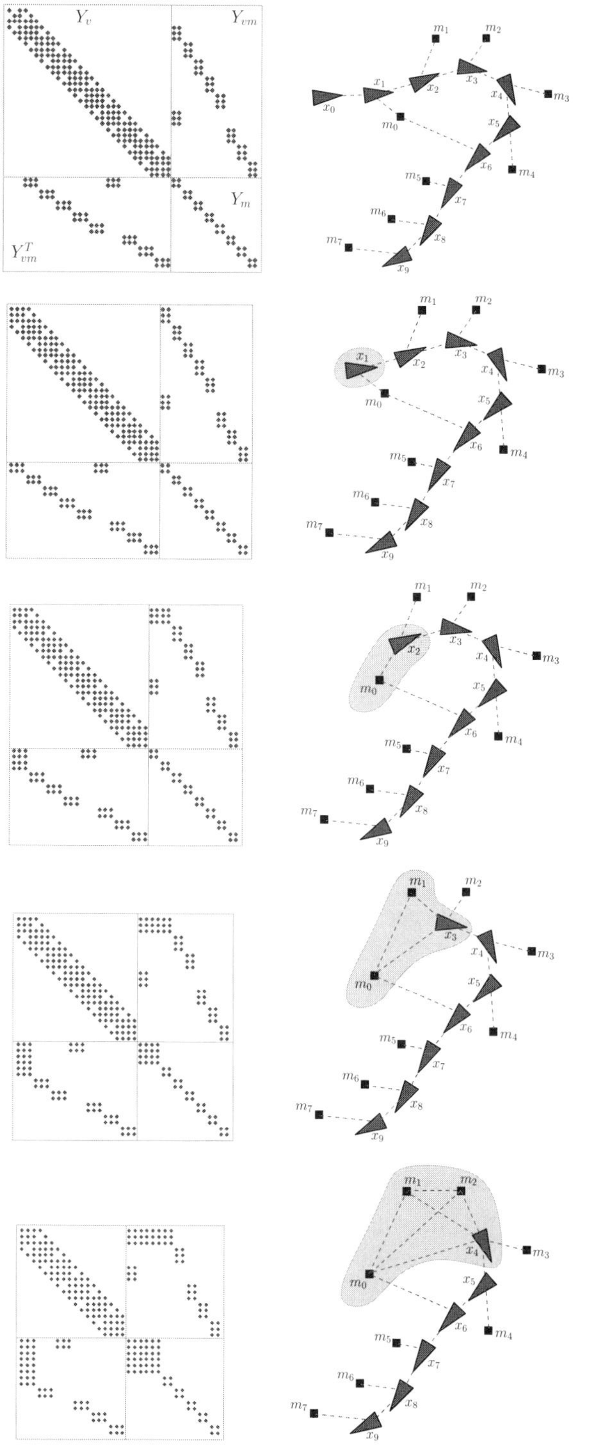

Fig. 2. Illustrates how the information matrix (left column) and MRF (right column) change as poses $\mathbf{x}_0$ (row 2), $\mathbf{x}_1$ (row 3), $\mathbf{x}_2$ (row 4) and $\mathbf{x}_3$ (row 5) are marginalised out. The light grey region indicates the Markov blanket that corresponds to the prior term (11) in Equation (10).

III. Sliding Window SLAM

Each iteration of "standard" EKF-based SLAM provides a maximum a posteriori estimate for the state at the current time step. Modern understanding of this recognises that previous poses have been marginalised out. In contrast, Full-SLAM [2] finds the maximum a posteriori estimate of the entire pose history. This is akin to bundle-adjustment techniques from photogrammetry [8], and has the advantage that more accurate solutions can be found since optimisation is performed over past and future data. It does of course suffer from the problem of growth without bound in the state size. Recently [17] proposed *Sliding Window SLAM*; where optimisation is carried out over a time window of length τ. This aims to capture the advantages of Full-SLAM, but rather than retaining the entire trajectory history, poses older than τ are marginalised out. In our work we take advantage of the optimisation over the trajectory history not only to improve the pose estimates, but crucially in order to allow reversible data association and model selection to take place. Details of this are presented in section IV, but first we review the two governing equations of sliding window SLAM which are: (i) an optimisation step

$$\{\hat{\mathbf{x}}_{t-\tau:t}, \hat{\mathbf{M}}\} = \underset{\{\mathbf{X}_{t-\tau:t}, \mathbf{M}\}}{\arg\max} \left(P(\mathbf{x}_{t-\tau}, \mathbf{M} | \mathbf{z}_{1:t-\tau}, \mathbf{u}_{1:t-\tau}) \times \prod_{j=t-\tau+1}^{t} P(\mathbf{z}_j | \mathbf{x}_j, \mathbf{M}) P(\mathbf{x}_j | \mathbf{x}_{j-1}, \mathbf{u}_j) \right) \quad (10)$$

and (ii) a marginalisation step

$$P(\mathbf{x}_{t-\tau+1}, \mathbf{M} | \mathbf{z}_{1:t-\tau+1}, \mathbf{u}_{1:t-\tau+1}) = \int P(\mathbf{x}_{t-\tau+1}, \mathbf{x}_{t-\tau}, \mathbf{M} | \mathbf{z}_{1:t-\tau+1}, \mathbf{u}_{1:t-\tau+1}) \mathbf{d}\mathbf{x}_{t-\tau}. \quad (11)$$

The term $P(\mathbf{x}_{t-\tau}, \mathbf{M} | \mathbf{z}_{1:t-\tau}, \mathbf{u}_{1:t-\tau})$ is the prior used at time t and is just the posterior at time $t-\tau$ i.e. the distribution of vehicle pose and map at the beginning of the sliding window. In practice this is only recalculated when $t-\tau > 0$, before that time the distribution of initial vehicle pose $P(\mathbf{x}_0)$ is used and the marginalisation step is left out.

A. Marginalisation in Information Form

It is well known that it is possible to decouple states $\mathbf{y}_1$ from a system of equations of the form:

$$\begin{bmatrix} \mathbf{A} & \mathbf{B} \\ \mathbf{B}^T & \mathbf{D} \end{bmatrix} \begin{bmatrix} \mathbf{y}_1 \\ \mathbf{y}_2 \end{bmatrix} = \begin{bmatrix} \mathbf{b}_1 \\ \mathbf{b}_2 \end{bmatrix} \quad (12)$$

using the Schur Complement [8] method. The idea is to pre multiply both sides of the equation with the matrix $[\mathbf{I} \quad \mathbf{0}; -\mathbf{B}^T\mathbf{A}^{-1} \quad \mathbf{I}]$, which results in a system of equations where $\mathbf{y}_2$ can be solved independently of $\mathbf{y}_1$ i.e.

$$\begin{bmatrix} \mathbf{A} & \mathbf{B} \\ \mathbf{0} & \mathbf{D} - \mathbf{B}^T\mathbf{A}^{-1}\mathbf{B} \end{bmatrix} \begin{bmatrix} \mathbf{y}_1 \\ \mathbf{y}_2 \end{bmatrix} = \begin{bmatrix} \mathbf{b}_1 \\ \mathbf{b}_2 - \mathbf{B}^T\mathbf{A}^{-1}\mathbf{b}_1 \end{bmatrix} \quad (13)$$

The term $\mathbf{D} - \mathbf{B}^T\mathbf{A}^{-1}\mathbf{B}$ is known as the Schur Complement and corresponds to the information matrix for the decoupled system. If this system of equations represents a least-squares problem as described in Section II then this is equivalent to marginalising out the random variables $\mathbf{y}_1$. Let us now consider what happens to the structure of $\mathbf{D} - \mathbf{B}^T\mathbf{A}^{-1}\mathbf{B}$ as old poses are marginalised out. Figure 2 shows the effect of marginalising out poses one by one. The first row shows the situation before any marginalisation. The second row corresponds to marginalising out $\mathbf{x}_0$, which results in no change of structure in the information matrix (because no features were observed from this pose) but does introduce a prior on the vehicle state $\mathbf{x}_1$. Then in the third row $\mathbf{x}_1$ has been marginalised out and a link has been introduced between $\mathbf{x}_2$ and $\mathbf{m}_0$, which is also seen in the $\mathbf{Y}_{vm}$ block of the information matrix (this extra link and the prior on $\mathbf{x}_2$ and $\mathbf{m}_0$ is explained by the prior term in Equation (10)). As poses $\mathbf{x}_2$ and $\mathbf{x}_3$ are marginalised out more links are again introduced between the oldest pose and the landmarks; links are also introduced between landmarks that are no longer observed from the oldest pose. In practice we use the prior term (3) in our least-squares formulation to represent the prior term (11) in the sliding window; where we replace $\tilde{\mathbf{x}}_0$ with $\mathbf{y}_2$, $\mathbf{x}_0$ with the poses and landmarks that have a prior and $\mathbf{P}_0$ with $(\mathbf{D} - \mathbf{B}^T\mathbf{A}^{-1}\mathbf{B})^{-1}$. *To maintain probabilistic correctness only equations containing the pose being marginalised out should be included in the system* (12) *and then the modified system* (13) *should be solved for* $\mathbf{y}_2$.

IV. Reversible Data Association

So far the previous sections have covered the necessary background knowledge, let us now introduce the first of our two algorithms. We first relax our assumption of known data association and introduce integer data association parameters $\mathbf{D} \triangleq \{d_1, \ldots, d_t\}$, which assign measurement $\mathbf{z}_t$ to landmark $\mathbf{m}_{d_t}$. By combining sliding window estimation and least-squares with generalised expectation maximisation we can estimate both the continuous state estimates $\{\hat{\mathbf{X}}, \hat{\mathbf{M}}\}$ and the discrete data association parameters $\mathbf{D}$.

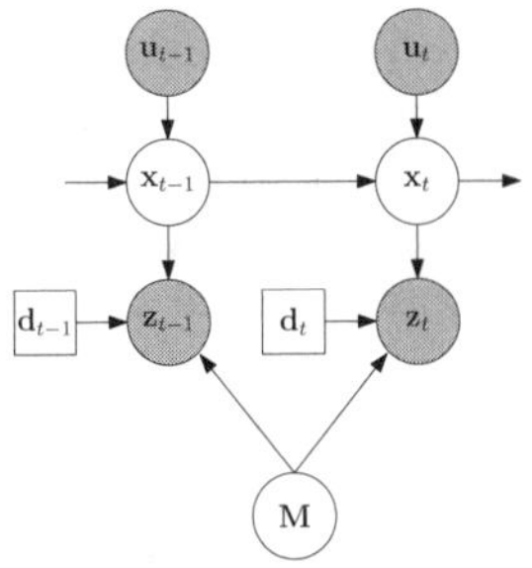

Fig. 3. A Bayesian network representing SLAM with reversible data association (Note:- Square boxes indicate discrete variables).

Figure 3 illustrates the Bayesian network that corresponds to the joint distribution of the relaxed problem:

$$P(\mathbf{X}, \mathbf{M}, \mathbf{D}, \mathbf{U}, \mathbf{Z}) = \quad (14)$$

$$P(\mathbf{x}_0) P(\mathbf{M}) P(\mathbf{D}) \prod_{t=1}^{T} P(\mathbf{z}_t | \mathbf{x}_t, \mathbf{M}, d_t) P(\mathbf{x}_t | \mathbf{x}_{t-1}, \mathbf{u}_t) \quad . \quad (15)$$

What we are really after is the MAP estimate $P(\mathbf{X}, \mathbf{M}|\mathbf{U}, \mathbf{Z})$, whereas what we have is $P(\mathbf{X}, \mathbf{M}, \mathbf{D}, \mathbf{U}, \mathbf{Z})$ where $\mathbf{D}$ is considered a nuisance parameter. One solution would be to marginalise $\mathbf{D}$ out i.e.

$$P(\mathbf{X}, \mathbf{M}|\mathbf{U}, \mathbf{Z}) = \int_{\mathbf{D}} P(\mathbf{X}, \mathbf{M}, \mathbf{D}|\mathbf{U}, \mathbf{Z})d\mathbf{D}.$$

Unfortunately in practice this integral is computationally intractable because the number of permutations of $\mathbf{D}$ grows exponentially with the length of the sliding window. A more tractable solution is to use the expectation maximisation algorithm [10], [5] to estimate $P(\mathbf{X}, \mathbf{M}|\mathbf{U}, \mathbf{Z})$. If we let $\mathbf{\Theta} = \{\mathbf{X}, \mathbf{M}\}$ and $\mathbf{\Psi} = \{\mathbf{U}, \mathbf{Z}\}$ then we would like $P(\mathbf{\Theta}|\mathbf{\Psi})$ as opposed to $P(\mathbf{\Theta}, \mathbf{D}|\mathbf{\Psi})$. The expectation maximisation algorithm achieves this by recursively applying the following two steps:

- E-Step: calculate $P(\mathbf{D}|\mathbf{\Theta}^k, \mathbf{\Psi})$.
- M-Step: $\mathbf{\Theta}^{k+1} = \arg\max_{\mathbf{\Theta}} \left(\int_{\mathbf{D}} P(\mathbf{D}|\mathbf{\Theta}^k, \mathbf{\Psi}) \log P(\mathbf{\Theta}|\mathbf{D}, \mathbf{\Psi})d\mathbf{D}\right)$.

A common simplification that is often applied to make the M-Step even more tractable is the 'winner-take-all' approach also known as classification expectation maximisation [3], [12], which assumes $P(\mathbf{D}|\mathbf{\Theta}^k, \mathbf{\Psi})$ to be a delta function centered on the best value of $\mathbf{D}$, reducing the algorithm to:

- E-Step: $\mathbf{D}^{k+1} = \arg\max_{\mathbf{D}} P(\mathbf{D}|\mathbf{\Theta}^k, \mathbf{\Psi})$.
- M-Step: $\mathbf{\Theta}^{k+1} = \arg\max_{\mathbf{\Theta}} P(\mathbf{\Theta}|\mathbf{D}^{k+1}, \mathbf{\Psi})$.

Finally, it has been shown that it is not necessary to complete the maximisation but that a single step where $P(\mathbf{\Theta}^{k+1}|\mathbf{D}^{k+1}, \mathbf{\Psi}) >= P(\mathbf{\Theta}^k|\mathbf{D}^k, \mathbf{\Psi})$ is not only sufficient for convergence but often improves the rate of convergence [13]. For probabilistic correctness it is necessary to use the joint distribution over landmarks and a single pose during the E-Step i.e. JCBB. In practice this is very slow for large numbers of measurements and so we also include in our results an implementation which makes an extra assumption of landmark independence during the E-Step i.e. chi-squared NN. In practice this method gives a significant improvement over other methods (which do not use reversible data association) without the full cost of JCBB. It is also interesting at this point to draw on the similarity between this approach and iterative closest point; the significant difference is that our method uses the underlying probability distribution (Mahalanobis distances) to find the most likely correspondences as opposed to the closest in a euclidean sense. Algorithm 1 gives a summary of our method.

V. SLAM in Dynamic Environments

We will now introduce our method for SLAM in dynamic environments (SLAMIDE). Let us start by relaxing the problem even further by: (i) introducing model selection parameters $\mathbf{V}_T \triangleq \{v_T^0, \ldots, v_T^k\}$, which consist of a binary indicator variable per landmark taking the value **stationary** or **dynamic** with probability p, $1-p$ respectively; (ii) extending the state vector for each landmark to include velocities $\dot{x}$, $\dot{y}$ and (iii) using the estimated landmark velocities as observations to

Algorithm 1: SLAM with reversible data association.

$\mathbf{P} = \mathbf{P}_0; \mathbf{x}_0 = \tilde{\mathbf{x}}_0; \mathbf{M} = []; \mathbf{D} = [];$
for *t=[0:T]* **do**
  DoVehiclePrediction();
  while $|\delta|_\infty > \epsilon$ **do**
    $\hat{\mathbf{D}}$ = DoDataAssociation();
    AddAnyNewLandmarks();
    Compute $\mathbf{A}$, $\mathbf{\Sigma}$ and $\mathbf{b}$;
    Solve for δ in $\mathbf{A}^T\mathbf{\Sigma}^{-1}\mathbf{A}\delta = \mathbf{A}^T\mathbf{\Sigma}^{-1}\mathbf{b}$;
    $\{\hat{\mathbf{x}}_{t-\tau:t}, \hat{\mathbf{M}}\} = \{\hat{\mathbf{x}}_{t-\tau:t}, \hat{\mathbf{M}}\} + \delta$;
    Compute $\mathbf{P}$ using triangular solve;
  end
  if $t - \tau > 0$ **then**
    Compute $\mathbf{y}_2$ and $\mathbf{D} - \mathbf{B}^T\mathbf{A}^{-1}\mathbf{B}$ using Schur Complement method (see Section III-A);
  end
end

a Hidden Markov Model (HMM) that estimates the probability of an object being dynamic or stationary. Figure 4 is a Bayesian network that shows our formulation of the SLAMIDE problem, where the most significant changes from normal SLAM are:

- The map becomes time dependent $\mathbf{M} \triangleq \{\mathbf{M}_0, \ldots, \mathbf{M}_t\}$.
- Model selection parameters $\mathbf{V}_T \triangleq \{v_T^0, \ldots, v_T^k\}$ are introduced.
- Data association parameters $\mathbf{D} \triangleq \{d_1, \ldots, d_t\}$ are introduced.

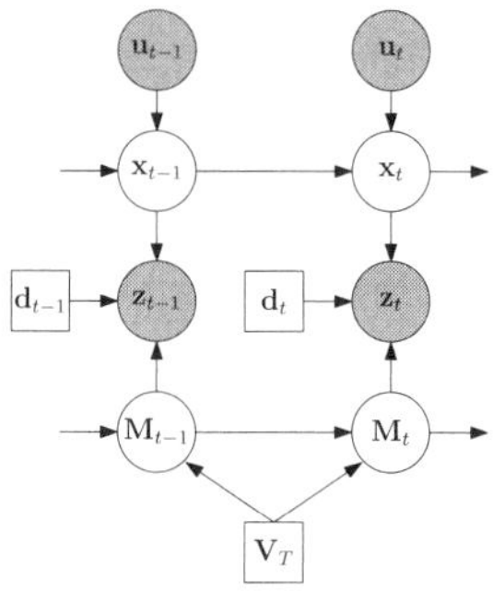

Fig. 4. A Bayesian network representing SLAMIDE (Note:- Square boxes indicate discrete variables).

The corresponding joint distribution $P(\mathbf{X}, \mathbf{M}, \mathbf{D}, \mathbf{V}, \mathbf{U}, \mathbf{Z})$ from Figure 4 is:

$$P(\mathbf{X}, \mathbf{M}, \mathbf{D}, \mathbf{V}, \mathbf{U}, \mathbf{Z}) = P(\mathbf{x}_0)P(\mathbf{M}_0)P(\mathbf{D})P(\mathbf{V}_T)\times \prod_{t=1}^{T} P(\mathbf{z}_t|\mathbf{x}_t, \mathbf{M}_t, d_t)P(\mathbf{x}_t|\mathbf{x}_{t-1}, \mathbf{u}_t)P(\mathbf{M}_t|\mathbf{M}_{t-1}, \mathbf{V}_T), \quad (16)$$

where:

- $P(\mathbf{V}_T)$ is a prior on the model selection parameters.

- $P(\mathbf{M}_t|\mathbf{M}_{t-1}, \mathbf{V}_T)$ is the motion model for the map given the current estimate of the model selection parameters. We use constant position for stationary landmarks and constant velocity with noise in $\dot{x}$ and $\dot{y}$ for dynamic landmarks.

Following the same principles used in our reversible data association method and including the extra nuisance parameter $\mathbf{V}$ we propose the following five steps to solve the optimisation:

1) $\mathbf{D}^{k+1} = \arg\max_{\mathbf{D}} P(\mathbf{D}|\mathbf{\Theta}^k, \mathbf{V}^k, \mathbf{\Psi})$
2) $\mathbf{\Theta}^{k+1} = \arg\max_{\mathbf{\Theta}} P(\mathbf{\Theta}|\mathbf{D}^{k+1}, \mathbf{V}^k, \mathbf{\Psi})$
3) $\mathbf{M}^{k+1'} = \arg\max_{\mathbf{M}} P(\mathbf{M}|\mathbf{X}^{k+1}, \mathbf{D}^{k+1}, \mathbf{V} = \mathbf{dyn}, \mathbf{\Psi})$
4) $\mathbf{V}^{k+1} = \arg\max_{\mathbf{V}} P(\mathbf{V}|\mathbf{X}^{k+1}, \mathbf{M}^{k+1'}, \mathbf{D}^{k+1}, \mathbf{\Psi})$
5) $\mathbf{\Theta}^{k+1} = \arg\max_{\mathbf{\Theta}} P(\mathbf{\Theta}|\mathbf{D}^{k+1}, \mathbf{V}^{k+1}, \mathbf{\Psi})$

Step 1: performs the data association using either NN or JCBB. In practice this is actually also computed at every iteration in steps 2, 3 and 5.

Step 2: is a least-squares optimisation for the vehicle poses and landmark states using the new data association. The main purpose of this optimisation is to refine the predicted vehicle and landmark locations using the new measurements. In practice this step is particularly important if the vehicle prediction is poor (large odometry noise), because large vehicle uncertainty gives rise to an ambiguous situation where it is hard to differentiate between vehicle and landmark motion.

Step 3: optimises for the landmark states assuming all landmarks are dynamic whilst holding the vehicle poses constant. The reason the vehicle poses are held constant is to remove any ambiguity between vehicle and landmark motion. This is reasonable if most of the landmarks maintain their model selection between time steps and hence the $\mathbf{X}^{k+1}$ given by Step 2 is close to the optimal answer.

Step 4: takes the answer from Step 3 and computes the next step of the HMM using a recursive Bayesian filter; where the likelihood model is a Gaussian on the average velocity with σ=2.0m/s and μ=0 and the prior $P(\mathbf{V}_T)$ for landmark j is:

$$P(v_T^j = \mathbf{stationary}) = \begin{cases} 0.6 & \text{if } v_{T-1}^j = \mathbf{stationary}, \\ 0.4 & \text{if } v_{T-1}^j = \mathbf{dynamic}, \end{cases}$$

which is based on $\mathbf{V}_{T-1}$ the model selection parameters chosen at the last time step. Given that the probability $P(v_T^j = \mathbf{stationary}) = p$ and $P(v_T^j = \mathbf{dynamic}) = 1 - p$ we threshold at 0.5 to get a discrete decision on which model to use.

Step 5: is a least-squares optimisation with the new model selection and data association parameters (using the answer from Step 2 as the starting point for optimisation). This step refines the estimate from Step 2 taking into account any changes in model selection to give the final estimate for this time step.

In practice this whole process only requires a few least-squares iterations typically: two in Step 2; one in Step 3 and two or three in Step 5; Step 1 and Step 4 are solved directly.

Map Management: When adding a new landmark we initialise its model selection probability to 0.5 (to reflect the uncertainty in whether it is dynamic or stationary) and add a very weak prior of zero initial velocity; this weak prior is essential to make sure that a landmark's velocity is always observable and hence our system of equations is positive definite. We also remove any dynamic landmarks that are not observed within the sliding window, this is done for two reasons: (i) real world objects do not obey a motion model exactly and so errors accumulate if you predict for too long and (ii) if you continue predicting a dynamic landmark and hence adding noise, then at some point measurements begin to get incorrectly associated to it due to the Mahalanobis test.

VI. Results

We use two simple 2D environments, which cover 400m by 400m, one with 15 landmarks and the other with 20 landmarks. In both environments the vehicle moves between three waypoints at 5m/s using proportional heading control (max. yaw rate 5°/sec) and provides rate-of-turn, forwards velocity and slip as odometry with covariance $\mathbf{Q}$. It has a range bearing sensor with a 360° field-of-view, 400m range and zero mean Gaussian noise added with covariance $\mathbf{R}$. The second environment is used for the dynamic object experiment where we progressively change stationary landmarks to dynamic landmarks, which move between waypoints using the same control scheme, speed and rate-of-turn as the vehicle.

We compare our method using either NN or JCBB for data association against an IEKF with either NN or JCBB. All experiments are 60 time steps long and use: a chi-squared threshold of $\boldsymbol{v}^T\mathbf{S}^{-1}\boldsymbol{v} < 16$; a sliding window length of 6 time steps; odometry noise (all noise quotes are for 1σ) of 0.1m/s on forwards velocity and 0.01m/s on slip; measurement noise of 1m for range and 0.5° for bearing; a maximum number of 8 iterations; the same stopping condition and the same initial vehicle uncertainty. The reason we use such a large chi-squared threshold is because for any significant angular uncertainty linearising the prediction covariance can cause all measurements to fall outside their data association gates.

In order to compare the performance of the algorithms we use two metrics: (i) the percentage of correct data association, which we define to be the ratio of correct associations between time steps w.r.t the number of potential correct associations (×100) and (ii) the percentage of consistent runs where we use the following test to determine whether a run is consistent: compute the Normalised Estimation Error Squared (NEES), which is defined as $D_t^2 = (\mathbf{x}_t - \hat{\mathbf{x}}_t)^T\mathbf{P}_t^{-1}(\mathbf{x}_t - \hat{\mathbf{x}}_t)$ and then for each time step perform the corresponding chi-squared test $D_t^2 \leq \chi^2_{r,1-\alpha}$ where r is the dimension of $\mathbf{x}_t$ and α is a threshold. We take the threshold α to be 0.05 and can then compute the probability of this test failing k times out of n from the binomial distribution $B(n, \alpha)$, which we use to threshold on the number of times the test can fail before we are 99% certain that a run is inconsistent.

We have carried out three Monte-Carlo simulation experiments (where each point on the graphs has been generated from 100 runs):

Figure 5 - Noise in rate-of-turn odometry: Performance was tested without clutter against increasing noise in rate-of-turn odometry with a variance of 1° to 60°. The IEKFJCBB and our RDJCBB both perform perfectly with data association but start becoming inconsistent more often for higher noise levels. As expected our RDNN outperforms the IEKFNN and matches the performance of IEKFJCBB up to around 25° of noise; this is interesting because it shows the RDNN could be used as a faster alternative to IEKFJCBB for medium noise problems.
Figure 6 - Number of clutter measurements: Performance was tested with a noise of 1° for rate-of-turn odometry against increasing clutter from 0 to 100 clutter measurements within the sensor range. This is where the real benefit of reversible data association becomes apparent. All algorithms tested use the same map management scheme, which is to remove landmarks that are not observed for three consecutive time steps after they have been added to the map. In the traditional IEKF this is done by simply removing them from the state vector and covariance matrix (marginalisation); whereas with our scheme if the information is removed before marginalisation i.e. the sliding window is longer than the time required to carry out map management then there is no effect on the estimate. This is clear from Figure 6 as both of our methods maintain their consistency with increasing clutter as opposed to the IEKF based methods which tail off.
Figure 7 - Percentage of dynamic objects: Performance was tested without clutter and with a noise of 1° for rate-of-turn odometry against an increasing percentage of dynamic objects from 0 to 100 percent. The figure clearly shows that using SLAMIDE to include dynamic objects allows us to navigate in regions with dynamic objects. We maintain a good level of consistency up to 90% of dynamic objects at which point the performance degrades until at 100% every run is inconsistent, which is because the system is no longer observable; i.e. there are ambiguities between vehicle and landmark motion.
Timing Results: With 20 measurements per time step and a sliding window length of 6 time steps on a 3.6GHz Pentium 4 the IEKF and SLAM with reversible data association run at approximately 30Hz and SLAMIDE runs at about 3Hz. We believe this can be significantly improved upon as we have yet to fully optimise the code, for instance we currently do a dense solve for $\mathbf{P} = \mathbf{Y}^{-1}$ which is a bottleneck (this could be heavily optimised or possibly avoided completely). Also, once a landmark has been created and passed the map management test, it always remains in the estimate; however, sliding window estimation is constant time if you choose to marginalise out landmarks i.e. maintain a constant state size.

VII. Conclusions and Future Work

We have proposed a method that combines sliding window optimisation and least-squares together with generalised expectation maximisation to do reversible model selection and data association. This allows us to include dynamic objects directly into the SLAM estimate, as opposed to other techniques which typically detect dynamic objects and then either treat them as outliers [20][1] or track them separately [18].

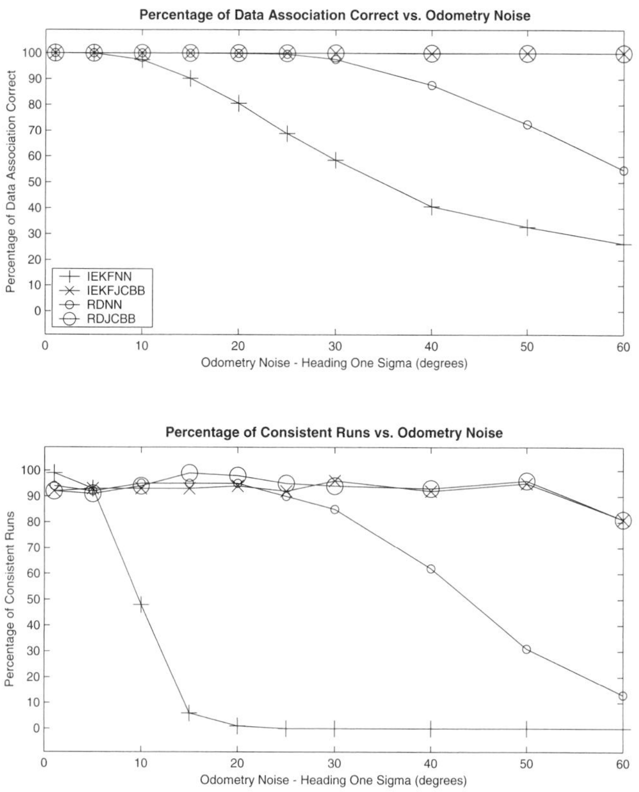

Fig. 5. Comparison for increasing odometry (rate-of-turn) noise.

Our initial simulation results show that: (i) our SLAMIDE algorithm significantly outperforms other methods which treat dynamic objects as clutter; (ii) our method for computing reversible data association remains consistent when other data association methods fail and (iii) our reversible data association provides excellent performance when clutter is present. Aside from simulation we have also successfully run our algorithms on a small set of real radar data with very promising initial results.

The first thing we would like to improve is the use of a Gaussian on average velocity for the likelihood model in our model selection. In practice this works well, however, a technique that does not introduce an extra threshold would be preferable, ideally it would work directly on the estimated distribution and the innovation sequence over the sliding window. Secondly, we have observed that the length of the sliding window in our experiments is often longer than it needs to be. We are currently considering an active threshold based on the convergence of parameters within the sliding window to select an appropriate length on the fly.

In summary, we have developed a method for robustly including dynamic objects directly into the SLAM estimate as opposed to treating them as outliers. The benefit of including dynamic objects in a single framework is clear for navigation and path planning; interestingly it also helps with localisation in highly dynamic environments, especially during short peri-

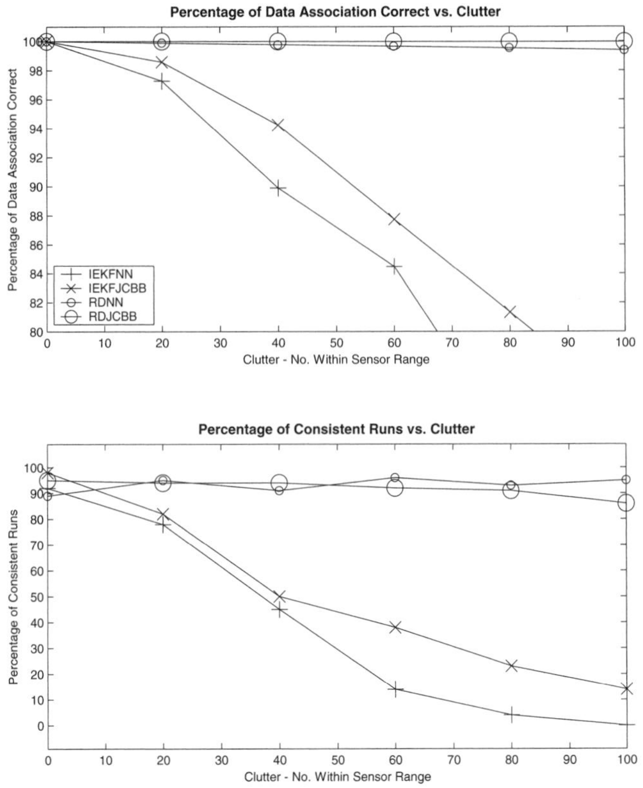

Fig. 6. Comparison for increasing clutter.

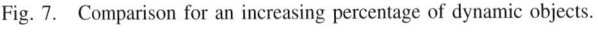

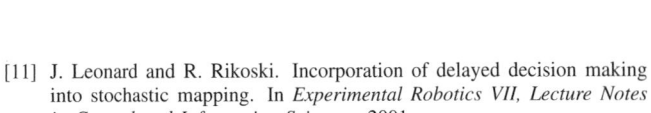

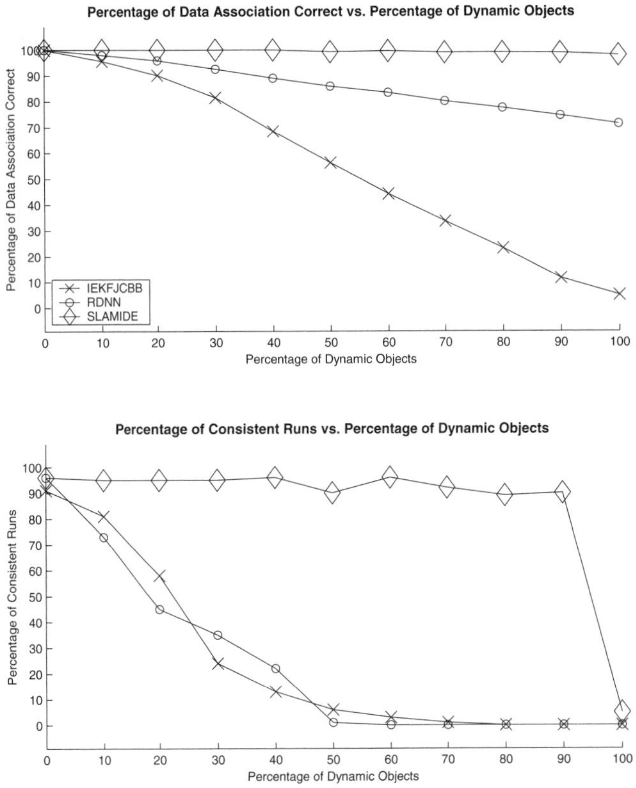

Fig. 7. Comparison for an increasing percentage of dynamic objects.

ods of time without any stationary landmark observations. Our longer term goal is to build a navigational aid using sliding window SLAM with reversible data association and reversible model selection to fuse data from a high performance pan tilt camera, marine radar, GPS and compass.

References

[1] D. Hähnel, R. Triebel, W. Burgard, and S. Thrun. Map building with mobile robots in dynamic environments. In *Proceedings of the IEEE International Conference on Robotics and Automation (ICRA)*, 2003.

[2] S. Thrun, W. Burgard and D. Fox. *Probabilistic Robotics.* MIT Press, 2005.

[3] G. Celeux and G. Govaert. A classification EM algorithm for clustering and two stochastic versions. *Comput. Stat. Data Anal.*, 14(3):315–332, 1992.

[4] T. Davis. *Direct Methods for Sparse Linear Systems.* SIAM, 2006.

[5] F. Dellaert. The expectation maximization algorithm. Technical Report GIT-GVU-02-20, Georgia Institute of Technology, 2002.

[6] F. Dellaert and M. Kaess. Square root SAM. *International Journal of Robotics Research*, 2006.

[7] W. E. L. Grimson. *Object Recognition by Computer: The Role of Geometric Constraints.* MIT Press, Cambridge MA, 1990.

[8] B. Triggs, P. McLauchlan, R. Hartley and A. Fitzgibbon. Bundle adjustment – a modern synthesis. In *Vision Algorithms: Theory and Practice*, volume 1883 of *Lecture Notes in Computer Science*, pages 298–372. Springer-Verlag, 2000.

[9] Y. Bar-Shalom, T. Kirubarajan and X. R. Li. *Estimation with Applications to Tracking and Navigation.* John Wiley & Sons, Inc., New York, NY, USA, 2002.

[10] A. P. Dempster, N. M. Laird and D. B. Rubin. Maximum likelihood from incomplete data via the EM algorithm. *J. Royal Stat. Society*, 1977.

[11] J. Leonard and R. Rikoski. Incorporation of delayed decision making into stochastic mapping. In *Experimental Robotics VII, Lecture Notes in Control and Information Sciences*, 2001.

[12] M. Meila and D. Heckerman. An experimental comparison of several clustering and initialization methods. In *Proceedings of Fourteenth Conference on Uncertainty in Artificial Intelligence*, 1998.

[13] R. Neal and G. Hinton. A view of the EM algorithm that justifies incremental, sparse, and other variants. In M. I. Jordan, editor, *Learning in Graphical Models.* Kluwer, 1998.

[14] J. Neira and J. Tardos. Data association in stochastic mapping using the joint compatibility test. In *IEEE Trans. on Robotics and Automation*, 2001.

[15] W. S. Wijesoma, L. D. L. Perera and M. D. Adams. Toward multidimensional assignment data association in robot localization and mapping. In *IEEE Transactions on Robotics*, April 2006.

[16] L. L. Scharf and L. T. McWhorter. Geometry of the Cramer-Rao bound. *Signal Process.*, 31(3):301–311, 1993.

[17] G. Sibley, G. S. Sukhatme and Larry Matthies. Constant time sliding window filter SLAM as a basis for metric visual perception. In *Proceedings of the IEEE International Conference on Robotics and Automation (ICRA)*, April 2007.

[18] C. C. Wang, C. Thorpe and S. Thrun. Online simultaneous localization and mapping with detection and tracking of moving objects: Theory and results from a ground vehicle in crowded urban areas. In *Proceedings of the IEEE International Conference on Robotics and Automation (ICRA)*, Taipei, Taiwan, September 2003.

[19] R. Eustice, H. Singh, J. Leonard, M. Walter and R. Ballard. Visually navigating the RMS Titanic with SLAM information filters. In *Proceedings of Robotics: Science and Systems*, Cambridge, USA, June 2005.

[20] D. Wolf and G. S. Sukhatme. Online simultaneous localization and mapping in dynamic environments. In *Proceedings of the IEEE International Conference on Robotics and Automation (ICRA)*, New Orleans, Louisiana, April 2004.

This work is sponsored by Servowatch Systems Ltd.

Sliding Mode Formation Tracking Control of a Tractor and Trailer - Car System

Fabio Morbidi
Dipartimento di Ingegneria dell'Informazione
University of Siena
Via Roma 56, 53100 Siena, Italy
Email: morbidi@dii.unisi.it

Domenico Prattichizzo
Dipartimento di Ingegneria dell'Informazione
University of Siena
Via Roma 56, 53100 Siena, Italy
Email: prattichizzo@dii.unisi.it

Abstract— **In this paper a new leader-follower formation of nonholonomic mobile robots is studied. The follower is a car-like vehicle and the leader is a tractor pulling a trailer. The leader moves along assigned trajectories and the follower is to maintain a desired distance and orientation to the trailer. A sliding mode control scheme is proposed for asymptotically stabilizing the vehicles to a time-varying desired formation. The attitude angles of the follower and the tractor are estimated via global exponential observers based on the invariant manifold technique. Simulation experiments illustrate the theory and show the effectiveness of the proposed formation controller and nonlinear observers.**

I. Introduction

Last years have seen a growing interest on control of formations of autonomous robots. This trend has been supported by the recent technological advances on computation and communication capabilities and by the observation that multiagent systems can perform tasks beyond the ability of individual vehicles. By formation control we simply mean the problem of controlling the relative position and orientation of the robots in a group while allowing the group to move as a whole. In this respect, research dealt with ground vehicles [9], [11], [18], surface and underwater autonomous vehicles (AUVs) [10], [12], unmanned aerial vehicles (UAVs) [4], [16], microsatellite clusters [2], [24].

One of the most important approaches to formation control is *leader-following*. A robot of the formation, designed as the leader, moves along a predefined trajectory while the other robots, the followers, are to maintain a desired posture (distance and orientation) to the leader.
Leader-follower architectures are known to have poor disturbance rejection properties. In addition, the over-reliance on a single agent for achieving the goal may be undesirable especially in adverse conditions. Nevertheless this approach is extremely simple since a reference trajectory is clearly defined by the leader and the internal formation stability is induced by stability of the individual vehicles' control laws.

The following common features can be identified in the literature concerning leader-follower formation control of mobile robots:

Kinematic models: Unicycle models are the most common in the literature [6], [29]. Car-like vehicles are frequent in papers dealing with vehicular systems and control of automated road vehicles (see, e.g. [22] and the references therein).

Formation shape: Rigid formations (i.e., formations where the inter-vehicle desired parameters are constant in time) are frequent in the literature (see the references above). Switching between different rigid formations has been studied recently in [7], [25].

Formation control: The most common formation control strategies are feedback linearization [7], [20], dynamic feedback linearization [19], [30], backstepping [17].

State estimation: The state of the formation is frequently supposed to be known. Otherwise standard nonlinear observers are used: e.g. the extended Kalman filter [7], [20] and recently the unscented Kalman filter [19].

In this paper a new challenging leader-follower formation of nonholonomic mobile robots is studied. The follower is a car-like vehicle and the leader is a multibody mobile robot: a tractor (a car-like vehicle) pulling a trailer. The leader moves along an assigned trajectory while the follower is to maintain a desired distance and orientation to the trailer.
The choice of this specific formation was motivated by a possible real-world application: the control of truck-trailer/car platoons in automated highway systems (AHSs) [1], [26]. Besides the applicative example, it is expected that the proposed multi-robot scheme could be of theoretical interest in nonholonomic systems research.

During the last three decades since [27], variable structure systems (VSS) and sliding mode control (SMC) (that plays a dominant role in VSS theory), have attracted the control research community worldwide (see e.g., [8], [14], [21] and the references therein). One of the distinguishing features of sliding mode is the discontinuous nature of the control action whose primary function is to switch between two different system structures such that a new type of system motion, called sliding mode, exists in a manifold. This peculiar characteristic results in excellent system performance which includes insensitivity to parametric uncertainty and external disturbances.
Sliding mode control has been largely used in the industrial electronic area (induction motors, electric drives [28]). Recently, it has been applied to trajectory tracking of nonholonomic vehicles [5], [31]. Nevertheless, only few

papers [13], [23], dealt with sliding mode control for robot formations.

In the present paper we propose a globally asymptotically stable sliding mode formation tracking controller for the tractor and trailer - car system. Differently from the literature, the desired leader-follower formation is allowed to vary in time arbitrarily. Actually, any continuous distance and orientation functions can be used to define the formation shape.

Finally, according to the invariant manifold technique proposed in [15], global exponential observers of the attitude angle of the follower and the tractor are designed. These observers are among the first applications of the theory developed by Karagiannis and Astolfi. They revealed simple to implement and they exhibited good performances in the simulation experiments, thus confirming that the invariant manifold technique is a viable alternative to standard nonlinear observer design strategies.

The rest of the paper is organized as follows. Section II is devoted to the problem formulation. In Section III the sliding mode formation tracking controller is designed. In Section IV the nonlinear observers of the attitude angles are presented. In Section V simulation experiments with noisy data illustrate the theory and show the closed-loop system performance. In Section VI the major contributions of the paper are summarized and future research lines are highlighted.

Notation: The following notation is used through the paper: $\mathbb{R}^+ = \{x \in \mathbb{R} \mid x > 0\}$, $\mathbb{R}_0^+ = \{x \in \mathbb{R} \mid x \geq 0\}$.
$\forall x \in \mathbb{R}$, $\text{sign}(x) = 1$ if $x > 0$, $\text{sign}(x) = 0$ if $x = 0$ and $\text{sign}(x) = -1$ if $x < 0$.
$\forall \mathbf{s} = (s_1, s_2)^T \in \mathbb{R}^2$, $\text{sign}(\mathbf{s}) = (\text{sign}(s_1), \text{sign}(s_2))^T$, $\|\mathbf{s}\| = \sqrt{s_1^2 + s_2^2}$, $\text{diag}(s_1, s_2) = \begin{pmatrix} s_1 & 0 \\ 0 & s_2 \end{pmatrix}$.

II. Problem Formulation

The leader-follower setup considered in the paper is presented in Fig. 1. The *follower* F is a car with rear wheels aligned with the vehicle and front wheels allowed to spin about the vertical axis. The kinematic model is,

$$\begin{cases} \dot{x}_F = v_F \cos\theta_F \\ \dot{y}_F = v_F \sin\theta_F \\ \dot{\theta}_F = \dfrac{v_F}{\ell_1} \tan\alpha_F \\ \dot{\alpha}_F = \omega_F \end{cases} \tag{1}$$

where (x_F, y_F) are the coordinates of the midpoint of the rear driving wheels, θ_F is the angle of the car body with respect to the x-axis, α_F the steering angle with respect to the car body, ℓ_1 the wheelbase of the car and v_F and ω_F are respectively the forward velocity of the rear wheels and the steering velocity of the car. The *leader* L is an articulated vehicle, a tractor (car) pulling a trailer.

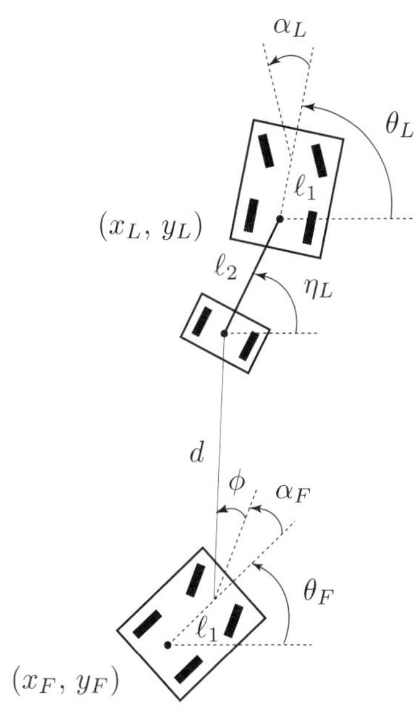

Fig. 1. Leader-follower setup.

The kinematic model is,

$$\begin{cases} \dot{x}_L = v_L \cos\theta_L \\ \dot{y}_L = v_L \sin\theta_L \\ \dot{\theta}_L = \dfrac{v_L}{\ell_1} \tan\alpha_L \\ \dot{\eta}_L = \dfrac{v_L}{\ell_2} \sin(\theta_L - \eta_L) \\ \dot{\alpha}_L = \omega_L \end{cases} \tag{2}$$

where x_L, y_L, θ_L and α_L are the car part variables of the leader, η_L is the angle of the trailer with respect to the x-axis and ℓ_2 is the distance between the rear wheels of the tractor and the wheels of the trailer. Hereafter we will assume that:

Hypothesis 1: The steering angle of the vehicles is bounded, namely $\alpha_F, \alpha_L \in [-\pi/3, \pi/3]$.

Hypothesis 2: The possible configurations of the tractor and trailer are restricted by a mechanical stop on the revolute joint connecting the two bodies, hence $\eta_L \in (0, \pi)$.

With reference to Fig. 1, the following definition introduces the notion of leader-follower formation used in this paper.

Definition 1: Let $d(t) : \mathbb{R}_0^+ \to \mathbb{R}^+$, $\phi(t) : \mathbb{R}_0^+ \to (-\pi/2, \pi/2)$ and $d(t), \phi(t) \in \mathcal{C}^1$, be two given functions. We say that L and F make a (d, ϕ)-formation with leader L at time t, if,

$$\begin{aligned} \begin{pmatrix} x_L - \ell_2 \cos\eta_L \\ y_L - \ell_2 \sin\eta_L \end{pmatrix} - \begin{pmatrix} x_F + \ell_1 \cos\theta_F \\ y_F + \ell_1 \sin\theta_F \end{pmatrix} &= \\ = d \begin{pmatrix} \cos(\theta_F + \alpha_F + \phi) \\ \sin(\theta_F + \alpha_F + \phi) \end{pmatrix} \end{aligned} \tag{3}$$

and, simply, that L and F make a (d,ϕ)-formation (with leader L) if (3) holds for any $t \geq 0$.
d is the distance between the midpoint of the wheels of the trailer and the midpoint of the front wheels of the follower and ϕ is the visual angle of the trailer from the follower, minus the steering angle α_F (see Fig. 1).

Note that in order to avoid collisions between the vehicles, a sufficient condition is that $d(t) \geq 2\ell_1 + \ell_2$ for any $t \geq 0$.

III. Sliding mode formation tracking control

In this section we design a sliding mode controller for globally asymptotically stabilizing the vehicles to the desired (d,ϕ)-formation. We here recall that the leader moves along an assigned trajectory and consequently the vector $(v_L,\, \omega_L)^T$ is known.

Proposition 1: According to Hypotheses 1, 2 and Definition 1, introduce the error vector

$$\begin{pmatrix} e_1 \\ e_2 \end{pmatrix} = \begin{pmatrix} x_L - \ell_2 \cos\eta_L \\ y_L - \ell_2 \sin\eta_L \end{pmatrix} - \begin{pmatrix} x_F + \ell_1 \cos\theta_F \\ y_F + \ell_1 \sin\theta_F \end{pmatrix} - d \begin{pmatrix} \cos(\theta_F + \alpha_F + \phi) \\ \sin(\theta_F + \alpha_F + \phi) \end{pmatrix} \quad (4)$$

and define the sliding surfaces

$$\begin{aligned} s_1 &= \dot e_1 + k_1\, e_1 \\ s_2 &= \dot e_2 + k_2\, e_2 \end{aligned} \quad (5)$$

where $k_1,\, k_2$ are positive constants. The control law

$$\begin{pmatrix} \dot v_F \\ \dot\omega_F \end{pmatrix} = \mathbf{A}^{-1}\Big(-\mathbf{B}\begin{pmatrix} \dot v_L \\ v_L^2 \\ \dot v_L \end{pmatrix} - \mathbf{C}\begin{pmatrix} v_F \\ \omega_F \\ v_F^2 \\ \omega_F^2 \\ v_F\,\omega_F \end{pmatrix} - \mathbf{D} - \mathbf{F}\,\mathrm{sign}(\mathbf{s})\Big) \quad (6)$$

globally asymptotically stabilizes the sliding surfaces to zero. The matrices **A**, **B**, **C**, **F** and the vector **D** are defined in the proof given below.
Proof: Define $\lambda_F = \theta_F + \alpha_F + \phi$ and $\gamma_L = \theta_L - \eta_L$.
Let compute $\dot s_1$ and $\dot s_2$ from (5),

$$\begin{aligned} \dot s_1 = {} & \ddot x_L + \ell_2 \cos\eta_L\, \dot\eta_L^2 + \ell_2 \sin\eta_L\, \ddot\eta_L - \ddot x_F + \ell_1 \cos\theta_F\, \dot\theta_F^2 \\ & + \ell_1 \sin\theta_F\, \ddot\theta_F - \ddot d \cos\lambda_F + 2\dot d \sin\lambda_F\, \dot\lambda_F + d\cos\lambda_F\, \dot\lambda_F^2 \\ & + d\sin\lambda_F\, \ddot\lambda_F + k_1\big(\dot x_L + \ell_2 \sin\eta_L\, \dot\eta_L - \dot x_F + \ell_1 \sin\theta_F\, \dot\theta_F \\ & - \dot d \cos\lambda_F + d \sin\lambda_F\, \dot\lambda_F\big), \end{aligned}$$

$$\begin{aligned} \dot s_2 = {} & \ddot y_L + \ell_2 \sin\eta_L\, \dot\eta_L^2 - \ell_2 \cos\eta_L\, \ddot\eta_L - \ddot y_F + \ell_1 \sin\theta_F\, \dot\theta_F^2 \\ & - \ell_1 \cos\theta_F\, \ddot\theta_F - \ddot d \sin\lambda_F - 2\dot d \cos\lambda_F\, \dot\lambda_F + d \sin\lambda_F\, \dot\lambda_F^2 \\ & - d\cos\lambda_F\, \ddot\lambda_F + k_2\big(\dot y_L - \ell_2 \cos\eta_L\, \dot\eta_L - \dot y_F - \ell_1 \cos\theta_F\, \dot\theta_F \\ & - \dot d \sin\lambda_F - d\cos\lambda_F\, \dot\lambda_F\big). \end{aligned}$$

Replacing the derivatives according to (1) and (2) and writing the resulting expression in a matrix form, we obtain,

$$\dot{\mathbf{s}} = \mathbf{A}\begin{pmatrix} \dot v_F \\ \dot\omega_F \end{pmatrix} + \mathbf{B}\begin{pmatrix} \dot v_L \\ v_L^2 \\ \dot v_L \end{pmatrix} + \mathbf{C}\begin{pmatrix} v_F \\ \omega_F \\ v_F^2 \\ \omega_F^2 \\ v_F\,\omega_F \end{pmatrix} + \mathbf{D}$$

where

$$\mathbf{A} = \begin{pmatrix} -\cos\theta_F + \tan\alpha_F\,(\sin\theta_F + \frac{d}{\ell_1}\sin\lambda_F) & d\sin\lambda_F \\ -\sin\theta_F - \tan\alpha_F\,(\cos\theta_F + \frac{d}{\ell_1}\cos\lambda_F) & -d\cos\lambda_F \end{pmatrix}.$$

The components of the matrices $\mathbf{B} \in \mathbb{R}^{2\times 3}$ and $\mathbf{C} \in \mathbb{R}^{2\times 5}$ are

$\mathbf{B}_{11} = k_1 \cos(\theta_L - \gamma_L)\cos\gamma_L$

$\mathbf{B}_{12} = \frac{\tan\alpha_L}{\ell_1}\cos(\theta_L - \gamma_L)\sin(2\,\theta_L - \gamma_L) + \frac{\sin\gamma_L}{\ell_2}\sin(2\gamma_L - \theta_L)$

$\mathbf{B}_{13} = \cos(\theta_L - \gamma_L)\cos\gamma_L$

$\mathbf{B}_{21} = k_2 \sin(\theta_L - \gamma_L)\cos\gamma_L$

$\mathbf{B}_{22} = -\frac{\tan\alpha_L}{\ell_1}\sin(\theta_L - \gamma_L)\sin\gamma_L + \frac{\sin\gamma_L}{\ell_2}\cos(2\gamma_L - \theta_L)$

$\mathbf{B}_{23} = \sin(\theta_L - \gamma_L)\cos\gamma_L,$

$\mathbf{C}_{11} = \tan\alpha_F\big(\frac{\sin\lambda_F(2\dot d + k_1 d) + 2d\dot\phi\cos\lambda_F}{\ell_1} + k_1\sin\theta_F\big) - k_1\cos\theta_F$

$\mathbf{C}_{12} = \sin\lambda_F(2\dot d + dk_1) + 2d\dot\phi\cos\lambda_F$

$\mathbf{C}_{13} = \frac{\sin\theta_F\tan\alpha_F}{\ell_1} + \frac{\tan^2\alpha_F}{\ell_1^2}(\ell_1\cos\theta_F + d\cos\lambda_F)$

$\mathbf{C}_{14} = d\cos\lambda_F$

$\mathbf{C}_{15} = \frac{1}{\ell_1\cos^2\alpha_F}(\ell_1\sin\theta_F + d\sin\lambda_F) + \frac{2d}{\ell_1}\cos\lambda_F\tan\alpha_F$

$\mathbf{C}_{21} = -\tan\alpha_F\big(\frac{\cos\lambda_F(2\dot d + k_2 d) - 2d\dot\phi\sin\lambda_F}{\ell_1} + k_2\cos\theta_F\big) - k_2\sin\theta_F$

$\mathbf{C}_{22} = -\cos\lambda_F(2\dot d + dk_2) + 2d\,\dot\phi\sin\lambda_F$

$\mathbf{C}_{23} = -\frac{\cos\theta_F\tan\alpha_F}{\ell_1} + \frac{\tan^2\alpha_F}{\ell_1^2}(\ell_1\sin\theta_F + d\sin\lambda_F)$

$\mathbf{C}_{24} = d\sin\lambda_F$

$\mathbf{C}_{25} = \frac{1}{\ell_1\cos^2\alpha_F}(\ell_1\cos\theta_F - d\cos\lambda_F) + \frac{2d}{\ell_1}\sin\lambda_F\tan\alpha_F$

and the vector **D** is given by,

$$\mathbf{D} = \begin{pmatrix} -\ddot d + d\dot\phi^2 - k_1\dot d & 2\dot d\dot\phi + d\ddot\phi + k_1 d\dot\phi \\ -2\dot d\dot\phi - d\ddot\phi - k_2 d\dot\phi & -\ddot d + d\dot\phi^2 - k_1\dot d \end{pmatrix}\begin{pmatrix} \cos\lambda_F \\ \sin\lambda_F \end{pmatrix}.$$

Consider the candidate Lyapunov function

$$V = \frac{1}{2}\mathbf{s}^T\mathbf{s}$$

whose derivative along the system trajectories is

$$\dot V = \mathbf{s}^T\Big(\mathbf{A}\begin{pmatrix} \dot v_F \\ \dot\omega_F \end{pmatrix} + \mathbf{B}\begin{pmatrix} \dot v_L \\ v_L^2 \\ \dot v_L \end{pmatrix} + \mathbf{C}\begin{pmatrix} v_F \\ \omega_F \\ v_F^2 \\ \omega_F^2 \\ v_F\,\omega_F \end{pmatrix} + \mathbf{D}\Big). \quad (7)$$

Substituting the control law (6) in (7), with $\mathbf{F} = \mathrm{diag}(f_1, f_2)$, $f_1, f_2 \in \mathbb{R}^+$, we obtain,

$$\dot{V} = -f_1|s_1| - f_2|s_2|$$

that is strictly less than zero for all $\mathbf{s} \neq \mathbf{0}$. Therefore (6) globally asymptotically stabilizes the sliding surfaces to zero. ■

Remark 1: The proposed control law (6) becomes singular when

$$\det(\mathbf{A}) = \frac{d\cos\phi}{\cos\alpha_F} = 0\,.$$

From Hypothesis 1 and the assumptions in Definition 1, $\det(\mathbf{A}) \neq 0$ for any $t \geq 0$, and consequently the proposed control law has no singularities.

Remark 2: Note that since control (6) explicitly depends on trailer's parameters (i.e., ℓ_2, η_L), the trajectory of the follower does not coincide in general with that of car following another car (the tractor). The two paths become similar when the curvature of leader's trajectory is small or the trailer is rigidly connected to the tractor.

IV. Observers Design

In this section, according to the invariant manifold technique introduced in [15], we design global exponential observers of the attitude angles θ_F and θ_L based on the measurement vectors $(x_F, y_F, \alpha_F)^T$ and $(x_L, y_L, \alpha_L)^T$, respectively.
An alternative observer of θ_L only based on $\gamma_L = \theta_L - \eta_L$, the angle between the cable connecting the tractor to the trailer and the trailer longitudinal axis, is also proposed.

A. Estimation of the angle θ_F

Proposition 2: Consider model (1) and assume that the set $\mathcal{Q} = \{t \in \mathbb{R}_0^+ : v_F(t) = 0\}$ is finite. Define the system,

$$\begin{aligned}\dot{\xi}_1 = &-(\xi_2 + \Lambda\, y_F\, v_F)\,\frac{v_F}{\ell_1}\,\tan\alpha_F - \\ &- \Lambda\, v_F^2\,(\xi_1 + \Lambda\, x_F\, v_F) - \Lambda\, x_F\,\dot{v}_F\end{aligned} \tag{8}$$

$$\begin{aligned}\dot{\xi}_2 = &(\xi_1 + \Lambda\, x_F\, v_F)\,\frac{v_F}{\ell_1}\,\tan\alpha_F - \\ &- \Lambda\, v_F^2\,(\xi_2 + \Lambda\, y_F\, v_F) - \Lambda\, y_F\,\dot{v}_F\end{aligned} \tag{9}$$

where Λ is a positive gain. Then, for any initial condition $(\xi_1(0),\, \xi_2(0))^T$,

$$\hat{\theta}_F = \arctan\left(\frac{\xi_2 + \Lambda\, y_F\, v_F}{\xi_1 + \Lambda\, x_F\, v_F}\right) \tag{10}$$

exponentially converges to the actual attitude angle θ_F.
Proof: Define the error variables,

$$z_1 = -\xi_1 - \Lambda\, x_F\, v_F + \cos\theta_F$$
$$z_2 = -\xi_2 - \Lambda\, y_F\, v_F + \sin\theta_F.$$

Their derivatives are,

$$\begin{aligned}\dot{z}_1 &= -\tfrac{v_F}{\ell_1}\sin\theta_F\tan\alpha_F - \dot{\xi}_1 - \Lambda\, v_F^2\cos\theta_F - \Lambda\, x_F\,\dot{v}_F \\ \dot{z}_2 &= \tfrac{v_F}{\ell_1}\cos\theta_F\tan\alpha_F - \dot{\xi}_2 - \Lambda\, v_F^2\sin\theta_F - \Lambda\, y_F\,\dot{v}_F.\end{aligned} \tag{11}$$

Substituting (8) and (9) in (11), we obtain,

$$\begin{aligned}\dot{z}_1 &= -\tfrac{v_F}{\ell_1}\, z_2\tan\alpha_F - \Lambda\, v_F^2\, z_1 \\ \dot{z}_2 &= \tfrac{v_F}{\ell_1}\, z_1\tan\alpha_F - \Lambda\, v_F^2\, z_2.\end{aligned} \tag{12}$$

Define the candidate Lyapunov function,

$$V = \frac{1}{2}\,(z_1^2 + z_2^2)$$

whose time derivative along (12) satisfies,

$$\dot{V} = -2\,\Lambda\, v_F^2\, V.$$

Since $\mathcal{Q}$ is finite, then necessarily there is a time instant $\bar{t}$ such that $\dot{V}(t) \leq -\varepsilon(t)\,V(t)$, $\varepsilon(t) \in \mathbb{R}^+$, $\forall t > \bar{t}$ and this concludes the proof. ■

B. Estimation of the angle θ_L

In order to estimate θ_L, an analogous observer can be designed by simply referring equations (8), (9), (10) to the leader's kinematics. This observer will be used in the simulation experiments.

An alternative observer of the angle θ_L is here introduced. It is not an exponential estimator of θ_L, but since it depends on a single measured variable, γ_L, is possibly less sensitive to disturbances.

Proposition 3: Consider model (2) and the system,

$$\dot{\xi}_3 = -\frac{v_L}{\ell_2}\,\xi_4\,\sin\gamma_L \tag{13}$$

$$\dot{\xi}_4 = \frac{v_L}{\ell_2}\,\xi_3\,\sin\gamma_L. \tag{14}$$

Then, for any $(\xi_3(0),\, \xi_4(0))^T$ such that $\hat{\eta}_L(0) = \eta_L(0)$,

$$\hat{\theta}_L = \gamma_L + \arctan\left(\frac{\xi_4}{\xi_3}\right)$$

gives an estimate of the angle θ_L.
Proof: The proof is analogous to that of Proposition 2. Define the error variables,

$$z_3 = -\xi_3 + \cos\eta_L$$
$$z_4 = -\xi_4 + \sin\eta_L.$$

Their derivatives are,

$$\begin{aligned}\dot{z}_3 &= -\tfrac{v_L}{\ell_2}\sin\eta_L\sin\gamma_L - \dot{\xi}_3 \\ \dot{z}_4 &= \tfrac{v_L}{\ell_2}\cos\eta_L\sin\gamma_L - \dot{\xi}_4.\end{aligned} \tag{15}$$

Substituting (13) and (14) in (15), we obtain,

$$\begin{aligned}\dot{z}_3 &= -\tfrac{v_L}{\ell_2}\, z_4\sin\gamma_L \\ \dot{z}_4 &= \tfrac{v_L}{\ell_2}\, z_3\sin\gamma_L.\end{aligned} \tag{16}$$

Define the candidate Lyapunov function,

$$V = \frac{1}{2}\,(z_3^2 + z_4^2).$$

Its derivative along (16) is $\dot{V} = 0$ from which the result follows. ■

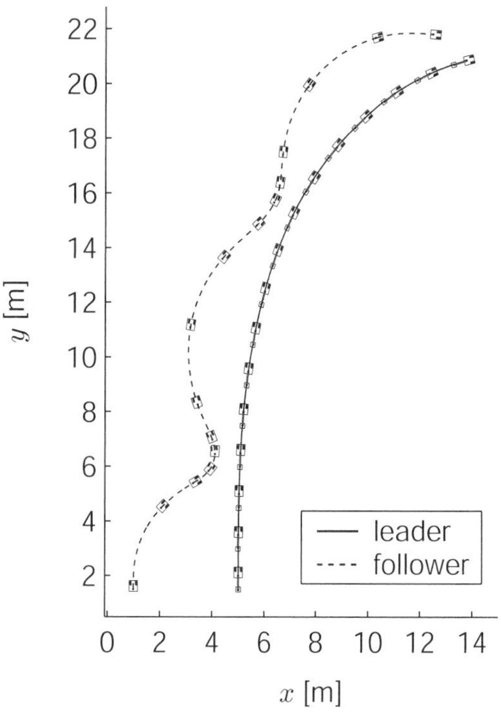

(a)

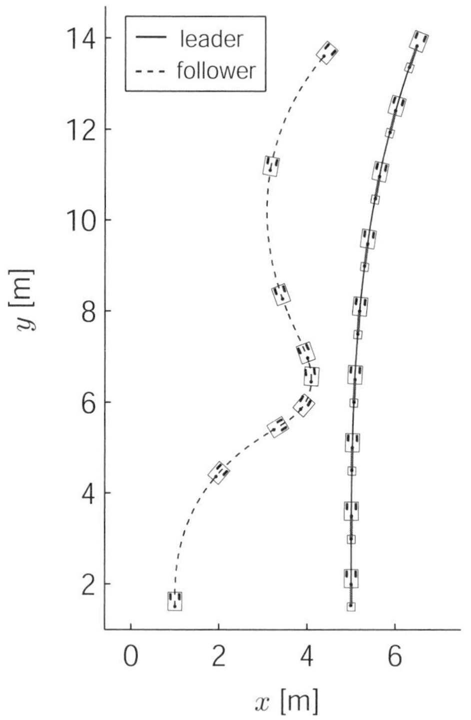

(b)

Fig. 2. Simulation I (ideal measurements): (a) Trajectory of the leader and the follower; (b) Trajectory of the vehicles in the first 8 seconds of the simulation.

Remark 3: Note that the exponential observer of θ_L does not use the angle γ_L. Nevertheless γ_L needs to be measured in order to compute the sliding surfaces s_1, s_2 and matrix $\mathbf{B}$ in (6). The same conclusions hold for the observer in Proposition 3 and vector $(x_L, y_L, \alpha_L)^T$.

V. Simulation Results

Two simulation experiments were carried out to evaluate the performance of the sliding mode controller and the nonlinear observers presented in Sections III and IV. The first simulation refers to the case of ideal measurements. In the second simulation the vectors $(x_F, y_F, \alpha_F)^T$, $(x_L, y_L, \alpha_L)^T$ and γ_L are supposed to be corrupted by zero mean gaussian noise with variance 0.5×10^{-3}. The initial conditions of the leader and the follower are, $x_L(0) = 5\,\mathrm{m}$, $y_L(0) = 2\,\mathrm{m}$, $\theta_L(0) = \pi/2\,\mathrm{rad}$, $\alpha_L(0) = 0\,\mathrm{rad}$, $\eta_L(0) = \pi/2\,\mathrm{rad}$, $x_F(0) = 1\,\mathrm{m}$, $y_F(0) = 1.5\,\mathrm{m}$, $\theta_F(0) = \pi/2\,\mathrm{rad}$, $\alpha_F(0) = 0\,\mathrm{rad}$.

In order to compute the actuating control input, equation (6) needs to be integrated and some initial values $v_F(0)$, $\omega_F(0)$ to be fixed. Although the stability of the control does not depend on these values, a common procedure to initialize the velocities in order to get good performances, is the following. Compute the derivative of (4) for $t = 0$ and equal it to zero,

$$\begin{aligned}\dot e_1(0) &= v_L(0)\cos\eta_L(0)\cos\gamma_L(0) + v_F(0)\big(-\cos\theta_F(0)\\ &\quad+\sin\theta_F(0)\tan\alpha_F(0) + \tfrac{d(0)}{\ell_1}\sin\lambda_F(0)\tan\alpha_F(0)\big)\\ &\quad+\omega_F(0)\,d(0)\sin\lambda_F(0) - \dot d(0)\cos\lambda_F(0)\\ &\quad+ d(0)\,\dot\phi(0)\sin\lambda_F(0) = 0\end{aligned}$$

$$\begin{aligned}\dot e_2(0) &= v_L(0)\sin\eta_L(0)\cos\gamma_L(0) - v_F(0)\big(\sin\theta_F(0)\\ &\quad+\cos\theta_F(0)\tan\alpha_F(0) + \tfrac{d(0)}{\ell_1}\cos\lambda_F(0)\tan\alpha_F(0)\big)\\ &\quad-\omega_F(0)\,d(0)\cos\lambda_F(0) - \dot d(0)\sin\lambda_F(0)\\ &\quad- d(0)\,\dot\phi(0)\cos\lambda_F(0) = 0.\end{aligned}$$

Solving the above equations with respect to $v_F(0)$, $\omega_F(0)$ with $\lambda_F(0) \in (0, \pi)$, we finally obtain,

$$v_F(0) = \frac{\cos\alpha_F(0)}{\cos\phi(0)}\big(v_L(0)\cos\gamma_L(0)\cos(\lambda_F(0)-\eta_L(0)) - \dot d(0)\big)$$

$$\begin{aligned}\omega_F(0) &= \frac{1}{d(0)\sin\lambda_F(0)}\Big(-v_L(0)\cos\eta_L(0)\cos\gamma_L(0)\\ &\quad+\frac{v_L(0)\cos\gamma_L(0)\cos(\lambda_F(0)-\eta_L(0)) - \dot d(0)}{\cos\phi(0)}\\ &\quad\cdot\big(\cos(\alpha_F(0)+\theta_F(0)) - \tfrac{d(0)}{\ell_1}\sin\lambda_F(0)\sin\alpha_F(0)\big)\\ &\quad+\dot d(0)\cos\lambda_F(0) - d(0)\,\dot\phi(0)\sin\lambda_F(0)\Big).\end{aligned}$$

We set $\ell_1 = 0.2\,\mathrm{m}$, $\ell_2 = 0.5\,\mathrm{m}$ and we chose the following parameters for the controller and the observers: $k_1 = k_2 = 1$, $f_1 = f_2 = 2$ and $\Lambda = 5$, $\xi_1(0) = \xi_2(0) = 0$. Table I shows leader's velocity and functions $d(t)$ and $\phi(t)$ used in Simulation I and II. The steering velocity of the leader in Simulation II (see Table I) is $\Omega(t) = \pi/600$ if $t < 3\,\mathrm{s}$, $\Omega(t) = -\pi/600$ if $t \geq 3\,\mathrm{s}$. Figs. 2 and 3 are relative to Simulation I.

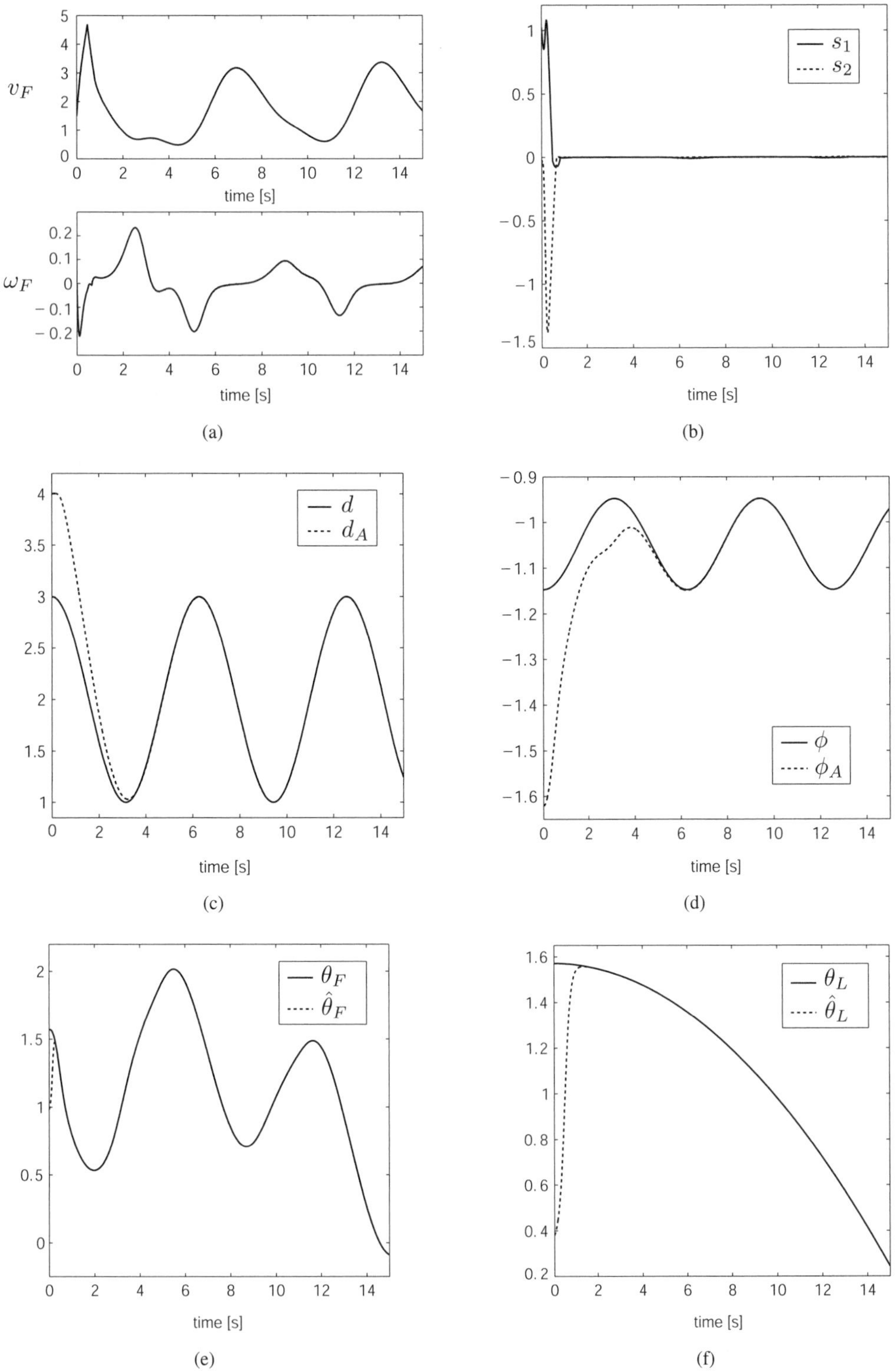

Fig. 3. Simulation I (ideal measurements): (a) Control inputs v_F and ω_F; (b) Sliding surfaces s_1 and s_2; (c) Desired distance d and actual distance d_A; (d) Desired orientation ϕ and actual orientation ϕ_A; (e) θ_F and $\hat{\theta}_F$; (f) θ_L and $\hat{\theta}_L$.

Fig. 2(a) shows the trajectory of the leader and the follower. In order to have a temporal reference in the figure the robots are drawn each second: the rectangles represent the tractor and the follower, while the small squares denote the trailer. Fig. 2(b) shows the trajectory of the vehicles in the first 8 seconds of the simulation. In Fig. 3(a) the control inputs v_F and ω_F are reported. In Fig. 3(b) the sliding surfaces s_1 and s_2 asymptotically converge to zero. Fig. 3(c) shows the

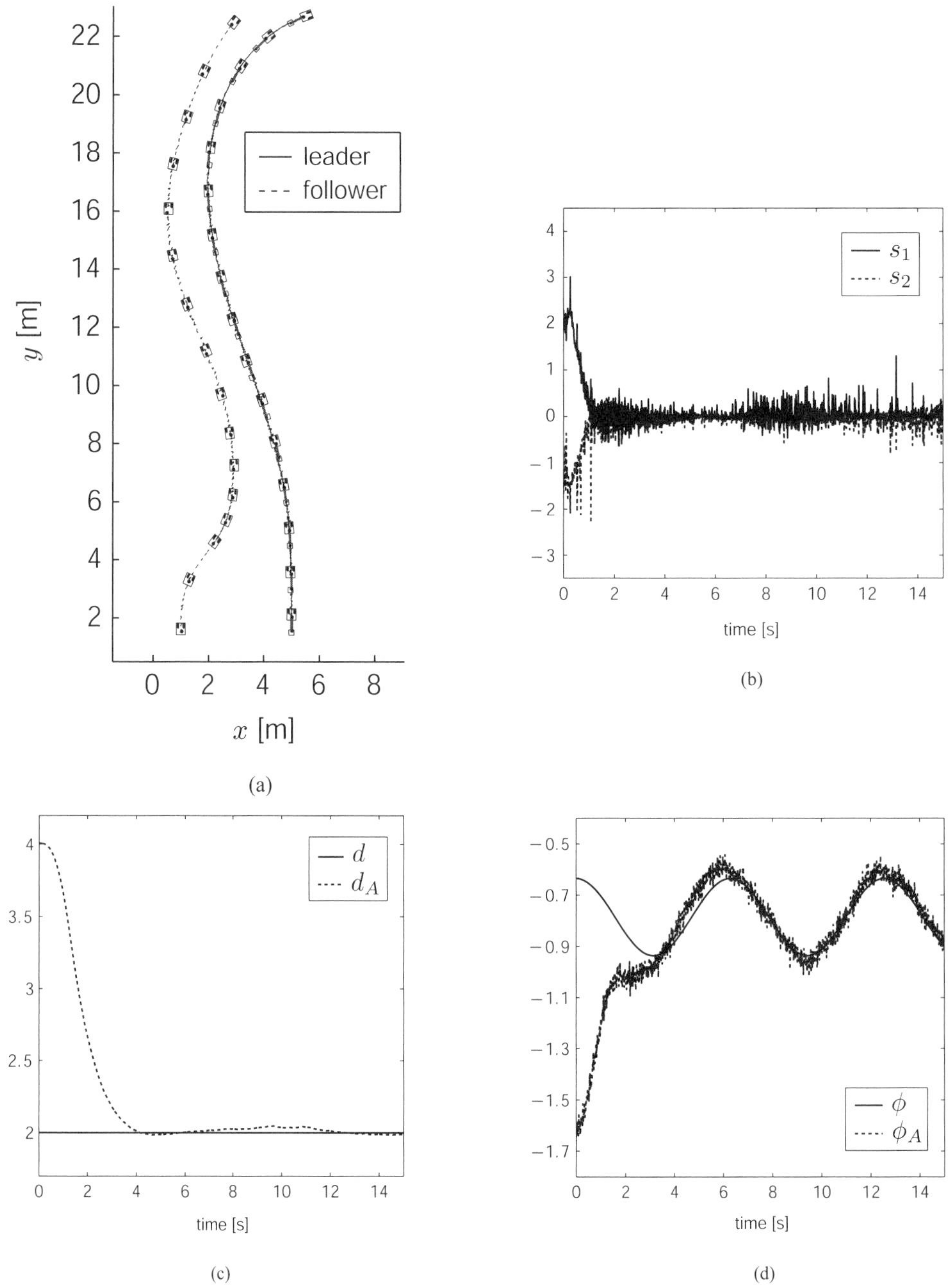

Fig. 4. Simulation II (noisy measurements): (a) Trajectory of the leader and the follower; (b) Sliding surfaces s_1 and s_2; (c) Desired distance d and actual distance d_A; (d) Desired orientation ϕ and actual orientation ϕ_A.

	Simulation I	Simulation II
$v_L(t)$ [m/s]	1.5	1.5
$\omega_L(t)$ [rad/s]	$-\pi/2000$	$\Omega(t)$
$d(t)$ [m]	$\cos(t)+2$	2
$\phi(t)$ [rad]	$-0.1\cos(t)-\pi/3$	$0.15\cos(t)-\pi/4$

TABLE I

LEADER'S VELOCITY AND DESIRED FUNCTIONS IN THE SIMULATIONS.

desired distance d and the actual distance

$$d_A = \left\| \begin{array}{c} x_L - \ell_2 \cos\eta_L - x_F - \ell_1 \cos\theta_F \\ y_L - \ell_2 \sin\eta_L - y_F - \ell_1 \sin\theta_F \end{array} \right\|.$$

Analogously, Fig. 3(d) shows the desired orientation ϕ and the actual orientation

$$\phi_A = \arctan\left(\frac{y_L - \ell_2 \sin\eta_L - y_F - \ell_1 \sin\theta_F}{x_L - \ell_2 \cos\eta_L - x_F - \ell_1 \cos\theta_F} \right) - \theta_F - \alpha_F.$$

d_A, ϕ_A and d, ϕ coincide after about 5 seconds. Finally, Figs. 3(e)-(f) show the time histories of θ_F, $\hat{\theta}_F$ and θ_L, $\hat{\theta}_L$.

The results of Simulation II are given in Fig. 4. Fig. 4(a) shows the trajectory of the leader and the follower, Fig. 4(b) the sliding surfaces and Figs. 4(c)-(d) the desired and actual distance and orientation. In spite of the noisy measurements, as in Simulation I, d_A, ϕ_A converge to d, ϕ and the sliding surfaces converge to zero. The control inputs v_F, ω_F and the time histories of θ_F, $\hat{\theta}_F$ and θ_L, $\hat{\theta}_L$ are not shown for lack of space. From the simulation experiments we observed that the estimate $\hat{\theta}_F$ (and analogously $\hat{\theta}_L$) is much more sensitive to position than angular disturbances.

VI. Conclusion

In this paper we propose a new leader-follower formation of nonholonomic mobile robots. The follower is a car and the leader is an articulated vehicle, a tractor pulling a trailer. The desired formation, defined by two parameters (a distance and an orientation function) is allowed to vary in time. A sliding mode formation tracking control scheme and nonlinear observers for the estimation of the attitude angles of the follower and the tractor, are designed. These observers are based on the invariant manifold technique recently proposed in [15]. The effectiveness of the proposed designs has been validated via simulation experiments.

Future research lines include the experimental validation of our control scheme and the extension of our results to vehicles with more involved kinematics (e.g. the fire truck model [3] could be considered for the leader). For the sake of simplicity in the present paper a single-leader, single-follower formation has been considered. Future investigations will cover the more general case of multi-leader/multi-follower formations (see, e.g. [6]).

Acknowledgments

The authors would like to thank the anonymous reviewers and conference attendees for their useful suggestions and constructive comments to improve the quality of paper.

References

[1] J.G. Bender. An overview of systems studies of automated highway systems. *IEEE Transactions on Vehicular Technology*, 40(1):82–99, 1991.

[2] R. Burns, C.A. McLaughlin, J. Leitner, and M. Martin. TechSat21: Formation Design, Control and Simulation. In *Proc. IEEE Aerospace Conference*, volume 7, pages 19–25, 2000.

[3] L.G. Bushnell, D.M. Tilbury, and S.S. Sastry. Steering Three-Input Nonholonomic Systems: The Fire Truck Example. *International Journal of Robotics Research*, 14(4):366–381, 1995.

[4] L.E. Buzogany, M. Pachter, and J.J. D'Azzo. Automated Control of Aircraft in Formation Flight. In *Proc. AIAA Guidance, Navigation, and Control Conference*, pages 1349–1370, 1993.

[5] D. Chwa. Sliding-Mode Tracking Control of Nonholonomic Wheeled Mobile Robots in Polar Coordinates. *IEEE Transactions on Control Systems Technology*, 12(4):637–644, 2004.

[6] L. Consolini, F. Morbidi, D. Prattichizzo, and M. Tosques. A Geometric Characterization of Leader-Follower Formation Control. In *Proc. IEEE International Conference on Robotics and Automation*, pages 2397–2402, 2007.

[7] A.K. Das, R. Fierro, V. Kumar, J.P. Ostrowsky, J. Spletzer, and C. Taylor. A Vision-Based Formation Control Framework. *IEEE Transaction on Robotics and Automation*, 18(5):813–825, 2002.

[8] R.A. DeCarlo, S.H. Zak, and G.P. Matthews. Variable Structure Control of Nonlinear Multivariable Systems: A Tutorial. *Proc. of the IEEE*, 76(3):212–232, 1988.

[9] K.D. Do and J. Pan. Nonlinear formation control of unicycle-type mobile robots. *Robotics and Autonomous Systems*, 55:191–204, 2007.

[10] D.B. Edwards, T.A. Bean, D.L. Odell, and M.J. Anderson. A leader-follower algorithm for multiple AUV formations. In *Proc. IEEE/OES Autonomous Underwater Vehicles*, pages 40–46, 2004.

[11] J.A. Fax and R.M. Murray. Information Flow and Cooperative Control of Vehicle Formations. *IEEE Transactions on Automatic Control*, 49(9):1465–1476, 2004.

[12] T.I. Fossen. *Guidance and Control of Ocean Vehicles*. John Wiley and Sons, New York, 1994.

[13] V. Gazi. Swarm Aggregations Using Artificial Potentials and Sliding-Mode Control. *IEEE Transactions on Robotics*, 21(6):1208–1214, 2005.

[14] J.Y. Hung, W. Gao, and J.C. Hung. Variable Structure Control: A Survey. *IEEE Transactions on Industrial Electronics*, 40(1):2–22, 1993.

[15] D. Karagiannis and A. Astolfi. Nonlinear observer design using invariant manifolds and applications. In *Proc. 44th IEEE Conference on Decision and Control*, pages 7775–7780, 2005.

[16] T.J. Koo and S.M. Shahruz. Formation of a group of unmanned aerial vehicles (UAVs). In *Proc. American Control Conference*, volume 1, pages 69–74, 2001.

[17] X. Li, J. Xiao, and Z. Cai. Backstepping based multiple mobile robots formation control. In *Proc. IEEE/RSJ International Conference on Intelligent Robots and Systems*, pages 887–892, 2005.

[18] Z. Lin, B.A. Francis, and M. Maggiore. Necessary and Sufficient Graphical Conditions for Formation Control of Unicycles. *IEEE Transactions on Automatic Control*, 50(1):121–127, 2005.

[19] G.L. Mariottini, F. Morbidi, D. Prattichizzo, G.J. Pappas, and K. Daniilidis. Leader-Follower Formations: Uncalibrated Vision-Based Localization and Control. In *Proc. IEEE International Conference on Robotics and Automation*, pages 2403–2408, 2007.

[20] G.L. Mariottini, G.J. Pappas, D. Prattichizzo, and K. Daniilidis. Vision-based Localization of Leader-Follower Formations. In *Proc. 44th IEEE Conference on Decision and Control*, pages 635–640, 2005.

[21] W. Perruquetti and J.P. Barbot. *Sliding mode control in engineering*. Control Engineering Series. Marcel Dekker Inc., 2002.

[22] M. Pham and D. Wang. A Unified Nonlinear Controller for a Platoon of Car-Like Vehicles. In *Proc. American Control Conference*, volume 3, pages 2350–2355, 2004.

[23] J. Sànchez and R. Fierro. Sliding Mode Control for Robot Formations. In *Proc. IEEE International Symposium on Intelligent Control*, pages 438–443, 2003.

[24] H. Schaub, S.R. Vadali, J.L. Junkins, and K.T. Alfriend. Spacecraft Formation Flying Control Using Mean Orbit Elements. *Journal of the Astronautical Sciences*, 48(1):69–87, 2000.

[25] J. Shao, G. Xie, J. Yu, and L. Wang. Leader-Following Formation Control of Multiple Mobile Robots. In *Proc. IEEE/RSJ International Symposium on Intelligent Control*, pages 808–813, 2005.

[26] D. Swaroop and J.K. Hedrick. Constant spacing strategies for platooning in Automated Highway systems. *Journal of dynamic systems, measurement, and control*, 121(3):462–470, 1999.

[27] V.I. Utkin. Variable structure systems with sliding modes. *IEEE Transactions on Automatic Control*, 22(2):212–222, 1977.

[28] V.I. Utkin. Sliding Mode Control Design Principles and Applications to Electric Drives. *IEEE Transactions on Industrial Electronics*, 40(1):23–36, 1993.

[29] R. Vidal, O. Shakernia, and S. Sastry. Following the Flock: Distributed Formation Control with Omnidirectional Vision-Based Motion Segmentation and Visual Servoing. *IEEE Robotics and Automation Magazine*, 11(4):14–20, 2004.

[30] E. Yang, D. Gu, and H. Hu. Nonsingular formation control of cooperative mobile robots via feedback linearization. In *Proc. IEEE/RSJ International Conference on Intelligent Robots and Systems*, pages 826–831, 2005.

[31] J.-M. Yang and J.-H. Kim. Sliding Mode Control for Trajectory Tracking of Nonholonomic Wheeled Mobile Robots. *IEEE Transactions on Robotics and Automation*, 15(3):578–587, 1999.

Map-Based Precision Vehicle Localization in Urban Environments

Jesse Levinson, Michael Montemerlo, Sebastian Thrun
Stanford Artificial Intelligence Laboratory
{jessel,mmde,thrun}@stanford.edu

***Abstract*— Many urban navigation applications (e.g., autonomous navigation, driver assistance systems) can benefit greatly from localization with centimeter accuracy. Yet such accuracy cannot be achieved reliably with GPS-based inertial guidance systems, specifically in urban settings.**

We propose a technique for high-accuracy localization of moving vehicles that utilizes maps of urban environments. Our approach integrates GPS, IMU, wheel odometry, and LIDAR data acquired by an instrumented vehicle, to generate high-resolution environment maps. Offline relaxation techniques similar to recent SLAM methods [2, 10, 13, 14, 21, 30] are employed to bring the map into alignment at intersections and other regions of self-overlap. By reducing the final map to the flat road surface, imprints of other vehicles are removed. The result is a 2-D surface image of ground reflectivity in the infrared spectrum with 5cm pixel resolution.

To localize a moving vehicle relative to these maps, we present a particle filter method for correlating LIDAR measurements with this map. As we show by experimentation, the resulting relative accuracies exceed that of conventional GPS-IMU-odometry-based methods by more than an order of magnitude. Specifically, we show that our algorithm is effective in urban environments, achieving reliable real-time localization with accuracy in the 10-centimeter range. Experimental results are provided for localization in GPS-denied environments, during bad weather, and in dense traffic. The proposed approach has been used successfully for steering a car through narrow, dynamic urban roads.

I. Introduction

In recent years, there has been enormous interest in making vehicles smarter, both in the quest for autonomously driving cars [4, 8, 20, 28] and for assistance systems that enhance the awareness and safety of human drivers. In the U.S., some 42,000 people die every year in traffic, mostly because of human error [34]. Driver assistance systems such as *adaptive cruise control*, *lane departure warning* and *lane change assistant* all enhance driver safety. Of equal interest in this paper are recent advances in autonomous driving. DARPA recently created the "Urban Challenge" program [5], to meet the congressional goal of having a third of all combat ground vehicles unmanned in 2015. In civilian applications, autonomous vehicles could free drivers of the burden of driving, while significantly enhancing vehicle safety.

Figure 1. The acquisition vehicle is equipped with a tightly integrated inertial navigation system which uses GPS, IMU, and wheel odometry for localization. It also possesses laser range finders for road mapping and localization.

All of these applications can greatly benefit from precision localization. For example, in order for an autonomous robot to follow a road, it needs to know where the road is. To stay in a specific lane, it needs to know where the lane is. For an *autonomous* robot to stay in a lane, the localization requirements are in the order of decimeters. As became painfully clear in the 2004 DARPA Grand Challenge [33], GPS alone is insufficient and does not meet these requirements. Here the leading robot failed after driving off the road because of GPS error.

The core idea of this paper is to augment inertial navigation by learning a detailed map of the environment, and then to use a vehicle's LIDAR sensor to localize relative to this map. In the present paper, maps are 2-D overhead views of the road surface, taken in the infrared spectrum. Such maps capture a multitude of textures in the environment that may be useful for localization, such as lane markings, tire marks, pavement, and vegetating near the road (e.g., grass). The maps are acquired by a vehicle equipped with a state-of-the-art inertial navigation system (with GPS) and multiple SICK laser range finders.

The problem of environment mapping falls into the realm of SLAM (simultaneous localization and mapping), hence there exists a huge body of related work on which our approach builds [3, 11, 12, 24, 22, 26]. SLAM addresses the problem of building consistent environment maps from a moving robotic vehicle, while simultaneously localizing the vehicle relative to these maps. SLAM has been at the core of a number of successful autonomous robot systems [1, 29].

At first glance, building urban maps is significantly easier than SLAM, thanks to the availability of GPS most of the time. While GPS is not accurate enough for mapping—as we show in this paper—it nevertheless bounds the localization error, thereby sidestepping hard SLAM problems such as the loop closure problem with unbounded error [17, 18, 31]. However, a key problem rarely addressed in the SLAM literature is that of dynamic environments. Urban environments are dynamic, and for localization to succeed one has to distinguish static aspects of the world (e.g., the road surface) relative to dynamic aspects

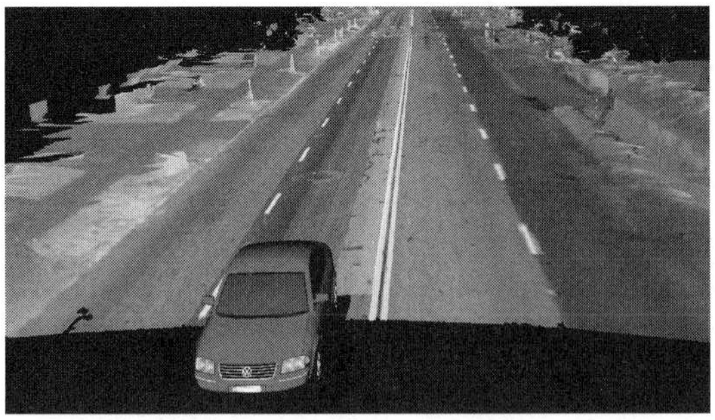

Figure 2. Visualization of the scanning process: the LIDAR scanner acquires range data *and* infrared ground reflectivity. The resulting maps therefore are 3-D infrared images of the ground reflectivity. Notice that lane markings have much higher reflectivity than pavement.

Figure 3. Example of ground plane extraction. Only measurements that coincide with the ground plane are retained; all others are discarded (shown in green here). As a result, moving objects such as car (and even parked cars) are not included in the map. This makes our approach robust in dynamic environments.

(e.g., cars driving by). Previous work on SLAM in dynamic environments [16, 19, 35] has mostly focused on tracking or removing objects that move at the time of mapping, which might not be the case here.

Our approach addresses the problem of environment dynamics, by reducing the map to features that with very high likelihood are static. In particular, using 3-D LIDAR information, our approach only retains the flat road surface; thereby removing the imprints of potentially dynamic objects (e.g., other cars, even if parked). The resulting map is then simply an overhead image of the road surface, where the image brightness corresponds to the infrared reflectivity.

Once a map has been built, our approach uses a particle filter to localize a vehicle in real-time [6, 9, 25, 27]. The particle filter analyzes range data in order to extract the ground plane underneath the vehicle. It then correlates via the *Pearson product-moment correlation* the measured infrared reflectivity with the map. Particles are projected forward through time via the velocity outputs from a tightly-coupled inertial navigation system, which relies on wheel odometry, an IMU and a GPS system for determining vehicle velocity. Empirically, we find that the our system very reliably tracks the location of the vehicle, with relative accuracy of ~10 cm. A dynamic map management algorithm is used to swap parts of maps in and out of memory, scaling precision localization to environments too large to be held in main memory.

Both the mapping and localization processes are robust to dynamic and hilly environments. Provided that the local road surface remains approximately laterally planar in the neighborhood of the vehicle (within ~10m), slopes do not pose a problem for our algorithms.

II. Road Mapping with GraphSLAM

Our mapping method is a version of the GraphSLAM algorithm and related constraint relaxation methods [2, 10, 13, 14, 21, 22, 30].

A. Modeling Motion

The vehicle transitions through a sequence of poses. In urban mapping, poses are five-dimensional vectors, comprising the x-y coordinates of the robot, along with its heading direction (yaw) and the roll and pitch angle of the vehicle (the elevation z is irrelevant for this problem). Let x_t denote the pose at time t. Poses are linked together through relative odometry data, acquired from the vehicle's inertial guidance system.

$$x_t \quad = \quad g(u_t, x_{t-1}) + \epsilon_t \qquad (1)$$

Here g is the non-linear kinematic function which accepts as input a pose x_{t-1} and a motion vector u_t, and outputs a projected new pose x_t. The variable ϵ_t is a Gaussian noise variable with zero mean and covariance R_t. Vehicle dynamics are discussed in detail in [15].

In log-likelihood form, each motion step induces a nonlinear quadratic constraint of the form

$$(x_t - g(u_t, x_{t-1}))^T \; R_t^{-1} \; (x_t - g(u_t, x_{t-1})) \qquad (2)$$

As in [10, 30], these constraints can be thought of as edges in a sparse Markov graph.

B. Map Representation

The map is a 2-D grid which assigns to each x-y location in the environment an infrared reflectivity value. Thus, we can treat the ground map as a orthographic infrared photograph of the ground.

To acquire such a map, multiple laser range finders are mounted on a vehicle, pointing downwards at the road surface (see Fig. 1). In addition to returning the range to a sampling of points on the ground, these lasers also return a measure of infrared reflectivity. By texturing this reflectivity data onto the 3-D range data, the result is a dense infrared reflectivity image of the ground surface, as illustrated in Fig. 2 (this map is similar to vision-based ground maps in [32]).

To eliminate the effect of non-stationary objects in the map on subsequent localization, our approach fits a ground plane to each laser scan, and only retains measurements that coincide with this ground plane. The ability to remove vertical objects is a key advantage of using LIDAR sensors over conventional cameras. As a result, only the flat ground is mapped, and other vehicles are automatically discarded from the data. Fig. 3 illustrates this process. The color labeling in this image illustrates the data selection: all green areas protrude above the ground plane and are discarded. Maps like these can be acquired even at night, as the LIDAR system does not rely on external light. This makes he mapping result much less dependent on ambient lighting than is the case for passive cameras.

For any pose x_t and any (fixed and known) laser angle relative to the vehicle coordinate frame α_i, the expected infrared reflectivity can easily be calculated. Let $h_i(m, x_t)$ be this function, which calculates the expected laser reflectivity for a given map m, a robot pose x_t, and a laser angle α_i. We model the observation process as follows

$$z_t^i = h_i(m, x_t) + \delta_t^i \tag{3}$$

Here δ_t^i is a Gaussian noise variable with mean zero and noise covariance Q_t.

In log-likelihood form, this provides a new set of constraints, which are of the form

$$(z_t^i - h_i(m, x_t))^T \, Q_t^{-1} \, (z_t^i - h_i(m, x_t))$$

The unknowns in this function are the poses $\{x_t\}$ and the map m.

C. Latent Variable Extension for GPS

In outdoor environments, a vehicle can use GPS for localization. GPS offers the convenient advantage that its error is usually limited to a few meters. Let y_t denote the GPS signal for time t. (For notational convenience, we treat y_t as a 3-D vector, with yaw estimate simply set to zero and the corresponding noise covariance in the measurement model set to infinity).

At first glance, one might integrate GPS through an additional constraint in the objective function J. The resulting constraints could be of the form

$$\sum_t (x_t - y_t)^T \, \Gamma_t^{-1} \, (x_t - y_t) \tag{4}$$

where Γ_t is the noise covariance of the GPS signal. However, this form assumes that GPS noise is *independent.* In practice, GPS is subject to *systematic* noise. because GPS is affected through atmospheric properties, which tend to change slowly with time.

Our approach models the systematic nature of the noise through a Markov chain, which uses GPS bias term b_t as a latent variable. The assumption is that the actual GPS measurement is corrupted by an additive bias b_t, which cannot be observed (hence is latent but can be inferred from data). This model yields constraints of the form

$$\sum_t (x_t - (y_t + b_t))^T \, \Gamma_t^{-1} \, (x_t - (y_t + b_t)) \tag{5}$$

In this model, the latent bias variables b_t are subject to a random walk of the form

$$b_t = \gamma \, b_{t-1} + \beta_t \tag{6}$$

Here β_t is a Gaussian noise variable with zero mean and covariance S_t. The constant $\gamma < 1$ slowly pulls the bias b_t towards zero (e.g., $\gamma = 0.999999$).

D. The Extended GraphSLAM Objective Function

Putting this all together, we obtain the goal function

$$\begin{aligned} J = {} & \sum_t (x_t - g(u_t, x_{t-1}))^T \, R_t^{-1} \, (x_t - g(u_t, x_{t-1})) \\ & + \sum_{t,i} (z_t^i - h_i(m, x_t))^T \, Q_t^{-1} \, (z_t^i - h_i(m, x_t)) \\ & + \sum_t (x_t - (y_t + b_t))^T \, \Gamma_t^{-1} \, (x_t - (y_t + b_t)) \\ & + \sum_t (b_t - \gamma b_{t-1})^T \, S_t^{-1} \, (b_t - \gamma b_{t-1}) \end{aligned} \tag{7}$$

Unfortunately, this expression cannot be optimized directly, since it involves many millions of variables (poses and map pixels).

E. Integrating Out the Map

Unfortunately, optimizing J directly is computationally infeasible. A key step in GraphSLAM, which we adopt here, is to first integrate out the map variables. In particular, instead of optimizing J over all variables $\{x_t\}$, $\{b_t\}$, and m, we first optimize a modified version of J that contains only poses $\{x_t\}$ and biases $\{b_t\}$, and then compute the most likely map. This is motivated by the fact that, as shown in [30], the map variables can be integrated out in the SLAM joint posterior. Since nearly all unknowns in the system are with the map, this simplification makes the problem of optimizing J much easier and allows it to be solved efficiently.

In cases where a specific surface patch is only seen once during mapping, our approach can simply ignore the corresponding map variables during the pose alignment process, because such patches have no bearing on the pose estimation. Consequently, we can safely remove the associated constraints from the goal function J, without altering the goal function for the poses.

Of concern, however, are places that are seen more than once. Those *do* create constraint between pose variables from which those places were seen. These constraints correspond to the famous loop closure problem in SLAM [17, 18, 31]. To integrate those map variables out, our approach uses an effective approximation known as map matching [21]. Map matching compares local submaps, in order to find the best alignment.

Our approach implements map matching by first identifying regions of overlap, which will then form the local maps. A region of overlap is the result of driving over the same terrain twice. Formally it is defined as two disjoint sequences of time indices, $t_1, t_2, \ldots$ and $s_1, s_2, \ldots$, such that the corresponding grid cells in the map show an overlap that exceeds a given threshold θ.

Once such a region is found, our approach builds two separate maps, one using only data from $t_1, t_2, \ldots$, and the other only with data from $s_1, s_2, \ldots$. It then searches for the alignment that maximizes the measurement probability, assuming that both adhere to a single maximum likelihood infrared reflectivity map in the area of overlap.

Specifically, a linear correlation field is computed between these maps, for different x-y offsets between these images.

Since the poses prior to alignment have already been post-processed from GPS and IMU data, we find that rotational error between matches is insignificant. Because one or both maps may be incomplete, our approach only computes correlation coefficients from elements whose infrared reflectivity value is known. In cases where the alignment is unique, we find a single peak in this correlation field. The peak of this correlation field is then assumed to be the best estimate for the local alignment. The relative shift is then labeled δ_{st}, and the resulting constraint of the form above is added to the objective J.

This map matching step leads to the introduction of the following constraint in J:

$$(x_t + \delta_{st} - x_s)^T \, L_{st} \, (x_t + \delta_{st} - x_s) \qquad (8)$$

Here d δ_{st} is the local shift between the poses x_s and x_t, and L_{st} is the strength of this constraint (an inverse covariance). Replacing the map variables in J with this new constraint is clearly approximate; however, it makes the resulting optimization problem tractable.

The new goal function J' then replaces the many terms with the measurement model by a small number of between-pose constraints. It is of the form:

$$\begin{aligned} J' &= \sum_t (x_t - g(u_t, x_{t-1}))^T \, R_t^{-1} \, (x_t - g(u_t, x_{t-1})) \\ &\quad + \sum_t (x_t - (y_t + b_t))^T \, \Gamma_t^{-1} \, (x_t - (y_t + b_t)) \\ &\quad + \sum_t (b_t - \gamma b_{t-1})^T \, S_t^{-1} \, (b_t - \gamma b_{t-1}) \\ &\quad + \sum_t (x_t + \delta_{st} - x_s)^T \, L_{st} \, (x_t + \delta_{st} - x_s) \qquad (9) \end{aligned}$$

This function contains no map variables, and thus there are orders of magnitude fewer variables in J' than in J. J' is then easily optimized using CG. For the type of maps shown in this paper, the optimization takes significantly less time on a PC than is required for acquiring the data. Most of the time is spent in the computation of local map links; the optimization of J' only takes a few seconds. The result is an adjusted robot path that resolves inconsistencies at intersections.

F. Computing the Map

To obtain the map, our approach needs to simply fill in all map values for which one or more measurements are available. In grid cells for which more than one measurement is available, the average infrared reflectivity minimizes the joint data likelihood function.

We notice that this is equivalent to optimize the missing constraints in J, denoted J'':

$$J'' = \sum_{t,i} (z_t^i - h_i(m, x_t))^T \, Q_t^{-1} \, (z_t^i - h_i(m, x_t)) \qquad (10)$$

under the assumption that the poses $\{x_t\}$ are known. Once again, for the type of maps shown in this paper, the computation of the aligned poses requires only a few seconds.

When rendering a map for localization or display, our implementation utilizes hardware accelerated OpenGL to render smoothly interpolated polygons whose vertices are based on the distances and intensities returned by the three lasers as the robot traverses its trajectory. Even with a low-end graphics card, this process is faster than real-time.

(a) Map acquired on a sunny day.

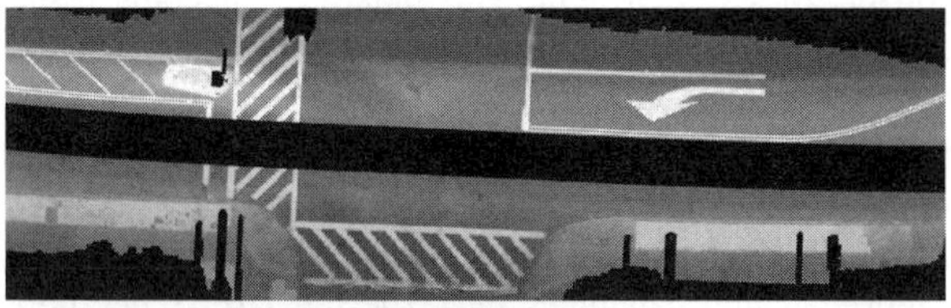

(b) Same road segment on a rainy day.

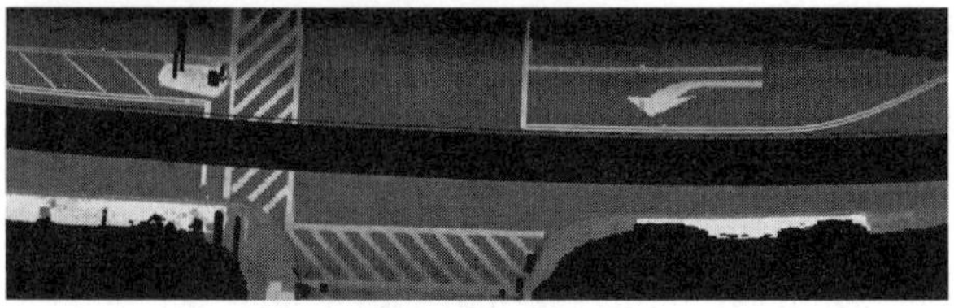

Figure 4. Patch of the map acquired in bright sunlight on a sunny day (top), and at night in heavy rain (middle). By correlating scans with the map, instead of taking absolute differences, the weather-related brightness variation has almost no effect on localization.

III. Online Localization

A. Localization with Particle Filters

Localization uses the map in order to localize the vehicle relative to the map. Localization takes place in real-time, with a 200 Hz motion update (measurements arrive at 75 Hz per laser).

Our approach utilizes the same mathematical variable model discussed in the previous section. However, to achieve real-time performance, we utilize a particle filter, known in robotics as Monte Carlo localizer [7]. The particle filter maintains a three-dimensional pose vector (x, y, and yaw); roll and pitch are assumed to be sufficiently accurate as is. The motion prediction in the particle filter is based on inertial velocity measurements, as stated in Sect. II-A.

Measurements are integrated in the usual way, by affecting the importance weight that sets the resampling probability. As in the mapping step, a local ground plane analysis removes measurement that correspond to non-ground objects. Further, measurements are only incorporated for which the corresponding map is defined, that is, for which a prior reflectivity value is available in the map. For each measurement update, the importance weight of the k-th particle is given by

$$\begin{aligned} w_t^{[k]} &= \exp\{-\frac{1}{2} \, (x_t^{[k]} - y_t)^T \, \Gamma_t^{-1} \, (x_t^{[k]} - y_t)\} \, \cdot \qquad (11) \\ &\quad \left(\operatorname{corr} \left[\begin{pmatrix} h_1(m, x_t^{[k]}) \\ \vdots \\ h_{180}(m, x_t^{[k]}) \end{pmatrix}, \begin{pmatrix} z_t^1 \\ \vdots \\ z_t^{180} \end{pmatrix} \right] + 1 \right) \end{aligned}$$

Here, corr() is the Pearson product-moment correlation. The actual resampling takes place at a much lower frequency than the measurement/pose updates, to reduce variance. As usual, importance weight updates are integrated multiplicatively. To avoid catastrophic localization errors, a small number of particles are continuously drawn from current GPS pose estimate.

This "sensor resetting" trick was proposed in [23]. GPS, when available, is also used in the calculation of the measurement likelihood to reduce the danger of particles moving too far away from the GPS location.

One complicating factor in vehicle localization is weather. As illustrated in Fig. 4, the appearance of the road surface is affected by rain, in that wet surfaces tend to reflect less infrared laser light than do dry ones. To adjust for this effect, the particle filter normalizes the brightness and standard deviation for each individual range scan, and also for the corresponding local map stripes. This normalization transforms the least squares difference method into the computation of the Pearson product-moment correlation with missing variables (empty grid cells in the map). Empirically, we find that this local normalization step is essential for robust performance .

We also note that the particle filter can be run entirely without GPS, where the only reference to the environment is the map. Some of our experiments are carried out in the absence of GPS, to illustrate the robustness of our approach in situations where conventional GPS-based localization is plainly inapplicable.

B. Data Management

Maps of large environments at 5-cm resolution occupy a significant amount of memory. We have implemented two methods to reduce the size of the maps and to allow relevant data to fit into main memory.

When acquiring data in a moving vehicle, the rectangular area which circumscribes the resulting laser scans grows quadratically with distance, despite that the data itself grows only linearly. In order to avoid a quadratic space requirement, we break the rectangular area into a square grid, and only save squares for which there is data. When losslessly compressed, the grid images require approximately 10MB per mile of road at 5-cm resolution. This would allow a 200GB hard drive to hold 20,000 miles of data.

Although the data for a large urban environment can fit on hard drive, it may not all be able to fit into main memory. Our particle filter maintains a cache of image squares near the vehicle, and thus requires a constant amount of memory regardless of the size of the overall map.

IV. Experimental Results

We conducted extensive experiments with the vehicle shown in Fig. 1. This vehicle is equipped with a state-of-the-art inertial navigation system (GPS, inertial sensors, and wheel odometry), and three down-facing laser range finders: one facing the left side, one facing the right side, and one facing the rear. In all experiments, we use a pixel resolution of 5cm.

A. Mapping

We tested the mapping algorithm successfully on a variety of urban roads. One of our testing environments is an urban area, shown in Fig. 5. This image shows an aerial view of the testing environment, with the map overlayed in red. To generate this map, our algorithm automatically identified and aligned several hundreds match points in a total of 32 loops. It corrected the trajectory and output consistent imagery at 5-cm resolution. Fig. 7 shows a close-up of this map in a residential area lined with parked cars, whose positions changed between different data runs.

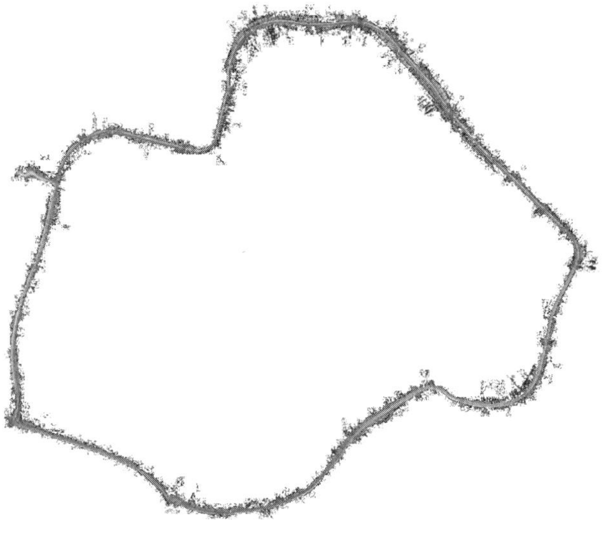

Figure 6. Campus Dr., a loop around Stanford 4 miles in circumference. To acquire this map, we drove the same loop twice, filling in gaps that are occluded by moving vehicles.

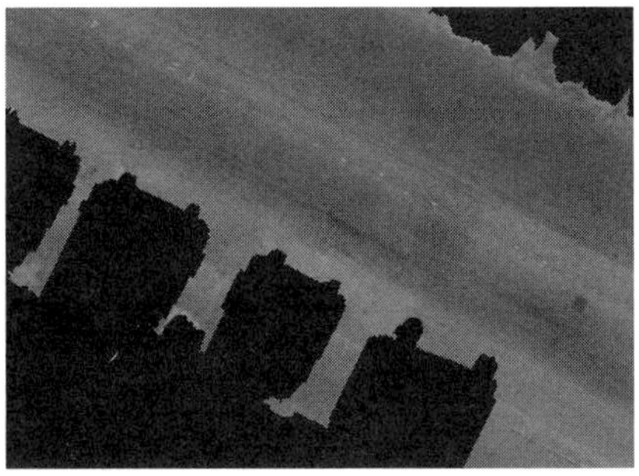

Figure 7. Map with many parked cars. Notice the absence of lane markers.

One of the key results of the map alignment process is the removal of the so-called "ghosting" effects. Ghosting occurs in the absence of the map alignment step, where the location of measurements are set by GPS alone. Fig. 8a shows an example of ghosting, in which features occur twice in the map. Clearly, such a map is too inaccurate for precision localization. Fig. 8b shows the adjusted map, generated with the approach described in this paper. Here the ghosting effect has disappeared, and the map is locally consistent.

Figure 6 shows a large loop which is 4 miles in circumference. Our algorithm automatically identified and aligned 148 match points during the second traversal of the loop. In this experiment, the average magnitude of the corrections calculated by the matching algorithm for these two loops is 35 cm, indicating that without corrections, the two loops are quite poorly aligned, even with post-processed poses.

An interesting result is that areas missing in the first traversal can be filled in during the second. Fig. 9 shows three local maps. The map from the first traversal has a hole caused by oncoming traffic. The center map in this figure is the GPS

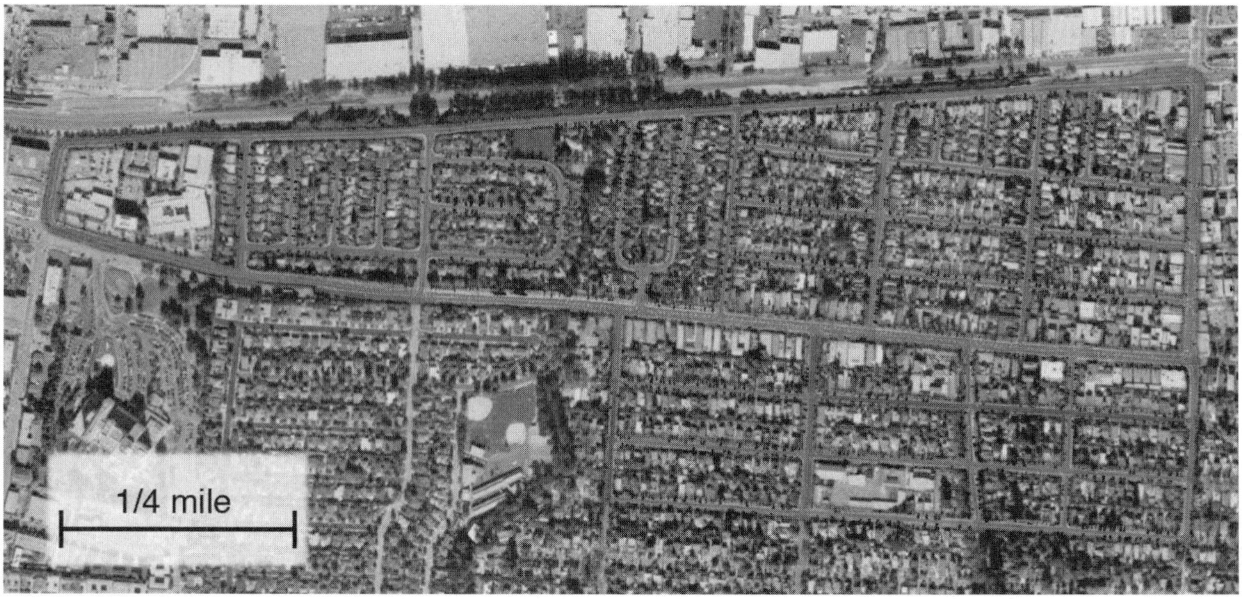

Figure 5. (Best viewed in color) Aerial view of Burlingame, CA. Regions of overlap have been adjusted according to the methods described in this paper. Maps of this size tend not to fit into main memory, but are swapped in from disk automatically during driving.

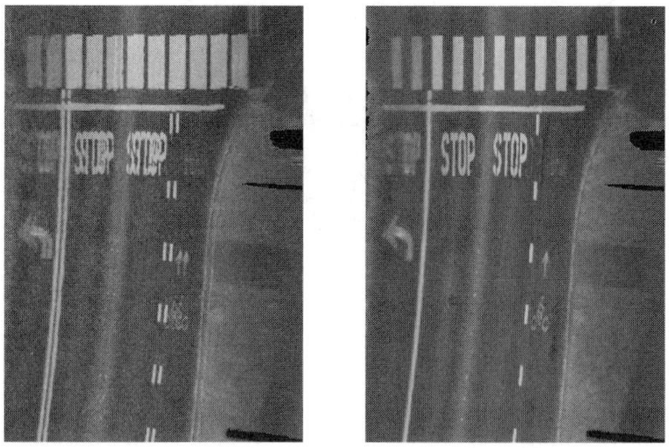

Figure 8. Infrared reflectivity ground map before and after SLAM optimization. Residual GPS drift can be seen in the ghost images of the road markings (left). After optimization, all ghost images have been removed (right).

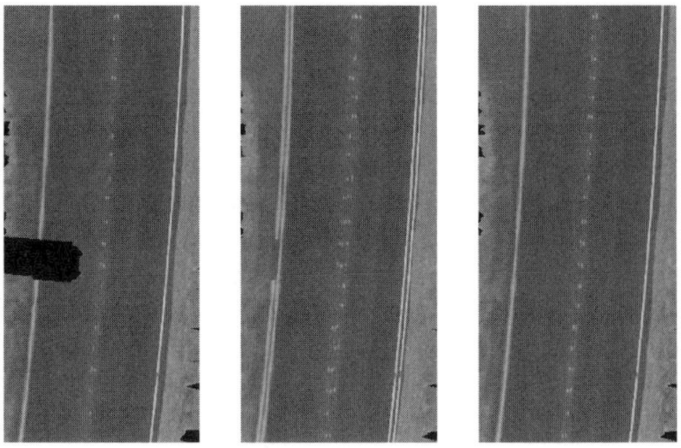

Figure 9. Filtering dynamic objects from the map leaves holes (left). These holes are often filled if a second pass is made over the road, but ghost images remain (center). After SLAM, the hole is filled and the ghost image is removed.

aligned map with the familiar ghosting effect. On the right is the fused map with our alignment procedure, which possesses no hole and exhibits no ghosting.

The examples shown here are representative of all the maps we have acquired so far. We find that even in dense urban traffic, the alignment process is robust to other cars and non-stationary obstacles. No tests were conducted in open featureless terrain (e.g., airfields without markings or texture), as we believe those are of low relevance for the stated goal of urban localization.

B. Localization

Our particle filter is adaptable to a variety of conditions. In the ideal case, the vehicle contains an integrated GPS/IMU system, which provides locally consistent velocity estimates and global pose accuracy to within roughly a meter. In our tests in a variety of urban roads, our GPS-equipped vehicle was able to localize in real-time relative to our previously created maps with errors of less than 10 cm, far exceeding the accuracy with GPS alone.

We ran a series of experiments to test localization relative to the learned map. All localization experiments used separate data from the mapping data; hence the environment was subject to change that, at times, was substantial. In fact, the map in Fig. 5 was acquired at night, but all localization experiments took place during daylight. In our experimentation, we used between 200 and 300 particles.

Fig. 10 shows an example path that corresponds to 20 minutes of driving and localization. During this run, the average disagreement between real-time GPS pose and our method was 66cm. Manual alignment of the map and the

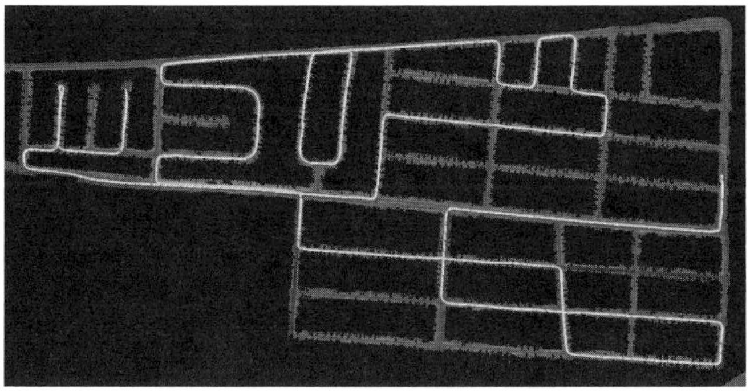

Figure 10. (Best viewed in color) Typical driving path during localization shown in green, and overlayed on a previously built map, acquired during 20 minutes of driving. For this and other paths, we find that the particle filter reliably localizes the vehicle.

(a) GPS localization induces ≥1 meter of error.

(b) No noticeable error in particle filter localization.

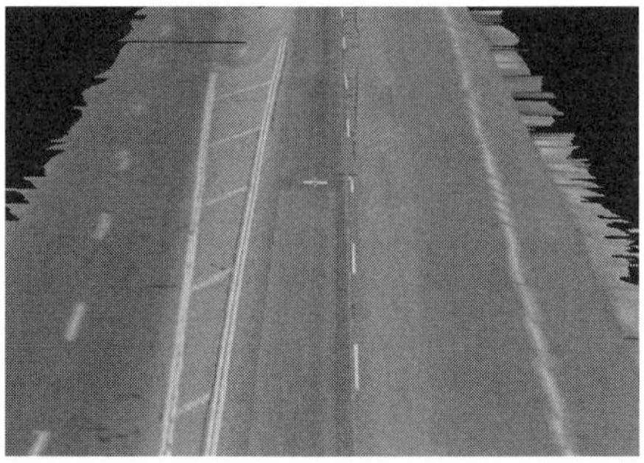

Figure 11. (a) GPS localization is prone to error, even (as shown here) with a high-end integrated inertial system and differential GPS using a nearby stationary antenna. (b) The particle filter result shows no noticeable error.

incoming LIDAR measurements suggests that the lateral error was almost always within 10cm. Occasionally, errors were larger; when turning, errors were sometimes as large as 30cm. However, this is still well below the error rate of the GPS-based inertial system.

A typical situation is shown in Fig. 11. Fig. 11a superimposes data acquired online by the vehicle and the previously acquired map. The error here is greater than one meter. Fig. 11b shows the result with our approach, where the error is below the map resolution. The red line in both figures corresponds to the path of the localizing vehicle.

One of the key aspects of the localization method is its ability to localize entirely in the absence of GPS. For the following experiments, we switched off the GPS receiver,

Stanford Ave.		
Distance Traveled (m)	Our Error (cm)	Odometry Error (cm)
50	7	98
100	3	149
150	35	0
200	13	8
250	4	133
300	22	272
350	8	428
400	23	589
450	13	783
499	10	924

Figure 12. This table compares the accuracy of pose estimation in the absence of GPS or IMU data. The right column is obtained by odometry only; the center by particle filter localization relative to the map. Clearly, odometry alone accumulates error. Our approach localizes reliably without any GPS or IMU.

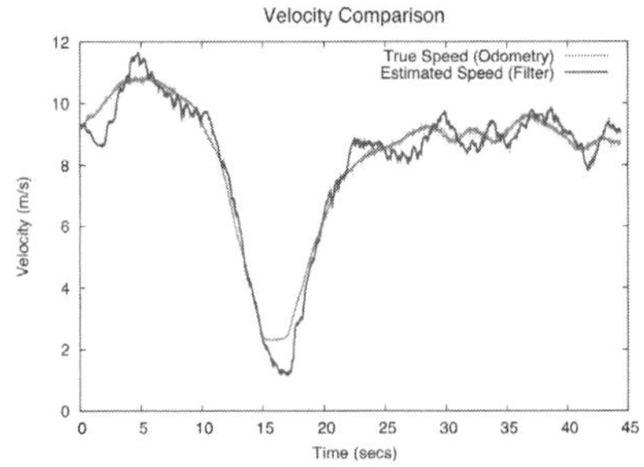

Figure 13. Estimated velocity from odometry and our particle filter, not using odometry or GPS. The strong correspondence illustrates that a particle filter fed with laser data alone can determine the vehicle velocity, assuming a previously acquired map. Note that in this case, accurate velocity estimation subsumes accurate position estimation.

and exclusively used wheel odometry and steering angle for localization. Clearly, odometry alone eventually diverges from the true position. Nevertheless, our localization method reliable tracks the location of the car.

This is illustrated by the result in Fig. 12, which quantitatively compares the accuracy of the particle filter localizer with odometry data. In both cases, the filters are initialized with the correct pose. Error is measured along 10 hand-labeled reference points in 50 meter increments.

Not surprisingly, the odometry estimates grows quickly as the vehicle moves, whereas our method results in a small error. The error is slightly larger than in the experiments above, due to the absence of GPS in the forward projection step of the particles. Nevertheless, this experiment illustrates successful tracking in the absence of GPS.

Finally, an extreme test of our software's ability to localize using laser data was performed by disallowing any motion or position data whatsoever in the motion update. Thus odometry, IMU, and GPS measurements were all ignored. In this case, the particles' state vector included x and y position, yaw, steering angle, velocity, and acceleration. The particles were initialized near the true position, and reasonable values were assumed for the rate of change of the control parameters. Remarkably, our system was able to track the position and the velocity of the vehicle using nothing but laser data (see Fig. 13).

C. Autonomous Driving

In a series of experiments, we used our approach for semi-autonomous urban driving. In most of these experiments, gas and brakes were manually operated, but steering was controlled by a computer. The vehicle followed a fixed reference trajectory on a university campus. In some cases, the lane width exceeded the vehicle width by less than 2 meters.

Using the localization technique described here, the vehicle was able to follow this trajectory on ten out of ten attempts without error; never did our method fail to provide sufficient localization accuracy. Similar experiments using only GPS consistently failed within meters, illustrating that GPS alone is insufficient for autonomous urban driving. We view the ability to use our localization method to steer a vehicle in an urban environment as a success of our approach.

V. Conclusions

Localization is a key enabling factor for urban robotic operation. With accurate localization, autonomous cars can perform accurate lane keeping and obey traffic laws (e.g., stop at stop signs). Although nearly all outdoor localization work is based on GPS, GPS alone is insufficient to provide the accuracy necessary for urban autonomous operation.

This paper presented a localization method that uses an environment map. The map is acquired by an instrumented vehicle that uses infrared LIDAR to measure the 3-D structure and infrared reflectivity of the environment. A SLAM-style relaxation algorithm transforms this data into a globally consistent environment model with non-ground objects removed. A particle filter localizer was discussed that enables a moving vehicle to localize in real-time, at 200 Hz.

Extensive experiments in urban environments suggest that the localization method surpasses GPS in two critical dimensions: accuracy and availability. The proposed method is significantly more accurate than GPS; its relative lateral localization error to the previously recorded map is mostly within 5cm, whereas GPS errors are often in the order of 1 meter. Second, our method succeeds in GPS-denied environments, where GPS-based systems fail. This is of importance to navigation in urban canyons, tunnels, and other structures that block or deflect GPS signals.

The biggest disadvantage of our approach is its reliance on maps. While road surfaces are relatively constant over time, they still may change. In extreme cases, our localization technique may fail. While we acknowledge this limitation, the present work does not address it directly. Possible extensions would involve filters that monitor the sensor likelihood, to detect sequences of "surprising" measurements that may be indicative of a changed road surface. Alternatively, it may be possible to compare GPS and map-based localization estimates to spot positioning errors. Finally, it may be desirable to extend our approach to incorporate 3-D environment models beyond the road surface for improved reliability and accuracy, especially on unusually featureless roads.

References

[1] C. Baker, A. Morris, D. Ferguson, S. Thayer, C. Whittaker, Z. Omohundro, C. Reverte, W. Whittaker, D. Hähnel, and S. Thrun. A campaign in autonomous mine mapping. In ICRA 2004.

[2] M. Bosse, P. Newman, J. Leonard, M. Soika, W. Feiten, and S. Teller. Simultaneous localization and map building in large-scale cyclic environments using the atlas framework. *IJRR*, 23(12), 2004.

[3] P. Cheeseman and P. Smith. On the representation and estimation of spatial uncertainty. *IJR*, 5, 1986.

[4] DARPA. DARPA Grand Challenge rulebook, 2004. On the Web at http://www.darpa.mil/grandchallenge05/Rules_8oct04.pdf.

[5] DARPA. DARPA Urban Challenge rulebook, 2006. On the Web at http://www.darpa.mil/grandchallenge/docs/Urban_Challenge_Rules_121106.pdf.

[6] F. Dellaert, W. Burgard, D. Fox, and S. Thrun. Using the condensation algorithm for robust, vision-based mobile robot localization. ICVP 1999.

[7] F. Dellaert, D. Fox, W. Burgard, and S. Thrun. Monte Carlo localization for mobile robots. ICRA 1999.

[8] E.D. Dickmanns. Vision for ground vehicles: history and prospects. *IJVAS*, 1(1) 2002.

[9] A Doucet. On sequential simulation-based methods for Bayesian filtering. TR CUED/F-INFENG/TR 310, Cambridge University, 1998.

[10] T. Duckett, S. Marsland, and J. Shapiro. Learning globally consistent maps by relaxation. ICRA 2000.

[11] H.F. Durrant-Whyte. Uncertain geometry in robotics. *IEEE TRA*, 4(1), 1988.

[12] A. Eliazar and R. Parr. DP-SLAM: Fast, robust simultaneous localization and mapping without predetermined landmarks. IJCAI 2003.

[13] J. Folkesson and H. I. Christensen. Robust SLAM. ISAV 2004.

[14] U. Frese, P. Larsson, and T. Duckett. A multigrid algorithm for simultaneous localization and mapping. *IEEE Transactions on Robotics*, 2005.

[15] T.D. Gillespie. *Fundamentals of Vehicle Dynamics*. SAE Publications, 1992.

[16] J. Guivant, E. Nebot, and S. Baiker. Autonomous navigation and map building using laser range sensors in outdoor applications. *JRS*, 17(10), 2000.

[17] J.-S. Gutmann and K. Konolige. Incremental mapping of large cyclic environments. CIRA 2000.

[18] D. Hähnel, D. Fox, W. Burgard, and S. Thrun. A highly efficient FastSLAM algorithm for generating cyclic maps of large-scale environments from raw laser range measurements. IROS 2003.

[19] D. Hähnel, D. Schulz, and W. Burgard. Map building with mobile robots in populated environments. IROS 2002.

[20] M. Hebert, C. Thorpe, and A. Stentz. *Intelligent Unmanned Ground Vehicles*. Kluwer, 1997.

[21] K. Konolige. Large-scale map-making. AAAI, 2004.

[22] B. Kuipers and Y.-T. Byun. A robot exploration and mapping strategy based on a semantic hierarchy of spatial representations. *JRAS*, 8, 1991.

[23] S. Lenser and M. Veloso. Sensor resetting localization for poorly modelled mobile robots. ICRA 2000.

[24] J. Leonard, J.D. Tardós, S. Thrun, and H. Choset, editors. ICRA Workshop W4, 2002.

[25] J. Liu and R. Chen. Sequential monte carlo methods for dynamic systems. *Journal of the American Statistical Association*, 93. 1998.

[26] M.A. Paskin. Thin junction tree filters for simultaneous localization and mapping. IJCAI 2003.

[27] M. Pitt and N. Shephard. Filtering via simulation: auxiliary particle filter. *Journal of the American Statistical Association*, 94, 1999.

[28] D. Pomerleau. *Neural Network Perception for Mobile Robot Guidance*. Kluwer, 1993.

[29] C. Thorpe and H. Durrant-Whyte. Field robots. ISRR 2001

[30] S. Thrun and M. Montemerlo. The GraphSLAM algorithm with applications to large-scale mapping of urban structures. *IJRR*, 25(5/6), 2005.

[31] N. Tomatis, I. Nourbakhsh, and R. Siegwart. Hybrid simultaneous localization and map building: closing the loop with multi-hypothesis tracking. ICRA 2002.

[32] I. Ulrich and I. Nourbakhsh. Appearance-based obstacle detection with monocular color vision. AAAI 2000.

[33] C. Urmson, J. Anhalt, M. Clark, T. Galatali, J.P. Gonzalez, J. Gowdy, A. Gutierrez, S. Harbaugh, M. Johnson-Roberson, H. Kato, P.L. Koon, K. Peterson, B.K. Smith, S. Spiker, E. Tryzelaar, and W.L. Whittaker. High speed navigation of unrehearsed terrain: Red Team technology for the Grand Challenge 2004. TR CMU-RI-TR-04-37, 2004.

[34] Bureau of Transportation Statistics U.S. Department of Transportation. Annual report, 2005.

[35] C.-C. Wang, C. Thorpe, and S. Thrun. Online simultaneous localization and mapping with detection and tracking of moving objects: Theory and results from a ground vehicle in crowded urban areas. ICRA 2003.

Dense Mapping for Range Sensors:
Efficient Algorithms and Sparse Representations

Manuel Yguel, Christopher Tay Meng Keat, Christophe Braillon,
Christian Laugier, Olivier Aycard

Abstract— **This paper focuses on efficient occupancy grid building based on wavelet occupancy grids, a new sparse grid representation and on a new update algorithm for range sensors. The update algorithm takes advantage of the natural multiscale properties of the wavelet expansion to update only parts of the environement that are modified by the sensor measurements and at the proper scale. The sparse wavelet representation coupled with an efficient algorithm presented in this paper provides efficient and fast updating of occupancy grids. It leads to real-time results especially in 2D grids and for the first time in 3D grids. Experiments and results are discussed for both real and simulated data.**

I. Introduction and Previous Work

The Simultaneous Localization And Mapping (SLAM) issue has found very convincing solutions in the past few years, especially in 2 dimensions. Thanks to a number of contributions [1] [2] [3] [4], it is now feasible to navigate and build a map while maintaining an estimate of the robot's position in an unknown 2D indoor environment on a planar floor. In these 2D conditions, the problem is theoretically and practically solved even in a populated environment [5]. Some of the most impressive approaches are based on grid-based fast-slam algorithms [6] [7] [3], which offer a unified framework for landmark registration and pose calculation thanks to occupancy grids (OG) [8]. This approach yields several advantages: it provides robots with the ability to build an accurate dense map of the static environment, which keeps track of all possible landmarks and represents open spaces at the same time. Only a simple update mechanism, which filters moving obstacles naturally and performs sensor fusion, is required. In contrast to other methods, there is no need to perform landmark extraction as the raw data from range measurements are sufficient. One of the benefits is accurate self positioning, which is particularly visible in the accuracy of the angle estimate. However, the major drawback is the amount of data required to store and process the grid, as a grid that represents the environment has an exponential memory cost as the number of dimensions increases. In 2D SLAM, this drawback is overcome by the sheer power of the computer and its huge memory. But this issue cannot be avoided for 3D SLAM even with today's desktop computing capabilities. Recently, methods to deal with the 3D instance of the SLAM problem, in undulating terrains [4] have used landmark extraction, clustering and a special algorithm for spurious data detection. However, this map framework does not handle out-of-date data; therefore the extra cost of removing or updating data coming from past poses of moving objects is not considered.

In this paper, we choose to use OGs in a hierarchical manner related to those of [9] but embedded in the wavelet theory which allows us to present a new algorithm called wavelet hierarchical rasterization that hierarchically updates the wavelet occupancy grid in the relevant area of the environment. It does not require, as with the previous approach [10], any intermediate representation for adding observations in the wavelet grid. This leads to real-time dense mapping in 2D and we propose a special instance of this algorithm that performs well enough for real-time 3D grid modelling. This method is intrinsically multi-scale and thus one of its major advantages is that the mapping could be performed at any resolution in an anytime fashion.

There exists a large panel of dense mapping techniques: amongst the other popular representations of dense 3D data are raw data points [4], triangle mesh [11] [12], elevation maps [13], [14] or 2^d-tree based representations [15], [9]. However there are major drawbacks in using such representations. With clouds of points it is not easy to generalize: roughly speaking, there is no simple mechanism to fill in the holes. Moreover, the clouds of points are generated by the successive records of range measurements, thus the amount of data is prohibitive after a few hours of recording. The triangle mesh representation is a kind of $2\frac{1}{2}$-D map and the space representation is also incomplete. In simple elevation maps [11], for the same reasons, holes in the environment such as tunnels are not part of the set of representable objects. This problem is overcome in [14] since there is a little number of vertical steps for each part of the map. The most serious point is that most of these methods lack a straightforward data fusion mechanism. In particular, it is rarely simple to include information on the absence of features. Triangle mesh [12] and elevation maps [14] suffer most from this problem. Therefore most of the time these representations are obtained as a batch processing or for a little environment.

For range sensors OGs represent the probability for the presence of a reflective surface at any world location. Therefore the ability to update the map for both the presence and the absence of data is a major advantage, which we call the evolution property. With OGs this property does not come from a batch process but is part of the probabilistic map model definition. The cost is that a huge amount of memory is needed to cope with the map discretization.

In [10], a wavelet grid-based approach was introduced, which makes it possible to represent grids in a compact but flexible format. In pyramid maps representations [8], [9], information is stored at each scale and there is a lot of redundancy but multiscale information is available. Conversely, probabilistic 2^d-trees record data at the leaves [15] [9] and the whole depth of the tree must be traversed to update the representation (fig. 3). Wavelet occupancy grids synthesize the advantages of both approaches: there is no redundancy and they make it possible to have multiscale editing by storing at finer scale only the differences with the previous coarser scale (see section II-B). Furthermore this representation allows compression by the elimination of redundant information where there are no significant additional details such as for empty or uncertain spaces with a theoretical analysis of information loss. In addition, this paper describe a real-time algorithm for hierarchical updating of the occupancy representation in 3D for the first time.

In order to build the map, a standard approach, [10], will use an intermediate standard grid representation on which a wavelet transform will be performed. Even if a 2D wavelet transform can be performed in real-time, the extension to the case of a 3D transform in real-time is not apparent. So for a reasonable field of view, it makes the previous method unfeasible for 3D data. Our algorithm overcomes this difficulty with a hierarchical strategy that updates only the relevant areas of the environment and at the proper scale. In a first section, we will present the wavelet framework and the data structure while underlining differences with probabilistic 2^d-trees. In a second section the sensor model within the occupancy grid framework for the wavelet space is described. Next, we present the wavelet hierarchical rasterization algorithm. Lastly, we present our results in 2D on real data and in simulated 3D data where correct localisation is provided. Although in all the paper the algorithm is described for any kind of range sensor, the implementation and the experimental section are with laser data only.

II. Wavelets

In this paper, the occupancy state is represented as a spatial function. Our main contribution is an occupancy updating technique that can be performed in a compact manner. At the heart of the method is wavelet representation, which is a popular tool in image compression. Indeed, there exists a similarity between OGs and images [8]. The wavelet transform known as the Mallat algorithm successively averages each scale, starting from the finest scale (fig. 1, from right to left). This produces an oracle predicting the information stored in finer cells, then only differences from the oracle are encoded. This averaging produces the next coarser scale and differences with neighboring samples at the fine scale gives the associated so called detail coefficients. There is no loss of information in that process since the information contained in the finer scale can be recovered from its average and detail coefficients. Since two neighboring samples are often similar, a large number of the detail coefficients turn out to be very small in magnitude. Truncating or removing these small coefficients from the representation introduces only small errors in the reconstructed signal, giving a form of lossy signal compression. Lossless compression is obtained by removing only zero coefficients. In this paper wavelets are just used as a special kind of vector space basis that allows good compression. It is beyond the scope of this paper to give details about wavelet theory; references can be found in [16] [17] [18].

A. Notations

Wavelets are built from two sets of functions: scaling and detail functions (also known as wavelet functions). Scaling functions, $\Phi(x)$, capture the average or lower frequency information and a scaling coefficient is noted s_t^l. Detail functions, $\Psi(x)$, capture the higher frequency information and a detail coefficient for a detail function f is noted $d_{t,f}^l$. The set of wavelet basis functions can be constructed by the translation and dilation of the scaling and detail functions. Thus each of the basis functions or coefficients is indexed by a scale l and a translation index t. Moreover a detail function is indexed by its type f. In this paper, the non-standard Haar wavelet basis is used. For non-standard Haar wavelet basis, there is only one mother scaling function and $2^d - 1$ mother wavelet functions, where d is the dimension of the signal. Expanding a function O in the Haar wavelet basis is described as:

$$O(x) = s_0^{-N}\Phi_0^{-N} + \sum_{l=-N}^{l=0} \sum_t \sum_f d_{t,f}^l \Psi_{t,f}^l, \qquad (1)$$

where f is an index from 1 to $2^d - 1$, and N the level such that the whole grid appears as one cell. As can be seen in eq. 1, only one scaling coefficient and one scaling function are required in the expansion of any function $O(x)$. As shown in fig. 1, the scaling coefficients at other levels are computed as part of the decompression (from left to right) or compression (from right to left) processes.

The scaling coefficient for a certain level l and translation t holds the average of values contained in the support of the scaling function. The support of any Haar basis function in dimension d is a d-cube *e.g.* a square in 2D and a cube in 3D. If the finest level is 0 and coarser levels are indexed by decreasing negative integers, the side of such a d-cube is 2^{-l} where the unit is in number of samples at level 0.

B. Tree structure

The key step in a wavelet decomposition is the passage from one scale to another. The support of a Haar wavelet function at level l is exactly partitioned by the support of the 2^d wavelet functions at level $l+1$, (see Fig. 1 for dimension 1). This leads to a quadtree for the case of a 2D space or an octree for a 3D space. Each representation hierarchically maps the whole explored space. A node of the tree stores $2^d - 1$ detail coefficients and potentially 2^d children that encode finer details if they are necessary to reconstruct the expanded function. The key step of a node creation is described in fig. 2. Only 3 coefficients in 2D remain at each leaf while at each node is recorded the mean occupancy of the underlying area. In a standard quadtree, 4 coefficients are necessary and, for example in [9], the storage of the mean is redundant, whereas in the wavelet representation it is not.

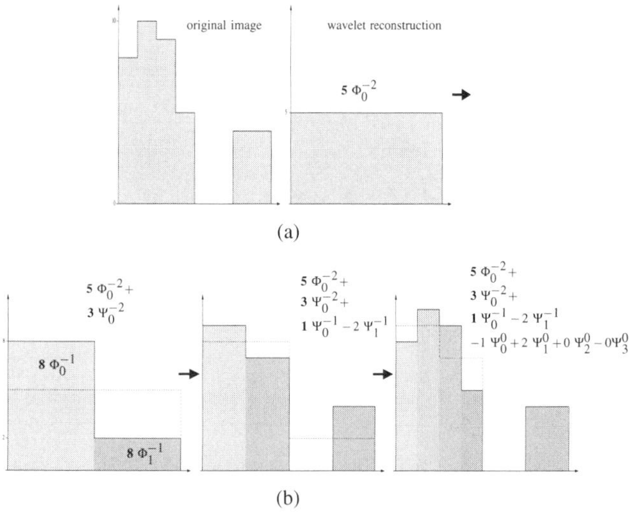

Fig. 1. The 1D image (upper, left) is: $[8,10,9,5,0,0,4,4]$, and its unnormalized (used here because it is simpler to display) Haar representation is: $[5,3,1,-2,-1,2,0,0]$. The image is then reconstructed one level at a time as follows: $[5] \rightarrow [5+3,5-3]=[8,2] \rightarrow [8+1,8-1,2-2,2+2]=[9,7,0,4]$ and so on. Here 0 is the finest scale index or the scale where data is gathered and -2 is the coarsest scale. As in one dimension there is only one kind of detail function, the subscripts refers only to translation (t) indices of eq. (1).

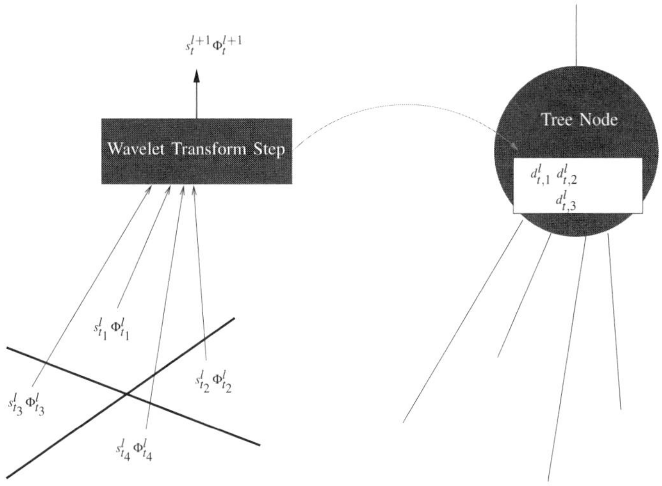

Fig. 2. A key step in a Haar wavelet transform in 2D. Four scaling samples at scale l generate 1 coarser scaling coefficient at scale $l+1$ and 3 detail coefficients at scale l that are stored in a wavelet tree node. In general the tree node has 4 children that describe finer resolutions for each space subdivision. But if each child is a leaf and has only zero-detail coefficients then all the child branches can be pruned without information loss. And the tree node becomes a leaf.

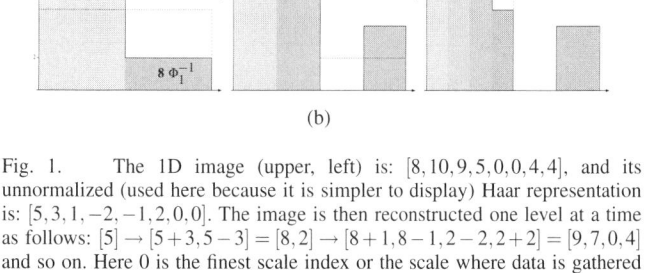

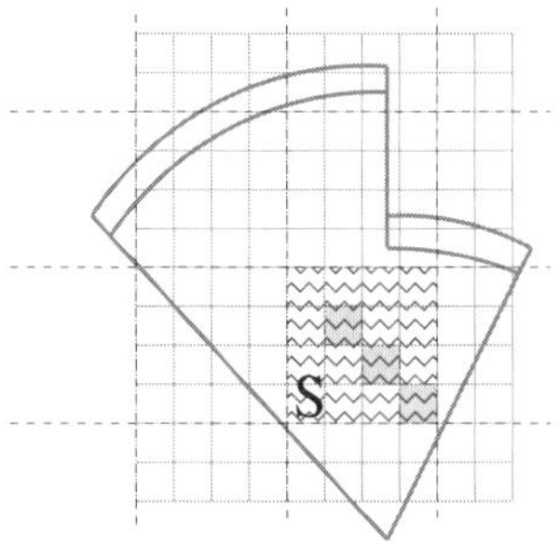

Fig. 3. Hierarchical updates are possible with wavelets: in a probabilistic quadtree all the pale/yellow squares are to be updated since they fall into the empty space. With wavelets the scale of the large square with waves is the last one that requires updating; therefore, the computation is efficient. The frontier of the coarse cells of the quadtree are marked in dashed/black lines.

The Haar wavelet data structure is exactly a 2^d-tree for its topology but not for the encoded data. Therefore the indexing of a cell with wavelet OG is as fast as with probabilistic 2^d-trees, however the retrieving of occupancy needs a small number of inverse wavelet transform operations[1]. Furthermore it not only stores spatially organized data, but also summarizes the data at different resolutions, which enables hierarchical updating. For example, fig. 3, the occupancy of all finest squares inside the empty area (fig. 6) decreases of the same amount, thus in the coarsest cell of the quadtree with waves the update is a constant. Also, the mean of the update over the waved square equals the value of the update for each finer cell, so all finer wavelet coefficients of the update are zero. As the update process is just the sum of the map wavelet representation with the update wavelet representation (section III), it produces efficient updates in areas that are coherent in the observation. Coherent areas are those that are adjacent to each other and have the same occupancy.

At the top of the structure, the root of the tree stores the scaling coefficient at the coarsest level and the support of the corresponding scaling function includes all the spatial locations of the signal data or the bounding box of the observed places.

III. Occupancy Grids and Range Sensor Models

OG is a very general framework for environment modelling associated with range sensors such as laser range-finders, sonar, radar or stereoscopic video camera. Each measurement of the sensor consists of the range to the nearest obstacle for a certain direction. Thus a range measurement divides the space into three areas: an *empty* space before the obstacle, an *occupied* space at the obstacle location and the *unknown* space everywhere else. In this context, an OG is a stochastic tessellated representation of spatial information that maintains probabilistic estimates of the occupancy state of each cell in a lattice [8]. In this framework, every cell are independently updated for each sensor measurement, and the only difference between cells is their positions in the grid. The distance which we are interested in, so as to define cell occupancy, is the relative position of the cell with respect to the sensor location. In the next subsection, the Bayesian equations for cell occupancy update are specified with cell positions relative to the sensor.

A. Bayesian cell occupancy update.

a) Probabilistic variable definitions:

[1]The number of operations is a small constant (4 sums and 2 multiplications in 2D, 7 sums and 3 multiplications in 3D) per scale and the number of scales is the depth of the tree which is logarithmic ($\ln_4$ in 2D and $\ln_8$ in 3D) in the number of cells.

- Z a random variable[2] for the sensor range measurements in the set $\mathscr{Z}$.
- $O_{x,y} \in \mathscr{O} \equiv \{\text{occ}, \text{emp}\}$. $O_{x,y}$ is the state of the cell (x,y), where $(x,y) \in \mathbb{Z}^2$. $\mathbb{Z}^2$ is the set of indexes of all the cells in the monitored area.

b) Joint probabilistic distribution: the lattice of cells is a type of Markov field and in this article the sensor model assumes cell independence. This leads to the following expression for the joint distribution for each cell.

$$P(O_{x,y}, Z) = P(O_{x,y})P(Z|O_{x,y}) \tag{2}$$

Given a sensor measurement z we apply the Bayes rule to derive the probability for cell (x,y) to be occupied 4:

$$p(o_{x,y}|z) = \frac{p(o_{x,y})p(z|o_{x,y})}{p(\text{occ})p(z|\text{occ}) + p(\text{emp})p(z|\text{emp})} \tag{3}$$

The two conditional distributions $P(Z|\text{occ})$ and $P(Z|\text{emp})$ must be specified in order to process cell occupancy update. Defining these functions is an important part of many works ([8], [19]). The results in [20] prove that for certain choices of parameters[3] these functions are piecewise constants:

$$p(z|[O_{x,y} = \text{occ}]) = \begin{cases} c_1 \text{ if } z < \rho \\ c_2 \text{ if } z = \rho \\ c_3 \text{ otherwise.} \end{cases} \tag{4}$$

$$p(z|[O_{x,y} = \text{emp}]) = \begin{cases} c_1 \text{ if } z < \rho \\ c_4 \text{ if } z = \rho \\ c_5 \text{ otherwise.} \end{cases} \tag{5}$$

where ρ is the range of the cell (x,y).

As explained in [10], the cell update requires operations that are not part of the set of wavelet vector operations[4] (product and quotient). Thus a better form is necessary to operate updates on the wavelet form of occupancy functions.

B. Log-ratio form of occupancy update

As occupancy is a binary variable, a quotient between the likelihoods of the two states of the variable is sufficient to describe the binary distribution. The new representation used is:

$$\text{log-odd}(O_{x,y}) = \log \frac{p([O_{x,y} = \text{occ}])}{p([O_{x,y} = \text{emp}])} \tag{6}$$

In the Bayesian update of the occupancy, the quotient makes the marginalization term disappear and thanks to a logarithm transformation, sums are sufficient for the inference:

[2]For a certain variable V we will note in upper case the variable, in lower case v its realization, and we will note $p(v)$ for $P([V = v])$ the probability of a realization of the variable.

[3]The sensor model failure rate, the sensor range discretization and the prior occupancy probability are the parameters. Prior occupancy is chosen very low, the world being assumed nearly empty. Only the last parameter is relevant for establishing that the functions are piece-wise constant [20].

[4]The product and the quotient operators are not base inner operators of a vector space.

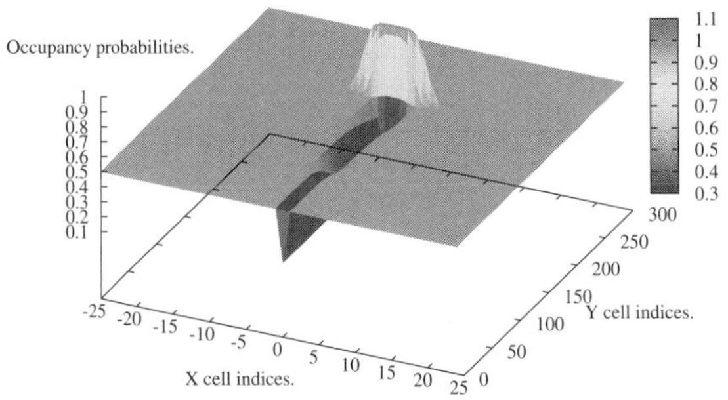

Fig. 4. Update of a 2D OG after a sensor reading, initially each cell occupancy was unknown, *i.e.* 0.5 probability. The sensor beam has an aperture of 7 degrees. The sensor is positioned in (0,0).

$$\log \frac{p(\text{occ}|z)}{p(\text{emp}|z)} = \log \frac{p(\text{occ})}{p(\text{emp})} + \log \frac{p(z|\text{occ})}{p(z|\text{emp})} = \text{log-odd}_0 + \text{log-odd}(z) \tag{7}$$

Therefore the vector space generated by the wavelet basis with its sum inner operator is sufficient to represent and update OGs. This inference with sums was originally proposed by Elfes and Moravec [8], but only for performance reasons. Here it is also necessary to allow inference to be performed within the compressed data.

C. Log-ratio form of sensor model functions

It is straightforward to derive from eq. 4 and 5 the sensor model equations in log-ratio form that we note thus:

$$\text{log-odd}(z) = \begin{cases} 0 \text{ if } z < \rho \\ \log(c_2/c_4) = \text{log-odd}_{\text{occ}} \text{ if } z = \rho \\ \log(c_3/c_5) = \text{log-odd}_{\text{emp}} \text{ otherwise.} \end{cases} \tag{8}$$

where ρ is the range of the cell (x,y), way to define each constant is given in [20][5]. One can notice that the update term is zero if the cell is beyond the sensor readings, thus no update is required in this case.

IV. Hierarchical rasterization of polygon or polyhedron

This section describe the main contribution of this article which consists of a fast algorithm for updating an occupancy grid expanded as a non-standard Haar wavelet series from a set of range measurements.

A. Problem statement

Given the sensor position, the beam geometry and the measured ranges, it is possible to define the polygon (fig. 6) or polyhedron viewed by the sensor within the grid. Each time the sensor position changes or measured ranges change a new

[5]in the experiments: $\text{log-odd}_{\text{emp}} = -5.46$ and $\text{log-odd}_{\text{occ}} = 16.97$

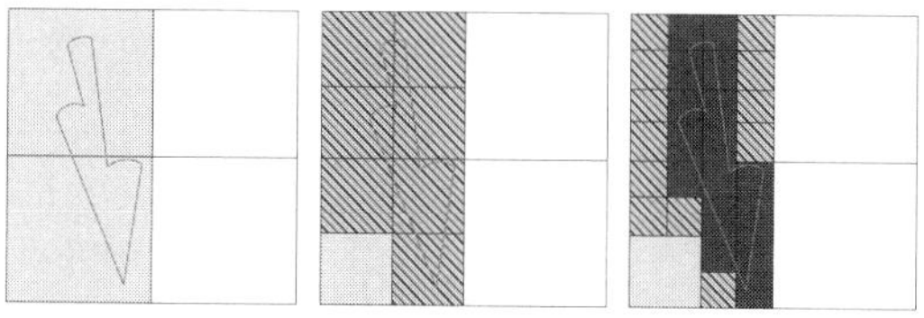

Fig. 5. The hierarchical process of updating the grid: from the coarsest scale to the finest. To save computing time, areas that are outside the polygon of view or totally included inside the areas classified as empty are detected and processed early in the hierarchy.

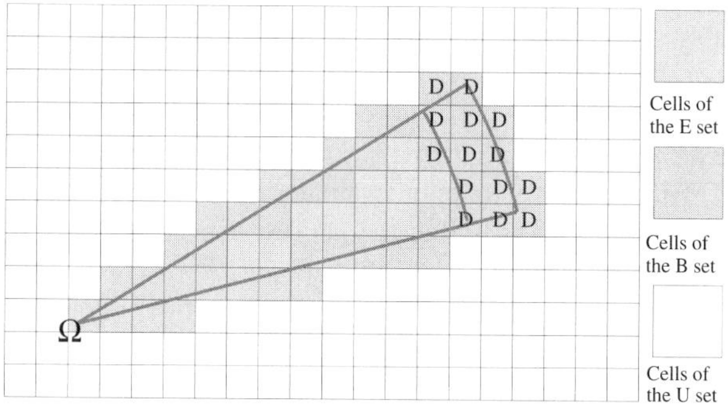

Fig. 6. A range-finder beam. The range finder is located at Ω and its field of view is surrounded by red/thick boundaries. It defines the three kinds of cell types. The band within the obstacle lies is at the top right end of the field of view. Thus the cells marked with a "D" stand for cells where a detection event occurs.

relative position of the polygon or polyhedron and the grid must be computed in order to update the grid. The standard approach for updating occupancy grids, in the context of laser sensors, will be to traverse the cells along each laser sensor beam and update the cells. This traversal method induces difficulties in calculating the coverage area for each laser sensor beam in order to avoid inaccuracies such as aliasing. An easier alternative will be to traverse every cell of the grid and for each cell, perform a simple test to determine the state of the cell. In this case, with a grid size of 1024 cells per dimension, a 2D square grid contains more than 1 million cells and a 3D cubic grid contains more than 1 billion. Even if real-time performance can be obtained in 2D, it does not seem to be the case in 3D. Therefore the problem is to find a method that efficiently updates the grid without traversing every cell of the grid. As shown in fig. 6 and eq. 8, a range measurement defines three sets of cells. The first set, E, contains cells that are observed as empty. The second set, U, contains cells that are considered as unknown. The third set, B (for boundaries), contains cells that are partially empty, unknown or occupied. The elements of the third set are mainly found at the boundaries formed by the sensor beams at its two extreme angles and at the neighborhood of an obstacle. Section III-C states that the U set can be avoided in the update process. Therefore an update step must iterate through the cells that intersect either the polygon in 2D or the polyhedron in 3D that describe the sensor beam boundaries (fig. 5). The following describes an algorithm that performs the correct iteration through the grid in an efficient manner through the use of wavelets.

B. Exhaustive hierarchical space exploration

The key idea in the exploration of the grid space (fig. 5) is to define a predicate, *existIntersection*, which is true if a given set of grid cells intersect the volume defined by the field of view of the sensor beams (blue/dark gray plus red/medium gray cells in fig. 6). The absence of intersection indicates that the given set of cells are outside the sensor field of view and do not need updating. When *existIntersection* returns true, a special sub case needs to be considered in addition: if the set of cells is totally included in the sensor field of view, all the cells belong to E (blue/dark gray cells in fig. 6) and their occupancy is decreased by the same amount of log-odd$_{\text{emp}}$, eq. 7.

As the algorithm is able to detect uniform regions recursively, the grid representation should allow to update those regions, and wavelets provide a natural mechanism for doing so. In this first version of the algorithm, the grid is traversed hierarchically following the Haar wavelet support partition. For each grid area, the *existIntersection* predicate guides the search. If there is an intersection, the traversal reaches deeper into the grid hierarchy, *i.e.* exploring finer scales. Otherwise it stops at the current node. Then the wavelet transform is performed recursively beginning from this last node as described in fig. 2 for the 2D case.

Algorithm 1 HierarchicalWavRaster(subspace S, sensor beam B)

for each subspace i of S: $i = 0, \ldots, n$ **do**
  if sizeof(i) = minResolution **then**
    v_i = evalOccupancy(i)
  else if existIntersection(i, B) **then**
    if $i \in E$ **then**
      $v_i = \text{log-odd}_{\text{emp}}$ /*eq. 8*/
    else
      v_i = HierarchicalWavRaster(i, B)
    end if
  else
    $v_i = 0$ /*$i \in U$*/
  end if
end for
$\{s_S^{l+1,obs}, d_{f_1,S}^{l,obs}, \cdots, d_{f_n,S}^{l,obs}\}$ =waveletTransform($\{v_0, \cdots, v_n\}$)
for each $d_{f,S}^l$: **do**
  $d_{f,S}^l \leftarrow d_{f,S}^l + d_{f,S}^{l,obs}$ /*update inference*/
end for
returns the scaling coefficient $s_S^{l+1,obs}$

Algorithm 1 gives the pseudo-code of the exhaustive hierarchical grid traversal. Here n is the maximum index of the space subdivisions that a node contains at one finer scale *i.e.* 3 for 2D and 7 for 3D Haar wavelet transforms. The algorithm is recursive and begins with the whole grid as the first subspace defined by the root of the wavelet tree. Its result

is used to update the mean of the wavelet tree which is also the coefficient of the scaling function at the coarsest level. The *sizeof* function gets the resolution of the subspace i and *minResolution* represents the resolution of a cell in the grid. The *evalOccupancy* function evaluates the occupancy of a cell; it can proceed by sampling the cell occupancy.
Such an algorithm is very efficient in 2D but as it refines every area on the sensor beam boundaries it explores at least the whole perimeter of the polygon of view in 2D (red/medium gray cells in fig. 6). Equivalently in 3D, the explored part is all the surface of the polyhedron of view and it is far too large to be explored in real-time. That is why a better algorithm is required.

C. Improved hierarchical space exploration

Most of the space where a robot is to move about is largely empty. Thus it is not efficient to begin with a map initialized with a probability of 0.5 since this probability will decrease almost everywhere toward the minimum probability p_{emp}. Equivalently, since each boundary between an area observed as an empty one and an area outside the sensor field of view separates cells that are almost all empty, updating occupancy along this boundary is useless. Following this remark algorithm 1 is modified in a lazy algorithm that investigates finer iterations through the grid only if an update is required.

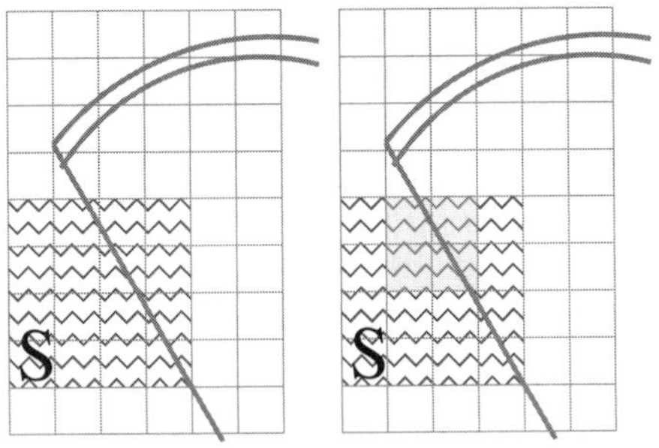

(a) (b)

Fig. 7. Two different cases for the iteration along a boundary of the field of view that separates the E set and the U set. Fig. 7(a) artificial separation, S (with waves) was totally empty and the observation of a part of its interior (on the right of the red boundary) does not bring any information gain. Fig. 7(b) the separation brings information about the state of a part of an obstacle: the yellow/pale square area that is inside the field of view (on the right of the red/thick boundary).

An update is almost always required for cells that are in the obstacle neighborhood (cells marked with 'D' in fig. 6) so iteration is always performed in areas that contain such a cell. But for boundaries that separate cells that belong to the U set and to the E set, iteration is required only if the E set corrects the knowledge in the grid (fig. 7(b)); otherwise the iterations can stop early in the hierarchy (fig. 7(a)).

In algorithm 2 three main differences appear: first an inverse wavelet transform is performed to retrieve the information about the current state of the traversed subspace (line $1-2$). Second, line 7, the intersection function returns *OCCUPIED*

Algorithm 2 HierarchicalWavRaster(subspace S, mean occupancy of subspace S: s_S^{l+1}, empty bound p_{emp}, sensor beam B)

```
1:  {v_0^g, ..., v_n^g} = inverse
2:      WaveletTransform({s_S^{l+1}, d^l_{f1,S}, ..., d^l_{fn,S}})
3:  for each subspace i of S: i = 0,...,n do
4:    if sizeof(i) = minResolution then
5:      v_i = evalOccupancy(i)
6:    else
7:      spaceState = existIntersection( i, B )
8:      if spaceState is UNKNOWN then
9:        v_i = 0
10:     else if spaceState is OCCUPIED then
11:       v_i = HierarchicalWavRaster( i, B )
12:     else if spaceState is EMPTY and v_i^g > p_emp then
13:       v_i = HierarchicalWavRaster( i, B )
14:     else if spaceState is EMPTY then
15:       v_i = log-odd_emp          /*eq. 8*/
16:     end if
17:   end if
18:   v_i^g <- v_i^g + v_i         /*update inference*/
19: end for
20: {δ_S^{l+1}, d^l_{f1,S}, ..., d^l_{fn,S}} = waveletTransform({v_0^g, ..., v_n^g})
21: returns the scaling coefficient s_S^{l+1,obs} = s_S^{l+1} - δ_S^{l+1}
```

only if the subspace intersects an obstacle and it returns *EMPTY* if the subspace is included in $E \cup U$. Third the value of the minimum possible occupancy p_{emp} is a parameter of the algorithm in order to compare the state of the traversed subspace with information gain brought by the sensor observations (line 12).
The major difference between the maps produced by the first and the second algorithm is that in the second algorithm there is no *a priori* unknown area. Thus it is no longer possible to store the position of the unexplored parts of the world. This could be a problem if one wants to drive the robot toward *terra incognita*. Nevertheless the information concerning unknown areas is used all the same in the definition of the polygon of view, so that occlusions are handled when observations are processed.
One of the most important parts of the previous algorithms are the intersection queries: the definition of *existIntersection*. These functions must be really optimized in order to retrieve fast algorithms. Each kind of range sensor requires its own implementation of *existIntersection*. A simple implementation of such a function is easy to write since it involves only geometric intersection primitives, therefore we will not describe one extensively here for lack of space. In our own implementation we have used an explicit representation of a polygon or polyhedron of the sensor view with vertices and edges and implicit representation of a grid cell with its index. Then the polygon-polygon or the polyhedron-polyhedron intersection is computed, if this test fails an inclusion test is performed to test if one object is included in the other.

The beams where a max-range reading occur, which could be produced by non-reflective surfaces, are safely replaced by totally unknown area. Therefore in presence of such readings the polygon/polyhedron is splitted into several parts connected by the laser-scanner origine.

V. Experiments

We performed experiments[6] on 2D real data sets and on 3D simulated data with noise. In the 3D simulations a rotating sick was used. For all the experiments the position of the laser were given. For real experiments, corrected data sets were used[7], whereas for simulated ones[8], the simulator provided us with the correct sensor position. For 2D and 3D both, we use data sets that contain only static obstacles and data sets that contain moving obstacles also. We test first and second algorithm on 2D data sets and only second algorithm on 3D data sets.

A. Methodology

c) Memory: for all experiments we give the memory used by an OG, a probabilistic-tree with and without mean stored and a wavelet OG at the end of the mapping. As probabilistic-tree and wavelet OG cover grids with size of power of two, we do not use OG with equivalent size to compare which would have been unfair. Instead, we compute the bounding box of all the range measurements and use an OG of the same size of that bounding box.

d) Computing time: in 2D we compare two things: first the use of the polygon rasterization algorithm against ray tracing with a Bresenham algorithm and OG rasterization using inverse sampling with a standard OG, second the hierarchical algorithms with probabilistic-trees and wavelet OG. We do not give results for other algorithms than polygon rasterization on hierarchical representations. Since all the cell accesses with the other algorithms are random accesses witch is totally inefficient with hierarchical rasterization, the results are not of interest. In 3D we only present the results with the wavelet OG representation. The mean, min and max of the update time per scan are computed.

e) Map quality: : in 2D, we evaluate map quality by comparing the resulting map with the one obtained by the OG rasterization using inverse sampling by computing the l_2 norm of the difference of the 2 grids. In 3D the quality is just visually estimated.

B. Results

f) Memory: These results show that the wavelet and probabilistic-tree performs the same concerning memory saving, witch follows the theory. As predicted, the Probabilistic-trees with the mean are however a bit more expansive. Both representations saves, in average, about 91% for 2D grids and 94% for 3D grids of the required memory compared with a classic OG representation. The amount of memory saved is larger in 3D than in 2D because the proportion of empty space is far more important.

g) Computing time: For the comparison of update algorithm on the same OG representation, polygon rasterization and Bresenham performs almost the same witch is interesting since Bresenham does not handled beam width and is therefore far less accurate than polygon rasterization. They both performs far better than the inverse sampling. The second algorithm performs better for both representations: probabilistic-tree and wavelet OG, even if an inverse wavelet transform is computed in the last case (10 times faster in 2D, and 20 times faster in 3D). The probabilistic-tree performs better on static environments, although the difference is not of importance. Probabilistic-tree with mean performs only slightly better on static environments than wavelet OG, since, as wavelet OGs, they compute the mean. Concerning a dynamic environment, wavelet OG is slightly faster, which we reward the true multi-scale nature of this representation.

h) Quality: For 2D and 3D grids, comparisons with a standard grid construction algorithm show that there are no significant differences. In the 3D results (fig. 8), part of the ground (on the right of the map) is not entirely mapped because the density of measurements is not uniform but depends on the vehicle velocity. As the map is considered empty *a priori* unseen parts of the ground appear as holes. Thus it would be interesting to use a ground model to initialize the map in the future works. Then, ground measurements would only correct the *a priori* and that would save a lot of computing time, as the ground is the main obstacle.

VI. Conclusions and Future Works

A. Conclusions

The objective of this work is to present new algorithms that make OG building in wavelet space feasible. We show that wavelet space is naturally a good space to represent huge functions such as occupancy functions in 3D. In contrast to previous works, we do not need intermediate representation to build and fuse OGs in wavelet space with the new wavelet hierarchical rasterization algorithms. Thanks to the hierarchical organization of the computation, computing time is definitely sufficient for real-time in 2D and enough for real-time in 3D. With that achievement, the main contribution of this work is to present an OG updating algorithm which is also useful in 3D. The use of Haar wavelets bring no significant computation speed-up or pittfall compare to probabilistic-tree with mean but is slightly better in memory saving. Our long-term objective is to use the unified grid-based fast-slam framework in 3D environments. The requirements for an environment representation suitable for fast-slam are:

1) fast updating and scan matching to construct the map and calculate the current robot's pose in real time,
2) a hierarchical grid representation to handle multiple maps in multiple particles efficiently,
3) a small amount of memory per grid to ensure efficiency in the previously stated conditions.

The last two requirements and half of the first one are fulfilled by this work; thus it is now possible to consider very powerful

[6] All experiments were done with an Intel$^{(R)}$ Pentium$^{(R)}$ IV CPU 3.00*GHz*.
[7] CSAIL (MIT), FR-Campus, FR079, FR101, thanks to Cyrill Stachniss [21]
[8] Inria static parking, Inria parking with a moving car

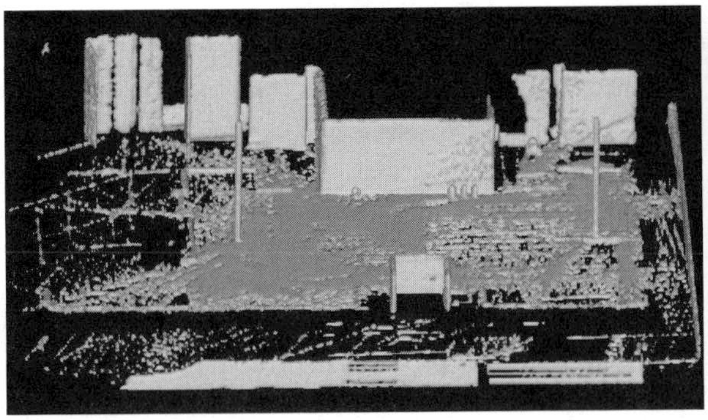

(a)

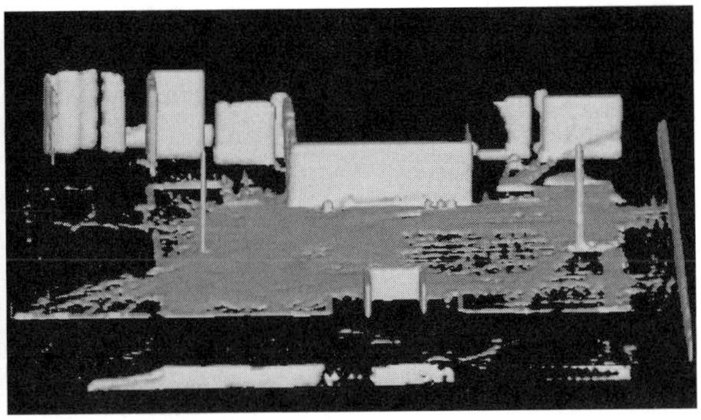

(b)

Fig. 8. The wavelet OG obtained from a simulation of 3D data gathering with a rotating laser range-finder. Fig. 8(a) and 8(b) provide two views of the reconstruction of the grid from the wavelet grid at scale −1 (cell side of $0.20m$) and scale −2 (cell side of $0.40m$). It is noticeable that salient details as the lamp-post or the 4 pole landmarks before the wall are accurately mapped. The wall is interestingly smooth too, and that is a feature obtained by the oracle of the scaling view: details appear at finer views.

algorithms such as a navigation grid-based fast-slam or grid-based multiple-target tracking in 3D based upon wavelet occupancy grids.

B. Future Works

In the future we will explore several major areas of improvement. As the intersection query is the most time-consuming part of the algorithm, we plan to work first on optimizing this part of the algorithm. Another area of improvement is the kind of wavelets that is used to compress the map. Haar wavelets are the poorest kind of wavelets for compression properties, so it will be interesting to work with higher order wavelets that are able to compress much more complex functions such as quadrics because it will approximate a map with locally Gaussian occupancy density in a far better way for example. Finally, the tree structure of the data allows parallel traversal of the environment and we plan to develop parallel instances of the hierarchical rasterization algorithm. The proposed algorithm is a general one and its validity area is theoretically the set of all range sensors. We plan to apply this algorithm using other kinds of range sensors such as a stereo camera. However, our main objective is now to derive a localization algorithm based on this grid representation to obtain a complete grid-based slam algorithm in 3D.

REFERENCES

[1] J. Gutmann and K. Konolige, "Incremental mapping of large cyclic environments," in *Proc. of the IEEE International Conference on Robotics and Automation (ICRA)*, Monterey, California, November 1999, pp. 318–325.

[2] M. Bosse, P. Newman, J. Leonard, M. Soika, W. Feiten, and S. Teller, "An atlas framework for scalable mapping," in *Proc. of the IEEE International Conference on Robotics and Automation (ICRA)*, 2003.

[3] G. Grisetti, C. Stachniss, and W. Burgard, "Improving grid-based slam with rao-blackwellized particle filters by adaptive proposals and selective resampling," in *Proc. of the IEEE International Conference on Robotics and Automation (ICRA)*, 2005, pp. 2443–2448.

[4] D. Cole and P. Newman, "Using laser range data for 3d slam in outdoor environments," in *Proc. of the IEEE International Conference on Robotics and Automation (ICRA)*, Florida, 2006.

[5] D. Hähnel, D. Schulz, and W. Burgard, "Map building with mobile robots in populated environments," in *Proc. of the IEEE/RSJ International Conference on Intelligent Robots and Systems (IROS)*, 2002.

[6] K. P. Murphy, "Bayesian map learning in dynamic environments," in *NIPS*, 1999, pp. 1015–1021.

[7] A. Eliazar and R. Parr, "DP-SLAM: Fast, robust simultaneous localization and mapping without predetermined landmarks," in *Proc. of the Int. Conf. on Artificial Intelligence (IJCAI)*, Acapulco, Mexico, 2003, pp. 1135–1142.

[8] A. Elfes, "Occupancy grids: a probabilistic framework for robot perception and navigation," Ph.D. dissertation, Carnegie Mellon University, 1989.

[9] G. K. Kraetzschmar, G. Pagès Gassull, and K. Uhl, "Probabilistic quadtrees for variable-resolution mapping of large environments," in *Proceedings of the 5th IFAC/EURON Symposium on Intelligent Autonomous Vehicles*, M. I. Ribeiro and J. Santos Victor, Eds., July 2004.

[10] M. Yguel, O. Aycard, and C. Laugier, "Wavelet occupancy grids: a method for compact map building," in *Proc. of the Int. Conf. on Field and Service Robotics*, 2005.

[11] M. Levoy, K. Pulli, B. Curless, S. Rusinkiewicz, D. Koller, L. Pereira, M. Ginzton, S. Anderson, J. Davis, J. Ginsberg, J. Shade, and D. Fulk, "The digital michelangelo project: 3D scanning of large statues," in *Siggraph 2000, Computer Graphics Proceedings*, ser. Annual Conference Series, K. Akeley, Ed. ACM Press / ACM SIGGRAPH / Addison Wesley Longman, 2000, pp. 131–144.

[12] S. Thrun, C. Martin, Y. Liu, D. Hähnel, R. Emery Montemerlo, C. Deepayan, and W. Burgard, "A real-time expectation maximization algorithm for acquiring multi-planar maps of indoor environments with mobile robots," *IEEE Transactions on Robotics and Automation*, vol. 20, no. 3, pp. 433–442, 2004.

[13] M. Hebert, C. Caillas, E. Krotkov, I. S. Kweon, and T. Kanade, "Terrain mapping for a roving planetary explorer," in *Proc. of the IEEE International Conference on Robotics and Automation (ICRA)*, vol. 2, May 1989, pp. 997–1002.

[14] R. Triebel, P. Pfaff, and W. Burgard, "Multi-level surface maps for outdoor terrain mapping and loop closing," in *"Proc. of the International Conference on Intelligent Robots and Systems (IROS)"*, 2006.

[15] P. Payeur, P. Hébert, D. Laurendeau, and C. Gosselin, "Probabilistic octree modeling of a 3-d dynamic environment," in *Proc. IEEE ICRA 97*, Albuquerque, NM, Apr. 20-25 1997, pp. 1289–1296.

[16] I. Daubechies, *Ten Lectures on Wavelets*, ser. CBMS-NSF Series in Applied Mathematics. Philadelphia: SIAM Publications, 1992, no. 61.

[17] S. Mallat, *A Wavelet Tour of Signal Processing*. San Diego: Academic Press, 1998.

[18] E. J. Stollnitz, T. D. Derose, and D. H. Salesin, *Wavelets for Computer Graphics: Theory and Applications*. Morgan Kaufmann, 1996.

[19] S. Thrun, "Learning occupancy grids with forward sensor models," *Autonomous Robots,*, vol. 15, pp. 111–127, 2003.

[20] M. Yguel, O. Aycard, and C. Laugier, "Efficient gpu-based construction of occupancy grids using several laser range-finders," *Int. J. on Vehicle Autonomous System*, to appear in 2007.

[21] C. Stachniss, "Corrected robotic log-files." http://www.informatik.uni-freiburg.de/ stachnis/datasets.html.

Gaussian Beam Processes: A Nonparametric Bayesian Measurement Model for Range Finders

Christian Plagemann Kristian Kersting Patrick Pfaff Wolfram Burgard
Albert-Ludwigs-University of Freiburg, Department for Computer Science, 79110 Freiburg, Germany
{plagem,kersting,pfaff,burgard}@informatik.uni-freiburg.de

***Abstract*— In probabilistic mobile robotics, the development of measurement models plays a crucial role as it directly influences the efficiency and the robustness of the robot's performance in a great variety of tasks including localization, tracking, and map building. In this paper, we present a novel probabilistic measurement model for range finders, called Gaussian beam processes, which treats the measurement modeling task as a nonparametric Bayesian regression problem and solves it using Gaussian processes. The major benefit of our approach is its ability to generalize over entire range scans directly. This way, we can learn the distributions of range measurements for whole regions of the robot's configuration space from only few recorded or simulated range scans. Especially in approximative approaches to state estimation like particle filtering or histogram filtering, this leads to a better approximation of the true likelihood function. Experiments on real world and synthetic data show that Gaussian beam processes combine the advantages of two popular measurement models.**

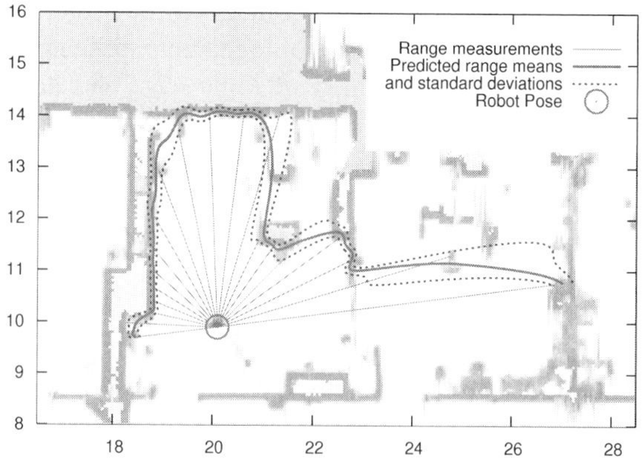

Fig. 1. The predictive distribution of range measurements for an uncertain robot pose within an office environment (heading angle varies by $\pm 5°$). Scales are given in meters. The straight, red lines depict one possible range scan in this setting.

I. Introduction

Acquiring, interpreting, and manipulating information from sensors is one of the fundamental tasks within mobile robotics. For instance, based on models for the robot's kinematics and perception, a robot might be asked to perform tasks such as building a map of the environment, determining its precise location within a map, or navigating to a particular place. In designing robots to operate in the real world, one cannot avoid dealing with the issue of uncertainty. Uncertainty arises from sensor limitations, noise, and the fact that most complex environments can only be represented and perceived in a limited way. It is therefore not surprising that state-of-the-art approaches build on probabilistic foundations and model the robot's perception as a probability density $p(\mathbf{z}|\mathbf{x})$, where $\mathbf{z}$ is an observation and $\mathbf{x}$ denotes the state of the robot, e.g., its position relative to the environment.

Among the most widely used types of sensors are range sensors such as laser range finders and sonar sensors. The popularity of range sensors in research and industry is due to the fact that spatial information about the environment can directly be acquired and that state-of-the-art sensor hardware is quite accurate and reliable. Range sensors measure distances r_i to nearby objects along certain directions $\boldsymbol{\alpha}_i$ (possibly multivariate bearing angles) relative to the sensor. Hence, for a vector $\mathbf{r} = (r_1, \dots, r_m)$ of distance measurements with corresponding bearing angles $\mathcal{A} = (\boldsymbol{\alpha}_1, \dots, \boldsymbol{\alpha}_m)$, the likelihood function $p(\mathbf{z}|\mathbf{x})$ can be rewritten as $p(\mathbf{r}|\mathcal{A}, \mathbf{x})$.

In this paper, we propose a novel, generative model for $p(\mathbf{r}|\mathcal{A}, \mathbf{x})$, called Gaussian beam processes (GBP), which takes a nonparametric Bayesian regression view on the measurement modeling task. We treat the measurements $\mathbf{z} = (r_i, \boldsymbol{\alpha}_i)_{i=1}^m$ as a set of samples from a stochastic process $p(r|\boldsymbol{\alpha})$ and assume the process to be a Gaussian process [1], i.e., any finite collection of random variables r_j is assumed to have a joint Gaussian distribution. Learning in this framework means recording or simulating a training set of range scans and adjusting a predefined covariance function accordingly. This can be done online while the robot is operating or offline.

We put special emphasis on the application of GBPs to mobile robot localization, but nevertheless present the model in a general form that should be useful in many other applications for range sensor. A major benefit of our model in localization tasks is that it naturally also allows to estimate the distribution $p(\mathbf{r}|\mathcal{A}, \mathcal{X})$ of range measurements for a region $\mathcal{X}$ of the pose space. As an example, consider Figure 1. It shows the predictive distribution of range measurements for a mobile robot with an uncertain heading angle, i.e., its location is fixed, the orientation angle, however, is known only up to $\pm 5°$. It can be clearly seen from the visualized standard deviations of the range predictions that the model accurately identifies the distances that can be predicted with high confidence despite of the angular uncertainty in the sensor pose. The ability to

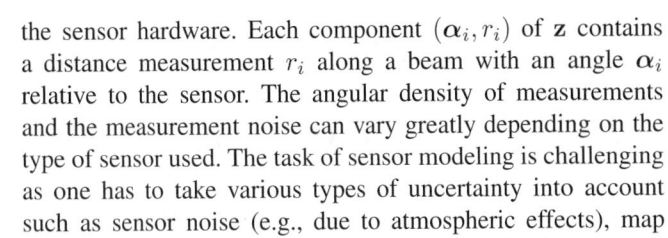

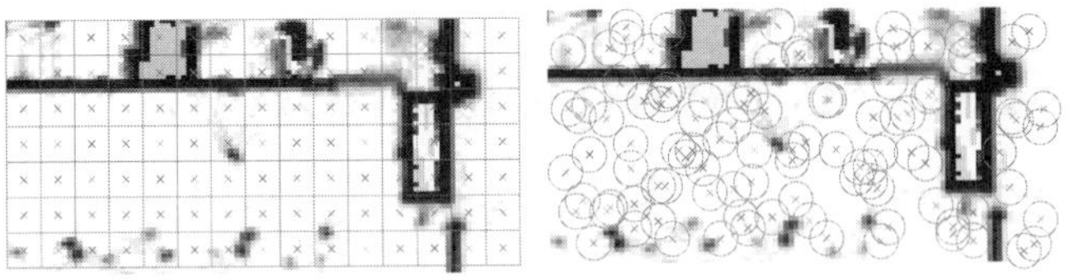

Fig. 2. The posterior distribution of robot poses is typically approximated by discretizing the pose space (left) or by sampling it (right).

learn and represent such distributions is of special value in applications in which the posterior is approximated using a discrete set of pose hypotheses. In histogram filtering, for example, the pose space is partitioned into a finite set of grid cells (see the left diagram of Figure 2). With the GBP model, we can estimate the observation likelihood $p(\mathbf{z}|\mathcal{X})$ for a whole grid cell $\mathcal{X}$ directly rather than having to numerically approximate $\frac{1}{|\mathcal{X}|}\int_{\mathcal{X}} p(\mathbf{z}|\mathbf{x})\, d\mathbf{x}$ from point estimates $p(\mathbf{z}|\mathbf{x})$, $\mathbf{x} \in \mathcal{X}$, of the likelihood function. This ability is also useful for particle filtering in which the posterior is represented by a finite set of weighted samples. It is a well-known fact that highly peaked likelihood functions have to be regularized in practical applications, because the number of particles is limited. A popular way of doing this is to locally average the likelihood function in the vicinity of each particle (see the right diagram of Figure 2), rather than taking point estimates at the exact particle locations only. Additionally, the Gaussian process treatment offers the following distinct benefits:

- The model is fully predictive as it is able to predict ranges at arbitrary bearing angles, i.e., also for angles in between two beams of an actual scan and for beams that have been classified as erroneous. For such predictions, the model also yields the predictive uncertainties.
- Neither the number of range measurements per scan nor their bearing angles have to be fixed beforehand.
- By representing correlations between adjacent beams using parameterized covariance functions, only few recorded or simulated range scans $(\mathcal{A}_j, \mathbf{r}_j)$ are required to learn an accurate model.
- Gaussian processes are mathematically well-established. There exists a great pool of methods for learning, likelihood evaluation, and prediction.

The paper is organized as follows. In Section II, we discuss related work. Section III presents our Gaussian beam process model. In Section IV, we describe its application to Monte Carlo localization and present the results of extensive evaluations on real world and synthetic data.

II. Related Work

Probabilistic measurement models (or observation models) are conditional probability distributions $p(\mathbf{z}|\mathbf{x})$ that characterize the distribution of possible sensor measurements $\mathbf{z}$ given the state $\mathbf{x}$ of the system. In the context of mobile robot localization with laser range finders, for instance, $\mathbf{x}$ denotes the three-dimensional pose (2D location and orientation) of the robot and $\mathbf{z}$ stands for a vector of range readings received from the sensor hardware. Each component $(\boldsymbol{\alpha}_i, r_i)$ of $\mathbf{z}$ contains a distance measurement r_i along a beam with an angle $\boldsymbol{\alpha}_i$ relative to the sensor. The angular density of measurements and the measurement noise can vary greatly depending on the type of sensor used. The task of sensor modeling is challenging as one has to take various types of uncertainty into account such as sensor noise (e.g., due to atmospheric effects), map and pose uncertainty (e.g., caused by discretization errors), and environmental dynamics like object displacements.

Feature-based approaches typically extract a set of features from the range scan $\mathbf{z}$ and match them to features contained in an environmental model in order to obtain $p(\mathbf{z}|\mathbf{x})$. Whereas such approaches have been proven to be robust in various applications, they assume that the features are known beforehand and that they can be extracted reliably, which might be hard in unstructured or cluttered environments. Alternative approaches directly operate on the dense measurements and therefore are applicable even in situations in which the relevant features are unknown.

Beam-based models consider each value r_i of the measurement vector $\mathbf{z}$ as a separate range measurement and represent its one-dimensional distribution by a parametric function depending on the expected range measurement in the respective beam direction (see Fox *et al.* [2] for example). Such models are closely linked to the geometry and the physics involved in the measurement process. In the remainder of this paper, we will also denote such models as ray cast models because they rely on ray casting operations within an environmental model, e.g., an occupancy grid map, to calculate the expected beam lengths. As a major drawback, the traditional approach assumes independent beams, which leads to overly peaked likelihood functions when one increases the number of beams per measurement (e.g., to increase the spatial resolution). In practice, this problem is dealt with by sub-sampling of measurements, by introducing minimal likelihoods for beams, by inflating the measurement uncertainty [3], or by other means of regularization of the resulting likelihoods (see, e.g., Arulampalam *et al.* [4]).

Correlation-based methods typically build local maps from consecutive scans and correlate them with a global map [5, 6]. A simple and effective approach that is also associated to this class of models is the so-called likelihood fields model or end point model [7]. Here, the likelihood of a single range measurement is a function of the distance of the respective end point of a beam to the closest obstacle in the environment. Like in the ray cast model, each beam is treated independently. This model lacks a physical explanation, as it can basically "see through walls", but it is more efficient than ray cast models and works well in practice.

Work that specifically dealt with peaked measurement models include Pfaff *et al.* [8], who adapt the smoothness of the likelihood model depending on the region covered by the individual particles, Fox *et al.* [9], and Kwok *et al.* [10], who adapt the number of particles depending on the progress of the localization process and computational power. These approaches have been developed independently from specific measurement models and should be directly applicable to GBPs as well. Finally, GBPs are related to Gutierrez-Osuna *et al.*'s [11] neural networks approach to modeling the measurement of an ultrasonic range sensor, which in contrast to GBPs assumes scans of fixed size.

Our GBP model, which is detailed in the following section, seeks to combine the advantages of the above mentioned approaches. It represents correlations between adjacent beams using covariance functions, it respects the geometry and physics of the measurement process, and it provides a natural and intuitive way of regularizing the likelihood function.

Gaussian processes have already received considerable attention within the robotics community. Schwaighofer *et al.* [12] introduced a positioning system for cellular networks based on Gaussian processes. Brooks *et al.*[13] proposed a Gaussian process model in the context of appearance-based localization with an omni-directional camera. Ferris *et al.*[14] applied Gaussian processes to locate a mobile robot from wireless signal strength. Plagemann *et al.* [15] used Gaussian processes to detect failures on a mobile robot. Indeed, Bayesian (regression) approaches have been also followed for example by Ting *et al.*[16] to identify rigid body dynamics and Grimes *et al.*[17] to learn imitative whole-body motions.

GBPs are also related to the theory and application of Gaussian processes. Most Gaussian processes methods rely on the assumption that the noise level is uniform throughout the domain [18], or at least, its functional dependency is known beforehand [19]. Gaussian beam processes contribute a novel way of treating input-dependent noise. In this respect, it is most closely related to Goldberg's *et al.* [20] approach and models the variance using a second Gaussian process in addition to a Gaussian process governing the noise-free output value. In contrast to Goldberg *et al.*, however, we do not use a time-consuming Markov chain Monte Carlo method to approximate the posterior noise variance but a fast most-likely-noise approach. This has the additional advantage that our approach fully stays within the Gaussian process framework so that more advanced Gaussian process techniques such as online learning, depending outputs, non-stationary covariance functions, and sparse approximations can easily be adapted.

III. Gaussian Beam Processes

We pose the task of estimating $p(\mathbf{r}|\mathcal{A},\mathbf{x})$ as a regression problem and model the function that maps beam angles $\boldsymbol{\alpha}$ to range measurements r as a stochastic process. In other words, we regard the individual measurements r_i as a collection of random variables indexed by the respective beam angles $\boldsymbol{\alpha}_i$. By placing a Gaussian process prior over this function, we get a simple yet powerful model for likelihood estimation of range measurements as well as for prediction. Concretely, for mobile robot localization, we propose to build GBP models online for all robot pose hypotheses $\mathbf{x}$. The training set $\mathcal{D} = \{(\boldsymbol{\alpha}_i, r_i)\}_{i=1}^n$ for learning such a model is simulated using ray casting operations relative to $\mathbf{x}$ using a metric map of the environment. For certain applications, one needs to estimate $p(\mathbf{r}|\mathcal{A},\mathcal{X})$, i.e. the distribution of range measurements for a region $\mathcal{X}$ in pose space. In this case, the training set $\mathcal{D}$ is simply built by sampling poses $\mathbf{x}$ from $\mathcal{X}$ and simulating the corresponding range scans. In the following, we will derive the general model for d-dimensional angular indices $\boldsymbol{\alpha}_i$ (e.g., $d=1$ for planar sensing devices, $d=2$ for 3D sensors).

Given a training set $\mathcal{D}$ of range and bearing samples, we want to learn a model for the non-linear and noisy functional dependency $r_i = f(\boldsymbol{\alpha}_i) + \epsilon_i$ with independent, normally distributed error terms ϵ_i. The idea of Gaussian processes is to view all target values r_i as jointly Gaussian distributed $p(r_1,\ldots,r_n|\boldsymbol{\alpha}_1,\ldots,\boldsymbol{\alpha}_n) \sim \mathcal{N}(\boldsymbol{\mu},K)$ with a mean $\boldsymbol{\mu}$ and covariance matrix K.

The mean $\boldsymbol{\mu}$ is typically assumed $\mathbf{0}$ and K is defined by $k_{ij} := k(\boldsymbol{\alpha}_i,\boldsymbol{\alpha}_j) + \sigma_n^2\delta_{ij}$, depending on a covariance function k and the global noise variance parameter σ_n. The covariance function represents the prior knowledge about the underlying function f and does not depend on the target values $\mathbf{r}$ of $\mathcal{D}$. Common choices, that we also employ throughout this work, are the squared exponential covariance function

$$k_{SE}(\boldsymbol{\alpha}_i,\boldsymbol{\alpha}_j) = \sigma_f^2 \exp\left(-\frac{\Delta_{ij}^2}{2\ell^2}\right), \tag{1}$$

with $\Delta_{ij} = \|\boldsymbol{\alpha}_i - \boldsymbol{\alpha}_j\|$, which has a relatively strong smoothing effect, and a variant of the Matern type of covariance function $k_M(\boldsymbol{\alpha}_i,\boldsymbol{\alpha}_j) =$

$$\sigma_f^2\left(1 + \frac{\sqrt{5}\Delta_{ij}}{\ell} + \frac{\sqrt{5}\Delta_{ij}^2}{3\ell^2}\right)\cdot\exp\left(-\frac{\sqrt{5}\Delta_{ij}}{\ell}\right). \tag{2}$$

These two covariance functions are called *stationary*, since they only depend on the distance Δ_{ij} between input locations $\boldsymbol{\alpha}_i$ and $\boldsymbol{\alpha}_j$. In the definitions above, σ_f denotes the amplitude (or signal variance) and ℓ is the characteristic length-scale, see [21] for a detailed discussion. These parameters plus the global noise variance σ_n are called hyper-parameters of the process. They are typically denoted as $\boldsymbol{\Theta} = (\sigma_f,\ell,\sigma_n)$. Since any set of samples from the process are jointly Gaussian distributed, predictions of m new range values $\mathbf{r}^* = (r_1^*,\ldots,r_m^*)$, at given angles $\mathcal{A}^* = (\boldsymbol{\alpha}_1^*,\ldots,\boldsymbol{\alpha}_m^*)$ can be performed by conditioning the $n+m$-dimensional joint Gaussian on the known target values of the training set $\mathcal{D}$. This yields an m-dimensional predictive normal distribution $\mathbf{r}^* \sim \mathcal{N}(\boldsymbol{\mu}^*,\Sigma^*)$

$$\boldsymbol{\mu}^* = E(\mathbf{r}^*) = K^*\left(K + \sigma_n^2 I\right)^{-1}\mathbf{r} \tag{3}$$

$$\Sigma^* = V(\mathbf{r}^*) = K^{**} + \sigma_n^2 I - K^*\left(K + \sigma_n^2 I\right)^{-1} K^{*T} \tag{4}$$

with the covariance matrices $K \in \mathbb{R}^{n\times n}$, $K_{ij} = k(\boldsymbol{\alpha}_i,\boldsymbol{\alpha}_j)$, $K^* \in \mathbb{R}^{m\times n}$, $K_{ij}^* = k(\boldsymbol{\alpha}_i^*,\boldsymbol{\alpha}_j)$, and $K^{**} \in \mathbb{R}^{m\times m}$, $K_{ij}^{**} = k(\boldsymbol{\alpha}_i^*,\boldsymbol{\alpha}_j^*)$, and the training targets $\mathbf{r} \in \mathbb{R}^n$. The hyper-parameters of the Gaussian process can either be learned by maximizing the likelihood of the given data points or, for fully Bayesian treatment, can be integrated over using parameter-specific prior distributions. In this work, we adapt the hyper-parameters by maximizing the marginal likelihood of $\mathcal{D}$ using the hybrid Monte-Carle approach described in [1].

So far, we have introduced the standard Gaussian processes framework for regression problems. In the following, we

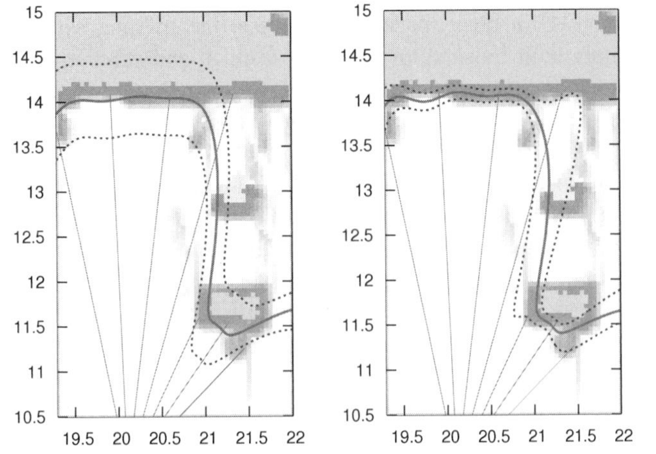

Fig. 3. The effect of modeling non-constant noise on a data set of range measurements simulated for the case of an uncertain sensor orientation ($\pm 5°$). Standard Gaussian process regression (left) assumes constant noise for all bearing angles. Modeling heteroscedasticity (our model, on the right) yields lower predictive uncertainties at places with low expected noise levels such as the wall in front. The straight, red lines depict one possible range scan in this setting.

describe a novel way of treating input-dependent noise, which leads to more accurate models in our application domain.

A. Modeling Non-Constant Noise

Gaussian processes as introduced above assume a constant noise term, i.e., identically distributed error terms ϵ_i over the domain. For modeling range sensor measurements, however, the variance of range values in each beam direction is, along with its mean value, an important feature of the sought-after distribution of range measurements. To overcome this, we extended the standard Gaussian process framework to deal with heteroscedasticity, i.e., non-constant noise. Figure 3 illustrates the effect of this treatment on the predictive distribution for range values. The left diagram depicts the standard procedure that assumes a constant noise term for all bearings $\boldsymbol{\alpha}$. Our heteroscedastic treatment, depicted in the right diagram, achieves a significantly better fit to the data set while still not over-fitting to the individual samples.

To deal with the heteroscedasticity inherent in our problem domain, we basically follow the approach of Goldberg *et al.* [20], who condition a standard Gaussian processes $\mathcal{G}_c$ on latent noise variables sampled from a separate noise process $\mathcal{G}_n$. Let $\mathbf{v} \in \mathbb{R}^n$ be such noise variances at the n given data points and $\mathbf{v}^* \in \mathbb{R}^m$ those for the m locations to be predicted, then the predictive distribution changes to

$$\boldsymbol{\mu}^* = K^* (K + K_v)^{-1} \mathbf{r} , \quad (5)$$

$$\Sigma^* = K^{**} + K_v^* - K^* (K + K_v)^{-1} K^{*T} , \quad (6)$$

where $K_v = \text{diag}(\mathbf{v})$ and $K_v^* = \text{diag}(\mathbf{v}^*)$. Now, as the noise variances $\mathbf{v}$ and $\mathbf{v}^*$ cannot be known a-priori, they have to be integrated over for predicting $\mathbf{r}^*$

$$p(\mathbf{r}^*|\mathcal{A}^*, \mathcal{D}) \quad (7)$$

$$= \int \underbrace{p(\mathbf{r}^*|\mathcal{A}^*, \mathbf{v}, \mathbf{v}^*, \mathcal{D})}_{p_r} \cdot \underbrace{p(\mathbf{v}, \mathbf{v}^*|\mathcal{A}^*, \mathcal{D})}_{p_v} \, d\mathbf{v}d\mathbf{v}^* .$$

Given the variances $\mathbf{v}$ and $\mathbf{v}^*$, the prediction p_r in Eq. (7) is a Gaussian with mean and variance as discussed above. The problematic term is indeed p_v as it makes the integral difficult to handle analytically. Therefore, Goldberg *et al.* [20] proposed a Monte Carlo approximation that alternately samples from p_r and p_v to fit both curve and noise rates. The sampling is quite time consuming and the expectation can be approximated by the most likely noise levels $\tilde{\mathbf{v}}$ and $\tilde{\mathbf{v}}^*$. That is, we approximate the predictive distribution as

$$p(\mathbf{r}^*|\mathcal{A}^*, \mathcal{D}) \approx p(\mathbf{r}^*|\mathcal{A}^*, \tilde{\mathbf{v}}, \tilde{\mathbf{v}}^*, \mathcal{D}) , \quad (8)$$

where $(\tilde{\mathbf{v}}, \tilde{\mathbf{v}}^*) = \arg\max_{(\tilde{\mathbf{v}}, \tilde{\mathbf{v}}^*)} p(\tilde{\mathbf{v}}, \tilde{\mathbf{v}}^*|\mathcal{A}^*, \mathcal{D})$. This will be a good approximation, if most of the probability mass of $p(\tilde{\mathbf{v}}, \tilde{\mathbf{v}}^*|\mathcal{A}^*, \mathcal{D})$ is concentrated around $(\tilde{\mathbf{v}}, \tilde{\mathbf{v}}^*)$. Moreover, the noise levels can be modeled using a standard Gaussian process. Thus, we have two interacting processes: $\mathcal{G}_n$ predicts the noise levels and $\mathcal{G}_c$ uses the predicted noise levels in (5) and (6). To learn the hyperparameters of both processes, we basically follow an alternating learning scheme in the spirit of the Expectation-Maximization algorithm: (1) fix the noise levels and learn $\mathcal{G}_c$ using a standard maximum likelihood estimator; (2) fix $\mathcal{G}_c$, estimate the empirical noise levels of $\mathcal{G}_c$ on the training data and estimated $\mathcal{G}_n$ using them as target data. Initially, the noise levels are set to the empirical noise levels of a constant-noise Gaussian process induced on the training data.

As covariance functions, we use the Matern type as stated in Equation 2 for the range process and the squared exponential one for the noise process. This matches the intuition that the noise process should exhibit more smoothness than the range process, which was also supported by our experiments. This, however, is not a mandatory choice. With properly learned hyperparameters, using the squared exponential function for both processes yields a nearly as high performance in our application.

B. Evaluating the Joint Data Likelihood of Observations

For m new range measurements $\mathbf{z} = \{(\boldsymbol{\alpha}_i, r_i)\}_{i=1}^m$ indexed by the beam orientations $\boldsymbol{\alpha}_i$, the model has to estimate the data likelihood $p(\mathbf{z}|\mathcal{D}, \boldsymbol{\Theta})$ given the training data $\mathcal{D}$ and the learned covariance parameters $\boldsymbol{\Theta}$. We solve this by considering the predictive distribution for range measurements $\mathbf{r}^*$ at the very same beam orientations $\boldsymbol{\alpha}_1^*, \ldots, \boldsymbol{\alpha}_m^*$, which is an m-dimensional Gaussian distribution as defined by (5) and (6). As this predictive distribution is a multivariate Gaussian, we can directly calculate the observation likelihood for the data vector $\mathbf{z}$ by evaluating the density function

$$p(\mathbf{z}|\boldsymbol{\mu}^*, \Sigma^*) = \left[(2\pi)^{\frac{m}{2}} |\Sigma^*|^{\frac{1}{2}}\right]^{-1} \cdot \quad (9)$$

$$\exp\left(-\frac{1}{2}(\mathbf{z} - \boldsymbol{\mu}^*)^T {\Sigma^*}^{-1} (\mathbf{z} - \boldsymbol{\mu}^*)\right)$$

or, in a more convenient form

$$\log p(\mathbf{z}|\boldsymbol{\mu}^*, \Sigma^*) = -\frac{1}{2}(\mathbf{z}-\boldsymbol{\mu}^*)^T \Sigma^{*-1}(\mathbf{z}-\boldsymbol{\mu}^*) -\frac{1}{2}\log|\Sigma^*| - \frac{m}{2}\log(2\pi) \ . \quad (10)$$

C. Regression over Periodic Spaces

In our application, we have to account for the fact that our input vectors $\boldsymbol{\alpha}_i$ are angular quantities rather than unconstrained real valued vectors. This means that an angular distance metric has to be incorporated into the covariance function to avoid discontinuities at $\pm\pi$. For the one dimensional case (for planar sensing devices), we use

$$\|\alpha, \beta\|_a := \begin{cases} |\alpha - \beta| & \text{if } |\alpha-\beta| \leq \pi \\ 2\pi - |\alpha - \beta| & \text{otherwise} \ . \end{cases} \quad (11)$$

Indeed, we also have to adapt the covariance functions themselves to the *periodic* structure of the input space. For example, a periodic variant of the squared exponential covariance function on the unit circle is

$$k(\alpha_i, \alpha_j) = \sigma_f^2 \sum_{p=-\infty}^{\infty} \exp\left(-\frac{|(\alpha_i + 2\pi p) - \alpha_j|^2}{2\ell^2}\right) \ , \quad (12)$$

which takes infinitively many influences of a data point on itself into account. The squared exponential covariance function, however, has a strong locality for relevant values of σ_f^2 and ℓ. All summands with $|\alpha_i - \alpha_j| >= 2\pi$ in Eq. (12) cannot even be represented using double precision, because their value is too close to zero. We can therefore safely ignore the periodicity in practice and only use the standard covariance function with the modified distance metric of Eq. (11).

D. Efficient Inference by Exploiting Locality

The covariance functions employed in this work are stationary, i.e., they assign small covariance values to those pairs of input points which lie far apart. With the given machine precision, this implies that the resulting covariance matrices are effectively band limited and only have non-zero entries close to the diagonal. This property can be exploited to speed up the computations by using optimized algorithms for sparse matrix operations. In this work, we used the UMFPACK package [22], an optimized solver for unsymmetric sparse linear systems, which resulted in significantly reduced computation times as shown in Figure 4. The run-times are given in seconds for a full iteration of simulating the scan, building the heteroscedastic model, and evaluating the observation likelihood for a given scan with 31 beams. The gain in speed depicted in this figure is due to the sparsity induced by the limitations of machine precision only. In addition to this, the covariances could be truncated actively to much tighter bounds before a notable loss of precision occurs.

IV. Experimental Results

In this section, we will report on our experimental evaluation. The intention of this evaluation is to determine how well

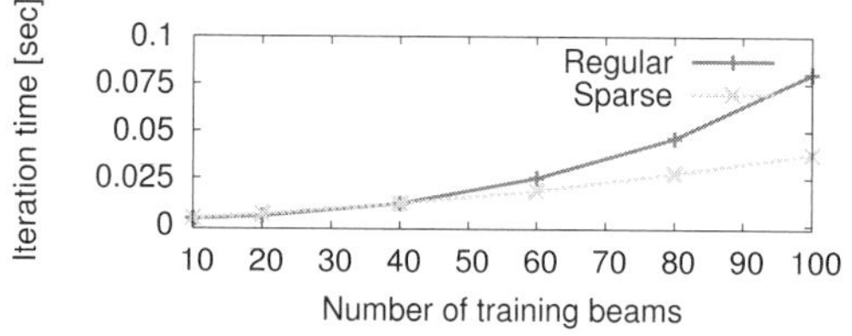

Fig. 4. The gain in speed due to sparse matrix calculations without a loss of precision. Exploiting sparsity reduces the iteration times drastically, especially for larger problem sizes.

the proposed GBP model performs compared to state-of-the-art probabilistic measurement models for laser range finders. To this aim, we applied our measurement model to one of the classical problems of robotics, mobile robot localization and tracking. We implemented our approach in C++. The platform used was a Pioneer PII DX8+ robot equipped with a laser range finder in a typical office environment. As the primary sensor is a planar sensing device, we only have to deal with one-dimensional bearing angles. To ensure a fair comparison, we independently optimized the parameters of all models using different data in all experiments.

We will proceed as follows. First, we briefly review the Monte Carlo localization scheme and discuss how the GBP model fits in there. Then, we present tracking and localization results with a real robot and, finally, we present results on simulated data to demonstrate the main benefits of our approach.

A. GBPs for Monte Carlo Localization

The task of mobile robot localization is to sequentially estimate the pose $\mathbf{x}$ of a moving robot in its environment. The key idea of Monte Carlo localization (MCL) [23], which belongs to the class of particle filtering algorithms, is to maintain a sampled approximation of the probability density $p(\mathbf{x})$ of the robot's own location. This belief distribution is updated sequentially according to

$$p(\mathbf{x}^{[t]}|\mathbf{z}^{[1:t]}, \mathbf{u}^{[0:t-1]}) = \eta \cdot p(\mathbf{z}^{[t]}|\mathbf{x}^{[t]}) \cdot \quad (13)$$
$$\int p(\mathbf{x}^{[t]}|\mathbf{u}^{[t-1]}, \mathbf{x}^{[t-1]}) \cdot p(\mathbf{x}^{[t-1]}|\mathbf{z}^{[1:t-1]}, \mathbf{u}^{[0:t-2]}) \, d\mathbf{x}^{[t-1]} \ .$$

Here, η is a normalization constant containing the prior observation likelihood $p(\mathbf{z}^{[t]})$, which is equal for the whole sample set and can therefore be neglected. The term $p(\mathbf{x}^{[t]}|\mathbf{u}^{[t-1]}, \mathbf{x}^{[t-1]})$ describes the probability that the robot is at position $\mathbf{x}^{[t]}$ given it executed the action $\mathbf{u}^{[t-1]}$ at position $\mathbf{x}^{[t-1]}$. Furthermore, $p(\mathbf{z}^{[t]}|\mathbf{x}^{[t]})$ denotes the probability of making observation $\mathbf{z}^{[t]}$ given the robot's current location is $\mathbf{x}^{[t]}$. The appropriate estimation of this quantity is the subject of this paper. Concretely, the update of the belief is realized by the following two alternating steps:

1) In the **prediction step**, we propagate each sample to a new location according to the robot's dynamics model $p(\mathbf{x}_t|\mathbf{u}_{t-1}, \mathbf{x}_{t-1})$ given the action $\mathbf{u}_{t-1}$ executed since the previous update.

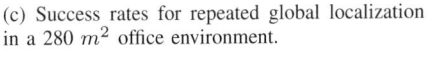

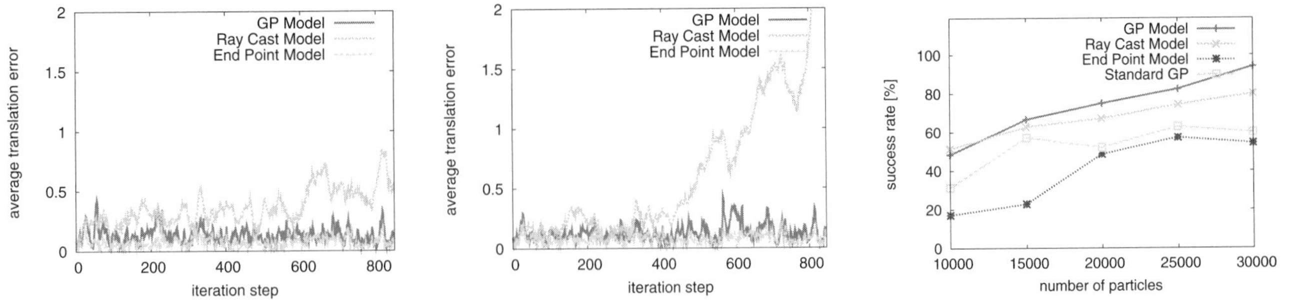

(a) Pose tracking errors for the different measurement models using 31 laser beams.

(b) With 61 laser beams, the filter using the ray cast model diverges after 400 steps.

(c) Success rates for repeated global localization in a 280 m^2 office environment.

Fig. 5. Pose tracking and localization results with a real robot in an office environment using a 180 degrees field of view. Subfigures (a) and (b) give the tracking displacement error (y-axis) in meters for an increasing number of iterations. The errors are averaged over 25 runs on the same trajectory. Subfigure (c) shows the global localization performances.

2) In the **correction step**, the new observation $\mathbf{z}_t$ is integrated into the sample set. This is done by adjusting the weight of each sample according to the likelihood $p(\mathbf{z}_t|\mathbf{x}_t)$ of sensing $\mathbf{z}_t$ given the robot pose $\mathbf{x}_t$.

The measurement model $p(\mathbf{z}|\mathbf{x})$ plays a crucial role in the correction step of the particle filter. Typically, very peaked models require a large number of particles and induce a high risk of filter divergence. Even when the particles populate the state space densely, the likelihoods of an observation might differ by several orders of magnitude. As the particles are drawn proportionally to the importance weights, which themselves are calculated as the likelihood of $\mathbf{z}_t$ given the pose $\mathbf{x}_t$ of the corresponding particle, a minor difference in $\mathbf{x}_t$ can already result in a large difference in likelihoods. This, in turn, would result in the depletion of such a particle in the re-sampling step. As a consequence, the "peakedness" of a measurement model should depend on the number of particles available and the size of the space to be covered. The ray cast model (RC) as well as the end point model (EP) have parameters for controlling their smoothness, which can be optimized for specific scenarios. In the following, we describe how GBPs can be applied to MCL and how the smoothness of the model can be defined in terms of an easy to interpret parameter.

As mentioned in Section III, we estimate $p(\mathbf{z}|\mathbf{x})$ by building a GBP model for the robot pose $\mathbf{x}$ online and evaluating the data likelihood of $\mathbf{z}$ according to Section III-B. For building the GBP model, we construct a training set $\mathcal{D}$ of simulated range measurements relative to $\overline{\mathbf{x}} \sim \mathcal{N}(\mathbf{x}, \boldsymbol{\sigma}_\mathbf{x})$ using an occupancy grid map of the environment. The random perturbations added to $\mathbf{x}$ account for the desired smoothness of the model as motivated above. Indeed, the pose variance parameter $\boldsymbol{\sigma}_\mathbf{x}$ introduced here, more naturally quantifies the level of regularization of GBPs compared to other models, as it is directly specified in the space of robot locations. Note that no sensory information is discarded at this point. For sufficiently high sampling densities, one could set $\sigma_\mathbf{x} = 0$ to get the fully peaked model.

The MCL measurement update step for the whole filter using GBPs can be summarized as follows:

Algorithm 1 GBPs-based Measurement Update for MCL

for all particles $\mathbf{x}$ **do**
  Generate $\mathcal{D}$ using ray casting in the given map at robot locations sampled from $\mathcal{N}(\mathbf{x}, \boldsymbol{\sigma}_\mathbf{x})$.
  Build local GBPs using $\mathcal{D}$ and the global covariance C.
  Compute all $\log p(\mathbf{z}|GBP)$ and weight the particles.
end for

B. Results on Tracking and Localizing a Real Robot

To demonstrate the tracking and global localization performance of the GBP approach, we implemented Algorithm 1 and evaluated it using real data acquired with a Pioneer PII DX8+ robot equipped with a laser range scanner in a typical office environment. The experiments described here are designed to investigate how well our GBP approach performs in comparison to the widely used ray cast model and the end point model. While our approach is computationally more demanding than the alternative ones, it still runs close to real-time for mobile robot tracking. Here, a full iteration including scan simulation and model building takes approximately 0.011 seconds.

In the first set of experiments, we assess the position tracking performance of the MCL filter using the different measurement models. The robot started in the corridor of an office environment and traversed a path through all adjacent rooms. Figure 5(a) depicts the average localization error for this experiment with 31 laser beams. As can be seen, the GBP model and the end point model show similar, good localization performance and both outperform the ray cast model. When using more beams for the same task, the difference to the ray cast model gets even more pronounced, see Figure 5(b). Due to the ray cast model's inability to deal with dependencies between beams, the risk of filter divergence increases with a growing number of beams used. In another experiment with 181 beams, the GBP model and the end point model showed

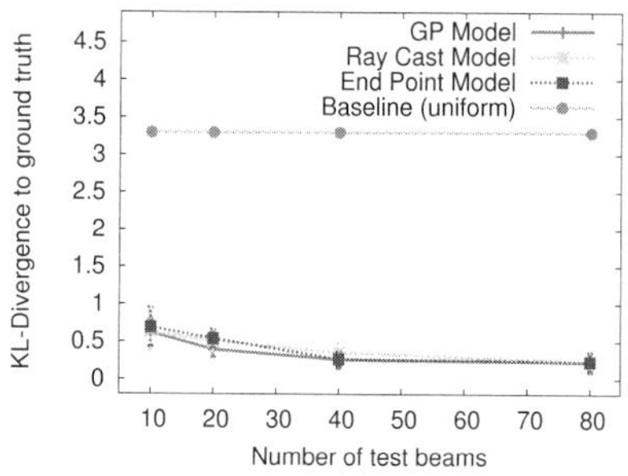

(a) Office room environment: All methods show similar performances and outperform the uniform model (pink) which assigns the same likelihood value to all grid cells. Lower KL-D values are better.

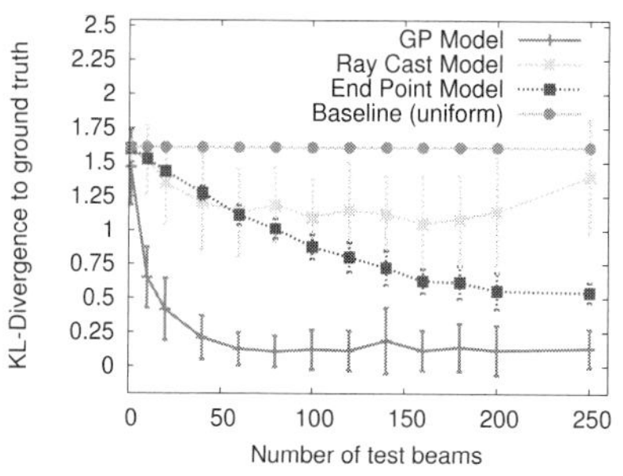

(b) Highly cluttered environment: The ray cast model (green) performs better than uniform (pink) but less well than the end point model (blue). Our GBP model (red) significantly outperforms the other methods in this environment.

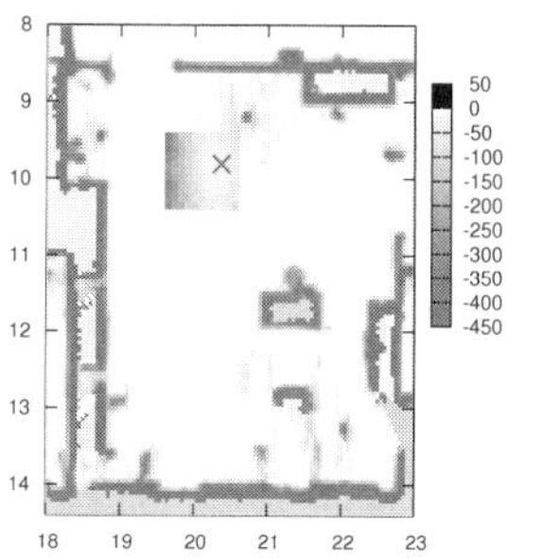

(c) Observation log likelihood induced by the GBP model for a single simulated scan on a discretized pose space in a typical office room. It shows a convex shape and the true robot pose (cross) is in the area of maximal values.

Fig. 6. Localization performance for a mobile robot in terms of the Kullback-Leibler divergence (KLD), which measures the distance of the discretized pose belief distribution to the known ground truth (lower values are better). The experiments were simulated in an office (a) and in a cluttered environment (b). The KLD (y-axis) is shown for a varying number of laser beams (x-axis). The baseline model (uniform) assigns the same log likelihoods to all grid cells, i.e., it does not use any sensor information at all. Subfigure (c) shows a typical likelihood function induced by our Gaussian beam process (GBP) model.

a similar behavior as before. The ray cast model, however, diverged even earlier then with 61 beams.

In a second set of experiments we investigated the robustness of our GBP approach for global localization. Here, the task is to find the pose of a moving robot within an environment using a stream of wheel encoder and laser measurements. The environment used consists of a long corridor and 8 rooms containing chairs, tables and other pieces of furniture. In total, the map is 20 meters long and 14 meters wide. The results are summarized in Figure 5(c), which shows the number of successful localizations after 8 integrations of measurements for the three measurement models and for different numbers of particles used. In the experiment, we assumed that the localization was achieved when more than 95 percent of the particles differed in average at most 30 cm from the true location of the robot. As can be seen from the diagram, the GBP model performs slightly better than the ray cast model and both outperform the end point model.

C. Results on Simulated Data in a Static Setting

In the previous section, we have evaluated the measurement models in the standard way for mobile robots, i.e., we have evaluated their performances in real-world tracking and localization tasks. Although this is closest to the actual application of the models (and should therefore round off any other evaluation strategy), it has also one major drawback: several external factors influence the evaluation, such as the choice of filtering algorithm, the sampling and resampling strategies, and the order in which places are visited along a trajectory. To investigate the strengths and weaknesses of the measurement models independently from specific tracking algorithms, we ran a different set of experiments in a static setting. Here, we use the Kullback-Leibler divergence (KL-D) on a discretized pose space to measure how well the different models are able to reconstruct a pose distribution given just the corresponding laser measurements. More precisely, for each measurement model, we

- discretize the space of robot poses using a three dimensional grid (2D location and heading) and let each grid cell represent one pose hypothesis $\mathcal{X}_i$,
- select a cell index t to contain the true robot pose $\mathbf{x}_t \in \mathcal{X}_t$,
- randomly draw m test poses within this cell t and simulate corresponding range measurement vectors $\mathbf{z}_m$ using a given occupancy grid map.
- Now, we evaluate the observation likelihoods $p(\mathbf{z}_m|\mathcal{X}_i)$ for each grid cell and each test observation and sum up the individual observation likelihoods per cell.
- Finally, we normalize the whole likelihood grid and compute the KL-D $D_{KL} = \sum_i p(\mathbf{z}_m|\mathcal{X}_i) \cdot \log\left(\frac{p(\mathbf{z}_m|\mathcal{X}_i)}{\delta_{i=t}}\right)$ to a binary ground truth grid, where all likelihood mass is concentrated at the cell t, i.e., the true robot pose.

For computing the KL-D measure, we employ the standard trick of adding an extremely small value to each cell for dealing with empty cells. The specific choice of this value did not have a notable influence on the measure.

Figure 6(c) depicts such a likelihood grid for the GBP model as well as the true robot location in an office environment. It can be seen that the observation likelihood is nicely peaked around the true robot pose and that the GBP model yields a smooth likelihood function. The KL-D results for this room are given in Figure 6(a). The diagram shows that all three models achieve comparable good performances when recovering the pose distribution in this situation. Additionally, we plot the KL-D for the uniform model taken as a baseline, which assigns the same, constant likelihood value to all cells. In highly cluttered environments such as a laboratory room with many chairs and tables, however, the GBP model clearly outperforms the other two models. As shown in Figure 6(b), the KL-D is always significantly lower and decreases with a high number of laser beams. The ray cast model shows even an increasing KL-D with increasing numbers of laser beams due to its lack of smoothness. In both experiments, we used a relatively coarse

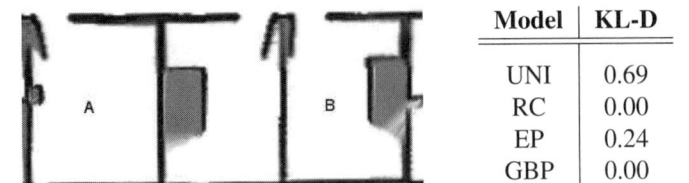

Model	KL-D
UNI	0.69
RC	0.00
EP	0.24
GBP	0.00

Fig. 7. In contrast to RC and GBP, the end point model (EP) cannot reliably distinguish between rooms A and B of the SDR building (left image). Since it only considers the end points of scans, it ignores the solid obstacle on the right-hand side of room B. Consequently, the resulting pose belief distribution shows a high KL-divergence to the ground truth (right) while RC and GBP achieve near optimal performances.

grid with grid cell areas of approximately $0.037m^2$.

Recall from the beginning that the end point model ignores the obstacles along the beam and replaces the ray casting operation of beam models by a nearest neighbor search. Figure 7 depicts a real situation in which this can lead to divergence of the localization process. To more quantitatively investigate this situation, we evaluated the different approaches in their capabilities to correctly classify laser scans simulated in room A as belonging to this room. As can be seen from the table on the right side of Figure 7, which contains the KL-D on a two cell grid (one cell for each room), the end point model produces a larger error than the ray cast model and the GBP model. This is mainly due to the fact that the end points of beams from room A also fit room B. The KL-D has been calculated for 100 simulated scans on a two cell grid, where each grid cell spans $1m^2$ in the center of room A respectively room B.

V. Conclusions

In this paper, we introduced Gaussian beam processes as a novel probabilistic measurement model for range finder sensors. The key idea of our approach is to view the measurement modeling task as a Bayesian regression problem and to solve it using Gaussian processes. As our experiments with real and simulated data demonstrate, Gaussian beam processes provide superior robustness compared to the ray cast model and the end point model by combining the advantages of both.

The model's ability to perform justified range predictions in arbitrary directions without having to consult a map is a promising starting point for future research. This might, for instance, allow to interpolate between measurements in settings where sensors only provide a sparse coverage of the environment.

Acknowledgments

We would like to thank Andreas Klöckner for his invaluable help in optimizing the sparse matrix calculations as well as Andrew Howard for providing the SDR building data set through the Radish repository [24]. This work has been supported by the EC under contract number FP6-004250-CoSy, by the German Federal Ministry of Education and Research (BMBF) under contract number 01IMEO1F (project DESIRE), and by the German Research Foundation (DFG) within the Research Training Group 1103.

References

[1] C. Williams and C. Rasmussen, "Gaussian processes for regression," in *Proc. of the Conf. on Neural Information Processing Systems (NIPS)*. MIT Press, 1995, pp. 514–520.

[2] D. Fox, W. Burgard, and S. Thrun, "Markov localization for mobile robots in dynamic environments," *Journal of Artificial Intelligence Research*, vol. 11, pp. 391–427, 1999.

[3] A. Petrovskaya, O. Khatib, S. Thrun, and A. Ng, "Bayesian estimation for autonomous object manipulation based on tactile sensors," in *Proc. of the IEEE Int. Conf. on Robotics & Automation (ICRA)*, Orlando, Florida, 2006.

[4] S. Arulampalam, S. Maskell, N. Gordon, and T. Clapp, "A tutorial on particle filters for on-line non-linear/non-gaussian bayesian tracking," in *IEEE Transactions on Signal Processing*, vol. 50, no. 2, February 2002, pp. 174–188.

[5] B. Schiele and J. Crowley, "A comparison of position estimation techniques using occupancy grids," in *Proc. of the IEEE Int. Conf. on Robotics & Automation (ICRA)*, 1994, pp. 1628–1634.

[6] K. Konolige and K. Chou, "Markov localization using correlation," in *Proc. of the Int. Conf. on Artificial Intelligence (IJCAI)*, 1999.

[7] S. Thrun, "A probabilistic online mapping algorithm for teams of mobile robots," *International Journal of Robotics Research*, vol. 20, no. 5, pp. 335–363, 2001.

[8] P. Pfaff, W. Burgard, and D. Fox, "Robust monte-carlo localization using adaptive likelihood models," in *European Robotics Symposium*, Palermo, Italy, 2006.

[9] D. Fox, "Kld-sampling: Adaptive particle filters," in *Proc. of the Conf. on Neural Information Processing Systems (NIPS)*, 2001.

[10] C. Kwok, D. Fox, and M. Meila, "Adaptive real-time particle filters for robot localization," in *Proc. of the IEEE International Conference on Robotics & Automation*, 2003.

[11] R. Gutierrez-Osuna, J. Janet, and R. Luo, "Modeling of ultrasonic range sensors for localization of autonomousmobile robots," in *IEEE Transactions on Industrial Electronics*, vol. 45, no. 4, August 1998, pp. 654–662.

[12] A. Schwaighofer, M. Grigoras, V. Tresp, and C. Hoffmann, "Gpps: A gaussian process positioning system for cellular networks." in *NIPS*, 2003.

[13] A. Brooks, A. Makarenko, and B. Upcroft, "Gaussian process models for sensor-centric robot localisation," in *ICRA*, 2006.

[14] B. Ferris, D. Haehnel, and D. Fox, "Gaussian processes for signal strength-based location estimation," in *Proceedings of Robotics: Science and Systems*, Philadelphia, USA, August 2006.

[15] C. Plagemann, D. Fox, and W. Burgard, "Efficient failure detection on mobile robots using particle filters with gaussian process proposals," in *Proc. of the Twentieth International Joint Conference on Artificial Intelligence (IJCAI)*, Hyderabad, India, 2007.

[16] J. Ting, M. Mistry, J. Peters, S. Schaal, and J. Nakanishi, "A bayesian approach to nonlinear parameter identification for rigid body dynamics," in *Proceedings of Robotics: Science and Systems*, Philadelphia, USA, August 2006.

[17] D. Grimes, R. Chalodhorn, and R. Rao, "Dynamic imitation in a humanoid robot through nonparametric probabilistic inference," in *Proceedings of Robotics: Science and Systems*, Philadelphia, USA, 2006.

[18] C. Williams, "Prediction with gaussian processes: From linear regression to linear prediction and beyond," in *Learning and inference in graphical models*, M. I. Jordan, Ed. Kluwer Acadamic, 1998, pp. 599–621.

[19] B. Schoelkopf, A. J. Smola, R. C. Williamson, and P. L. Bartlett, "New support vector algorithms." *Neural Computation*, vol. 12, pp. 1207–1245, 2000.

[20] P. Goldberg, C. Williams, and C. Bishop, "Regression with input-dependent noise: A gaussian process treatment," in *Proc. of the Conf. on Neural Information Processing Systems (NIPS)*, vol. 10, 1998.

[21] C. Rasmussen and C. Williams, *Gaussian Processes for Machine Learning*. MIT Press, 2006.

[22] T. A. Davis, "A column pre-ordering strategy for the unsymmetric-pattern multifrontal method," *ACM Trans. Math. Softw.*, vol. 30, no. 2, pp. 165–195, 2004.

[23] S. Thrun, D. Fox, W. Burgard, and F. Dellaert, "Robust Monte Carlo localization for mobile robots," *Artificial Intelligence*, vol. 128, no. 1–2, pp. 99–141, 2000.

[24] A. Howard and N. Roy, "The robotics data set repository (radish)," 2003. [Online]. Available: http://radish.sourceforge.net/

Vision-Aided Inertial Navigation for Precise Planetary Landing: Analysis and Experiments

Anastasios I. Mourikis, Nikolas Trawny, Stergios I. Roumeliotis
Dept. of Computer Science & Engineering,
University of Minnesota, Minneapolis, MN 55455
Email: {mourikis|trawny|stergios}@cs.umn.edu

Andrew Johnson and Larry Matthies
Jet Propulsion Laboratory,
Pasadena, CA 91125.
Email: {aej|lhm}@jpl.nasa.gov

***Abstract*—In this paper, we present the analysis and experimental validation of a vision-aided inertial navigation algorithm for planetary landing applications. The system employs tight integration of inertial and visual feature measurements to compute accurate estimates of the lander's terrain-relative position, attitude, and velocity in real time. Two types of features are considered: mapped landmarks, i.e., features whose global 3D positions can be determined from a surface map, and opportunistic features, i.e., features that can be tracked in consecutive images, but whose 3D positions are not known. Both types of features are processed in an extended Kalman filter (EKF) estimator and are optimally fused with measurements from an inertial measurement unit (IMU). Results from a sounding rocket test, covering the dynamic profile of typical planetary landing scenarios, show estimation errors of magnitude 0.16 m/s in velocity and 6.4 m in position at touchdown. These results vastly improve current state of the art for non-vision based EDL navigation, and meet the requirements of future planetary exploration missions.**

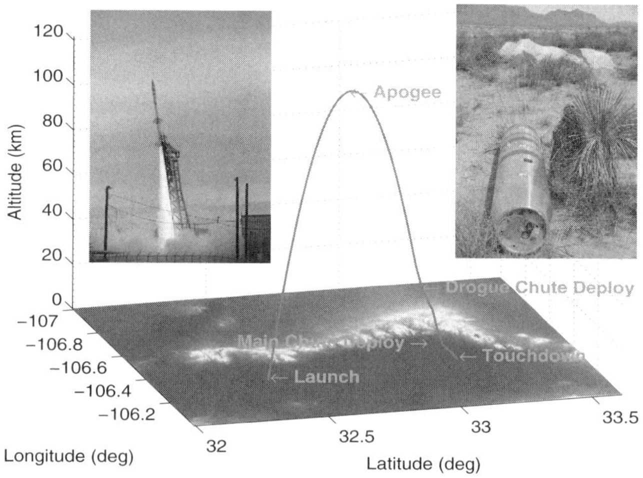

Fig. 1. Sounding Rocket Experiment 41.068

I. Introduction

Space missions involving Entry, Descent and Landing (EDL) maneuvers require high-accuracy position and attitude (pose) determination to enable precise trajectory control. On solar system bodies other than Earth, this is challenging due to the absence of artificial navigation aids such as GPS or radio-beacons. To date, robotic lander missions have used integration of acceleration and rotational velocity measurements from an inertial measurement unit (IMU), augmented by velocity and altitude information from Doppler radar. In these cases, integration of noise and biases, as well as errors in initialization, result in relatively large errors in the touchdown position estimate (e.g., for Mars Pathfinder and the Mars Exploration Rovers in the order of 100 km). However, several future planetary exploration missions will require meter-level landing accuracy, to facilitate the study of geological features of scientific interest [1]. In the absence of radiometric navigation beacons, the most attractive option for increasing the navigation accuracy during EDL is to use camera measurements. Cameras operate in almost any environment, are small, lightweight, and consume little power, and are therefore ideal sensors for space applications.

The Extended Kalman Filter (EKF)-based estimation algorithm presented in this paper combines inertial measurements from an IMU with camera observations of two types of features: (i) *Mapped Landmarks* (MLs), i.e., features whose global 3D coordinates can be determined from a map of the landing site [2], and (ii) *Opportunistic Features* (OFs), i.e., features that can be reliably detected in image sequences, but not in a map of the planet's surface [3]. We note that the term "map" refers to a co-registered pair of a digital elevation map (DEM) and satellite image, which is available *a priori*.

For precise vision-based EDL, observations of MLs are necessary, since they provide absolute pose information. However, due to the finite resolution of satellite imagery, MLs cannot typically be extracted during the last stages of the descent, when the lander is very close to the ground. To address this problem, in our work we employ OFs for improving the accuracy of pose estimation in these stages. Even though OFs cannot provide absolute pose information, they can be thought of as providing velocity information. Intuitively, viewing a static feature from multiple camera poses provides geometric constraints involving all these poses. The proposed EKF-based estimation algorithm (cf. Section IV-E) utilizes all the constraints from the OF and ML measurements optimally, while maintaining computational complexity only linear in the number of features. These characteristics enable high-accuracy pose estimation in real-time. The results from a sounding-rocket experiment (cf. Fig. 1), which are presented in Section V, demonstrate the filter's robustness and superior accuracy compared to existing state-of-the-art EDL navigation algorithms, which do not utilize vision information.

II. Related Work

Only a few recursive estimation approaches that utilize measurements of *a priori* known features have been proposed in the literature. In [4], a statistical (zero-acceleration) model is employed for propagating the pose estimate between ML observations. However, the use of a statistical model (rather than inertial measurements), limits the applicability of such an approach to maneuvers with *slow dynamics* that occur, for example, during spacecraft rendezvous and docking. In [5], [6], inertial measurements are fused with observations of artificial rectangular targets, and with heading measurements from a magnetometer. In their work, the authors use measurements both of the coordinates of a target's projection *and* of the *area* of this projection. Area measurements, though, may be difficult or unreliable when dealing with real planetary imagery, where visual features are less structured. Finally, in [7] inertial measurements are fused with bearing measurements to MLs, but the spacecraft's attitude is assumed to be perfectly known, which is not a valid assumption for EDL.

In addition to processing measurements to MLs, in this work we also utilize measurements to features whose position in the global frame is *not* known in advance (OFs). The standard method of treating such features is to include their positions in the state vector, and to estimate them along with the camera trajectory. This is the well-known Simultaneous Localization and Mapping (SLAM) problem, for which numerous approaches that employ vision and inertial sensing have recently been proposed (e.g., [8] and references therein). However, the need to maintain the landmark estimates in SLAM results in increased computational complexity (quadratic in the number of features for EKF-SLAM). Moreover, the main benefit of performing SLAM is the ability to achieve "loop closing" when revisiting an area. Due to the nature of EDL trajectories loop closing is not an important consideration, and thus the quadratic computational complexity of SLAM does not appear to be justified in the context of EDL. In contrast, our algorithm attains complexity only *linear* in the number of OFs.

In our work, OF measurements are employed for imposing constraints between multiple camera poses. In a similar spirit, in [7], [9], a history of only the latest *two* camera poses is considered, and visual measurements are utilized for imposing constraints between them. Moreover, pairwise relative-pose constraints are employed for pose estimation in several approaches that maintain a history of multiple camera poses (e.g., [10] and references therein). Contrary to that, the proposed algorithm does not use the measurements for deriving relative-pose estimates. This reduces the computational burden, avoids possible correlations between the displacement measurements [11], and is more robust to non-linearities. Finally, a sliding window of poses is also maintained in the VSDF algorithm [12]. However, the VSDF is tailored for cases where no motion model is available (in EDL the IMU measurements provide such a model), and its computational complexity is at least quadratic in the number of features.

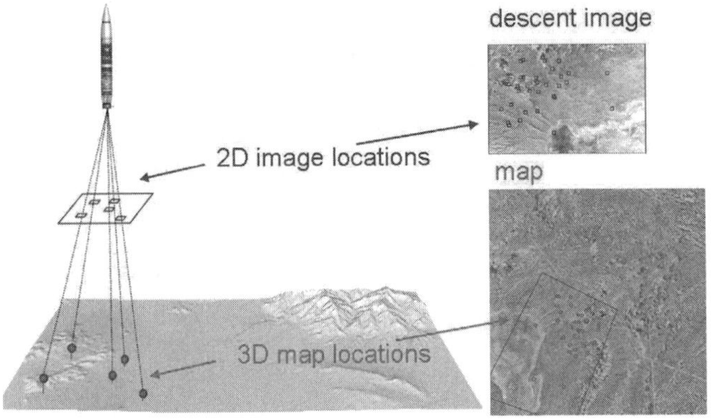

Fig. 2. ML algorithm concept: by matching templates between descent images and a visual map of the landing site, the algorithm produces measurements of the image projections of features with known 3D coordinates (i.e., MLs).

III. Image Processing

Several techniques can be applied for feature extraction in planetary landing applications. For example, Cheng *et al.* [13] propose using craters as landmarks for navigation. Craters are abundant on most bodies of interest in our solar system, and this makes them a very useful feature. However, there exist sites (e.g., in the polar regions of Mars) where craters are not present. In this case, more general feature types can be used, such as SIFT keypoints [14], and Harris corners [15]. In our work, the image processing module relies on Harris corners and normalized correlation, because (i) corner features can be extracted more efficiently than SIFT keys, since they do not require a search over image scale, (ii) Harris corners have been shown to be more robust to illumination changes than SIFT [16], and (iii) image correlation is a field-tested technology, which has already been employed in EDL applications [17].

The ML algorithm applies normalized correlation to match templates from the descent images to the map (cf. Fig. 2). Each selected template is warped prior to correlation, so as to have the same scale and orientation as the map. This enables us to reduce matching to a 2-D search, which can be carried out very efficiently. We note that the size of the search windows for correlation depends on the accuracy of the camera pose estimate. When the camera pose is very uncertain (e.g., when the first images are recorded), the search windows are very large, and directly applying correlation search for all features is computationally expensive. In that case, an FFT map-matching step is employed prior to ML matching. During this step, a rough alignment of the image with the map is obtained, by frequency-domain correlation of a single large template. Using this initial estimate, the dimensions of the search areas for ML matching are significantly reduced. For OF tracking, we perform pairwise correlation matching between images. To reduce the effects of the changing camera viewpoint, the homography that is induced by the camera motion between consecutive images is accounted for during correlation. For a detailed description of the image-processing algorithms, the reader is referred to [18].

Algorithm 1 Vision-aided Inertial Navigation Algorithm

Propagation: For each IMU measurement received, propagate the state and covariance estimates (cf. Section IV-B).

Image acquisition: Every time a new image is recorded,

- augment the state and covariance matrix with a copy of the current camera pose estimate (cf. Section IV-A).
- image processing module begins processing the new image.

Update: When the ML and OF measurements of a given image become available, perform an EKF update using

- all ML measurements of this image (cf. Section IV-D)
- the OFs that are no longer detected in the image sequence (cf. Section IV-E).

IV. Estimator Description

The proposed estimation algorithm (cf. Algorithm 1) employs an EKF to estimate the 3D pose of the *body frame* $\{B\}$, which is affixed on the spacecraft's IMU, with respect to a *global frame* of reference $\{G\}$. In this work, $\{G\}$ is selected as a planet-centered, planet-fixed frame of reference that rotates with the planet. The IMU measurements are processed immediately as they become available, for propagating the EKF state and covariance estimates, as discussed in Section IV-B. On the other hand, each time an image is recorded, the current camera pose estimate is appended to the state vector, and the covariance matrix is appropriately augmented. State augmentation is necessary for two reasons: First, due to the processing delays introduced by the image processing module, the camera measurements cannot be processed immediately[1]. Second, maintaining a window of camera poses enables the processing of OF measurements (cf. Section IV-E). Therefore, at any time instant the EKF state vector comprises (i) the *evolving* state, $\mathbf{X}_E$, which describes the current state of the spacecraft, and (ii) copies of up to N past poses of the camera. The maximum length of the camera pose history, N, is selected by pre-flight testing, and is chosen to be equal to the maximum number of images through which an OF can be tracked.

Every time the image processing module produces ML and/or OF measurements, an EKF update takes place. In our implementation, the ML measurements of the latest image are processed immediately as they become available. On the other hand, OF updates occur whenever an OF that has been tracked in a number of images is no longer detected in the latest image. At that time, all the measurements of this feature are processed using the method presented in Section IV-E. In the following sections we present the various steps of the algorithm in detail.

[1]Consider that at time-step k an image is recorded, and that the image measurements become available at time-step $k+d$. During the time interval $[k, k+d]$ IMU measurements are processed normally for state propagation. When, at time step $k+d$, the measurements that occurred at time-step k become available, applying an EKF update is possible, because the camera pose at time-step k is included in the state vector.

A. Structure of the EKF state vector

The evolving state of the EKF is described by the vector:

$$\mathbf{X}_E = \begin{bmatrix} {}^B_G\bar{q}^T & \mathbf{b}_g{}^T & {}^G\mathbf{v}_B{}^T & \mathbf{b}_a{}^T & {}^G\mathbf{p}_B^T \end{bmatrix}^T \quad (1)$$

where ${}^B_G\bar{q}$ is the unit quaternion [19] describing the rotation from the global frame to the body frame, ${}^G\mathbf{p}_B$ and ${}^G\mathbf{v}_B$ are the position and velocity of the body expressed with respect to the global frame, and finally $\mathbf{b}_g$ and $\mathbf{b}_a$ are 3×1 vectors that describe the biases affecting the gyroscope and accelerometer measurements, respectively. The IMU biases are modeled as random walk processes, driven by the white Gaussian noise vectors $\mathbf{n}_{wg}$ and $\mathbf{n}_{wa}$, respectively.

Given the definition of the evolving state in Eq. (1), the error-state vector for $\mathbf{X}_E$ is defined accordingly, as:

$$\widetilde{\mathbf{X}}_E = \begin{bmatrix} \boldsymbol{\delta\theta}_B^T & \widetilde{\mathbf{b}}_g^T & {}^G\widetilde{\mathbf{v}}_B^T & \widetilde{\mathbf{b}}_a^T & {}^G\widetilde{\mathbf{p}}_B^T \end{bmatrix}^T \quad (2)$$

For the position, velocity, and biases, the standard additive error definition is used (i.e., the error in the estimate $\hat{x}$ of a quantity x is defined as $\widetilde{x} = x - \hat{x}$). However, for the quaternion a different error definition is employed. In particular, if $\hat{\bar{q}}$ is the estimated value of the quaternion $\bar{q}$, then the orientation error is described by the *error quaternion* $\delta\bar{q}$, which is defined by the relation $\bar{q} = \delta\bar{q} \otimes \hat{\bar{q}}$. In this expression, the symbol $\otimes$ denotes quaternion multiplication. The error quaternion is

$$\delta\bar{q} \simeq \begin{bmatrix} \tfrac{1}{2}\boldsymbol{\delta\theta}^T & 1 \end{bmatrix}^T \quad (3)$$

Since attitude corresponds to 3 degrees of freedom, using $\boldsymbol{\delta\theta}$ to describe the attitude errors results in a minimal representation.

Assuming that N camera poses are included in the EKF state vector at time-step k, this vector has the following form:

$$\hat{\mathbf{X}}_k = \begin{bmatrix} \hat{\mathbf{X}}_{E_k}^T & {}^{C_1}_G\hat{\bar{q}}^T & {}^G\hat{\mathbf{p}}_{C_1}^T & \ldots & {}^{C_N}_G\hat{\bar{q}}^T & {}^G\hat{\mathbf{p}}_{C_N}^T \end{bmatrix}^T \quad (4)$$

where ${}^{C_i}_G\hat{\bar{q}}$ and ${}^G\hat{\mathbf{p}}_{C_i}$, $i = 1 \ldots N$ are the estimates of the camera attitude and position, respectively. The EKF error-state vector is defined accordingly:

$$\widetilde{\mathbf{X}}_k = \begin{bmatrix} \widetilde{\mathbf{X}}_{E_k}^T & \boldsymbol{\delta\theta}_{C_1}^T & {}^G\widetilde{\mathbf{p}}_{C_1}^T & \ldots & \boldsymbol{\delta\theta}_{C_N}^T & {}^G\widetilde{\mathbf{p}}_{C_N}^T \end{bmatrix}^T \quad (5)$$

B. Propagation

To derive the filter propagation equations, we employ discretization of the continuous-time IMU system model, as outlined in the following:

1) Continuous-time system modeling: The system model describing the time evolution of the evolving state is [20]:

$$\begin{aligned} {}^B_G\dot{\bar{q}}(t) &= \tfrac{1}{2}\boldsymbol{\Omega}(\boldsymbol{\omega}(t))\,{}^B_G\bar{q}(t), \quad \dot{\mathbf{b}}_g(t) = \mathbf{n}_{wg}(t) \\ {}^G\dot{\mathbf{v}}_B(t) &= {}^G\mathbf{a}(t), \quad \dot{\mathbf{b}}_a(t) = \mathbf{n}_{wa}(t), \quad {}^G\dot{\mathbf{p}}_B(t) = {}^G\mathbf{v}_B(t) \end{aligned} \quad (6)$$

In these expressions ${}^G\mathbf{a}$ is the body acceleration in the global frame, $\boldsymbol{\omega} = [\omega_x \;\; \omega_y \;\; \omega_z]^T$ is the body rotational velocity expressed in the body frame, and

$$\boldsymbol{\Omega}(\boldsymbol{\omega}) = \begin{bmatrix} -\lfloor \boldsymbol{\omega}\times \rfloor & \boldsymbol{\omega} \\ -\boldsymbol{\omega}^T & 0 \end{bmatrix}, \quad \lfloor \boldsymbol{\omega}\times \rfloor = \begin{bmatrix} 0 & -\omega_z & \omega_y \\ \omega_z & 0 & -\omega_x \\ -\omega_y & \omega_x & 0 \end{bmatrix}$$

The gyroscope and accelerometer measurements, $\boldsymbol{\omega}_m$ and $\mathbf{a}_m$ respectively, are given by:

$$\begin{aligned}\boldsymbol{\omega}_m &= \boldsymbol{\omega} + \mathbf{C}({}^B_G\bar{q})\boldsymbol{\omega}_G + \mathbf{b}_g + \mathbf{n}_g \\ \mathbf{a}_m &= \mathbf{C}({}^B_G\bar{q})({}^G\mathbf{a} - {}^G\mathbf{g} + 2\lfloor\boldsymbol{\omega}_G \times\rfloor {}^G\mathbf{v}_B + \lfloor\boldsymbol{\omega}_G \times\rfloor^2\, {}^G\mathbf{p}_B) \\ &\quad + \mathbf{b}_a + \mathbf{n}_a\end{aligned}$$

where $\mathbf{C}(\cdot)$ denotes the rotational matrix corresponding to the quaternion argument, and $\mathbf{n}_g$ and $\mathbf{n}_a$ are zero-mean, white Gaussian noise processes. It is important to note that, since the frame $\{G\}$ is not inertial, but rather planet-fixed, the IMU measurements incorporate the effects of the planet's rotation, $\boldsymbol{\omega}_G$. Moreover, the accelerometer measurements include the gravitational acceleration, ${}^G\mathbf{g}$, expressed in the local frame.

Applying the expectation operator in the state propagation equations (Eq. (6)) we obtain the equations for propagating the *estimates* of the evolving state:

$$\begin{aligned}{}^B_G\dot{\hat{\bar{q}}} &= \tfrac{1}{2}\boldsymbol{\Omega}(\hat{\boldsymbol{\omega}}){}^B_G\hat{\bar{q}}, \quad \dot{\hat{\mathbf{b}}}_g = \mathbf{0}_{3\times 1}, \\ {}^G\dot{\hat{\mathbf{v}}}_B &= \mathbf{C}^T_{\hat{q}}\hat{\mathbf{a}} - 2\lfloor\boldsymbol{\omega}_G \times\rfloor {}^G\hat{\mathbf{v}}_B - \lfloor\boldsymbol{\omega}_G \times\rfloor^2\, {}^G\hat{\mathbf{p}}_B + {}^G\mathbf{g} \qquad (7) \\ \dot{\hat{\mathbf{b}}}_a &= \mathbf{0}_{3\times 1}, \quad {}^G\dot{\hat{\mathbf{p}}}_B = {}^G\hat{\mathbf{v}}_B\end{aligned}$$

where for brevity we have denoted $\mathbf{C}_{\hat{q}} = \mathbf{C}({}^B_G\hat{\bar{q}})$, $\hat{\mathbf{a}} = \mathbf{a}_m - \hat{\mathbf{b}}_a$ and $\hat{\omega} = \boldsymbol{\omega}_m - \hat{\mathbf{b}}_g - \mathbf{C}_{\hat{q}}\boldsymbol{\omega}_G$. The linearized continuous-time model for the evolving error state is given by:

$$\dot{\widetilde{\mathbf{X}}}_E = \mathbf{F}_E\widetilde{\mathbf{X}}_E + \mathbf{G}_E\mathbf{n}_{\mathrm{IMU}} \qquad (8)$$

where $\mathbf{n}_{\mathrm{IMU}} = \begin{bmatrix}\mathbf{n}_g^T & \mathbf{n}_{wg}^T & \mathbf{n}_a^T & \mathbf{n}_{wa}^T\end{bmatrix}^T$ is the system noise. The covariance matrix of $\mathbf{n}_{\mathrm{IMU}}$, $\mathbf{Q}_{\mathrm{IMU}}$, depends on the IMU noise characteristics and is computed off-line during sensor calibration. Finally, the values of the jacobians $\mathbf{F}_E$ and $\mathbf{G}_E$, which appear in Eq. (8), are given in [18].

2) Discrete-time implementation: The IMU samples the signals $\boldsymbol{\omega}_m$ and $\mathbf{a}_m$ with a period T, and these measurements are used for state propagation in the EKF. Every time a new IMU measurement is received, the IMU state estimate is propagated using 5th order Runge-Kutta numerical integration of Eqs. (7). Moreover, the covariance matrix of the EKF state has to be propagated. For this purpose, we introduce the following partitioning for the covariance matrix:

$$\mathbf{P}_{k|k} = \begin{bmatrix}\mathbf{P}_{EE_{k|k}} & \mathbf{P}_{EC_{k|k}} \\ \mathbf{P}^T_{EC_{k|k}} & \mathbf{P}_{CC_{k|k}}\end{bmatrix} \qquad (9)$$

where $\mathbf{P}_{EE_{k|k}}$ is the 15×15 covariance matrix of the evolving state, $\mathbf{P}_{CC_{k|k}}$ is the $6N\times 6N$ covariance matrix of the camera pose estimates, and $\mathbf{P}_{EC_{k|k}}$ is the correlation between the errors in the evolving state and the camera pose estimates. With this notation, the covariance matrix of the propagated state is given by:

$$\mathbf{P}_{k+1|k} = \begin{bmatrix}\mathbf{P}_{EE_{k+1|k}} & \boldsymbol{\Phi}(t_k+T,t_k)\mathbf{P}_{EC_{k|k}} \\ \mathbf{P}^T_{EC_{k|k}}\boldsymbol{\Phi}(t_k+T,t_k)^T & \mathbf{P}_{CC_{k|k}}\end{bmatrix}$$

where the covariance of the evolving state at time-step $k+1$ is computed by numerical integration of the Lyapunov equation:

$$\dot{\mathbf{P}}_{EE} = \mathbf{F}_E\mathbf{P}_{EE} + \mathbf{P}_{EE}\mathbf{F}_E^T + \mathbf{G}_E\mathbf{Q}_{\mathrm{IMU}}\mathbf{G}_E^T \qquad (10)$$

Numerical integration is carried out for the time interval (t_k, t_k+T), with initial condition $\mathbf{P}_{EE_{k|k}}$. The state transition matrix $\boldsymbol{\Phi}(t_k+T, t_k)$ is similarly computed by numerical integration of the differential equation

$$\dot{\boldsymbol{\Phi}}(t_k+\tau, t_k) = \mathbf{F}_E\boldsymbol{\Phi}(t_k+\tau, t_k), \qquad \tau \in [0, T] \qquad (11)$$

with initial condition $\boldsymbol{\Phi}(t_k, t_k) = \mathbf{I}_{15}$.

C. State Augmentation

When a new image is recorded, the camera pose estimate is computed from the body pose estimate as follows:

$$\begin{aligned}{}^C_G\hat{\bar{q}} &= {}^C_B\bar{q} \otimes {}^B_G\hat{\bar{q}} \qquad (12) \\ {}^G\hat{\mathbf{p}}_C &= {}^G\hat{\mathbf{p}}_B + \mathbf{C}^T_{\hat{q}}\,{}^B\mathbf{p}_C \qquad (13)\end{aligned}$$

where ${}^C_B\bar{q}$ is the quaternion expressing the rotation between the body and camera frames, and ${}^B\mathbf{p}_C$ is the position of the origin of the camera frame with respect to $\{B\}$, both of which are known. This camera pose estimate is appended to the state vector, and the covariance matrix of the EKF is augmented accordingly [3].

D. Measurement Model for ML Observations

We now describe the EKF measurement model for treating visual observations of MLs. Consider that feature j, whose global coordinates are known *a priori*, is observed from the i-th camera pose included in the EKF state vector. In normalized image coordinates, this observation is described by the equation:

$$\mathbf{z}_i^{(j)} = \frac{1}{{}^{C_i}Z_j}\begin{bmatrix}{}^{C_i}X_j \\ {}^{C_i}Y_j\end{bmatrix} + \mathbf{n}_i^{(j)} \qquad (14)$$

where $\mathbf{n}_i^{(j)}$ is the 2×1 image noise vector, with covariance matrix $\mathbf{R}_i^{(j)} = \sigma^2_{\mathrm{im}}\mathbf{I}_2$. The feature position expressed in the camera frame, ${}^{C_i}\mathbf{p}_j$, is given by:

$${}^{C_i}\mathbf{p}_j = \begin{bmatrix}{}^{C_i}X_j \\ {}^{C_i}Y_j \\ {}^{C_i}Z_j\end{bmatrix} = \mathbf{C}({}^{C_i}_G\bar{q})({}^G\mathbf{p}_{\ell_j} - {}^G\mathbf{p}_{C_i}) \qquad (15)$$

where ${}^G\mathbf{p}_{\ell_j}$ is the *known* coordinate of the landmark in the global frame. The expected value of the measurement $\mathbf{z}_i^{(j)}$ is computed using the state estimates:

$$\hat{\mathbf{z}}_i^{(j)} = \frac{1}{{}^{C_i}\hat{Z}_j}\begin{bmatrix}{}^{C_i}\hat{X}_j \\ {}^{C_i}\hat{Y}_j\end{bmatrix} \text{ with } \begin{bmatrix}{}^{C_i}\hat{X}_j \\ {}^{C_i}\hat{Y}_j \\ {}^{C_i}\hat{Z}_j\end{bmatrix} = \mathbf{C}({}^{C_i}_G\hat{\bar{q}})({}^G\mathbf{p}_{\ell_j} - {}^G\hat{\mathbf{p}}_{C_i}) \qquad (16)$$

From Eqs. (14) and (16) we can compute the *residual* of this ML measurement, $\mathbf{r}_{\mathrm{ML}_i}^{(j)} = \mathbf{z}_i^{(j)} - \hat{\mathbf{z}}_i^{(j)}$. By linearization of Eq. (14), $\mathbf{r}_{\mathrm{ML}_i}^{(j)}$ is written as:

$$\mathbf{r}_{\mathrm{ML}_i}^{(j)} \simeq \mathbf{H}_{\delta\boldsymbol{\theta}_i}^{(j)}\delta\boldsymbol{\theta}_{C_i} + \mathbf{H}_{\mathbf{p}_i}^{(j)}\,{}^G\widetilde{\mathbf{p}}_{C_i} + \mathbf{n}_i^{(j)} = \mathbf{H}_{\mathrm{ML}_i}^{(j)}\widetilde{\mathbf{X}} + \mathbf{n}_i^{(j)} \qquad (17)$$

where

$$\mathbf{H}_{\delta\boldsymbol{\theta}_i}^{(j)} = \frac{1}{{}^{C_i}\hat{Z}_j}\begin{bmatrix}\mathbf{I}_2 & -\hat{\mathbf{z}}_i^{(j)}\end{bmatrix}\lfloor\mathbf{C}({}^{C_i}_G\hat{\bar{q}})({}^G\mathbf{p}_{\ell_j} - {}^G\hat{\mathbf{p}}_{C_i})\times\rfloor$$

$$\mathbf{H}_{\mathbf{p}_i}^{(j)} = -\frac{1}{{}^{C_i}\hat{Z}_j} \begin{bmatrix} \mathbf{I}_2 & -\hat{\mathbf{z}}_i^{(j)} \end{bmatrix} \mathbf{C}({}^{C_i}_{G}\hat{\bar{q}})$$

$$\mathbf{H}_{\mathrm{ML}_i}^{(j)} = \begin{bmatrix} \mathbf{0}_{3\times 15} & \mathbf{0}_{3\times 6} & \cdots & \underbrace{[\mathbf{H}_{\delta\boldsymbol{\theta}_i}^{(j)} \ \ \mathbf{H}_{\mathbf{p}_i}^{(j)}]}_{i-\text{th camera block}} & \cdots & \mathbf{0}_{3\times 6} \end{bmatrix}$$

The residual defined in Eq. (17) is employed for performing EKF updates, as described in Section IV-F.

E. Measurement Model for OF Observations

We present the OF measurement model for the case of a *single* feature, f_j, that is observed from a set of M_j poses, $\mathcal{S}_j$. Each observation of this feature is described by the measurement model of Eq. (14). Since the global coordinates of f_j are *not* known in advance, in order to compute the expected value of the measurements, we obtain an estimate of the position of the observed feature, ${}^G\hat{\mathbf{p}}_{\ell_j}$, by employing a least-squares minimization algorithm. Once this estimate has been computed, the expected value of each of the feature measurements can be evaluated, similarly to Eq. (16), with the sole difference that the *estimate* of the landmark position is used, instead of an *a priori* known value.

Linearization yields the following expression for the residual, $\mathbf{r}_i^{(j)} = \mathbf{z}_i^{(j)} - \hat{\mathbf{z}}_i^{(j)}$, of the i-th measurement:

$$\begin{aligned} \mathbf{r}_i^{(j)} &\simeq \mathbf{H}_{\delta\boldsymbol{\theta}_i}^{(j)} \delta\boldsymbol{\theta}_{C_i} + \mathbf{H}_{\mathbf{p}_i}^{(j)} {}^G\widetilde{\mathbf{p}}_{C_i} + \mathbf{H}_{f_i}^{(j)} {}^G\widetilde{\mathbf{p}}_{\ell_j} + \mathbf{n}_i^{(j)} \\ &= \mathbf{H}_{\mathbf{X}_i}^{(j)} \widetilde{\mathbf{X}} + \mathbf{H}_{f_i}^{(j)} {}^G\widetilde{\mathbf{p}}_{\ell_j} + \mathbf{n}_i^{(j)} \end{aligned}$$

Note that, in contrast to the case of ML observations, the measurement residual in this case is also affected by the error in the estimate of the landmark position. In the last expression, $\mathbf{H}_{f_i}^{(j)} = -\mathbf{H}_{\mathbf{p}_i}^{(j)}$ is the Jacobian of the residual with respect to ${}^G\hat{\mathbf{p}}_{\ell_j}$, and

$$\mathbf{H}_{\mathbf{X}_i}^{(j)} = \begin{bmatrix} \mathbf{0}_{3\times 15} & \mathbf{0}_{3\times 6} & \cdots & \underbrace{[\mathbf{H}_{\delta\boldsymbol{\theta}_i}^{(j)} \ \ \mathbf{H}_{\mathbf{p}_i}^{(j)}]}_{i-\text{th camera block}} & \cdots & \mathbf{0}_{3\times 6} \end{bmatrix}$$

By stacking the residuals corresponding to all the observations of this feature, we obtain:

$$\mathbf{r}^{(j)} \simeq \mathbf{H}_{\mathbf{X}}^{(j)} \widetilde{\mathbf{X}} + \mathbf{H}_{f}^{(j)} {}^G\widetilde{\mathbf{p}}_{\ell_j} + \mathbf{n}^{(j)} \qquad (18)$$

where $\mathbf{r}^{(j)}$, $\mathbf{H}_{\mathbf{X}}^{(j)}$, $\mathbf{H}_{f}^{(j)}$, and $\mathbf{n}^{(j)}$ are block vectors or matrices with elements $\mathbf{r}_i^{(j)}$, $\mathbf{H}_{\mathbf{X}_i}^{(j)}$, $\mathbf{H}_{f_i}^{(j)}$, and $\mathbf{n}_i^{(j)}$, for $i \in \mathcal{S}_j$. Assuming that feature observations in different images are independent, the covariance matrix of the noise vector $\mathbf{n}^{(j)}$ is $\mathbf{R}^{(j)} = \sigma_{\mathrm{im}}^2 \mathbf{I}_{2M_j}$.

It should be clear that the residual derived in Eq. (18) *cannot* be directly used for performing EKF updates, since the landmark position error, ${}^G\widetilde{\mathbf{p}}_{\ell_j}$, is correlated with the state errors (recall that ${}^G\hat{\mathbf{p}}_{\ell_j}$ is computed using the state estimates and the measurements $\mathbf{z}_i^{(j)}$ in a least-squares minimization routine). To overcome this problem, we define a residual $\mathbf{r}_{\mathrm{OF}}^{(j)}$, by *projecting* $\mathbf{r}^{(j)}$ on the left nullspace of the matrix $\mathbf{H}_f^{(j)}$. Specifically, if we let $\mathbf{U}$ denote the unitary matrix whose columns form the basis of the left nullspace of $\mathbf{H}_f^{(j)}$, we obtain:

$$\mathbf{r}_{\mathrm{OF}}^{(j)} = \mathbf{U}^T(\mathbf{z}^{(j)} - \hat{\mathbf{z}}^{(j)})$$

$$\simeq \mathbf{U}^T \mathbf{H}_{\mathbf{X}}^{(j)} \widetilde{\mathbf{X}} + \mathbf{U}^T \mathbf{n}^{(j)} = \mathbf{H}_{\mathrm{OF}}^{(j)} \widetilde{\mathbf{X}} + \mathbf{n}_o^{(j)} \qquad (19)$$

It is worth noting that $\mathbf{r}_{\mathrm{OF}}^{(j)}$ and $\mathbf{H}_{\mathrm{OF}}^{(j)}$, can be computed without explicitly evaluating $\mathbf{U}$. Instead, these projections of $\mathbf{r}$ and $\mathbf{H}_{\mathbf{X}}^{(j)}$ on the nullspace of $\mathbf{H}_f^{(j)}$ can be computed very efficiently using Givens rotations [21]. The covariance matrix of the noise vector $\mathbf{n}_o^{(j)}$ can be easily shown to be equal to $\sigma_{\mathrm{im}}^2 \mathbf{I}_{2M_j-3}$.

The residual $\mathbf{r}_o^{(j)}$ is *independent* of the errors in the feature coordinates, and thus EKF updates can be performed based on it. Eq. (19) defines a *linearized* constraint between all the camera poses from which the feature f_j was observed. This residual expresses all the available information that the measurements $\mathbf{z}_i^{(j)}$ provide for the M_j states, and thus the resulting EKF update is optimal, except for the inaccuracies caused by linearization.

F. EKF Updates

In the preceding sections we presented the measurement models that we employ for treating ML and OF observations. Once all the ML and OF measurements that must be processed at a given time-step are determined (as described in Algorithm 1), the corresponding residual vectors and measurement Jacobians (Eqs. (17) and (19)) are created. Stacking all these together yields the following residual vector:

$$\mathbf{r} = \mathbf{H}\widetilde{\mathbf{X}} + \mathbf{n} \qquad (20)$$

where $\mathbf{r}$ is a block vector with elements $\mathbf{r}_{\mathrm{ML}_i}^{(j)}$ and $\mathbf{r}_{\mathrm{OF}}^{(j)}$, $\mathbf{H}$ is a block matrix with elements $\mathbf{H}_{\mathrm{ML}_i}^{(j)}$ and $\mathbf{H}_{\mathrm{OF}}^{(j)}$, and $\mathbf{n}$ is a noise vector of dimension L (equal to the length of $\mathbf{r}$), with covariance matrix $\mathbf{R} = \sigma_{\mathrm{im}}^2 \mathbf{I}_L$. Once the residual, $\mathbf{r}$, and the measurement Jacobian matrix, $\mathbf{H}$ of Eq. (20) have been computed, the EKF update proceeds according to the standard equations [22]. In our work, we employ the QR decomposition of the matrix $\mathbf{H}$ to reduce the computational complexity of EKF updates [7]. At the end of the update step, the oldest camera pose is marginalized out of the EKF state, to allow for the inclusion of the next one.

V. Experiments

In order to validate the algorithm's performance in conditions as close to actual planetary landing as possible, a sounding rocket experiment was conducted in April 2006, at White Sands Missile Range (WSMR), New Mexico.

A. Hardware Description

A commercially available analog camera (Pulnix TM-9701) was added to an existing mission payload consisting of a GLN-MAC IMU and a Thales G12 GPS, onboard a Terrier Orion Sounding Rocket (cf. Fig. 1 for the experimental setup). The nadir-pointing camera provided descent imagery from parachute deployment to landing at 30 frames/s with a resolution of 768×484 pixels, 8 bits/pixel, and a field of view (FOV) of $38° \times 24°$. A common GPS time tag from a commercial timecode generator was used to synchronize 50 Hz IMU, 10 Hz GPS, and 30 Hz image data. Images, IMU data, and GPS measurements were downlinked in real-time

Parameter	Sounding Rocket	Mars Landing
Parachute Deploy Alt.	4200 m above ground	2000 m above ground
Vertical Velocity	10 m/s at touchdown	1 m/s at td.
Horizontal Velocity	3 m/s at touchdown	< 0.5 m/s at td.
Off nadir angle	$\leq 12°$	$< 20°$
Off nadir angular rate	≤ 19 °/s	< 60 °/s
Roll Rate	≤ 360 °/s	< 60 °/s

TABLE I
DYNAMICS COMPARISON BETWEEN SOUNDING ROCKET AND MARS EDL.

during flight over an S-band telemetry channel and recorded on the ground.

The data collected during this experiment were processed off-line. We should note, however, that our algorithm is capable of real-time operation. The FFT correlation-based feature matching is predicted to run at 5-20 Hz in an FPGA-based implementation currently under development at JPL, and the current C++ implementation of the pose estimator runs at 20 Hz on a 2 GHz CPU, with the number of stored poses set to $N = 20$.

B. *Experiment Profile and Relevance*

The rocket reached an apogee altitude of 123 km, followed by drogue and main parachute opening at 28 km and 4.2 km altitude, respectively. After a total flight time of 805 s, 376 s of which on the parachute, the vehicle landed 78 km downrange from the launch pad. The dynamics encountered during the parachuted phase of the sounding rocket flight are similar to those during an actual Mars landing, as shown by the comparison in Table I.

Fig. 1 shows the rocket's trajectory superimposed on a 3D map of the area. A zoomed-in view of the flight path after main parachute deployment is depicted in Fig. 3. Pure integration of the IMU measurements (blue dashed line) yielded fairly accurate results until right after the deployment of the main parachute, but then quickly diverged. The opening of the parachute caused the rocket's motion to be extremely jerky for several seconds. Integrating the large acceleration measurements recorded in this period, using the attitude estimates that had error accumulated over the preceding 431 s of flight, resulted in large position errors. Note that up to this point no images were available. Once the first few images were processed, the VISINAV algorithm corrected the accumulated error, and the estimated trajectory became virtually indistinguishable from that measured by the GPS.

As shown in Table II, ML measurements were processed during two separate phases of flight, one between 3800 m and 3100 m, using images at 3 Hz, and the second between 1600 m and 230 m above ground, processing only one frame per second (1 Hz), and yielding up to 80 MLs per image (cf. Fig. 4). The artificial gap, during which the filter had to rely on open-loop IMU integration, was introduced to emulate a temporary failure of ML detection. In EDL, this could arise, for example, due to large off-nadir viewing angles

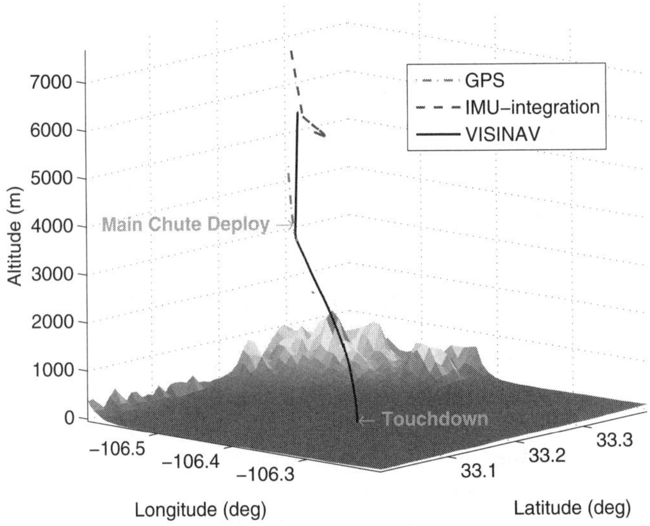

Fig. 3. Zoomed-in view of trajectory after main parachute deployment.

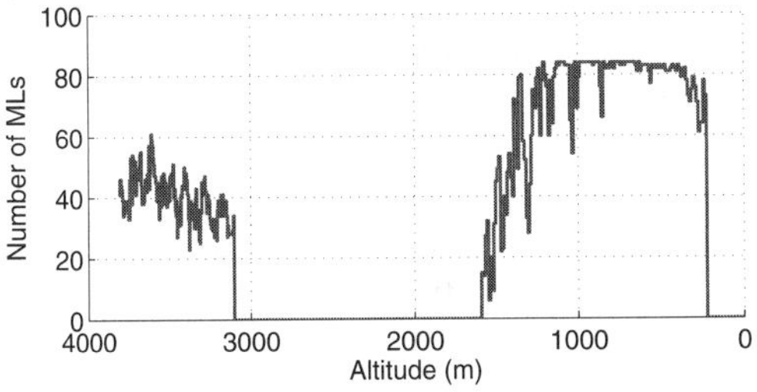

Fig. 4. Number of detected MLs vs. altitude.

caused by a pose correction maneuver during powered descent. Once images become available after this intermediate open-loop phase, a significant change in image scale has occurred. Despite this, the ML algorithm is capable of successfully matching MLs, and the EKF estimates' uncertainty drops sharply within a few seconds after resuming ML updates (cf. Figs. 5, 7 at 1600 m altitude).

The 3D ground coordinates for the MLs were obtained from USGS 1 Meter Digital Orthoimagery Quarter Quadrangles taken in 2001, combined with 1 arc second finished elevation data from the Shuttle Radar Topography Mission [23]. For feature matching in the first set, the entire 7×8 km map image was resampled to ~ 7 m/pixel, while for the second set the map was cropped to 2×2 km and used at its base resolution of 1 m/pixel.

At some point during descent, the number of features within the FOV becomes too small, and the difference in resolution between camera and map too significant to allow successful ML correspondences to be established. In this experiment, we emulated this behavior by stopping ML updates at 230 m altitude. To compensate, starting at 330 m, the filter began to perform OF updates at a frequency of 3 Hz. Even though OFs do not make the system observable (as opposed to images containing at least three MLs), they allow for precise estimation of linear and rotational velocity, resulting in very small error growth during the final 230 m of descent.

C. Algorithm Performance

Ground-truth for position and velocity was obtained from GPS measurements. Figs. 5 and 7 show the resulting errors and the corresponding 3σ-bounds for velocity and position in the local North-East-Down (NED) frame. Table II gives the error norms for the position and velocity estimates at the beginning and end of the different update phases.

a) First ML set: When processing the first MLs, the algorithm converges within about 5 seconds from the large error ($\sim$ 2700 m) accumulated during IMU integration to within 18 m of GPS ground-truth (cf. Figs. 3, 7). During the open-loop integration phase between 3100 m to 1600 m altitude, the pose uncertainty is again increasing.

b) Second ML set and OFs: The pose estimates almost instantaneously converge close to ground truth once ML updates resume at 1600 m. ML- and OF-updates (the latter starting at 330 m) reduce the position uncertainty bounds to approximately ± 4 m, and the velocity uncertainty to less than ± 0.25 m/s along each axis (3σ). Notice that the actual error at the beginning of OF updates is smaller than at the end of ML processing ten seconds later. This can be attributed to the already decreasing number of detected MLs within the FOV at this altitude, due to the difference in resolution between satellite and camera image (cf. Fig. 4). With the discontinuation of ML processing at 230 m altitude, the pose uncertainty begins to increase even more, although still at a very low rate (cf. the zoomed-in view of the errors for the final 300 m in Figs. 6 and 8). As predicted, this is the result of the system becoming unobservable. Table II shows the final velocity and position error magnitudes at touchdown, which are approximately 6.4 m for position and 0.16 m/s for velocity.

Similar to position and velocity, the attitude uncertainty bounds were decreased to $\pm 0.15^\circ$ accuracy along each axis (3σ) during processing of ML updates, with a temporary increase to $\pm 0.9^\circ$ during open-loop IMU integration between 3100 m and 1600 m altitude. Due to the lack of ground-truth, the attitude uncertainty was determined from the EKF estimates of the covariance matrix. The figures for position and velocity errors (cf. Figs. 5-8) show that the filter estimates for these variables are consistent, indicating that the attitude estimates are also consistent. The filter attitude estimate was further verified through an independent measurement of the final attitude at touchdown using a compass.

VI. Conclusion

In this paper, we have presented the analysis and experimental validation of a vision-aided inertial navigation algorithm for planetary landing applications. Results from a sounding rocket test, covering the dynamic profile of typical planetary EDL scenarios, showed estimation errors of magnitude 0.16 m/s in velocity and 6.4 m in position at touchdown. These results vastly improve current state of the art for EDL navigation without vision, and meet the requirements of future planetary exploration missions [1]. The algorithm tightly couples IMU and camera measurements of mapped landmarks (MLs) and opportunistic features (OFs), in a resource-adaptive and hence real-time capable fashion. It is thus able to provide very accurate, high-bandwidth estimates for precision guidance and

	Altitude (m)	Time (s)	Position Error (m)	Velocity Error (m/s)
Beg. 1st ML set	3800	454	2722.3	10.45
End 1st ML set	3100	514	16.9	0.18
Beg. 2nd ML set	1600	647	78.2	1.38
End 2nd ML set	230	781	5.1	0.23
Beg. OFs	330	771	3.7	0.15
Touchdown	0	805	6.4	0.16
Touchdown (IMU-only)	0	805	9169.5	32.70

TABLE II
CONDITIONS FOR THE DIFFERENT EKF UPDATE PHASES.

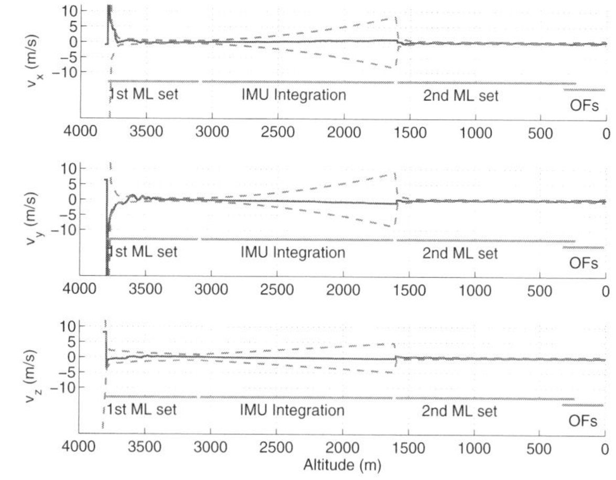

Fig. 5. Velocity error expressed in NED frame (blue solid line) and corresponding 3σ-bounds (red dashed line). Note that x-y-z in the plots corresponds to N-E-D.

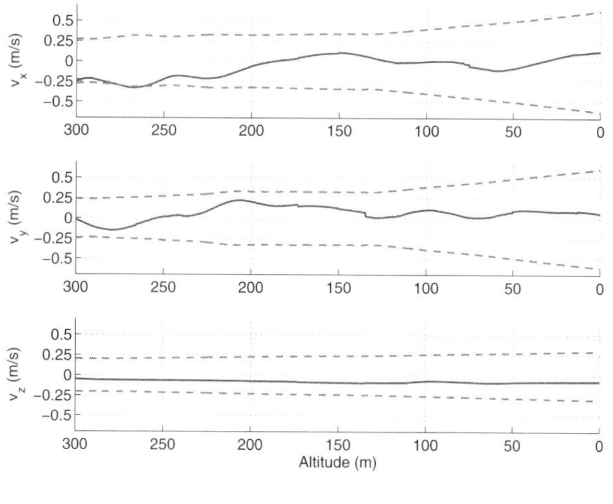

Fig. 6. Velocity error in NED frame (zoomed-in view of Fig. 5 before touchdown).

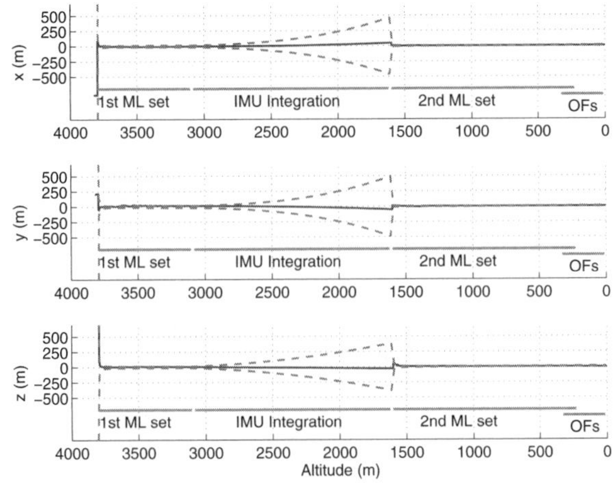

Fig. 7. Position error expressed in NED frame (blue solid line) and corresponding 3σ-bounds (red dashed line).

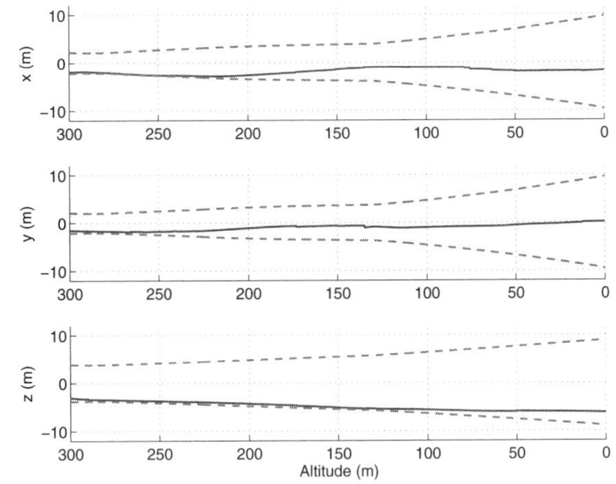

Fig. 8. Position error in NED frame (zoomed-in view of Fig. 7 before touchdown).

control. It should be pointed out that the proposed estimator is modular and easily extendable to incorporate additional sensor information (e.g., doppler radar), as well as different image processing algorithms that provide unit vector measurements to corresponding features. Future work includes optimal selection of a subset of the most informative image features to track. This would allow efficient use of camera information in the presence of limited computational resources.

ACKNOWLEDGEMENTS

This work was supported by the University of Minnesota (DTC), the NASA Mars Technology Program (MTP-1263201), and the National Science Foundation (EIA-0324864, IIS-0643680).

REFERENCES

[1] NASA, "Solar system exploration roadmap," http://solarsystem.nasa.gov/multimedia/downloads/SSE_RoadMap_2006_Report_FC-A_med.pdf, Sep. 2006.

[2] N. Trawny, A. I. Mourikis, S. I. Roumeliotis, A. E. Johnson, and J. Montgomery, "Vision-aided inertial navigation for pin-point landing using observations of mapped landmarks," *Journal of Field Robotics*, vol. 24, no. 5, pp. 357–378, May 2007.

[3] A. I. Mourikis and S. I. Roumeliotis, "A multi-state constraint Kalman filter for vision-aided inertial navigation," in *Proceedings of the IEEE International Conference on Robotics and Automation*, Rome, Italy, April 2007, pp. 3565–3572.

[4] J. L. Crassidis, R. Alonso, and J. L. Junkins, "Optimal attitude and position determination from line-of-sight measurements," *The Journal of the Astronautical Sciences*, vol. 48, no. 2–3, pp. 391–408, April–Sept. 2000.

[5] A. Wu, E. Johnson, and A. Proctor, "Vision-aided inertial navigation for flight control," in *Proceedings of the AIAA Guidance, Navigation, and Control Conference*, no. AIAA 2005-5998, San Francisco, CA, Aug. 2005.

[6] G. F. Ivey and E. Johnson, "Investigation of methods for simultaneous localization and mapping using vision sensors," in *Proceedings of the AIAA Guidance, Navigation, and Control Conference*, no. AIAA 2006-6578, Keystone, CO, Aug. 2006.

[7] D. S. Bayard and P. B. Brugarolas, "An estimation algorithm for vision-based exploration of small bodies in space," in *Proceedings of the 2005 American Control Conference*, vol. 7, Portland, Oregon, Jun. 2005, pp. 4589–4595.

[8] D. Strelow, "Motion estimation from image and inertial measurements," Ph.D. dissertation, Carnegie Mellon University, Nov. 2004.

[9] S. I. Roumeliotis, A. E. Johnson, and J. F. Montgomery, "Augmenting inertial navigation with image-based motion estimation," in *IEEE International Conference on Robotics and Automation (ICRA)*, Washington D.C., 2002, pp. 4326–33.

[10] R. M. Eustice, H. Singh, and J. J. Leonard, "Exactly sparse delayed-state filters for view-based SLAM," *IEEE Transactions on Robotics*, vol. 22, no. 6, pp. 1100–1114, Dec. 2006.

[11] A. I. Mourikis and S. I. Roumeliotis, "On the treatment of relative-pose measurements for mobile robot localization," in *Proc. IEEE Int. Conf. on Robotics and Automation*, Orlando, FL, May 15-19 2006, pp. 2277 – 2284.

[12] P. McLauchlan, "The variable state dimension filter," Centre for Vision, Speech and Signal Processing, University of Surrey, UK, Tech. Rep., 1999.

[13] Y. Cheng and A. Ansar, "Landmark based position estimation for pin-point landing on mars," in *Proceedings of the 2005 IEEE International Conference on Robotics and Automation (ICRA)*, Barcelona, Spain, Apr. 2005, pp. 4470–4475.

[14] D. G. Lowe, "Distinctive image features from scale-invariant keypoints," *International Journal of Computer Vision*, vol. 2, no. 60, pp. 91–110, 2004.

[15] C. Harris and M. Stephens, "A combined corner and edge detector," in *Proceedings of the 4th Alvey Vision Conference*, 1988, pp. 147–151.

[16] A. Ansar, "2004 small body GN&C research report: Feature recognition algorithms," in *Small Body Guidance Navigation and Control FY 2004 RTD Annual Report (Internal Document)*. Pasadena, CA: Jet Propulsion Laboratory, 2004, no. D-30282 / D-30714, pp. 151–171.

[17] A. Johnson, R. Willson, Y. Cheng, J. Goguen, C. Leger, M. SanMartin, and L. Matthies, "Design through operation of an image-based velocity estimation system for Mars landing," *International Journal of Computer Vision*, vol. 74, no. 3, pp. 319–341, September 2007.

[18] A. I. Mourikis, N. Trawny, S. I. Roumeliotis, A. Johnson, A. Ansar, and L. Matthies, "Vision-aided inertial navigation for spacecraft entry, descent, and landing," *Journal of Guidance, Control, and Dynamics*, 2007, submitted.

[19] W. G. Breckenridge, "Quaternions proposed standard conventions," Jet Propulsion Laboratory, Pasadena, CA, Interoffice Memorandum IOM 343-79-1199, 1999.

[20] A. B. Chatfield, *Fundamentals of High Accuracy Inertial Navigation*. Reston, VA: American Institute of Aeronautics and Astronautics, Inc., 1997.

[21] G. Golub and C. van Loan, *Matrix computations*. The Johns Hopkins University Press, London, 1996.

[22] P. S. Maybeck, *Stochastic Models, Estimation and Control*. New York: Academic Press, 1979, vol. 1+2.

[23] U.S. Geological Survey, "Seamless data distribution system," http://seamless.usgs.gov/index.asp, 2006.

Optimal Kinodynamic Motion Planning for 2D Reconfiguration of Self-Reconfigurable Robots

John Reif
Department of Computer Science, Duke University

Sam Slee
Department of Computer Science, Duke University

***Abstract*— A self-reconfigurable (SR) robot is one composed of many small modules that autonomously act to change the shape and structure of the robot. In this paper we consider a general class of SR robot modules that have rectilinear shape that can be adjusted between fixed dimensions, can transmit forces to their neighbors, and can apply additional forces of unit maximum magnitude to their neighbors. We present a kinodynamically optimal algorithm for general reconfiguration between any two distinct, 2D connected configurations of n SR robot modules. The algorithm uses a third dimension as workspace during reconfiguration. This entire movement is achieved within $O(\sqrt{n})$ movement time in the worst case, which is the asymptotically optimal time bound. The only prior reconfiguration algorithm achieving this time bound was restricted to linearly arrayed start and finish configurations (known as the "x-axis to y-axis problem"). All other prior work on SR robots assumed a constant velocity bound on module movement and so required at least time linear in n to do the reconfiguration.**

I. Introduction

The dynamic nature of self-reconfigurable robots makes them ideally suited for numerous environments with challenging terrain or unknown surroundings. Applications are apparent in exploration, search and rescue, and even medical settings where several specialized tools are required for a single task. The ability to efficiently reconfigure between the many shapes and structures of which an SR robot is capable is critical to fulfilling this potential.

In this paper we present an $O(\sqrt{n})$ movement time algorithm for reconfiguration between general, connected 2D configurations of modules. Our algorithm accomplishes this by transforming any 2D configuration into a vertical column in a third dimension in $O(\sqrt{n})$ movement time. By the Principle of Time Reversal mentioned in [4], we may then reconfigure from that vertical column to any other 2D configuration in the same movement time. Thus, we have an algorithm for reconfiguration between general 2D configurations by going through this intermediate column state.

We begin with a general, connected 2D configuration in the x/y-axis plane, with the z-axis dimension used as our workspace. The configuration is represented as a graph with a node for each module and an edge between nodes in the graph for pairs of modules that are directly connected. The robot modules are then reconfigured in O(1) movement time so that they remain in the x/y-axis plane, but a Hamiltonian Cycle (a cycle that visits each module exactly once) is formed through the corresponding graph and is known.

Using this cycle we create a Hamiltonian Path which may be reconfigured in $O(\sqrt{n})$ movement time so that each module has a unique location in the z-axis direction. It is in this stage that our kinodynamic formulation of the problem is necessary to achieve the desired movement time bound. Finally, in two further stages all modules will be condensed to the same x-axis location and y-axis location in $O(\sqrt{n})$ movement time. This forms the intermediate configuration of a z-axis column of modules. Reversing this process allows general reconfiguration between connected, 2D configurations.

Organization The rest of the paper is organized as follows. Work related to the results given in this paper is summarized in Section II. In Section III the problem considered in this paper is more rigorously defined and the notation that will be used in the remainder of the paper is given. In Section IV we transform a general, connected 2D configuration of modules into a 2D configuration with a known Hamiltonian Cycle through its modules in O(1) movement time. In Section V that Hamiltonian Cycle is used to allow reconfiguration into the z-axis column intermediate stage. Finally, we conclude with Section VI.

II. Related Work

The results given in this paper primarily apply to lattice or substrate style SR robots. In this form, SR robot modules attach to each other at discrete locations to form lattice-like structures. Reconfiguration is then achieved by individual modules rolling, walking, or climbing along the surfaces of the larger robotic structure formed by other modules in the system. Several abstract models have been developed by various research groups to describe this type of SR robot.

In the work of [7] these robots were referred to as metamorphic robots (another common term for SR robots [2, 3, 8]). In this model the modules were identical 2D, independently controlled hexagons. Each hexagon module had rigid bars for its edges and bendable joints at the 6 vertices. Modules would move by deforming and "rolling" along the surfaces formed by other, stationary modules in the system. In [1] the sliding cube model was presented to represent lattice-style modules. Here each module was a 3D cube that was capable of sliding along the flat surfaces formed by other cube modules. In addition, cube modules could make convex or concave transitions between adjacent perpendicular surfaces.

Most recently, in [4] an abstract model was developed that set explicit bounds on the physical properties and abilities of robot modules. According to those restrictions, matching upper and lower bounds of $O(\sqrt{n})$ movement time for an example worst case reconfiguration problem were given. The algorithm for that problem, the *x-axis to y-axis problem*, was designed only for that example case and left open the question of finding a general reconfiguration algorithm. This paper gives an $O(\sqrt{n})$ movement time algorithm for reconfiguration between general, connected 2D configurations. This algorithm satisfies the abstract model requirements of [4] and matches the lower bound shown in that paper.

While the results of this paper are theoretical, the abstract modules used do closely resemble the compressible unit or expanding cube hardware design. Here an individual module can expand or contract its length in any direction by a constant factor dependent on the implementation. Modules then move about in the larger system by having neighboring modules push or pull them in the desired direction using coordinated expanding and contracting actions. Instances of this design include the Crystal robot by Rus et. al. [5] and the Telecube module design by Yim et. al. [6]. In [5, 6] algorithms are given that require time at least linear in the number of modules.

III. Notation and Problem Formulation

Bounds and Equations We assume that each module has unit mass, unit length sides, and can move with acceleration magnitude upper bounded by 1. This acceleration is created by exerting a unit-bounded force on neighboring modules so that one module slides relative to its neighbors. Each module in any initial configuration begins with 0 velocity in all directions. We also assume that each module may exert force to contract itself from unit length sides to 1/2 unit length or expand up to 3 units in length in any axis direction in O(1) time. These assumptions match the abstract module requirements stated in [4]. Friction and gravitational forces are ignored. Our analysis will make use of the following physics equations:

$$F_i = m_i a_i \quad (1)$$

$$x_i(t) = x_i(0) + v_i(0)t + \frac{1}{2}a_i t^2 \quad (2)$$

$$v_i(t) = v_i(0) + a_i t\ . \quad (3)$$

In these equations F_i is force applied to a module i having mass m_i and acceleration a_i. Similarly, $x_i(t)$ and $v_i(t)$ are module i's position and velocity after moving for time t.

Problem Formulation Define the coordinate location of a module in a given axis direction as being the smallest global coordinate of any point on that module. This must be the location of some face of the module since all modules are assumed to be rectangles with faces aligned with the 3 coordinate axes. Let A and B be two connected configurations, each of n modules with each module having a z-axis coordinate of 0 and unit-length dimensions. Let at least one module δ in A and one module β in B have the same coordinate location. The desired operation is reconfiguration from A to B while satisfying the requirements of the abstract model given in [4].

IV. Creating a Hamiltonian Cycle

We begin to lay the groundwork for our algorithm by first showing that a Hamiltonian Cycle may be formed through a collection of modules in a general, connected 2D configuration along the x/y-axis plane. We consider this configuration as a graph, with the modules represented as nodes and connections between touching, neighboring modules represented as edges in the graph. First a spanning tree for this graph will be found. Then the modules will be reconfigured to form a Hamiltonian Cycle through them that essentially traces around the original spanning tree.

A. Finding A Spanning Tree

To facilitate the creation of such a cycle, we will form a spanning tree by alternating stages of adding rows and stages of adding columns of modules to the tree. We later show how this method of creating a spanning tree is useful. We define a *connected row* of modules as a row of modules $i = 1, \ldots, k$ each having the same y-axis coordinate and with face-to-face connections between adjacent modules i and $i+1$ along that row. A *connected column* is defined in the same way as a connected row, but with each module having the same x-axis coordinate. An *endpoint module* has only one adjacent neighbor in the spanning tree.

Define a *free module* as one that has not yet been added to the spanning tree. Consider a case where we have a partially formed spanning tree T and a given module p that has been added to T. Let the *free-connected row* about p be the longest possible connected row of modules which includes p and is composed only of modules, other than p, that are free. Similarly, the *free-connected column* about p is the longest possible connected column of modules which includes p and is composed only of free modules (with p possibly not free). These definitions are used in the algorithm below and examples are given in Figure 1.

CREATE_SPANNING_TREE(Module m)

Input: A module m in the 2D configuration.
Output: A spanning tree of all configuration modules.
Initialize: Add m to spanning tree T.
Add m to queue R.
Add m to queue C.

While: R is nonempty or C is nonempty.
Repeat:
For-each module x in R
  Remove x from queue R.
  For-each module y in x's free-connected row
    Add y to tree T as part of that row.
    Add y to queue C.
  End-for-each
End-for-each

For-each module x in C
  Remove x from queue C.
  For-each y in x's free-connected column
    Add y to tree T as part of that column.
    Add y to queue R.
  End-for-each
End-for-each

The algorithm above simply adds modules to the spanning tree through alternating stages of adding free-connected rows and columns of modules. The **For-each** loops are executed synchronously so rows and columns are added one at a time (i.e. no time overlap). The fact that we add these synchronously and that we alternate between row stages and column stages will be used in later proofs.

Spanning Tree Property 1: *If two modules a and b have a face-to-face connection, the same y-axis (x-axis) coordinate, and were each added to the spanning tree as part of rows (columns), then a and b were added as part of the same free-connected row (column).*

An example run of this tree creation algorithm is given in the top row of pictures in Figure 1. The bottom row of pictures shows how the spanning tree may be converted to a *double-stranded* spanning tree, which will soon be discussed.

Lemma 1: *The algorithm described above finds a spanning tree through a given 2D connected configuration along the x/y plane and satisfies Spanning Tree Property 1.*

Proof: To prove that a spanning tree is formed we must show that all modules are added to the tree and that no cycles are created. First, consider a case where the algorithm for creating the spanning tree has finished, but one or more modules have not been added to the spanning tree. Since the given 2D configuration is assumed to be connected, there must be some module q in the group, not yet added to the spanning tree, that has a face-to-face connection to another module d which has been added to the spanning tree.

Yet, in this case q would be a part of either module d's free connected row or its free connected column. So q would have been added to the spanning tree when d was added or in the next stage when the free connected row or column about d was added. Hence, we cannot have that the algorithm has finished and module q is not a part of the spanning tree. Thus, all modules must be part of the spanning tree.

Furthermore, with the exception of the first module added to the tree (module p_0) modules are added to the spanning tree through either a free connected column or a free connected row. Then each module m has at most one 'parent' module in the spanning tree: the module preceding m along the row or column with which m was added. Hence, no cycles are formed and a complete spanning tree has been created. Also, each module is added to the spanning tree exactly once and is considered for inclusion in the spanning tree at most 4 times (once when each of its 4 possible neighbors are added). Thus, the algorithm terminates and, from a centralized controller viewpoint, it takes $O(n)$ computation time.

Finally, to prove that Spanning Tree Property 1 holds, consider two modules a and b that were added to the spanning tree from different rows r_1 and r_2, respectively, but share a face-to-face connection and have the same y-axis coordinate. Since rows are added synchronously one at a time, either r_1 or r_2 was added first. Without loss of generality let it be r_1. Then at the time that module a was added as part of r_1, module b would have been free. Thus, module b would have been included in r_1, the same free-connected row as module

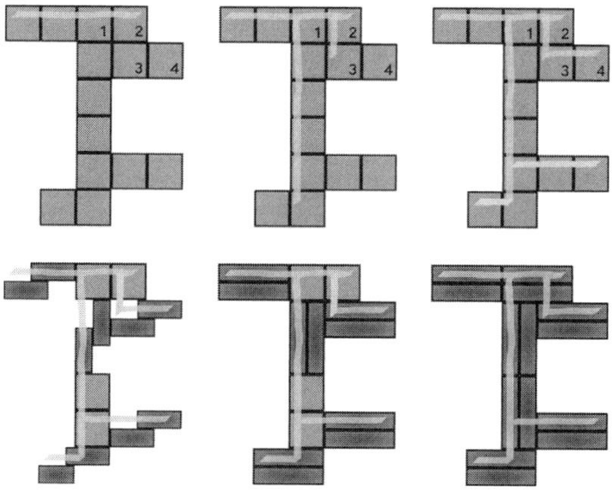

Fig. 1. **Top Row:** Creating a spanning tree by successive rows and columns. **Bottom Row:** Transforming into a double-stranded spanning tree.

a, instead of r_2. Thus, Spanning Tree Property 1 holds for rows. A matching argument for columns also holds (replacing rows r_1 and r_2 with columns c_1 and c_2 and letting a and b have the same x-axis coordinate instead of the same y-axis coordinate). Thus, Spanning Tree Property 1 holds for both rows and columns. □

B. Doubling the Spanning Tree

With a spanning tree through our given 2D configuration now found, we may focus on turning that into a Hamiltonian Cycle. A basic step for this is to divide the modules along the spanning tree into adjacent pairs of modules. We then reconfigure each of these pairs of consecutive 1×1 dimension modules along that spanning tree into $\frac{1}{2} \times 2$ dimension modules that run parallel to each other in the direction of the original spanning tree. An example is shown in the 3 pictures along the bottom row of Figure 1.

This reconfiguration is done so that each pair of modules maintains the same 1×2 length dimension bounding box around those modules. So, all such reconfigurations may be done locally and simultaneously in two stages. Using two stages allows alternating pairs of modules to be kept stationary so that the overall configuration does not become disconnected. At all times adjacent modules along the spanning tree will share at least a corner connection from the 2D 'bird's eye' view (which is really an edge connection since each module has a 3D rectilinear shape).

The effect of this reconfiguration is to transform our original tree into a "double-stranded" spanning tree. That is, a spanning tree which has two parallel sets of modules along each edge of the tree, but with single modules still permitted at the endpoints of the tree. This allows traversal of the tree by using one module from each pair to travel 'down' a tree branch, and the other module in the pair to travel back 'up' the branch. Further reconfiguration work will transform this intuition into an actual Hamiltonian Cycle.

Lemma 2: *Given the spanning tree T formed in Lemma 1, the modules in that tree may be paired such that the only modules remaining single are endpoint modules in that tree.*

Proof Sketch: Begin with the root module p_0 and pair

modules while moving away from p_0. At the end of any row/column if a module p is left single it is either: (1) an endpoint of the spanning tree, or (2) included in some other column/row where pairs have not yet been assigned. In case (1) no more work is needed and in case (2) recursively solve the subtree that has module p as its root. □

Now that we have the modules in our spanning tree paired together, the next step is to reconfigure those pairs so that the desired double-stranded spanning tree is formed. This reconfiguration and its proof are similar to the *confined cubes swapping problem* introduced in [4].

Lemma 3: *Consider a row of n modules in the x/y plane, labeled $i = 1, \ldots, n$ along the row, each having unit length dimensions. Each pair of adjacent modules along the row may be reconfigured in O(1) movement time so that each has $1/2{\times}2$ unit dimensions in the x/y plane while maintaining the same 1×2 unit dimension bounding box around each pair in that plane throughout reconfiguration.*

Proof Sketch: Same as given in the *c. c. s. problem* in [4]. □

With the reconfiguration step proven for a single row, we may now state that the entire double-stranded spanning tree may be formed in $O(1)$ movement time.

Lemma 4: *Given the spanning tree formed in Lemma 1, and the pairing of modules along that tree given in Lemma 2, that spanning tree may be reconfigured into a double-stranded spanning tree in O(1) movement time.*

Proof Sketch: Module pairs reconfigure just as in Lemma 3, but here all pairs reconfigure simultaneously in one of two stages. Placing adjacent module pairs into different stages keeps the total configuration connected throughout. □

C. Forming a Hamiltonian Cycle

With the lemmas above we have shown that any connected, 2D configuration of modules may be reconfigured into a double-stranded (DS) spanning tree in O(1) movement time. The next step is to form a Hamiltonian Cycle. Note that a single DS module pair, or a module left single, trivially forms a "local" cycle of 2 or 1 module, respectively. Thus, all that remains is to merge local cycles that are adjacent along the DS spanning tree. This will form a Hamiltonian Cycle that effectively traces around the original tree.

Single modules only occur at the endpoints of the spanning tree. So, we only have 2 types of adjacent local cycle merge cases to consider: (1) DS module pair adjacent to another DS pair, and (2) DS pair adjacent to a single module. Thus, it is sufficient to consider these merge cases from the viewpoint of a given DS pair. Figure 2 illustrates an example DS pair and the 6 possible locations, A-F, of adjacent local cycles. These 6 neighbor locations have 2 types: (1) at the endpoint of the DS pair (locations A and D) and (2) along the side of the DS pair (B, C, E, and F).

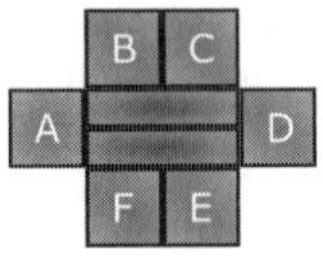

Fig. 2. The 6 possible locations for neighbors adjacent to a double-stranded module pair.

Since the original spanning tree was made with straight rows and columns, adjacent DS pairs along the tree will typically form straight lines or make perpendicular row-column connections. The lone exception is a short row-column-row or column-row-column sequence where the middle section is only 2 modules long. This can sometimes cause adjacent DS pairs that are parallel but slightly offset. An example of this situation, which we refer to as the *kink case*, occurred in the example in Figure 1 (modules 1,2,3 and 4). Later it is also shown more clearly in Figure 4.

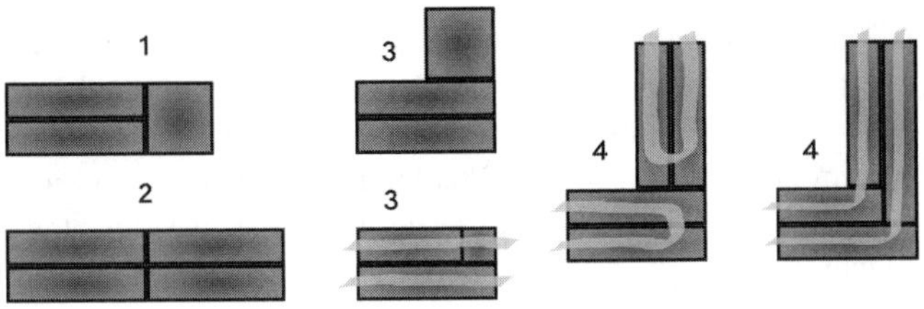

Fig. 3. **(1,2):** Adjacent neighbors occurring at 'endpoint' locations A or D. **(3,4):** Adjacent neighbors occurring at 'side' locations B, C, F, or E.

We now have only 5 types of local cycle merges to handle: (1) pair-single merge at the end of the pair, (2) pair-pair merge at the end of each pair, (3) pair-single merge at the side of the pair, (4) pair-pair merge at the side of 1 pair and end of the other pair, and (5) pair-pair merge at the side of each pair (kink case). A given DS pair must be able to simultaneously merge with all neighboring local cycles adjacent to it along the DS spanning tree. In order to merge adjacent cycles, there needs to be sufficient face-to-face module connections between the 2 cycles to allow one cycle to be "inserted" into the other cycle. In the 2D viewpoint of this paper's figures, this means shared edges between modules rather than just a shared point. We now present 5 reconfiguration rules for handling all 5 cases as well as 4 further rules for resolving potential conflicts between the first 5 rules. Typically, a rule first shows the cycles to be merged and then shows the reconfigured modules after their cycles have been merged in $O(1)$ movement time. In the following descriptions, a module's *length* refers to its longest dimension and a module's *width* refers to its shortest dimension in the x/y-plane.

Rules 1, 2: The leftmost two pictures of Figure 3 show examples of merge types 1 and 2: neighbors at the ends of the DS pair. No reconfiguration of the modules is necessary as the adjacent modules already share sufficient edges to allow cycle merges. Thus, no conflict can arise between multiple applications of rules 1 and 2.

Rules 3, 4: In Figure 3 rules are also given for handling merge types 3 and 4. The two middle pictures in Figure 3 show how to insert a single module into a DS pair while the two rightmost pictures show how to insert another DS pair. In reconfigurations for this rule modules travel O(1) distance and do not exceed the original bounding boxes of the modules shown. Thus, reconfiguration takes O(1) movement time and does not interfere with other modules in the system. Note that rules 3 and 4 may simultaneously be applied at each of the four possible side locations along this DS pair without conflict.

The only possible conflict occurs with the module end that was extended in rule 4, when this same module end is also faced with another rule 4 application along its side (a rule 3 application at this same side location is not a problem). This

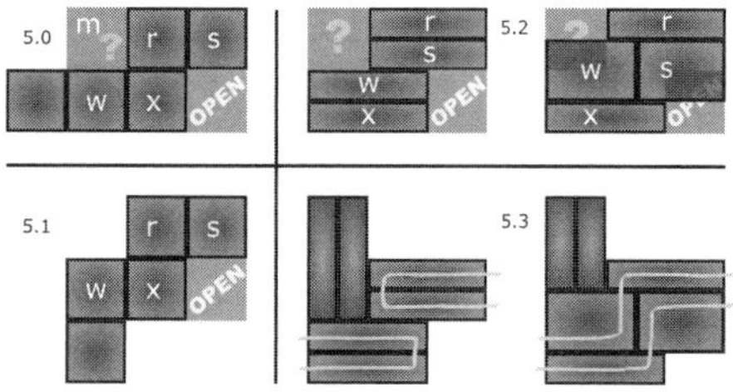

Fig. 4. **Rule 5:** Double-stranded module pairs occurring in a 'kink' case.

second rule would require the module end to contract, creating a direct conflict with the first rule. In this case the conflicted module does contract and the void it was supposed to extend and fill will be handled later by reconfiguration rule 8.

Rule 5: In Figure 4 we present a reconfiguration rule for handling the last type of local cycle merge, the type 5 'kink case'. Dark blue modules are those adjacent to each other along the kink case. Light red modules are from a different part of the spanning tree not directly connected to any of the blue modules. In this example the blue modules were added in a left-to-right, row-column-row order sequence. We now prove two small properties about this kink case arrangement.

Proposition 1: *The green square labeled "open" is not occupied by any module in the original 2D configuration.*

Proof: Since the blue modules in this kink case were added to the original spanning tree in a left-to-right order, modules w and x were added first as part of a left-to-right row (before r and s were added). If the green square were occupied by some module g then it would have been added either as part of a row or part of a column. If added as part of a row then by Spanning Tree Property 1 module g would be part of the same row as modules w and x and thus would be added before modules r and s. Yet, in this case in the next part of the sequence for creating the kink case, modules r and s would have been added as parts of different columns and therefore could not have been paired together as shown.

So, if module g exists it must have been added to the spanning tree as part of a column. In this case, the column would have to be added before modules w and x were added as part of a row. Otherwise module g would be part of that row instead. Yet, this means that module s would have been part of the same column. Thus, modules r and s could not have been paired together as shown and follow modules w and x as part of a left-to-right kink case. Thus, module g could not have been added as part of a column and hence could not exist. Thus, the green square must be unoccupied. □

Proposition 2: *The red 'question mark' square, if occupied by a 'red' module, must have been added to the spanning tree as part of a column and must not have a neighbor to its left adjacent to it in the spanning tree.*

Proof Sketch: Let the supposed red module be labeled m. If m is added to the tree before module w then it would be part of the same column as w or part of the same row as r. Either would prevent the kink case shown. Also, m cannot be added after module s because then it would be part of a column coming up from w – meaning it would not be a red module. So m is added to the tree after w but before s, and thus must be part of a column. Also, it cannot have a red neighbor to its left (added from a row) because such a module would first be added as part of a blue column. □

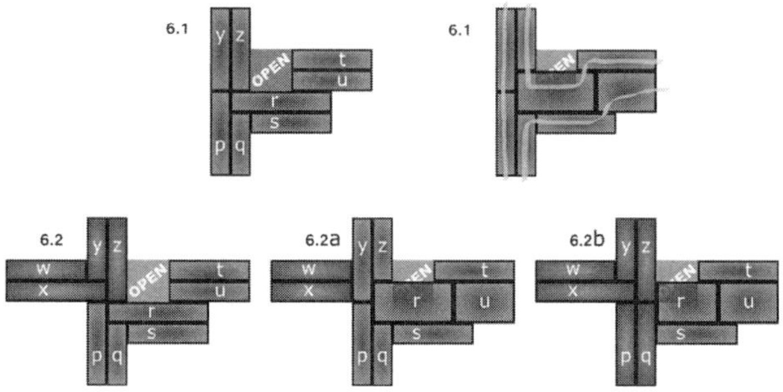

Fig. 5. **Rule 6:** Correcting a conflict between a rule 4 and a rule 5 case.

These properties allow rule 5 to work. The green square in the bottom-right corner is open, so module s can expand into that space. If the upper-left corner (question mark space m) is filled by a blue module, by rules 3 and 4 it will first be inserted to the left of module w and will not be a problem. Otherwise that corner is filled by a red module which may be pushed upwards as shown in picture 5.3.

Only two minor issues remain: (1) rule case 4 requires a module with width $1/2$ to extend its length and fill a void, but modules w and s in the kink case have width 1, and (2) module s expands its width into an open space below it, but what if another module from a different kink case also expands into that space? These conflicts are handled by rules 6 and 7.

Rule 6: Figure 5 gives a reconfiguration rule for handling one of the possible conflicts caused by the previous kink case. In all pictures shown the green square marked "open" must be unoccupied. If the kink case in Figure 5 (modules r, s, t and u) was formed right-to-left then by Proposition 1 that green square must have been unoccupied for the kink case to form. Otherwise the kink case is formed left-to-right. Consider if some module m did occupy the green space. Module m cannot be part of a row because by Spanning Tree Property 1 it would be part of the same row as modules t and u, preventing the kink case from forming.

The only remaining possibility is that module m was added to the tree as part of a column. This column must have been added after the bottom row of the kink case. Otherwise either module r or s would have been part of that column and not part of the kink case shown. Thus, r and s are added to the tree before any module in the green space (module m). Yet, then in the same stage when r and s are added as part a row (left-to-right), module m would have been added as part of a row along with modules t and u (left-to-right) since the column to its left was added before modules r and s. Again, this prevents the kink case from forming and so is not possible. Hence, the green square must be unoccupied in all pictures shown. If we only have blue modules (all adjacent along the spanning tree) then reconfiguration may proceed as in 6.1.

The only other conflict from combining the kink case with rule 4 is shown in Figure 5, picture 6.2, and is handled by either 6.2a or 6.2b. In this arrangement, the conflict occurs if we require DS pair (w, x) to be adjacent to DS pair (y, z) in the spanning tree. It is not important that these module pairs be adjacent, just that each module pair be attached to the tree at exactly one location. So, the tactic used is split and steal: break apart adjacent DS pairs and add one pair, along with its subtree, to a different portion of the spanning tree. First, consider the case where pair (w, x) is the parent of pair (y, z) (i.e. (w, x) precedes (y, z) in the DS tree). In this case split off pair (y, z) and insert it into the 'blue' portion of the spanning tree as shown in picture 6.2a.

In the opposite scenario, pair (y, z) is the parent of pair (w, x). In this case, the blue kink case must have been formed from right to left. Note that if the kink case was formed from left to right, modules p and q would have been added as part of a column. Yet, since pair (y, z) is the parent of pair (w, x), modules y and z would have been added as part of a column. By Spanning Tree Property 1, this means that y, z, p, and q would have to be part of the same column. This is not the case as shown in picture 6.2. Therefore, the kink case must have been formed right to left and pair (r, s) must be the parent of pair (p, q). Thus, we can split off pair (p, q) and add it to the red portion of the spanning tree as shown in 6.2b.

In both 6.2a and 6.2b, the DS module pair chosen to move to a new location was the child pair rather than the parent. Therefore, when the parent and child are split, the parent can stay at its old location. Meanwhile, the child pair may be successfully added to its new location or may be stolen by a third spanning tree location not shown. In this way, no conflicts will arise between multiple applications of Rule 6.

Rule 7: Picture 7.0 in Figure 6 depicts a case where two different kink cases will attempt to expand into the same open space. Picture 7.1 depicts the resulting collision. To resolve this conflict, consider the case where module pair (p, q) is the parent of pair (y, z). In this case we can, once again, apply the "split and steal" tactic and join module pair (y, z) to the blue portion of the spanning tree. Figures 7.2 and 7.3 depict how this can happen depending on whether module y has a neighbor to its left in the blue question mark area shown in picture 7.3. Alternatively, if module pair (a, b) was the parent of pair (w, x), then the same reconfiguration could be applied to add module pair (w, x) to the red portion of the spanning tree instead. If neither case occurs, then pair (y, z) is the parent of pair (p, q), and pair (w, x) is the parent of pair (a, b). In this case, we can add modules y, z, p, and q to the blue portion of the spanning tree as shown in picture 7.4.

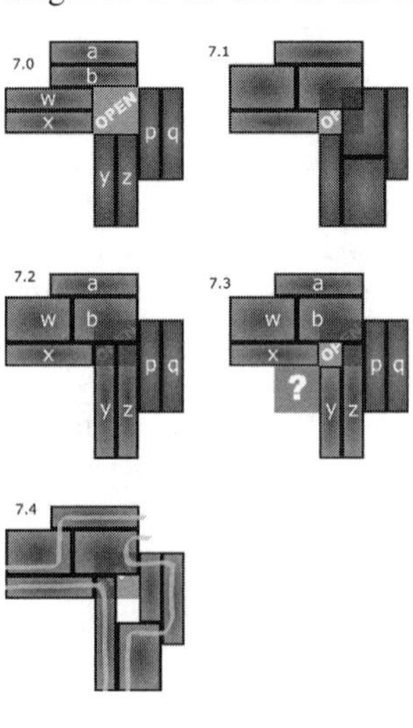

Fig. 6. **Rule 7:** Correcting a conflict between two rule 5 cases.

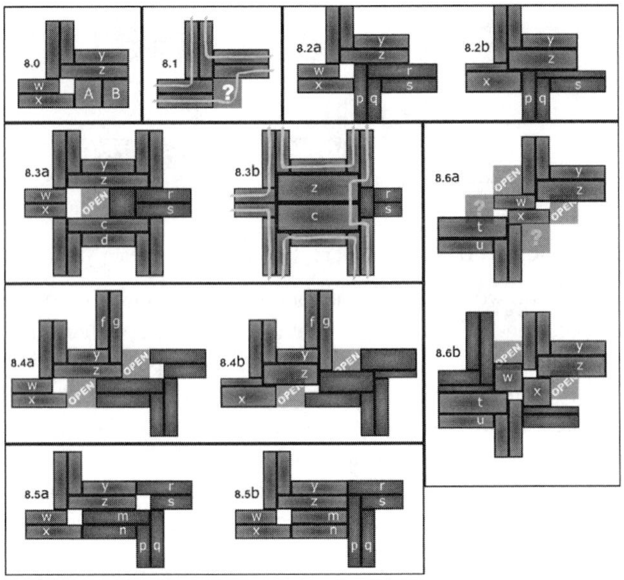

Fig. 7. **Rule 8:** Correcting a conflict between two rule 4 cases.

Rule 8: In Figure 7 rules are given to resolve a conflict between two simultaneous rule 4 applications. The initial state of this module arrangement is given with the blue modules in picture 8.0. The green squares labeled A and B are given simply to label the two important spaces beneath module z. For example, if space A is occupied by any blue module — one directly adjacent either to pair (w, x) or pair (y, z) on the spanning tree — then no reconfiguration is necessary and the cycles may be merged as shown in picture 8.1. This is true regardless of what type of blue module occupies space A.

Alternatively, if space A is not occupied by a blue module then this conflict case can typically be resolved by the reconfigurations shown in the transition from picture 8.2a to picture 8.2b. Here module z expands downward to make a connection with module x, but module x also expands upward to meet z. To allow this, w must contract its width (shortest dimension in x/y-plane) to $1/4$ and potentially red modules occupying spaces A or B must also contract. Red modules p, q, r and s represent such a case. Note that r contracts but s does not, leaving the boundary between the two modules at the same location. Thus, module r maintains any connections to other modules it previously had (though with smaller area).

The reconfiguration from 8.2a to 8.2b allows red modules in spaces A or B may contract and maintain prior connections. However, conflicts can occur between two simultaneous applications of rule 8. In pictures 8.3a and 8.3b a case is shown where expanding module z must connect not to module x but to another expanding module c instead. Since modules w and x have unlabeled blue modules above and below in these pictures they must have been added by a row. Therefore by Spanning Tree Property 1 there must be an open space between them and any red modules filling the space between z and c. Because of this, any such red modules are surrounded on 3 sides by a blue module or open space and these red modules can be contracted out of the way. An example is given with r, s, and the red single module shown. Now modules z and c can expand until they make contact (a width of 1 for each).

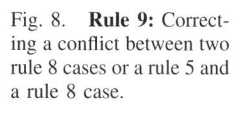

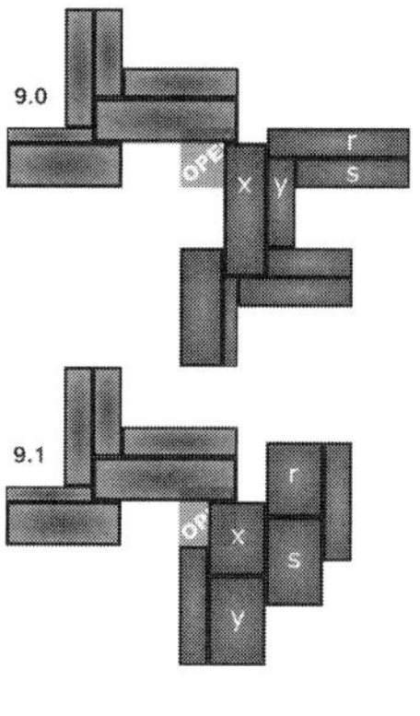

Fig. 8. **Rule 9:** Correcting a conflict between two rule 8 cases or a rule 5 and a rule 8 case.

Two more possible conflicts between rule 8 applications are handled in the transition from picture 8.4a to 8.4b and in going from 8.5a to 8.5b. Here spaces A and B are partially or completely filled by another module facing a rule 8 application. In the case where only space B is filled (8.4a/8.4b) then module z can have its length contracted before expanding its width. A potential connecting module g may then expand its length to maintain the connection. The corresponding red modules do the same. In these pictures it can be proven that the green spaced marked "open" are in fact unoccupied, but the reconfigurations shown work even if these spaces were occupied. If spaces A and B are both occupied by red modules in a rule 8 case, then the conflicted modules are stolen to become blue modules just as was done for rule 7. In 8.5a and 8.5b these modules are m and n and by changing them from the red to the blue portion of the spanning tree this conflict is easily handled.

With this fix, all possibilities have been checked to ensure that blue module z can expand downward if needed. The last remaining problem is that module z must connect to horizontal DS pair module x as in picture 8.2b. This requires module w to contract its width so x can expand. A second rule 8 application could require the opposite thing. This case is depicted in picture 8.6a and the resulting reconfiguration fix is given in 8.6b. Note that here modules w and x achieve the required connections (w to t and x to z) while exceeding their original bounding box by a width of 1/4 unit. Thus, any module previously occupying that space can be pushed out of the way without further conflict (as with the red module given below x in picture 8.6b). If the blue portion of the spanning tree in these pictures began in the lower-left corner (either with pair (t, u) or the vertical blue DS pair next to it) then it can be shown that the green squares marked "open" must be unoccupied. Again, the reconfigurations shown work without creating further conflicts even if these spaces are filled.

Rule 9: When the modules in rule 8 expand outside their original bounding boxes any modules in those locations are pushed out of the way without conflict. Rule 9, depicted in Figure 8, shows that this is true even if those "previous" occupants were modules that underwent a rule 5 or another rule 8 application. With all potential conflicts between rules now resolved, we have a complete rule set for transforming the DS spanning tree into a Hamiltonian Cycle.

Lemma 5: *Reconfiguration Rules 1-9 successfully transform a DS spanning tree into a known Hamiltonian Cycle through the modules in O(1) movement time.*

Proof Sketch: A cycle is formed by merging adjacent local cycles. Five types of merges are needed. Rules 1-5 handle these types. Rules 6-9 resolve any conflicts. Rules 1-4 are used in unison, then rules 5-9, in $O(1)$ total move time. □

V. General 2D Reconfiguration

In the previous section we showed that any given 2D configuration of modules along the x/y plane could be reconfigured in O(1) movement time so that a known Hamiltonian Cycle through those modules was formed. Now, given such a 2D Hamiltonian Cycle, in this section we will show how to transform that into a single column of modules in the z-axis direction in O($\sqrt{n}$) movement time. By using this z-axis column as an intermediate step, we will have the ability to perform general reconfiguration between any 2 configurations of modules along the x/y plane in O($\sqrt{n}$) movement time.

Suppose that we have an initial configuration of n modules with unit-length dimensions in a single row along the x-axis. Let the modules be labeled $i = 0, \ldots, n$ along this row so that module i has its leftmost edge at x-axis coordinate $x_i = i$. If z_i is used to denote the z-axis coordinate of module i's bottom face, then each module initially has $z_i = 0$. Recall that the coordinate of module i along some axis direction is the lowest global coordinate of any point on i.

Our goal is to reconfigure these n modules, starting at rest, into an x/z-axis diagonal with length n in each direction and final velocity 0 for each module in all directions. Note that this may be achieved by moving each module i a distance of 1 unit-length upward in the z-axis direction relative to its adjacent neighbor $i-1$. Module $i = 0$ is stationary. This reconfiguration will be completed in movement time $T = 2\sqrt{n-1}$ while meeting the requirements of the SRK model.

All modules must begin and end with 0 velocity in all directions and remain connected to the overall system throughout reconfiguration. The effects of friction and gravity are ignored to simplify calculations, though the bottom face of module $i = 0$ is assumed to be attached to an immovable base to give a foundation for reconfiguration movements.

Lemma 6: *An x-axis row of n modules with unit length dimensions in the z-axis, may be reconfigured into an x/z-axis diagonal with length n in the z-axis direction in total movement time $T = 2\sqrt{n-1}$.*

Proof Sketch: Reconfiguration may be completed by each module traveling distance $x_i = 1$ relative to its neighbor on one side while lifting $\leq n-1$ modules on its other side. Each module has mass $m = 1$ and exerts force $F = 1 = ma$. So each module may accelerate by $\alpha = 1/(n-1)$. Since we need $x_i(T) = \frac{1}{2}\alpha_i T^2 = 1$, then $T = \sqrt{2/\alpha_i} \leq 2\sqrt{n-1}$. □

Corollary 7: *Any 2D configuration of modules in the x/y-axis plane with a Hamiltonian Path through it may be reconfigured in movement time $T = 2\sqrt{n-1}$ such that each module $i = 0, \ldots, n-1$ along that path finishes with a unique z-axis coordinate $z_i(T) = i$.*

Proof Sketch: Same as in Lemma 6. A module's neighbors are those preceding/following it along a Hamiltonian Path (from breaking a single connection in the Hamiltonian Cycle). □

In Section IV we showed that any connected 2D configuration of unit-length dimension modules along the x/y-axis plane could be reconfigured into a Hamiltonian Path of those modules in O(1) movement time. From corollary 7 we have shown that such a Hamiltonian Path may be reconfigured so that each module $i = 0, \ldots, n-1$ along that path has its own unique z-axis coordinate $z_i = i$. This required movement time $T = 2\sqrt{n-1}$ and created a winding stair step or "spiral" of modules. To reconfigure this new spiral configuration into a z-axis column, all that remains is to move the modules so that they all have the same x/y-plane coordinates.

Performing this contraction motion for each axis direction is similar to the reconfiguration proven in Lemma 6. The difference is that here all modules finish with the same coordinate value rather than begin with the same value. Also, the modules may have different lengths in the x-axis and y-axis directions as a result of reconfigurations used to create the initial Hamiltonian Path. However, those lengths are still no less than 1/2 unit and no more that 3 unit lengths and so we may use the same relative position and velocity analysis as was used in the proof for Lemma 6.

We will first show that all n modules may be reconfigured so that they have the same x-axis coordinate in total movement time $T \leq 4\sqrt{(n-1)}$. In particular, since the bottom module $i = 0$ is assumed to have its bottom face attached to an immovable base, all modules will finish with the same x-axis coordinate as module $i = 0$. Let the relative x-axis location and velocity of module i at time t be denoted as $x_i(t)$ and $v_i(t)$, respectively. These values for module $i = 1, \ldots, n-1$ are relative to the module $i-1$ directly below module i.

Note that if module i has relative location $x_i(t) = 0$ then at time t module i has the same x-axis coordinate as module $i-1$ below it. Module $i = 0$ does not move throughout reconfiguration. Thus, having $x_i(T) = 0$ for each module i would give all modules the same global x-axis coordinate. Finally, let a z-axis "spiral" configuration of n modules be defined as one where for modules $i = 0, \ldots, n-1$ module i has unit length in the z-axis direction, a z-axis coordinate of $z_i = i$ and has either an edge-to-edge or a face-to-face connection with module $i-1$, if $i > 0$, and with module $i+1$, if $i < n-1$. All modules are assumed to begin at rest and we will show that the desired reconfiguration may be completed in movement time $T = 4\sqrt{(n-1)}$ while meeting the requirements of the SRK model.

Lemma 8: *A z-axis spiral configuration of n modules may be reconfigured in movement time $T = 4\sqrt{(n-1)}$ such that all modules have final relative x-axis coordinate $x_i(T) = 0$.*

Proof Sketch: Same approach as in Corollary 7, but modules are condensed rather than expanded. Slightly more time is required as modules may have x-axis length up to 3. □

From the result just proven, we may conclude that the same reconfiguration may be taken in the y-axis direction.

Corollary 9: *A z-axis spiral configuration of n modules may be reconfigured in movement time $T = 4\sqrt{(n-1)}$ so that all modules have final relative y-axis coordinate $y_i(T) = 0$.*

Proof Sketch: Same as for Lemma 8. □

Theorem: *Let A and B be two connected configurations of n modules, with each module having a z-axis coordinate of 0 and unit-length dimensions. Let modules δ in A and β in B have the same coordinate location. Reconfiguration from A to B may be completed in $O(\sqrt{n})$ movement time while satisfying the requirements of the SRK model.*

Proof Sketch: From configuration A, Lemma's 1-5 create a Hamiltonian Cycle in O(1) time. Lemma 6 – Corollary 9 gives a z-axis column in $O(\sqrt{n})$ time. We break the cycle so that δ is the base of the column. Reversing the process gives B. □

VI. Conclusion

In this paper we presented a novel algorithm for general reconfiguration between 2D configurations of expanding cube-style self-reconfigurable robots. This algorithm requires $O(\sqrt{n})$ movement time, met the requirements of the SRK model given in [4], and is asymptotically optimal as it matches the lower bound on general reconfiguration given in [4]. In addition, it was also shown that a known Hamiltonian Cycle could be formed in any 2D configuration of expanding cube-style modules in O(1) movement time.

There are a number of open problems remaining. For simplicity, in this paper we assumed that all reconfigurations of modules were executed by a centralized controller to permit synchronous movements. In the future, asynchronous control would be preferable. Local, distributed control of reconfiguration movements is also a topic of future interest. In this paper we have ignored the effects of friction and gravity in the SRK model, but frictional and gravitational models should be included in future work. Note that this makes the Principle of Time Reversal invalid. Furthermore, the general problem of reconfiguration between two arbitrarily connected 3D configurations remains open, although we have developed an algorithm for certain types of configurations. Our current work focuses on extending this, and developing simulation software to implement these algorithms with the SRK model.

VII. Acknowledgement

This work has been supported by grants from NSF CCF-0432038 and CCF-0523555.

References

[1] K. Kotay and D. Rus. Generic distributed assembly and repair algorithms for self-reconfiguring robots. In *Proc. of IEEE Intl. Conf. on Intelligent Robots and Systems*, 2004.

[2] A. Pamecha, C. Chiang, D. Stein, and G. Chirikjian. Design and implementation of metamorphic robots. In *Proceedings of the 1996 ASME Design Engineering Technical Conference and Computers in Engineering Conference*, 1996.

[3] A. Pamecha, I. Ebert-Uphoff, and G. Chirikjian. Useful metrics for modular robot motion planning. In *IEEE Trans. Robot. Automat.*, pages 531–545, 1997.

[4] J. H. Reif and S. Slee. Asymptotically optimal kinodynamic motion planning for self-reconfigurable robots. In *Seventh International Workshop on the Algorithmic Foundations of Robotics (WAFR2006)*, July 16-18 2006.

[5] D. Rus and M. Vona. Crystalline robots: Self-reconfiguration with unit-compressible modules. *Autonomus Robots*, 10(1):107–124, 2001.

[6] S. Vassilvitskii, J. Suh, and M. Yim. A complete, local and parallel reconfiguration algorithm for cube style modular robots. In *Proc. of the IEEE Int. Conf. on Robotics and Automation*, 2002.

[7] J. Walter, B. Tsai, and N. Amato. Choosing good paths for fast distributed reconfiguration of hexagonal metamorphic robots. In *Proc. of the IEEE Intl. Conf. on Robotics and Automation*, pages 102–109, 2002.

[8] Jennifer E. Walter, Jennifer L. Welch, and Nancy M. Amato. Distributed reconfiguration of metamorphic robot chains. In *PODC '00*, pages 171–180, 2000.

A Discrete Geometric Optimal Control Framework for Systems with Symmetries

Marin Kobilarov
USC

Mathieu Desbrun
Caltech

Jerrold E. Marsden
Caltech

Gaurav S. Sukhatme
USC

Abstract— **This paper studies the optimal motion control of mechanical systems through a discrete geometric approach. At the core of our formulation is a discrete *Lagrange-d'Alembert-Pontryagin* variational principle, from which are derived discrete equations of motion that serve as constraints in our optimization framework. We apply this discrete mechanical approach to holonomic systems with symmetries and, as a result, geometric structure and motion invariants are preserved. We illustrate our method by computing optimal trajectories for a simple model of an air vehicle flying through a digital terrain elevation map, and point out some of the numerical benefits that ensue.**

I. Introduction

The goal of this paper is to design *optimal motion control* algorithms for robotic systems with *symmetries*. That is, we consider the problem of computing the controls $f(t)$ necessary to move a finite-dimensional mechanical system with configuration space Q from its initial state $(q(0) = q_i, \dot{q}(0) = \dot{q}_i)$ to a goal state $(q(T) = q_f, \dot{q}(T) = \dot{q}_f)$, while minimizing a cost function of the form:

$$J(q, \dot{q}, f) = \int_0^T C(q(t), \dot{q}(t), f(t))\mathrm{d}t. \quad (1)$$

Minimum control effort problems can be implemented using $C(q(t), \dot{q}(t), f(t)) = \|f(t)\|^2$, while minimum-time problems involve $C(q(t), \dot{q}(t), f(t)) = 1$. Additional nonlinear equality or inequality constraints on the configuration (such as obstacle avoidance in the environment) and velocity variables can be imposed as well. Systems of interest captured by this formulation include autonomous vehicles such as unmanned helicopters, micro-air vehicles or underwater gliders.

A. Related work

Trajectory design and motion control of robotic systems have been studied from many different perspectives. Of particular interest are geometric approaches [1, 2, 3] that use symmetry and reduction techniques [4, 5]. Reduction by symmetry can be used to greatly simplify the optimal control problem and provide a framework to compute motions for general nonholonomic systems [6]. A related approach, applied to an underwater eel-like robot, involves finding approximate solutions using truncated basis of cyclic input functions [7]. There are a number of successful methods for motion planning with obstacles—see [8] for references.

While standard optimization methods are based on shooting, multiple shooting, or collocation techniques, recent work on Discrete Mechanics and Optimal Control (DMOC, see [9, 10, 11]) proposes a different discretization strategy. At the core of DMOC is the use of *variational integrators* [12] that are derived from the discretization of variational principles such as Hamilton's principle for conservative systems or Lagrange-D'Alembert for dissipative systems. Unlike other existing variational approaches [6, 13] where the continuous equations of motion are enforced as constraints and *subsequently* discretized, DMOC *first* discretizes the variational principles underlying the mechanical system dynamics; the resulting discrete equations are then used as constraints along with a discrete cost function to form the control problem. Because the discrete equations of motion result from a discrete variational principle, momenta preservation and symplecticity are automatically enforced, avoiding numerical issues (like numerical dissipation) that generic algorithms often possess.

B. Contributions

In this paper, we extend the generalized variational principle of [14, 15, 16] to the DMOC framework to derive optimization algorithms based on structure-preserving, discrete-mechanical integrators. In particular, we employ a discrete *Lagrange-d'Alembert-Pontryagin* principle to characterize mechanical systems with symmetries and external forces. We use this new discrete geometric optimal control framework for holonomic systems (possibly underactuated and/or with symmetries) and illustrate the implemented algorithms with a simulated example of a simplified helicopter flying through a canyon.

The numerical benefits of our discrete geometric approach are numerous. First, it automatically preserves motion invariants and geometric structures of the continuous system, exhibits good energy behavior, and respects the work-energy balance due to its variational nature. Such properties are often crucial for numerical accuracy and stability, in particular for holonomic systems such as underwater gliders traveling at low energies along ocean currents. Second, it benefits from an exact reconstruction of curves in the Lie group configuration space from curves in its Lie algebra. Thus, numerical drift, for example associated with enforcing rotation frame orthogonality constraints, is avoided. Third, the simplicity of the variational principle allows flexibility of implementation.

Finally, this framework is flexible enough to strike a balance between a desired order of accuracy and runtime efficiency.

In addition to these well-documented advantages of discrete variational methods, there is growing evidence that DMOC methods are especially well suited for optimization problems. In particular, their discrete variational nature seems to offer very good trajectory approximation even at low temporal resolutions. This stability vis-a-vis resolution is particularly suitable for design and exploration purposes as well as for hierarchical optimizations, as it leads to faster convergence towards optimal solutions.

It is also important to note that non-holonomic constraints can also be imposed in our framework. We refer to [17] for details on rolling constraints and Chaplygin systems. However, in this paper we focus solely on holonomic systems with symmetries.

II. Overview of Mechanical Integrators

A mechanical integrator integrates a dynamical system forward in time. The construction of such numerical algorithms usually involves some form of discretization or Taylor expansion that results in either implicit or explicit equations to compute the next state in time. In an optimal control setting, these equations are then used as constraints.

Instead, the integrators employed in this paper are based on the discretization of variational principles, i.e. *variational integrators*. In essence, they ensure the optimality (in the sense of Hamilton's principle, for instance) of the discrete path of the mechanical system in space-time. In addition, certain systems have group structure and symmetries that can be factored out directly in order to obtain more accurate and efficient integrators, e.g. *Lie group integrators*. After giving a brief overview of such integrators below we present a variational principle to derive more general integrators that account for symmetries.

A. Variational Integrators

Variational integrators [12] are derived from a variational principle (e.g., Hamilton's principle) using a discrete Lagrangian. Unlike standard integration methods, variational integrators can preserve momenta, energy, and symplectic structure (i.e., a symplectic 2-form in phase space) for conservative systems; in the presence of forces and/or dissipation, they compute the change in these quantities with remarkable accuracy. Such features are obviously desirable for accurate dynamics simulation. The underlying theory has discrete analogs of Noether's theorem and the Legendre transform, and a Lagrange-d'Alembert principle to handle non-conservative forces and constraints. *Discrete mechanics*, therefore, stands as a self-contained theory similar to Hamiltonian or Lagrangian mechanics [15] and has already been applied to several domains: nonsmooth variational collision integration [18], elasticity simulation in computer graphics [14], satellite formation trajectory design [19], optimal control of rigid bodies [11], of articulated bodies in fluid [10, 20], and optimal control of wheeled robots [21].

In the variational integration setting, the state space TQ is replaced by a product of two manifolds $Q \times Q$ [12]. Thus, a velocity vector $(q, \dot{q}) \in TQ$ is represented by a pair of points $(q_0, q_1) \in Q \times Q$. A path $q : [0, T] \to Q$ is replaced by a discrete path $q_d : \{kh\}_{k=0}^{N} \to Q$ ($q_d = \{q_0, ..., q_N\}$, $q_k = q(kh)$), $Nh = T$. One formulates a discrete version of Hamilton's principle (i.e. $\delta \int_0^T L(q, \dot{q})dt = 0$) by approximating the integral of the Lagrangian $L : TQ \to \mathbb{R}$ between q_k and q_{k+1} by a discrete Lagrangian $L_d : Q \times Q \to \mathbb{R}$

$$L_d(q_k, q_{k+1}) \approx \int_{kh}^{(k+1)h} L(q(t), \dot{q}(t))\mathrm{d}t.$$

The discrete principle then requires that

$$\delta \sum_{k=0}^{N-1} L_d(q_k, q_{k+1}) = 0,$$

where variations are taken with respect to each position q_k along the path, and the resulting equations of motion become

$$D_2 L_d(q_{k-1}, q_k) + D_1 L_d(q_k, q_{k+1}) = 0.$$

Example: For example, consider a Lagrangian of the form $L(q, \dot{q}) = \frac{1}{2}\dot{q}^T M \dot{q} - V(q)$ and define the discrete Lagrangian $L_d(q_k, q_{k-1}) = hL\left(q_{k+\frac{1}{2}}, (q_{k+1} - q_k)/h\right)$, using the notation $q_{k+\frac{1}{2}} := (q_k + q_{k+1})/2$. The resulting equations are

$$M \frac{q_{k+1} - 2q_k + q_{k-1}}{h^2} = -\frac{1}{2}(\nabla V(q_{k-\frac{1}{2}}) + \nabla V(q_{k+\frac{1}{2}})),$$

which is a discrete analog of Newton's law $M\ddot{q} = -\nabla V(q)$. For controlled (i.e., non conservative) systems, forces can be added using a discrete version of Lagrange-d'Alembert principle and discrete virtual work in a similar manner.

B. Lie Group Integrators

Lie group integrators preserve symmetry and group structure for systems with motion invariants. Consider a system on configuration manifold $Q = G \times M$ where G is a Lie group (with Lie algebra $\mathfrak{g}$) whose action leaves the system invariant, i.e., it preserves the induced momentum map. For example, $G = SE(3)$ can represent the group of rigid body motions of a free-floating articulated body while M is a space of internal variables describing the joints of the body. The idea is to transform the system equations from the original state space TQ into equations on the *reduced* space $\mathfrak{g} \times TM$ (elements of TG are translated to the origin and expressed in the algebra $\mathfrak{g}$) which is a linear space where standard integration methods can be used. The inverse of this transformation is then used to map curves in the algebra variables back to the group. Two standards maps have been commonly used to achieve this transformation for any Lie group G:

- Exponential map $\exp : \mathfrak{g} \to G$, defined by $\exp(\xi) = \gamma(1)$, with $\gamma : \mathbb{R} \to G$ is the integral curve through the identity of the left invariant vector field associated with $\xi \in \mathfrak{g}$ (hence, with $\dot{\gamma}(0) = \xi$);

- Canonical coordinates of the second kind $\mathrm{ccsk} : \mathfrak{g} \to G$, $\mathrm{ccsk}(\xi) = \exp(\xi^1 e_1) \cdot \exp(\xi^2 e_2) \cdot \ldots \cdot \exp(\xi^n e_n)$, where $\{e_i\}$ is the Lie algebra basis.

A third choice, valid only for certain *quadratic* matrix groups [22] (which include the rigid motion groups $SO(3)$, $SE(2)$, and $SE(3)$), is the Cayley map $\mathrm{cay} : \mathfrak{g} \to G$, $\mathrm{cay}(\xi) = (e - \xi/2)^{-1}(e + \xi/2)$. Although this last map provides only an approximation to the integral curve defined by exp, we include it as one of our choices since it is very easy to compute and thus results in a more efficient implementation. Other approaches are also possible, e.g., using retraction and other commutator-free methods; we will however limit our exposition to the three aforementioned maps in the formulation of the discrete reduced principle presented in the next section.

C. Unified View

The optimal control algorithms in this paper are based on a discrete version of the Lagrange-d'Alembert-Pontryagin (LDAP) principle [16]. The LDAP viewpoint unifies the Lagrangian and Hamiltonian descriptions of mechanics [15] and extends to systems with symmetries and constraints. The discrete version of this principle yields integration schemes that generalize both the variational and Lie group integrators mentioned above.

The Lagrange-d'Alembert-Pontryagin Principle: We briefly recall the general formulation of the continuous LDAP principle for a system with Lagrangian $L : TQ \to \mathbb{R}$ and control force[1] $f : [0, T] \to T^*Q$. For a curve $(q(t), v(t), p(t))$ in $TQ \oplus T^*Q$, $t \in [0, T]$ the principle states that

$$\delta \int_0^T \{L(q, v) + p \cdot (\dot{q} - v)\}dt + \int_0^T f(t) \cdot \delta q(t) dt = 0, \tag{2}$$

for variations that vanish at the endpoints. The curve $v(t)$ describes the velocity determined from the dynamics of the system. In view of the formulation, v does not necessarily correspond to the rate of change of the configuration q. The additional variable p, though, indirectly enforces this dependence and corresponds to both Lagrange multipliers and the momenta of the system. Thus 2 generalizes the Lagrange-d'Alembert principle and is linked to the Pontryagin maximum principle of optimal control.

The LDAP principle is conceptually equivalent to the Lagrange-d'Alembert principle. Nevertheless, in the discrete setting, the LDAP principle provides a more powerful framework for designing mechanical integrators. One notable benefit lies in the ability to derive higher-order integrators and, in the case of systems with symmetries, to tailor the algorithm structure to achieve a desired accuracy or efficiency [15]. While in this paper we do not explore higher order approximations and, in essence, our formulation in Sec. III-B could be alternatively derived using the discrete Euler-Poincaré (DEP) approach [23], we follow the LDAP formulation because of its greater flexibility. Another benefit appears in the discretization of systems with nonholonomic constraints. In particular, the optimal control method proposed in this paper is extended to nonholonomic systems of Chaplygin type in [17] (with the general case soon to follow) in a unified variational formulation.

[1] In the Lagrangian setting a force is an element of the cotangent bundle T^*Q, i.e. a one-form $\langle f, \cdot \rangle$ that pairs with velocity vectors to produce the total work $\int_0^T \langle f, \dot{q} \rangle dt$ done by the force along a path between $q(0)$ and $q(T)$.

III. Systems with Symmetries

In this section we develop the optimal control formulation for mechanical systems with symmetries. Assume that the configuration space is an n-dimensional Lie group G with algebra $\mathfrak{g}$ and Lagrangian $L : TG \to \mathbb{R}$ that is left invariant under the action of G. Using the invariance we can reduce such systems by introducing the *body-fixed* velocity $\xi \in \mathfrak{g}$ defined by translation to the origin $\xi = TL_{g^{-1}}\dot{g}$ and the reduced Lagrangian $\ell : TG/G \to \mathbb{R}$ such that $\ell(\xi) = L(g^{-1}g, g^{-1}\dot{g}) = L(e, \xi)$. The system is controlled using a control vector $u : [0, T] \to \mathbb{U}$, where $\mathbb{U} \subset \mathbb{R}^c$, $c \leq n$, is the set of controls applied with respect to a body-fixed basis $\{F^1, ..., F^c\}$, $F^i : [0, T] \to \mathfrak{g}^*$.

A. The Continuous System

The continuous equations of motion are derived from the reduced Hamilton (or Lagrange-d'Alembert in the presence of forces) principle [4, 5] and have the standard form

$$\dot{g} = g\xi, \tag{3}$$

$$\mu = \ell'(\xi), \tag{4}$$

$$\dot{\mu} = \mathrm{ad}^*_\xi \mu + u_i F^i. \tag{5}$$

Eq. (5) are the forced Euler-Poincaré equations, with $\mu \in \mathfrak{g}^*$ denoting the system momentum, and (3) are the reconstruction equations. Note that $\mathrm{ad}^*_\xi \mu$ is defined by $\langle \mathrm{ad}^*_\xi \mu, \eta \rangle = \langle \mu, \mathrm{ad}_\xi \eta \rangle$, where $\mathrm{ad}_\xi \eta = [\xi, \eta]$ for $\eta \in \mathfrak{g}$.

Example: A Simple Helicopter

Consider the following simplistic model of a helicopter-like vehicle (Fig. 1). The vehicle is modeled as a single underactuated rigid body with mass m and principal moments of inertia I_1, I_2, I_3 (the inertia matrix is denoted $\mathbb{I} = \mathrm{diag}(I_1, I_2, I_3)$). The vehicle is controlled through a *collective* u_c (lift produced by the main rotor) and a *yaw* u_ψ (force produced by the rear rotor), while the direction of the lift is controlled by tilting the main blades forward or backward through a *pitch* α_p and a sideways *roll* α_r. The configuration space is $Q = SO(3) \times \mathbb{R}^3$ with $(R, p) \in Q$ denoting orientation and position. Ignoring aerodynamic effects, we treat the rotational dynamics separately from the translational dynamics.

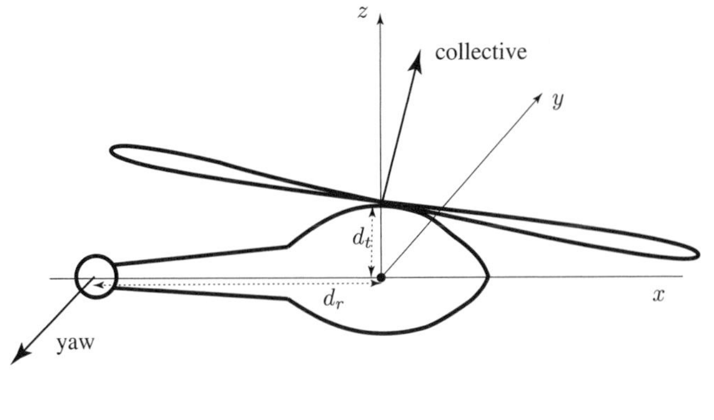

Fig. 1. Air vehicle model.

Rotations: Let $\Omega \in \mathbb{R}^3 \sim \mathfrak{so}(3)$ denote the body angular velocity. The Lagrangian of the rotational dynamics is $\ell(\Omega) = \frac{1}{2}\Omega^T \mathbb{I}\,\Omega$. In the absence of forces the system is invariant under rotations and we can set $G = SO(3)$ as the symmetry group. The torque control input basis can be written in matrix form as

$$F_\Omega(\alpha_p, \alpha_r) := \begin{bmatrix} d_t \sin\alpha_r & 0 \\ d_t \sin\alpha_p \cos\alpha_r & 0 \\ 0 & d_r \end{bmatrix},$$

where d_t and d_r are the distances from the top and rear blades to the center of mass of the helicopter. The reduced equations of the rotational dynamics corresponding to (3)-(5) are

$$\dot{R} = R\Omega, \tag{6}$$

$$\Pi = \mathbb{I}\,\Omega, \tag{7}$$

$$\dot{\Pi} = \Pi \times \Omega + F_\Omega(\alpha)u, \tag{8}$$

where $\Pi \in \mathbb{R}^3 \sim \mathfrak{so}(3)^*$ denotes angular momentum, $\alpha = (\alpha_p, \alpha_r)$, and $u = (u_c, u_\psi)$.

Translations: The translational dynamics is described by

$$m\ddot{p} = RF_V(\alpha)u + f^g,$$

where $f^g = (0, 0, -9.8m)$ denotes gravity, and

$$F_V(\alpha_p, \alpha_r) := \begin{bmatrix} \sin\alpha_p \cos\alpha_r & 0 \\ -\sin\alpha_r & -1 \\ \cos\alpha_p \cos\alpha_r & 0 \end{bmatrix}.$$

Next we present the derivation of the corresponding discrete equations of motion and describe how to use them as constraints in the optimization problem.

B. Discrete Reduced LDAP Principle

The *discrete* reduced Hamilton-Pontryagin principle for conservative systems was introduced in [15]. We now propose a simple extension to systems with internal forces. The principle is formulated using the *discrete reduced path* denoted by $(g, \xi, \mu)_d : \{t_k\}_{k=0}^N \to G \times \mathfrak{g} \times \mathfrak{g}^*$, where $g_d = \{g_0, ..., g_N\}$ and ξ_d, μ_d analogously defined. The discrete control force which approximates a continuous control force is defined as $f_d : \{t_k\}_{k=0}^N \to \mathfrak{g}^*$. The map $TL^*_{g(t)^{-1}} : \mathfrak{g}^* \to TG^*$ transforms the body-fixed momentum or force back to the cotangent space at point $g(t)$. The reduced LDAP principle results from expressing the principle (2) in terms of the reduced variables, i.e. by substituting $L(g, \dot{g}) \Rightarrow \ell(\xi)$, $v \Rightarrow TL_g\xi$, $p \Rightarrow TL^*_{g^{-1}}\mu$ in (2). The discrete principle is then obtained by approximating the integrals in (2) according to:

$$\delta \sum_{k=0}^{N-1} h\left[\ell(\xi_k) + \left\langle \mu_k, \tau^{-1}(g_k^{-1}g_{k+1})/h - \xi_k \right\rangle\right] + \sum_{k=0}^{N-1}\left[TL^*_{g_k^{-1}} f_k^- \cdot \delta g_k + TL^*_{g_{k+1}^{-1}} f_k^+ \cdot \delta g_{k+1}\right] = 0, \tag{9}$$

where the map $\tau : \mathfrak{g} \to G$ defines the *difference* between two configurations g_k and g_{k+1} on the group through an element in the Lie algebra. τ is selected as a local diffeomorphism [15] such that $\tau(\xi)\cdot\tau(-\xi) = e$. The *left* (resp., *right*) discrete force $f_k^- \in \mathfrak{g}^*$ (resp., $f_k^+ \in \mathfrak{g}^*$, as shown below) is such that the work done by f along each discrete segment is approximated using the work done at the beginning (resp., the end) of the segment; that is, as described in [12], these two discrete forces provide an approximation of the continuous forcing through:

$$\int_{kh}^{(k+1)h} TL^*_{g(t)^{-1}} f(t) \cdot \delta g(t) dt \approx TL^*_{g_k^{-1}} f_k^- \cdot \delta g_k + TL^*_{g_{k+1}^{-1}} f_k^+ \cdot \delta g_{k+1}.$$

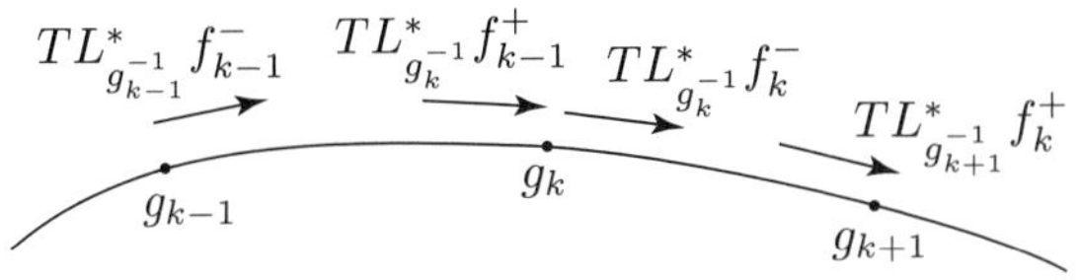

After taking variations in (9) we obtain the following discrete equations of motion (see Sec. 4.2 in [15] for details).

$$g_k^{-1}g_{k+1} = \tau(h\xi_k), \tag{10}$$

$$\mu_k = \ell'(\xi_k), \tag{11}$$

$$(d\tau^{-1}_{h\xi_k})^*\mu_k - (d\tau^{-1}_{-h\xi_{k-1}})^*\mu_{k-1} = f_{k-1}^+ + f_k^-, \tag{12}$$

where $d\tau_\xi : \mathfrak{g} \to \mathfrak{g}$ is the *right-trivialized tangent* of $\tau(\xi)$ defined by $\mathrm{D}\,\tau(\xi)\cdot\delta = TR_{\tau(\xi)}(d\tau_\xi \cdot \delta)$ and $d\tau_\xi^{-1} : \mathfrak{g} \to \mathfrak{g}$ is its inverse[2]. Equations (10)-(12) can be understood as a discrete approximation of equations (3)-(5). They define a second-order accurate integrator that is part of the *variational Euler* family of methods [15].

The exact form of (12) depends on the choice of τ. It is important to point out that this choice will influence the computational efficiency of the optimization framework when the equalities above are enforced as constraints. There are several choices commonly used for integration on Lie groups. We give details of two particularly representative examples: the exponential map (exp), and the Cayley map (cay)—see Sec. II-B for their definitions. Note that other maps, such as canonical coordinates of the second kind (ccsk) (also based on the exponential map), can be similarly derived.

[2]D is the standard derivative map (here taken in the direction δ)

Exponential map: The right-trivialized derivative of the map exp and its inverse are defined as

$$\begin{aligned} \operatorname{dexp}(x)y &= \sum_{j=0}^{\infty} \frac{1}{(j+1)!} \operatorname{ad}_x^j y, \\ \operatorname{dexp}^{-1}(x)y &= \sum_{j=0}^{\infty} \frac{B_j}{j!} \operatorname{ad}_x^j y, \end{aligned} \tag{13}$$

where B_j are the Bernoulli numbers. Typically, these expressions are truncated in order to achieve a desired order of accuracy. The first few Bernoulli numbers are $B_0 = 1$, $B_1 = -1/2$, $B_2 = 1/6$, $B_3 = 0$ (see [22, 24] for details). Setting $\tau = \exp$, (12) becomes

$$\operatorname{dexp}^{-1}(h\xi_k)^*\mu_k - \operatorname{dexp}^{-1}(-h\xi_{k-1})^*\mu_{k-1} = f_{k-1}^+ + f_k^-.$$

Cayley map: The derivative maps of cay (see Sec.IV.8.3 in [24] for derivation) are

$$\begin{aligned} \operatorname{dcay}(x)y &= \left(e - \frac{x}{2}\right)^{-1} y \left(e + \frac{x}{2}\right)^{-1}, \\ \operatorname{dcay}^{-1}(x)y &= \left(e - \frac{x}{2}\right) y \left(e + \frac{x}{2}\right). \end{aligned} \tag{14}$$

Using $\tau = \operatorname{cay}$ (see also [15]) (12) simplifies to

$$\begin{aligned} &\mu_k - \mu_{k-1} - \frac{h}{2}\left(\operatorname{ad}^*_{\xi_k} \mu_k + \operatorname{ad}^*_{\xi_{k-1}} \mu_{k-1}\right) \\ &- \frac{h^2}{4}\left(\xi_k^* \mu_k \xi_k^* - \xi_{k-1}^* \mu_{k-1} \xi_{k-1}^*\right) = f_{k-1}^+ + f_k^-. \end{aligned} \tag{15}$$

The Cayley map provides a coarser approximation than the exponential map exp, but its simple form is suitable for efficient implementation.

Discrete Forces: There are various ways to construct valid discrete forces f^+ and f^-. A simple approach, in the spirit of the midpoint rule, is to assume that the left and right discrete forces at each segment are equal, and defined as the average of the forces applied in the beginning and the end of the segment:

$$f_k^- = f_k^+ = \frac{h}{2}\left[\frac{f_k + f_{k+1}}{2}\right].$$

Example: Assume that we use the Cayley map and the midpoint rule to construct a variational integrator for the air vehicle model defined in Sec. III-A. The discrete equations of rotational motion (corresponding to eqs. (10)-(12), and making use of (15)) become

$$\begin{aligned} &R_k^T R_{k+1} = \operatorname{cay}(h\widehat{\Omega_k}), \\ &\Pi_k = \mathbb{I}\,\Omega_k, \\ &\Pi_k = \Pi_{k-1} + \frac{h}{2}(\Pi_{k-1} \times \Omega_{k-1} + \Pi_k \times \Omega_k) \\ &\quad + \frac{h^2}{4}((\Omega_{k-1}^T \Pi_{k-1})\Omega_{k-1} - (\Omega_k^T \Pi_k)\Omega_k) \\ &\quad + \frac{h}{4}(F_\Omega(\alpha_{k-1})u_{k-1} + 2F_\Omega(\alpha_k)u_k + F_\Omega(\alpha_{k+1})u_{k+1}), \end{aligned} \tag{16}$$

where the map $\widehat{\cdot} : \mathbb{R}^3 \to \mathfrak{so}(3)$ is defined by

$$\widehat{\Omega} = \begin{bmatrix} 0 & -\Omega^3 & \Omega^2 \\ \Omega^3 & 0 & -\Omega^1 \\ -\Omega^2 & \Omega^1 & 0 \end{bmatrix}.$$

The translational motion is derived using a standard variational integrator (see the example in Sec. II-A) and, with the addition of forces, becomes

$$\begin{aligned} m\left[\frac{p_{k+1} - 2p_k + p_{k-1}}{h^2}\right] &= \frac{1}{4}\left[R_{k-1}F_V(\alpha_{k-1})u_{k-1}\right. \\ &\left.+ 2R_k F_V(\alpha_k)u_k + R_{k+1}F_V(\alpha_{k+1})u_{k+1}\right] + f_g. \end{aligned} \tag{17}$$

IV. Direct Optimal Control Formulation

A straightforward way to find a numerical solution to the optimal control problem is to formulate a nonlinear program that minimizes the cost function over all discrete configurations, velocities, and forces, while satisfying the boundary conditions and the discrete equations of motion enforced as equality constraints. Additional equality or inequality constraints can also be enforced.

A. Problem Formulation

The optimal control problem can be directly formulated as

Compute: g_d, ξ_d, f_d, h

minimizing $\displaystyle\sum_{k=0}^{N-1} C_d(g_k, \xi_k, f_k^{\pm}, h)$

subject to:

$$\begin{cases} g_0 = g_i,\ \xi_0 = \xi_i,\ g_N = g_f,\ \xi_{N-1} = \xi_f, \\ \text{Equations } (10) - (12) \quad \text{for} \quad k = 0, ..., N-1, \\ H(g_d, \xi_d, h) \geq 0, \\ \xi_k \in [\xi_l, \xi_u], f_k \in [f_l, f_u], h \in [h_l, h_u], \end{cases} \tag{18}$$

where C_d is a discrete approximation of C defined in (1) and and $H : G \times \mathfrak{g} \times \mathbb{R} \to \mathbb{R}^p$ are inequality constraints. The formulation allows time to vary and the last constraint places bounds on the time variation as well as bounds on all other variables.

Remarks

Average Velocity: The variables denoted ξ_N and μ_N have no effect on the trajectory g_d so we can treat these last points as irrelevant to the optimization. This is coherent with thinking of each velocity ξ_k as the average body-fixed velocity along the kth path segment between configurations g_k and g_{k+1}.

Velocity at the boundary: A second remark must be made regarding velocity boundary conditions. For simplicity, we work with the boundary conditions $\xi_0 = \xi_i$ and $\xi_{N-1} = \xi_f$ which are not exact, since according to the above assumption ξ_k represents an average velocity. A proper treatment of the exact velocity boundary conditions given by $\xi(0)$ and $\xi(T)$ requires further constraints such as

$$\begin{aligned} &(d\tau^{-1}_{h/2\xi(0)})^* \ell'(\xi(0)) - (d\tau^{-1}_{-h/2\xi_0})^* \mu_0 = f_0^-, \\ &(d\tau^{-1}_{h/2\xi_{N-1}})^* \mu_{N-1} - (d\tau^{-1}_{-h/2\xi(T)})^* \ell'(\xi(T)) = f_{N-1}^+. \end{aligned}$$

However, for simplicity and computational efficient, we assume that the boundary condition is in terms of the initial and final *average* velocity.

B. Algorithm Construction

Midpoint Rule: The discrete cost function can be constructed using the midpoint rule as

$$C_d(g_k, \xi_k, f_k^{\pm}, h) = hC\left(g_{k+\frac{1}{2}}, \xi_k, \frac{f_k + f_{k+1}}{2}\right) \quad (19)$$

where $g_{k+\frac{1}{2}} = g_k\tau(\frac{h}{2}\xi_k)$, i.e. the midpoint along the flow defined by τ. There are other choices besides the midpoint rule that can lead to integrators of arbitrary high order of accuracy, e.g., composition methods and symplectic Runga-Kutta methods [12]. The midpoint rule is a particularly pertinent choice for optimization problems since it provides a good balance between accuracy and efficiency.

Implementation: The optimal control formulation (18) can be solved using a standard constrained optimization technique such as sequential quadratic programming (SQP). A continuous control curve f can be obtained from a discrete solution curve f_d using linear interpolation of f_d (in case of the midpoint rule) or some higher order interpolation consistent with the order of accuracy of the chosen discretization.

V. Application

We have implemented our framework for solving general optimal control problems of the form (18). It is written in C++ and uses the sparse SQP solver SNOPT [25]. The system is used to compute a control-effort minimizing trajectory for the simulated helicopter between two zero-velocity states in an artificial canyon. We use the discrete system described in the example of Sec.III-B with equations of motion defined by (16) and (17) and a discrete cost function defined by (19). Fig. 2 shows a typical output of our system and the resulting trajectory and control curves are shown on Fig. 5.

Controllability: In order to establish the controllability of the system one can use the good symmetric products [5] of the two vectors obtained from the columns of the matrix

$$\begin{bmatrix} \mathbb{I}^{-1}F_{\Omega}(\alpha) \\ \frac{1}{m}F_V(\alpha) \end{bmatrix},$$

and show that the system is locally configuration controllable at zero velocity. Since the actuator inputs have bounds, one has to design an algorithm which allows time to vary in order to accommodate these limits. In our implementation, the time-step h (and hence the final time T) is part of the optimization state vector and is allowed to vary within some prescribed bounds.

Optimization Variables: For efficiency the components of the matrices R_k are not part of the optimization state vector and the trajectory R_d is reconstructed from the trajectory Ω_d internally during optimization. Alternatively, one could parametrize the rotations (e.g., using quaternions) and optimize over these additional coordinates as well. In our experience, both approaches perform similarly well.

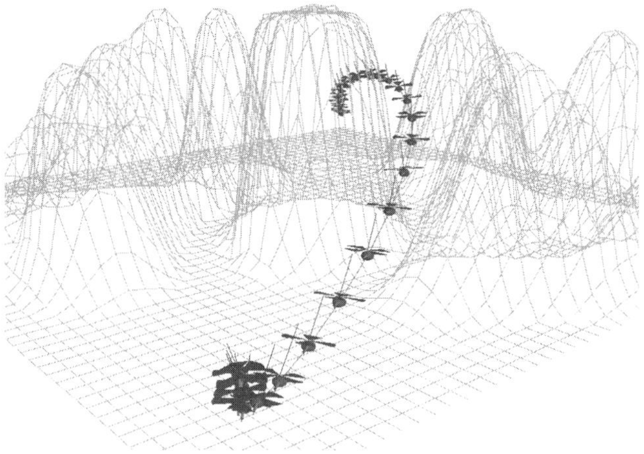

Fig. 2. Example of an optimized trajectory in a complex environment: a helicopter path through an outdoor canyon.

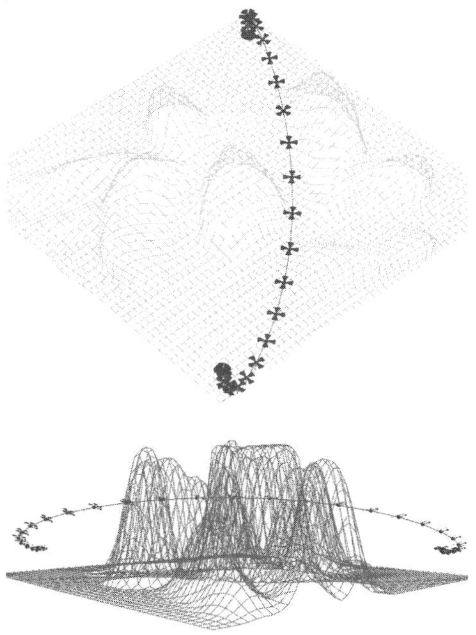

Fig. 3. Top and side views of the helicopter trajectory shown in Fig. 2

Obstacles: A robot can be required to stay away from obstacles by enforcing inequality constraints $H_i(R,p) = \text{dist}(\mathcal{A}(R,p), \mathcal{O}_i) - d_s$, where $\mathcal{A} \subset \mathbb{R}^3$ is the region occupied by the robot, $\mathcal{O}_i \subset \mathbb{R}^3$ represent the static obstacles, and d_s is some safety distance. The function dist computes the minimum distance between two rigid objects. In our implementation both the canyon and the helicopter are triangulated surfaces and we use the Proximity Query Package (PQP) to compute dist.

The optimization runs efficiently in the absence of obstacles or with simple (smooth and convex) obstacles (taking in the order of a few seconds for most tests). On the other hand, complex rough terrains can slow down the system significantly.

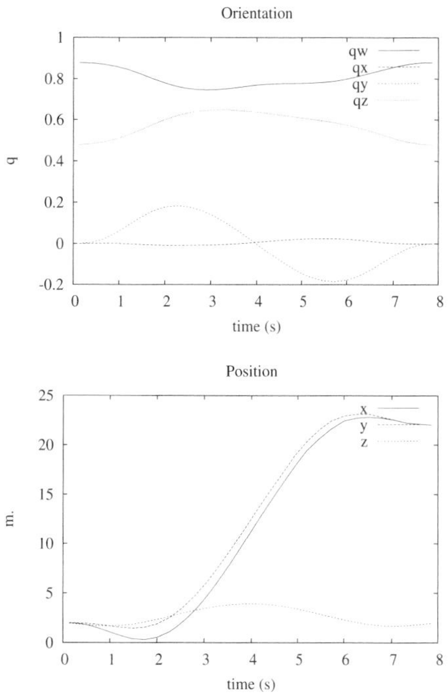

Fig. 4. The orientation and position as functions of time for the helicopter trajectory shown in Fig. 2.

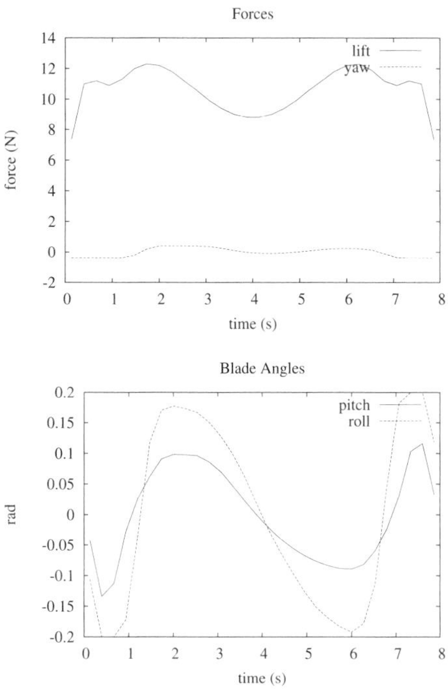

Fig. 5. The control forces and blade angles as functions of time for the helicopter trajectory shown in Fig. 2.

Note also that our simple implementation faces the same robustness issues as many optimization procedures: a bad choice for the initial trajectory may lead to a local minima. One way to speedup convergence is to start with a good initial obstacle-free path computed, for example, using a local obstacle avoidance method (e.g. [26]). Nevertheless, as already pointed out by several similar works cited in our references, this approach should be used to produce small-scale local solutions that are combined by more *global* methods in a hierarchical fashion (e.g. see [8] regarding incremental and roadmap planners, as well as [27, 28] for planning using primitives). An obvious application is also the refinement of trajectories produced by discrete or sampling-based planners. Finally, in order to assess the exact numerical benefits of this method, a detailed performance analysis and comparison to related methods is needed and is currently under way.

VI. Conclusion

There are many ways to solve optimal control problems, as confirmed by the rich literature on this subject. We have presented one approach that focuses on a proper discretization of the system dynamics through the extremization of a discrete action. An optimal control framework for systems with symmetries is then derived and applied to a simple robotic model. It will be useful to further optimize our approach using ideas from, e.g., [6, 7, 29]. A comparison with the closely related method [11] is also an obvious research direction. Since the optimization algorithm is inherently local (as are most gradient based methods) our method will be most effective in a global hierarchical optimization framework. This is a central issue related to the nature of discrete mechanics. In that respect, discrete mechanics can be used to perform provably correct coarse-to-fine discretization of time and space which is linked to global convergence of trajectories.

Future Directions and Related Work

There are several promising directions of potential benefit to robotics, in which the discrete geometric approach is currently being extended. One of these is to further exploit the simplicity of the approach as well as its hierarchical benefits. Another is the discretization and optimal control of nonholonomic systems. Initial results with applications to Chaplygin systems and car-like robots can be found in [17]. Yet another direction is the inclusion of environment constraints such as obstacles or interaction constraints in groups of vehicles directly as part of the discrete optimization framework. Coarse-to-fine methods can then be used to jointly deform such constraints along with solution trajectories to produce more efficient methods with provable convergence properties. For example, multi-resolution extensions of the optimization methods presented in this paper already exhibit real-time performance when applied to rigid body systems. Finally, hierarchical and decentralized (e.g. [19]) DMOC methods are under development to address multi-vehicle or other complex systems in a scalable and robust manner.

As has been indicated, there are several important issues that need additional attention. One is a detailed comparison to other optimization, both in terms of accuracy and in terms of hierarchical and coarse-to-fine benefits. A preliminary and encouraging study was done in [21], but this needs systematic and detailed analysis.

ACKNOWLEDGMENT

We are grateful to Eva Kanso, Nawaf Bou-Rabee, Sina Ober-Blöbaum, and Sigrid Leyendecker for their interest and helpful comments. This work was supported in part by NSF (CCR-0120778, CCR-0503786, IIS-0133947 and ITR DMS-0453145), DOE (DE-FG02-04ER25657), the Caltech Center for Mathematics of Information and AFOSR Contract FA9550-05-1-0343.

REFERENCES

[1] S. Kelly and R. Murray, "Geometric phases and robotic locomotion," *Journal of Robotic Systems*, vol. 12, no. 6, pp. 417–431, 1995.

[2] J. Ostrowski, "Computing reduced equations for robotic systems with constraints and symmetries," *IEEE Transactions on Robotics and Automation*, pp. 111–123, 1999.

[3] E. A. Shammas, H. Choset, and A. A. Rizzi, "Motion planning for dynamic variable inertia mechanical systems with non-holonomic constraints," in *International Workshop on the Algorithmic Foundations of Robotics*, 2006.

[4] J. E. Marsden and T. S. Ratiu, *Introduction to Mechanics and Symmetry*. Springer, 1999.

[5] F. Bullo and A. Lewis, *Geometric Control of Mechanical Systems*. Springer, 2004.

[6] J. P. Ostrowski, J. P. Desai, and V. Kumar, "Optimal gait selection for nonholonomic locomotion systems," *The International Journal of Robotics Research*, vol. 19, no. 3, pp. 225–237, 2000.

[7] J. Cortes, S. Martinez, J. P. Ostrowski, and K. A. McIsaac, "Optimal gaits for dynamic robotic locomotion," *The International Journal of Robotics Research*, vol. 20, no. 9, pp. 707–728, 2001.

[8] S. M. LaValle, *Planning Algorithms*. Cambridge University Press, Cambridge, U.K., 2006.

[9] O. Junge, J. Marsden, and S. Ober-Blöbaum, "Discrete mechanics and optimal control," in *Proccedings of the 16th IFAC World Congress*, 2005.

[10] E. Kanso and J. Marsden, "Optimal motion of an articulated body in a perfect fluid," in *IEEE Conference on Decision and Control*, 2005, pp. 2511–2516.

[11] T. Lee, N. McClamroch, and M. Leok, "Optimal control of a rigid body using geometrically exact computations on SE(3)," in *Proc. IEEE Conf. on Decision and Control*, 2006.

[12] J. Marsden and M. West, "Discrete mechanics and variational integrators," *Acta Numerica*, vol. 10, pp. 357–514, 2001.

[13] J. P. Desai and V. Kumar, "Motion planning for cooperating mobile manipulators," *Journal of Robotic Systems*, vol. 16, no. 10, pp. 557–579, 1999.

[14] L. Kharevych, Weiwei, Y. Tong, E. Kanso, J.E. Marsden, P. Schroder, and M.Desbrun, "Geometric, variational integrators for computer animation," in *Eurographics/ACM SIGGRAPH Symposium on Computer Animation*, 2006, pp. 43–51.

[15] N. Bou-Rabee and J. E. Marsden, "Reduced Hamilton-Pontryagin variational integrators," preprint.

[16] H. Yoshimura and J. Marsden, "Dirac structures in Lagrangian mechanics part ii: Variational structures," *Journal of Geometry and Physics*, vol. 57, pp. 209–250, dec 2006.

[17] M. Kobilarov, M. Desbrun, J. Marsden, and G. S. Sukhatme, "Optimal control using d'alembert-pontryagin nonholonomic integrators," Center for Robotics and Embedded Systems, University of Southern California, Tech. Rep., 2007.

[18] R. Fetecau, J. Marsden, M. Ortiz, and M. West, "Nonsmooth Lagrangian mechanics and variational collision integrators," *SIAM Journal on Applied Dynamical Systems*, vol. 2, no. 3, pp. 381–416, 2003.

[19] O. Junge, J. Marsden, and S. Ober-Blöbaum, "Optimal reconfiguration of formation flying spacecraft - a decentralized approach," in *45th IEEE Conference on Decision and Control*, 2006, pp. 5210–5215.

[20] S. Ross, "Optimal flapping strokes for self-propulsion in a perfect fluid," in *American Control Conference*, 2006.

[21] M. Kobilarov and G. S. Sukhatme, "Optimal control using nonholonomic integrators," in *IEEE International Conference on Robotics and Automation*, Apr 2007, pp. 1832–1837.

[22] E. Celledoni and B. Owren, "Lie group methods for rigid body dynamics and time integration on manifolds," *Comput. meth. in Appl. Mech. and Eng.*, vol. 19, no. 3,4, pp. 421–438, 2003.

[23] J.E. Marsden and S. Pekarsky and S. Shkoller, "Discrete Euler-Poincaré and Lie-Poisson equations," *Nonlinearity*, vol. 12, p. 1647–1662, 1999.

[24] E. Hairer, C. Lubich, and G. Wanner, *Geometric Numerical Integration*, ser. Springer Series in Computational Mathematics. Springer-Verlag, 2006, no. 31.

[25] P. E. Gill, W. Murray, and M. A. Saunders, "SNOPT: An SQP algorithm for large-scale constrained optimization," *SIAM J. on Optimization*, vol. 12, no. 4, pp. 979–1006, 2002.

[26] D. E. Chang, S. Shadden, J. E. Marsden, and R. Olfati-Saber, "Collision avoidance for multiple agent systems," in *IEEE Conference on Decision and Control*, vol. 42, 2003, pp. 539–543.

[27] E. Frazzoli, M. A. Dahleh, and E. Feron, "Maneuver-based motion planning for nonlinear systems with symmetries," *IEEE Transactions on Robotics*, vol. 21, no. 6, pp. 1077–1091, dec 2005.

[28] C. Dever, B. Mettler, E. Feron, and J. Popovic, "Nonlinear trajectory generation for autonomous vehicles via parameterized maneuver classes," *Journal of Guidance, Control, and Dynamics*, vol. 29, no. 2, pp. 289–302, 2006.

[29] M. B. Milam, K. Mushambi, and R. M. Murray, "A new computational approach to real-time trajectory generation for constrained mechanical systems," in *IEEE Conference on Decision and Control*, vol. 1, 2000, pp. 845–851.

BS-SLAM: Shaping the World

Luis Pedraza*, Gamini Dissanayake†, Jaime Valls Miro†, Diego Rodriguez-Losada*, and Fernando Matia*
*Universidad Politecnica de Madrid (UPM)
C/ Jose Gutierrez Abascal, 2. 28006, Madrid, Spain
Email: luis.pedraza@ieee.org
†Mechatronics and Intelligent Systems Group. University of Technology Sydney (UTS)
NSW2007, Australia

Abstract— **This paper presents BS-SLAM, a simultaneous localization and mapping algorithm for use in unstructured environments that is effective regardless of whether features correspond to simple geometric primitives such as lines or not. The coordinates of the control points defining a set of B-splines are used to form a complete and compact description of the environment, thus making it feasible to use an extended Kalman filter based SLAM algorithm. The proposed method is the first known EKF-SLAM implementation capable of describing both straight and curve features in a parametric way. Appropriate observation equation that allows the exploitation of virtually all observations from a range sensor such as the ubiquitous laser range finder is developed. Efficient strategies for computing the relevant Jacobians, perform data association, initialization and expanding the map are presented. The effectiveness of the algorithms is demonstrated using experimental data.**

I. Introduction

Developing an appropriate parameterization to represent the map is the key challenge for simultaneous localization and mapping in unstructured environments. While a substantial body of literature exists in methods for representing unstructured environments during robot mapping, most of these are unsuitable for use with the Kalman filter based simultaneous localization and mapping. For example, the use of popular occupancy grid approach [13], based on dividing the environment into small cells and classify these as occupied or unoccupied, and its variants such as those using quad-trees would result in an impractically large state vector.

In much of the early SLAM work the map is described by a set of points [6], [9], [11]. While this simplifies the formulation and the complexity of the SLAM estimator, there are two main disadvantages in relying solely on point features. The obvious problem arises if the environment does not have sufficient structure to be able to robustly extract point features, for example in an underground mine [12]. The more significant issue is the fact that much of the information acquired from a typical sensor, such as a laser range finder, does not correspond to point features in the environment. Therefore, the raw information from the sensor needs to be analysed and observations corresponding to stable point features extracted. During this process, typically more than 95% of the observations are discarded and the information contained in these observations wasted. One strategy to exploit this additional information is to model the environment using alternative geometric primitives such as line segments [3], [15] or polylines [18]. While this has been successful in many indoor environments, presence of curved elements can create significant problems. In particular, attempting to interpret information from a sensor using an incorrect model is one of the major causes of failure of many estimation algorithms. Some efforts have been made in the past when circle features are available [20], but that's still a big simplification. Thus a more generic representation of the environment can potentially improve the robustness of the SLAM implementations.

Alternative to using a specific environmental model is to use information from sensor observations from different robot poses to obtain an accurate relationship between these poses. While this strategy has been successfully used to generate accurate visualizations of complex structures [7] and detailed maps of indoor environments [8], it can not exploit the inherent information gain that occur in traditional SLAM due to the improvement of map quality.

In this paper, it is proposed to use B-splines to represent the boundary between occupied and unoccupied regions of a complex environment. B-splines provide naturally compact descriptions consisting of both lines and curves. Vertices of the control polygons that describe a set of B-splines are used to represent a complex environment by a state vector. Computationally efficient strategies for (a) initializing and extending the state vector when new parts of the environment are discovered, (b) formulating a suitable observation equation that allows the exploitation of virtually all the information gathered from a laser range finder, and (c) evaluation of appropriate Jacobians for easy implementation of Kalman filter equations are presented in detail. The algorithms proposed are evaluated for effectiveness using data gathered from real environments.

The paper is organized as follows. Section II introduces some general concepts and properties of B-spline curves. Section III shows how these powerful tools fit into the EKF-SLAM framework. Finally, some experimental results and conclusions are presented in sections IV and V.

II. Splines Fundamentals

In this section, a brief introduction to the fundamental concepts of the B-splines theory is presented. The term *spline* is used to refer to a wide class of functions that are used in applications requiring interpolation or smoothing of data in a flexible and computationally efficient way. A spline of degree k (order $k-1$) is a piecewise polynomial curve; i.e. a curve divided into several pieces, where each of these pieces is described by a polynomial of degree k, and the different pieces accomplish with certain continuity properties at the joint points or knots. Their most common representation is based on the utilization of B-splines (where B stands for *basic*).

A. B-Splines Definition

Letting $\mathbf{s}(t)$ be the position vector along the curve as a function of the parameter t, a spline curve of order k, with control points $\mathbf{x}_i \ (i = 0 \ldots n)$ and knot vector $\Xi = \{\xi_0, \ldots, \xi_{n+k}\}$ can be expressed as:

$$\mathbf{s}(t) = \sum_{i=0}^{n} \mathbf{x}_i \beta_{i,k}(t) \tag{1}$$

where $\beta_{i,k}(t)$ are the normalized B-spline basis functions of order k which are defined by the Cox-de Boor recursion formulas [16], [4]:

$$\beta_{i,1}(t) = \begin{cases} 1 & \text{if } \xi_i \le t \le \xi_{i+1} \\ 0 & \text{otherwise} \end{cases} \tag{2}$$

and

$$\beta_{i,k}(t) = \frac{(t-\xi_i)}{\xi_{i+k-1}-\xi_i}\beta_{i,k-1}(t) + \frac{(\xi_{i+k}-t)}{\xi_{i+k}-\xi_{i+1}}\beta_{i+1,k-1}(t) \tag{3}$$

The knot vector is any nondecreasing sequence of real numbers ($\xi_i \le \xi_{i+1}$ for $i = 0, \ldots, n+k-1$) and can be defined in two different ways: *clamped*, when the multiplicity of the extreme knot values is equal to the order k of the B-spline, and *unclamped* [16, 14]. When clamped knot vectors are used, first and last control polygon points define the beginning and end of the spline curve.

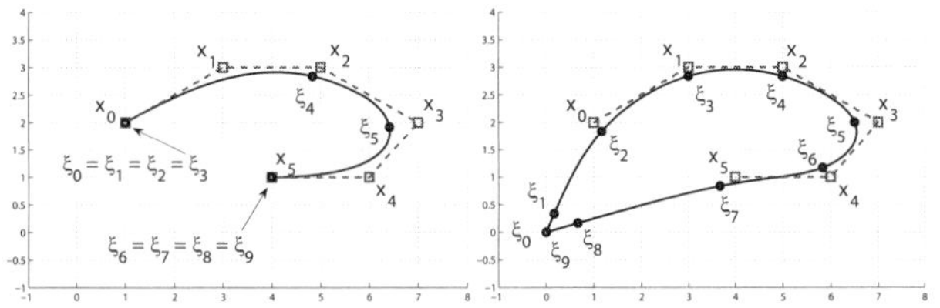

Fig. 1. Examples of splines in B-spline form. **a)** Cubic spline with clamped knot vector ($\Xi_c : \xi_0 = \ldots = \xi_3 \le \ldots \le \xi_6 = \ldots = \xi_9$). **b)** Cubic spline with unclamped knot vector ($\Xi_c : \xi_0 \le \ldots \le \xi_9$).

B. Properties of Spline Curves

In this section, some properties of spline curves, defined as linear combination of B-splines, are summarized.

1) The curve generally follows the shape of the control polygon.
2) Any affine transformation is applied to the curve by applying it to the control polygon vertices.
3) Each basis function $\beta_{i,k}(t)$ is a piecewise polynomial of order k with breaks $\xi_i, \ldots, \xi_{i+k}$, vanishes outside the interval $[\xi_i, \xi_{i+k})$ and is positive on the interior of that interval:

$$\beta_{i,k}(t) > 0 \quad \Leftrightarrow \quad \xi_i \le t < \xi_{i+k} \tag{4}$$

4) As a consequence of property 3, the value of $\mathbf{s}(t)$ at a site $\xi_j \le t \le \xi_{j+1}$ for some $j \in \{k-1, \ldots, n\}$ depends only on k of the coefficients:

$$\mathbf{s}(t) = \sum_{i=j-k+1}^{j} \mathbf{x}_i \beta_{i,k}(t) \tag{5}$$

5) The sum of all the B-spline basis functions for any value of the parameter t is 1:

$$\sum_{i=0}^{n} \beta_{i,k}(t) = 1 \tag{6}$$

6) The derivative of a B-spline of order k is a spline of one order $k-1$, whose coefficients can be obtained differencing the original ones [4].

$$\frac{d\mathbf{s}(t)}{dt} = \mathbf{s}'(t) = (k-1)\sum_{i=0}^{n} \frac{\mathbf{x}_i - \mathbf{x}_{i-1}}{\xi_{i+k-1} - \xi_i}\beta_{i,k-1}(t) \tag{7}$$

For further information and justification of the previous properties, please see [4], [14], [1] and [16].

C. Curve Fitting

Here we consider the problem of obtaining a spline curve that fits a set of data points $\mathbf{d}_j$, $j = 0 \ldots m$. If a data point lies on the B-spline curve, then it must satisfy equation 1:

$$\mathbf{d}_j = \beta_{0,k}(t_j)\mathbf{x}_0 + \ldots + \beta_{n,k}(t_j)\mathbf{x}_n, \quad j = 0 \ldots m$$

system of equations which can be more compactly written as

$$\mathbf{d} = \mathbf{B}\mathbf{x} \left\{ \begin{array}{lcl} \mathbf{d} & = & \begin{bmatrix} \mathbf{d}_0 & \mathbf{d}_1 & \ldots & \mathbf{d}_m \end{bmatrix}^T \\ \mathbf{x} & = & \begin{bmatrix} \mathbf{x}_0 & \mathbf{x}_1 & \ldots & \mathbf{x}_n \end{bmatrix}^T \\ \mathbf{B} & = & \begin{bmatrix} \beta_{0,k}(t_0) & \ldots & \beta_{n,k}(t_0) \\ \vdots & \ddots & \vdots \\ \beta_{0,k}(t_m) & \ldots & \beta_{n,k}(t_m) \end{bmatrix} \end{array} \right. \tag{8}$$

Matrix $\mathbf{B}$, usually referred to as the *collocation matrix*, has for each of its rows at most k non-null values. The parameter value t_j defines the position of each data point $\mathbf{d}_j$ along the B-spline curve, and can be approximated by the chord length between data points:

$$\left. \begin{array}{lcl} t_0 & = & 0 \\ t_j & = & \sum_{s=1}^{j} |\mathbf{d}_s - \mathbf{d}_{s-1}|, \quad j \ge 1 \end{array} \right\} \tag{9}$$

being the total length of the curve

$$\ell = \sum_{s=1}^{m} |\mathbf{d}_s - \mathbf{d}_{s-1}| \tag{10}$$

which is taken as the maximum value of the knot vector.

The most general case occurs when $2 \le k \le n+1 < m+1$; the problem is over specified and can only be solved in a mean sense. A least squares solution can be computed making use of the pseudoinverse matrix of $\mathbf{B}$:

$$\mathbf{x} = \left[\mathbf{B}^T\mathbf{B}\right]^{-1}\mathbf{B}^T\mathbf{d} = \mathbf{\Phi}\mathbf{d} \tag{11}$$

Once the order of the B-spline bases k is predefined, the number of control polygon vertices $n+1$, and the parameter values along the curve are known (as calculated from equation 9), the basis functions $\beta_{i,k}(t_j)$ and hence the matrix $\mathbf{B}$ can be obtained.

In the work here presented, clamped knot vectors are generated taking the total length of the curve ℓ (equation 10), and defining a knot density which depends on the complexity of the environment. Recall that knots are the joints of the polynomial pieces a spline is composed of, so complex environments need a high knot density, while segments can be described properly using longer polynomial pieces (lower knot density). The interested reader is referred to [5] and [2], where more information about curve fitting methods can be found.

III. SLAM WITH B-SPLINES

A. Data Management

A commonly used sensor in mobile robotics is the laser range-finder. Whenever a robots makes an observation of its environment with such a device, a set of m data points $\mathbf{d}_i \in \Re^2$ is obtained (we're considering here the 2D scenario). In this section, methods for extracting parametric splines representing the physical detected objects are presented. It is also shown the way of performing data association, establishing comparisons between the detected splines and the map splines.

1) Obtaining Splines: When a new data set is obtained from an observation of the environment, splines are extracted in a three-stages process (see Fig. 2):

- **Primary segmentation**: Data stream is split into pieces separated by measurements out of range, if any. A set of N_1 data vectors is obtained, called primary segmentation objects, and denoted $F_{1,1}, F_{1,2}, \ldots, F_{1,N_1}$ (see Fig. 2.c).
- **Secondary segmentation**: An analysis of the relative positions of consecutive data points is performed. The aim is to detect points close enough as to belong to the same feature, and also detect corners, but allowing points not to lie on the same segment (Fig. 2.f). A set of N_2 secondary segmentation features is obtained (Fig. 2.d).
- **Fitting**: Each of the secondary segmentation objects is fitted by a cubic spline, as described in section II-C.

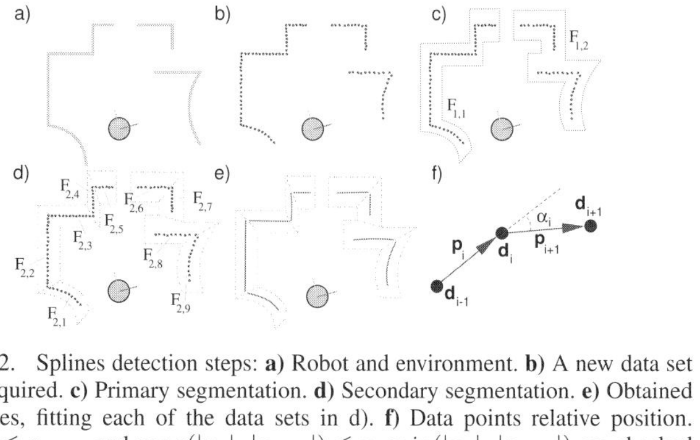

Fig. 2. Splines detection steps: **a)** Robot and environment. **b)** A new data set is acquired. **c)** Primary segmentation. **d)** Secondary segmentation. **e)** Obtained splines, fitting each of the data sets in d). **f)** Data points relative position. $|\alpha_i| \leq \alpha_{max}$ and $max(|\mathbf{p}_i|, |\mathbf{p}_{i+1}|) \leq \eta \cdot min(|\mathbf{p}_i|, |\mathbf{p}_{i+1}|)$ are checked (typically $\alpha_{max} \in [0, \pi/4]$ and $\eta \in [1.5, 2]$).

2) Splines Association: In this section, the data association process is described. At each sampling time, k, a new set of splines is obtained, being necessary to achieve a good feature association between the curves that are already contained in the map $(\mathbf{s}_{m,1}, \ldots, \mathbf{s}_{m,N})$, and the new ones that have just been detected $(\mathbf{s}_{o,1}, \ldots, \mathbf{s}_{o,N_2})$.

The process is explained with an example. Take the situation displayed in figure 3. Figure 3.a shows the initial configuration: a map containing one spline $\mathbf{s}_{m,1}(t)$, and two new features which have just been detected, $\mathbf{s}_{o,1}(u)$ and $\mathbf{s}_{o,2}(v)$ (notice the lack of correspondence between the parameters of different curves). A rough measure of the distances between the splines can be obtained by measuring the distances between the control points of each spline (recall here property 1). In order to improve this search, the map can be simplified choosing only the splines that are close to the robot's position.

At each sampling time, the position of the observed splines is calculated taking into account the best available estimation for the robot's position and orientation. Then, the distances from the control points of each of the detected splines to the control points of the splines contained in the map are calculated, and each of the observed splines is matched to the closest one on the map, following this criterion:

$$min\left(dist\left(\mathbf{x}_{m,i}, \mathbf{x}_{o,j}\right)\right) \leq d_{min}, \left\{ \begin{array}{l} i = 1 \ldots n_m \\ j = 1 \ldots n_o \end{array} \right. \quad (12)$$

where $\mathbf{x}_{m,i}$ are the control points of the map spline, $\mathbf{x}_{o,j}$ are the control points of the observed spline, n_m and n_o are, respectively, the number of control points of the spline in the map and the detected spline, and $dist\left(\mathbf{x}_{m,i}, \mathbf{x}_{o,j}\right)$ represents the euclidean distance between two control points.

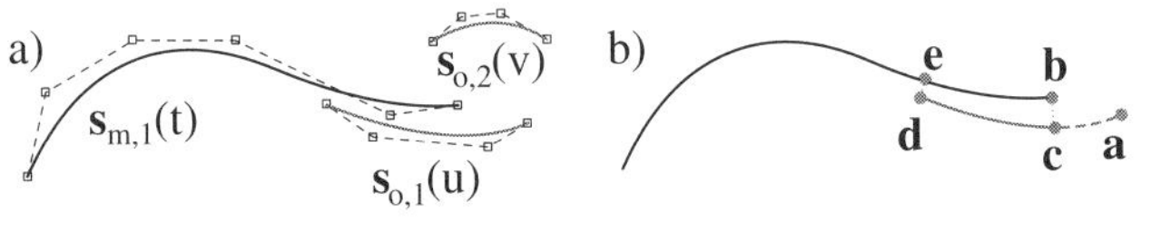

Fig. 3. Data association process. Comparison of control points positions (a) and parameters correspondence (b).

If no spline in the map is close enough to a detected spline (as occurs with $\mathbf{s}_{o,2}$ in figure 3.a) then, this new object is added to the map once its position has been updated. If a spline is associated with a map feature, then it is necessary to obtain a matching between their points, as depicted in figure 3.b. This matching provides information about the correspondence between the parameterizations of both curves, and is very useful when a spline extension is required. The process is as follows:

- One of the extreme points of the observed spline is considered (point $\mathbf{a}$)
- The closest point [19] on the map spline to the point $\mathbf{a}$ is calculated (point $\mathbf{b}$)
- If $\mathbf{b}$ is one of the extreme points of the map's spline, then, the closest point to $\mathbf{b}$ on the observed spline is calculated (point $\mathbf{c}$). Else, point $\mathbf{a}$ is matched with point $\mathbf{b}$.

- The process is repeated taking as starting point the other extreme of the observed spline (point $\mathbf{d}$ in the picture), which is associated with the point $\mathbf{e}$ on the map spline.

At the end of this process, not only correspondent pairs of points $(\mathbf{c}, \mathbf{b})$ and $(\mathbf{d}, \mathbf{e})$ are obtained, but also a correspondence between the map spline parameter t and the observed spline parameter u: (u_{ini}, t_{ini}) and (u_{fin}, t_{fin}), given that

$$\begin{array}{ll} \mathbf{c} = \mathbf{s}_{o,1}(u_{ini}) & \mathbf{b} = \mathbf{s}_{m,1}(t_{ini}) \\ \mathbf{d} = \mathbf{s}_{o,1}(u_{fin}) & \mathbf{e} = \mathbf{s}_{m,1}(t_{fin}) \end{array} \quad (13)$$

The simple data association process described in this section, though quite simple and based upon euclidean distance metric, has performed very robustly in our experiments. However, benefits of parametric representation are not yet fully exploited, and further research is being undertaken in this sense.

B. The State Model

The state of the system is composed by the robot (the only mobile element) and all the map features, modeled in the work here presented by cubic splines. When these splines are expressed as linear combination of B-splines, the state of each of them can be represented by the positions of their control polygon vertices, given a fixed and known knot vector which generates a basis of B-splines for each of the map features.

Referring all the positions and orientations to a global reference system $\{\mathbf{u}_W, \mathbf{v}_W\}$, and assuming the robot as the first feature in the map (F_0) the following expressions describe the state of the system at a certain time k:

$$\mathbf{x}_{F_0} = \mathbf{x}_r = [x_r, y_r, \phi_r]^T \quad (14)$$

$$\mathbf{x}_{F_i} = \mathbf{x}_{s_i} = [x_{i,0}, \ldots, x_{i,n_i}, y_{i,0}, \ldots, y_{i,n_i}]^T \quad (15)$$
$$i = 1, \ldots, N$$

and finally

$$\mathbf{x} = \left[\mathbf{x}_r^T, \mathbf{x}_{s_1}^T, \ldots, \mathbf{x}_{s_N}^T\right]^T \quad (16)$$

In the previous equations, N is the number of map static elements (map features) and n_i is the number of control points for each of them. Note that the number of control points for each of the splines contained in the map can be variable, as features are progressively extended as new areas of the environment are explored.

$$\mathbf{x}(k) \sim N(\hat{\mathbf{x}}(k|k), \mathbf{P}(k|k)) \quad (17)$$

where

$$\hat{\mathbf{x}}(k|k) = \left[\begin{array}{cccc} \hat{\mathbf{x}}_r(k|k) & \hat{\mathbf{x}}_{s_1}(k|k) & \ldots & \hat{\mathbf{x}}_{s_N}(k|k) \end{array}\right] \quad (18)$$

and

$$\mathbf{P}(k|k) = \begin{bmatrix} \mathbf{P}_{rr}(k|k) & \mathbf{P}_{rs_1}(k|k) & \ldots & \mathbf{P}_{rs_N}(k|k) \\ \mathbf{P}_{s_1 r}(k|k) & \mathbf{P}_{s_1 s_2}(k|k) & \ldots & \mathbf{P}_{s_1 s_N}(k|k) \\ \vdots & \vdots & \ddots & \vdots \\ \mathbf{P}_{s_N s_1}(k|k) & \mathbf{P}_{s_N s_2}(k|k) & \ldots & \mathbf{P}_{s_N s_N}(k|k) \end{bmatrix} \quad (19)$$

C. The Observation Model

The implementation of a EKF-SLAM algorithm requires of an observation model; i.e. some expression which allows to predict the measurements that are likely to be obtained by the robot sensors given the robot pose and the current knowledge of the environment. This measurement model can be understood, in our particular case, as the calculation of the intersection of the straight line defined by a laser beam (for each position across the laser angular range) with the splines contained in the map (Fig. 4). Unfortunately, calculating the intersection of a straight line with a parametric curve, in the form $\mathbf{s}(t) = [s_x(t), s_y(t)]^T$ is not suitable for an explicit mathematical formulation. There is a whole field of research regarding this complex problem known as *ray tracing* [17].

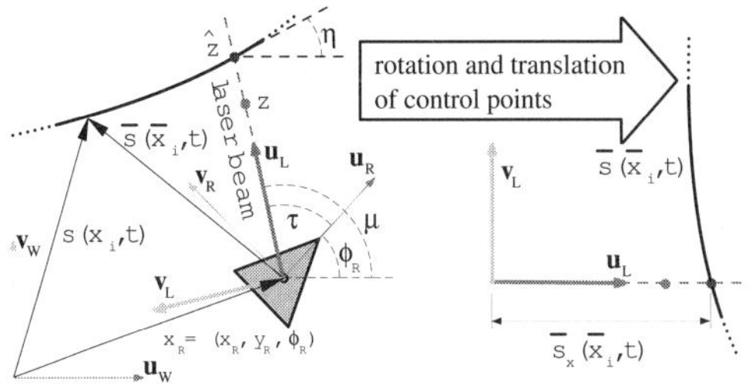

Fig. 4. Observation model

In this document, the predicted measurement is calculated in an iterative way, making use of property 2 in section II-B and the Newton-Raphson method for calculating the zeros of a function. The first step is to define an orthonormal reference system $\{\mathbf{u}_L, \mathbf{v}_L\}$, centered in the robot reference system $\{\mathbf{u}_R, \mathbf{v}_R\}$ and with the abscissas axis $\mathbf{u}_L$ defined by the laser beam direction and orientation (see Fig. 4). Let $\bar{\mathbf{s}}(\bar{\mathbf{x}}_i(\mathbf{x}_i, \mathbf{x}_r), t)$ be the position vector along a spline curve expressed in such a reference system (we're making here explicit the functional dependency between a spline and its control points). The relationship between control points $\mathbf{x}_i = [x_i, y_i]^T$ and $\bar{\mathbf{x}}_i = [\bar{x}_i, \bar{y}_i]^T$, $i = 0 \ldots n$, is given by:

$$\begin{bmatrix} \bar{x}_i \\ \bar{y}_i \end{bmatrix} = \begin{bmatrix} cos\mu & sin\mu \\ -sin\mu & cos\mu \end{bmatrix} \begin{bmatrix} x_i - x_r \\ y_i - y_r \end{bmatrix} \quad (20)$$

being μ the angle of the considered laser beam in the global reference system; i.e. given the laser orientation in the robot reference system, τ:

$$\mu = \phi_r + \tau \quad (21)$$

In this context, the measurement prediction $\hat{z} = h(\mathbf{x}_i, \mathbf{x}_r)$ is given by the value of $\bar{s}_x(\bar{x}_i(\mathbf{x}_i, \mathbf{x}_r), t^*)$, where t^* is the value of the parameter t that makes $\bar{s}_y(\bar{y}_i(\mathbf{x}_i, \mathbf{x}_r), t^*) = 0$. As only a reduced set of k control points affect the shape of the curve for each value of the parameter t, a small number of this points need to be rotated and translated to the new reference system. The initial guess for each map spline associated with an observation can be obtained directly from the data association stage, and for each of the laser beams positions the solution obtained in the previous prediction can be used as initial guess.

During the experiments, a maximum of two iterations were needed for this calculation, with a precision of 0.0001 m.

Despite the lack of an explicit observation model, it is still possible to calculate the derivatives of this expected measurements with respect to the map state in an approximate way. Once calculated the value t^* which makes $\bar{s}_y(t^*) = 0$, the expected measurement in the nearness of this parameter location, assuming small perturbations in the state vector, can be approximated by:

$$h(\mathbf{x}_i, \mathbf{x}_r) = \bar{s}_x \left(\bar{x}_i(\mathbf{x}_i, \mathbf{x}_r), t^* - \frac{\bar{s}_y(\bar{y}_i(\mathbf{x}_i, \mathbf{x}_r), t^*)}{\bar{s}'_y(\bar{y}_i(\mathbf{x}_i, \mathbf{x}_r), t^*)} \right) \quad (22)$$

and derivating with respect to the control points positions:

$$\frac{\partial h}{\partial \mathbf{x}_i} = \frac{\partial \bar{s}_x}{\partial \bar{x}_i}\frac{\partial \bar{x}_i}{\partial \mathbf{x}_i} + \bar{s}'_x(t^*)\frac{\frac{\partial \bar{s}'_y}{\partial \bar{y}_i}\frac{\partial \bar{y}_i}{\partial \mathbf{x}_i}\bar{s}_y(t^*) - \frac{\partial \bar{s}_y}{\partial \bar{y}_i}\frac{\partial \bar{y}_i}{\partial \mathbf{x}_i}\bar{s}'_y(t^*)}{\left[\bar{s}'_y(t^*)\right]^2}$$
$$= \frac{\partial \bar{s}_x}{\partial \bar{x}_i}\frac{\partial \bar{x}_i}{\partial \mathbf{x}_i} - \frac{1}{\tan(\eta - \mu)}\frac{\partial \bar{s}_y}{\partial \bar{y}_i}\frac{\partial \bar{y}_i}{\partial \mathbf{x}_i} \quad (23)$$

The partial derivatives in equation 23 can be easily obtained looking at equations 1 and 20. For example from equation 1 we can obtain:

$$\frac{\partial \bar{s}_x}{\partial \bar{x}_i} = \frac{\partial \bar{s}_y}{\partial \bar{y}_i} = \beta_{i,k}(t) \quad (24)$$

So, finally, we can write:

$$\frac{\partial h}{\partial x_i} = \beta_{i,k}(t^*)\left[cos\mu + \frac{sin\mu}{\tan(\eta - \mu)}\right] \quad (25)$$
$$\frac{\partial h}{\partial y_i} = \beta_{i,k}(t^*)\left[sin\mu - \frac{cos\mu}{\tan(\eta - \mu)}\right] \quad (26)$$

Similarly, making use of property 5, and equations 1 and 20, we can obtain:

$$\frac{\partial h}{\partial x_r} = -cos\mu - \frac{sin\mu}{\tan(\eta - \mu)} \quad (27)$$
$$\frac{\partial h}{\partial y_r} = -sin\mu + \frac{cos\mu}{\tan(\eta - \mu)} \quad (28)$$
$$\frac{\partial h}{\partial \phi_r} = \frac{\hat{z}}{\tan(\eta - \mu)} \quad (29)$$

These equations will allow the efficient calculation of the relevant Jacobians in the following sections.

D. Applying the EKF

In this section, results obtained in previous sections are put together and combined in the working frame of the Extended Kalman Filter with the aim of incrementally building a map of an environment modeled with cubic splines.

1) Kalman Filter Prediction: Between the times k and $k+1$ the robot makes a relative movement, given by the vector

$$\mathbf{u}(k+1) \sim N(\hat{\mathbf{u}}(k+1), \mathbf{Q}(k+1)) \quad (30)$$

Under the hypothesis that the only moving object in the map is the robot, the a priori estimation of the state at time $k+1$ is given by:

$$\hat{\mathbf{x}}_r(k+1|k) = \mathbf{f}_r(\hat{\mathbf{x}}_r(k|k), \hat{\mathbf{u}}(k+1)) \quad (31)$$
$$\hat{\mathbf{x}}_{s_i}(k+1|k) = \hat{\mathbf{x}}_{s_i}(k|k) \quad (32)$$

and its covariance:

$$\mathbf{P}(k+1|k) = \mathbf{F}_x(k+1)\mathbf{P}(k|k)\mathbf{F}_x^T(k+1) + $$
$$+\mathbf{F}_u(k+1)\mathbf{Q}(k+1)\mathbf{F}_u^T(k+1) \quad (33)$$

The Jacobian matrices are

$$\mathbf{F}_x(k+1) = \begin{bmatrix} \left.\frac{\partial \mathbf{f}_r}{\partial \mathbf{x}_r}\right|_{\hat{\mathbf{x}}_r(k|k),\hat{\mathbf{u}}(k+1)} & \mathbf{0} & \ldots & \mathbf{0} \\ \mathbf{0} & \mathbf{I}_{n_1} & \ldots & \mathbf{0} \\ \vdots & \vdots & \ddots & \vdots \\ \mathbf{0} & \mathbf{0} & \ldots & \mathbf{I}_{n_N} \end{bmatrix} \quad (34)$$

$$\mathbf{F}_u(k+1) = \begin{bmatrix} \left.\frac{\partial \mathbf{f}_r}{\partial \mathbf{u}}\right|_{\hat{\mathbf{x}}_r(k|k),\hat{\mathbf{u}}(k+1)} \\ \mathbf{0} \\ \vdots \\ \mathbf{0} \end{bmatrix} \quad (35)$$

where $\mathbf{f}_r$ depends on the mobile platform being considered.

2) Kalman Filter Update: Once obtained the expected measurements for each of the laser beams positions of an observation associated with a map spline, the innovation covariance matrix is given by [6]:

$$\mathbf{S}(k+1) = \mathbf{H}_\mathbf{x}(k+1)\mathbf{P}(k+1|k)\mathbf{H}_\mathbf{x}^T(k+1) + \mathbf{R}(k+1) \quad (36)$$

where the Jacobian is:

$$\mathbf{H}_\mathbf{x}(k+1) = \left[\frac{\partial \mathbf{h}}{\partial \mathbf{x}_r}\ \mathbf{0} \ldots \mathbf{0}\ \frac{\partial \mathbf{h}}{\partial \mathbf{x}_{s_i}}\ \mathbf{0} \ldots \mathbf{0}\right] \quad (37)$$

In the previous equation, the term $\frac{\partial \mathbf{h}}{\partial \mathbf{x}_r}$ is calculated making use of equations 27, 28 and 29, and term $\frac{\partial \mathbf{h}}{\partial \mathbf{x}_{s_i}}$ is calculated from 25 and 26. The gain matrix is calculated as follows

$$\mathbf{W}(k+1) = \mathbf{P}(k+1|k)\mathbf{H}_\mathbf{x}^T(k+1)\mathbf{S}^{-1}(k+1) \quad (38)$$

Finally, the state estimation and its covariance are updated according to:

$$\hat{\mathbf{x}}(k+1|k+1) = \hat{\mathbf{x}}(k+1|k) + \mathbf{W}(k+1)\mathbf{h}(k+1) \quad (39)$$

$$\mathbf{P}(k+1|k+1) = [\mathbf{I} - \mathbf{W}(k+1)\mathbf{H}_x(k+1)]\mathbf{P}(k+1|k) \quad (40)$$

E. Extending the Map

The map is incrementally built in two ways: adding new objects, and extending objects already contained in the map.

1) Adding New Objects to the Map: Whenever a new observation is not associated with any of the objects already contained in the map, it is considered as a new object, and the spline which defines its shape is added to the map. Given a map containing N static features, and a set of measurements $z_i, i = p \ldots p+q$ corresponding to a new detected feature, the augmented state vector is:

$$\mathbf{x}^a = \mathbf{g}(\mathbf{x}, \mathbf{z}) \Leftrightarrow \begin{cases} \mathbf{x}_r^a = \mathbf{x}_r \\ \mathbf{x}_{s_i}^a = \mathbf{x}_{s_i} \quad ,i = 1, \ldots, N \\ \mathbf{x}_{s_{N+1}}^a = \mathbf{g}_{s_{N+1}}(\mathbf{x}_r, \mathbf{z}) \end{cases} \quad (41)$$

where $\mathbf{g}_{s_{N+1}}(\mathbf{x}_r, \mathbf{z})$ is defined by the fitting of the q new data points as described in section II-C:

$$\begin{bmatrix} x_{N+1,0} \\ \vdots \\ x_{N+1,n_N} \end{bmatrix} = \mathbf{\Phi} \begin{bmatrix} x_r + z_p \cos(\phi_r + \tau_p) \\ \vdots \\ x_r + z_{p+q} \cos(\phi_r + \tau_{p+q}) \end{bmatrix} \quad (42)$$

$$\begin{bmatrix} y_{N+1,0} \\ \vdots \\ y_{N+1,n_N} \end{bmatrix} = \mathbf{\Phi} \begin{bmatrix} y_r + z_p \sin(\phi_r + \tau_p) \\ \vdots \\ y_r + z_{p+q} \sin(\phi_r + \tau_{p+q}) \end{bmatrix} \quad (43)$$

The new covariance matrix for the augmented state vector is:

$$\mathbf{P}^a = \mathbf{G_x P G_x}^T + \mathbf{G_z R G_z}^T \quad (44)$$

and the Jacobians $\mathbf{G_x} = \frac{\partial \mathbf{g}}{\partial \mathbf{x}}$ and $\mathbf{G_z} = \frac{\partial \mathbf{g}}{\partial \mathbf{z}}$:

$$\mathbf{G_x} = \begin{bmatrix} \mathbf{I}_r & \mathbf{0} & \dots & \mathbf{0} \\ \mathbf{0} & \mathbf{I}_{n_1} & \dots & \mathbf{0} \\ \vdots & \vdots & \ddots & \vdots \\ \mathbf{0} & \mathbf{0} & \dots & \mathbf{I}_{n_N} \\ \hline \frac{\partial \mathbf{g}_{s_{N+1}}}{\partial \mathbf{x}_r} & \mathbf{0} & \dots & \mathbf{0} \end{bmatrix}, \mathbf{G_z} = \begin{bmatrix} \mathbf{0} \\ \mathbf{0} \\ \vdots \\ \mathbf{0} \\ \hline \frac{\partial \mathbf{g}_{s_{N+1}}}{\partial \mathbf{z}} \end{bmatrix} \quad (45)$$

with

$$\frac{\partial \mathbf{g}_{s_{N+1}}}{\partial \mathbf{x}_r} = \begin{bmatrix} \mathbf{\Phi} \begin{bmatrix} 1 & 0 & -z_p \sin\mu_p \\ \vdots & \vdots & \vdots \\ 1 & 0 & -z_{p+q} \sin\mu_{p+q} \end{bmatrix} \\ \hline \mathbf{\Phi} \begin{bmatrix} 0 & 1 & z_p \cos\mu_p \\ \vdots & \vdots & \vdots \\ 0 & 1 & z_{p+q} \sin\mu_{p+q} \end{bmatrix} \end{bmatrix} \quad (46)$$

$$\frac{\partial \mathbf{g}_{s_{N+1}}}{\partial \mathbf{z}} = \begin{bmatrix} \mathbf{\Phi} \begin{bmatrix} \cos\mu_p & \dots & 0 \\ \vdots & \ddots & \vdots \\ 0 & \dots & \cos\mu_{p+q} \end{bmatrix} \\ \hline \mathbf{\Phi} \begin{bmatrix} \sin\mu_p & \dots & 0 \\ \vdots & \ddots & \vdots \\ 0 & \dots & \sin\mu_{p+q} \end{bmatrix} \end{bmatrix} \quad (47)$$

2) Extending Map Objects: Frequently, observations are only partially associated with a map feature (as in figure 3). This means that a new unexplored part of a map object is being detected and, consequently, this spline must be extended. Take for instance the situation displayed in figure 5, where the j-th map spline has been partially associated with an observation, and the information contained in a new set of data points

$$\mathbf{d_i} = \begin{bmatrix} d_i^x \\ d_i^y \end{bmatrix} = \begin{bmatrix} x_r + z_i \cos\mu_i \\ y_r + z_i \sin\mu_i \end{bmatrix}, \quad i = q, \dots, q+m \quad (48)$$

must be integrated into the map feature. The extended state vector will be:

$$\mathbf{x}^e = \mathbf{g_e}(\mathbf{x}, \mathbf{z}) \Leftrightarrow \begin{cases} \mathbf{x}_r^e &= \mathbf{x}_r \\ \mathbf{x}_{s_i}^e &= \mathbf{x}_{s_i}, \quad i \neq j \\ \mathbf{x}_{s_j}^e &= \mathbf{g}_{s_j}^e(\mathbf{x}_r, \mathbf{x}_j, \mathbf{z}) \end{cases} \quad (49)$$

With the objective of calculating $\mathbf{g}_{s_j}^e(\mathbf{x}_r, \mathbf{x}_j, \mathbf{z})$, a similar scheme to the one used during the data approximation is followed.

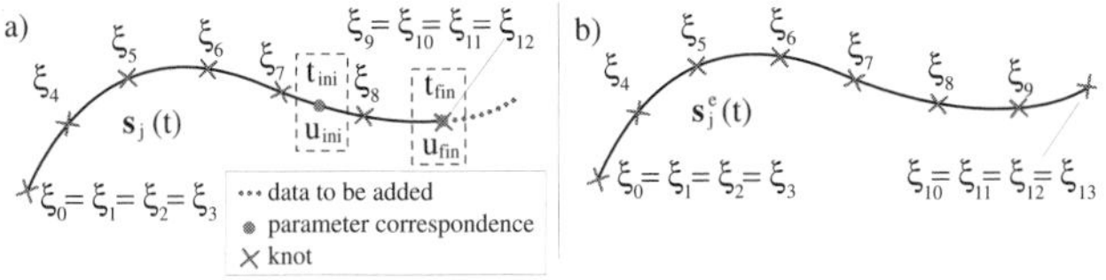

Fig. 5. Map spline extension with new data after Kalman filter update.

In [14] an unclamping algorithm is proposed, which is successfully applied in [10] with the goal of extending a B-spline curve to a single given point. Our problem is to extend a map spline, given a new set of measured data points. The following equations show how, combining the unclamping algorithm with the approximation scheme previously proposed, it is possible to extend the map features as new measurements are obtained. Given a spline curve, defined by a set of control points $\mathbf{x}_i$ and clamped knot vector in the form:

$$\Xi : \underbrace{\xi_0 = \dots = \xi_{k-1}}_{k} \leq \xi_k \leq \dots \leq \xi_n \leq \underbrace{\xi_{n+1} = \dots = \xi_{k+n}}_{k}$$

the unclamping algorithm proposed in [14] calculates the new control points corresponding to the unclamped knot vectors:

$$\Xi_r\!: \underbrace{\xi_0 = \dots = \xi_{k-1}}_{k} \leq \xi_k \leq \dots \leq \xi_n \leq \bar{\xi}_{n+1} \leq \dots \leq \bar{\xi}_{k+n}$$

$$\Xi_l\!: \bar{\xi}_0 \leq \dots \leq \bar{\xi}_{k-1} \leq \xi_k \leq \dots \leq \xi_n \leq \underbrace{\xi_{n+1} = \dots = \xi_{k+n}}_{k}$$

The right-unclamping algorithm (on the side corresponding to higher values for the parameter t) of a spline of order k, converting the knot vector Ξ into Ξ_r, and obtaining the new control points $\mathbf{x}_i^r$ is as follows:

$$\mathbf{x}_i^r = \begin{cases} \mathbf{x}_i, & i = 0, 1, \dots, n-k+1 \\ \mathbf{x}_i^{k-2}, & i = n-k+2, n-k+3, \dots, n \end{cases} \quad (50)$$

where

$$\mathbf{x}_i^j = \begin{cases} \mathbf{x}_i^{j-1}, & i = n-k+2, n-k+3, \dots, n-j \\ \frac{\mathbf{x}_i^{j-1} - (1-\alpha_i^j)\mathbf{x}_{i-1}^j}{\alpha_i^j}, & i = n-j+1, \dots, n \end{cases} \quad (51)$$

being

$$\alpha_i^j = \frac{\bar{\xi}_{n+1} - \bar{\xi}_i}{\bar{\xi}_{i+j+1} - \bar{\xi}_i} \quad (52)$$

and the initial values are set to

$$\mathbf{x}_i^0 = \mathbf{x}_i, \quad i = n-k+2, n-k+3, \dots, n \quad (53)$$

If all the splines are cubic B-splines ($k = 4$), the previous expressions can be reduced to:

$$\left.\begin{aligned} \mathbf{x}_i^r &= \mathbf{x}_i, \quad i = 0, \dots, n-2 \\ \mathbf{x}_{n-1}^r &= -\Gamma_{n-1}^2 \mathbf{x}_{n-2} + \frac{1}{\gamma_{n-1}^2} \mathbf{x}_{n-1} \\ \mathbf{x}_n^r &= \Gamma_n^2 \Gamma_{n-1}^2 \mathbf{x}_{n-2} - \left(\frac{\Gamma_n^2}{\gamma_{n-1}^2} + \frac{\Gamma_n^1}{\gamma_n^2} \right) \mathbf{x}_{n-1} + \frac{1}{\gamma_n^1 \gamma_n^2} \mathbf{x}_n \end{aligned}\right\} \quad (54)$$

being

$$\gamma_i^j = \frac{\bar{\xi}_{n+1} - \bar{\xi}_i}{\bar{\xi}_{i+j+1} - \bar{\xi}_i} \quad \text{and} \quad \Gamma_i^j = \frac{1-\gamma_i^j}{\gamma_i^j} \quad (55)$$

Similar results can be obtained when converting a clamped knot vector Ξ into a left-unclamped one Ξ_l:

$$\left.\begin{array}{l} \mathbf{x}_0^l = \frac{1}{\omega_0^1\omega_0^2}\mathbf{x}_0 - \left(\frac{\Omega_0^2}{\omega_1^2} + \frac{\Omega_0^1}{\omega_0^2}\right)\mathbf{x}_1 + \Omega_0^2\Omega_0^1\mathbf{x}_2 \\ \mathbf{x}_1^l = \frac{1}{\omega_1^2}\mathbf{x}_1 - \Omega_1^2\mathbf{x}_2 \\ \mathbf{x}_i^l = \mathbf{x}_i, \quad i = 2,\dots,n \end{array}\right\} \tag{56}$$

with

$$\omega_i^j = \frac{\bar{\xi}_{k-1}-\bar{\xi}_{i+k}}{\bar{\xi}_{i+k-j-1}-\bar{\xi}_{i+k}} \text{ and } \Omega_i^j = \frac{1-\omega_i^j}{\omega_i^j} \tag{57}$$

These results can be combined with the methodology proposed in section II-C for obtaining new splines as new data is acquired, being aware of the following considerations:

- A parametrization for the new measurements to be integrated in the spline, congruent with the map spline parametrization, can be obtained easily from the data association stage.
- The knot vector needs to be unclamped and might need to be extended with p extra knots in order to make room for the new span being added. New knots number and spacing are chosen according to the specified knot density, and as many new control points as new knots need to be added.

This way, the system of equations 8 is written for the new data points, extended with previous equations 54 and/or 56, and its least-squares solution provides a matrix-form linear relationship between the old control points $\mathbf{x}_j$ and the sampled data $\mathbf{d}_i$, and the new control points $\mathbf{x}_j^e$:

$$\begin{bmatrix} x_{j,0}^e \\ \vdots \\ x_{j,n_j+p}^e \end{bmatrix} = \mathbf{\Phi^e} \begin{bmatrix} d_q^x \\ \vdots \\ d_{q+m}^x \\ \hline x_{j,0} \\ \vdots \\ x_{j,n_j} \end{bmatrix}, \begin{bmatrix} y_{j,0}^e \\ \vdots \\ y_{j,n_j+p}^e \end{bmatrix} = \mathbf{\Phi^e} \begin{bmatrix} d_q^y \\ \vdots \\ d_{q+m}^y \\ \hline y_{j,0} \\ \vdots \\ y_{j,n_j} \end{bmatrix} \tag{58}$$

Note that once chosen an unclamped knot vector Ξ^e for the extended spline, Φ^e can be considered a constant matrix. The new covariance matrix after extending the j-th spline is:

$$\mathbf{P}^e = \mathbf{G}_\mathbf{x}^e\mathbf{P}\mathbf{G}_\mathbf{x}^{e\,T} + \mathbf{G}_\mathbf{z}^e\mathbf{R}\mathbf{G}_\mathbf{z}^{e\,T} \tag{59}$$

where the involved Jacobians $\mathbf{G}_\mathbf{x}^e = \frac{\partial \mathbf{g}^e}{\partial \mathbf{x}}$ and $\mathbf{G}_\mathbf{z}^e = \frac{\partial \mathbf{g}^e}{\partial \mathbf{z}}$ have the following appearance:

$$\mathbf{G}_\mathbf{x}^e = \begin{bmatrix} \mathbf{I}_r & \mathbf{0} & \dots & \mathbf{0} & \dots & \mathbf{0} \\ \mathbf{0} & \mathbf{I}_{n_1} & \dots & \mathbf{0} & \dots & \mathbf{0} \\ \vdots & \vdots & \ddots & \vdots & \ddots & \vdots \\ \frac{\mathbf{g}_{s_j}^e}{\partial \mathbf{x}_r} & \vdots & \ddots & \frac{\partial \mathbf{g}_{s_j}^e}{\partial \mathbf{x}_{s_j}} & \ddots & \vdots \\ \vdots & \vdots & \ddots & \vdots & \ddots & \vdots \\ \mathbf{0} & \mathbf{0} & \dots & \mathbf{0} & \dots & \mathbf{I}_{n_N} \end{bmatrix}, \mathbf{G}_\mathbf{z}^e = \begin{bmatrix} \mathbf{0} \\ \mathbf{0} \\ \vdots \\ \frac{\partial \mathbf{g}_{s_j}^e}{\partial \mathbf{z}} \\ \vdots \\ \mathbf{0} \end{bmatrix} \tag{60}$$

being

$$\frac{\partial \mathbf{g}_{s_j}^e}{\partial \mathbf{x}_r} = \begin{bmatrix} \mathbf{\Phi}^e \begin{bmatrix} 1 & 0 & -z_p \sin\mu_p \\ \vdots & \vdots & \vdots \\ 1 & 0 & -z_{p+q}\sin\mu_{p+q} \\ \hline \mathbf{0} & \mathbf{0} & \mathbf{0} \end{bmatrix} \\ \hline \mathbf{\Phi}^e \begin{bmatrix} 0 & 1 & z_p \cos\mu_p \\ \vdots & \vdots & \vdots \\ 0 & 1 & z_{p+q}\sin\mu_{p+q} \\ \hline \mathbf{0} & \mathbf{0} & \mathbf{0} \end{bmatrix} \end{bmatrix}, \frac{\partial \mathbf{g}_{s_j}^e}{\partial \mathbf{x}_{s_j}} = \begin{bmatrix} \mathbf{\Phi}^e \begin{bmatrix} \mathbf{0} \\ \hline \mathbf{I} \end{bmatrix} \\ \hline \mathbf{\Phi}^e \begin{bmatrix} \mathbf{0} \\ \hline \mathbf{I} \end{bmatrix} \end{bmatrix}$$

and

$$\frac{\partial \mathbf{g}_{s_j}^e}{\partial \mathbf{z}} = \begin{bmatrix} \mathbf{\Phi}^e \begin{bmatrix} \cos\mu_p & \dots & 0 \\ \vdots & \ddots & \vdots \\ 0 & \dots & \cos\mu_{p+q} \\ \hline \mathbf{0} & \dots & \mathbf{0} \end{bmatrix} \\ \hline \mathbf{\Phi}^e \begin{bmatrix} \sin\mu_p & \dots & 0 \\ \vdots & \ddots & \vdots \\ 0 & \dots & \sin\mu_{p+q} \\ \hline \mathbf{0} & \dots & \mathbf{0} \end{bmatrix} \end{bmatrix}$$

IV. Experimental Results

Several experiments have been performed with real data in order to validate the results here presented. Figure 6 shows a map representing an environment with predominant flat features (segments), with 81 splines defined by 332 control points. Figure 7 depicts the map of a bigger and more

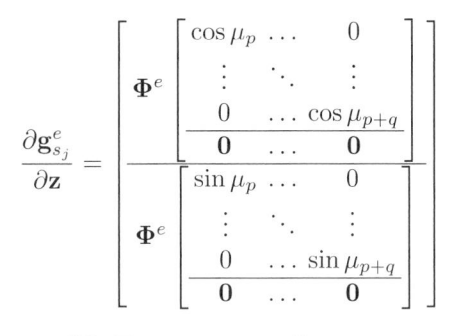

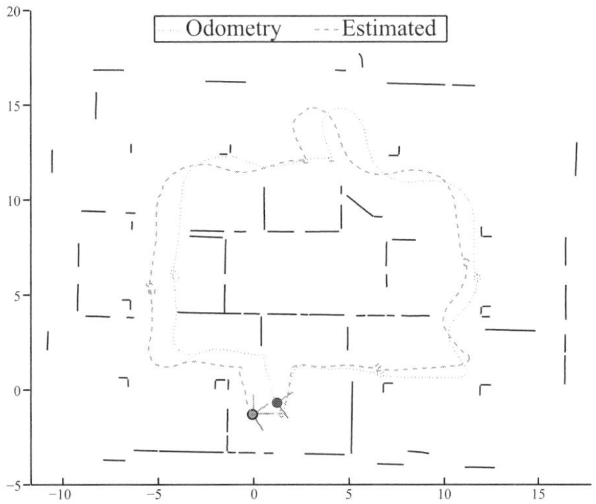

Fig. 6. Map of a career fair held at the School of Industrial Engineering (UPM) in Madrid.

complex environment with a mixture of both straight and curved features. In this case 461 control points defining a total of 96 splines are necessary. In both experiments, a B21r robot with a SICK laser was used for the data acquisition with a sampling frequency of 5 Hz, and density of 2 knots/m was used for knot vectors generation. No geometric constraints have been used in the map building process.

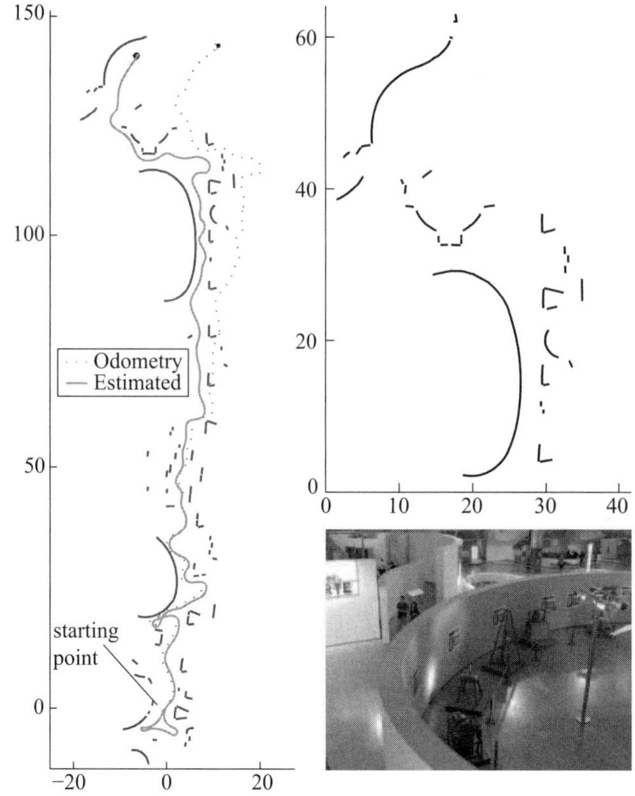

Fig. 7. Map of the Museum of Science "Principe Felipe" (Valencia, Spain) (left). Detail view and photography (right).

V. Conclusion

A new methodology for simultaneous localization and mapping in arbitrary complex environments has been proposed. For the first time, it has been shown how the powerful and computationally efficient mathematical tools that splines are, fit into the EKF-SLAM framework allowing the representation of complex structures in a parametric way. Despite the apparent difficulty that this symbiosis between EKF-SLAM and splines theory could involve, simple and easily programmable matrix-form expressions have been obtained. Specifically, the new ray tracing method here proposed, constitutes a mathematical artifice that not only provides an effective observation model, but also makes easier the obtaining of the Jacobians for the Kalman filter update. It seems clear that, when simple descriptions of the environment are insufficient or unfeasible, any other SLAM algorithm could benefit from the versatility of this new parametric representation.

No information is lost, as control points incrementally encapsulate the uncertainties of laser measurements, thanks to the application of the new matrix-form algorithms here proposed for splines enlargement. Moreover, these concepts and techniques are suitable for a 3D extension. Finally, this new representation constitutes a big step forward, compared to current SLAM techniques based on geometric maps, allowing the mathematical interpretation and reasoning over the maps being built regardless the shape of the contained features.

Acknowledgement

The first author thanks to the "Consejería de Educación de la Comunidad de Madrid" for the PhD scholarship being granted to him, and to the "ACR Centre of Excellence for Autonomous Systems" for having him as visiting postgraduate research student during some months in 2006.

References

[1] J. H. Ahlberg, E. N. Nilson, and J. L. Walsh. *The Theory of Splines and their Applications*. Academic Press, New York, USA, 1967.

[2] A. Atieg and G. A. Watson. A class of methods for fitting a curve or surface to data by minimizing the sum of squares of orthogonal distances. *J. Comput. Appl. Math.*, 158(2):277–296, 2003.

[3] J. A. Castellanos, J. M. M. Montiel, J. Neira, and J. D. Tardós. The SPmap: A probabilistic framework for simultaneous localization and map building. *IEEE Trans. Robot. Automat.*, 15(5):948–952, October 1999.

[4] C. de Boor. *A Practical Guide to Splines*. Springer, 1978.

[5] C. de Boor and J. Rice. Least squares cubic spline approximation I – fixed knots. Technical Report CSD TR 20, Dept. of Computer Sciences, Purdue University, 1968.

[6] M. W. M. G. Dissanayake, P. M. Newman, H. F. Durrant-Whyte, S. Clark, and M. Csorba. A solution to the simultaneous localization and map building (SLAM) problem. *IEEE Trans. Robot. Automat.*, 17(3):229–241, 2001.

[7] R. Eustice, H. Singh, and J. Leonard. Exactly sparse delayed-state filters. In *Proc. of the IEEE International Conference on Robotics and Automation (ICRA)*, pages 2428–2435, Barcelona, Spain, April 2005.

[8] G. Grisetti, C. Stachniss, and W. Burgard. Improving grid-based slam with rao-blackwellized particle filters by adaptive proposals and selective resampling. In *Proc. of the IEEE International Conference on Robotics and Automation (ICRA)*, pages 2443–2448, 2005.

[9] J. Guivant, E. Nebot, and H. Durrant-Whyte. Simultaneous localization and map building using natural features in outdoor environments. *Intelligent Autonomous Systems 6*, 1:581–586, July 2000.

[10] S.-M. Hu, C.-L. Tai, and S.-H. Zhang. An extension algorithm for B-Splines by curve unclamping. *Computer-Aided Design*, 34(5):415–419, April 2002.

[11] J. J. Leonard, H. F. Durrant-Whyte, and I. J. Cox. Dynamic map building for an autonomous mobile robot. *Int. J. Rob. Res.*, 11(4):286–298, 1992.

[12] R. Madhavan, G. Dissanayake, and H. F. Durrant-Whyte. Autonomous underground navigation of an LHD using a combined ICP–EKF approach. In *Proc. of the IEEE International Conference on Robotics and Automation (ICRA)*, volume 4, pages 3703–3708, 1998.

[13] H. Moravec and A. Elfes. High resolution maps from wide angle sonar. In *Proc. of the IEEE International Conference on Robotics and Automation (ICRA)*, volume 2, pages 116–121, March 1985.

[14] L. Piegl and W. Tiller. *The NURBS Book (2nd ed.)*. Springer-Verlag New York, Inc., New York, NY, USA, 1997.

[15] D. Rodriguez-Losada, F. Matia, and R. Galan. Building geometric feature based maps for indoor service robots. *Robotics and Autonomous Systems*, 54(7):546–558, July 2006.

[16] D. F. Rogers. *An introduction to NURBS: With historical perspective*. Morgan Kaufmann Publishers Inc., San Francisco, CA, USA, 2001.

[17] A. J. Sweeney and R. H. Barrels. Ray-tracing free-form B-spline surfaces. *IEEE Computer Graphics & Applications*, 6(2):41–49, February 1986.

[18] M. Veeck and W. Burgard. Learning polyline maps from range scan data acquired with mobile robots. In *Proc. of the IEEE/RSJ International Conference on Intelligent Robots and Systems (IROS)*, volume 2, pages 1065–1070, 2004.

[19] H. Wang, J. Kearney, and K. Atkinson. Robust and efficient computation of the closest point on a spline curve. In *Proc. of the 5th International Conference on Curves and Surfaces*, pages 397–406, San Malo, France, 2002.

[20] S. Zhang, L. Xie, M. Adams, and F. Tang. Geometrical feature extraction using 2d range scanner. In *Proc. of the 4th International Conference on Control and Automation (ICCA'03)*, pages 901–905, 2003.

An Implicit Time-Stepping Method for Multibody Systems with Intermittent Contact

Nilanjan Chakraborty, Stephen Berard, Srinivas Akella, and Jeff Trinkle
Department of Computer Science
Rensselaer Polytechnic Institute
Troy, New York 12180
Email: {chakrn2, sberard, sakella, trink}@cs.rpi.edu

Abstract— **In this paper we present an implicit time-stepping scheme for multibody systems with intermittent contact by incorporating the contact constraints as a set of complementarity and algebraic equations within the dynamics model. Two primary sources of stability and accuracy problems in prior time stepping schemes for differential complementarity models of multibody systems are the use of polyhedral representations of smooth bodies and the approximation of the distance function (arising from the decoupling of collision detection from the solution of the dynamic time-stepping subproblem). Even the simple example of a disc rolling on a table without slip encounters these problems. We assume each object to be a convex object described by an intersection of convex inequalities. We write the contact constraints as complementarity constraints between the contact force and a distance function dependent on the closest points on the objects. The closest points satisfy a set of algebraic constraints obtained from the KKT conditions of the minimum distance problem. These algebraic equations and the complementarity constraints taken together ensure satisfaction of the contact constraints. This enables us to formulate a *geometrically* implicit time-stepping scheme (*i.e.*, we do not need to approximate the distance function) as a nonlinear complementarity problem (NCP). The resulting time-stepper is therefore more accurate; further it is the first geometrically implicit time-stepper that does not rely on a closed form expression for the distance function. We demonstrate through example simulations the fidelity of this approach to analytical solutions and previously described simulation results.**

I. Introduction

To automatically plan and execute tasks involving intermittent contact, one must be able to accurately predict the object motions in such systems. Applications include haptic interactions, collaborative human-robot manipulation, such as rearranging the furniture in a house, as well as industrial automation, such as simulation of parts feeders. Due to the intermittency of contact and the presence of stick-slip frictional behavior, dynamic models of such multibody systems are inherently (mathematically) nonsmooth, and are thus difficult to integrate accurately. In fact, commercially available software systems such as Adams, have a difficult time simulating any system with unilateral contacts. Users expect to spend considerable effort in a trial-and-error search for good simulation parameters to obtain believable, not necessarily accurate, results. Even the seemingly simple problem of a sphere rolling on a horizontal plane under only the influence of gravity is challenging for commercial simulators.

The primary sources of stability and accuracy problems are polyhedral approximations of smooth bodies, the decoupling of collision detection from the solution of the dynamic time-stepping subproblem, and approximations to the quadratic Coulomb friction model. This paper focuses on the development of *geometrically implicit* optimization-based time-stepper for dynamic simulation. More specifically, state-of-the-art time-steppers [16, 15, 9] use geometric information obtained from a collision detection algorithm at the current time, and the state of the system at the end of the time step is computed (by solving a dynamics time step subproblem) without modifying this information. Thus, state-of-the-art time-steppers can be viewed as *explicit* methods with respect to geometric information. We develop the first time-stepping method that is implicit in the geometric information (when the distance function is not available in closed form) by incorporating body geometry in the dynamic time-stepping subproblem. In other words, our formulation solves the collision detection and dynamic stepping problem in the same time-step, which allows us to satisfy contact constraints at the end of the time step. The resulting subproblem at each time-step will be a mixed nonlinear complementarity problem and we call our time-stepping scheme a *geometrically implicit* time-stepping scheme.

To illustrate the effects of geometric approximation, consider the simple planar problem of a uniform disc rolling on a horizontal support surface. For this problem, the exact solution is known, *i.e.,* the disc will roll at constant speed *ad infinitum.* However, when the disc is approximated by a uniform regular polygon, energy is lost a) due to collisions between the vertices and the support surface, b) due to contact sliding that is resisted by friction and c) due to artificial impulses generated by the approximate distance function that is to be satisfied at the end of the time-step. We simulated this example in dVC [3] using the Stewart-Trinkle time-stepping algorithm [16]. The parametric plots in Figure 1 show the reduction of kinetic energy over time caused by the accumulation of these effects. The top plot shows that increasing the number of edges, with the step-size fixed, decreases the energy loss; the energy loss approaches a limit determined by the size of the time-step. The bottom plot shows reducing energy loss with decreasing step size, with the number of vertices fixed at 1000. However, even with the decrease in time-step an energy loss limit is reached.

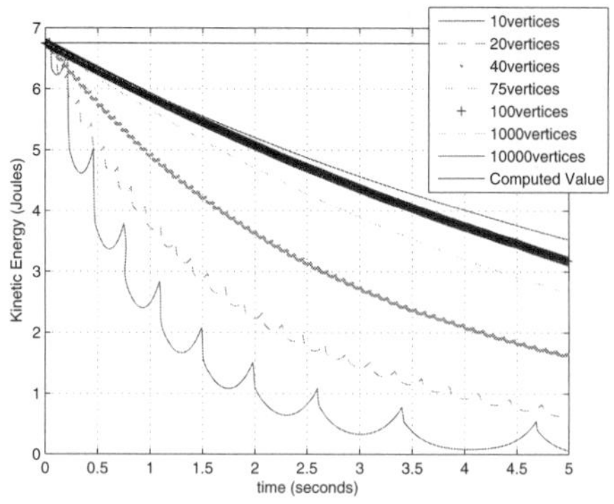

(a) As the number of edges of the "rolling" polygon increases, the energy loss decreases. The computed value obtained by our time-stepper using an implicit surface description of the disc is the horizontal line at the top. The time step used is 0.01 seconds.

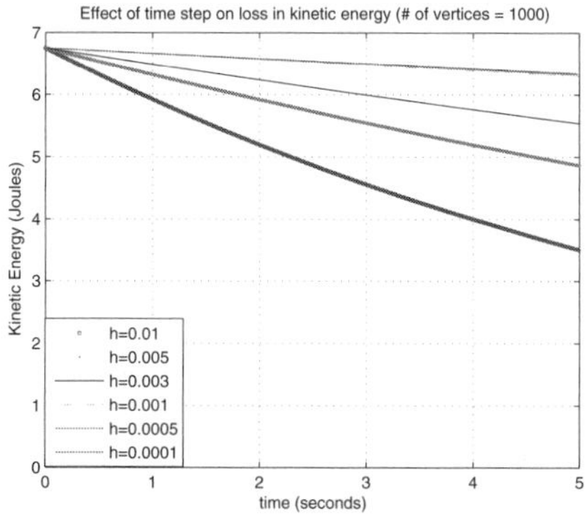

(b) For a given number of edges, the energy loss decreases with decreasing step size, up to a limit. In this case, the limit is approximately 0.001 seconds (the plots for 0.001, 0.0005, and 0.0001 are indistinguishable).

Fig. 1. For a disc rolling on a surface, plots of the reduction of kinetic energy over time caused by approximating the disc as a uniform regular polygon.

These plots make it clear that the discretization of geometry and linearization of the distance function lead to the artifact of loss in energy in the simulations.

To address these issues and related problems that we have encountered (*e.g.*, parts feeding), we present a highly accurate geometrically implicit time-stepping method for convex objects described as an intersection of implicit surfaces. This method also takes into consideration other important nonlinear elements such as quadratic Coulomb friction. This method will provide a baseline for understanding and quantifying the errors incurred when using a geometrically explicit method and when making various linearizing approximations. Our ultimate goal is to develop techniques for automatically selecting the appropriate method for a given application, and to guide method switching, step size adjustment, and model approximations on the fly.

Our paper is organized as follows. In Section II we survey the relevant literature. Section III presents the dynamics model for multi-rigid-body systems with an elliptic dry friction law. In Section IV we develop a new formulation of the contact constraints. In Section V, we give examples that validate and elucidate our time-stepping scheme. Finally we present our conclusions and lay out the future work.

II. Related Work

Dynamics of multibody systems with unilateral contacts can be modeled as differential algebraic equations (DAE) [7] if the contact interactions (sliding, rolling, or separating) at each contact are known. However, in general, the contact interactions are not known *a priori*, but rather must be discovered as part of the solution process. To handle the many possibilities in a rigorous theoretical and computational framework, the model is formulated as a differential complementarity problem (DCP) [4, 17]. The differential complementarity problem is solved using a time-stepping scheme and the resultant system of equations to be solved at each step is a mixed (linear/nonlinear) complementarity problem. Let $\mathbf{u} \in \mathbb{R}^{n_1}$, $\mathbf{v} \in \mathbb{R}^{n_2}$ and let $g : \mathbb{R}^{n_1} \times \mathbb{R}^{n_2} \rightarrow \mathbb{R}^{n_1}$, $f : \mathbb{R}^{n_1} \times \mathbb{R}^{n_2} \rightarrow \mathbb{R}^{n_2}$ be two vector functions and the notation $\mathbf{0} \leq \mathbf{x} \perp \mathbf{y} \geq \mathbf{0}$ imply that $\mathbf{x}$ is orthogonal to $\mathbf{y}$ and each component of the vectors are non-negative.

Definition 1: The differential (or dynamic) complementarity problem is to find $\mathbf{u}$ and $\mathbf{v}$ satisfying

$$\dot{\mathbf{u}} = g(\mathbf{u}, \mathbf{v}), \qquad \mathbf{u}, \text{free}$$
$$0 \leq \mathbf{v} \ \perp \ f(\mathbf{u}, \mathbf{v}) \geq 0$$

Definition 2: The mixed complementarity problem is to find $\mathbf{u}$ and $\mathbf{v}$ satisfying

$$g(\mathbf{u}, \mathbf{v}) = 0, \qquad \mathbf{u}, \text{free}$$
$$0 \leq \mathbf{v} \ \perp \ f(\mathbf{u}, \mathbf{v}) \geq 0$$

The three primary modeling approaches for multibody systems with unilateral contacts are based on three different assumptions about the flexibility of the bodies. The assumptions from most to least realistic (and most to least computationally complex) are: 1) the bodies are fully deformable, 2) the bodies have rigid cores surrounded by compliant material, 3) the bodies are fully rigid. The first assumption leads to finite element approaches, for which one must solve very large difficult complementarity problems or variational inequalities at each time step. The second assumption leads to smaller subproblems that can be solved more easily [15, 13], but suitable values of the parameters of the compliant layer are difficult to determine. The assumption of rigid bodies leads to the smallest subproblems and avoids the latter problem of determining material deformation properties.

Independent of which underlying assumption is made, the methods developed to date have one problem in common that fundamentally limits their accuracy – they are not implicit with respect to the relevant geometric information. For example, at the current state, a collision detection routine is called to determine separation or penetration distances between the bodies, but this information is not incorporated as a function of the unknown future state at the end of the current time step. A goal of a typical time-stepping scheme is to guarantee consistency of the dynamic equations and all model constraints at the end of each time step. However, since the geometric information at the end of the current time step is approximated from that at the start of the time step, the solution will be in error.

Early time-steppers used linear approximations of the local geometry at the current time [16, 1]. Thus each contact was treated as a point on a plane or a line on a (non-parallel) line and these entities were assumed constant for the duration of the time step. Besides being insufficiently accurate in some applications, some unanticipated side-effects arose [5].

Increased accuracy can be obtained in explicit schemes by including curvature information. This was done by Liu and Wang [9] and Pfeiffer and Glocker [14] by incorporating kinematic differential equations of rolling contact (Montana [12]). Outside the complementarity formalism, Kry and Pai [8] and Baraff [2] also make use of the contact kinematics equations in dynamic simulations of parametric and implicit surface objects respectively.

The method of Tzitzouris [19] is the only geometrically implicit method developed to date, but unfortunately it requires that the distance function between the bodies and two levels of derivatives be available in closed form. However, it is possible that this method could run successfully after replacing the closed-form distance functions with calls to collision detection algorithms and replacing derivatives with difference approximations from multiple collision detection calls, but polyhedral approximations common to most collision detection packages would generate very noisy derivatives. To our knowledge, such an implementation has not been attempted. One other problem with Tzitzouris' method is that it adapts its step size to precisely locate every collision time. While this is a good way to avoid interpenetration at the end of a time step, it has the undesirable side-effect of forcing the step size to be unreasonably small when there are many interacting bodies [10]. The method we propose does not suffer from this problem.

III. Dynamic Model for Multibody Systems

In complementarity methods, the instantaneous equations of motion of a rigid multi-body system consist of five parts: (a) Newton-Euler equations, (b) Kinematic map relating the generalized velocities to the linear and angular velocities, (c) Equality constraints to model joints, (d) Normal contact condition to model intermittent contact, and (e) Friction law. Parts (a) and (b) form a system of ordinary differential equations, (c) is a system of (nonlinear) algebraic equations, (d) is a system of complementarity constraints, and (e) can be written as a system of complementarity constraints for any friction law that obeys the maximum work dissipation principle. In this paper we use an elliptic dry friction law [18]. Thus, the dynamic model is a *differential complementarity* problem. To solve this system of equations, we have to set up a discrete time-stepping scheme and solve a complementarity problem at each time step. We present below the continuous formulation as well as a Euler time-stepping scheme for discretizing the system. To simplify the exposition, we ignore the presence of joints or bilateral constraints in the following discussion. However, all of the discussion below holds in the presence of bilateral constraints.

To describe the dynamic model mathematically, we first introduce some notation. Let $\mathbf{q}_j$ be the position and orientation of body j in an inertial frame and $\boldsymbol{\nu}_j$ be the concatenated vector of linear velocities $\mathbf{v}$ and angular velocities $\boldsymbol{\omega}$. The generalized coordinates, $\mathbf{q}$, and generalized velocity, $\boldsymbol{\nu}$ of the whole system are formed by concatenating $\mathbf{q}_j$ and $\boldsymbol{\nu}_j$ respectively. Let λ_{in} be the normal contact force at the ith contact and $\boldsymbol{\lambda}_{\mathrm{n}}$ be the concatenated vector of the normal contact forces. Let λ_{it} and λ_{io} be the orthogonal components of the friction force on the tangential plane at the ith contact and $\boldsymbol{\lambda}_{\mathrm{t}}$, $\boldsymbol{\lambda}_{\mathrm{o}}$ be the respective concatenated vectors. Let λ_{ir} be the frictional moment about the ith contact normal and $\boldsymbol{\lambda}_{\mathrm{r}}$ be the concatenated vector of the frictional moments. The instantaneous dynamic model can then be written as follows.

Newton-Euler equations of motion:

$$\mathbf{M}(\mathbf{q})\dot{\boldsymbol{\nu}} = \mathbf{W}_{\mathrm{n}}\boldsymbol{\lambda}_{\mathrm{n}}+\mathbf{W}_{\mathrm{t}}\boldsymbol{\lambda}_{\mathrm{t}}+\mathbf{W}_{\mathrm{o}}\boldsymbol{\lambda}_{\mathrm{o}}+\mathbf{W}_{\mathrm{r}}\boldsymbol{\lambda}_{\mathrm{r}}+\boldsymbol{\lambda}_{\mathrm{app}}+\boldsymbol{\lambda}_{\mathrm{vp}} \quad (1)$$

where $\mathbf{M}(\mathbf{q})$ is the inertia tensor, $\boldsymbol{\lambda}_{\mathrm{app}}$ is the vector of external forces, $\boldsymbol{\lambda}_{\mathrm{vp}}$ is the vector of Coriolis and centripetal forces, $\mathbf{W}_{\mathrm{n}}$, $\mathbf{W}_{\mathrm{t}}$, $\mathbf{W}_{\mathrm{o}}$, and $\mathbf{W}_{\mathrm{r}}$ are dependent on $\mathbf{q}$ and map the normal contact forces, frictional contact forces, and frictional moments to the body reference frame.

Kinematic Map:

$$\dot{\mathbf{q}} = \mathbf{G}(\mathbf{q})\boldsymbol{\nu} \quad (2)$$

where $\mathbf{G}$ is the matrix mapping the generalized velocity of the body to the time derivative of the position and orientation. The Jacobian $\mathbf{G}$ may be a non-square matrix (*e.g.*, using a unit quaternion to represent orientation) but $\mathbf{G}^T\mathbf{G} = \mathbf{I}$.

Nonpenetration Constraints: The normal contact constraint for the ith contact is

$$0 \leq \lambda_{in} \perp \psi_{in}(\mathbf{q}, t) \geq 0 \quad (3)$$

where ψ_{in} is a signed distance function or *gap function* for the ith contact with the property $\psi_{in}(\mathbf{q}, t) > 0$ for separation, $\psi_{in}(\mathbf{q}, t) = 0$ for touching, and $\psi_{in}(\mathbf{q}, t) < 0$ for interpenetration. The above gap function is defined in the configuration space of the system. Note that there is usually no closed form expression for $\psi_{in}(\mathbf{q}, t)$.

Friction Model:

$$(\lambda_{it},\ \lambda_{io},\ \lambda_{ir}) \in \text{argmax}\{-(v_{it}\lambda'_{it} + v_{io}\lambda'_{io} + v_{ir}\lambda'_{ir}) : (\lambda'_{it},\ \lambda'_{io}, \lambda'_{ir}) \in \mathbf{F}_i(\lambda_{in}, \mu_i)\}$$

where $\mathbf{v}_i$ is the relative velocity at contact i and the friction ellipsoid is defined by $\mathbf{F}_i(\lambda_{in}, \mu_i) = \{(\lambda_{it},\ \lambda_{io},\ \lambda_{ir}) : \left(\frac{\lambda_{it}}{e_{it}}\right)^2 + \left(\frac{\lambda_{io}}{e_{io}}\right)^2 + \left(\frac{\lambda_{ir}}{e_{ir}}\right)^2 \leq \mu_i^2\lambda_{in}^2\}$ where e_{it}, e_{io} and e_{ir} are given positive constants defining the friction ellipsoid and μ_i is the coefficient of friction at the ith contact.

We use a velocity-level formulation and an Euler time-stepping scheme to discretize the above system of equations. Let t_ℓ denote the current time, and h be the time step.

Use the superscripts ℓ and $\ell+1$ to denote quantities at the beginning and end of the ℓth time step respectively. Using $\dot{\nu} \approx (\boldsymbol{\nu}^{\ell+1} - \boldsymbol{\nu}^{\ell})/h$ and $\dot{\mathbf{q}} \approx (\mathbf{q}^{\ell+1} - \mathbf{q}^{\ell})/h$, and writing in terms of the impulses we get the following discrete time system.

$$\begin{aligned}
\mathbf{M}\boldsymbol{\nu}^{\ell+1} &= \mathbf{M}\boldsymbol{\nu}^{\ell} + h(\mathbf{W}_{\rm n}\boldsymbol{\lambda}_{\rm n}^{\ell+1} + \mathbf{W}_{\rm t}\boldsymbol{\lambda}_{\rm t}^{\ell+1} + \mathbf{W}_{\rm o}\boldsymbol{\lambda}_{\rm o}^{\ell+1} \\
&\quad + \mathbf{W}_{\rm r}\boldsymbol{\lambda}_{\rm r}^{\ell+1} + \boldsymbol{\lambda}_{\rm app}^{\ell} + \boldsymbol{\lambda}_{\rm vp}) \\
\mathbf{q}^{\ell+1} &= \mathbf{q}^{\ell} + h\mathbf{G}\boldsymbol{\nu}^{\ell+1} \\
0 \le h\boldsymbol{\lambda}_{\rm n}^{\ell+1} &\perp \psi_{\rm n}(\mathbf{q}^{\ell+1}) \ge 0 \\
h(\lambda_{it}^{\ell+1}, \lambda_{io}^{\ell+1}, \lambda_{ir}^{\ell+1}) &\in \operatorname{argmax}\{-((v_{it}^{\ell+1})\lambda'_{it} + (v_{io}^{\ell+1})\lambda'_{io} \\
&\quad + (v_{ir}^{\ell+1})\lambda'_{ir}) \\
&\quad : h(\lambda_{it}^{\ell+1}, \lambda_{io}^{\ell+1}, \lambda_{ir}^{\ell+1}) \in \mathbf{F}_i(h\lambda_{in}, \mu_i)\}
\end{aligned} \tag{4}$$

The argmax formulation of the friction law has a useful alternative formulation obtained from the Fritz-John optimality conditions [18]:

$$\begin{aligned}
\mathbf{E}_{\rm t}^2\mathbf{U}\mathbf{p}_{\rm n} \circ \mathbf{W}_{\rm t}^T\boldsymbol{\nu}^{\ell+1} + \mathbf{p}_{\rm t} \circ \boldsymbol{\sigma} &= 0 \\
\mathbf{E}_{\rm o}^2\mathbf{U}\mathbf{p}_{\rm n} \circ \mathbf{W}_{\rm o}^T\boldsymbol{\nu}^{\ell+1} + \mathbf{p}_{\rm o} \circ \boldsymbol{\sigma} &= 0 \\
\mathbf{E}_{\rm r}^2\mathbf{U}\mathbf{p}_{\rm n} \circ \mathbf{W}_{\rm r}^T\boldsymbol{\nu}^{\ell+1} + \mathbf{p}_{\rm r} \circ \boldsymbol{\sigma} &= 0 \\
(\mathbf{U}\mathbf{p}_{\rm n}) \circ (\mathbf{U}\mathbf{p}_{\rm n}) - (\mathbf{E}_{\rm t}^2)^{-1}(\mathbf{p}_{\rm t} \circ \mathbf{p}_{\rm t}) - (\mathbf{E}_{\rm o}^2)^{-1}(\mathbf{p}_{\rm o} \circ \mathbf{p}_{\rm o}) & \\
- (\mathbf{E}_{\rm r}^2)^{-1}(\mathbf{p}_{\rm r} \circ \mathbf{p}_{\rm r}) &\ge 0
\end{aligned} \tag{5}$$

where the impulse $\mathbf{p}_{(\cdot)} = h\boldsymbol{\lambda}_{(\cdot)}$, the matrices $\mathbf{E}_{\rm t}$, $\mathbf{E}_{\rm o}$, $\mathbf{E}_{\rm r}$, and $\mathbf{U}$ are diagonal with ith diagonal element equal to e_i, e_o, e_r, and μ_i respectively, $\boldsymbol{\sigma}$ is a concatenated vector of the Lagrange multipliers arising from the conversion from the argmax formulation and σ_i is equal to the magnitude of the slip velocity at contact i, and $\circ$ connotes the Hadamard product.

The above subproblem at a time step is either an LCP or an NCP depending on the time evaluation of $\mathbf{W}_{(\cdot)}$ the approximation used for $\psi_{\rm n}(\mathbf{q}^{\ell+1})$, and the representation of the friction model. If $\mathbf{W}_{(\cdot)}$ are evaluated at ℓ, and we use a first order Taylor series expansion for $\psi_{\rm n}(\mathbf{q})$ and a linearized representation of the friction ellipsoid, we have an LCP. However, the approximations involved introduce numerical artifacts as discussed in Section I. Moreover, the linear approximation of the friction ellipsoid also leads to certain artifacts. In contrast, if we evaluate $\mathbf{W}_{(\cdot)}$ at $\ell+1$, use a quadratic friction law (Equation (5)), and use $\psi_{\rm n}(\mathbf{q}^{\ell+1})$, we have an NCP. We call this formulation a *geometrically implicit* formulation because it ensures that the contact conditions are satisfied at the end of the time step. However, evaluating $\psi_{\rm n}(\mathbf{q}^{\ell+1})$ is possible only if we have a closed form expression for the distance function, which we do not have in general. Instead, we propose to define the gap function in terms of the closest points between the two objects and provide a set of algebraic equations for finding these closest points during the time step. The next section discusses this approach in detail and proves that the conditions will enforce satisfaction of contact constraints at the end of the time step.

IV. Contact Constraint

In this section we rewrite the contact condition (Equation 3) as a complementarity condition in the work space, combine it with an optimization problem to find the closest points, and prove that the resultant system of equations ensures that the contact constraints are satisfied. Let us consider the ith contact. For ease of exposition, we assume here that each object is a convex object described by a single implicit surface. A more general formulation where each object is described by an intersection of implicit surfaces is given in Appendix A. Let the two objects be defined by convex functions $f(\xi_1) \le 0$ and $g(\xi_2) \le 0$ respectively, where ξ_1 and ξ_2 are the coordinates of points in the two objects. Let $\mathbf{a}_1$ and $\mathbf{a}_2$ be the closest points on the two objects. The equation of an implicit surface has the property that for any point $\mathbf{x}$, the point lies inside the object for $f(\mathbf{x}) < 0$, on the object surface for $f(\mathbf{x}) = 0$, and outside the object for $f(\mathbf{x}) > 0$. Thus, we can define the gap function in work space as either $f(\mathbf{a}_2)$ or $g(\mathbf{a}_1)$ and write the complementarity conditions as either one of the following two conditions:

$$\begin{aligned}
0 \le \lambda_{in} \perp f(\mathbf{a}_2) \ge 0 \\
0 \le \lambda_{in} \perp g(\mathbf{a}_1) \ge 0
\end{aligned} \tag{6}$$

where $\mathbf{a}_1$ and $\mathbf{a}_2$ are given by

$$\operatorname{argmin} \ \{\|\xi_1 - \xi_2\|^2 : f(\xi_1) \le 0, \ \ g(\xi_2) \le 0\} \tag{7}$$

It can be shown easily from the Karush-Kuhn-Tucker (KKT) conditions of Equation 7 that $\mathbf{a}_1$ and $\mathbf{a}_2$ are the solutions of the following system of algebraic equations.

$$\begin{aligned}
\mathbf{a}_1 - \mathbf{a}_2 = -l_1\nabla f(\mathbf{a}_1) &= l_2\nabla g(\mathbf{a}_2) \\
f(\mathbf{a}_1) = 0, \qquad g(\mathbf{a}_2) &= 0
\end{aligned} \tag{8}$$

where l_1 and l_2 are the Lagrange multipliers. The geometric meaning of the first two equations is that the normals to the two surfaces at their closest points are aligned with the line joining the closest points. The solution to Equation 8 gives the closest point when the two objects are separate. However, when $\mathbf{a}_1 = \mathbf{a}_2$, the solution is either the touching point of the two surfaces or a point lying on the intersection curve of the two surfaces 2. Thus, as written, Equations 8 and 6 do not guarantee non-penetration. However, note that the distinction

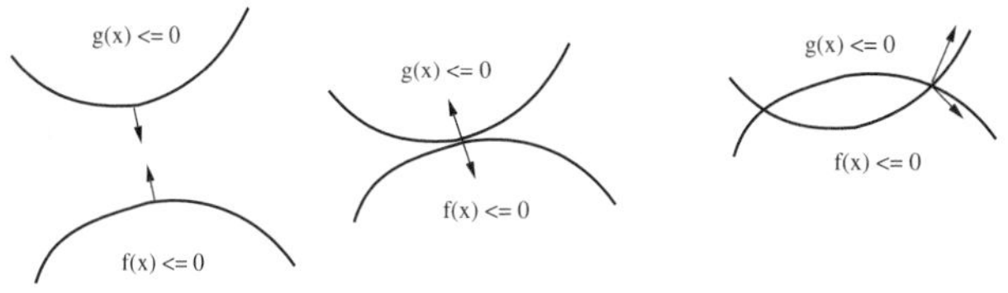

Fig. 2. Three Contact cases: (left) Objects are separate (middle) Objects are touching (right) Objects are intersecting.

between touching points and intersecting points is that the normals to the two surfaces at the touching points are aligned while it is not so for intersection points. When $\mathbf{a}_1 = \mathbf{a}_2$, we lose the geometric information that the normals at the two points are aligned if we write our equations in the form above.

Rewriting the above equations in terms of the unit vectors allows us to avoid this problem.

$$\begin{aligned} \mathbf{a}_1 - \mathbf{a}_2 &= -\|\mathbf{a}_1 - \mathbf{a}_2\| \frac{\nabla f(\mathbf{a}_1)}{\|\nabla f(\mathbf{a}_1)\|} \\ \frac{\nabla f(\mathbf{a}_1)}{\|\nabla f(\mathbf{a}_1)\|} &= -\frac{\nabla g(\mathbf{a}_2)}{\|\nabla g(\mathbf{a}_2)\|} \\ f(\mathbf{a}_1) = 0, &\qquad g(\mathbf{a}_2) = 0 \end{aligned} \tag{9}$$

Proposition: *Equation 6 and 9 together represent the contact constraints, i.e., the two objects will satisfy the contact constraints at the end of each time step if and only if Equation 6 and 9 hold together.*

Proof: As discussed above.

Note that since the first two vector equations are equating unit vectors, there are only two independent equations for each, and the above system has 6 independent equations in 6 variables. We can now formulate the mixed NCP for the geometrically-implicit time-stepper. The vector of unknowns $\mathbf{z}$ can be partitioned into $\mathbf{z} = [\mathbf{u},\ \mathbf{v}]$ where $\mathbf{u} = [\boldsymbol{\nu},\ \mathbf{a}_1,\ \mathbf{a}_2,\ \mathbf{p}_t,\ \mathbf{p}_o,\ \mathbf{p}_r]$ and $\mathbf{v} = [\mathbf{p}_n,\ \boldsymbol{\sigma}]$. The equality constraints in the mixed NCP are:

$$\begin{aligned} 0 &= -\mathbf{M}\boldsymbol{\nu}^{\ell+1} + \mathbf{M}\boldsymbol{\nu}^{\ell} + \mathbf{W}_{\mathrm{n}}^{\ell+1}\mathbf{p}_{\mathrm{n}}^{\ell+1} + \mathbf{W}_{\mathrm{t}}^{\ell+1}\mathbf{p}_{\mathrm{t}}^{\ell+1} \\ &\quad + \mathbf{W}_{\mathrm{o}}^{\ell+1}\mathbf{p}_{\mathrm{o}}^{\ell+1} + \mathbf{W}_{\mathrm{r}}^{\ell+1}\mathbf{p}_{\mathrm{r}}^{\ell+1} + \mathbf{p}_{\mathrm{app}}^{\ell} + \mathbf{p}_{\mathrm{vp}}^{\ell} \\ 0 &= (\mathbf{a}_1^{\ell+1} - \mathbf{a}_2^{\ell+1}) + \|\mathbf{a}_1^{\ell+1} - \mathbf{a}_2^{\ell+1}\| \frac{\nabla f(\mathbf{a}_1^{\ell+1})}{\|\nabla f(\mathbf{a}_1^{\ell+1})\|} \\ 0 &= \frac{\nabla f(\mathbf{a}_1^{\ell+1})}{\|\nabla f(\mathbf{a}_1^{\ell+1})\|} + \frac{\nabla g(\mathbf{a}_2^{\ell+1})}{\|\nabla g(\mathbf{a}_2^{\ell+1})\|} \\ 0 &= f(\mathbf{a}_1^{\ell+1}) \\ 0 &= g(\mathbf{a}_2^{\ell+1}) \\ 0 &= \mathbf{E}_{\mathrm{t}}^2\mathbf{U}\mathbf{p}_n^{\ell+1} \circ (\mathbf{W}_{\mathrm{t}}^T)^{\ell+1}\boldsymbol{\nu}^{\ell+1} + \mathbf{p}_{\mathrm{t}}^{\ell+1} \circ \boldsymbol{\sigma}^{\ell+1} \\ 0 &= \mathbf{E}_{\mathrm{o}}^2\mathbf{U}\mathbf{p}_n^{\ell+1} \circ (\mathbf{W}_{\mathrm{o}}^T)^{\ell+1}\boldsymbol{\nu}^{\ell+1} + \mathbf{p}_{\mathrm{o}}^{\ell+1} \circ \boldsymbol{\sigma}^{\ell+1} \\ 0 &= \mathbf{E}_{\mathrm{r}}^2\mathbf{U}\mathbf{p}_n^{\ell+1} \circ (\mathbf{W}_{\mathrm{r}}^T)^{\ell+1}\boldsymbol{\nu}^{\ell+1} + \mathbf{p}_{\mathrm{r}}^{\ell+1} \circ \boldsymbol{\sigma}^{\ell+1} \end{aligned} \tag{10}$$

The complementarity constraints on $\mathbf{v}$ are:

$$0 \le \begin{bmatrix} \mathbf{p}_{\mathrm{n}}^{\ell+1} \\ \boldsymbol{\sigma}^{\ell+1} \end{bmatrix} \perp \begin{bmatrix} f(\mathbf{a}_2^{\ell+1}) \\ \zeta \end{bmatrix} \ge 0 \tag{11}$$

where

$$\begin{aligned} \zeta = \mathbf{U}\mathbf{p}_{\mathrm{n}}^{\ell+1} \circ \mathbf{U}\mathbf{p}_{\mathrm{n}}^{\ell+1} &- \left(\mathbf{E}_{\mathrm{t}}^2\right)^{-1}\left(\mathbf{p}_{\mathrm{t}}^{\ell+1} \circ \mathbf{p}_{\mathrm{t}}^{\ell+1}\right) \\ - \left(\mathbf{E}_{\mathrm{o}}^2\right)^{-1}\left(\mathbf{p}_{\mathrm{o}}^{\ell+1} \circ \mathbf{p}_{\mathrm{o}}^{\ell+1}\right) &- \left(\mathbf{E}_{\mathrm{r}}^2\right)^{-1}\left(\mathbf{p}_{\mathrm{r}}^{\ell+1} \circ \mathbf{p}_{\mathrm{r}}^{\ell+1}\right) \end{aligned}$$

In the above formulation, we see $\mathbf{u} \in \mathbb{R}^{6n_b+9n_c}$, $\mathbf{v} \in \mathbb{R}^{2n_c}$, the vector function of equality constraints maps $[\mathbf{u}, \mathbf{v}]$ to $\mathbb{R}^{6n_b+9n_c}$ and the vector function of complementarity constraints maps $[\mathbf{u}, \mathbf{v}]$ to $\mathbb{R}^{2n_c}$ where n_b and n_c are the number of bodies and number of contacts respectively. If using convex bodies only, the number of contacts can be determined directly from the number of bodies, $n_c = \sum_{i=1}^{n_b-1} i$.

V. Illustrative Examples

In this section we present three examples to validate our technique against known analytical results and previous approaches. The first example is the same example of a disc rolling without slip on a plane that we studied in Section I. The second example, taken from [18], consists of a sphere spinning on a horizontal surface that comes to rest due to torsional friction. The time taken by the sphere to come to a complete stop is known analytically and we demonstrate that the results of our simulation agree with the analytical predictions. The final example consists of a small ball moving in contact with two larger fixed balls. We include it here to compare our solutions with those based on earlier approaches [18, 9]. All of our numerical results were obtained by PATH [6], a free solver that is one of the most robust complementarity problem solvers available.

A. *Example 1: Disc on a Plane*

In this example we revisit the unit disc example from Section I. For illustrative purposes, we explain the formulation of the full dynamic model in detail. The normal axis of the contact frame $\hat{n}$ always points in the inertial y-axis direction and tangential axis $\hat{t}$ always coincides with the x-direction. The mass matrix, $\mathbf{M}$ is constant and the only force acting on the body is due to gravity. The equation of the disc is given by $f_1(x, y) = (x - q_x)^2 + (y - q_y)^2 - 1$, where $\mathbf{q}$ is the location of the center of the disc in the inertial frame. Let $\mathbf{a}_1$ be the closest point on body 1 (the disc) to the x-axis. Similarly, let $\mathbf{a}_2$ be the closest point on the x-axis to body 1 ($a_{2_y} = 0$ and can be removed from the system of unknowns). Given this information, the matrices for this system can be shown to be: $\mathbf{M} = \mathrm{diag}(m, m, 0.5m)$ $\mathbf{p}_{\mathrm{app}} = [0, -9.81 \cdot m \cdot h, 0]^T$.

$$\mathbf{W_n} = \begin{bmatrix} \hat{n} \\ \mathbf{r}^{a_1} \otimes \hat{n} \end{bmatrix} \quad \mathbf{W_t} = \begin{bmatrix} \hat{t} \\ \mathbf{r}^{a_1} \otimes \hat{t} \end{bmatrix}$$

$$\nabla_{a_1} f_1(a_1^{\ell+1}) = \begin{bmatrix} 2(a_{1x}^{\ell+1} - q_x) \\ 2(a_{1y}^{\ell+1} - q_y) \end{bmatrix}$$

where $\mathbf{r}^{a_1}$ is the vector from the center of gravity of the disc to $\mathbf{a}_1$ and $\otimes$ connotes the 2D analog of the cross product.

There are 9 unknowns for this system: $\mathbf{z} = [\boldsymbol{\nu},\ \mathbf{a}_1,\ a_{2_x},\ p_{\mathrm{n}},\ p_{\mathrm{t}},\ \sigma]$ We can now formulate the entire system of equations for this simple model:

$$0 = -\mathbf{M}\boldsymbol{\nu}^{\ell+1} + \mathbf{M}\boldsymbol{\nu}^{\ell} + \mathbf{W}_{\mathrm{n}}^{\ell+1}p_{\mathrm{n}}^{\ell+1} + \mathbf{W}_{\mathrm{t}}^{\ell+1}p_{\mathrm{t}}^{\ell+1} + \mathbf{p}_{\mathrm{app}} \tag{12}$$

$$0 = \mathbf{a}_2^{\ell+1} - \mathbf{a}_1^{\ell+1} + \|\mathbf{a}_2^{\ell+1} - \mathbf{a}_1^{\ell+1}\|\hat{n} \tag{13}$$

$$0 = \frac{\nabla_{\mathbf{a}_1} f_1(\mathbf{a}_1^{\ell+1})}{\|\nabla_{\mathbf{a}_1} f_1(\mathbf{a}_1^{\ell+1})\|} + \hat{n} \tag{14}$$

$$0 = f_1(\mathbf{a}_1^{\ell+1}) \tag{15}$$

$$0 \le f_1(\mathbf{a}_2^{\ell+1}) \tag{16}$$

$$0 = \mu p_{\mathrm{n}}^{\ell+1}(\mathbf{W}_{\mathrm{t}}^{\mathbf{T}^{\ell+1}}\boldsymbol{\nu}^{\ell+1}) + \sigma^{\ell+1}p_{\mathrm{t}}^{\ell+1} \tag{17}$$

$$0 \le \mu^2 p_{\mathrm{n}}^{\ell+1}p_{\mathrm{n}}^{\ell+1} - p_{\mathrm{t}}^{\ell+1}p_{\mathrm{t}}^{\ell+1} \tag{18}$$

where equations 13 and 14 each provide one independent equation.

The initial configuration of the disc is $\mathbf{q} = [0, 1, 0]$, initial velocity is $\boldsymbol{\nu} = [-3, 0, 3]$, mass is $m = 1$, and $\mu = 0.4$.

Figure 1(a) shows the kinetic energy of the disc for our implicit representation along with the Stewart-Trinkle LCP implementation using various levels of discretization as it rolls along the horizontal surface. When using an implicit curve representation to model the disc and our formulation we get no energy loss (within the numerical tolerance of 10^{-6} used for our simulations) as seen by the horizontal line. When using the LCP formulation we have energy loss as discussed earlier.

B. Example 2: Sphere on a Plane

Here we consider a sphere spinning on a plane about the normal of the plane. The initial configuration of the sphere is $\mathbf{q} = [0,\ 0,\ 1,\ 1,\ 0,\ 0,\ 0]^T$ where the first three elements are the position of the center of the sphere and the last 4 are the unit quaternion representing the orientation. The initial generalized velocity is $\boldsymbol{\nu} = [0,\ 0,\ 0,\ 0,\ 0,\ 1.962]^T$, the first three elements are the linear terms, the last three elements the angular. The friction parameters used were $e_t = 1$, $e_o = 1$, $e_r = 0.4$ and $\mu = 0.2$. A step size of $h = 0.07$ seconds was chosen.

From the initial conditions given, the sphere should rotate in place with a constant deceleration due to the torsional friction. Figure 3 shows a plot of the velocities for the sphere given by the time-stepping formulation. The analytical solution for this problem predicts that all velocities except ω_z should be zero, and w_z should be decreasing linearly to 0 with a slope of -1.962, reaching 0 at exactly $t = 1$ seconds. The plot shows that we agree with the analytical solution except that we reach zero velocity at $t = 1.05$, since we are using a fixed time step and the time at which spinning stops is in the middle of the time step. The friction forces (Figure 4) also follow the analytical solution. The tangential component of friction force is 0. The tangential moment does not drop to 0 at 1.05 s, since we are using an impulse-based time-stepping formulation with a fixed time step and there is a torsional moment between 0.98 to 1 second which contributes to the impulse. Our results match those of the Tzitzouris-Pang and Stewart methods presented in [18].

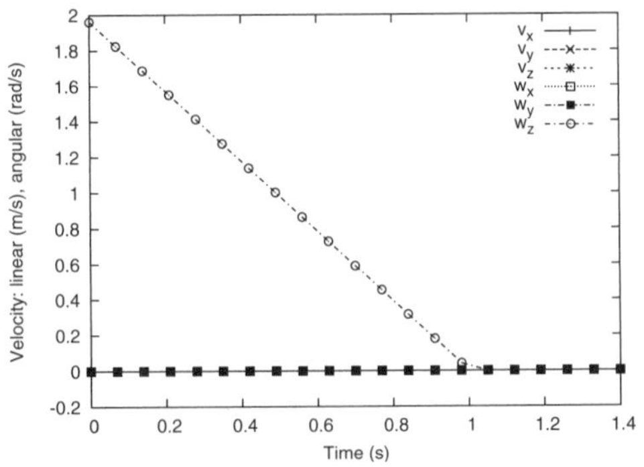

Fig. 3. Linear and angular velocities for Example 2. All velocities except ω_z are zero throughout the simulation.

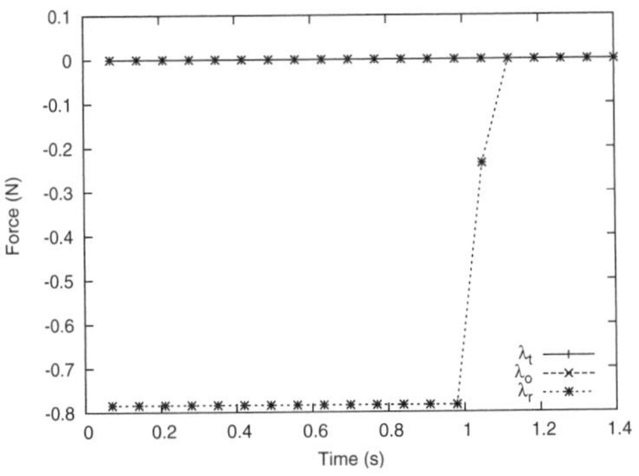

Fig. 4. Forces for Example 2. The tangential forces are both 0 for the entire simulation, and the torsional force transitions to zero when the sphere switches from a sliding contact to sticking.

C. Example 3: Sphere on Two Spheres

This example consists of a small ball rolling and sliding on two larger balls, and is chosen to compare our model with those presented in [18] and [9]. Figure 5 shows a small unit sphere in simultaneous contact with two larger fixed spheres. The sphere of radius 10 units is located at $(0,\ 0,\ 0)$ in the inertial frame and the sphere of radius 9 units is located at $(0,\ 11.4,\ 0)$. There is also a constant force of $\lambda_{app} = [1.0,\ 2.6,\ -9.81,\ 0,\ 0,\ 0]^T$ applied to the small sphere. With this force, the sphere initially has one of its contacts rolling while the other contact is simultaneously sliding, the rolling contact transitions to sliding, and both contacts eventually separate. It is important to emphasize that all these transitions are captured using a fixed time step implementation.

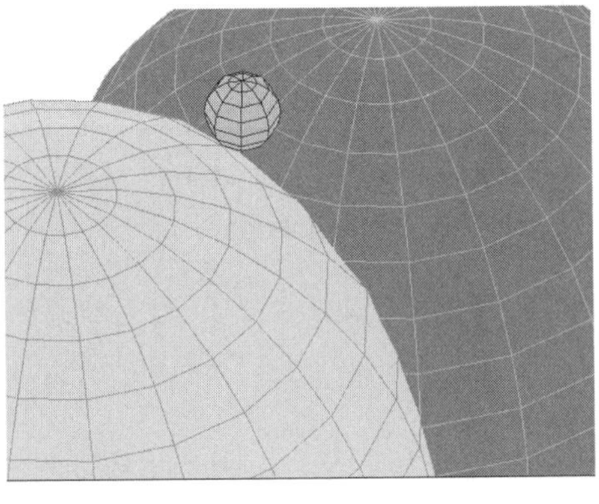

Fig. 5. A small sphere in contact with two large spheres.

The initial configuration of the small sphere is $\mathbf{q} = [0,\ 6.62105263157895,\ 8.78417110772903,\ 1,\ 0,\ 0,\ 0]^T$. The initial velocity is $\boldsymbol{\nu} = [\ 0,\ 0,\ 0,\ 0,\ 0,\ 0]$. The friction parameters are: $e_t = 1$, $e_o = 1$, $e_r = 0.3$, and $\mu = 0.2$. There were a total of 28 unknowns in our NCP formulation. We used a step size $h = 0.01$ (Tzitzouris-Pang use $h = 0.1$).

The generalized velocity of the sphere is shown in Figure 6. The smooth velocity profile agrees well with the nonlinear Tzitzouris-Pang formulation [18]. The Liu-Wang formulation [9] experienced non-smooth velocity jumps when the small sphere separated from the larger fixed spheres, which they attributed to an explicit time-stepping scheme. In the LCP Stewart-Trinkle implementation, the velocity profiles were very non-smooth. These results further confirm our belief that both linearization and explicit time-stepping lead to inaccuracies.

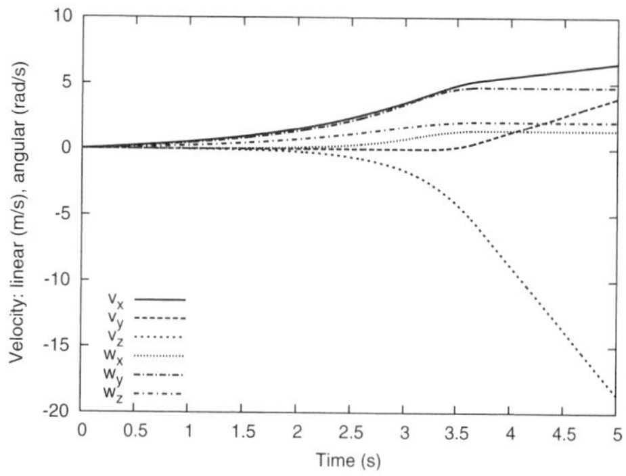

Fig. 6. Velocities of small moving sphere.

The fidelity of our method is further emphasized by Figures 7 and 8 that show the forces and sliding speed magnitudes at the two contacts. Contact 1 starts as a sliding contact and we see the sliding speed increases as the normal force decreases. Also, the magnitude of the friction force is equal to $\mu\lambda_{1n}$, consistent with our friction law for a sliding contact. At approximately 3.2 seconds, the small sphere separates from the large sphere at this contact, and all forces acting at contact 1 and the sliding speed drop to zero. Contact 2 on the other hand starts out as a rolling contact until approximately $t = 3$ seconds when it transitions to sliding. During the rolling phase the frictional magnitude is bounded by $\mu\lambda_{2n}$ as required by the friction law, and the sliding speed is 0. At the transition to sliding, the magnitude of the friction force becomes equal to $\mu\lambda_{2n}$ and the sliding speed begins to increase. Finally, at approximately $t = 3.6$ seconds, the contact breaks and all forces at this contact and the sliding speed drop to zero.

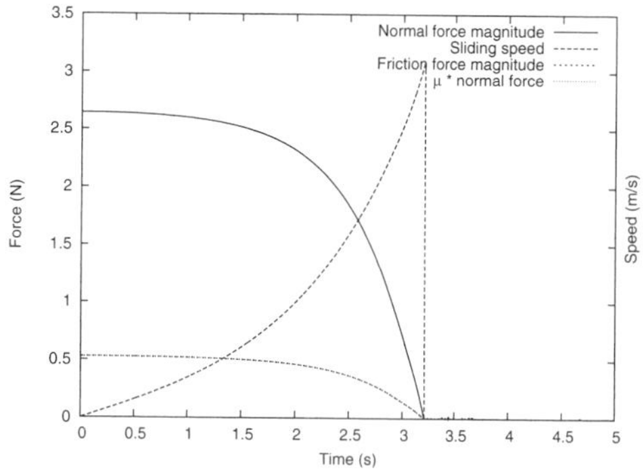

Fig. 7. Force and sliding speed at contact 1. Contact 1 is always sliding until separation, hence the μ normal force curve and friction magnitude curve overlap for the duration. The value of $\mu = 0.2$

Unlike the other approaches, we modeled the spheres as special cases of an ellipsoid:

$$f(x, y, z) = \left(\frac{x}{a}\right)^2 + \left(\frac{y}{b}\right)^2 + \left(\frac{z}{c}\right)^2 - 1$$

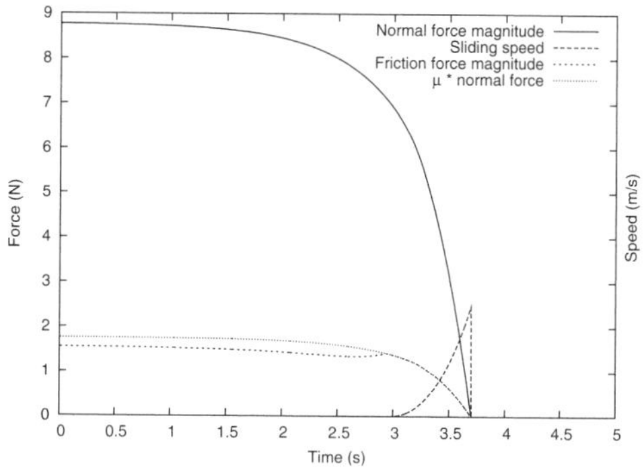

Fig. 8. Force and sliding speed at contact 2. The value of $\mu = 0.2$

where we set $a = b = c = 1$ for a unit sphere centered at the origin. This flexibility allows us to test various shapes by altering the 3 parameters, while the more difficult dynamical equations remain unaltered. For example, changing the a, b, c parameters would allow us to model ellipsoids rolling on ellipsoids; the Tzitzouris-Pang formulation cannot be applied since there is no analytical gap function.

VI. Conclusion

We presented the first geometrically implicit time-stepping scheme for objects described as intersections of convex inequalities. This approach overcomes stability and accuracy problems associated with polygonal approximations of smooth objects and approximation of the distance function for two objects in intermittent contact. We developed a new formulation for the contact constraints in the work space which enabled us to formulate a geometrically implicit time-stepping scheme as an NCP. We demonstrated through example simulations the fidelity of this approach to analytical solutions and previously described simulation results.

We see several directions for future work. We would like to address the question of existence and uniqueness of solutions of the NCP we formulate. We will perform more extensive numerical experimentation, and compare these solutions with solutions obtained when the closest distance is computed through a function call. We plan to precisely quantify the tradeoffs between the computation speed and physical accuracy of simulations for different object representations (*e.g.*, polyhedral, implicit, spline), friction approximations, and the choice between geometrically explicit or implicit methods. Although we have restricted our discussion to convex objects, we believe that this framework can be extended to non-convex objects described as unions of convex objects as well as parametric surfaces. Finally, we want to incorporate different impact laws to simulate a broader class of problems.

Acknowledgment

We wish to thank Todd Munson for his insightful discussions on NCPs. This work was supported by the National Science Foundation under grants IIS-0093233(CAREER), 0139701 (DMS-FRG), 0413227 (IIS-RCV), and 0420703

(MRI). Any opinions, findings, and conclusions or recommendations expressed in this material are those of the author(s) and do not necessarily reflect the views of the National Science Foundation.

REFERENCES

[1] M. Anitescu, J. F. Cremer, and F. A. Potra. Formulating 3D contact dynamics problems. *Mechanics of structures and machines*, 24(4):405, 1996.

[2] D. Baraff. Curved surfaces and coherence for non-penetrating rigid body simulation. *Computer Graphics*, 24(4):19–28, August 1990.

[3] S. Berard, J. Trinkle, B. Nguyen, B. Roghani, V. Kumar, and J. Fink. daVinci Code: A multi-model simulation and analysis tool for multi-body systems. In *IEEE International Conference on Robotics and Automation*, pages 2588–2593, Rome, Italy, Apr. 2007.

[4] R. W. Cottle, J. Pang, and R. E. Stone. *The Linear Complementarity Problem*. Academic Press, 1992.

[5] K. Egan, S. Berard, and J. Trinkle. Modeling nonconvex constraints using linear complementarity. Technical Report 03-13, Department of Computer Science, Rensselaer Polytechnic Institute, 2003.

[6] M. C. Ferris and T. S. Munson. Complementarity problems in GAMS and the PATH solver. *Journal of Economic Dynamics and Control*, 24(2):165–188, Feb. 2000.

[7] E. J. Haug, S. C. Wu, and S. M. Yang. Dynamics of mechanical systems with Coulomb friction, stiction, impact and constraint addition-deletion theory. *Mechanism and Machine Theory*, 21:365–446, 1986.

[8] P. G. Kry and D. K. Pai. Continuous contact simulation for smooth surfaces. *ACM Transactions on Graphics*, 22(1):106–129, Jan. 2003.

[9] T. Liu and M. Y. Wang. Computation of three-dimensional rigid-body dynamics with multiple unilateral contacts using time-stepping and Gauss-Seidel methods. *IEEE Transactions on Automation Science and Engineering*, 2(1):19–31, January 2005.

[10] B. Mirtich. *Impulse-based Dynamics Simulation of Rigid Body Systems*. PhD thesis, University of California, Berkley, 1996.

[11] B. Mirtich, Y. Zhuang, K. Goldberg, J. Craig, R. Zanutta, B. Carlisle, and J. Canny. Estimating pose statistics for robotic part feeders. In *IEEE International Conference on Robotics and Automation*, pages 1140–1146, Minneapolis, MN, Apr. 1996.

[12] D. J. Montana. The kinematics of contact and grasp. *International Journal of Robotics Research*, 7(3):17–32, June 1998.

[13] M. Pauly, D. K. Pai, and L. J. Guibas. Quasi-rigid objects in contact. In *ACM SIGGRAPH/Eurographics Symposium on Computer Animation*, August 2004.

[14] F. Pfeiffer and C. Glocker. *Multibody Dynamics with Unilateral Constraints*. John Wiley, New York, 1996.

[15] P. Song, J.-S. Pang, and V. Kumar. A semi-implicit time-stepping model for frictional compliant contact problems. *International Journal for Numerical Methods in Engineering*, 60:2231–2261, 2004.

[16] D. Stewart and J. Trinkle. An implicit time-stepping scheme for rigid body dynamics with inelastic collisions and Coulomb friction. *International Journal of Numerical Methods in Engineering*, 39:2673–2691, 1996.

[17] J. Trinkle, J. Pang, S. Sudarsky, and G. Lo. On dynamic multi-rigid-body contact problems with Coulomb friction. *Zeitschrift für Angewandte Mathematik und Mechanik*, 77(4):267–279, 1997.

[18] J. C. Trinkle, J. Tzitzouris, and J. S. Pang. Dynamic multi-rigid-body systems with concurrent distributed contacts: Theory and examples. *Philosophical Transactions on Mathematical, Physical, and Engineering Sciences, Series A*, 359(1789):2575–2593, Dec. 2001.

[19] J. E. Tzitzouris. *Numerical Resolution of Frictional Multi-Rigid-Body Systems via Fully Implicit Time-Stepping and Nonlinear Complementarity*. PhD thesis, Mathematics Department, Johns Hopkins University, 2001.

APPENDIX

CONTACT CONDITIONS FOR OBJECTS REPRESENTED AS INTERSECTIONS OF CONVEX INEQUALITIES

We present here the contact conditions for the general case where each convex object is defined as an intersection of convex inequalities. Let $f_j(\xi_1) \leq 0, j = 1, \dots, m$, $g_j(\xi_2) \leq 0, j = m+1, \dots, n$, be convex functions representing the two convex objects. Since the closest point is outside the object if it is outside at least one of the intersecting surfaces, the complementarity conditions for nonpenetration can be written as either one of the following two sets of conditions:

$$\begin{aligned} 0 \leq \lambda_{in} \perp \max\{f_j(\mathbf{a}_2)\} \geq 0 \quad j = 1, \dots m \\ 0 \leq \lambda_{in} \perp \max\{g_j(\mathbf{a}_1)\} \geq 0 \quad j = m+1, \dots n \end{aligned} \tag{19}$$

where $\mathbf{a}_1$ and $\mathbf{a}_2$ are the closest points on the two bodies and are given by the KKT conditions

$$\begin{aligned} \mathbf{a}_1 - \mathbf{a}_2 = -\sum_{i=1}^{m} l_i \nabla f_i(\mathbf{a}_1) = \sum_{j=m+1}^{n} l_j \nabla g_j(\mathbf{a}_2) \\ f_i(\mathbf{a}_1) + s_i = 0 \\ g_j(\mathbf{a}_2) + s_j = 0 \\ l_i s_i = l_j s_j = 0 \\ s_i \geq 0, \quad s_j \geq 0 \end{aligned} \tag{20}$$

where s_i, s_j are the slack variables. At the optimal solution only some of the constraints are active. Thus the optimality conditions can be written as the following set of nonlinear equations:

$$\begin{aligned} \mathbf{a}_1 - \mathbf{a}_2 = -\sum_{i \in \mathbb{I} \cap \{i\}} l_i \nabla f_i(\mathbf{a}_1) = \sum_{j \in \mathbb{I} \cap \{j\}} l_j \nabla g_j(\mathbf{a}_2) \\ f_k(\mathbf{a}_1) = 0 \quad k \in \mathbb{I} \cap \{i\} \\ g_k(\mathbf{a}_2) = 0 \quad k \in \mathbb{I} \cap \{j\} \end{aligned} \tag{21}$$

where $\mathbb{I}$ is the index set of active constraints. Equations 19 and 21 together represent the contact constraints as long as $\mathbf{a}_1 \neq \mathbf{a}_2$. Using arguments similar to the single surface case in Section IV we can see that it is not possible to distinguish between touching points and intersecting points using the above formulation. In this case also, we can rewrite Equation 21 suitably by equating unit vectors to eliminate the intersection point solutions. Without loss of generality, we can set one of the Lagrange multipliers to be 1 and scale the others and rewrite Equation 21 as

$$\begin{aligned} \mathbf{a}_1 - \mathbf{a}_2 = \|\mathbf{a}_1 - \mathbf{a}_2\| \Big(\frac{\nabla f_{k_1}(\mathbf{a}_1)}{\|\nabla f_{k_1}(\mathbf{a}_1)\|} \\ + \sum_{k \in \{\mathbb{I} \setminus k_1\} \cap \{i\}} l_k \frac{\nabla f_k(\mathbf{a}_1)}{\|\nabla f_k(\mathbf{a}_1)\|} \Big) \\ \frac{\nabla f_{k_1}(\mathbf{a}_1)}{\|\nabla f_{k_1}(\mathbf{a}_1)\|} + \sum_{k \in \{\mathbb{I} \setminus k_1\} \cap \{i\}} l_k \frac{\nabla f_k(\mathbf{a}_1)}{\|\nabla f_k(\mathbf{a}_1)\|} \\ = \frac{\nabla g_{k_2}(\mathbf{a}_2)}{\|\nabla g_{k_2}(\mathbf{a}_2)\|} + \sum_{k \in \{\mathbb{I} \setminus k_2\} \cap \{j\}} l_k \frac{\nabla g_k(\mathbf{a}_2)}{\|\nabla g_k(\mathbf{a}_1)\|} \\ f_k(\mathbf{a}_1) = 0 \quad k \in \mathbb{I} \cap \{i\} \\ g_k(\mathbf{a}_2) = 0 \quad k \in \mathbb{I} \cap \{j\} \end{aligned} \tag{22}$$

Proposition: *Equation 19 and 22 together represent the non-penetration constraints, i.e., the two objects will satisfy the contact constraints at the end of each time step if and only if Equation 19 and 22 hold together.*

Synthesis of Constrained nR Planar Robots to Reach Five Task Positions

Gim Song Soh
Robotics and Automation Laboratory
University of California
Irvine, California 92697-3975
Email: gsoh@uci.edu

J. Michael McCarthy
Department of Mechanical and Aerospace Engineering
University of California
Irvine, California 92697-3975
Email: jmmccart@uci.edu

***Abstract*—In this paper, we design planar nR serial chains that provide one degree-of-freedom movement for an end-effector through five arbitrarily specified task positions. These chains are useful for deployment linkages or the fingers of a mechanical hand.**

The trajectory of the end-effector pivot is controlled by n-1 sets of cables that are joined through a planetary gear system to two input variables. These two input variables are coupled by a four-bar linkage, and the movement of the end-effector around its end joint is driven by a second four-bar linkage. The result is one degree-of-freedom system.

The design of the cabling system allows control of the shape of the chain as it moves through the task positions. This combines techniques of deployable linkage design with mechanism synthesis to obtain specialized movement with a minimum number of actuators. Two example designs for a 6R planar chain are presented, one with a square initial configuration and a second with a hexagonal initial configuration.

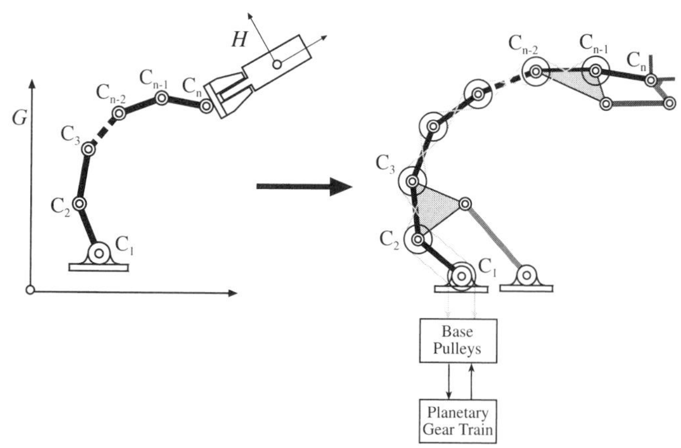

Fig. 1. The Design Problem. The figure on the left shows a schematic of planar nR serial robot, and the figure on the right shows a schematic of the constrained nR robot.

I. Introduction

In this paper, we present a methodology for the design of planar nR serial chains such that the end-effector reaches five arbitrary task positions with one degree-of-freedom. The method is general in that we may introduce mechanical constraints between m drive joints and $n-m$ driven joints. For the purposes of this paper, we consider $m = 1$, and show how to use cables combined with planetary gear trains to constrain the end-effector movement to three degrees-of-freedom. Then, we design four-bar chains attached at the base and at the end-effector to obtain a one degree-of-freedom system. See Figure 1. As an example, we obtain a one degree-of-freedom 6R planar chain that reaches five specified task positions. This work can be used to control the shape of deployable chains, as well as for the design of mechanical fingers for specialized grasping.

II. Literature Review

Cable systems are employed to reduce the inertia of a manipulator system by permitting the actuators to be located remotely, thus a higher power-to-weight ratio over a direct drive system, see Lee (1991) [1]. Recent work on cable-driven robotic systems includes Arsenault and Gosselin (2006) [2], who introduce a 2-dof tensegrity mechanism to increase the reachable workspace and task flexibility of a serial chain robot. Also, Lelieveld and Maeno (2006) [3] present a 4 DOF portable haptic device for the index finger driven by a cable-actuated rolling-link-mechanism (RLM). Palli and Melchiorri (2006) [4] explore a variation on cables using a tendon sheath system to drive a robotic hand.

Our work is inspired by Krovi et al. (2002) [5], who present a synthesis theory for nR planar serial chains driven by cable tendons. Their design methodology sizes the individual links and pulley ratios to obtain the desired end-effector movement used to obtain an innovative assistive device. Our approach borrows from deployable structures research Kwan and Pelligrino (1994) [6], and uses symmetry conditions to control the overall movement of the chain, while specialized RR cranks position the base and end-effector joints.

For our serial chain the designer can choose the location of the base joint $\mathbf{C}_1$, the attachment to the end-effector $\mathbf{C}_n$, and the dimensions of the individual links which for convenience we assume are the same size. See Figure 1. Our goal is to design the mechanical constraints that ensures the end-effector reaches each of five task positions, in response to an input drive variable. This research extends the ideas presented in Leaver and McCarthy (1987) [7] and Leaver et al. (1988) [8] using the tendon routing analysis of Lee (1991) [1]. Also see Tsai (1999) [9]. While our proposed system appears complicated, it has only one actuator.

III. The Planar nR Serial Chain Robot

Let the configuration of an nR serial chain be defined by the coordinates $\mathbf{C}_i = (x_i, y_i)$, $i = 1, \ldots, n$ to be each of its revolute joints. See Figure 1. The distances $a_{i,i+1} = |\mathbf{C}_{i+1} - \mathbf{C}_i|$ defined the lengths of each link. Attach a frame B_i to each of these links so its origin is located at $\mathbf{C}_i$ and its x axis is directed toward $\mathbf{C}_{i+1}$. The joints $\mathbf{C}_1$ and $\mathbf{C}_n$ are the attachments to the base frame $F = B_1$ and the moving frame $M = B_n$, respectively, and we assume they are the origins of these frames. The joint angles θ_i define the relative rotation about the joints $\mathbf{C}_i$.

Introduce a world frame G and task frame H so the kinematics equations of the nR chain are given by

$$[D] = [G][Z(\theta_1)][X(a_{12})] \ldots [X(a_{n-1,n})[Z(\theta_n)][H], \quad (1)$$

where $[Z(\theta_i)]$ and $[X(a_{i,i+1})]$ are the 3×3 homogeneous transforms that represent a rotation about the z-axis by θ_i, and a translation along the x-axis by $a_{i,i+1}$, repspectively. The transformation $[G]$ defines the position of the base of the chain relative to the world frame, and $[H]$ locates the task frame relative to the end-effector frame. The matrix $[D]$ defines the coordinate transformation from the world frame G to the task frame H.

IV. Kinematics of Tendon Driven Robots

The relationship between the rotation of the base pulley and the joint angles in an open-loop chain with (k+1) links can be described by:

$$\theta_k^* = \theta_1 \pm (\frac{r_{k,2}}{r_k})\theta_2 \pm \ldots (\frac{r_{k,i}}{r_k})\theta_i \pm \ldots (\frac{r_{k,k}}{r_k})\theta_k, \quad (2)$$

where θ_k^* denotes the base pulley angle of the kth tendon, θ_i denotes the joint angle of the ith joint, $r_{k,i}$ denotes the radius of the pulley of the ith joint of the kth tendon, and r_k denotes the radius of the base pulley of the kth tendon. The sign of each term, $(\frac{r_{k,i}}{r_k})\theta_i$, in (2) is to be determined by the number of cross-type routing preceding the ith joint axis. If the number of cross-type routing is even, the sign is positive, else it is negative. See the connection between $\mathbf{C}_1$ and $\mathbf{C}_2$ of Tendon 3 in Figure 2 for an example of cross-type routing. For derivation of this equation, refer to Lee (1991) [1].

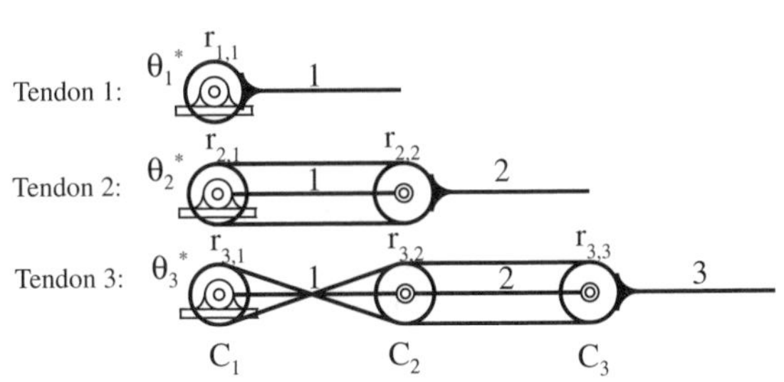

Fig. 2. Planar schematic for a three-DOF planar robot with structure matrix (5).

Writing (2) once for each of the tendon routing k, $k = 1, \ldots, n$ for an nR serial robot, yields the following linear transformation in matrix form:

$$\mathbf{\Theta}^* = [B][R]\mathbf{\Theta}. \quad (3)$$

where $\mathbf{\Theta}^* = (\theta_1^*, \theta_2^*, \ldots, \theta_n^*)^T$ denotes the base pulley angles and $\mathbf{\Theta} = (\theta_1, \theta_2, \ldots, \theta_n)^T$ denotes the joint angles. If we let $r_{k,j} = r_k$, $k = 1, \ldots, n$ such that all the pulleys have the same diameter, then $[B][R]$ would be reduced to a lower triangular $n \times n$ matrix with entries $+1$ or -1,

$$[B][R] = [B] = \begin{bmatrix} 1 & 0 & 0 & \ldots & 0 \\ 1 & \pm 1 & 0 & \ldots & 0 \\ \vdots & \vdots & \vdots & \ldots & \vdots \\ 1 & \pm 1 & \pm 1 & \ldots & \pm 1 \end{bmatrix}. \quad (4)$$

The matrix $[B]$ is known as the structure matrix, Lee (1991) [1]. It describes how the tendons are routed. The matrix $[R]$ is a diagonal scaling matrix with elements representing the ratio of the radii of the i-th joint pulley to the i-th base pulley. In this paper, we size the pulleys to be of equal lengths such that this matrix is the identity.

We illustrate with an example on constructing the tendons of a 3R serial robot from the structure matrix. Suppose the structure matrix, $[B]$ is defined to be

$$[B] = \begin{bmatrix} 1 & 0 & 0 \\ 1 & 1 & 0 \\ 1 & -1 & -1 \end{bmatrix}. \quad (5)$$

Each columns of the matrix $[B]$ corresponds to the routing prior to the joint pivots, and each row represent an independent tendon drive. The element in the first row is always $+1$ since we are driving the fixed pivot $\mathbf{C}_1$ directly with θ_1^*. For the second row, it represent a parallel-type routing between the moving pivot $\mathbf{C}_2$ and fix pivot $\mathbf{C}_1$ since the sign of the second element remains unchanged. For the last row, the second element changes sign, we get a cross-type between the moving pivot $\mathbf{C}_2$ and fix pivot $\mathbf{C}_1$. Also, the sign remains unchanged after the second element, parallel-type follows between the moving pivot $\mathbf{C}_3$ and $\mathbf{C}_2$. See Figure 2 for the planar schematic for the structure matrix (5).

V. Schematic of a Constrain nR Planar Robot

The schematic of a constrain nR Planar Robot consists of a nR serial chain, $N-1$ tendons, a planetary gear train, and two RR chains. See Figure 3. Our goal is to use the $N-1$ tendons, a planetary gear train, and two RR chains to mechanically constrain a specified nR serial chain that reaches five task position to one degree-of-freedom. The $N-1$ base pulleys were connected to the $N-1$ tendons to group the actuation point to the base for easy attachment to a planetary gear train.

The specified nR serial chain have n degrees of freedom. The attachment of the $N-1$ tendons to the nR serial chain does not change the system degrees of freedom, but it allows the control of the first $n-1$ joint angles from the base independently. If we add a planetary gear train to couple the various $n-1$ base pulleys as shown in Figure 5, we are able

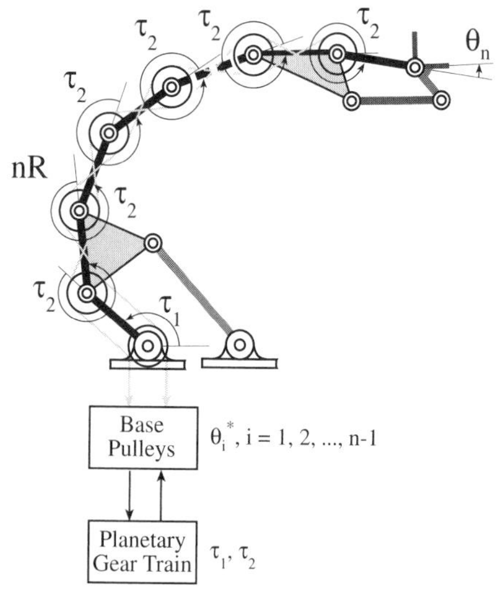

Fig. 3. Schematic for a constrain nR Planar Robot driven by cables, mechanically constraint by planetary gear train and RR cranks.

to reduce the tendon driven robot to a three degree-of-freedom system. Two RR constraints were then added to obtain a one degree-of-freedom system. The first RR chain serves to couple the planetary gear train inputs τ_1 and τ_2 together, and the purpose of the second RR chain is to couple the nth joint, θ_n, to τ_2 and hence τ_1.

As shown in Figure 3, if we denote τ_1 as the input that drive the first joint angle, we can see that τ_2 is related to the input τ_1 through the four bar linkage formed from the attachment of an RR constraint to the second link. The planetary gear train was designed in such a way that it impose symmetry condition on the joint angles so that $\theta_2 = \theta_3 = \cdots = \theta_{n-1} = \tau_2$. Hence τ_2 get reflected through the tendons to serve as the input to the floating four bar linkage from the RR attachment between the end-effector and the $(n-2)$th link. The output from this floating four-bar linkage is the joint angle θ_n. In other words, if we drive the system with input τ_1, τ_2 and θ_n will be driven by the input τ_1 through the mechanical coupling of the planetary gear train, and the two RR constraints.

VI. The Coupling between Joint Angles and Inputs

The desired coupling between joint angles, Θ, and the inputs, Γ, of a nR serial chain can be described by matrix $[A]$ as follows,

$$\Theta = [A]\Gamma. \tag{6}$$

Our goal here is to constrain the nR serial chain to a three degree-of-freedom system first using tendons and a planetary gear train before we add RR chainss to constrain it to one degree-of-freedom. Hence $\Gamma = (\tau_1, \tau_2)^T$ in particular if we ignore θ_n for now. The desired coupling between the inputs and the joint angles for a nR serial planar robot can be obtained by specifying the tendon routing, the size of the base pulleys, and the transmission coupling the input to the base pulleys. This is equivalent to selecting the matrices $[B]$, $[S]$, and $[C]$ so we obtain the desired coupling matrix [A] in the relation

$$[A] = [B]^{-1}[S][C]. \tag{7}$$

$[B]$ is a matrix that describes the relationship between the relative joint angles and the rotation of the base pulleys, see (3). $[S]$ is a diagonal matrix that scales the sum of the row elements of matrix $[C]$ to unity. This matrix was included to account for the effects of changing the size of the base pulleys and is the ratio of the ith base pulley to the ith joint pulley. $[C]$ is a matrix that describes the relationship between the rotation of the base pulleys to the inputs of the transmission system.

VII. The Transmission System

The transmission that couples the nR serial robot is composed of two stages. The bottom stage connects the controlled inputs to the base pulleys using simple and planetary gear trains and the top stage connects the base pulleys to the joint pulleys using tendons.

The relationship between the inputs, Γ and the base pulley rotations Θ^*, which defines the bottom stage of the transmission is given by:

$$\Theta^* = [C]\Gamma. \tag{8}$$

The matrix $[C]$ is called the mechanical coupling matrix. Its elements determine the gear train design which couples the inputs to the tendon drive pulleys. In this paper, our goal is to first design a transmission system that couples the nR serial chain to three degree-of-freedom, and then design two RR cranks to constraint the system to one degree-of-freedom. The equation relating the base pulley rotation, θ_i^* to the inputs at the transmission τ_1 and τ_2 is:

$$\theta_i^* = c_i\tau_1 + d_i\tau_2. \tag{9}$$

Three cases exist for (9). They correspond to a simple gear train if either c_i or $d_i = 0$, a planetary gear train if both c_i and d_i are non-zero, and un-driveable if both c_i and d_i are zero. An example of a planetary gear train with sum of $c_i = 1$ and $d_i = 3$ is as shown in Figure 4.

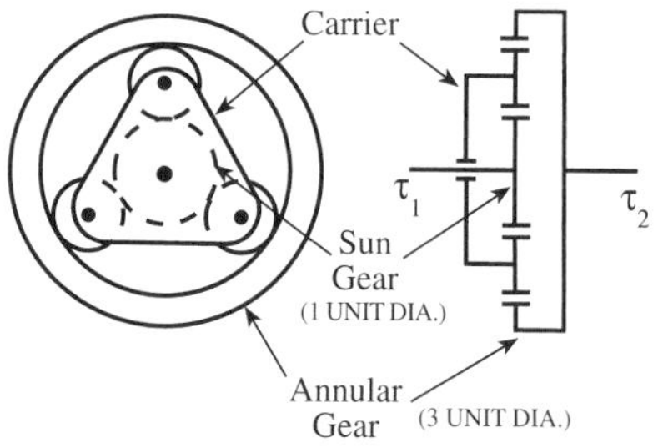

Fig. 4. Planetary Gear Train

VIII. nR Chain Design Methodology

Our design of constraining nR robots using tendon, planetary gear train and RR chainss to reach five specified task positions proceed in three steps. The following describes the design process.

A. Step 1 - Solving the Inverse Kinematics of the nR Serial Chain

Given five task positions $[T_j], j = 1, \ldots, 5$ of the end-effector of the nR chain, we seek to solve the equations

$$[D] = [T_j], \quad j = 1, \ldots, 5, \tag{10}$$

to determine the joint parameter vectors $\mathbf{\Theta}_j = (\theta_{1,j}, \theta_{2,j}, \ldots, \theta_{n,j})^T$, where $\theta_{i,j}$ represents the ith joint angle at the jth position. Because there are three independent constraints in this equation, we have free parameters when $n > 3$. In what follows, we show how to use these free parameters to facilitate the design of mechanical constraints on the joint movement so the end-effector moves smoothly through the five task positions.

In the design of serial chain robots, it is convenient to have near equal length links to reduce the size of workspace holes. For this reason, we assume that the link dimensions of our nR planar robot satisfy the relationship

$$a_{12} = a_{23} = \cdots = a_{n-1,n} = l. \tag{11}$$

This reduces the specification of the nR robot to the location of the base joint $\mathbf{C}_1$ in G, and the end-effector joint $\mathbf{C}_n$ relative to the task frame H. In this paper, we take the task frame H to be the identity matrix to simplify the problem even more. The designer is free to chose the location of the base joint $\mathbf{C}_1$ and the link dimensions l.

We define the input crank joint angle to be,

$$\theta_{1,j} = \sigma_j, \quad j = 1, \ldots, 5, \tag{12}$$

and if the planar nR robot has $n > 3$ joints, then we impose a symmetry condition

$$\theta_{2,j} = \theta_{3,j} = \cdots = \theta_{n-1,j} = \lambda_j, \quad j = 1, \ldots, 5, \tag{13}$$

in order to obtain a unique solution to the inverse kinematics equations.

B. Step 2 - Designing of a planetary gear train

The second step of our design methodology consists of designing a two-degree-of-freedom planetary gear train to drive the base pulleys to achieve the symmetry conditions listed in (13). Note that we ignore the degree of freedom of the nth joint at this stage. The gear train controls the shape of the chain through the two inputs. Proper selection of the routing of the tendons from their joints to their base pulley can simplify gear train design. We now generalize the routing of the $n-1$ tendon such that we only require a planetary gear train with inputs τ_1 and τ_2 to drive the first $n-1$ joint angles. Consider the following $(n-1) \times 2$ desired coupling matrix

$$[A] = \begin{bmatrix} 1 & 0 \\ 0 & 1 \\ \vdots & \vdots \\ 0 & 1 \end{bmatrix}. \tag{14}$$

If we chose the $(n-1) \times (n-1)$ structure matrix $[B]$ with elements such that starting from the second column, we have +1 on the even columns and -1 on the odd columns to get the following lower triangular matrix,

$$[B] = \begin{bmatrix} 1 & 0 & 0 & 0 & 0 & \ldots & 0 \\ 1 & 1 & 0 & 0 & 0 & \ldots & 0 \\ 1 & 1 & -1 & 0 & 0 & \ldots & 0 \\ 1 & 1 & -1 & 1 & 0 & \ldots & 0 \\ 1 & 1 & -1 & 1 & -1 & \ldots & 0 \\ \vdots & \vdots & \vdots & \vdots & \vdots & \ldots & \vdots \\ 1 & 1 & -1 & 1 & -1 & \ldots & \pm 1 \end{bmatrix}. \tag{15}$$

Refer to Figure 7 and 8 for a graph and schematic representation of such a routing system. Using (7), we get

$$[B][A] = [S][C] = \begin{bmatrix} 1 & 0 \\ 1 & 1 \\ 1 & 0 \\ 1 & 1 \\ 1 & 0 \\ \vdots & \vdots \\ c_{n-1,1} & c_{n-1,2} \end{bmatrix}. \tag{16}$$

If we introduce a $(n-1) \times (n-1)$ diagonal scaling matrix,

$$[S] = \begin{bmatrix} 1 & 0 & 0 & 0 & \ldots & 0 \\ 0 & 2 & 0 & 0 & \ldots & 0 \\ 0 & 0 & 1 & 0 & \ldots & 0 \\ 0 & 0 & 0 & 2 & \ldots & 0 \\ \vdots & \vdots & \vdots & \vdots & \ldots & \vdots \\ 0 & 0 & 0 & 0 & \ldots & s_{n-1} \end{bmatrix}, \tag{17}$$

with element $s_{n-1} = 1$ if n is even, and $s_{n-1} = 2$ if n is odd. It would result in a mechanical coupling matrix $[C]$, with row elements adding to unity, of the form

$$[C] = \begin{bmatrix} 1 & 0 \\ \frac{1}{2} & \frac{1}{2} \\ 1 & 0 \\ \frac{1}{2} & \frac{1}{2} \\ 1 & 0 \\ \vdots & \vdots \\ c_{n-1,1} & c_{n-1,2} \end{bmatrix}. \tag{18}$$

Note that the rows of matrix $[C]$ consists of elements $(1, 0)$ at odd rows and $(\frac{1}{2}, \frac{1}{2})$ at even rows. The values at the last row would depend on if n-1 is odd or even. Because of this special alternating structure, we only require a two degree-of-freedom planetary gear train to drive the n-1 tendon base pulleys.

The mechanical coupling matrix $[C]$ indicates that the planetary gear train used should have the property such that τ_1 drives odd base joints, and τ_1 and τ_2 drives even base joints. A planetary gear train that fits this characteristic would be a bevel gear differential since an ordinary planetary gear train would result in a locked state. Also, the scaling matrix indicates that the ratio of the even k^{th} base pulley to the even k^{th} joint pulley is 1 to 2. See Figure 5 and Figure 6 for an example of such a bevel gear differential and tendon routing on a 6R planar robot.

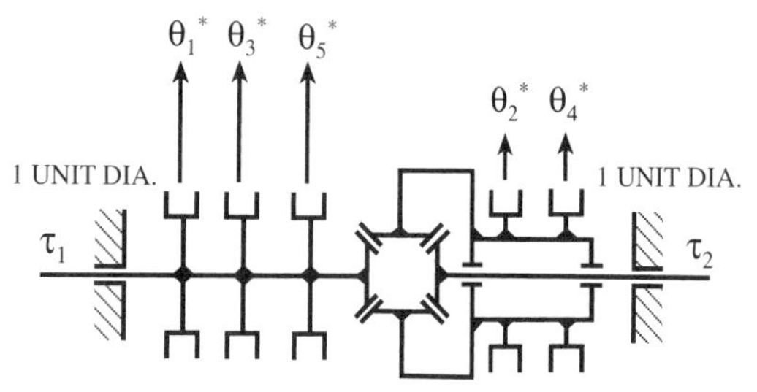

Fig. 5. Planetary Gear Train for a 6R Planar serial Robot

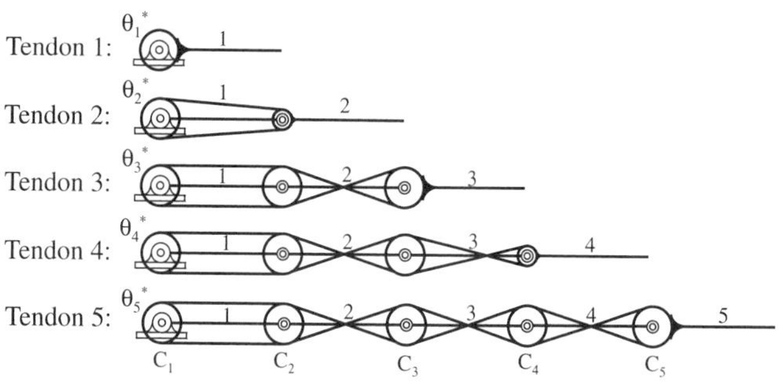

Fig. 6. Tendon Routing for a 6R Planar Robot

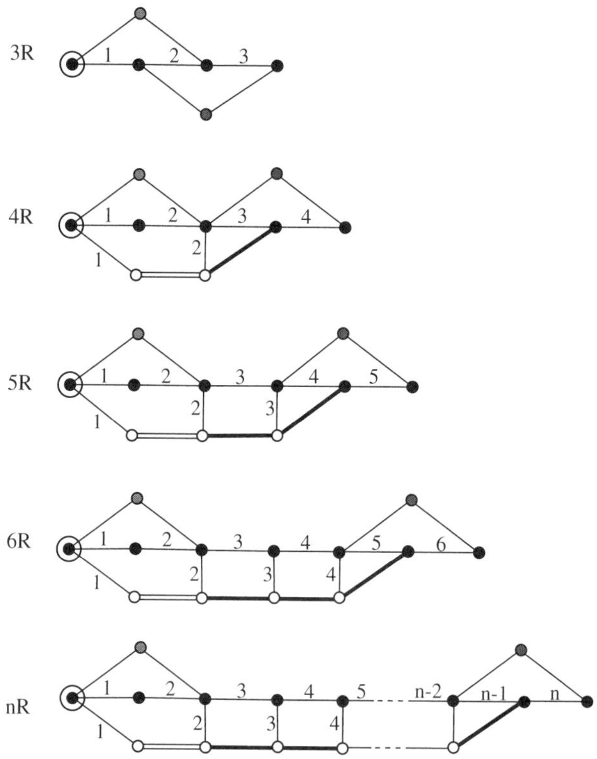

Fig. 7. This shows the graph representation of mechanically constrained serial chain. Each link is represented as a node and each R joint as an edge. Parallel type routing is denoted by a double-line edge and cross type routing is denoted by a heavy edge.

C. Step 3 - Synthesis of an RR Drive Crank

The last step of our design methodology consists of sizing two RR chainss that constrains the nR robot to one degree-of-freedom. Figure 7 and 8 shows how we can systematically attach two RR chainss to constraint the three degree-of-freedom nR cable driven robot to one degree-of-freedom. In this section, we expand the RR synthesis equations, Sandor and Erdman (1984) [10] and McCarthy (2000) [11], to apply to this situation.

Let $[B_{k-2,j}]$ be five position of the $(k-2)$th moving link, and $[B_{k,j}]$ be the five positions of the kth moving link measured in a world frame F, $j = 1, \ldots, 5$. Let $\mathbf{g}$ be the coordinates of the R-joint attached to the $(k-2)$th link measured in the link frame B_{k-2}, see Figure 9. Similarly, let $\mathbf{w}$ be the coordinates of the other R-joint measured in the link frame B_k. The five positions of these points as the two moving bodies move between the task configurations are given by

$$\mathbf{G}^j = [B_{k-2,j}]\mathbf{g} \quad \text{and} \quad \mathbf{W}^j = [B_{k,j}]\mathbf{w} \tag{19}$$

Now, introduce the relative displacements $[R_{1j}] = [B_{k-2,j}][B_{k-2,1}]^{-1}$ and $[S_{1j}] = [B_{k,j}][B_{k,1}]^{-1}$, so these equations become

$$\mathbf{G}^j = [R_{1j}]\mathbf{G}^1 \quad \text{and} \quad \mathbf{W}^j = [S_{1j}]\mathbf{W}^1 \tag{20}$$

where $[R_{11}] = [S_{11}] = [I]$ are the identity transformations.

The point $\mathbf{G}^j$ and $\mathbf{W}^j$ define the ends of a rigid link of length R, therefore we have the constraint equations

$$([S_{1j}]\mathbf{W}^1 - [R_{1j}]\mathbf{G}^1) \cdot ([S_{1j}]\mathbf{W}^1 - [R_{1j}]\mathbf{G}^1) = R^2 \tag{21}$$

These five equations can be solved to determine the five design parameters of the RR constraint, $\mathbf{G}^1 = (u, v, 1)$, $\mathbf{W}^1 = (x, y, 1)$ and R. We will refer to these equations as the *synthesis equations* for the RR link.

To solve the synthesis equations, it is convenient to introduce the displacements $[D_{1j}]$= $[R_{1j}]^{-1}[S_{1j}]$=$[B_{k-2,1}][B_{k-2,j}]^{-1}[B_{k,j}][B_{k,1}]^{-1}$, so these equations become

$$([D_{1j}]\mathbf{W}^1 - \mathbf{G}^1) \cdot ([D_{1j}]\mathbf{W}^1 - \mathbf{G}^1) = R^2 \tag{22}$$

which is the usual form of the synthesis equations for the RR crank in a planar four-bar linkage, see McCarthy (2000)[11]. Subtract the first of these equations from the remaining to cancel R^2 and the square terms in the variables u, v and x, y. The resulting four bilinear equations can be solved algebraically, or numerically using something equivalent to *Mathematica's* `Nsolve` function to obtain the desired pivots.

The RR chain imposes a constraint on the value of τ_2 as a function of τ_1 given by

$$\tau_2 = \arctan\left(\frac{b \sin\psi - a \sin\tau_1}{g + b\cos\psi - a\cos\tau_1}\right) + \beta, \tag{23}$$

where β is the joint angle between pivots $\mathbf{C}_2\mathbf{W}_1$ and $\mathbf{C}_2\mathbf{C}_3$,

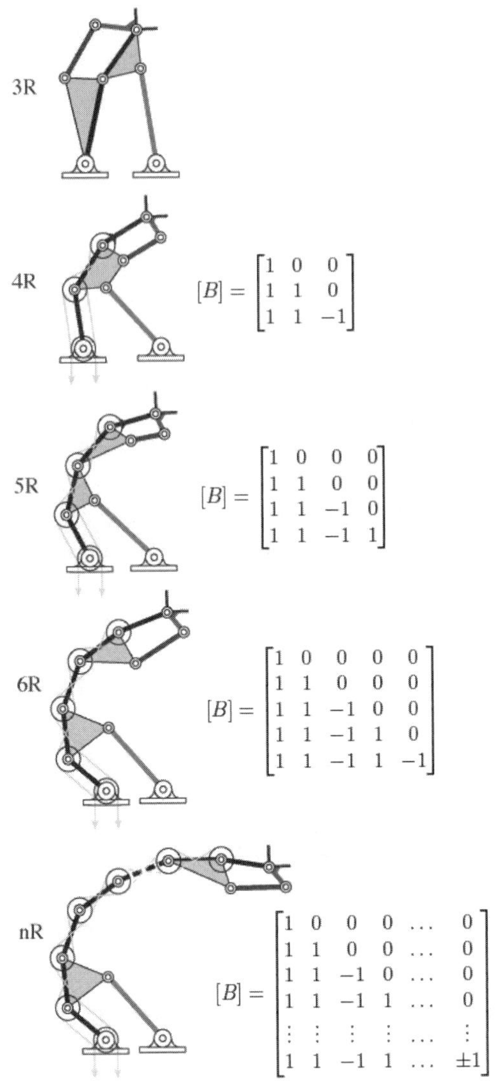

$$[B]=\begin{bmatrix}1&0&0\\1&1&0\\1&1&-1\end{bmatrix}$$

$$[B]=\begin{bmatrix}1&0&0&0\\1&1&0&0\\1&1&-1&0\\1&1&-1&1\end{bmatrix}$$

$$[B]=\begin{bmatrix}1&0&0&0&0\\1&1&0&0&0\\1&1&-1&0&0\\1&1&-1&1&0\\1&1&-1&1&-1\end{bmatrix}$$

$$[B]=\begin{bmatrix}1&0&0&0&\dots&0\\1&1&0&0&\dots&0\\1&1&-1&0&\dots&0\\1&1&-1&1&\dots&0\\\vdots&\vdots&\vdots&\vdots&\dots&\vdots\\1&1&-1&1&\dots&\pm1\end{bmatrix}$$

Fig. 8. This shows the kinematic structure of mechanically constrained serial chain. The matrix on the right is the structure matrix $[B]$, which describes how the tendons are routed. This show that the structure extends to any length of nR robot

and

$$\psi = \arctan(\frac{B}{A}) \pm \arccos(\frac{C}{\sqrt{A^2+B^2}})$$
$$A = 2ab\cos\tau_1 - 2gb$$
$$B = 2ab\sin\tau_1$$
$$C = g^2 + b^2 + a^2 - h^2 - 2ag\cos\tau_1. \quad (24)$$

a, b, g and h are the lengths of link $\mathbf{C}_1\mathbf{C}_2$, $\mathbf{G}_1\mathbf{W}_1$, $\mathbf{C}_1\mathbf{G}_1$ and $\mathbf{C}_2\mathbf{W}_1$ respectively, and ψ is the angle of the driven crank b. See Figure 9 for the various notation.

IX. Configuration Analysis

In order to evaluate and animate the movement of these linkage systems, we must analyze the system to determine its configuration for each value of the input angle θ_1. Figure 9 shows that these systems consist of two interconnected four bar linkages. Our approach uses the analysis procedure of four bar linkages from McCarthy (2000) [11]. Starting from frame K, we solve for $\mathbf{W}_1$ for a given θ_1. Then, we calculate the coupler angle and hence the joint angle θ_2 using (23). This joint angle value gets transmitted to the rest of the n-2 joints to drive the link $\mathbf{C}_{n-1}\mathbf{G}_n$. Next, we analyze the second four bar linkage in frame B_{n-1} to complete the analysis. The result is a complete analysis of the mechanically constrained nR planar robot.

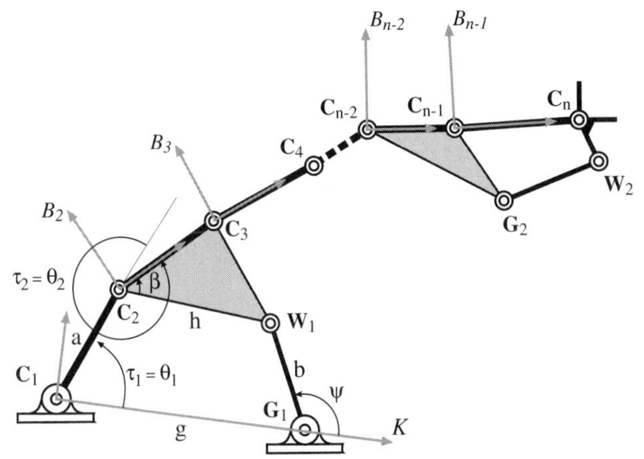

Fig. 9. This shows our conventions for the analysis of a mechanically constrained nR planar robot

TABLE I
FIVE TASK POSITIONS FOR THE END-EFFECTOR OF THE PLANAR 6R ROBOT WITH A SQUARE INITIAL CONFIGURATION

$Task$	$Position(\phi, x, y)$
1	$(0°, 0, -1)$
2	$(-39°, 0.25, 2)$
3	$(-61°, 0.8, 2.35)$
4	$(-94°, 1.75, 2.3)$
5	$(-117°, 2.25, 2.1)$

X. Design of a 6R Planar Chain

We illustrate how the design methodology can be used to control the shape of the deployable chain with two examples on a 6R planar robot, one with a square initial configuration and a second with a hexagonal initial configuration.

A. Example with a square initial configuration

Let the task positions of the planar robot in Table I be represented by $[T_j], j = 1\dots, 5$. These positions represent the movement of the manipulator of the 6R planar robot from a square folded state to its extended state. The forward kinematics of the 6R serial chain is given by

$$[D] = [G][Z(\theta_1)][X(a_{12})]\dots[X(a_{56})[Z(\theta_6)][H]. \quad (25)$$

To solve for the inverse kinematics of the 6R chain, we select $\mathbf{C}_1 = (0, 0)$ such that $[G]$ becomes the identity $[I]$, and the link dimensions of the 6R chain $a_{12} = a_{23} = a_{34} = a_{45} = a_{56} = 1m$. In addition, we assume $[H]$ to be the identity $[I]$,

TABLE II

INVERSE KINEMATICS OF THE 6R CHAIN AT EACH TASK POSITIONS WITH A SQUARE INITIAL CONFIGURATION

Task	σ_j	λ_j	$\theta_{6,j}$
1	270°	−90°	90°
2	181.26°	−49.19°	−23.49°
3	159.75°	−44.28°	−43.65°
4	132.50°	−39.88°	−66.97°
5	118.62°	−37.80°	−84.43°

TABLE III

SOLUTION OF $\mathbf{G}_1\mathbf{W}_1$ WITH A SQUARE INITIAL CONFIGURATION

Pivots	$\mathbf{G}_1$	$\mathbf{W}_1$
1	$(0,0)$	$(0,-1)$
2	$(\mathbf{0.557},\mathbf{0.134})$	$(\mathbf{0.888},-\mathbf{0.223})$
3	$(0.61-0.68i, 0.51+0.62i)$	$(1.17-0.64i, 0.37+0.17i)$
4	$(0.61+0.68i, 0.51-0.62i)$	$(1.17+0.64i, 0.37-0.17i)$

and apply symmetry conditions such that the joint angles at each of the task position, $\theta_{2,j} = \theta_{3,j} = \cdots = \theta_{5,j} = \lambda_j,\ j = 1,\ldots,5$. Also, if we let $\theta_{1,j} = \sigma_j,\ j = 1,\ldots,5$, we are able to solve for σ_j, λ_j, and $\theta_{6,j}$ at each of the task positions by equating the forward kinematics equation (25) with the task $[T_j], j = 1\ldots,5$. See Table II.

Once we solve the inverse kinematics, the positions of its links $B_{1,j}, B_{2,j},\ldots,B_{6,j},\ j = 1,\ldots,5$ in each of the task positions can be determined. This means that we can identify five positions $[T_j^{B_2}],\ j = 1,\ldots,5$, when the end-effector is in each of the specified task positions. These five positions become our task positions in computing the RR chain $\mathbf{G}_1\mathbf{W}_1$ using (22) to constrain B_2 to ground. Note that since B_0 is the ground, $[D_{1j}]$ reduces to $[B_{2,j}][B_{2,1}]^{-1},\ j = 1,\ldots,5$.

Similarly, we compute the RR chain $\mathbf{G}_2\mathbf{W}_2$ using (22) with $[D_{1j}] = [B_{4,1}][B_{4,j}]^{-1}[B_{6,j}][B_{6,1}]^{-1}$.

We used this synthesis methodology and the five task positions listed in Table I to compute the pivots of the RR chains. We obtained two real solutions for $\mathbf{G}_1\mathbf{W}_1$, and four real solutions for $\mathbf{G}_2\mathbf{W}_2$, which result in one of the existing link for each case. Table III, and IV show the solution of the various possible pivots. The required transmission design and tendon routing is shown in Figure 5 and 6 respectively.

TABLE IV

SOLUTION OF $\mathbf{G}_2\mathbf{W}_2$ WITH A SQUARE INITIAL CONFIGURATION

Pivots	$\mathbf{G}_2$	$\mathbf{W}_2$
1	$(0,0)$	$(0,-1)$
2	$(0.239,-0.639)$	$(0.097,-1.063)$
3	$(\mathbf{1.772},-\mathbf{0.298})$	$(\mathbf{0.506},-\mathbf{1.272})$
4	$(3.962, 17.757)$	$(0.243,-0.812)$

TABLE V

FIVE TASK POSITIONS FOR THE END-EFFECTOR OF THE PLANAR 6R ROBOT WITH A HEXAGON INITIAL CONFIGURATION

Task	*Position*(ϕ,x,y)
1	$(0°,-1,0)$
2	$(-39°,-0.15,2)$
3	$(-61°,0.8,2.35)$
4	$(-94°,1.75,2.3)$
5	$(-117°,2.25,2.1)$

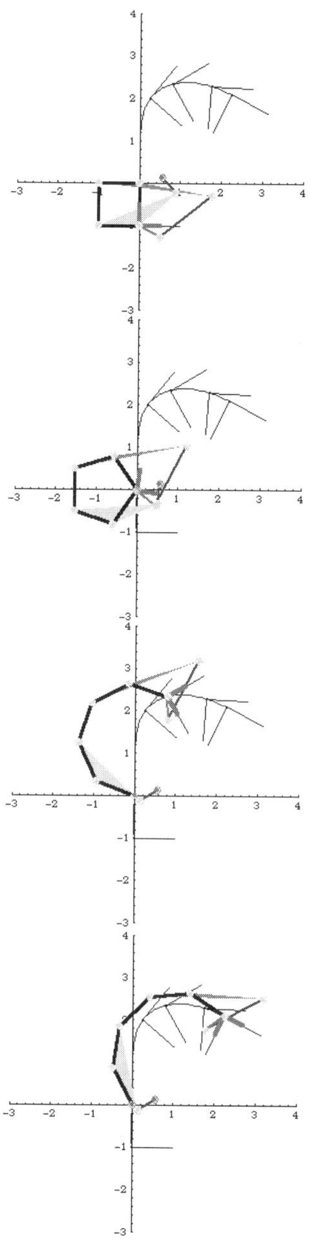

Fig. 10. This mechanically constrained 6R chain guides the end effector through five specified task positions with a square initial configuration

We follow the analysis procedure described in IX to animate the movement of the constrained 6R robot. See Figure 10.

B. Example with a hexagon initial configuration

We follow the procedure described in the earlier section to constrain the 6R chain with a hexagon initial configuration to one degree-of-freedom. Table V represents the movement of the manipulator of the 6R planar robot from a hexagon folded

TABLE VI

SOLUTION OF $\mathbf{G}_1\mathbf{W}_1$ WITH A HEXAGON INITIAL CONFIGURATION

Pivots	$\mathbf{G}_1$	$\mathbf{W}_1$
1	$(0,0)$	$(0.5,-0.866)$
2	$(-0.019,0.030)$	$(3.966,-24.384)$
3	$(0.015,-0.006)$	$(0.396,-0.442)$
4	$(\mathbf{0.179},\mathbf{-0.070})$	$(\mathbf{-0.170},\mathbf{0.141})$

TABLE VII

SOLUTION OF $\mathbf{G}_2\mathbf{W}_2$ WITH A HEXAGON INITIAL CONFIGURATION

Pivots	$\mathbf{G}_2$	$\mathbf{W}_2$
1	$(-1.5,-0.866)$	$(-1,0)$
2	$(\mathbf{-1.379},\mathbf{0.250})$	$(\mathbf{-1.218},\mathbf{0.131})$
3	$(-1.22-0.20i,0.05-0.39i)$	$(-1.04-0.04i,-0.01+0.05i)$
4	$(-1.22+0.20i,0.05+0.39i)$	$(-1.04+0.04i,-0.01-0.05i)$

state to its extended state. We solve the inverse kinematics of the 6R chain at each of the task positions by equating the forward kinematics equation (25). Once we solve the inverse kinematics, the positions of its links was used to design RR chain $\mathbf{G}_1\mathbf{W}_1$, and $\mathbf{G}_2\mathbf{W}_2$. We obtained four real solution for $\mathbf{G}_1\mathbf{W}_1$, and two real solutions for $\mathbf{G}_2\mathbf{W}_2$. Table VI, and VII show the various possible pivots. Again, we use the same transmission design and tendon routing as shown in Figure 5 and 6 respectively.

We follow the analysis procedure described in IX to animate the movement of the constrained 6R robot. See Figure 11

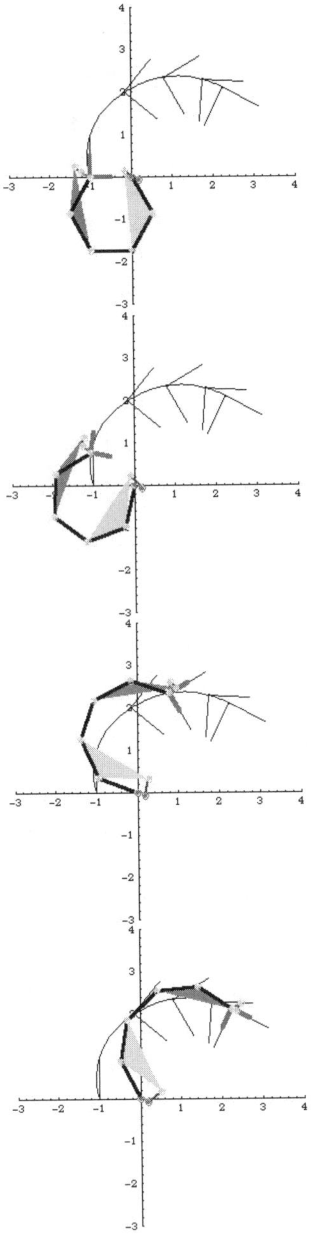

Fig. 11. This mechanically constrained 6R chain guides the end effector through five specified task positions with a hexagon initial configuration

XI. CONCLUSIONS

In this paper, we show that the design of a cable drive train integrated with four-bar linkage synthesis yields new opportunities for the constrained movement of planar serial chains. In particular, we obtain a 6R planar chain through five arbitrary task positions with one degree-of-freedom. The same theory applies to nR chains with more degrees of freedom. Two examples show that the end-effector moves through the specified positions, while the system deploys from square and hexagonal initial configurations.

REFERENCES

[1] J.J. Lee, *Tendon-Driven Manipulators: Analysis, Synthesis, and Control*, PhD. dissertation, Department of Mechanical Engineering, University of Maryland, College Park. MD; 1991.

[2] M. Arsenault and C. M. Gosselin, "Kinematic and Static Analysis of a Planar Modular 2-DoF Tensegrity Mechanism", *Proceedings of the Robotics and Automation*, Orlando, Florida, 2006, pp. 4193-4198.

[3] M. J. Lelieveld and T. Maeno, "Design and Development of a 4 DOF Portable Haptic Interface with Multi-Point Passive Force Feedback for the Index Finger", *Proceedings of the Robotics and Automation*, Orlando, Florida, 2006, pp. 3134-3139.

[4] G. Palli and C. Melchiorri, "Model and Control of Tendon-Sheath Transmission Systems", *Proceedings of the Robotics and Automation*, Orlando, Florida, 2006, pp. 988-993.

[5] Krovi, V., Ananthasuresh, G. K., and Kumar, V., "Kinematic and Kinetostatic Synthesis of Planar Coupled Serial Chain Mechanisms," *ASME Journal of Mechanical Design*, 124(2):301-312, 2002.

[6] A. S. K. Kwan and S. Pellegrino, "Matrix Formulation of Macro-Elements for Deployable Structures," *Computers and Structures*, 15(2):237-251, 1994.

[7] S. O. Leaver and J. M. McCarthy, "The Design of Three Jointed Two-Degree-of-Freedom Robot Fingers," *Proceedings of the Design Automation Conference*, vol. 10-2, pp. 127-134, 1987.

[8] S. O. Leaver, J. M. McCarthy and J. E. Bobrow, "The Design and Control of a Robot Finger for Tactile Sensing," *Journal of Robotic Systems*, 5(6):567-581, 1988.

[9] L.W. Tsai, *Robot Analysis: The Mechanics of Serial and Parallel Manipulators*, John Wiley & Sons Inc, NY; 1999.

[10] G.N. Sandor and A.G. Erdman, A., *Advanced Mechanism Design: Analysis and Synthesis, Vol. 2.* Prentice-Hall, Englewood Cliffs, NJ; 1984.

[11] J. M. McCarthy, *Geometric Design of Linkages*, Springer-Verlag, New York; 2000.

Automatic Scheduling for Parallel Forward Dynamics Computation of Open Kinematic Chains

Katsu Yamane
Department of Mechano-Informatics
University of Tokyo
Email: yamane@ynl.t.u-tokyo.ac.jp

Yoshihiko Nakamura
Department of Mechano-Informatics
University of Tokyo
Email: nakamura@ynl.t.u-tokyo.ac.jp

Abstract— **Recent progress in the algorithm as well as the processor power have made the dynamics simulation of complex kinematic chains more realistic in various fields such as human motion simulation and molecular dynamics. The computation can be further accelerated by employing parallel processing on multiple processors. In fact, parallel processing environment is becoming more affordable thanks to recent release of multiple-core processors. Although several parallel algorithms for the forward dynamics computation have been proposed in literature, there still remains the problem of automatic scheduling, or load distribution, for handling arbitrary kinematic chains on a given parallel processing environment. In this paper, we propose a method for finding the schedule that minimizes the computation time. We test the method using three human character models with different complexities and show that parallel processing on two processors reduces the computation time by 35–36%.**

I. Introduction

Dynamics simulation of kinematic chains is an essential tool in robotics to test a mechanism or controller before actually conducting hardware experiments. In graphics, such technique can be applied to synthesizing realistic motions of virtual characters and objects. Dynamics simulation of human body models has a number of applications in biomechanics and medical fields. Some algorithms have also been applied to molecular dynamics to estimate the three-dimensional structure of chemical materials such as protein. In spite of the recent progress in algorithms and processor power, realtime simulation of highly complex systems (over 100 links) is still a challenging research issue.

One of the possible ways to further improve the computation speed is to employ parallel processing. In fact, several parallel algorithms have been proposed for the forward dynamics computation of kinematic chains [1]–[4]. These algorithms have $O(N)$ complexity on fixed number of processors and $O(\log N)$ complexity if $O(N)$ processors are available. These algorithms therefore require huge number of processors to fully appreciate the power of parallel computation, and it was considered unrealistic to assume such computational resource for personal use.

The situation has changed by the recent release of multiple-core processors because they would significantly reduce the cost for constructing a parallel processing environment. In particular, the Cell^{TM} processor [5] used in PlayStation3 has 7–8 vector processing cores which function as parallel processors with distributed memory. By optimizing the parallel forward dynamics algorithms for this processor, we would be able to considerably accelerate the computation for dynamics simulation on low-cost hardware.

Another technical barrier towards practical parallel dynamics simulation is finding the optimal scheduling, or load distribution, for a given structure and number of processors. In general, the total amount of floating-point operations increases as the number of parallel processes increases. For example, the total computational cost of a schedule that divides the computation into four processes is greater than that of two processes, and the computation time will not be reduced unless the program runs on more than four processors. Another point to be considered is that waiting time of the processors should be kept minimum to maximize the effect of parallel processing. These facts imply that the optimal schedule depends on the number of available processors as well as the target structure. For a dynamics simulator to be practical for personal use, it should be able to automatically find the optimal schedule.

In this paper, we propose an automatic scheduling method that can find the optimal schedule for any given open kinematic chain and number of available processors. A* search algorithm is applied to obtaining the best schedule. Although the scheduling process only takes place during the initialization or when the link connectivity has changed, we employ several heuristics to find the solution in a reasonable time. Our method is also applicable to fairly wide range of parallel forward dynamics algorithms. Although we have only tested our method on Assembly-Disassembly Algorithm (ADA) proposed by the authors [1], the same method can be applied to Divide-and-Conquer Algorithm (DCA) [2] and Hybrid Direct-Iterative Algorithm (HDIA) [3] with small modifications.

The rest of the paper is organized as follows. We first provide the background information of this work in section II, including a brief summary of the parallel forward dynamics algorithm ADA [1]. Section III presents the automatic scheduling method, which is the main contribution of this paper. The performance of the scheduling method will be demonstrated in section IV, followed by the concluding remarks.

II. Parallel Forward Dynamics Computation

This section provides a brief summary of the parallel forward dynamics algorithm ADA [1], [6]. We first review

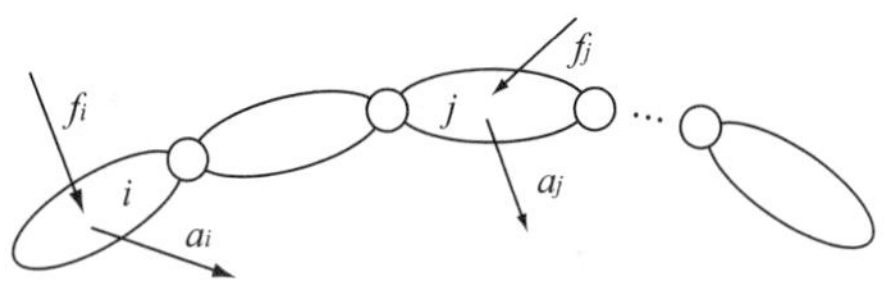

Fig. 1. The concept of Articulated Body Inertia.

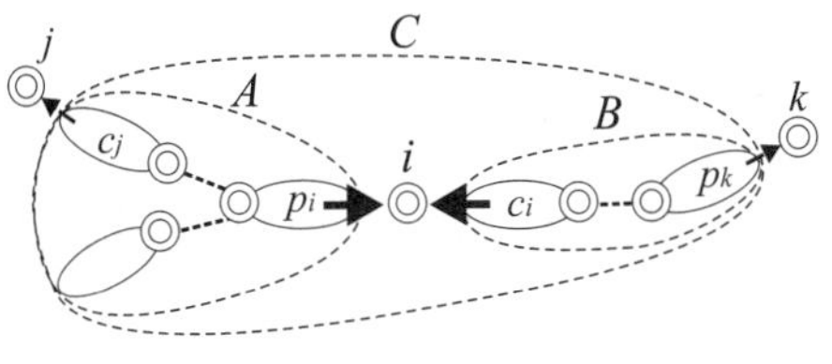

Fig. 2. Connecting partial chains A and B to assemble partial chain C.

the concept of Articulated-Body Inertia and its inverse [7] and then summarize the outline of ADA. After discussing the computational cost of the algorithm, we introduce the concept of *schedule tree* for representing a schedule to easily identify its parallelism and computation time. We also show the results of parallel processing experiments of a simple serial chain and demonstrate the importance of finding the optimal schedule.

A. Articulated-Body Inertia

The concept of Articulated-Body Inertia (ABI) was first proposed by Featherstone [7]. ABI is the apparent inertia of a collection of links connected by joints (articulated body) when a test force is applied to a link (handle) in the articulated body. The equation of motion of a kinematic chain can be written in a compact form by using ABI (see Fig. 1):

$$\boldsymbol{f}_i = \boldsymbol{I}_i^A \boldsymbol{a}_i + \boldsymbol{p}_i^A \tag{1}$$

where $\boldsymbol{I}_i^A$ is the ABI of the articulated body when link i is the handle, $\boldsymbol{f}_i$ is the test force, $\boldsymbol{p}_i^A$ is the bias force, and $\boldsymbol{a}_i$ is the acceleration of link i. Note that the ABI of a kinematic chain may change if a different link is chosen as the handle. All the equations in this paper will be represented by spatial notation [7].

Because ABI is symmetric and positive definite [7], Eq.(1) is equivalent to

$$\boldsymbol{a}_i = \boldsymbol{\Phi}_i \boldsymbol{f}_i + \boldsymbol{b}_i \tag{2}$$

where $\boldsymbol{\Phi}_i$ is the inverse of ABI and called Inverse Articulated Body Inertia (IABI) [7] and $\boldsymbol{b}_i$ is the bias acceleration.

The major advantage of using Eq.(2) instead of Eq.(1) is that we can attach multiple handles to an articulated body. For example, if we attach a new handle to link j in Fig. 1, the accelerations of two handles i and j is written as

$$\begin{aligned}\boldsymbol{a}_i &= \boldsymbol{\Phi}_i \boldsymbol{f}_i + \boldsymbol{\Phi}_{ij} \boldsymbol{f}_j + \boldsymbol{b}_i \\ \boldsymbol{a}_j &= \boldsymbol{\Phi}_{ji} \boldsymbol{f}_i + \boldsymbol{\Phi}_j \boldsymbol{f}_j + \boldsymbol{b}_j\end{aligned}$$

where $\boldsymbol{\Phi}_{ij}$ and $\boldsymbol{\Phi}_{ji}(= \boldsymbol{\Phi}_{ij}^T)$ are the IABIs representing the coupling between the handles.

B. Outline of ADA

ADA computes the joint acceleration by the following two steps:

1) assembly: starting from the individual links, recursively add a joint one by one and compute the IABI of the partial chains, and
2) disassembly: starting from the full chain, remove a joint one by one in the reverse order of step 1) and compute the joint acceleration of each removed joint.

Consider the case of Fig. 2, where partial chains A and B are connected to form chain C through joint i. Partial chain C will be connected to other partial chains through joints j and k in the subsequent assembly operations. In general, each partial chain can have any number of such joints and we shall denote the set by $\mathcal{O}_i$ and the size of $\mathcal{O}_i$ by N_{Oi}. $\mathcal{O}_i$ is empty for the joint lastly added in step 1). In the example of Fig. 2, $\mathcal{O}_i = \{j\ k\}$ and $N_{Oi} = 2$.

In step 1), assuming that we know the IABI of chains A and B, $\boldsymbol{\Phi}_{A**}$ and $\boldsymbol{\Phi}_{B**}$ respectively, as well as the bias accelerations $\boldsymbol{b}_{A*}$ and $\boldsymbol{b}_{B*}$, we compute the IABIs and bias accelerations of chain C by the following equations:

$$\begin{aligned}\boldsymbol{\Phi}_{Cj} &= \boldsymbol{\Phi}_{Aj} - \boldsymbol{\Phi}_{Aji} \boldsymbol{W}_i \boldsymbol{\Phi}_{Aij} && (3)\\ \boldsymbol{\Phi}_{Ck} &= \boldsymbol{\Phi}_{Bk} - \boldsymbol{\Phi}_{Bki} \boldsymbol{W}_i \boldsymbol{\Phi}_{Bik} && (4)\\ \boldsymbol{\Phi}_{Cjk} &= \boldsymbol{\Phi}_{Ckj}^T = \boldsymbol{\Phi}_{Aji} \boldsymbol{W}_i \boldsymbol{\Phi}_{Bjk} && (5)\\ \boldsymbol{b}_{Cj} &= \boldsymbol{b}_{Aj} - \boldsymbol{\Phi}_{Aji} \boldsymbol{\gamma}_i && (6)\\ \boldsymbol{b}_{Ck} &= \boldsymbol{b}_{Bk} + \boldsymbol{\Phi}_{Bki} \boldsymbol{\gamma}_i && (7)\end{aligned}$$

where

$$\begin{aligned}\boldsymbol{W}_i &= \boldsymbol{N}_i^T \boldsymbol{V}_i \boldsymbol{N}_i \\ \boldsymbol{\gamma}_i &= \boldsymbol{W}_i \boldsymbol{\beta}_i + \boldsymbol{R}_i^T \boldsymbol{\tau}_i.\end{aligned}$$

$\boldsymbol{N}_i \in \boldsymbol{R}^{6\times(6-n_i)}$ and $\boldsymbol{R}_i \in \boldsymbol{R}^{6\times n_i}$ are the matrices that represent the constraint and motion space of joint i respectively [8], $\boldsymbol{\tau}_i \in \boldsymbol{R}^{n_i}$ is the joint torque of joint i, n_i is the degrees of freedom of joint i, and

$$\begin{aligned}\boldsymbol{V}_i &= (\boldsymbol{N}_i(\boldsymbol{\Phi}_{Ai} + \boldsymbol{\Phi}_{Bi})\boldsymbol{N}_i^T)^{-1} \\ \boldsymbol{\beta}_i &= \boldsymbol{b}_{Ai} - \boldsymbol{b}_{Bi} - (\boldsymbol{\Phi}_{Ai} + \boldsymbol{\Phi}_{Bi})\boldsymbol{R}_i^T \boldsymbol{\tau}_i\end{aligned}$$

where we omit the terms including the time derivatives of $\boldsymbol{N}_i$ and $\boldsymbol{R}_i$ for clarity of representation. The initial values for IABI are the spatial inertia matrices of the rigid bodies.

In step 2), assuming that we know the forces applied joints j and k, we can compute the constraint force at joint i by

$$\boldsymbol{n}_i = \boldsymbol{V}_i \boldsymbol{N}_i (\boldsymbol{\Phi}_{Aij} \boldsymbol{f}_j - \boldsymbol{\Phi}_{Bik} \boldsymbol{f}_k + \boldsymbol{\beta}_i) \tag{8}$$

and then compute the joint acceleration by

$$\ddot{q}_i = \boldsymbol{R}_i((\boldsymbol{\Phi}_{Ai} + \boldsymbol{\Phi}_{Bi})\boldsymbol{N}_i^T \boldsymbol{n}_i - \boldsymbol{\Phi}_{Aij} \boldsymbol{f}_j + \boldsymbol{\Phi}_{Bik} \boldsymbol{f}_k - \boldsymbol{\beta}_i). \tag{9}$$

Finally $\boldsymbol{f}_i$, the total joint force at joint i which will be used in the subsequent computations, is computed by

$$\boldsymbol{f}_i = \boldsymbol{N}_i^T \boldsymbol{n}_i + \boldsymbol{R}_i^T \boldsymbol{\tau}_i. \tag{10}$$

The total computational cost for assembling and disassembling joint i in Fig. 2 depends on two factors: (1) n_i, the degrees of freedom of joint i, and (2) N_{Oi}, number of joints in $\mathcal{O}_i$. The cost can be approximated by the following formulae:

$$C_i(N_{Oi}, n_i) = \alpha N_{Oi}^2 + \beta N_{Oi} + \gamma n_i + \delta \qquad (11)$$

where the first term represents the cost for computing the N_{Oi}^2 IABIs, the second term for computing the N_{Oi} bias accelerations and one joint acceleration from N_{Oi} external forces, the third term for computing $\boldsymbol{W}_i$, the last term for computing the inertia matrices of the individual links, and α, β, γ and δ are constants. Because only N_{Oi} can be modified by changing the schedule, Eq.(11) indicates that smaller N_{Oi} would lead to less total floating-point operations.

The constants could be determined by counting the number of floating-point operations as a function of N_{Oi} and n_i; however, we chose to determine them by actually measuring the computation time for various mechanisms and schedules. The advantage of this method is that it can take into account the implementation-specific optimizations and that it would be possible to write a code to automatically determine the values for other forward dynamics algorithms.

C. Schedule Tree

ADA can be executed on multiple processes in parallel because of the following reasons:

- in the example shown in Fig. 2, the computations of the IABIs of partial chains A and B can be performed in parallel, and
- similarly, the accelerations of the joints included in partial chains A and B can be computed in parallel.

In addition, ADA allows any sequence of joints in step 1), and the parallelism and computational efficiency depends on the sequence. Scheduling can therefore be regarded as the problem of finding the best sequence of joints for step 1). However, it is difficult to identify the number of processes required to perform the schedule by looking only at the sequence.

We propose to represent a schedule by a binary tree where each node denotes a joint and every joint is included once. We refer to this tree as *schedule tree*. Figure 3 shows the three schedule trees derived from different assembly sequences (a)–(c) for a simple four-joint serial chain shown on the top of Fig. 3. The concept is similar to *assembly tree* [2] where each node of the tree represents a partial chain, while in our schedule tree it represents a joint to clarify the relationship between the joints and processors.

In a schedule tree, each node has zero to two direct descendants. If a node has two descendants as node 2 in schedules (b) and (c) of Fig. 3, the joint connects two partial chains composed of all the subsequent descendants of each direct descendant. A node with one descendant connects a rigid link and a partial chain, and a node without a descendant connects two rigid links. The recursive process to compute the IABIs starts from the leaves towards the root, while the process to compute the joint accelerations propagates from the root towards the leaves.

A schedule tree is useful for identifying which joints can be processed in parallel and estimating the total computation

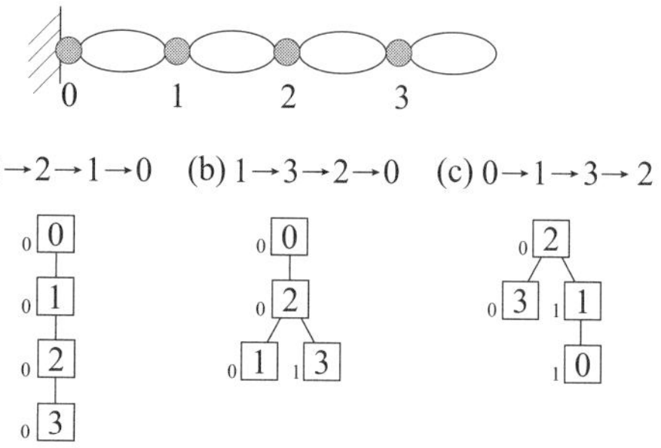

Fig. 3. Three schedule tree examples for different assembly orders of the four-joint serial chain on the top.

time. The two descendants of a node in a schedule tree can obviously be processed in parallel. For example, the numbers at the lower-left corner of each node in Fig. 3 indicates the process in which the node is handled when two processors numbered 0 and 1 are available. If the number of available processors is greater than the number of the leaves of the tree, the height of the schedule tree gives a rough estimate of the computation time. For example, the computation time for schedule (a) in Fig. 3 would be four times longer than that of processing a single joint, while those for schedules (b) and (c) would be approximately equivalent to processing three joints if more than two processors are available.

D. Parallel Processing Experiments

We implemented ADA for parallel processing using C++ programming language and MPICH2 [9] for inter-process communication. The codes were compiled by gcc version 4.0.2. All the examples, including the ones in section IV, were executed on a cluster consisting of dual Xeon 3.8GHz servers running Linux operating system.

Figure 4 shows the performance of parallel forward dynamics computation of a 200-joint serial chain with fixed root link. The solid line represents the computation times on two processors with various schedules. The schedule trees were constructed manually by first choosing a joint as the root, whose index is indicated by the horizontal axis of Fig. 4, and then sequentially appending the joints of each of the partial chains divided by the chosen joint, the joints next to the root being the direct descendants. The assembly process starts by assembling the joints at the both ends in parallel, and ends at the root of the schedule tree. Schedule (c) in Fig. 3 is an example of such schedule for the four-joint chain with root index 2.

The horizontal dashed line in Fig. 4 represents the computation time on single processor when the links were assembled sequentially from the end link to the root, which results in the minimum number of floating-point operation and therefore is the best schedule for serial computation.

As intuitively expected, parallel computation shows the best performance when the computational load is equally distributed to the two processors, which reduces the computation time by 33%. However, the performance degrades if inappro-

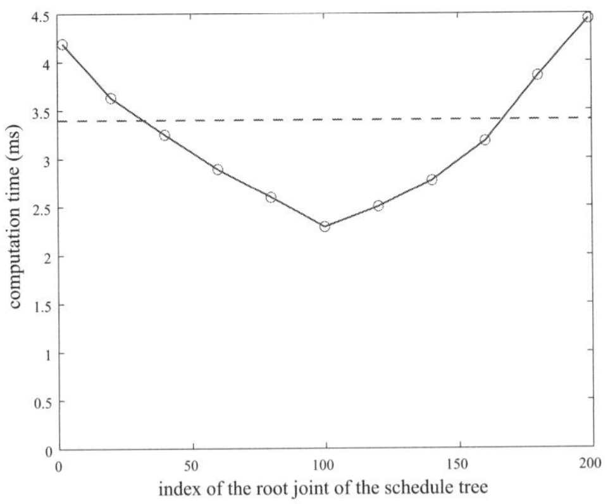

Fig. 4. Computation time of the forward dynamics of a 200-joint serial chain on two processors with various schedules. The dashed horizontal line represents the computation time on single processor.

priate schedule is used and becomes even worse than serial computation due to the communication cost. This experiment demonstrates the importance of selecting appropriate schedule to fully extract the advantage of parallel processing. We also need an algorithm for automatically finding the optimal schedule for more general cases, e.g. for branched chains or when more processors are available, because the optimal schedule is not trivial any more.

III. Automatic Scheduling

A. Overview

The purpose of the automatic scheduling process is to automatically find the best schedule, or the best assembly order, that minimizes the computation time for the given mechanism and number of available processes. As described in the previous section, the forward dynamics algorithms we consider in this paper allow any assembly order. The number of all possible orders therefore becomes as large as $N!$ for a mechanism with N joints. We employ several heuristics to avoid the search in such a huge space.

The first observation is that different assembly orders may lead to the same schedule tree. This fact can be easily illustrated by using the examples shown in Fig. 5 for the same serial chain as in Fig. 3. In Fig. 5, the two schedules (a) and (b) result in the conceptually equivalent schedule trees. The order of adding joints 1 and 3 obviously does not affect the schedule tree because the resulting partial chains are independent of each other. We can reduce the search space by eliminating such duplicated assembly orders.

The second observation is that, as described in section III-D, the best schedule can be determined without further running the search process if the number of processes assigned to a partial chain becomes one. This fact implies that the cost for the search depends more on the number of processors rather than the number of joints. We can considerably reduce the search time because practical parallel processing environments usually have far less processors than the number of joints.

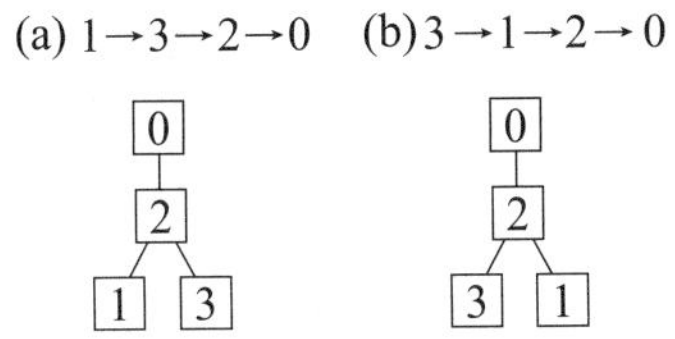

Fig. 5. Different assembly orders resulting in the same schedule tree.

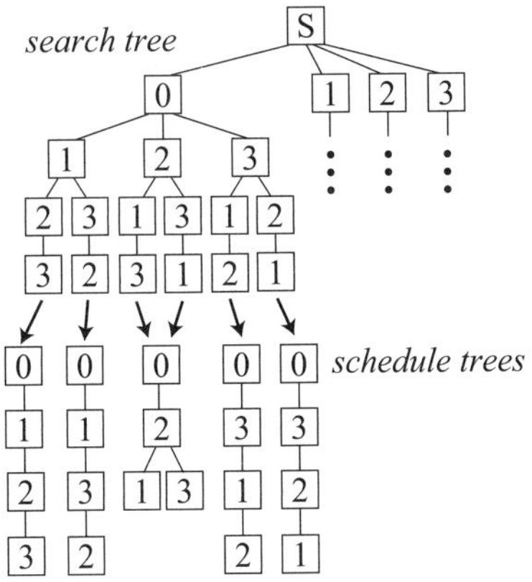

Fig. 6. Search and schedule trees.

B. Search Algorithm

The problem of finding the optimal schedule can be viewed as a travelling salesman problem where the joints are the cities to visit and the cost function is the time to cover all cities. The difference in multiple-processor case is that more than one salesmen are working in parallel and each city is visited by only one of them. We also have the constraint that a city should be visited later than some other cities due to the dependencies between the IABI of partial chains.

We apply A* search [10] to our problem of finding the optimal schedule. The general form of A* search is summarized in the Appendix. The following three functions for a node x have to be customized for our problem:

- $x.NextNodes()$: returns the set of descendant nodes of x
- $x.Cost()$: returns the cost to add x to the search tree
- $x.AstarCost()$: returns an underestimation of the cost from x to a goal

One point to note here is the relationship between the schedule tree and the tree constructed during the search (called the *search tree* hereafter). Each path from the root node to a leaf node of a search tree is associated with a schedule tree. We depict an example of the search tree and associated schedule trees in Fig. 6 for the simple serial chain in Fig. 3, where the node marked "S" is a dummy start node.

We describe the three functions customized for our problem in the rest of this subsection.

1) NextNodes(): A naive way to implement this function is to add all unvisited joints as the descendants. However, this approach may yield duplicated schedules as observed in the

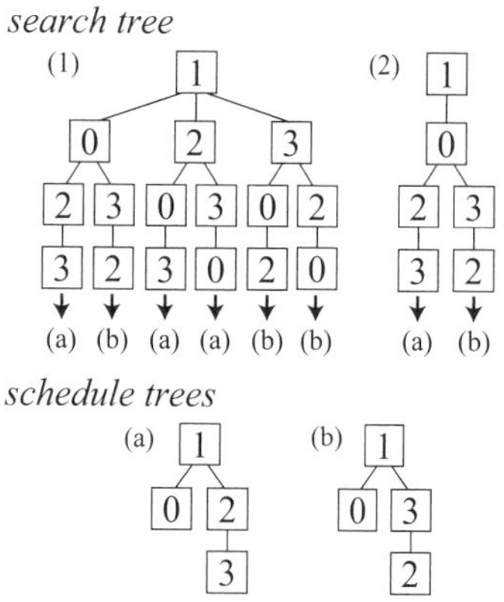

Fig. 7. Search tree by naive version of *NextNodes()* (1) and improved version (2).

previous subsection.

In order to reduce the number of nodes in the search tree, we select the descendants such that the corresponding schedule tree can be effectively extended. When we try to extend the search tree by adding descendants to node i associated with joint i, we construct the (incomplete) schedule tree derived from the joint sequence up to node i, and look for nodes which have only 0–1 descendants. Let us denote the index of such joint in the highest layer of the schedule tree by j. Cutting joint j generates two partial chains: one on the root side and the other on the end link side. If joint j has no descendants, the joints on the root side will be added as direct descendants of joint i in the search tree, which will eventually add the first descendant of joint j in the schedule tree. If joint j has one descendant, which means that the joints on the root side have already been added, we add the joints on the end link side.

Figure 7 shows the comparison between the naive version of *NextNode()* and our improved version, using the part of the search tree in Fig. 6 under node 1 directly below the start node. Although the naive version generates the large search tree (1), each of them results in one of the schedule trees (a) and (b). Using the improved version (2), on the other hand, the tree only includes necessary nodes.

The search tree (2) is constructed as follows. Suppose we are extending the search tree by adding descendant(s) to node 1. Because the corresponding schedule tree also has only one node associated with joint 1, we try to add the descendants to this node. Joint 1 divides the chain into two partial chains composed of the joints $\{0\}$ (towards the root) and $\{2, 3\}$ (towards the end). We therefore add joint 0 as the descendant of node 1 of the search tree. In the next step, because node 1 in the schedule tree has only one descendant, we try to find the second, which should be either joint 2 or 3. Two branches are therefore added to node 0 in the search tree.

2) Cost(): The cost of a node is defined as the total computation time increased by adding the joint. At each node in the search tree, we maintain the total active time of each process. The computation time for processing a node is added to the total active times of all the processes assigned to the node. The actual computation time is obtained by looking for the maximum active time among the processes. The cost of a node is then obtained by subtracting the total cost of its ascendant from its own total cost.

3) AstarCost(): We first compute the minimum costs to assemble the two partial chains generated by cutting the corresponding partial chain at the joint, using the method described in section III-D. We then take the larger cost and divide it by the number of processors available at the node. This is guaranteed to be an underestimate because a schedule for multiple processors results in more total floating-point operations than the one for single processor.

C. Assigning Processes

Once we have a schedule tree, we then assign the processes to each node. The point to be considered here is that the time for communication between the processes (in shared-memory environments, the time for waiting for the permission to access the memory) should be kept minimum. Strictly speaking, the amount of data to be passed also affects the communication time and should be considered in the scheduling process. However, we ignore the data size because it is usually small (288 bytes for each IABI) and the delay for invoking the communication is a more serious issue in most communication devices.

Let us denote the number of all available processes by N_P, which are numbered $0 \ldots N_P - 1$. Here we assume that N_P is a power of 2. The basic approach is to assign a group of processes to a node in the schedule tree, and divide the group into two if the node has two descendants. We denote the range of processes assigned to node i of the schedule tree by $[a_{Pi}, b_{Pi})$ which means that processes $a_{Pi}, a_{Pi}+1, \ldots b_{Pi}-1$ are assigned to node i. The actual computations for assembling and disassembling joint i take place at process a_{Pi}.

The procedure for assigning processes to the nodes is described as follows:

1) Initialize $a_{Pr} = 0$ and $b_{Pr} = N_P$ where r is the index of the root node.
2) At node i
   a) If $b_{Pi} - a_{Pi} = 1$, or if node i has one descendant, set the same range for the descendant(s).
   b) If node i has two descendants m and n, and $b_{Pi} - a_{Pi} > 1$, set $a_{Pm} = a_{Pi}, b_{Pm} = a_{Pn} = b_{Pi} + (b_{Pi} - a_{Pi})/2, b_{Pn} = b_{Pi}$.
3) Recursively call step 2) for all descendants.

The communication cost is minimized by ensuring that one of the descendants is handled at the same process, therefore requiring no communication.

D. Optimal Schedule for Single Process

The optimal schedule for single process is the one that minimizes the number of floating-point operations. As Eq.(11) and the subsequent discussion imply, we have to keep N_{Oi} at every joint as small as possible. To realize this, for each joint we divide the joints in $\mathcal{O}_i$ into two groups by checking whether they exists towards the root side or end side of the

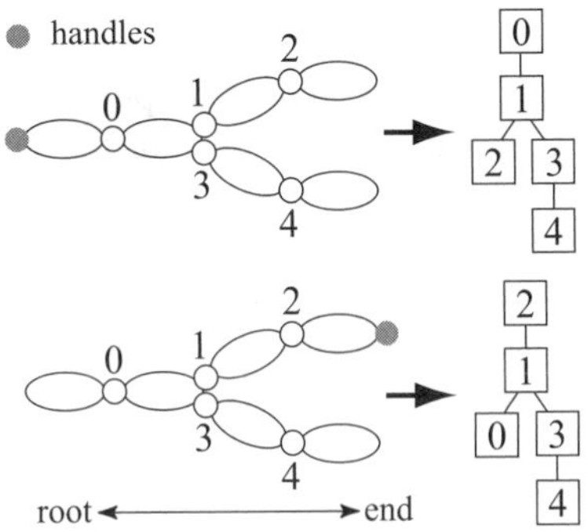

Fig. 8. Examples of partial chains and their optimal schedule for single process, with different $\mathcal{O}_i$.

joint. Let us denote the number of joints in the two groups by N^r_{Oi} and N^e_{Oi} respectively. If $N^r_{Oi} \leq N^e_{Oi}$ we put the joint at the lower layer of the schedule tree, i.e. processed earlier in the assembly step, because we can keep N_{Oi} smaller than in the reverse order. This principle defines the sequence of adding joints where there is no branches. At branches, we count the number of joints included in $\mathcal{O}_i$ in each branch and then process the branches in the ascending order of the number.

Figure 8 shows two examples of optimal schedules for a partial chain composed of joints 0–4, with different $\mathcal{O}_i$. In both cases, $\mathcal{O}_i$ includes one joint represented by the gray circle. The branch neighboring the joint in $\mathcal{O}_i$ should be processed later in the assembly process regardless of the hierarchy in the structure. The joints in that branch are therefore placed near the top of the schedule tree.

IV. Experiments

A. Setup

We used the same computer environment as in section II-D and determined the constants in Eq.(11) as $\alpha = 1.6, \beta = 1.0, \gamma = -1.0$, and $\delta = 14.4$, which gives the total computation time in µs for assembling and disassembling a joint. The negative value for γ is because the computation of $\boldsymbol{V}_i$ involves the inversion of a $(6 - n_i) \times (6 - n_i)$ matrix. The coefficients for n_i^3 and n_i^2 turned out to be too small to be identified by this method, and therefore practically negligible.

B. Serial Chains

We applied the method to finding the optimal schedule for handling serial chains on two- and four-processor environments. In the two-processor case, the optimal schedule is trivial as shown in Fig. 4: the root of the schedule tree is the joint at the middle of the chain, and other joints are added sequentially towards the ends. We therefore only confirm that the method can find the trivial solution. The four-processor case is no longer trivial. Dividing the chain into four equal-length partial chains and assigning one process to each partial chain is not the optimal schedule because the two partial chains in the middle requires more computations than those at the ends because of larger $N_{Oi}(= 2)$.

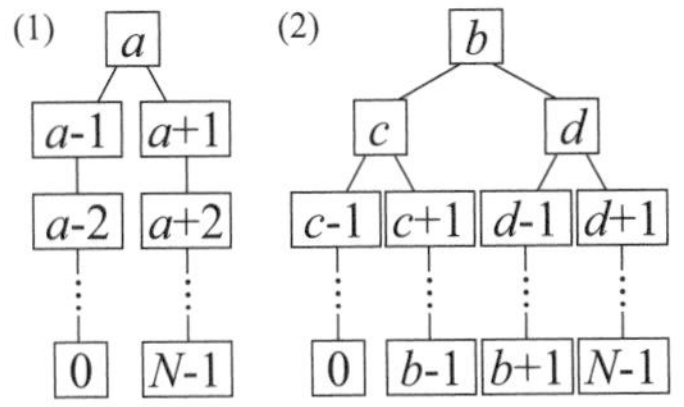

Fig. 9. Optimal schedules for handling N-joint serial chain on two processors (1) and four processors (2).

TABLE I

THE VALUES OF a, b, c AND d FOR THE SCHEDULE TREES IN FIG. 9.

DOF	2 processes	4 processes		
N	a	b	c	d
16	7	7	3	11
32	15	15	8	22
64	31	31	17	45

The scheduling algorithm derived the schedule trees for handling N-joint serial chain on two and four processors shown in Fig. 9, where the actual indices a, b, c and d were the values shown in Table I for $N = 16, 32$ and 64. The schedules for two processors match the trivial solutions. Those for four processors also reasonable because the internal two partial chains are slightly shorter than the others.

C. Branched Chains

We applied the method to finding the optimal schedule for handling branched chains on two processors. The scheduling is not trivial as in the case of serial chains. We used three human figures with different complexities (Fig. 10): 40 DOF, 52 DOF, and 161 DOF composed of 1 DOF rotational, 3 DOF spherical, and 6 DOF free joints. In all models, the method selected a joint in the backbone near the neck as the root of the schedule tree. Table II shows the computation time on one to four processors, and the ratios of speedup. Parallel computation reduced the computation time by 38–43% for two processors and 39–54% for four processors. The speedup gain is comparable to that of Fig. 4 which is the ideal load balance case. The effect of parallel computation was more prominent in complex models.

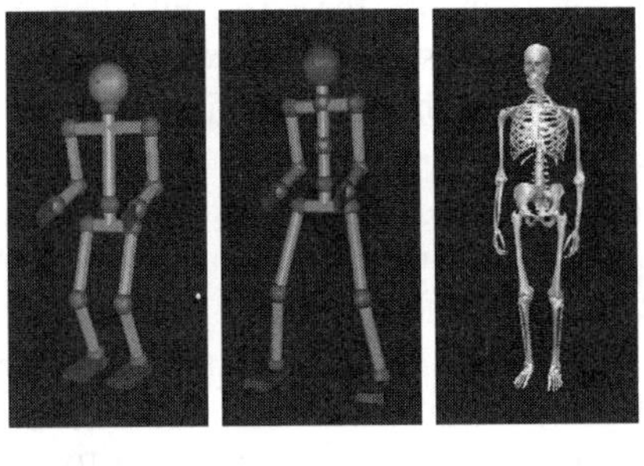

Fig. 10. Three human character models for test; from left to right: 40 DOF, 52 DOF, and 161 DOF.

TABLE II

COMPUTATION TIMES FOR SERIAL AND PARALLEL COMPUTATIONS (MS) AND RATIO OF SPEEDUP.

DOF	40	52	161
♯ of joints	15	19	53
1 process	0.249	0.301	0.773
2 processes (speedup)	0.155 (38%)	0.186 (38%)	0.443 (43%)
4 processes (speedup)	0.153 (39%)	0.177 (41%)	0.356 (54%)

V. CONCLUSION

In this paper, we proposed a method for automatically scheduling the parallel forward dynamics computation of open kinematic chains. The conclusion of the paper is summarized as follows:

1) We proposed an efficient method for applying A* search algorithm to the scheduling problem.
2) We proposed a systematic method for assigning processors to the partial chains considering the communication cost.
3) The method was applied to three human character models with different complexity and reduced the computation time by up to 43% on two processors and 54% on four processors.

The method is applicable to parallel forward dynamics algorithms which can be physically interpreted as successive connection of two partial chains described as schedule trees, e.g. [1]–[3]. The coefficients of Eq.(11) should be modified accordingly.

Future work includes extension to closed kinematic chains. This problem can be partially solved by the following procedure: (1) cut the loops by removing some joints and apply the method described in this paper to the resulting open chain, and (2) insert the removed joints to the root of the schedule tree, although the resulting schedule may not be optimal. Another issue is the relationship between the schedule and numerical accuracy. If the schedule affects the accuracy, it would be possible and beneficial to optimize the accuracy as well in the automatic scheduling process.

ACKNOWLEDGEMENTS

This work was supported by the Ministry of Education, Culture, Sports, Science and Technology (MEXT) and the New Energy and Industrial Technology Development Organization (NEDO), Japan.

REFERENCES

[1] K. Yamane and Y. Nakamura, "Efficient Parallel Dynamics Computation of Human Figures," in *Proceedings of the IEEE International Conference on Robotics and Automation*, May 2002, pp. 530–537.

[2] R. Featherstone, "A Divide-and-Conquer Articulated-Body Algorithm for Parallel $O(\log(n))$ Calculation of Rigid-Body Dynamics. Part1: Basic Algorithm," *International Journal of Robotics Research*, vol. 18, no. 9, pp. 867–875, September 1999.

[3] K. Anderson and S. Duan, "Highly Parallelizable Low-Order Dynamics Simulation Algorithm for Multi-Rigid-Body Systems," *AIAA Journal on Guidance, Control and Dynamics*, vol. 23, no. 2, pp. 355–364, March-April 2000.

[4] A. Fijany, I. Sharf, and G. D'Eleuterio, "Parallel $O(\log N)$ Algorithms for Computation of Manipulator Forward Dynamics," *IEEE Transactions on Robotics and Automation*, vol. 11, no. 3, pp. 389–400, 1995.

[5] IBM Research, "The Cell Project at IBM Research," online, http://www.research.ibm.com/cell/.

[6] K. Yamane and Y. Nakamura, "$O(N)$ Forward Dynamics Computation of Open Kinematic Chains Based on the Principle of Virtual Work," in *Proceedings of IEEE International Conference on Robotics and Automation*, 2001, pp. 2824–2831.

[7] R. Featherstone, *Robot Dynamics Algorithm*. Boston, MA: Kluwer Academic Publishers, 1987.

[8] K. Yamane and Y. Nakamura, "Parallel $O(\log N)$ Algorithm for Dynamics Simulation of Humanoid Robots," in *Proceedings of IEEE-RAS International Conference on Humanoid Robotics*, Genoa, Italy, December 2006, pp. 554–559.

[9] Mathematics and Computer Science Division, Argonne National Laboratory, "MPICH2 Home Page," online, http://www.mcs.anl.gov/mpi/mpich2/.

[10] Steven M. LaValle, *Planning Algorithms*. New York, NY: Cambridge University Press, 2006.

APPENDIX

Algorithm 1 shows the general form of A^* search [10], where T is the search tree, Q is a list of nodes sorted in the ascending order of $priority\_cost$ of each node. The following operations are predefined for list Q and search tree T:

- $Q.GetFirst()$: extract the first node in Q
- $Q.Insert(x)$: insert node x to Q such that the nodes are aligned in the ascending order of $x.priority\_cost$
- $T.SetRoot(x)$: set node x as the root of T
- $T.AddDescendant(x, x')$: add x' to T as a descendant of x

while the actions of the following methods for node x should be customized to fit the particular problem:

- $x.NextNodes()$: returns the set of descendant nodes of x
- $x.Cost()$: returns the cost to add x to the search tree
- $x.AstarCost()$: returns an underestimation of the cost from x to a goal

Algorithm 1 General A* Search

Require: the initial node x_I and set of goal nodes X_G

```
1:  x_I.total_cost ← 0
2:  x_I.priority_cost ← 0
3:  Q.Insert(x_I)
4:  T.SetRoot(x_I)
5:  while Q not empty do
6:    x ← Q.GetFirst()
7:    if x ∈ X_G then
8:      return x
9:    end if
10:   for all x' ∈ x.NextNodes() do
11:     T.AddDescendant(x, x')
12:     x'.total_cost ← x.total_cost + x'.Cost()
13:     x'.priority_cost ← x'.total_cost + x'.AstarCost()
14:     Q.Insert(x')
15:   end for
16: end while
17: return NULL
```

CRF-Matching: Conditional Random Fields for Feature-Based Scan Matching

Fabio Ramos* Dieter Fox† Hugh Durrant-Whyte*

* ARC Centre of Excellence for Autonomous Systems
Australian Centre for Field Robotics
The University of Sydney
Sydney, NSW, Australia

† Dept. of Computer Science & Engineering
University of Washington
Seattle, WA, USA

***Abstract*—Matching laser range scans observed at different points in time is a crucial component of many robotics tasks, including mobile robot localization and mapping. While existing techniques such as the Iterative Closest Point (ICP) algorithm perform well under many circumstances, they often fail when the initial estimate of the offset between scans is highly uncertain. This paper presents a novel approach to 2D laser scan matching. CRF-Matching generates a Condition Random Field (CRF) to reason about the joint association between the measurements of the two scans. The approach is able to consider arbitrary shape and appearance features in order to match laser scans. The model parameters are learned from labeled training data. Inference is performed efficiently using loopy belief propagation. Experiments using data collected by a car navigating through urban environments show that CRF-Matching is able to reliably and efficiently match laser scans even when no a priori knowledge about their offset is given. They additionally demonstrate that our approach can seamlessly integrate camera information, thereby further improving performance.**

I. Introduction

Many robotics tasks require the association of sensor data observed at different points time. For instance, in mobile robot mapping, a robot needs to be able to accurately determine the spatial relationship between different laser scans. While this task is rather straightforward if enough prior knowledge about the relative location of the scans is given, it becomes more challenging when no knowledge about the spatial relationship is available. Given two scans of sensor measurements, the matching problem can be defined as finding a transformation that best matches one scan to another. For example, when laser range finders are used in mobile robotics, such a transformation corresponds to the movement performed by the robot between scans. To find the transformation between two scans, it is necessary to associate the individual measurements in one scan with the corresponding measurements in the other scan.

The most widely used algorithm for matching range sensors in robotics is the Iterative Closest Point (ICP) algorithm [5]. ICP alternates between nearest neighbor association and least-squares optimization to compute the best transformation between the laser scans given the most recent association. Although ICP and its extensions are fast and in general produce good results, simple nearest neighbor association has a number of drawbacks. First, ICP frequently generates incorrect transformation estimates when the initial offset between the scans is large. This is in part caused by the association cost function that does not take into account higher-level information from the data, such as shape descriptions. Second, ICP does not provide adequate means of fusing data collected by multiple types of sensors to improve matching. Third, ICP provides only limited support for estimating the uncertainty of the resulting transformation. Such uncertainty estimates are important in the context of tasks such as robot localization or mapping using Bayes filters [21].

This paper presents CRF-Matching, an alternative procedure for laser scan matching based on Conditional Random Fields (CRFs) [8]. CRFs are undirected graphical models that are very powerful for modeling relational information (spatial data for example). By directly modeling the conditional probability of the hidden states given the observations rather than the joint probability, CRFs avoid the difficult task of specifying a generative model for observations, as necessary in techniques such as Hidden Markov Models (HMMs) or Markov Random Fields (MRFs). As a result, CRFs can handle arbitrary dependencies between observations, which gives them substantial flexibility in using high-dimensional feature vectors.

CRF-Matching focuses on the problem of data association between two laser scans. This is accomplished by converting the individual measurements of one laser scan into hidden nodes of a CRF. The states of each node range over all measurements in the other scan. The CRF models arbitrary information about local appearance and shape of the scans. Consistency of the association is achieved by connections between nodes in the CRF. CRF-Matching learns model parameters discriminatively from sets of aligned laser scans. When applied to a new pair of scans, maximum *a posteriori* estimation is used to determine the data association, which in turn specifies the spatial transformation between the scans.

Extensive experiments show that CRF-Matching significantly outperforms ICP when matching laser range-scans with large spatial offset. Furthermore, they show that our approach is able to reliably match scans without a priori information about their spatial transformation, and to incorporate visual information to further improve matching performance.

This paper is organized as follows. After discussing related work in Section II, we provide an overview of Conditional Random Fields in Section III. CRF-Matching is introduced in Section IV, followed by an experimental evaluation in Section V. We conclude in Section VI.

II. Related Work

ICP has been applied to robotics quite successfully, however, it does not explicitly account for sensor rotation since it uses the Euclidean distance to compute the nearest neighbor. To overcome this limitation, [12] combines the normal NN with angular constraints in the Iterative Dual Correspondence (IDC). The algorithm uses two types of correspondences (translation and rotation) and at each iteration performs two optimizations. [12] also proposes the interpolation of lines between laser points to improve robustness for large transformations. Although these extensions improve the basic ICP algorithm, they do not eliminate the chance the algorithm reaches a poor local minima.

Methods for computing the transformation uncertainty from ICP were also proposed [2], [17]. However, they do not take into account the association uncertainty between pairs of measurements. This can cause large errors in the uncertainty estimation for the transformation since weak associations can equally contribute to the overall estimate.

Shape matching has been a long-standing problem especially for the computer vision community (see [22] for a review). Various techniques have been applied to represent shape including Fourier descriptors [20], parametric curves [13], and geodesic distances [9]. We use some of these ideas in this work to encode shape and image properties.

A similar probabilistic model for 3D matching of non-rigid surfaces was proposed by [1]. The model is trained generatively and assumes a large number of range points in the objects for accurate results. The main feature employed is the geodesic distance which performs well when there is a well-defined object structure. However, in unstructured environments this assumption is not valid. For this reason we employ several features for shape description with the addition of image features when convenient.

Loop closure detection in outdoor environments was investigated in [15]. A combined laser and image matching was proposed with a vocabulary of features. In this paper, we also combine laser and camera information for loop closure detection as a possible application for our technique. The key benefit of our method is the deployment of a single probabilistic model able to fuse and provide uncertainty estimation in a natural manner.

III. Preliminaries

A. Conditional Random Fields (CRF)

Conditional random fields (CRFs) are undirected graphical models developed for labeling sequence data [8]. CRFs directly model $p(\mathbf{x}|\mathbf{z})$, the *conditional* distribution over the hidden variables $\mathbf{x}$ given observations $\mathbf{z}$. This is in contrast to generative models such as Hidden Markov Models or Markov Random Fields, which apply Bayes rule to infer hidden states [18]. Due to this structure, CRFs can handle arbitrary dependencies between the observations $\mathbf{z}$, which gives them substantial flexibility in using high-dimensional feature vectors.

The nodes in a CRF represent hidden states, denoted $\mathbf{x} = \langle \mathbf{x}_1, \mathbf{x}_2, \ldots, \mathbf{x}_n \rangle$, and data, denoted $\mathbf{z}$. The nodes $\mathbf{x}_i$, along with the connectivity structure represented by the undirected edges between them, define the conditional distribution $p(\mathbf{x}|\mathbf{z})$ over the hidden states $\mathbf{x}$. Let $\mathcal{C}$ be the set of cliques (fully connected subsets) in the graph of a CRF. Then, a CRF factorizes the conditional distribution into a product of *clique potentials* $\phi_c(\mathbf{z}, \mathbf{x}_c)$, where every $c \in \mathcal{C}$ is a clique in the graph and $\mathbf{z}$ and $\mathbf{x}_c$ are the observed data and the hidden nodes in the clique c, respectively. Clique potentials are functions that map variable configurations to non-negative numbers. Intuitively, a potential captures the "compatibility" among the variables in the clique: the larger the potential value, the more likely the configuration. Using clique potentials, the conditional distribution over hidden states is written as

$$p(\mathbf{x} \mid \mathbf{z}) = \frac{1}{Z(\mathbf{z})} \prod_{c \in \mathcal{C}} \phi_c(\mathbf{z}, \mathbf{x}_c), \qquad (1)$$

where $Z(\mathbf{z}) = \sum_{\mathbf{x}} \prod_{c \in \mathcal{C}} \phi_c(\mathbf{z}, \mathbf{x}_c)$ is the normalizing partition function. The computation of this partition function can be exponential in the size of $\mathbf{x}$. Hence, exact inference is possible for a limited class of CRF models only.

Potentials $\phi_c(\mathbf{z}, \mathbf{x}_c)$ are described by log-linear combinations of *feature functions* $\mathbf{f}_c$, *i.e.*,

$$\phi_c(\mathbf{z}, \mathbf{x}_c) = \exp\left(\mathbf{w}_c^T \cdot \mathbf{f}_c(\mathbf{z}, \mathbf{x}_c)\right), \qquad (2)$$

where $\mathbf{w}_c^T$ is a weight vector, and $\mathbf{f}_c(\mathbf{z}, \mathbf{x}_c)$ is a function that extracts a vector of features from the variable values. Using feature functions, we rewrite the conditional distribution (1) as

$$p(\mathbf{x} \mid \mathbf{z}) = \frac{1}{Z(\mathbf{z})} \exp\left\{ \sum_{c \in \mathcal{C}} \mathbf{w}_c^T \cdot \mathbf{f}_c(\mathbf{z}, \mathbf{x}_c) \right\} \qquad (3)$$

B. Inference

Inference in CRFs can estimate either the marginal distribution of each hidden variable $\mathbf{x}_i$ or the most likely configuration of all hidden variables $\mathbf{x}$ (*i.e.*, MAP estimation), as defined in (3). Both tasks can be solved using *belief propagation* (BP), which works by sending local messages through the graph structure of the model. Each node sends messages to its neighbors based on messages it receives and the clique potentials, which are defined via the observations and the neighborhood relation in the CRF.

BP generates exact results in graphs with no loops, such as trees or polytrees. However, since the models used in our approach contain various loops, we apply loopy belief propagation, an approximate inference algorithm that is not guaranteed to converge to the correct probability distribution [14]. In our experiments, we compute the MAP labeling of a scan match using max-product loopy BP. Fortunately, even when the algorithm failed to converge, our experiments showed reasonable results.

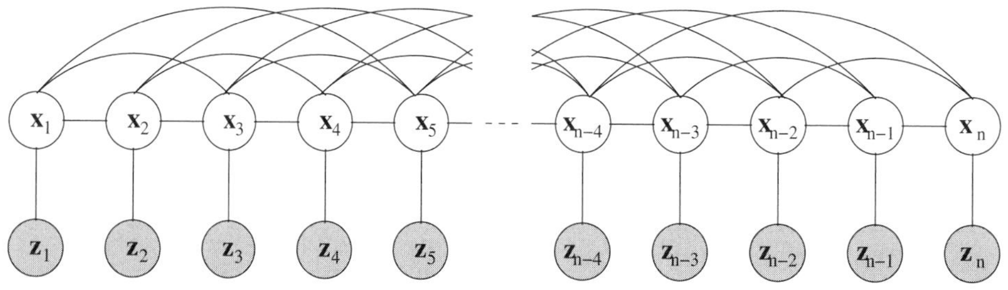

Fig. 1. Graphical representation of the CRF-Matching model. The hidden states $\mathbf{x}_i$ indicate the associations between the points in the different scans. The observations $\mathbf{z}_i$ corresponds to shape or visual appearance information extracted from the two laser scans.

C. Pseudo-Likelihood Parameter Learning

The goal of CRF parameter learning is to determine the weights of the feature functions used in the conditional likelihood (3). CRFs learn these weights discriminatively by maximizing the conditional likelihood of labeled training data. While there is no closed-form solution for optimizing (3), it can be shown that (3) is convex relative to the weights $\mathbf{w}_c$. Thus, the global optimum of (3) can be found using a numerical gradient algorithm. Unfortunately, this optimization runs an inference procedure at each iteration, which can be intractably inefficient in our case.

We therefore resort to maximizing the *pseudo-likelihood* of the training data, which is given by the sum of local likelihoods $p(\mathbf{x}_i \mid \text{MB}(\mathbf{x}_i))$, where $\text{MB}(\mathbf{x}_i)$ is the Markov blanket of variable $\mathbf{x}_i$: the set of the immediate neighbors of $\mathbf{x}_i$ in the CRF graph [4]. Optimization of this pseudo-likelihood is performed by minimizing the negative of its log, resulting in the following objective function:

$$L(\mathbf{w}) = -\sum_{i=1}^{n} \log p(\mathbf{x}_i \mid \text{MB}(\mathbf{x}_i), \mathbf{w}) + \frac{(\mathbf{w}-\widetilde{\mathbf{w}})^T(\mathbf{w}-\widetilde{\mathbf{w}})}{2\sigma^2} \quad (4)$$

Here, the terms in the summation correspond to the negative pseudo log-likelihood and the right term represents a Gaussian shrinkage prior with mean $\widetilde{\mathbf{w}}$ and variance σ^2. Without additional information, the prior mean is typically set to zero. In our approach, we use unconstrained L-BFGS [11], an efficient gradient descent method, to optimize (4). The key advantage of maximizing pseudo-likelihood rather than the likelihood (3) is that the gradient of (4) can be computed extremely efficiently, without running an inference algorithm. Learning by maximizing pseudo-likelihood has been shown to perform very well in different domains; see [7], [19], [6].

IV. CRF-Matching

A. Model Definition

In order to find the association between a laser scan A and another scan B, CRF-Matching generates a CRF that contains a hidden node $\mathbf{x}_i$ for each laser point in scan A. Such a CRF is shown in Figure 1. For now, we assume that each point in scan A can be associated to a point in scan B; outliers will be discussed at the end of this section. To reason about associations between the two scans, each hidden state $\mathbf{x}_i$ of the CRF ranges over all points in laser scan B.

The nodes $\mathbf{z}_i$ in Figure 1 correspond to features associated with the individual laser points. These features describe local appearance properties of the laser scans. In order to achieve global consistency between the individual data associations, each hidden node is connected to other hidden nodes in the network. We now describe the individual features used in the clique potentials of the CRF model.

B. Local Features

CRF-Matching can employ arbitrary local features to describe shape, image properties, or any particular aspect of the data. Since our focus is on associating points in scan A to similar points in scan B, our features describe *differences* between data points. The learning algorithm provides means to weight each of the resulting features to best associate the data. The local features are described as follows:

Spatial distance: This feature measures the distance between points in one scan w.r.t. points in the other scan. This is the basic feature used in ICP, which rather than representing the shape of the scan, accounts for small position transformations. If we denote the locations of individual points in scan A and B by $z_{A,i}$ and $z_{B,j}$, respectively, then the feature $\mathbf{f}_\text{d}$ for point i in scan A can be defined as

$$\mathbf{f}_\text{d}(i, j, z_A, z_B) = \frac{\|z_{A,i} - z_{B,j}\|^2}{\sigma^2}, \quad (5)$$

where σ^2 is the variance of the distances in the training data. Note that the value of this feature depends on the point j to which i is associated. The reader may also notice that this feature is only useful if the initial offset and rotation between the two scans is small.

Shape difference: These features capture how much the local shape of the laser scans differs for each possible association. While local shape can be captured by various types of features, we chose to implement very simple shape features measuring distance, angle, and geodesic distance along the scans.

To generate distance features, we compute for each point in scan A its distance to other points in scan A. These other points are chosen based on their relative indices in the scan. The same distances can be computed for points in scan B, and the resulting feature is the difference between the distance values in A and B. To see, let k be an index offset for the distance feature. The feature value corresponding to points i

and j is computed as follows:

$$\mathbf{f}_{\text{dist}}(i,j,k,z_A,z_B) = \frac{\left\| \|z_{A,i} - z_{A,i+k}\| - \|z_{B,j} - z_{B,j+k}\| \right\|^2}{\sigma^2}. \tag{6}$$

In our implementation this feature is computed for index offsets $k \in \{-1, 1, -3, 3, -5, 5\}$. The reader may notice that this feature is based on the assumption that the two scans have similar densities. However, even though this assumption is often violated, we found this feature to be very valuable to increase consistency of data associations.

Another way to consider local shape is by computing the difference between the angles of points in both scans w.r.t their neighbors. The angle of a point $z_{A,i}$ is defined as the angle between the segments connecting the point to its neighbors, where k indicates the index offset to the neighbor. This feature is defined as:

$$\mathbf{f}_{\text{angle}}(i,j,k,z_A,z_B) = \frac{\left\| \angle\left(\overline{z_{A,i-k}z_{A,i}}, \overline{z_{A,i}z_{A,i+k}}\right) - \angle\left(\overline{z_{B,j-k}z_{B,j}}, \overline{z_{B,j}z_{B,j+k}}\right) \right\|^2}{\sigma^2} \tag{7}$$

As with the distance feature, we compute the difference of angles for neighbors 1, 3 and 5 points apart.

The *geodesic distance* provides additional shape information as it is defined as the sum of Euclidean distances between points along the scan. As with the previous features, it can be calculated for different neighborhoods representing local or long-term shape information. Given points $z_{A,i}$ and $z_{B,j}$ and a neighborhood k, the geodesic distance feature is computed as:

$$\mathbf{f}_{\text{geo}}(i,j,k,z_A,z_B) = \frac{\left\| \sum_{l=i}^{i+k-1} \|z_{A,l+1} - z_{A,l}\| - \sum_{l=j}^{j+k-1} \|z_{B,l+1} - z_{B,l}\| \right\|}{\sigma^2} \tag{8}$$

[1] used such a feature for matching 3D laser scans.

Visual appearance: When camera data is available, it can be integrated with shape information from the scans to help with the association. The projection of laser points into the camera image is used to extract an image patch for each laser point. We use Principal Components Analysis to reduce the dimensionality of image features. The feature is then computed as the Euclidean distance between the principal components of a patch $z^I_{A,i}$ in scan A and $z^I_{B,j}$ in scan B:

$$\mathbf{f}_{\text{PCA}}(i,j,z^I_A,z^I_B) = \frac{\left\| z^I_{A,i} - z^I_{B,j} \right\|^2}{\sigma^2}. \tag{9}$$

All features described so far are *local features* in that they only depend on a single hidden state i in scan A (indices j and k in the features define nodes in scan B and neighborhood size). The main purpose of these features is to associate scan points that have similar appearance. However, in order to generate *consistent* associations it is necessary to define features that relate the hidden states in the CRF to each other.

C. Pairwise Features

The following features are used to define the clique potentials of nodes connected in the CRF.

Association: The main purpose of this feature is to ensure consistency (avoiding crossover labels for example) and enforce sequential labeling: If a measurement i in A is associated to measurement j in B, its neighbor $i+1$ has a high chance of being associated to $j+1$ in B. To measure consistency, we define a function $\Delta = \mathbf{x}_{i+k} - \mathbf{x}_i$ that determines the difference between the hidden states of two nodes $\mathbf{x}_i$ and $\mathbf{x}_{i+k}$. When $\Delta = k$, then the associations of $\mathbf{x}_i$ and $\mathbf{x}_{i+k}$ are fully consistent. We convert Δ into ten different binary features, corresponding to different Δ values.

Pairwise distance: This feature is very similar to $\mathbf{f}_{\text{dist}}$ described above. However, instead of being defined over a *single* hidden node $\mathbf{x}_i$ only, it measures the consistency between the associations of *two* hidden nodes $\mathbf{x}_i$ and $\mathbf{x}_j$:

$$\mathbf{f}_{\text{pair}}(i,j,m,n,z_A,z_B) = \frac{\left\| \|z_{A,i} - z_{A,j}\| - \|z_{B,m} - z_{B,n}\| \right\|^2}{\sigma^2}. \tag{10}$$

Here, i and j are the indices of points in scan A, and m and n are values of their hidden nodes. In essence, this feature assumes that the two scans have the same shape and scale. In contrast to $\mathbf{f}_{\text{dist}}$, however, it does not assume that the scans have the same spatial resolution. Such a feature has been shown to be very valuable for matching camera images [3].

D. Outlier Detection

So far we assumed that every point in scan A can be associated to a point in scan B. However, such an approach cannot handle changes in an environment or partially overlapping laser scans. In order to deal with such cases, the hidden state of each node contains one additional value that corresponds to "outlier". Unfortunately, most of the features discussed so far can and should not be computed if one of the involved hidden nodes is an outlier. For instance, the pairwise distance feature $\mathbf{f}_{\text{pair}}$ cannot be computed if the hidden value m is set to "outlier", since there is no associated laser point $z_{B,m}$ to which the distance can be computed. Fortunately, CRFs are extremely well suited for handling such cases. We simply define two additional binary feature functions that are true whenever the value of an involved hidden node is "outlier" (we get one function for local features and one function for pairwise features). Whenever such a feature is true, the values of all corresponding feature functions are set to zero. When learning the weights of the different feature values, the algorithm automatically learns weights for the binary outlier features that result in most consistent data associations.

E. Boosting Features

CRFs are able to directly incorporate the local, continuous features described above. However, in order to model more complex, nonlinear relationships between hidden states and feature values, it is advantageous to discretize the features. Recently, Friedman and colleagues [6] showed that it is possible to learn AdaBoost classifiers from labeled training data and to use the resulting classifier outputs as features in a CRF. We apply their approach by learning boosted classifiers that combine the local features described above.

To train Adaboost, each pair of laser points is labeled as "associated" or "not associated". Adaboost then generates a discretization of the continuous features and an initial classification of pairs of laser points (whether or not they should be associated). The output value of Adaboost, which is a non-linear combination of the learned features, is used as an observation in the CRF. Note that this "observation" is different for each possible hidden state in the CRF (each hidden state gives a different association of laser points). The CRF then learns a weight for the Adaboost output in combination with the other weights (pairwise feature weights and outlier feature weight). For the experiments Adaboost was trained with 50 decision stumps.

F. Position estimation

To compute the spatial transformation between two laser scans, CRF-Matching uses the uncertainty over the association obtained. The uncertainty is incorporated in a least squared optimization problem for the non-outlying points:

$$Err = \sum_{i=1}^{n} w_i \left(z_{A,i} R + T - z_{B,a(i)} \right)^2 ,$$

where $a(i)$ is the point in scan B associated with point i, and R and T are the rotation and translation matrices respectively. The weight w_i corresponds to the probability of the association for point i obtained with belief propagation. Using this approach, points that are more likely to have the correct association have more importance on position estimation. The same does not occur in ICP where each point has the same contribution.

G. Algorithm Overview and Performance

CRF-Matching converts the matching problem into a probabilistic inference procedure on a CRF where the hidden states correspond to association hypotheses. Outliers are handled explicitly by an additional “outlier” state. We defined various local shape and appearance features. Instead of using these features directly in the CRF, we first train boosted classifiers in order to model more complex relationships between the features. The weights of all local and pairwise features are learned from labeled training scans using pseudo-likelihood maximization. During testing, we apply loopy belief propagation (LBP) to compute the maximum a posteriori association for the complete scan. This association can then be used to determine the spatial transformation between the two scans.

The computational complexity of loopy belief propagation is linear in the number of hidden nodes (points in scan A), and the complexity of computing each message is quadratic in the number of hidden states (points in scan B). In our experiments LBP converges in 0.1 second in scans with 90 beams and 1 second in scans with 180 beams.

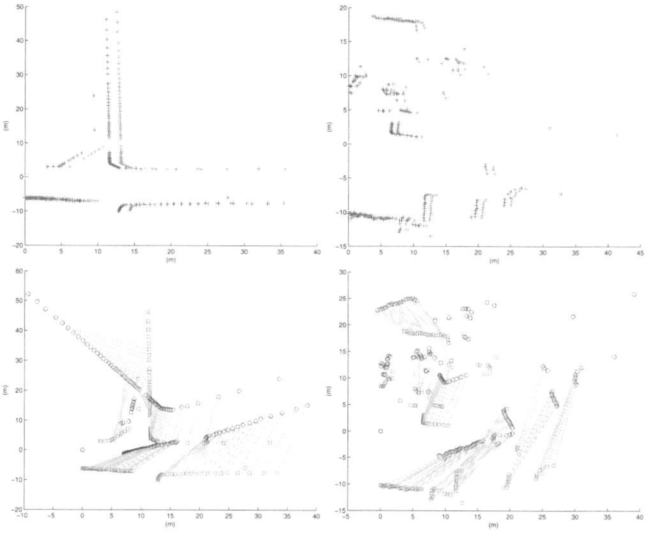

Fig. 2. Two examples of the scans used in the experiments. Original pairs of scans (top) are rotated and translated. CRF matching and ICP are applied to the new configuration. The bottom pictures show the MAP correspondence obtained with CRF-Matching.

Offset	1	2	3	4	5	6	7	8	9	10
X(m)	0.1	0.25	0.5	1.0	1.5	2.0	2.5	3.0	3.5	4.0
Y(m)	0.1	0.25	0.5	1.0	1.5	2.0	2.5	3.0	3.5	4.0
θ(Deg.)	1	5	10	15	20	25	30	40	50	60

TABLE I

TRANSLATION AND ROTATION OFFSETS USED IN THE EXPERIMENTS.

V. EXPERIMENTS

We performed experiments with outdoor data collected with a modified car travelling at 0 to 40 km/h. The car performed several loops around the university campus which has structured areas with buildings, walls and cars, and areas less structured with bush, trees and lawn fields. 20 pairs of scans obtained at different points of the trajectory were used for training and 50 different pairs were used for testing. The scans of each pair are taken approximately 0.25 second apart while the car is in movement. To further evaluate the robustness of the algorithm, the scans are translated and rotated using 10 different offsets. The translation offsets ranged from 0.1m to 4m in x and y direction, and the rotations ranged from 1° to 60° degrees. Table I shows the ten different offsets used. Since we know the additional translation and rotation, the ground truth can be estimated from the association over the original scans by running ICP multiple times for different outlier thresholds. For each offset, a CRF model is trained using the training set with 20 pairs of scans. In all experiments, 90 beams per laser scan are associated (results achieved with 180 beams are virtually identical).

A. Laser-based Matching

In this experiment, we compare the robustness of ICP and CRF for laser scan matching. We have also tested the IDC algorithm with mutual closest point and trimming with similar results to ICP. The experiment is illustrated in Figure 2: a

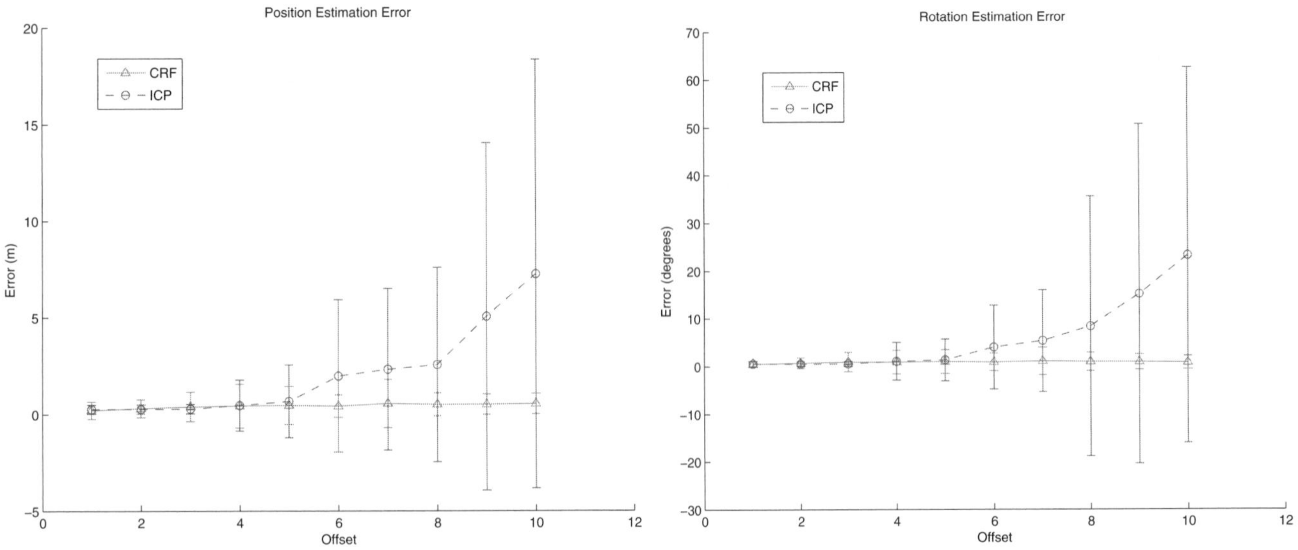

Fig. 4. Translation estimate error (top) and rotation estimate error (bottom) for 90-point scans. While ICP error increases significantly after the 5th offset, CRF matching error is constant for all configurations tested.

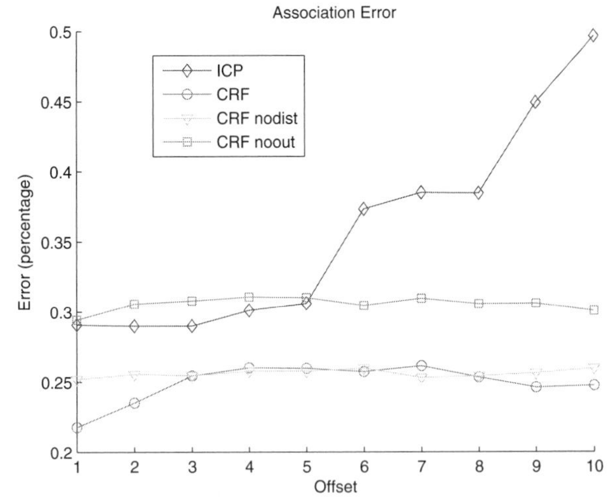

Fig. 3. Correspondence error for ICP and CRF matching for 90-point laser scans. The performance of ICP degrades after offset 5 while CRF matching remains constant.

pair of scans is obtained during the trajectory and artificially rotated and translated. ICP and CRF matching are then applied and compared. To guarantee ICP reaches its best output, the algorithm is computed twice, for a threshold of 20 metres and for a threshold of 5 metres (ICP treats nearest neighbor distances beyond the threshold as outliers). In each case 100 ICP iterations are performed.

Figure 3 summarizes the results of this experiment. For each offset, it shows the fraction of measurements that are not identical to the "ground truth" association computed with ICP on the non-shifted scans. It can be noticed that the performance of ICP degrades significantly for translations larger than 1.5 meter combined with rotation of 20 degrees (offset 5). In contrast, CRF matching keeps the same performance independently of the offset. This indicates the main benefit of our approach; the algorithm is able to associate points by shape when it is necessary or by distance when this feature is more relevant.

Figure 4 shows the spatial transformation errors. As can be seen, for small offsets, the translation and rotation errors of ICP and CRF-Matching are very low (graphs labeled ICP and CRF). While CRF-Matching maintains this low error for larger offsets, the performance of ICP degrades significantly. In order to evaluate the ability of CRF-Matching to align scans without prior knowledge about the relative locations of laser scans, we removed the distance feature defined in (5) from the feature set of CRF-Matching. The resulting graph, labeled "CRF nodist" in Figure 3, indicates that CRF-Matching can globally match laser scans. The graph labeled "CRF noout" shows the performance when the CRF does not contain the outlier state. In this case, the association error increases, which shows that explicit outlier handling is important for real outdoor laser data.

We also performed experiments removing classes of features to evaluate the importance of each feature for the overall performance. When we removed the local shape features or the image features the performance was 10% worse (measured by the association error). Removing the pairwise features and making the matching entirely based on the output of Adaboost resulted in error rates three times as high as CRF-Matching. This indicates the importance of the joint reasoning of CRFs. Overall, the most important component of the network is the pairwise distance feature $\mathbf{f}_{\mathrm{pair}}$, which models the rigidity of the transformation. Leaving this feature out results in 60% increase in the association error. Removing the pairwise association features resulted in a relative increase of 30%.

For trajectory estimation, the benefits of CRF matching are more evident. The key property of weighting each association

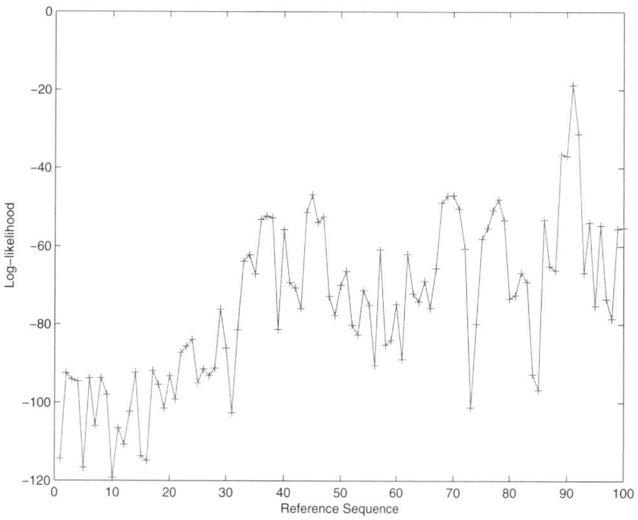

Fig. 5. Log-likelihood of the matching between a reference scene and scenes from a sequence used to detect loop closure. The log-likelihood was computed from the MAP association of CRF-Matching.

according to their matching probability significantly improves the performance. It also demonstrates that especially for rotation, ICP suffers a severe degradation for large offsets.

B. Loop Closure Detection with Laser and Camera Data

Loop closure detection is a key challenge for reliable simultaneous localization and map building (SLAM) [16], [15]. In this preliminary experiment, we show that CRF matching can be employed to detect loop closure in urban environments using both laser and camera. The data was acquired using an adapted car travelling at 30 km/h approximately. We use the approach described in [23] to compute the correct extrinsic calibration of these two sensors.

To detect the loop closure, a reference scene (camera and laser data) from the initial position of the vehicle is matched against a sequence of scenes. To compare different matches, the log-likelihood from the MAP association is computed. The maximum of this scoring function indicates the most probable location where the loop was closed. Figure 5 shows the log-likelihood for all scenes. The peak is clearly visible at scene 91 which indicates the best match for the reference scene. Figure 6 shows the reference image and image 91 where the loop closure was detected. It should be noted that even though the images look rather similar, the actual laser scans differ substantially between the two scenes (Figure 7).

VI. Conclusions and Future Work

We introduced CRF-Matching, a novel approach to matching laser range scans. In contrast to existing scan matching techniques such as ICP, our approach takes shape information into account in order to improve the matching quality. Additionally, CRF-Matching explicitly reasons about outliers and performs a joint estimation over the data associations of all beams in a laser scan. By using a Conditional Random Field as the underlying graphical model, our approach is able to incorporate arbitrary features describing the shape and visual appearance of laser scans. The parameters of the model are learned efficiently from labeled training data. Maximum a posteriori data association computed via loopy belief propagation is used to determine the transformation between laser scans.

Our experiments show that CRF-Matching clearly outperforms ICP on scan matching tasks that involve large uncertainties in the relative locations of scans. Our approach is able to consistently match laser scans in real time without any information about their relative locations. Additional experiments using camera information indicate that the performance of our approach increases as richer sensor data becomes available.

We consider these results extremely encouraging. In future work we will investigate various extensions to the basic CRF-Matching described here. In addition to developing different feature functions, we plan to integrate CRF-Matching into an outdoor SLAM technique. Since CRF-Matching computes full distributions over data associations, it is possible to estimate the uncertainty of the resulting spatial transformations (for instance by sampling data associations). This uncertainty can be used directly by a SLAM approach.

Our current approach considers outliers as the only alternative to matching a measurement. However, one can extend CRF-Matching to additionally model different types of objects such as cars, buildings, trees, and bushes. Such a model could also perform improved data association for moving obstacles. The key advantage of our technique is the development of a framework for performing all these reasoning steps within a single model, in contrast to existing techniques.

We will also investigate the application of CRF-Matching to the more challenging problem of underwater mapping. Here, the observations are camera images annotated with 3D depth information provided by sonar. We hope that the ability of our approach to incorporate arbitrary features and to jointly reason about complete data associations will allow it to perform significantly better than current approaches. To make CRF-Matching applicable to 3D scans with many thousands of points, we intend to develop a hierarchical technique that automatically selects increasingly large subsets of points for matching. Such an approach could include classifiers that are trained not to associate but to select good points for matching. Finally, we will investigate the application of Virtual Evidence Boosting, which has recently been shown to provide superior feature selection capabilities for CRFs [10].

Acknowledgments

This work is partly supported by the ARC Centres of Excellence programme funded by the Australian Research Council (ARC) and the New South Wales State Government, by the University of Sydney Visiting Collaborative Research Fellowship Scheme, and by DARPA's ASSIST and CALO Programmes (contract numbers: NBCH-C-05-0137, SRI subcontract 27-000968).

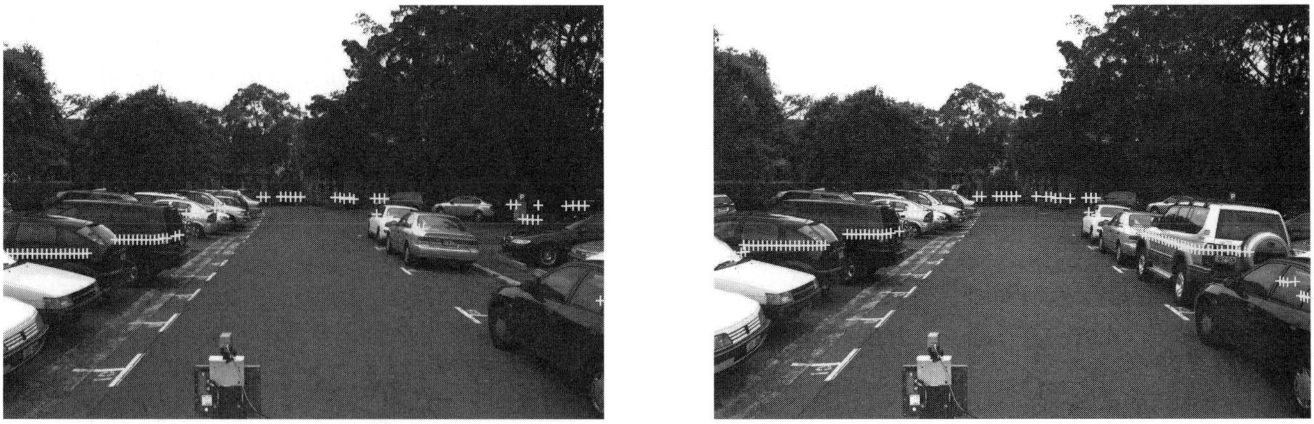

Fig. 6. Reference scene (left) and scene 91 (right) with projected laser point (gray/yellow crosses). CRF matching identifies this scene as the best match indicating the most probable position where the loop closure occurred. Note that the laser scans are substantially different.

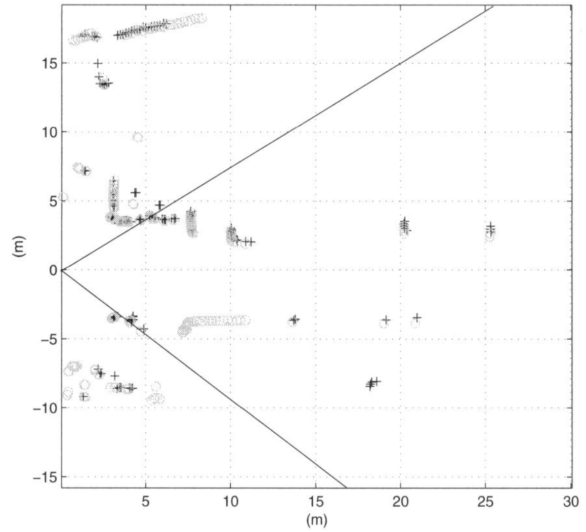

Fig. 7. Matched laser scans from Figure 6. The black crosses are points from the reference scene and the gray/green circles are points from scene 91. The dark lines indicate the camera field of view.

References

[1] D. Anguelov, P. Srinivasan, D. Koller, S. Thrun, J. Rodgers, and J. Davis. SCAPE: Shape completion and animation of people. *ACM Transactions on Graphics (Proc. of SIGGRAPH)*, 24(3), 2005.

[2] O. Bengtsson and A. J. Baerveldt. Localization in changing environments - estimation of a covariance matrix for the IDC algorithm. In *Proc. of the IEEE/RSJ International Conference on Intelligent Robots and Systems (IROS)*, 2001.

[3] A. Berg, T. Berg, and J. Malik. Shape matching and object recognition using low distortion correspondences. In *Proc. of the IEEE Computer Society Conference on Computer Vision and Pattern Recognition (CVPR)*, 2005.

[4] J. Besag. Statistical analysis of non-lattice data. *The Statistician*, 24, 1975.

[5] P. J. Besl and McKay N. D. A method for registration of 3-d shapes. *IEEE Transactions on Pattern Analysis and Machine Intelligence (PAMI)*, 14(2):239–256, 1992.

[6] S. Friedman, D. Fox, and H. Pasula. Voronoi random fields: Extracting the topological structure of indoor environments via place labeling. In *Proc. of the International Joint Conference on Artificial Intelligence (IJCAI)*, 2007.

[7] S. Kumar and M. Hebert. Discriminative random fields: A discriminative framework for contextual interaction in classification. In *Proc. of the International Conference on Computer Vision (ICCV)*, 2003.

[8] J. Lafferty, A. McCallum, and F. Pereira. Conditional random fields: Probabilistic models for segmenting and labeling sequence data. In *Proc. of the International Conference on Machine Learning (ICML)*, 2001.

[9] M. Leventon, E. Grimson, and O. Faugeras. Statistical shape influence in geodesic active contours. In *Proc. of the IEEE Computer Society Conference on Computer Vision and Pattern Recognition (CVPR)*, 2001.

[10] L. Liao, T. Choudhury, D. Fox, and H. Kautz. Training conditional random fields using virtual evidence boosting. In *Proc. of the International Joint Conference on Artificial Intelligence (IJCAI)*, 2007.

[11] D. Liu and J. Nocedal. On the limited memory BFGS method for large scale optimization. *Math. Programming*, 45(3, (Ser. B)), 1989.

[12] F. Lu and E. Milios. Robot pose estimation in unknown environments by matching 2D range scans. *Journal of Intelligent and Robotic Systems*, 18, 1997.

[13] D. Metaxas. *Physics-Based Deformable Models*. Kluwer Academic, 1996.

[14] K. Murphy, Y. Weiss, and M. Jordan. Loopy belief propagation for approximate inference: An empirical study. In *Proc. of the Conference on Uncertainty in Artificial Intelligence (UAI)*, 1999.

[15] P. Newman, D. Cole, and K. Ho. Outdoor SLAM using visual appearance and laser ranging. In *Proc. of the IEEE International Conference on Robotics & Automation (ICRA)*, Orlando, USA, 2006.

[16] P. Newman and K. Ho. SLAM - loop closing with visually salient features. In *Proc. of the IEEE International Conference on Robotics & Automation (ICRA)*, Barcelona, Spain, 2005.

[17] S. Pfister, K. L. Kriechbaum, S. I. Roumeliotis, and J. W. Burdick. Weighted range sensor matching algorithms for mobile robot displacement estimation. In *Proc. of the IEEE International Conference on Robotics & Automation (ICRA)*, 2002.

[18] L. R. Rabiner. A tutorial on hidden Markov models and selected applications in speech recognition. In *Proceedings of the IEEE*. IEEE, 1989. IEEE Log Number 8825949.

[19] M. Richardson and P. Domingos. Markov logic networks. *Machine Learning*, 62(1-2), 2006.

[20] L. H. Staib and J. S. Duncan. Boundary finding with parametrically deformable models. *IEEE Transactions on Pattern Analysis and Machine Intelligence (PAMI)*, 14(11):1061–1075, 1992.

[21] S. Thrun, W. Burgard, and D. Fox. *Probabilistic Robotics*. MIT Press, Cambridge, MA, September 2005. ISBN 0-262-20162-3.

[22] R. Veltkamp and M. Hagedoorn. State-of-the-art in shape matching. Technical Report UU-CS-1999-27, Utrecht University, Netherlands, 1999.

[23] Q. Zhang and R. Pless. Extrinsic calibration of a camera and laser range finder (improves camera calibration). In *Proc. of the IEEE/RSJ International Conference on Intelligent Robots and Systems (IROS)*, Sendai, Japan, 2004.

Control of Many Agents Using Few Instructions

Timothy Bretl
Department of Aerospace Engineering
University of Illinois at Urbana-Champaign
tbretl@uiuc.edu

Abstract— **This paper considers the problem of controlling a group of agents under the constraint that every agent must be given the same control input. This problem is relevant for the control of mobile micro-robots that all receive the same power and control signals through an underlying substrate. Despite this restriction, several examples in simulation demonstrate that it is possible to get a group of micro-robots to perform useful tasks. All of these tasks are derived by thinking about the relationships between robots, rather than about their individual states.**

I. Introduction

A growing number of applications require coordinated control of multiple agents. For example, applications that have received considerable interest from the robotics community include remote monitoring with mobile sensor networks, collision and congestion avoidance with automated air traffic control systems, and cooperative search and rescue with fleets of unmanned vehicles. Control architectures used in these applications vary widely—they might be deliberative or reactive, they might be computed online or offline, they might use global or local information, and they might be implemented using centralized or distributed processing. But regardless of the control architecture, we typically assume that each agent is capable either of acting independently or of following a distinct control input.

In this paper, we are interested in multi-agent systems in which agents are not capable of acting independently. In particular, we are interested in *multi-agent systems in which every agent must be given the same control input.*

We are motivated by recent work in the development of untethered, steerable micro-robots [8]. These robots consist of an untethered scratch drive actuator (used for forward locomotion) and a cantilevered steering arm (used to turn, through frictional contact with the substrate). They are globally controllable, their motion resembling that of unicycles and other car-like vehicles. However, these robots do not have onboard sensing or processing, do not receive commands on distinct communication channels, and do not act independently. Instead, every robot receives the same power and control signal through an underlying electrical grid.

Some work by [8] has focused on removing this restriction. Electromechanical hysteresis is already used to independently send "forward" and "turn" commands by snapping the cantilever up or down. By adding more cantilever arms and making the hysteresis more sophisticated, it is possible to create distinct robots that each respond to some sequences of commands but not to others. The limits of this technique, particularly in terms of how many robots can be controlled simultaneously, have yet to be defined.

In this paper we take a different approach, showing that it is possible to get a group of micro-robots to perform useful tasks even if every robot receives the same control signal.

We begin by considering point robots with simple kinematics (and without nonholonomic constraints). In Section II we show that it is impossible to move two point robots to arbitrary locations if the same control input is sent to both. In Section III we show that a small change in how the control input is interpreted (with respect to local coordinate frames rather than a global coordinate frame) makes it possible to change the relative location between two point robots, even if we still cannot move them to arbitrary locations. In particular, we show that the way in which we can change the relationship between the two robots has distinct structure—for example, we can direct them to meet at the same location, but this location is both unique and fixed. In Section IV, we apply these ideas to enable not just two, but a much larger number of point robots to perform useful tasks. In Section V, we return to the micro-robot example and consider a group of unicycles that must roll without slipping. Finally, in Section VI, we discuss related work and in particular point out several other applications of future interest (micro-robots are only one example).

It is important to understand that we are not advocating any particular control architecture or computational approach in this paper. Indeed, not only do we focus entirely on open-loop control (the steering or motion-planning problem), but we also neglect many phenomena that might influence control design, such as process noise, sensor uncertainty, and the presence of obstacles in the environment. These phenomena may even present alternative ways of differentiating state in response to a common input. However, it is difficult to determine how to take advantage of these phenomena in general—instead, we choose to focus on phenomena that exhibit distinct structure.

In particular, the key message here is that by thinking about *relationships* between robots (rather than about their individual states), we are led to consider tasks that are not only possible, but that admit algorithmic solutions with provable levels of performance. We frame several interesting motion planning problems as a result—for some, we present a method of solution; for others, we leave open questions for future work.

II. Uniform control input, uniform response

Consider two agents moving in a plane with no obstacles. Imagine that we can direct their motion by telling them to

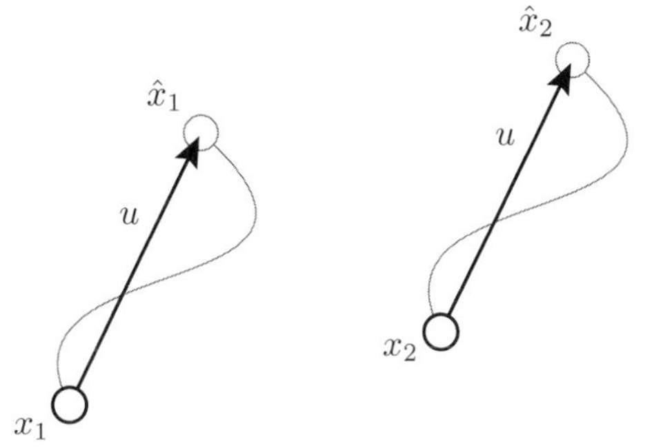

Fig. 1. Two agents are instructed to follow the same path, specified with respect to a global coordinate frame. The total displacement is represented by the vector u. The relative location between agents remains fixed.

move a certain distance north, south, east, or west (or in some arbitrary direction relative to these coordinates). We will assume that they follow our instructions exactly, without error. However, we will also assume that the same set of instructions is given to both agents—we cannot tell one to move east and the other west, for example. Under this restriction, is it possible to simultaneously direct both agents to arbitrary locations on the field? Alternatively, is it possible to direct both agents to meet at the same location?

We can easily show that the answer to both questions is *no*. In particular, let the position of each agent be $x_1, x_2 \in \mathbb{R}^2$. Let the vector $u \in \mathbb{R}^2$ represent our set of instructions, telling both agents to move a distance $\|u\|$ in a direction $u/\|u\|$ that is relative to a fixed coordinate frame. So the position of agent i after following our instructions is $x_i + u$. First, given arbitrary goal locations $\hat{x}_1$ and $\hat{x}_2$ in $\mathbb{R}^2$, we want to know if some u exists that satisfies

$$\hat{x}_1 = x_1 + u$$
$$\hat{x}_2 = x_2 + u.$$

Such a u exists only if $\hat{x}_1 - x_1 = \hat{x}_2 - x_2$. So the goal locations may not be chosen arbitrarily—they must be in the same place relative to both agents. Second, we want to know if some u exists that satisfies

$$x_1 + u = x_2 + u.$$

Such a u exists only if $x_1 = x_2$. So the two agents can be directed to meet at the same location only if they begin at the same location.

In short, it seems we are considering an example that is not interesting. The two agents are moving in formation—the relative location between them is fixed (see Fig. 1).

III. Uniform Control Input, Selective Response

Consider the same two agents moving in a plane with no obstacles. Again assume that we can direct their motion by telling them to move a certain distance in a certain direction, that they follow our instructions exactly, and that the same set of instructions is given to both agents. But now, rather than specify the direction of motion with respect to a global coordinate frame (north, south, east, west), imagine that we

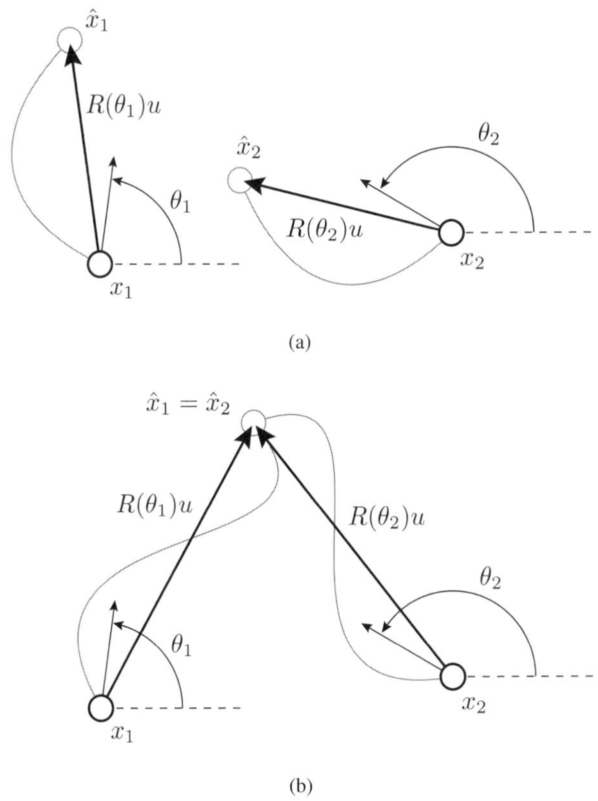

Fig. 2. Two agents are instructed to follow the same path, specified with respect to local coordinate frames. The total displacement is u in local coordinates, which corresponds to $R(\theta_1)u$ and $R(\theta_2)u$ in global coordinates. (a) The relative location between agents does not remain fixed. (b) In fact, the agents can be directed to meet at the same location.

can specify it with respect to a local coordinate frame (forward, back, right, left). So if we tell both agents to move forward, they may move in different directions, depending on which way they are initially facing. We assume that neither agent changes their orientation as they move, so that their local coordinate frames are fixed.

It is still impossible to simultaneously direct both agents to arbitrary locations. Let the position of each agent be $x_1, x_2 \in \mathbb{R}^2$. Let the initial (and fixed) orientation of each agent be $\theta_1, \theta_2 \in (-\pi, \pi]$. Let the vector $u \in \mathbb{R}^2$ represent our set of instructions, telling both agents to move a distance $\|u\|$ in a direction $u/\|u\|$ that is relative to their local coordinate frame. Define the rotation matrix

$$R(\theta) = \begin{bmatrix} \cos\theta & -\sin\theta \\ \sin\theta & \cos\theta \end{bmatrix}.$$

So the position of agent i after following our instructions is $x_i + R(\theta_i)u$. Given arbitrary goal locations $\hat{x}_1$ and $\hat{x}_2$ in $\mathbb{R}^2$, we want to know if some u exists that satisfies

$$\hat{x}_1 = x_1 + R(\theta_1)u$$
$$\hat{x}_2 = x_2 + R(\theta_2)u.$$

Such a u exists only if $R(\theta_1)^T(\hat{x}_1 - x_1) = R(\theta_2)^T(\hat{x}_2 - x_2)$. So again, the goal locations may not be chosen arbitrarily—they must be in the same place relative to both agents with respect to their local coordinate systems.

However, it is now possible to direct both agents to meet at the same location (as long as we do not care where that location is). We want to know if some u exists that satisfies

$$x_1 + R(\theta_1)u = x_2 + R(\theta_2)u.$$

Such a u is given by

$$u = -\left(R(\theta_2) - R(\theta_1)\right)^{-1}(x_2 - x_1).$$

The inverse $\left(R(\theta_2) - R(\theta_1)\right)^{-1}$ exists if and only if $\theta_1 \neq \theta_2$. So if each agent has a different initial orientation (or if both agents begin at the same location, the trivial case), then both agents can be directed to meet at the same location. Of course, this location may not be chosen arbitrarily. Instead, it is unique and is fixed by x_1, x_2, θ_1, and θ_2. We denote it by

$$u_{12} = -\left(R(\theta_2) - R(\theta_1)\right)^{-1}(x_2 - x_1)$$

in local coordinates and by

$$\begin{aligned}\hat{x}_{12} &= x_1 + R(\theta_1)u_{12}\\ &= x_2 + R(\theta_2)u_{12}\end{aligned}$$

in global coordinates. But although u_{12} is fixed, the path taken to reach u_{12} is not. For example, consider a set of instructions specified by an arbitrary continuous curve $u\colon [0,1] \mapsto \mathbb{R}^2$ such that $u(0) = 0$. Then for all $t \in [0,1]$ such that $u(t) = u_{12}$, the two agents are at the same location, meeting at $\hat{x}_{12}$. Conversely, for all $t \in [0,1]$ such that $u(t) \neq u_{12}$, the two agents are at different locations.

We have made only a small change to our example from Section II: interpreting instructions in local coordinate frames rather than in a global coordinate frame. But now, even though we still give the same set of instructions to two agents, we can get something interesting to happen—we can direct them to meet at the same location (see Fig. 2). This result may seem surprising, even though it was easy to prove.

IV. Control of Many Agents

In Sections II-III, we showed that it is possible to change the relative location between two agents (modeled as points in the plane) even if the same control input is sent to both. In particular, we showed that the way in which we can change the relationship between two agents has distinct structure—for example, we can direct them to meet at the same location, but this location is both unique and fixed. In this section we apply these ideas to enable a group of agents to perform useful tasks.

A. *Sequential information passing with no propagation*

Consider n agents moving in a plane with no obstacles. Again assume that we can direct their motion by telling them to move a certain distance in a certain direction, that they all follow our instructions exactly, and that the same set of instructions is given to everyone. Also, assume as in Section III that we specify the direction of motion with respect to a local coordinate frame (forward, back, right, left).

Imagine that one agent has an important piece of information. We would like this agent to pass the information on to everyone else. We assume that the only way this agent can pass information to another agent is by meeting them at the same location. So we would like to compute a set of instructions that directs the agent with information to meet everyone else while traveling the minimum distance possible. We call this task *sequential information passing with no propagation.*

Note that we are using the term "information" as an abstraction for something physical that is passed around. For example, one might imagine micro-robots propagating a marker (such as a colored dye or a chemical substance) or exchanging micro-scale construction materials. It is not assumed that the robots can communicate, or in fact do anything on their own other than follow a common control input.

Let the position and initial orientation of each agent be $x_i \in \mathbb{R}^2$ and $\theta_i \in (-\pi, \pi]$, respectively, for all $i \in \{1, \ldots, n\}$. We assume that $\theta_i \neq \theta_j$ for all $i, j \in \{1, \ldots, n\}$ such that $i \neq j$. Let the continuous curve $u\colon [0,1] \mapsto \mathbb{R}^2$ represent our set of instructions, followed with respect to each agent's local coordinate frame. Define the rotation matrix

$$R(\theta) = \begin{bmatrix}\cos\theta & -\sin\theta\\ \sin\theta & \cos\theta\end{bmatrix}.$$

So the position of agent i after following our instructions for time $t \in [0,1]$ is $x_i + R(\theta_i)u(t)$. Assume without loss of generality that agent 1 has the "important information" and must meet everyone else. Recall from Section III that agent 1 can only meet agent $i \neq 1$ at the fixed location

$$\begin{aligned}\hat{x}_i &= x_1 + R(\theta_1)u_i\\ &= x_i + R(\theta_i)u_i\end{aligned}$$

where

$$u_i = -\left(R(\theta_i) - R(\theta_1)\right)^{-1}(x_i - x_1).$$

We also define the initial location by $\hat{x}_1 = x_1$ and $u_1 = 0$. So agent 1 must visit $\hat{x}_i$ for all $i \in \{1, \ldots, n\}$. Since the total distance traveled is equal to the length of the curve u, our task is to find u of minimum length that passes through u_i for all $i \in \{1, \ldots, n\}$. The optimal u is clearly a sequence of straight lines between meeting points. It can be specified by an ordering $i(1), \ldots, i(n)$ of the set $\{1, \ldots, n\}$ and found by solving the following problem:

$$\begin{aligned}\text{minimize}\quad & \sum_{k=1}^{n-1} \|u_{i(k+1)} - u_{i(k)}\|\\ \text{subject to}\quad & k \in \{i(1), \ldots, i(n)\} \text{ for all } k \in \{1, \ldots, n\}\\ & i(k) \in \{1, \ldots, n\} \text{ for all } k \in \{1, \ldots, n\}\\ & i(1) = 1\end{aligned}$$

In fact, this problem can be recast as a traveling salesman problem. Define a graph by a vertex set $V = \{1, \ldots, n\}$ and an

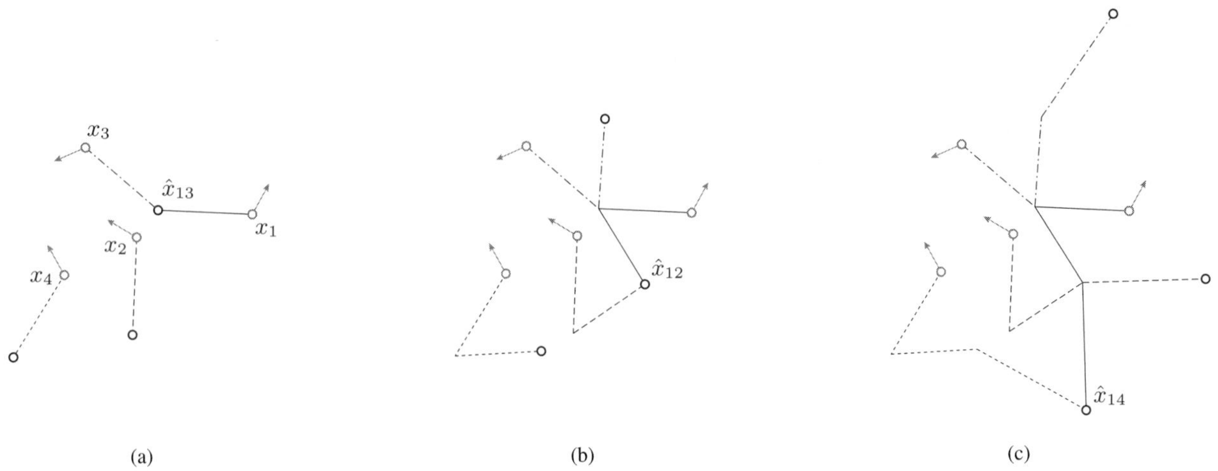

Fig. 3. *Information passing without propagation.* An optimal tour for one agent (beginning at x_1) to pass information to three other agents by meeting them each in turn, under the restriction that all agents must be instructed to follow the same path, specified with respect to local coordinate frames. In this tour, the first agent is directed sequentially to the meeting points $\hat{x}_{13}$, $\hat{x}_{12}$, and $\hat{x}_{14}$.

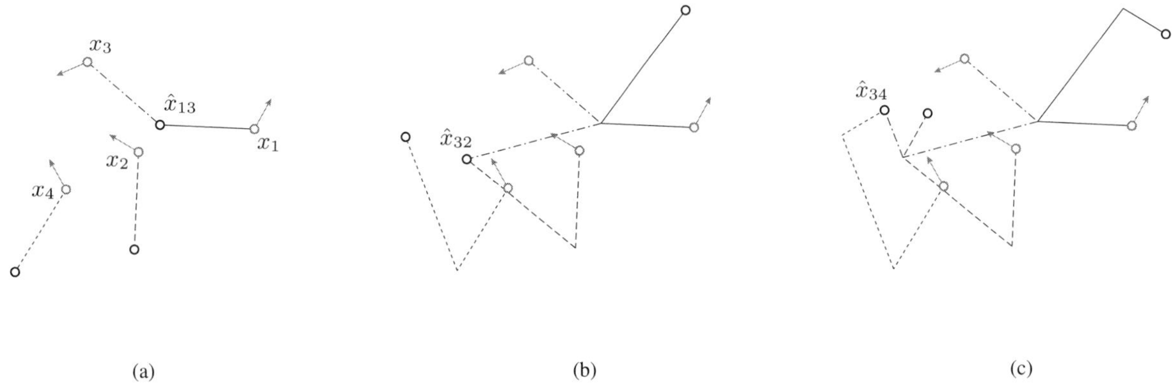

Fig. 4. *Information passing with propagation.* An optimal tour for the same example shown in Fig. 3, again under the restriction that all agents must be instructed to follow the same path, specified with respect to local coordinate frames. In this case, however, agents that have been given information are allowed to pass it along to other agents themselves. As a result, the agents are directed sequentially to different meeting points: $\hat{x}_{13}$, $\hat{x}_{32}$, and $\hat{x}_{34}$.

arc set $A = \{(i, j) \text{ for all } i, j \in V \text{ such that } i \neq j\}$. Define a cost matrix with elements

$$c_{ij} = \begin{cases} 0 & \text{if } j = 1 \\ \|u_i - u_j\| & \text{otherwise} \end{cases}$$

for all $i, j \in V$ such that $i \neq j$. Then the least-cost Hamiltonian circuit through the graph defined by V and A defines an optimal ordering $i(1), \ldots, i(n)$ of the set $\{1, \ldots, n\}$.

Figure 3 shows a simple example for $n = 4$ agents. Since each agent follows the same path (interpreted with respect to a local coordinate frame), it is impossible for them all to meet at the same location. So, the first agent meets each of the other three sequentially. The curve u consists of three straight lines: from $u_1 = 0$ to u_3, from u_3 to u_2, and from u_2 to u_4.

B. Sequential information passing with propagation

In the previous section we considered an information-passing task for a group of n agents, that required one agent to meet each of the others in turn. But what if that first agent is not the only one that can pass information to others? Indeed, it may be natural to assume that once a second agent is given information, they too are able to pass it along. So in this section, we would like to compute a set of instructions that directs each agent to be met by someone who has already received the information, while traveling the minimum distance possible (where as usual we assume that the same set of instructions is given to everyone and interpreted with respect to local coordinate frames). We call this task *sequential information passing with propagation.*

Again we denote each agent's position by x_i and initial orientation by θ_i, and assume that $\theta_i \neq \theta_j$ for all $i \neq j$. The continuous curve $u\colon [0, 1] \mapsto \mathbb{R}^2$ represents our set of instructions, followed with respect to each agent's local coordinate frame as represented by a rotation matrix $R(\theta_i)$. So as before, the position of agent i after following our instructions for time $t \in [0, 1]$ is $x_i + R(\theta_i)u(t)$. Assume that agent 1 is initially given the information to be passed on. Two agents i and $j \neq i$ can only meet at the fixed location

$$\begin{aligned}\hat{x}_{ij} &= x_i + R(\theta_i)u_{ij} \\ &= x_j + R(\theta_j)u_{ij}\end{aligned}$$

where

$$u_{ij} = -\left(R(\theta_j) - R(\theta_i)\right)^{-1}(x_j - x_i).$$

We also define initial locations by $\hat{x}_{ii} = x_{ii}$ and $u_{ii} = 0$ for all $i \in \{1, \ldots, n\}$. Although $\hat{x}_{ij} = \hat{x}_{ji}$ and $u_{ij} = u_{ji}$, by convention we will write $\hat{x}_{ij}$ and u_{ij} if information is being passed from agent i to agent j, and $\hat{x}_{ji}$ and u_{ji} if information is being passed from agent j to agent i. So for all $j \in \{1, \ldots, n\}$, each agent j must visit and receive information at $\hat{x}_{ij}$ (equivalently, u must pass through u_{ij}) from some agent $i \in \{1, \ldots, n\}$ that has already received it.

Since the total distance traveled is equal to the length of u, the optimal path is a sequence of straight lines between meeting points. It can be specified by orderings $i(1), \ldots, i(n)$ and $j(1), \ldots, j(n)$ of the set $\{1, \ldots, n\}$ and found by solving the following problem:

$$\begin{aligned}
\text{minimize} \quad & \sum_{k=1}^{n-1} \|u_{i(k+1)j(k+1)} - u_{i(k)j(k)}\| \\
\text{subject to} \quad & k \in \{j(1), \ldots, j(n)\} \text{ for all } k \in \{1, \ldots, n\} \\
& i(k) \in \{j(1), \ldots, j(k-1)\} \text{ for all } k \in \{2, \ldots, n\} \\
& i(k) \in \{1, \ldots, n\} \text{ for all } k \in \{1, \ldots, n\} \\
& j(k) \in \{1, \ldots, n\} \text{ for all } k \in \{1, \ldots, n\} \\
& i(1) = 1 \\
& j(1) = 1
\end{aligned}$$

The first constraint ensures that each agent is met; the second constraint ensures that they are met by someone who already has information. The other constraints set the variable domains and the initial conditions.

It seems that, as in Section IV-A, this problem might be recast as a traveling salesman problem. It resembles the "generalized" traveling salesman problem in which cities are divided into groups and the salesman must visit at least one city in each group. However, the ordering constraints in our information passing problem make this formulation difficult. In any case, this problem is harder to solve than the one in Section IV-A, having $((n-1)!)^2$ possible orderings rather than $(n-1)!$. On the other hand, this problem has distinct structure that may be useful: for example, denoting $R_{ij} = R(\theta_i) - R(\theta_j)$, we see that

$$R_{12}u_{12} + R_{23}u_{23} + R_{34}u_{34} + R_{41}u_{41} = 0.$$

Figure 4 shows a simple example for $n = 4$ agents. In fact, this is the same example as in Fig. 3. However, here we take advantage of the ability to propagate information (rather than requiring one agent to meet all the others) to make the total distance traveled smaller. After receiving the information, the third agent (not the first) passes it along to the second and fourth agents. The curve u consists of three straight lines:

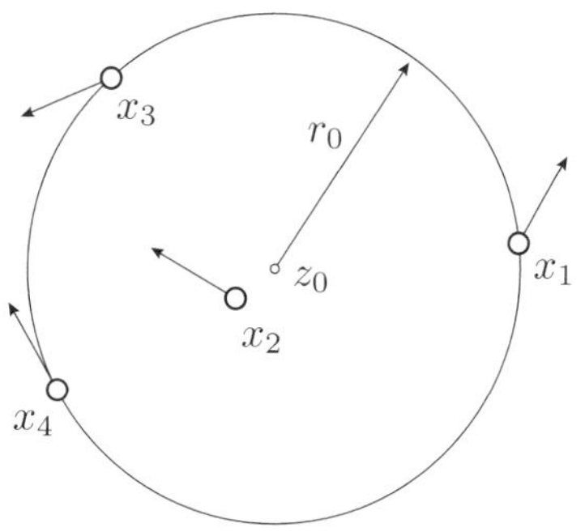

(a)

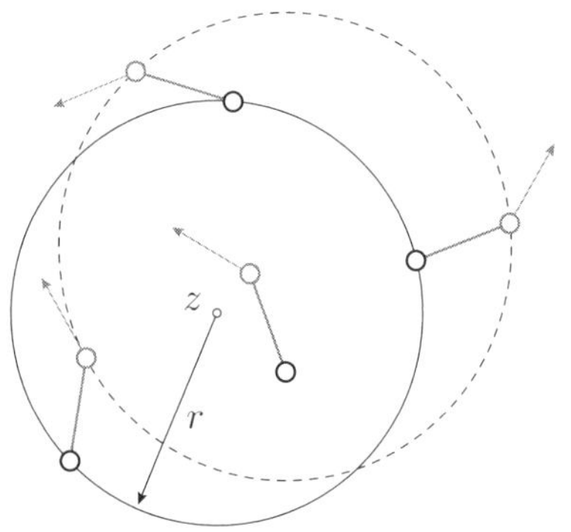

(b)

Fig. 5. Moving four agents inside a disc of minimum radius, for the same example shown in Figs. 3 and 4. As before, all agents follow the same path, interpreted with respect to local coordinate frames. In this case $r/r_0 \approx 0.9$.

from $u_{11} = 0$ to u_{13}, from u_{13} to u_{32}, and from u_{32} to u_{34}. This path is about 10% shorter than the one in Fig. 3.

C. Moving everyone close together

Again imagine that we are directing the motion of n agents in a plane. In previous sections, we have emphasized the fact that if everyone follows the same path with respect to their local coordinate frame, it is impossible for them all to meet at the same location at the same time (they can only meet pairwise). But what if we are interested in having everyone move close together rather than having everyone actually meet? In this section, we show that an optimal policy to accomplish this task can readily be computed.

As usual we denote each agent's position by x_i and initial orientation by θ_i. Let the vector $u \in \mathbb{R}^2$ represent our instructions, telling all agents to move a distance $\|u\|$ in a direction $u/\|u\|$. These instructions are followed with respect to each agent's local coordinate frame, as represented by a rotation matrix $R(\theta_i)$. So, the position of agent i after

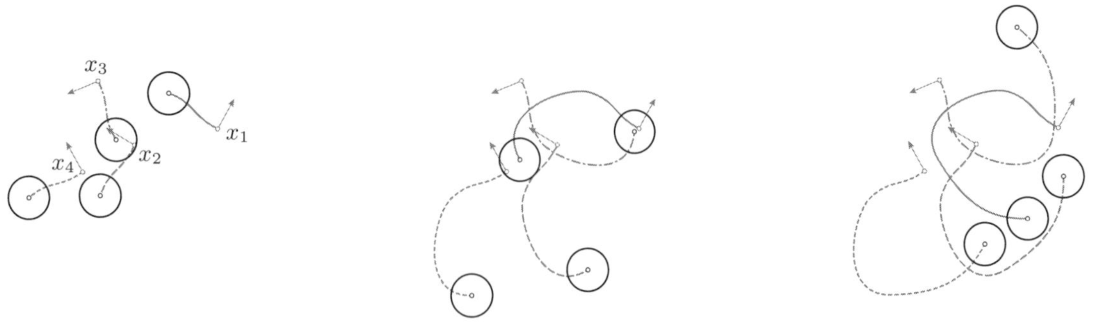

(a)

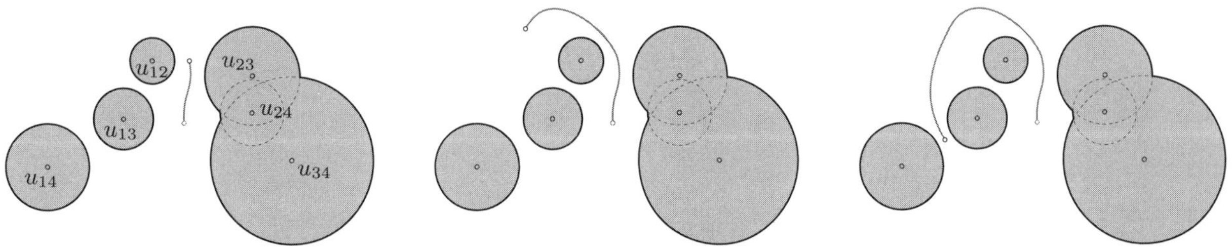

(b)

Fig. 6. Motion that avoids collision between four agents, for the same example shown in Figs. 3-5. Each agent has the same radius, and follows the same path (interpreted with respect to its local coordinate frame). (a) The paths $x_i + R(\theta_i)u(t)$ followed by each agent i in the workspace. (b) The curve $u(t)$ given as input. The shaded region is the set of all $u \in \mathbb{R}^2$ that would place two robots in collision.

following our instructions is $x_i + R(\theta_i)u$.

Subsequently, a disc with center $z \in \mathbb{R}^2$ and radius $r > 0$ contains the entire group of agents if $\|R(\theta_i)u + x_i - z\| \leq r$ for all $i = 1, \ldots, n$. We measure the *distribution* of the group by the radius of the smallest such disc (for any z). We would like to find an input u that minimizes this distribution. We can do this by solving the following second-order cone program (SOCP):

$$\begin{aligned} &\text{minimize} && r \\ &\text{subject to} && \|R(\theta_i)u + x_i - z\| \leq r \quad \text{for all } i = 1, \ldots, n. \end{aligned}$$

This SOCP is convex, and can be solved efficiently (by using an interior-point or primal-dual algorithm, for example [3]).

Figure 5 shows a simple example for $n = 4$ agents (the same agents as in Figs. 3 and 4). In this case, the optimal input u reduces the radius of the smallest disc containing all four agents by approximately 10%.

D. Avoiding collisions

Up to now, we have considered tasks that require agents to meet at the same location (in other words, to collide) or at least to move closer together. In this section, we consider the problem of *avoiding* collision. As before, assume that we are directing the motion of n agents in the plane by specifying a single path that each agent follows with respect to their local coordinate frame. But now, also assume that each agent has some physical shape—for example, we might model agents as discs of fixed radius. We would like to plan a collision-free path between any reachable start and goal configurations.

Denote each agent's position by x_i and initial orientation by θ_i. Assume that $\theta_i \neq \theta_j$ for all $i \neq j$. The continuous curve $u\colon [0,1] \mapsto \mathbb{R}^2$ represents our set of instructions, followed with respect to each agent's local coordinate frame as given by a rotation matrix $R(\theta_i)$. So, the position of agent i after following our instructions for time t is $x_i + R(\theta_i)u(t)$. Likewise, the distance between any two agents i and j in the workspace is

$$d_{ij}(u) = \|(R(\theta_i) - R(\theta_j))u + (x_i - x_j)\|$$

Level sets of d_{ij} as a function of $u \in \mathbb{R}^2$ are circles centered at

$$u_{ij} = -\left(R(\theta_j) - R(\theta_i)\right)^{-1}(x_j - x_i).$$

In particular, if we model each agent by a disc of radius r in the workspace, one can show that the set of all $u \in \mathbb{R}^2$ placing agents i and j in collision is a disc centered at u_{ij} with radius

$$\hat{r}_{ij} = \frac{2r}{\sqrt{2 - 2\cos(\theta_1 - \theta_2)}}.$$

Assume we are given $u(0)$ and $u(1)$, corresponding to reachable start and goal configurations $x_i + R(\theta_i)u(0)$ and $x_i + R(\theta_i)u(1)$, respectively, for each agent i. Then the problem of planning a collision-free path for all agents between start and goal is a *path planning* problem in a two-dimensional configuration space $\mathbb{R}^2$ with $n(n-1)/2$ circular obstacles, each a disc of radius $\hat{r}_{ij}$ centered at u_{ij} for some i and j. Many different algorithms are available to solve such problems [12], [6], [13].

Figure 6 shows a simple example for $n = 4$ agents (the same as in Figs. 3-5). The path of each agent in the workspace is shown in Fig. 6(a); the curve u (our set of instructions) is shown in Fig. 6(b). Since there are four agents, there are six circular obstacles—shown as shaded regions—that must be avoided by u. Notice that although the radius of each agent is the same, the radius of each obstacle can be arbitrarily large, depending on the difference $\theta_i - \theta_j$ between each pair of initial orientations.

V. THE MICRO-ROBOT EXAMPLE

Thus far we have assumed a simple kinematic model for each agent. However, one of the motivational applications with which we began this paper involved car-like micro-robots subject to a nonholonomic constraint. If we again assume that two agents are directed to follow the same path (interpreted with respect to local coordinate frames), but now assume that they roll without slipping, is it still possible to direct them both to meet at the same location? Is that location still a single fixed point? Is it still impossible to direct both agents to arbitrary locations?

In fact, if we model each agent as a unicycle (alternatively, a kinematic car or a differential drive vehicle), then the answer to each question is *yes*. This result follows simply from the invariance of unicycle trajectories under rotation and translation. In particular, consider a unicycle that moves forward at constant (unit) velocity, so we can direct its motion only by changing the turning rate $v(t) \in \mathbb{R}$ as a function of time. Denote the position and heading of the unicycle by $u \in \mathbb{R}^2$ and $\theta \in S^1$, respectively. Then we can write the equations of motion as

$$\dot{u} = \begin{bmatrix} \cos\theta \\ \sin\theta \end{bmatrix}$$
$$\dot{\theta} = v$$

For a given control input $v(t)$, these equations can be integrated to find $u(t)$ and $\theta(t)$. Assume that both $u(0) = 0$ and $\theta(0) = 0$. If we apply the same control input $v(t)$ to a second unicycle with states u' and θ' and with different initial

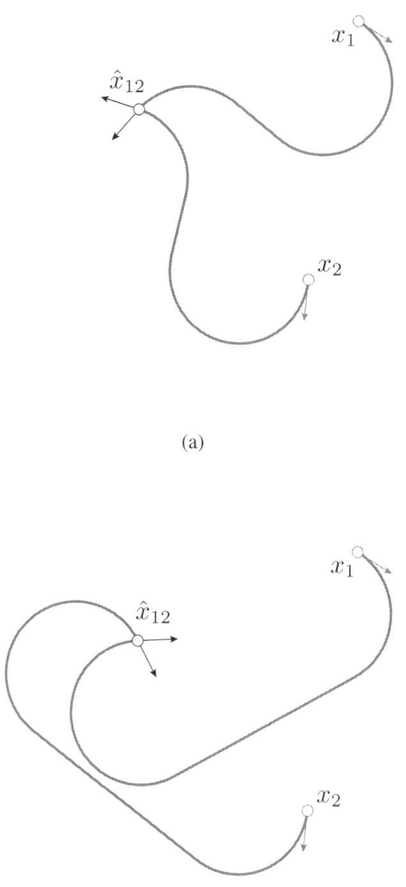

(a)

(b)

Fig. 7. Examples of different ways for two agents to meet at the same location, when their motion is subject to a nonholonomic constraint. Both agents are modeled as unicycles that can only drive forward and that have a minimum turning radius. Their final angle depends on the path taken; their meeting point does not.

conditions $u'(0) = u'_0$ and $\theta'(0) = \theta'_0$, then we can show that

$$u'(t) = u'_0 + R(\theta'_0)u(t)$$
$$\theta'(t) = \theta'_0 + \theta(t),$$

where $R(\theta)$ is defined as the rotation matrix

$$R(\theta) = \begin{bmatrix} \cos\theta & -\sin\theta \\ \sin\theta & \cos\theta \end{bmatrix}.$$

Consequently, if we model each agent i as a unicycle with initial position $x_i \in \mathbb{R}^2$ and orientation $\theta_i \in (-\pi, \pi]$, then its position after time t is $x_i + R(\theta_i)u(t)$, exactly as before. The only difference now is that the input $u(t)$, a curve in $\mathbb{R}^2$, must itself satisfy the equations of motion of a unicycle.

Figure 7 shows an example of directing two agents to meet at the same location (exactly as described in Section III), assuming that they are both unicycles with a minimum turning radius. Notice that the final heading of both agents depends on the path taken to reach the meeting point. So, despite the fact that our set of instructions is now only one-dimensional (turning rate as a function of time), we can still get a group of agents to perform useful tasks.

VI. Related work

A growing number of applications require the coordinated control of multiple agents. For many of these applications, it is possible to engineer the behavior of, and interactions between, individual agents. For example, one strategy for controlling large groups of mobile robots is to take advantage of flocking or swarming behavior. By following simple control rules based on local sensor measurements (such as regulating distance from nearest neighbors [20]), it is possible for a group of identical robots to accomplish complex global tasks such as movement synchronization [11], rendezvous [7], platooning [15], [10], and distributed manipulation [16]. Moreover, few external inputs are needed to influence this group behavior. For example, a formation can be controlled by commanding the location of its centroid or the variance of its distribution [1] or by directing the movement of a "virtual leader" [14].

In this paper, we are interested in applications for which it is impossible to engineer the behavior of individual agents. In particular, we have focused on one multi-agent system [8] in which every agent receives the same control input. A variety of other multi-agent systems are subject to similar constraints. For example, recent work has considered problems such as herding cattle with a robotic sheepdog [22] or with "virtual fences" [4], interacting with cockroaches using a mobile robot [5], guiding crowds during emergency egress [19], and even using groups of micro-organisms to manipulate small objects [18], [21]. These applications involve considerable uncertainty, so existing solution approaches seek to control the distribution of a group, modeled either by moments [9] or by a probability density function [17] using the formalism of stochastic optimal control. We have presented a different approach that is computationally more tractable but that does not yet address uncertainty. Our approach was based on thinking about *relationships* between agents, and deriving a set of useful tasks in that context.

VII. Conclusion

In this paper we considered the problem of controlling an entire team of agents using only a single control input. Initially we were motivated to pursue this problem by the development of untethered, steerable micro-robots [8], all of which receive the same power and control signals through an underlying electrical grid. We showed through several examples that, despite this restriction, it is possible to get the micro-robots to perform useful tasks. These tasks include meeting at the same location, sequentially passing "information," gathering close together, and avoiding collision.

There are several opportunities for future work. For example, in this paper we focused entirely on open-loop control. These results could be extended to closed-loop control, including a consideration of process noise, sensor uncertainty, and the presence of obstacles in the environment. These results could also be applied to other untethered micro-robots, such as [2], or to the other applications mentioned in Section VI. Finally, several of the motion planning problems raised in Section IV remain to be solved.

References

[1] G. Antonelli and S. Chiaverini. Kinematic control of platoons of autonomous vehicles. *IEEE Trans. Robot.*, 22(6):1285–1292, Dec. 2006.

[2] D. Bell, S. Leutenegger, K. Hammar, L. Dong, and B. Nelson. Flagella-like propulsion for microrobots using a nanocoil and a rotating electromagnetic field. In *Int. Conf. Rob. Aut.*, pages 1128–1133, Roma, Italy, Apr. 2007.

[3] S. Boyd and L. Vandenberghe. *Convex Optimization*. Cambridge University Press, 2004.

[4] Z. Butler, P. Corke, R. Peterson, and D. Rus. From robots to animals: virtual fences for controlling cattle. *Int. J. Rob. Res.*, 25(5-6):485–508, 2006.

[5] G. Caprari, A. Colot, R. Siegwart, J. Halloy, and J.-L. Deneubourg. Insbot: Design of an autonomous mini mobile robot able to interact with cockroaches. In *Int. Conf. Rob. Aut.*, 2004.

[6] H. Choset, K. Lynch, S. Hutchinson, G. Kanto, W. Burgard, L. Kavraki, and S. Thrun. *Principles of Robot Motion: Theory, Algorithms, and Implementations*. MIT Press, 2005.

[7] J. Cortés, S. Martínez, and F. Bullo. Robust rendezvous for mobile autonomous agents via proximity graphs in arbitrary dimensions. *IEEE Trans. Automat. Contr.*, 51(8):1289–1296, Aug. 2006.

[8] B. R. Donald, C. G. Levey, C. D. McGray, I. Paprotny, and D. Rus. An untethered, electrostatic, globally controllable MEMS micro-robot. *J. Microelectromech. Syst.*, 15(1):1–15, Feb. 2006.

[9] R. A. Freeman, P. Yang, and K. M. Lynch. Distributed estimation and control of swarm formation statistics. In *American Control Conference*, pages 749–755, Minneapolis, MN, June 2006.

[10] V. Gazi. Swarm aggregations using artificial potentials and sliding-mode control. *IEEE Trans. Robot.*, 21(6):1208–1214, Dec. 2005.

[11] A. Jadbabaie, J. Lin, and A. S. Morse. Coordination of groups of mobile autonomous agents using nearest neighbor rules. *IEEE Trans. Automat. Contr.*, 48(6):988–1001, June 2003.

[12] J.-C. Latombe. *Robot Motion Planning*. Kluwer Academic Publishers, Boston, MA, 1991.

[13] S. M. LaValle. *Planning algorithms*. Cambridge University Press, New York, NY, 2006.

[14] N. E. Leonard and E. Fiorelli. Virtual leaders, artificial potentials and coordinated control of groups. In *IEEE Conf. Dec. Cont.*, pages 2968–2973, Orlando, FL, Dec. 2001.

[15] Y. Liu, K. M. Passino, and M. M. Polycarpou. Stability analysis of m-dimensional asynchronous swarms with a fixed communication topology. *IEEE Trans. Automat. Contr.*, 48(1):76–95, Jan. 2003.

[16] A. Martinoli, K. Easton, and W. Agassounon. Modeling swarm robotic systems: a case study in collaborative distributed manipulation. *Int. J. Rob. Res.*, 23(4-5):415–436, 2004.

[17] D. Milutinović and P. Lima. Modeling and optimal centralized control of a large-size robotic population. *IEEE Trans. Robot.*, 22(6):1280–1285, Dec. 2006.

[18] N. Ogawa, H. Oku, K. Hashimoto, and M. Ishikawa. Microrobotic visual control of motile cells using high-speed tracking system. *IEEE Trans. Robot.*, 21(3):704–712, June 2005.

[19] X. Pan, C. S. Han, K. Dauber, and K. H. Law. Human and social behavior in computational modeling and analysis of egress. *Automation in Construction*, 15(4):448–461, July 2006.

[20] J. H. Reif and H. Wang. Social potential fields: A distributed behavioral control for autonomous robots. *Robotics and Autonomous Systems*, 27:171–194, 1999.

[21] K. Takahashi, N. Ogawa, H. Oku, and K. Hashimoto. Organized motion control of a lot of microorganisms using visual feedback. In *IEEE Int. Conf. Rob. Aut.*, pages 1408–1413, Orlando, FL, May 2006.

[22] R. Vaughan, N. Sumpter, A. Frost, and S. Cameron. Robot sheepdog project achieves automatic flock control. In *Int. Conf. on the Simulation of Adaptive Behaviour*, 1998.

Safety Evaluation of Physical Human-Robot Interaction via Crash-Testing

Sami Haddadin, Alin Albu-Schäffer, Gerd Hirzinger
Institute of Robotics and Mechatronics
DLR - German Aerospace Center
P.O. Box 1116, D-82230 Wessling, Germany
{sami.haddadin, alin.albu-schaeffer, gerd.hirzinger}@dlr.de

Abstract—**The light-weight robots developed at the German Aerospace Center (DLR) are characterized by their low inertial properties, torque sensing in each joint and a load to weight ratio similar to humans. These properties qualify them for applications requiring high mobility and direct interaction with human users or uncertain environments. An essential requirement for such a robot is that it must under no circumstances pose a threat to the human operator. To actually quantify the potential injury risk emanating from the manipulator, impact test were carried out using standard automobile crash-test facilities at the ADAC[1]. In our evaluation we focused on unexpected rigid frontal impacts, i.e. injuries e.g. caused by sharp edges are excluded. Several injury mechanisms and so called Severity Indices are evaluated and discussed with respect to their adaptability to physical human-robotic interaction.**

I. Introduction & Motivation

The desired coexistence of robotic systems and humans in the same physical domain, by sharing the same workspace and actually cooperating in a physical manner, poses the very fundamental problem of ensuring safety to the user and the robot.

Safety in terms of industrial robots usually consists of isolating the workspace of the manipulator from the one of human beings by a safety guard with locked safety doors or light barriers [1]. Once the safety door is opened or the light barrier is crossed, the robot is stopped immediately.

On the other hand an increasing interest has recently been observed in domestic and industrial service robots, characterized by desirable, and under certain circumstances even unavoidable physical interaction [2], [3], [4]. Therefore, a resulting essential requirement is to guarantee safety for human users in regular operation mode as well as in possible fault modes.

This requirement necessitates a quantification of potential danger by an objective measure of injury severity. Once it is possible to correlate the behavior of the robot with an estimation of the resulting injury, it has to be guaranteed that the actions of the robot cannot cause an exceedance of a safe maximum value if physical contact with the robot occurs.

According to ISO-10218, which defines new collaborative operation requirements for industrial robots [5], one of the following conditions always has to be fulfilled: The TCP[2]/flange velocity needs to be ≤ 0.25m/s, the maximum dynamic power ≤ 80W, or the maximum static force ≤ 150N. However, these values are not derived from real human impact experiments or any biomechanical analysis but base on heuristics, intending to give a human the possibility to actively avoid dangerous situations. Up to now there do not exist any injury indices tailored to the needs of robotics and those borrowed from other fields (e.g. the Head Injury Criterion (HIC) [6], [7]) were mainly evaluated in simulation so far. Further approaches are outlined in [8], [9], [10], but the importance of human biomechanics was barely investigated. To fill this gap we decided to measure the potential danger emanating from the DLR lightweight robot III (LWRIII) by impact tests at a certified crash-test facility. These tests were conducted at the Crash-Test Center of the German Automobile Club ADAC. The robot was commanded to move on a predefined trajectory and hit various dummy body parts at TCP velocities up to 2m/s. Based on the outcome of these experiments we can draw some general conclusions related to potential danger of robots, depending on their mass and velocity.

In our evaluation we concentrated on unexpected impacts of a smooth surface related to the three body regions head, neck, and chest. Injury mechanisms caused by sharp tools or similar injury sources were not taken into consideration, since these mechanisms cannot be measured with standard crash-test dummies. To evaluate the resulting injury severity the European testing protocol EuroNCAP was applied. The results of several injury criteria for head, neck, and chest were measured by the ADAC and will be presented in the paper. Because an overview of commonly used Severity Indices is missing in robotics, a short presentation on them will be given as well. The most prominent index for the head is the Head Injury Criterion [11] which was already introduced to robotics in [6], [7] and used as a basis for new actuation concepts [12], [13].

As mentioned above, work that has been carried out up to now in the field of physical human-robot interaction was mainly based on simulations. These contributions indicated high potential injury of humans by means of HIC, already at a robot speed of 1m/s. This also perfectly matched to the "common sense" expectation that a robot moving at maximal speed (e.g. due to malfunction) can cause high impact injury. In this sense the paper presents very surprising and striking

[1]German Automobile Club
[2]Tool Center Point

results.
Moreover, one of the main contributions of this paper is the first experimental evaluation of HIC in standard crash-test facilities. Additionally to the impact evaluation it will be shown that even with an ideally fast (physical) collision detection one is not able to react fast enough to a stiff collision (e.g. head) in order to decrease the effect of the contact forces for link inertias similar or larger to the ones of the LWRIII.

In Section II the evaluated injury limits and measures are defined and briefly explained, followed by the description of the testing setup in Section III. Consecutively, experimental results are presented in Section IV. The following evaluation and discussion lead to a number of surprising and quite general conclusions outlined in Section V.

II. Classifying Injury Severity

Before actually introducing the definition of evaluated Severity Indices, an intuitive and internationally established definition of injury level will be given.

A. The Abbreviated Injury Scale

AIS	SEVERITY	TYPE OF INJURY
0	None	None
1	Minor	Superficial Injury
2	Moderate	Recoverable
3	Serious	Possibly recoverable
4	Severe	Not fully recoverable without care
5	Critical	Not fully recoverable with care
6	Fatal	Unsurvivable

TABLE I
Definition of the Abbreviated Injury Scale.

A definition of injury level developed by the AAAM[3] and the AMA[4] is the Abbreviated Injury Scale (AIS) [14]. It subdivides the observed level of injury into seven categories from *none* to *fatal* and provides a very intuitive classification (see Tab.I). Of course this classification gives no hint *how* to measure possible injury, this is provided by so called Severity Indices.

B. EuroNCAP

The ADAC crash-tests are carried out according to the EuroNCAP[5] which is based on the Abbreviated Injury Scale. The EuroNCAP, inspired by the American NCAP, is a manufacturer independent crash-test program uniting the European ministries of transport, automobile clubs and underwriting associations with respect to their testing procedures and evaluations [15]. The outcome of the tests, specified in the program, is a scoring of the measured results via a sliding scale system. Upper and lower limits for the injury potentials are mostly defined such that they correlate to a certain probability of AIS ≥ 3 (see e.g. Fig.5,6). Between these two values the corresponding score (injury potential) is calculated by linear interpolation. A standardized color code indicates injury potential and is given in Tab.II.

Colorcode	Color	Injury potential
Red		Very high
Brown		High
Orange		Medium
Yellow		Low
Green		Very low

TABLE II
Injury Severity and Corresponding Color Code.

Since standard dummy equipment enables the measurement of Severity Indices for the head, neck and chest, those ones evaluated by our tests at the crash-test facilities will now be defined.

C. Injury Criteria For The Head

According to [16] most research carried out in connexions with automobile crash-testing distinguishes two types of head loadings:

1) *Direct Interaction:* An impact or blow involving a collision of the head with another solid object at appreciable velocity. This situation is generally characterized by large linear accelerations and small angular accelerations during the impact phase.
2) *Indirect Interaction:* An impulse loading including a sudden head motion without direct contact. The load is generally transmitted through the head-neck junction upon sudden changes in the motion of the torso and is associated with large angular accelerations of the head.

Since the potential danger is disproportionately higher by direct interaction, this work will concentrate on the first potential injury source.

Especially for the head quite many criteria for type 1 interactions are available. Their major theoretical basis is the so called WSTC[6], a fundamental experimental injury tolerance curve forming the underlying biomechanical data of all presented head criteria. The limit values of the following injury criteria are defined in the EuroNCAP protocol. For the head they represent a 5% probability of occurring AIS ≥ 3 injury.

1) Head Injury Criterion: The most frequently used head Severity Index is the Head Injury Criterion [11], defined as

$$\mathrm{HIC}_{36} = \max_{\Delta t}\left\{\Delta t\left(\frac{1}{\Delta t}\int_{t_1}^{t_2}||\ddot{\mathbf{x}}_H||_2 \mathrm{d}t\right)^{\left(\frac{5}{2}\right)}\right\} \leq 650 \qquad (1)$$
$$\Delta t = t_2 - t_1 \leq \Delta t_{\max} = 36\mathrm{ms}.$$

$||\ddot{\mathbf{x}}_H||$ is the resulting acceleration of the human head[7] and has to be measured in $g = 9.81\mathrm{m/s}^2$. The optimization

[3]Association for the Advancement of Automotive Medicine
[4]American Medical Association
[5]European National Car Assessment Protocol
[6]Wayne State Tolerance Curve
[7]$||\ddot{\mathbf{x}}||_2$ =Euclidean norm

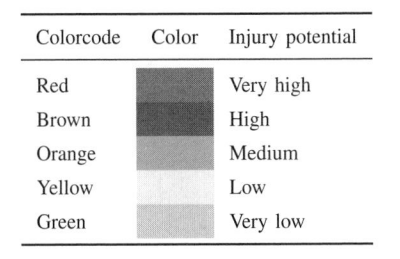

is done by varying t_1 and t_2, i.e. the start and stop time are both parameters of the optimization process. Intuitively speaking, the HIC weights the resulting head acceleration and impact duration, which makes allowance of the fact that the head can be exposed to quite high accelerations and is still intact as long as the impact duration is kept low. In addition to the HIC_{36} the identically defined HIC_{15} with $\Delta t_{\max} = 15\text{ms}$ exists. Comparing both likelihood distributions yields that corresponding injury probabilities for HIC_{15} are more restrictive than for the HIC_{36} (see Sec.II-C.3).

2) 3ms-Criterion:

$$a_{3\text{ms}} = \frac{1}{\Delta t}\int_{t_1}^{t_2+\Delta t} ||\ddot{\mathbf{x}}_H||_2 \mathrm{d}t \leq 72\text{g}, \quad \Delta t = 3\text{ms} \quad (2)$$

This criterion requires the maximum 3ms-average of the resulting acceleration to be less than 72g. Any shorter impact duration has only little effect on the brain.

3) Converting Severity Indices to the Abbreviated Injury Scale: Unfortunately, Severity Indices do usually not provide a direct scaling of injury but rather a limit between severe and non-severe injury. Furthermore, they are defined with respect to different physical domains and thus are not directly comparable to each other, nor can they be combined. In order to cope with this deficit, mappings were developed to translate a Severity Index to the Abbreviated Injury Scale. The NHTSA[8] specified the expanded Prasad/Mertz curves [17] for converting HIC_{15} values to the probability $p(AIS \geq i)$ of the corresponding AIS level i which are shown in Fig.1(left). In [18] a conversion from HIC_{36} to $p(AIS \geq 2,3,4)_{\text{HIC36}}$ is defined. Since the EuroNCAP underlays its injury risk level definition mainly on the $p(AIS \geq 3)$-level, the corresponding functions for both HICs are illustrated in Fig.1(right):

$$p(AIS \geq 3)_{\text{HIC15}} = \frac{1}{1+e^{3.39+\frac{200}{\text{HIC}_{15}}-0.00372\text{HIC}_{15}}} \quad (3)$$

$$p(AIS \geq 3)_{\text{HIC36}} = \Phi\left(\frac{\ln(\text{HIC}_{36})-\mu}{\sigma}\right), \quad (4)$$

with $\Phi(.)$ denoting the cumulative normal distribution with mean $\mu = 7.45231$ and standard deviation $\sigma = 0.73998$. For our very short impacts the evaluation of HIC_{15} and HIC_{36} lead to the same numerical value. Obviously the HIC_{15} indicates a higher risk level than the HIC_{36} for the same numerical value and is therefore more restrictive.

D. Injury Criteria For The Neck

In general, the injury mechanisms of the human neck are related to forces and bending torques acting on the spinal column. In the EuroNCAP the corresponding limits are defined with respect to the positive cumulative exceedance time as denoted in Tab.III. Between these values a linear interpolation is carried out. The corresponding taxonomy of the neck is illustrated in Fig.2, whereas the EuroNCAP specifies limit values only for the motions listed in Tab.III.

[8]National Highway Traffic Safety Administration

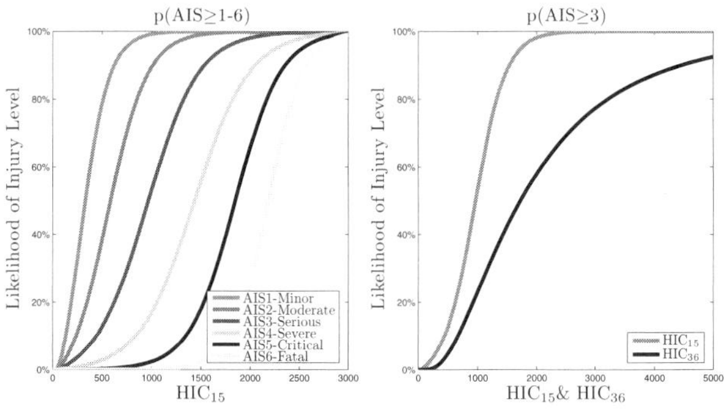

Fig. 1. Mapping HIC_{15} to the Abbreviated Injury Scale (left) and comparing $p(AIS \geq 3)_{\text{HIC15}}$ with $p(AIS \geq 3)_{\text{HIC36}}$ (right).

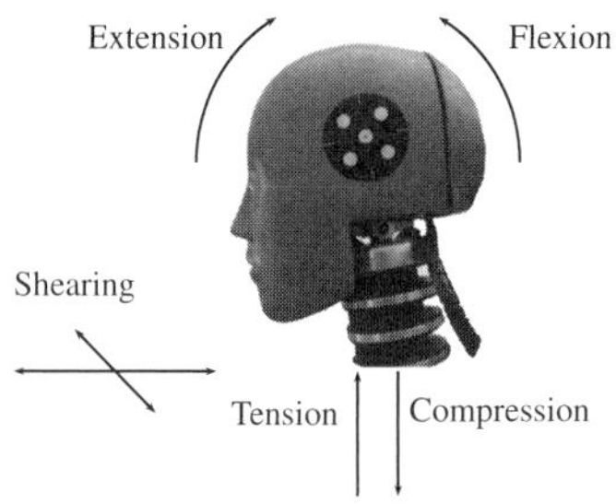

Fig. 2. Taxonomy of neck motions.

E. Injury Criteria For The Chest

1) Compression-Criterion: From evaluated cadaver experiments it was derived that acceleration and force criteria alone are intrinsically not able to predict the risk of internal injuries of the thorax which tend to be a greater threat to human survival than skeletal injury. Kroell analyzed a large data base of blunt thoracic impact experiments and realized that the Compression Criterion

$$CC\ (\ \ ||\text{p}\,\mathbf{x}_C||_2 \leq \text{HH\&\&} \quad (5)$$

is a superior indicator of chest injury severity, where p $\mathbf{x}_C$ is the chest deflection. Especially sternal impact was shown to cause compression of the chest until rib fractures occur ([19], [20]).

2) Viscous-Criterion: The second criterion for the chest is the Viscous Criterion (VC), which is also known as Soft Tissue Criterion [20],[21]. It can be formulated as

$$VC\ (\ \ c_c||\text{p}\,\dot{\mathbf{x}}_C||_2\frac{||\text{p}\,\mathbf{x}_C||_2}{l_c} \leq \text{u.I}\frac{\&}{s}, \quad (6)$$

defined as the product of compression velocity and the normalized thoracic deflection. The scaling factor c_c and the deformation constant (actually the initial torso thickness) l_c depend on the used dummy and are summarized in [22].

III. Experimental Setup

The instrumentation of the used HybridIII dummy is shown in Fig.3. It represents the standard equipment to measure the described injury criteria at a sampling frequency of HkHz. The signals are filtered according to [23].

Load	@0ms	@25 − 35ms	@45ms
Shearing: F_x, F_y	1.9/3.1kN	1.2/1.5kN	1.1/1.1kN
Tension: F_z	2.7/3.3kN	2.3/2.9kN	1.1/1.1kN
Extension: M_y	42/57Nm	42/57Nm	42/57Nm

TABLE III
HIGHER AND LOWER PERFORMANCE LIMIT SPECIFIED FOR THE HUMAN NECK (SEE AS WELL FIG.8).

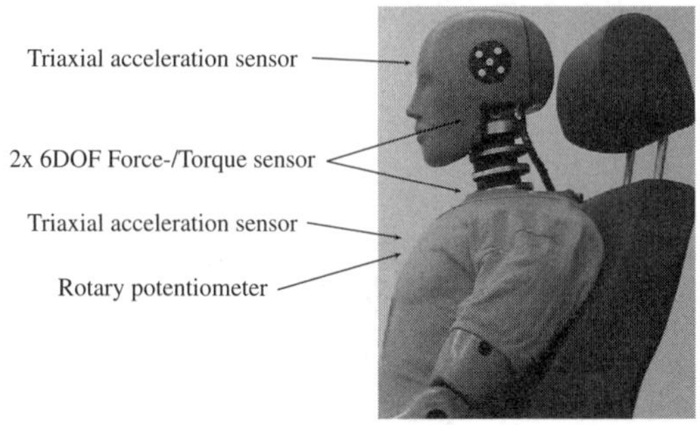

Fig. 3. HybridIII Dummy Instrumentation.

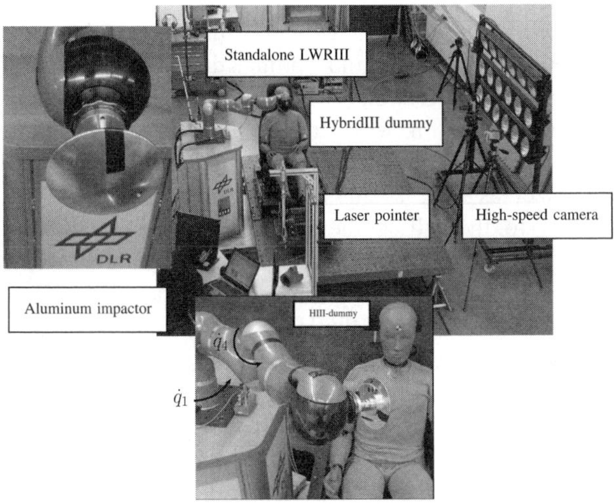

Fig. 4. Test setup for impact experiments.

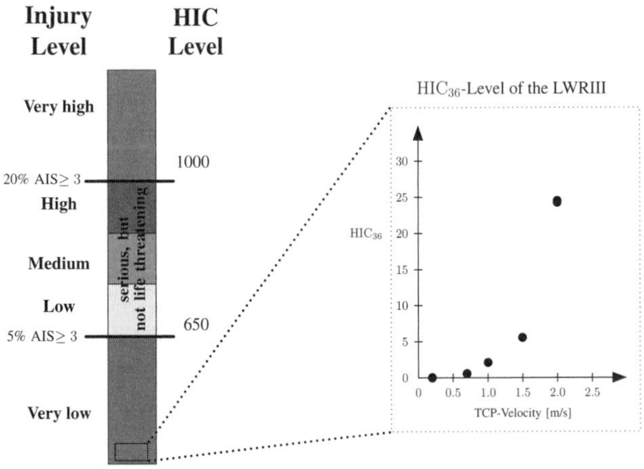

Fig. 5. Resulting HIC_{36} values for varying impact velocities, rated according to the EuroNCAP *Assessment Protocol And Biomechanical Limits*.

In Fig.4 the overall test setup is shown. It consists of the LWRIII, the full dummy equipment, a high-speed camera and a laser pointer to ensure a reproducibility of the tests. The 7DOF[9] flexible-joint robot has a weight of 14kg and a load to weight ratio ≈ 1. It is equipped with motor and link side position and torque sensors in each joint. All calculations of the indices were carried out by the ADAC, thus were done according to the EuroNCAP program. In order to ensure rigid and defined contact, a 1kg aluminum impactor was used which was equipped with high bandwidth force and acceleration sensors (see Fig.4). The desired trajectory[10] was a rest-to-rest motion which start and end configuration was given by

$$\mathbf{q}_{\text{start}} = (-45 \quad 90 \quad -90 \quad -45 \quad 0 \quad -90 \quad 147)^\circ$$
$$\mathbf{q}_{\text{end}} = (\ 45 \quad 90 \quad -90 \quad 45 \quad 0 \quad -90 \quad 147)^\circ.$$

In order to maximize the joint mass matrix (reflected inertia was ≈ 4kg at the TCP) the trajectory was selected such that the robot hits the dummy in outstretched position. Furthermore, high TCP velocities can be achieved in this impact configuration. In our experiments we chose the robot velocities to be $||\dot{\mathbf{x}}||_{\text{TCP}} \in \{0.2, 0.7, 1.0, 1.5, 2.0\}$m/s.

A TCP velocity of 2m/s is already relatively close to the maximal robot speed and, as will be pointed out later, poses in the case of impact a potential threat to the mechanics of the robot. Of course this position can be further optimized towards the absolute worst-case, but the described configuration seemed to be a reasonable first guess.

IV. EXPERIMENTAL RESULTS

A. Results for the Head

In Fig. 5 the resulting HIC_{36} values are plotted with respect to the impact velocity of the robot. The corresponding injury classification was described in Sec.II-B. In order to classify an impact into the *green* labeled region, the occurring HIC_{36} must not exceed 650, which corresponds to a resulting 5%-probability of serious injury ($\text{AIS} \geq 3$). This value originates from [24], [25] and differs only slightly from the one obtained by the fitting function (4).

As indicated in Fig.5, the HIC_{36} caused by the LWRIII is below 25 at 2m/s which corresponds to a *very low* injury level. The resulting probability of injury severity obtained by (3) and (4) is $\approx 0\%$ for all categories. Another aspect that clearly can be extracted from Fig.5 is that the HIC_{36} is rapidly increasing with robot velocity.

Similar to the results of the HIC_{36}, very low potential danger is indicated by the 3ms-Criterion. Even at a tip velocity of 2m/s less than 20% of the lower limit of 72g are reached (see Fig.6).

B. Results for the Neck

The resulting neck force $F_{\text{res}}^{\text{Neck}}$ for varying robot velocities caused by head impacts is illustrated in Fig.7. The actual impact is characterized by a very short peak which duration and maximum value depend on the impact velocity. For fast impacts a low level safety feature of the robot activates and

[9]Degrees Of Freedom
[10]See also the video on www.robotic.dlr.de/safe-robot

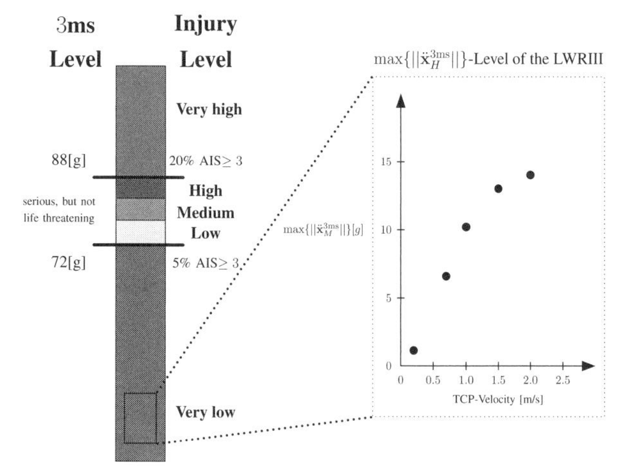

Fig. 6. Resulting 3ms-Criterion values for varying impact velocities, rated according to the EuroNCAP *Assessment Protocol And Biomechanical Limits.*

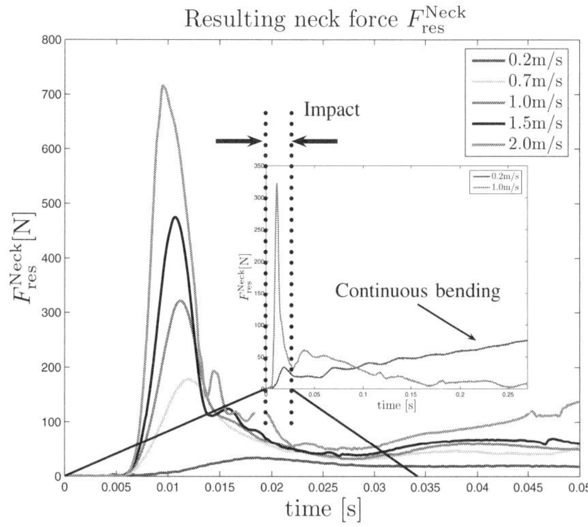

Fig. 7. Resulting impact force during head impacts.

stops it because the specified maximum joint torques are exceeded. Therefore, the *maximum* neck force/torque during the entire collision is determined by this *peak* force/torque occurring within the first $5-20$ms of the impact. On the other hand, if the impact velocity is very low (0.2m/s), the *impact* force is reduced dramatically and does not trigger the low-level stopping mechanism. Consequently, steadily growing neck bending can take place, increasing neck forces to even larger values than the ones caused by the original impact because the robot still follows its desired trajectory. This becomes clear if the neck forces for the impact velocities 0.2m/s and 1.0m/s are plotted for a longer time period (see Fig.7): After $\approx$ 20ms both *impact* maxima are over and at 1m/s the low-level stop of the robot is triggered because the impact forces (up to 2kN were measured at the aluminum impactor) cause extremely high joint torques. In contrast at 0.2m/s the neck force is steadily increasing and might become even larger than impact forces at higher velocities.

In Fig.8 the occurring upper neck shearing and tension/compression forces are plotted with respect to the positive cumulative exceedance time. Actually, only tension limits are

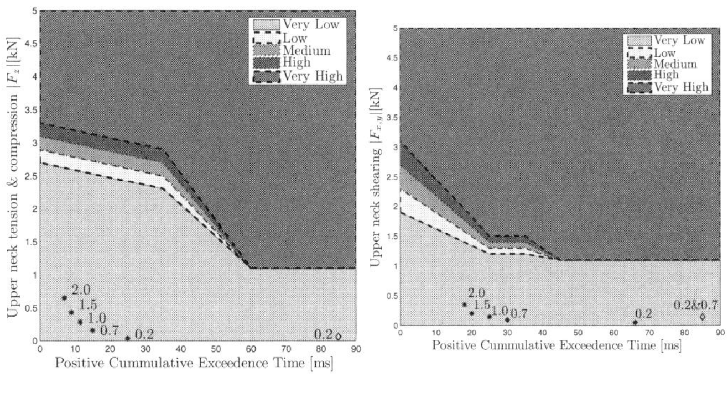

Fig. 8. Resulting $F_{x,y}$ and F_z values for varying impact velocities, rated according to the EuroNCAP *Assessment Protocol And Biomechanical Limits.*

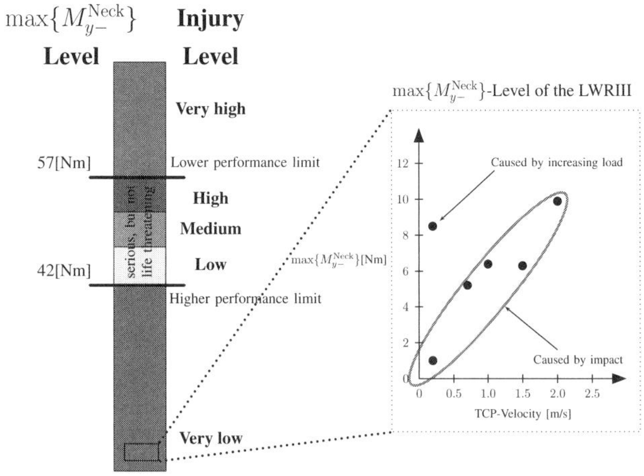

Fig. 9. Resulting $\max\{M_{y-}^{Neck}\}$ values for varying impact velocities, rated according to the EuroNCAP *Assessment Protocol And Biomechanical Limits.*

specified in the EuroNCAP, but according to [26] tension is more critical than compression and thus applying available limits to both, tension and compression seems to be a reasonable choice.

The tolerance values for neck forces are not constant, but a function of the exceedance time (see Sec.II-D). The resulting forces are labeled with the corresponding TCP velocity. A * indicates the forces caused by the impact and $\diamondsuit$ the ones by continuous bending, if they were finally larger than the impact forces. In order not to break the dummy neck the robot stopped a predefined distance after the collision occurred. This of course limits the bending forces & torques which otherwise would further increase. In Fig.9 the results of the extension torque are visualized. As for the previous head Severity Indices, the occurring neck forces/torques are totally subcritical, i.e. pose no threat to the human.

C. Results for the Chest

According to [27] a 5%-probability of serious chest injury (AIS$\geq$ 3) corresponds to a compression of 22mm and 50% to 50mm. In Fig.10 the resulting compression values are plotted with respect to the impact velocity of the robot. Again, the injury potential is very low, as the values range in the lowest quarter of the *green* area.

The results of the Viscous Criterion are not presented

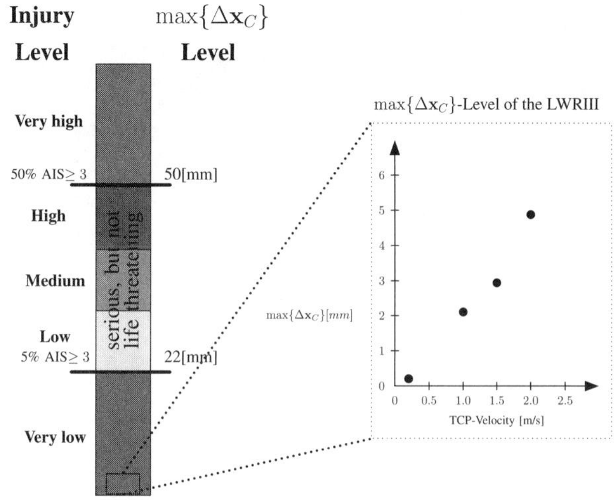

Fig. 10. Resulting max{$\Delta \mathbf{x}_C$} values for varying impact velocities, rated according to the EuroNCAP *Assessment Protocol And Biomechanical Limits*.

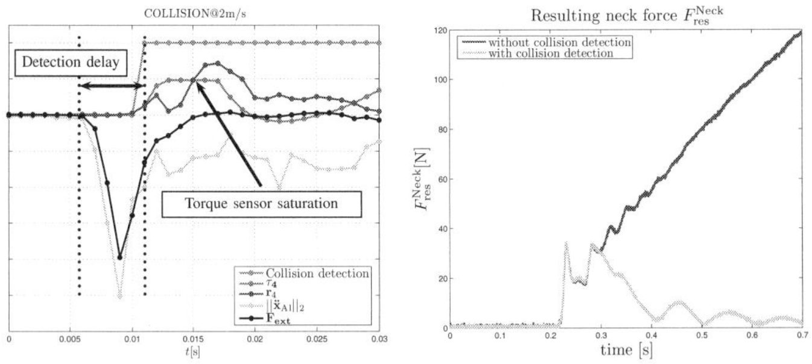

Fig. 11. Impact characteristics at 2m/s. All values are scaled to fit into one plot (left). The plot is mainly to show the timing of the signals: While acceleration and impact force are simultaneous, the joint torque and the additional external torque estimation react delayed to the impact. Resulting neck force with and without collision detection and reaction strategy (right).

because the resulting values were located within the range of noise and thus this criterion is not well suited, nor sensitive enough for our evaluation. This is related to the relatively low velocities, compared to the ones encountered in automotive crashes.

V. Evaluation & Discussion

All evaluated severity measures range within the lowest quarter of the *green* indicated area, i.e. the potential danger emanating from the LWRIII is intrinsically very low by means of injury measurable by automobile crash-test facilities such as the ADAC. These are surprising and gratifying results and to our knowledge, they represent the first systematic experimental evaluation of possible injuries during robot-human impacts using standardized testing facilities.

The low values of the Severity Indices are a direct consequence of the much lower speeds of the robot tip, compared to velocities of cars involved in crash-tests. There, impact tests at velocities starting at 10m/s, which is equivalent to 36km/h, are carried out. At such high velocities Severity Indices possibly exceed the defined limits.

Apart from the directly measured results one is able to draw fundamental implications to physical human-robot interaction, which now shall be outlined.

A. Typical Impact Characteristics

In order to illustrate certain problems and first implications resulting from the dynamics of a rigid head impact, a dummy head crash at 2m/s shall be evaluated. In Fig.11(left) the measured joint torque of the 4th axis $\boldsymbol{\tau}_4$, the contact force $\boldsymbol{F}_{\mathbf{ext}}$, and impactor acceleration $\ddot{\boldsymbol{x}}_{\mathbf{A1}}$ are visualized. Such a fast impact is characterized by a very high acceleration/force peak, lasting $6-10$ms and highly depending on the impact velocity. The maximum measured contact force and acceleration of the aluminum impactor were 2kN and 35g, respectively. Actually, this very short part of the impact is the relevant one for the evaluation of head Severity Indices, since here they reach their maximum value.

One clearly can see that before the joint torque starts increasing, the relevant force/acceleration peak period is practically over. Thus, during this particular time interval motor and link inertia are decoupled by the intrinsic joint elasticity, and only the link inertia is involved into the impact. Therefore, decreasing joint stiffness e.g. via antagonistic actuation would not have any effect on a hard contact head impact with link inertias similar or higher than the ones of the LWRIII. For collisions with softer body parts (e.g. the arm) the impact duration is lower and decreasing joint stiffness might reduce contact forces.

A collision detection and reaction scheme, based on the disturbance observer developed in [28], is used and indicated in Fig.11(left). It utilizes only the proprioceptive capabilities of the robot (joint torque and motor position) and provides a filtered version of the external torque $\boldsymbol{\tau}_{\mathbf{ext}}$. As soon as a collision has been detected (after a detection delay of $\approx$ 6ms) the robot switches within one cycle time of 1ms from position to torque control with gravitation compensation [29]. In other words, the commanded torque is $\boldsymbol{\tau}_{\mathbf{d}}$ x $\hat{\boldsymbol{g}}(\boldsymbol{\theta})$, where $\hat{\boldsymbol{g}}$ is a gravitation compensation based on the motor position $\boldsymbol{\theta}$.

The disturbance observer highly depends on the joint torque measurement. This implies that the collision detection cannot be used to decrease the resulting injury severity caused by rigid impacts at high robot velocities, since the impact itself is passed before the joint torques start increasing significantly. Thus, using this collision detection mechanism does not clarify whether one could potentially decrease the resulting injury by a faster detection mechanism. Therefore, the acceleration signal of the impactor was used as well to trigger the reaction strategy of the robot. Although one could now detect the collision within 1ms, the resulting injury criteria did not differ from the ones obtained with the previous collision detection scheme or even without any reaction strategy. This is explained by the fact that, even if a collision is detected timely, the motors cannot revert their motion sufficiently fast.

Another fundamental injury source is quasistatic loading occurring at lower velocities (see Sec.IV-B), or if no low-level stop in case of maximum joint torque exceedance would occur and the robot continued to follow its desired trajectory after the impact. Especially if a human body part is clamped or lacks somehow in mobility, this could become a very dangerous injury source. In Fig.11(right) the effect of the collision detection is visualized which proves to be a very fast and efficient way to handle quasistatic loadings.

B. Protecting The Robot

Another very important observation, already made at an impact velocity of 1m/s, is that the specified maximum joint torques of the robot were exceeded for several milliseconds during the impact (see Fig.11(left)).

A mechanical end stop in the robot limits the deflection range of the torque sensor which then goes into saturation. As already mentioned, a low-level emergency stop is initialized as soon as this event is triggered. It has been explained that even an ideally fast collision detection is not able to reduce the impact forces. Therefore, such dynamic loads pose a serious threat, potentially causing mechanical damage to the robot. This necessitates to think about how to prevent the robot from being damaged during such an impact, directly leading to the requirement of reducing the robot speed to subcritical values from the robot's point of view. One could even say that a rigid impact with the LWRIII poses a disproportionally higher threat to the robot than to the human.

C. Influence of Robot Mass & Velocity

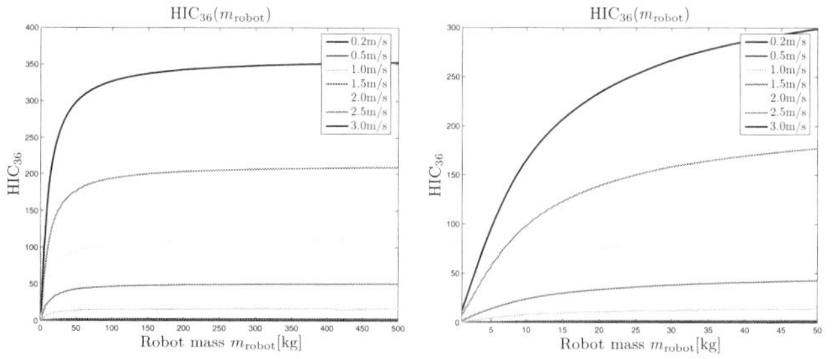

Fig. 12. HIC values resulting from 1DOF impact simulations between a robot and a dummy head model obtained by data fitting to real impact measurements. The curves show the dependency of HIC on the robot mass and are parameterized by the impact velocity.

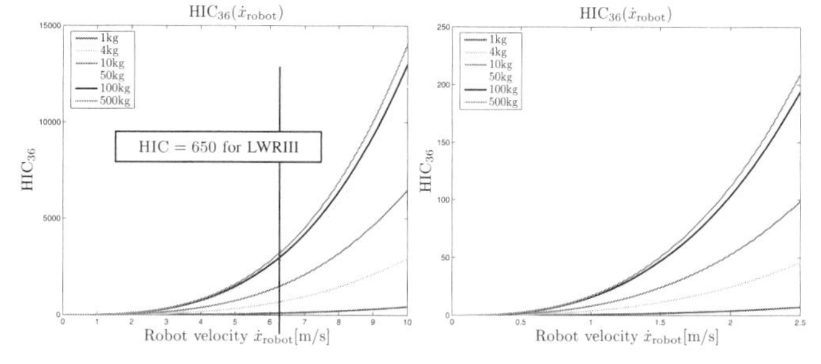

Fig. 13. HIC values resulting from 1DOF impact simulations between a robot and a dummy head model obtained by data fitting to real impact measurements. The curves show the dependency of HIC on the impact velocity and are parameterized by the robot mass.

In Fig.12,13 the simulation results of a robot colliding with a dummy head model, which was extracted from the real impact data, are plotted. A Hunt-Crossley model [30] was used to cope with the discontinuity in the classical mass spring damper approach caused by the damping element at the moment of impact. Although some variations between real experiments and this simulation may exist, very interesting and useful implications can be extracted from it. In Fig.12 the Head Injury Criterion was evaluated for robot masses up to 500kg and graphs were obtained for impact velocities of $||\dot{\mathbf{x}}||_{\mathrm{TCP}} \in \{0.2, 0.5, 1.0, 1.5, 2.0, 2.5, 3.0\}$m/s. They show that the HIC saturates for increasing robot mass at each impact velocity. This on the other hand indicates that at some point increasing robot mass does not result in higher HIC. In Fig.13 the vast effect of impact velocity is shown and additionally one can see the decreasing robot mass dependency. The impacts were simulated with robot velocities up to 10m/s and the graphs were obtained for reflected robot inertias of $\{1, 4, 10, 50, 100, 500\}$kg.

Additionally, a very intuitive and afterwards obvious interpretation of the saturation effect can be drawn: If we think of a very massive industrial robot, colliding at 2m/s with a human head it is nearly the same as if the human runs with 2m/s, which is equivalent to 7.2km/h, against a rigid wall. This is at least true for the short duration of the impact which is relevant for the HIC. Already intuitively it becomes clear that one would not be seriously injured by such an impact at walking speed.

Consequently, no robot whatever mass it has could become dangerous at 2m/s by means of the presented criteria as long as clamping and impacts with sharp surfaces can be excluded. In fact, even in case of clamping the Head Injury Criterion and the 3ms-Criterion would not indicate any potential injury. This is because they are determined within the first $5 - 10$ms of the impact and during this extremely short time interval the head does not move noticeable.

Generally speaking increasing robot mass and loads pose a threat if there is any chance a human body part could be clamped. Because of the higher inertias it would take the robot longer to decelerate and thus fatal injuries, e.g. exceeding chest deflection limits could be the consequence. This yields to another advantage of the LWRIII: By its lightweight design the inertias are low enough not to cause this kind of injury. In order to verify that statement, impact tests with industrial robots are going to be carried out, which probably yield similar HIC values as the ones obtained with the LWRIII, but the quasistatic contact force and penetration depth are going to increase significantly.

D. Consequences for Standards & Norms

The presented results imply that typical Severity Indices, such as the Head Injury Criterion are not applicable to robot-human impacts occurring at much lower velocities than the ones evaluated in the automobile industry. Actually, it has been shown in [31] that the mechanical response of a dummy indicates even higher injury severity than would occur in reality. Therefore, new criteria focusing on other injury mechanisms, such as clamping, lacerations and fractures have to be investigated. These statements do not directly apply to robots operating at much higher velocities but it is questionable anyway whether it is desirable for a human to **interact** with a robot moving at velocities considerably higher than 2m/s. Of course, another potential injury source still has to be investigated: Tools mounted on the robot can be arbitrarily dangerous and need to be investigated separately (what if the robot moves a knife).

In the beginning of this paper the ISO-10218 was introduced, defining new collaborative operation requirements. In our experiments we were able to drive eight times faster and cause thirteen times higher dynamical contact forces than are suggested by the norm for the static case (Severity Indices are usually defined for impacts in the range of milliseconds but static tolerance is usually higher than dynamic [26]). Still our impact experiments yielded such low injury risks raising the question whether this standard is not too conservative.

VI. Conclusion & Outlook

The motivation of this work was to investigate the potential danger emanating from the LWRIII with respect to physical human-robot interaction. Severity Indices were introduced, which are biomechanically motivated quantities indicating the injury severity of various human body parts. In this paper we focused on the blunt unexpected impact with the human standing still at the moment of impact. The important issue of tools still has to be investigated.

Impact tests were carried out using standard crash-test facilities, whereas numerous Severity Indices were evaluated according to the standardized EuroNCAP. The resulting values proved that a blunt impact between a human and the LWRIII does not cause serious harm[11]. Additionally it has been shown, that the results concerning the HIC can be generalized to robots of arbitrary mass.

Another major conclusion is that classical Severity Indices, established in the automobile industry cannot be transferred to the field of robotics because operating velocities are basically far too low to have any significant impact on them. Therefore, new criteria have to be proposed focusing on relevant injury mechanisms. Based on our results the adjustment or even redefinition of the automotive injury scaling system to robotic purposes will be of major importance.

A **video** illustrating and supporting the key aspects proposed and explained in the paper can be found at www.robotic.dlr.de/safe-robot

Acknowledgment

This work has been partially funded by the European Commission's Sixth Framework Programme as part of the projects *SMERobot*™ under grant no. 011838 and PHRIENDS under grant no. 045359.
The authors would like to thank Andreas Ratzek and all involved members of the ADAC team for the fruitful and enjoyable cooperation. Furthermore, we would like to express our appreciation to Professor Dr. Dimitrios Kallieris for his helpful advise.

References

[1] EN953, "Safety of Machinery - Guards - General Requirements for the Design and Construction of fixed and movable Guards," 1997.
[2] "http://www.phriends.eu/."
[3] "http://www.smerobot.org/."

[11] By means of presented criteria. Furthermore, the presented evaluation is carried out for average male dummies and not for females or children. For such an analysis further tests with specially designed female or child dummies would be necessary.

[4] Ch. Ott, O. Eiberger, W. Friedl, B. Bäuml, U. Hillenbrand, Ch. Borst, A. Albu-Schäffer, B. Brunner, H. Hirschmüller, S. Kielhöfer, R. Konietschke, M. Suppa, T. Wimböck, F. Zacharias, and G. Hirzinger, "A Humanoid Two-Arm System for Dexterous Manipulation," in *IEEE-RAS International Conference on Humanoid Robots*, 2006.
[5] ISO10218, "Robots for industrial environments - Safety requirements - Part 1: Robot," 2006.
[6] M. Zinn, O. Khatib, B. Roth, and J.K.Salisbury, "Playing It Safe - Human-Friendly Robots," *IEEE Robotics and Automation Magazine*, vol. 11, pp. 12–21, 2002.
[7] A. Bicchi and G. Tonietti, "Fast and Soft Arm Tactics: Dealing with the Safety-Performance Trade-Off in Robot Arms Design and Control," *IEEE Robotics and Automation Magazine*, vol. 11, pp. 22–33, 2004.
[8] H.-O. Lim and K. Tanie, "Human Safety Mechanisms of Human-Friendly Robots: Passive Viscoelastic Trunk and Passively Movable Base," *International Journal of Robotic Research*, vol. 19, no. 4, pp. 307–335, 2000.
[9] J. Heinzmann and A. Zelinsky, "Quantitative Safety Guarantees for Physical Human-Robot Interaction," *International Journal of Robotic Research*, vol. 22, no. 7-8, pp. 479–504, 2003.
[10] K. Ikuta, H. Ishii, and M. Nokata, "Safety Evaluation Method of Design and Control for Human-Care Robots," *International Journal of Robotic Research*, vol. 22, no. 5, pp. 281–298, 2003.
[11] J. Versace, "A Review of the Severity Index," *Proc 15th Stapp Conference*, vol. SAE Paper No.710881, 1971.
[12] M. Zinn, O. Khatib, and B. Roth, "A New Actuation Approach for Human Friendly Robot Design," *The International Journal of Robotics Research*, vol. 23, pp. 379–398, 2004.
[13] G. Tonietti, R. Schiavi, and A. Bicchi, *Optimal Mechanical/Control Design for Safe and Fast Robotics*, ser. Springer Tracts in Advanced Robotics, O. K. Marcelo H. Ang, Ed. Springer Berlin / Heidelberg, 2006, vol. 21.
[14] A. for the Advancement of Automotive medicine, *The Abbreviated Injury Scale (1990) Revision Update 1998*. Des Plaines/IL, 1998.
[15] EuroNCAP, "European Protocol New Assessment Programme - Assessment Protocol and Biomechanical Limits," 2003.
[16] J. McElhaney, R. Stalnaker, and V. Roberts, "Biomechanical Aspects of Head Injury," *Human Impact Response - Measurement and Simulation*, 1972.
[17] NHTSA, "Actions to Reduce the Adverse Effects of Air Bags," *FMVSS No. 208*, 1997.
[18] S. Kuppa, "Injury Criteria for Side Impact Dummies," *NHTSA*, 2004.
[19] I. Lau and D. Viano, "Role of Impact Velocity and Chest Compression in Thoracic Injury," *Avia. Space Environ. Med.*, vol. 56, 1983.
[20] ——, "The Viscous Criterion - Bases and Applications of an Injury Severity Index for Soft Tissues," *Proceedings of 30th Stapp Car Crash Conference*, vol. SAE Technical Paper No. 861882, 1986.
[21] ——, "Thoracic Impact: A Viscous Tolerance Criterion," *Tenth Experimental Safety Vehicle Conference*, 1985.
[22] D. P. V. S. Workgroup, "Crash Analysis Criteria Version 1.6.1," 2004.
[23] EuroNCAP, "European Protocol New Assessment Programme - Frontal Impact Testing Protocol," 2004.
[24] P. Prasad and H. Mertz, "The Position of the US Delegation to the ISO Working Group 6 on the Use of HIC in Automotive Environment," *SAE Paper 851246*, 1985.
[25] P. Prasad, H. Mertz, and G. Nusholtz, "Head Injury Risk Assessment for Forehead Impacts," *SAE Paper 960099*, 1985.
[26] D. Kallieris, *Handbuch gerichtliche Medizin*, B. Brinkmann and B. Madea, Eds. Springer Verlag, 2004.
[27] P. Prasad and H. Mertz, "Hybrid III sternal Deflection associated with Thoracic Injury Severities on Occupants restrained with Force-limiting Shoulder Belts," *SAE Paper 910812*, 1991.
[28] A. D. Luca, A. Albu-Schäffer, S. Haddadin, and G. Hirzinger, "Collision Detection and Safe Reaction with the DLR-III Lightweight Manipulator Arm," *IEEE/RSJ International Conference on Intelligent Robots and Systems (IROS)*, pp. 1623–1630, 2006.
[29] C. Ott, A. Albu-Schäffer, A. Kugi, S. Stramigioli, and G. Hirzinger, "A Passivity Based Cartesian Impedance Controller for Flexible Joint Robots - Part I: Torque Feedback and Gravity Compensation," *International Conference on Robotics and Automation (ICRA)*, pp. 2666–2672, 2004.
[30] K. Hunt and F. Crossley, "Coefficient of Restitution Interpreted as Damping in Vibroimpact," *ASME Journal of Applied Mechanics*, pp. 440–445, 1975.
[31] A. Rizzetti, D. Kallieris, P. Schiemann, and P. Mattern, "Response and Injury of the Head-Neck Unit during a Low Velocity Head Impact," *IRCOBI*, pp. 194–207, 1997.

Dimensionality Reduction Using Automatic Supervision for Vision-Based Terrain Learning

Anelia Angelova
Computer Science Dept.
California Institute of Technology
Email: anelia@vision.caltech.edu

Larry Matthies, Daniel Helmick
Jet Propulsion Laboratory
California Institute of Technology
lhm, dhelmick@jpl.nasa.gov

Pietro Perona
Electrical Engineering Dept.
California Institute of Technology
perona@vision.caltech.edu

***Abstract*— This paper considers the problem of learning to recognize different terrains from color imagery in a fully automatic fashion, using the robot's mechanical sensors as supervision. We present a probabilistic framework in which the visual information and the mechanical supervision interact to learn the available terrain types. Within this framework, a novel supervised dimensionality reduction method is proposed, in which the automatic supervision provided by the robot helps select better lower dimensional representations, more suitable for the discrimination task at hand. Incorporating supervision into the dimensionality reduction process is important, as some terrains might be visually similar but induce very different robot mobility. Therefore, choosing a lower dimensional visual representation adequately is expected to improve the vision-based terrain learning and the final classification performance. This is the first work that proposes *automatically supervised* dimensionality reduction in a probabilistic framework using the supervision coming from the robot's sensors. The proposed method stands in between methods for reasoning under uncertainty using probabilistic models and methods for learning the underlying structure of the data.**

The proposed approach has been tested on field test data collected by an autonomous robot while driving on soil, gravel and asphalt. Although the supervision might be ambiguous or noisy, our experiments show that it helps build a more appropriate lower dimensional visual representation and achieves improved terrain recognition performance compared to unsupervised learning methods.

I. Introduction

We consider the problem of learning to recognize terrain types from color imagery for the purposes of autonomous navigation. This is necessary because different terrains induce different mobility limitations on the vehicle. For example, the robot might get stuck in sand or mud, so it has to learn to avoid such terrains. Visual information is used as a forward-looking sensor to determine the terrain type *prior* to the robot entering the terrain, so that a better planning can be done. In this paper the robot learns *automatically* using its own mechanical measurements while traversing the terrains. In particular, the amount of robot slip is used as supervision for learning different terrain types and the robot's mobility on them.

Learning fully automatically is important, because in the context of autonomous navigation huge amounts of data are available and providing manual supervision is impractical. To avoid manual labeling, the so-called *self-supervised* learning methods have proposed to use the vehicle's sensors as supervision for learning [4], [11], [13], [16], [19]. The key idea of self-supervised learning is that one of the sensors can provide the ground truth for learning with another sensor and the underlying assumption is that the former sensor can be reliably clustered or thresholded [4], [11], [13], [16].

However, some signals obtained from the robot do not necessarily provide a unique clustering into well separable classes, but can be still useful for providing supervision. For example, different terrain types might induce similar robot mobility, i.e. the supervision might be *ambiguous*. In the particular case of slip, which is slope dependent, the robot can have the same slip on flat ground but different slip when traversing slopes. Our previous work [3] proposed a unified learning framework for this case, but its limitation is that the visual representation is low dimensional and the method can become numerically brittle or require prohibitive amounts of training data for higher dimensional inputs. Robotics applications often need to process data obtained from multiple sensors which is high dimensional. In particular, feature representations of visual data are typically of high dimensionality, especially if fine distinctions between terrains need to be done or a lot of intra-class variability has to be accommodated.

To cope with high dimensional input spaces, we propose to use the supervision, automatically obtained by the robot, to affect the dimensionality reduction process. The intuition is that two visually similar terrains which are not normally discriminated in the visual space, and are mapped to the same cluster in the lower dimensional space, might be discriminated properly after introducing the supervision. In our case the mechanical supervision is in the form of robot slip and might be ambiguous or noisy. To solve the problem in this setup, we present a probabilistic framework in which the mechanical supervision provided by the robot is used to learn the representation and classification of terrain types in the visual space automatically. This essentially means having the supervision help choose more appropriate and meaningful, with respect to the learning task, low dimensional projections of the initial visual data. Most previous dimensionality reduction techniques are completely unsupervised [17], [21], whereas here we propose to learn a more useful lower dimensional visual representation which at the same time allows for better discrimination of terrains determined to be different by the automatic mechanical supervision from the robot. The significance of the approach is that a fully automatic learning and recognition of terrain types can be performed *without* using human supervision for data labeling. Moreover, the method allows the supervision signal obtained by the robot to be noisy or ambiguous, i.e. it might not have a one-to-one correspondence to the visual data.

II. Previous work

Learning to recognize terrains from vision and to determine their characteristics regarding traversability or robot mobility has been widely applied for autonomous vehicles [11], [16], [24]. However, current methods are not automated enough and human supervision or some other heuristics are still needed to determine traversability [9], [16]. Recently, the concept of learning from the vehicle's sensors, referred to as *learning from proprioception* [16], or *self-supervised learning* [4], [13], [19], has emerged. This idea has proved to be particularly useful for extending the perception range [4], [9], [16], [19] which is crucial to increasing the speed and efficiency of the robot [4]. Self-supervised learning approaches require good separability in the space of sensor responses, so that a unique terrain class assignment for each example is obtained. The latter is not always possible, e.g. driving at slower speed cannot produce definitive enough vibration patterns to discriminate terrains [6].

Dimensionality reduction techniques have also become very popular in robotics applications, because the input visual data is of high dimensionality and more efficient representations are needed [8], [12], [22]. Most previous dimensionality reduction methods are unsupervised [7], [17], [21], as they have been intended for data representation. However, in our robotics application, where additional mechanical sensor measurements are available, it is more rational to use them as supervision in selecting better lower dimensional data representation. Some recent work has proposed to include prior information into the dimensionality reduction framework, for example, by using known class labels [20] or by assuming the projections of some examples are given [25]. In our case, the supervision, i.e. the knowledge about class-membership, is fairly weak and neither of these approaches can be applied.

This work extends the probabilistic formulation for dimensionality reduction using Mixture of Factor Analyzers (MoFA) [7], [12], [17] with the major distinction that additional measurements, obtained independently by the robot, are used as supervision in the dimensionality reduction process. Moreover, in [17], [12] the lower dimensionality representation is observed (obtained by applying the unsupervised dimensionality reduction algorithm Isomap [21] prior to learning), whereas here it is unknown and needs to be learned. The particular application addresses recognizing terrain types and inherent mobility related to robot slip using visual input, similar to [2], with the difference that learning is done with automatic supervision, provided by the robot, and does not need manual labeling of terrain types, as in [2]. Being able to predict certain mechanical terrain properties remotely from only visual information and other sensors onboard the vehicle has significant importance in autonomous navigation applications, because more intelligent planning could be done [16], [24].

III. Problem formulation

Consider the problem of predicting the mobility characteristics Z of the robot in each map cell of the forthcoming terrain using as input the visual information $\mathbf{x} \in \Omega$ in the cell and some information about the terrain geometry $\mathbf{y} \in \Phi$, e.g. local terrain slope (Ω is the visual space, Φ is the space of terrain slopes). The input variables $\mathbf{x}$ and $\mathbf{y}$ can be obtained by the robot from a distance, which will allow the prediction of the output variable from a distance too. Let us denote the function that needs to be evaluated as $Z = F(\mathbf{x}, \mathbf{y})$.

This problem can be reduced to recognizing the terrain type from visual information. That is, we can assume that there are a limited number (K) of terrain types that can be encountered and that on each terrain type the robot experiences different behavior (e.g. mobility):

$$F(\mathbf{x}, \mathbf{y}) = f_j(\mathbf{y}), \qquad \text{if } \mathbf{x} \in \Omega_j \tag{1}$$

where $\Omega_j \in \Omega$ are different subsets in the visual space, $\Omega_i \cap \Omega_j = \emptyset, i \neq j$ and $f_j(\mathbf{y})$ are (nonlinear) functions which work in the domain Φ and which change their behavior depending on the terrain. In other words, different mobility behaviors occur on different terrain types which are determined by visual information. Now the question is how to learn the mapping $Z = F(\mathbf{x}, \mathbf{y})$ from training data $D = \{(\mathbf{x}_i, \mathbf{y}_i), z_i\}_{i=1}^N$, where $\mathbf{x}_i$ are the visual representations of patches from the observed terrain, $\mathbf{y}_i$ are the terrain slopes, and z_i are the slip measurements when the robot traverses that terrain.

The input space X, representing the visual data, can be of a very high dimension, which impedes working with it. Instead, we work with a lower dimensional embedding U of the input space X. For that purpose we need to learn the embedding $R : X \rightarrow U$ itself. As the learning of this mapping requires prohibitive amount of data whenever the input is high dimensional, we assume, similar to [7], [12], that it takes a particular form. Namely:

$$\mathbf{x} = \Lambda_j \mathbf{u}_j + \mathbf{v}_j \qquad \text{for } \mathbf{x} \in \Omega_j \tag{2}$$

where Λ_j is the projection matrix and $\mathbf{u}_j$, $\mathbf{v}_j$ are normally distributed: $\mathbf{u}_j \sim \mathcal{N}(\mu_j, \Sigma_j)$, $\mathbf{v}_j \sim \mathcal{N}(\eta_j, \Psi_j)$. That is, we assume that a locally linear mapping is a good enough approximation for patches that belong to the same terrain class.

Figure 1 visualizes the problem when measurements of slip as a function of terrain slope are used as supervision. Robot slip is a measure of the lack of progress and is essentially the complement of robot mobility [2]. The measurements in Figure 1 are obtained from actual robot traversals and are computed as the difference between Visual Odometry (VO) based pose estimates [15] and the pose estimates from the kinematic model of the robot. The mechanical slip measurements are received completely automatically, as only the vehicle's sensors are needed to compute slip. A nonlinear model can approximate the slip behavior as a function of

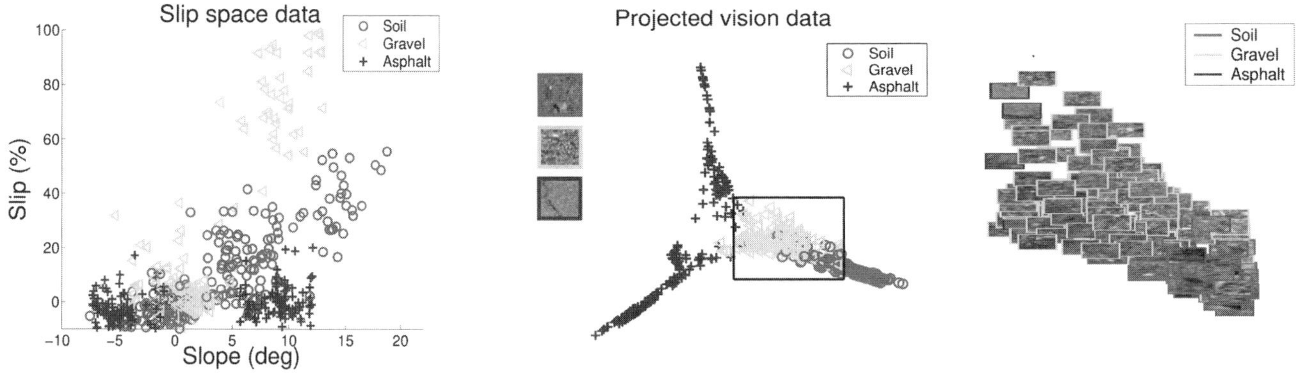

Fig. 1. Left: Slip measurements to be used as automatic supervision in our learning setup. Each training example consists of an image patch represented as a high dimensional point and a corresponding slip measurement represented as a function of the estimated slope angle. Middle: Lower dimensional projections of the visual data, obtained by the unsupervised dimensionality reduction algorithm Isomap [21]. The rectangle is expanded to the right and visualizes the original image patches. The ground truth terrain types in this figure are provided by human labeling, but our system works without human supervision and relies on the goodness-of-fit of nonlinear slip models to the slip measurements as automatic supervision to learn the terrain representation (dimensionality reduction), terrain classification, and the nonlinear slip models from the available training data.

slope for each terrain type. These models essentially act as supervision, but they are unknown and have to be learned from the data. The slopes can be easily estimated by the robot remotely using range data from stereo, ladar, etc., and a tilt sensor on the robot, which is readily available from the IMU, for example. We consider only the slip in the forward motion direction as dependent on the longitudinal slope, similar to slip measurements done for the Mars Exploration Rover [14], which is a simpler and more straightforward representation of slip than in [2]. This representation is also more convenient for using the slip measurements as supervision during learning. After the robot has learned how to visually discriminate the terrains, it is conceivable to learn more complex slip models using additional input variables (e.g. both longitudinal and lateral slopes, roughness, etc.), as in [2].

Figure 1 also shows the vision part of the input data, represented as described in Section V-B, projected into 2D by using the unsupervised dimensionality reduction algorithm Isomap [21]. As seen, there is a significant overlap between terrain classes which have visually similar patches. Because of the overlap, performing unsupervised, purely vision-based classification is not sufficient. So, to be able to learn to correctly discriminate these terrains and predict a potentially different mobility behavior on them, some form of supervision is needed. The key idea is that the dimensionality reduction process can also take advantage of the supervision information obtained from additional mechanical sensors.

The main problem in our formulation is that the slip signal to be used as supervision can be of very weak form and using slip measurements as supervision cannot be reduced to supervised learning, as in [4], [11]. In particular, because of the nonlinearity of the slip models $f_i(\mathbf{y})$, it is possible that some of the models overlap in parts of their domain (i.e. for some $i, j, i \neq j$, $f_i(\mathbf{y}) \equiv f_j(\mathbf{y})$, for $\mathbf{y} \in \Phi_0$, for some $\Phi_0 \subseteq \Phi$). For example, several terrains might exhibit the same slip for $\sim 0°$ slope, as seen in Figure 1, or simply two visually different terrain types might have the same slip behavior. Since some of the supervision (for some of the training examples) is inherently ambiguous, the slip supervision signals cannot be directly clustered into well separable classes. However, if two terrains exhibit different slip behavior for any slope range, the supervision should still be able to force a better discrimination in the visual space, even though not all examples can definitively exercise supervision. The intuition is that examples for which the supervision signal is strong will propagate it to the examples of ambiguous supervision in the same class through their visual similarity. Finally, as the supervision is collected automatically by the robot's mechanical sensors, it is rather noisy. To cope with noisy and ambiguous supervision signals necessitates a framework which allows reasoning under uncertainty.

To summarize, our goal is to learn the function $Z = F(\mathbf{x}, \mathbf{y})$ from the available training data $D = \{\mathbf{x}_i, \mathbf{y}_i, z_i\}_{i=1}^N$. Thus, after learning, the mechanical behavior z for some query input example $(\mathbf{x}_q, \mathbf{y}_q)$ will be predicted as $z = F(\mathbf{x}_q, \mathbf{y}_q)$. We do not want to use manual labeling of the terrain types during training, so the slip measurements z_i, which are assumed to have come from one of several unknown nonlinear models, act as the only supervision to the whole system. The main problem is that using the mechanical measurements as the only ground truth, or supervision, we have to learn both the terrain classification and the unknown nonlinear functions for each terrain. In particular, a combinatorial enumeration problem needs to be solved as a subproblem, which is known to be computationally intractable [10]. Furthermore, the supervision is noisy and ambiguous.

IV. Probabilistic Framework for Dimensionality Reduction Using Supervision

To solve the problem defined in Section III, we propose a probabilistic framework (Section IV-C) which performs dimensionality reduction and terrain classification by using automatic supervision and which can cope with both noisy and ambiguous supervision. A maximum likelihood estimation

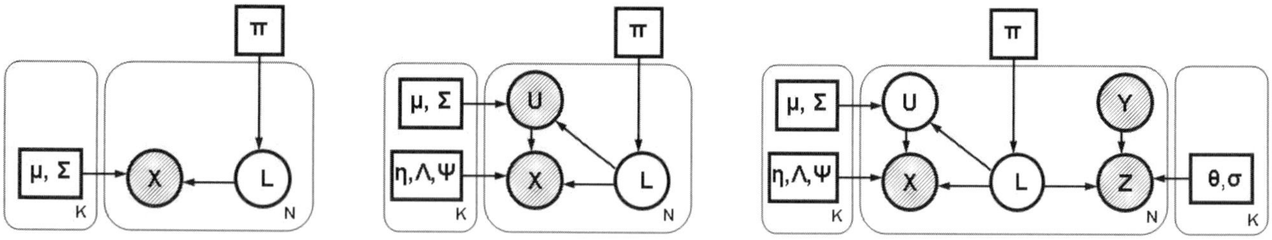

Fig. 2. Left: Graphical model of unsupervised clustering in the initial visual space. Middle: Graphical model of unsupervised dimensionality reduction based on MoFA [12], [17] (see also [7]). Right: Graphical model of automatically supervised dimensionality reduction in which mechanical measurements obtained automatically from the robot are used as supervision (proposed in this paper). The automatic supervision influences the selection of appropriate low dimensional representations and helps learn the distinction between different terrain types. The observed random variables are displayed in shaded circles.

will be done in this framework. To ease the exposition, we first describe two related probabilistic models.

A. Unsupervised clustering

The most straightforward approach to learn to classify examples corresponding to different terrains is to apply unsupervised learning (clustering). The corresponding graphical model is shown in Figure 2, left. The parameters μ_j, Σ_j are the means and covariances of each of the K clusters of visual data X and π_j are the prior probabilities of each class. The indicator variables L are *latent*, i.e. hidden, and are added to simplify the inference process; they define the class-membership of each training example, i.e. $L_{ij} = 1$ if the ith training example $\mathbf{x}_i$ belongs to the jth class. The model is used to learn the parameters of each class and the classification boundaries between them. However, inference in high dimensional spaces is numerically brittle and is limited by the amount and the diversity of the available training data.

B. Unsupervised dimensionality reduction

As operating in high dimensional spaces is not desirable, we wish to find a lower dimensional representation U of the initial visual space X. As previously shown [7], dimensionality reduction can be done using Mixture of Factor Analyzers (MoFA), which can be expressed probabilistically as follows:

$$P(X,U) = \sum_{j=1}^{K} P(X|U,C=j)P(U|C=j)P(C=j) \quad (3)$$

in which it is assumed that $\{X|U,C=j\} \sim \mathcal{N}(\Lambda_j U + \eta_j, \Psi_j)$ and $U \sim \mathcal{N}(\mu_j, \Sigma_j)$. In other words, the joint probability of X and U is assumed to be modeled as a mixture of K local linear projections, or factors (see Equation (2)) [7], [17]. In this paper we assume that U are latent variables. This is a more general case than both [7] and [17]. After introducing auxiliary latent variables L_{ij}, as above, we can write Equation (3) in the following way (which corresponds to the graphical model in Figure 2, middle):

$$P(X,U,L|\Theta_0) = P(X|U,L,\Theta_0)P(U|L,\Theta_0)P(L|\Theta_0),$$

where $\Theta_0 = \{\mu_j, \Sigma_j, \Lambda_j, \eta_j, \Psi_j, \pi_j\}_{j=1}^{K}$ contains the unknown parameters of the model. Because of the particular assumptions about the model, made in Equation (2), the probability of a data point $\mathbf{x}_i$ belonging to a terrain class j, given a latent representation $\mathbf{u}_i$, and the probability of the latent representation $\mathbf{u}_i$, given the class j, are expressed as:

$$P(\mathbf{x}_i|\mathbf{u}_i, L_{ij}=1) = \frac{e^{-\frac{1}{2}(\mathbf{x}_i - \Lambda_j \mathbf{u}_i - \eta_j)^T \Psi_j^{-1} (\mathbf{x}_i - \Lambda_j \mathbf{u}_i - \eta_j)}}{(2\pi)^{D/2}|\Psi_j|^{1/2}}$$

$$P(\mathbf{u}_i|L_{ij}=1) = \frac{1}{(2\pi)^{d/2}|\Sigma_j|^{1/2}} e^{-\frac{1}{2}(\mathbf{u}_i - \mu_j)^T \Sigma_j^{-1} (\mathbf{u}_i - \mu_j)},$$

where D and d are the dimensionalities of the initial visual space and the projected representation, respectively. Those distributions are modeled, so that a tractable solution to the maximum likelihood estimation problem is achieved.

C. Automatically supervised dimensionality reduction

Previous approaches have assumed the projections U of the data are known [12], [17] or have obtained them by unsupervised learning [7]. In this work we wish to have the automatic supervision influence which projections are chosen to best represent and consequently discriminate the visual classes. For that purpose we introduce supervision into the whole maximum likelihood framework, thus solving the initial problem in Equation (1), considering all the data available to the system. That is, the ambiguous mechanical supervision also takes part in the maximum likelihood decision.

In particular, we have two parts, a vision part, in which dimensionality reduction is done, and a mechanical behavior part, in which the slip measurements act as supervision. They are linked through the fact that they refer to the same terrain type, so they both give some information about this terrain. In other words, during learning, we can use visual information to learn something about the nonlinear mechanical models, and conversely, the mechanical feedback to supervise the vision based dimensionality reduction and terrain classification. Our goal is to make those two different sets of information interact.

The main problem is that the decision about the terrain types and learning of their mechanical behavior are not directly related (i.e. they are done in different, decoupled spaces) but they do refer to the same terrains. We can do that decoupling by using again the hidden variables L which define the class-membership of each training example (here $L_{ij} = 1$ if the ith training example $(\mathbf{x}_i, \mathbf{y}_i, z_i)$ has been generated by the jth nonlinear slip model and belongs to the jth terrain class). As

Input: Training data $\{\mathbf{x}_i, \mathbf{y}_i, z_i\}_{i=1}^N$, where $\mathbf{x}_i$ are the vision domain data, $\mathbf{y}_i$ are the geometry domain data, z_i are the mechanical supervision measurements. **Output:** Estimated parameters $\boldsymbol{\Theta}$ of the system.

Algorithm: Initialize the unknown parameters $\boldsymbol{\Theta}^0$. Set $t = 0$. Repeat until convergence:

1. (E-step) Estimate the expected values of L_{ij}, $\mathbf{u}_{ij}$ (we denote $\mathbf{u}_{ij} = E(\mathbf{u}|\mathbf{x}_i, L_{ij} = 1)$):

$L_{ij}^{t+1} = \frac{P(\mathbf{x}_i|L_{ij}=1,\boldsymbol{\Theta}^t)P(\mathbf{y}_i,z_i|L_{ij}=1,\boldsymbol{\Theta}^t)\pi_j^t}{\sum_{k=1}^K P(\mathbf{x}_i|L_{ik}=1,\boldsymbol{\Theta}^t)P(\mathbf{y}_i,z_i|L_{ik}=1,\boldsymbol{\Theta}^t)\pi_k^t}$, where $\mathbf{x}_i \sim \mathcal{N}(\Lambda_j^t \mu_j^t + \eta_j^t, \Psi_j^t + \Lambda_j^t \Sigma_j^t (\Lambda_j^t)')$

$\mathbf{u}_{ij}^{t+1} = \Upsilon[(\Lambda_j^t)'(\Psi_j^t)^{-1}(\mathbf{x}_i - \eta_j^t) + (\Sigma_j^t)^{-1}\mu_j^t]$, where $\Upsilon = [(\Sigma_j^t)^{-1} + (\Lambda_j^t)'(\Psi_j^t)^{-1}\Lambda_j^t]^{-1}$.

2. (M-step) Select the parameters Θ^{t+1} to maximize $CL(X, U, Y, Z, L|\Theta^t)$. Let $l_{ij}^{t+1} = L_{ij}^{t+1} / \sum_{r=1}^N L_{rj}^{t+1}$:

$\mu_j^{t+1} = \sum_{i=1}^N l_{ij}^{t+1}\mathbf{u}_{ij}^{t+1}$; $\quad \boldsymbol{\Sigma}_j^{t+1} = \sum_{i=1}^N l_{ij}^{t+1}\mathbf{u}_{ij}^{t+1}(\mathbf{u}_{ij}^{t+1})' - \mu_j^{t+1}(\mu_j^{t+1})' + \Upsilon$; $\quad \eta_j^{t+1} = \sum_{i=1}^N l_{ij}^{t+1}(\mathbf{x}_i - \Lambda_j^t \mathbf{u}_{ij}^{t+1})$

$\boldsymbol{\Lambda}_j^{t+1} = [\sum_{i=1}^N L_{ij}^{t+1}(\mathbf{x}_i - \eta_j^t)(\mathbf{u}_{ij}^{t+1})'][\sum_{i=1}^N L_{ij}^{t+1}(\mathbf{u}_{ij}^{t+1}(\mathbf{u}_{ij}^{t+1})' + \Upsilon)]^{-1}$; $\quad \boldsymbol{\Psi}_j^{t+1} = \sum_{i=1}^N l_{ij}^{t+1}(\mathbf{x}_i - \eta_j^{t+1} - \Lambda_j^{t+1}\mathbf{u}_{ij}^{t+1})(\mathbf{x}_i - \eta_j^{t+1})'$

$\theta_j^{t+1} = (G' L_j^{t+1} G)^{-1} G' L_j^{t+1} Z$; $\quad (\sigma_j^2)^{t+1} = \sum_{i=1}^N l_{ij}^{t+1}(z_i - G(\mathbf{y}_i, \theta_j^{t+1}))^2$; $\quad \pi_j^{t+1} = \sum_{i=1}^N L_{ij}^{t+1}/N$.

3. $t = t + 1$

Fig. 3. EM algorithm updates (see [1] for details).

an additional step, a dimensionality reduction of the visual part of the data is done, so now the supervision can affect the parameters related to the dimensionality reduction too. This essentially means preferring projections which fit the data, and therefore also the supervision, well. Now, given the labeling of an example is known, the slip supervision measurements and the visual information are independent. So, the complete likelihood factors as follows:

$$P(X, U, Y, Z, L|\Theta) = \underbrace{P(X|U, L, \Theta)P(U|L, \Theta)}_{\textit{Vision part, dim. red.}} \underbrace{P(Y, Z|L, \Theta)}_{\textit{Autom. supervision}} \underbrace{P(L|\Theta)}_{\textit{Prior}}$$

where $\Theta = \{\mu_j, \Sigma_j, \Lambda_j, \eta_j, \Psi_j, \theta_j, \sigma_j, \pi_j\}_{j=1}^K$ contains all the parameters that need to be estimated in the system. θ_j are the parameters of the nonlinear fit of the slip data and σ_j are their covariances (here they are the standard deviations, as the final measurement is one dimensional). The graphical model corresponding to this case is shown in Figure 2, right. This model allows the automatically obtained mechanical supervision to affect both the dimensionality reduction and the clustering process, thus improving a purely unsupervised learning for the purposes of the task at hand. Note that here the lower dimensional representation is hidden and that the supervision part can influence the visual learning and the dimensionality reduction through the latent variables L_{ij}.

The supervision part is as follows. The mechanical measurement data are assumed to have come from a nonlinear fit, which is modeled as a General Linear Regression (GLR) [18]. GLR is appropriate for expressing nonlinear behavior and is convenient for computation because it is linear in terms of the parameters to be estimated. For each terrain type j, the regression function $\tilde{Z}(Y) = E(Z|Y)$ is assumed to have come from a GLR with Gaussian noise: $f_j(Y) \equiv Z(Y) = \tilde{Z}(Y) + \epsilon_j$, where $\tilde{Z}(Y) = \theta_j^0 + \sum_{r=1}^R \theta_j^r g_r(Y)$, $\epsilon_j \sim \mathcal{N}(0, \sigma_j)$, and g_r are several nonlinear functions selected before the learning has started. Some example nonlinear functions to be used as building blocks for slip approximation are: x, x^2, e^x, $\log x$, $\tanh x$ (those functions are used later on in our experiments with the difference that the input parameter is scaled first). The parameters $\theta_j^0, ..., \theta_j^R, \sigma_j$ are to be learned for each model j. We assume the following probability model for z_i belonging to the jth nonlinear model conditioned on $\mathbf{y}_i$:

$$P(z_i|\mathbf{y}_i, L_{ij} = 1, \theta_j, \sigma_j) = \frac{1}{(2\pi)^{1/2}\sigma_j} e^{-\frac{1}{2\sigma_j^2}(z_i - G(\mathbf{y}_i, \theta_j))^2},$$

where $G(\mathbf{y}, \theta_j) = \theta_j^0 + \sum_{r=1}^R \theta_j^r g_r(\mathbf{y})$ and $\theta_j = (\theta_j^0, \theta_j^1, ..., \theta_j^R)$. $P(\mathbf{y}_i)$ is given an uninformative prior (here, uniform over a range of slopes).

With the help of the hidden variables L, the complete log likelihood function (CL) can be written as:

$$CL(X, U, Y, Z, L|\Theta) = \sum_{i=1}^N \sum_{j=1}^K L_{ij}[\log P(\mathbf{x}_i|\mathbf{u}_i, L_{ij} = 1, \Lambda_j, \eta_j, \Psi_j) + \log P(\mathbf{u}_i|L_{ij} = 1, \mu_j, \Sigma_j) + \log P(\mathbf{y}_i, z_i|L_{ij} = 1, \theta_j, \sigma_j) + \log \pi_j]$$

The introduction of the hidden variables L is crucial to simplifying the problem and allows for it to be solved efficiently with the Expectation Maximization (EM) algorithm [5], which tries to maximize the complete log likelihood (CL). The EM algorithm updates applied to our formulation of the problem are shown in Figure 3 (the detailed derivations of the updates are provided in [1]). In brief, the algorithm performs the following steps until convergence. In the E-step, the expected values of the unobserved variables $\mathbf{u}_{ij}$ and label assignments L_{ij} are estimated. In the M-step, the parameters for both the vision and the mechanical supervision side are selected, so as to maximize the complete log likelihood. In other words, at each iteration better parameters Θ are selected, in a sense that they increase the likelihood of the available data. As the two views are conditionally independent, the parameters for the vision and the mechanical side are updated independently of one another in the M-step. Note that it is through the variable L that the visual data and the mechanical supervision interact and that the automatic supervision can affect the local projections defining the dimensionality reduction through the variable U. The interaction happens in the E-step of each iteration, by updating the expected values of L and U which depend on *both* the visual data and the supervision. The new variables introduced in Figure 3 are defined as follows: L_j^t is a diagonal NxN matrix which has $L_{1j}^t, ...L_{Nj}^t$ on its diagonal, G is a Nx$(R+1)$ matrix such that $G_{ir} = g_r(\mathbf{y}_i)$, $G_{i(R+1)} = 1$, and Z is a Nx1 vector containing the measurements z_i [1].

D. Discussion

The main difference from previous approaches [7], [12], [17] is that we have incorporated automatic supervision into the framework, which directly affects the lower dimensionality projections and the terrain classification. Furthermore, the variables U corresponding to the low dimensional representation are *latent* (unlike [12], where they are known and obtained from Isomap, prior to learning) and can have arbitrary means and covariances which are learned (unlike [7], where they are assumed to be zero mean and unit variance). This is an important point, because it is through the latent variables U that the supervision can influence the dimensionality reduction process during learning.

The proposed maximum likelihood approach solves the abovementioned combinatorial enumeration problem [10] approximately by producing a solution which is guaranteed to be a local maximum only. Indeed, the EM solution is prone to getting stuck in a local maximum. For example, one can imagine creating adversarial mechanical models to contradict the clustering in visual space. In practice, for the autonomous navigation problem we are addressing, our intuition is that the mechanical measurements are correlated to a large extent with the vision input and will be only improving the vision based classification. This is seen later in the experiments.

V. Experimental Evaluation

In this section we apply the proposed automatically supervised dimensionality reduction algorithm to vision-based learning of different terrain types, using slip supervision obtained by the robot.

The learning setup is as follows. The robot collects data by building a map of the environment and obtaining geometry and appearance information for each map cell. When a particular cell is traversed, the robot measures the amount of slippage occurring and saves a training example composed of a visual feature vector (corresponding to a terrain patch), geometry feature vector (here only the slope angle), and the corresponding slip. The collected training examples are used for learning of the mapping between the input visual and geometric features and the output slip. This strategy is commonly applied to learning traversability or other terrain properties from vision [2], [11], [24]. VO [15] is used for robot localization.

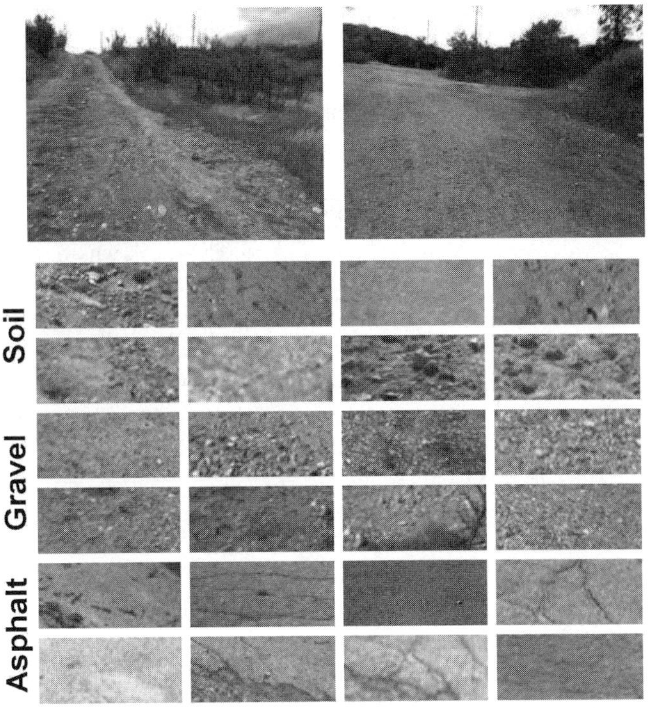

Fig. 4. Top: Example frames from driving on soil (left) and on gravel (right). Bottom: Patches from the classes in our dataset. The variability in texture appearance is one of the challenges present in our application domain. The dataset is collected under various weather conditions.

A. Dataset

The dataset has been collected by an autonomous LAGR[1] robot while driving on three terrains with different mobility in a natural park: soil, gravel and asphalt. Figure 4 shows example patches from the terrains and Figure 1 shows the collected slip measurements in the dataset. It is not known to the algorithm which terrain classes the input examples belong to: the slip and slope measurements (Figure 1) are the only information to be used for automatic supervision. The dataset is quite challenging as it is obtained in outdoor, off-road environments. In particular, a lot of intra-class variability can be observed in the appearance of the terrain patches and the mechanical slip measurements are very noisy.

B. Visual representation

Each terrain patch is represented as the frequency of occurrence (i.e. a histogram) of visual features, called textons, within a patch [23]. The textons are collected by using k-means of 5x5 pixel neighborhoods extracted at random from a pool of training images coming from all the classes (see [23] for details). In this case, 5 textons are selected for each terrain class in the data, constructing a 15-dimensional input feature vector. This representation, based on both color and texture, has been shown to achieve satisfactory classification results for generic textures [23], as well as for natural off-road terrains [2].

[1]LAGR stands for Learning Applied to Ground Robots and is an experimental all-terrain vehicle program funded by DARPA

C. Mechanical supervision

Robot slip is defined as the difference between the commanded velocity of the robot, obtained from its kinematics model and wheel encoder sensors, and its actual velocity between two consecutive steps [2]. The VO algorithm [15], running onboard the robot, is used to compute its actual velocity. Thus, the slip-based supervision is measured fully automatically by the robot. In these experiments we focus on slip in the forward motion direction as dependent on the longitudinal slope. The terrain slope is retrieved by performing a least-mean-squares plane fit on the average elevations of the map cells in a 4x4 cell neighborhood.

D. Experimental results

In this section we present experimental results of the dimensionality reduction with automatic supervision. We quantitatively evaluate the performance of the proposed algorithm for automatically supervised learning (Figure 2, right) compared to both unsupervised learning (Figure 2, middle [7]) and human supervised learning. While testing, terrain classification is performed first to find the most likely class index j^* given the input data X (let us denote $P(L_j) = P(C = j)$):

$$j^* = argmax_j P(C = j|X) \propto P(X|C = j)P(C = j) =$$
$$\int_u P(X|u, L_j)P(u|L_j)du P(L_j) \approx P(X|U_{ML}, L_j)P(L_j),$$

in which we approximate the integral by using the maximum likelihood lower dimensional projection (U_{ML}). Note that only the visual input is used to make this decision. Then, the expected slip is predicted by evaluating the j^*-th learned slip model $f_{j^*}(Y) = \theta^0_{j^*} + \sum_{r=1}^{R} \theta^r_{j^*} g_r(Y)$ for the given slope Y.

The average terrain classification and slip prediction errors and their standard deviations across 50 independent runs are shown in Figure 5. We compare learning and dimensionality reduction without supervision, with automatic supervision, and with human supervision. We have about $\sim$1000 examples which are split randomly into 70% training and 30% test sets in each run. As the correct slip models are not known, the ultimate test of performance is by comparing the predicted slip to the actual measured slip on a test set (not used in training). Slip prediction error is computed as: Err=$\sum_{i=1}^{N} |F(\mathbf{x}_i, \mathbf{y}_i) - z_i|/N$, where $F(\mathbf{x}_i, \mathbf{y}_i)$ is the predicted and z_i is the target slip for a test example $(\mathbf{x}_i, \mathbf{y}_i)$. The terrain classification results are evaluated by comparing to human labeled terrains. When using human supervision, the class-membership of each example is known, but the parameters of each class need to be estimated. The latter is equivalent to doing Factor Analysis in each class independently. Due to some overlap between the classes in the original visual space, the classification with human supervision can still incur some nonzero test error in terrain classification. To reflect the monotonic nature of slip, an additional constraint ($\theta_j \geq 0$) is imposed (see [1] for details).

As seen in Figure 5, learning with automatically supervised dimensionality reduction outperforms the unsupervised learning method and decreases the gap to learning with human

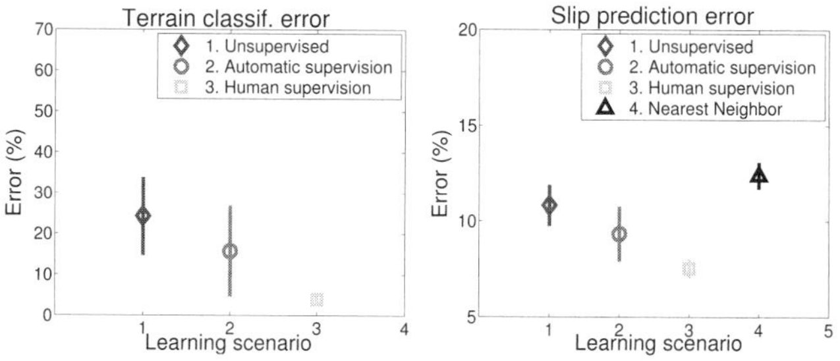

Fig. 5. Average test results for terrain recognition (left) and slip prediction (right). Comparison to a baseline nonlinear regression method is also shown.

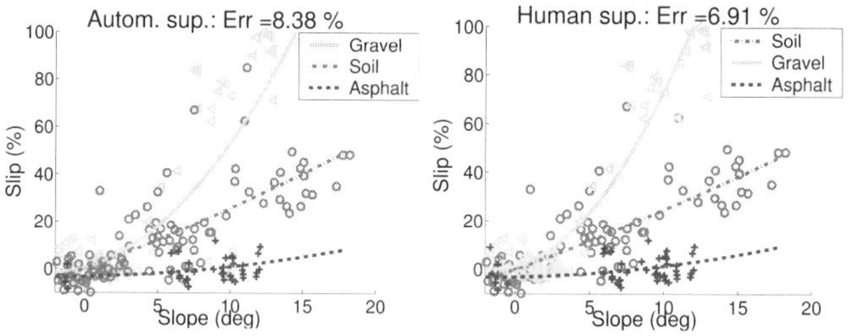

Fig. 6. The learned slip models and the classification of the test examples when learning with automatic supervision (left) and learning with human superivsion (right). The examples are marked according to their predicted terrain class labels (the colors and markers are consistent with Figure 1).

supervision. More precisely, learning with automatic supervision achieves about 42% and 45% of the possible margin for improvement between the unsupervised and the human supervised learning for terrain classification and slip prediction, respectively. Naturally, for the type of supervision used in these experiments (Figure 1), we cannot expect to fully close the gap to human supervision, because the supervision signals are not sufficiently well separable. The improved performance of the supervised dimensionality reduction compared to the unsupervised one is due to selecting more appropriate low dimensional visual representations which provide for better discrimination among the terrain classes and respectively for learning of more accurate slip models for each terrain. Comparing the results to [3] we can see that working with more descriptive high dimensional representations is instrumental to achieving better performance. At the same time, as the representation is more powerful, there is a smaller margin for improvement between the unsupervised and the human supervised learning. We also compared the results to a baseline nonlinear regression method, k-Nearest Neighbor, which learns directly the mapping from the inputs (visual features $\mathbf{x}$ and slope $\mathbf{y}$) to the output (slip z) and does not apply dimensionality reduction as an intermediate step. Note that directly learning the desired outputs, as is with k-Nearest Neighbor, important information about the structure of the problem, namely that there are several underlying terrain types on which potentially different slip behaviors occur, is ignored. As seen in Figure 5, the k-Nearest Neighbor is outperformed by the other three methods.

The learned nonlinear models for one of the runs are shown

in Figure 6. The resultant slip models when learning with automatic supervision are very similar to the ones generated by human supervision, which is due to having learned the correct terrain classification in the visual space. Note that, although the correct slip models have been learned, there are still examples which are misclassified for both learning scenarios because only the visual information is used during testing. The slip model used here has less inputs than in [2] and its main purpose is to act as supervision rather than achieve a good approximation of the slip signal. Now, given that the robot has *automatically* learned how to visually discriminate terrains by using the slip signals as supervision, the final slip prediction results can be further improved by applying a more advanced slip learning algorithm, e.g. by taking into consideration more inputs [2].

Our results show that using additional, automatically obtained, signals as supervision is worthwhile: it outperforms purely unsupervised vision-based learning and has the potential to substitute the expensive, tedious, and inefficient human labeling in applications related to autonomous navigation. Secondly, as more descriptive high dimensional feature representations are crucial to achieving better recognition performance, performing dimensionality reduction and utilizing the automatic supervision in the process is more advantageous than working with simpler lower dimensional representations.

VI. Conclusions and Future Work

We have proposed a novel probabilistic framework for dimensionality reduction which takes advantage of ambiguous and noisy supervision obtained automatically from the robot's onboard sensors. As a result, simultaneous learning of the lower dimensional representation, the terrain classification, and the nonlinear slip behavior on each terrain is done by using only automatically obtained measurements. The proposed method stands in between reasoning under uncertainty using probabilistic models and retrieving the underlying structure of the data (i.e. dimensionality reduction). The impact of the proposed method of automatically supervised dimensionality reduction is that: 1) a better visual representation can be created by utilizing the supervision from the robot, or the task at hand; 2) the robot can learn about terrains and their visual representation by using its own sensors as supervision; 3) after the learning has completed, the expected mobility behavior on different terrains can be predicted remotely.

We have shown experiments on a dataset collected while driving in the field, in which different terrain types are learned better from both vision and slip supervision than from vision alone and unsupervised dimensionality reduction. Significant improvements, currently under investigation, can be done by introducing temporal/spatial continuity to the consecutive/neighboring terrain measurements. Extending the method to online learning is an important future direction, in which the main challenges are determining which examples to keep in memory and estimating the number of terrains.

Acknowledgment: This research was carried out by the Jet Propulsion Laboratory, California Institute of Technology, under a contract with NASA, with funding from the Mars Technology Program. We thank Navid Serrano and the anonymous reviewers for their very useful comments on the paper.

References

[1] A. Angelova. EM algorithm updates for dimensionality reduction using automatic supervision. *Technical report*, 2007. http://www.vision.caltech.edu/anelia/publications/DimRedTR.pdf.

[2] A. Angelova, L. Matthies, D. Helmick, and P. Perona. Slip prediction using visual information. *Robotics: Science and Systems Conf.*, 2006.

[3] A. Angelova, L. Matthies, D. Helmick, and P. Perona. Learning slip behavior using automatic mechanical supervision. *International Conference on Robotics and Automation*, 2007.

[4] H. Dahlkamp, A. Kaehler, D. Stavens, S. Thrun, and G. Bradski. Self-supervised monocular road detection in desert terrain. *Robotics: Science and Systems Conference*, 2006.

[5] A. Dempster, N. Laird, and D. Rubin. Maximum likelihood from incomplete data via the EM algorithm. *Journal of the Royal Statistical Society, Series B*, 39(1):1–37, 1977.

[6] E. DuPont, R. Roberts, C. Moore, M. Selekwa, and E. Collins. Online terrain classification for mobile robots. *International Mechanical Engineering Congress and Exposition Conference*, 2005.

[7] Z. Ghahramani and G. Hinton. The EM algorithm for mixtures of factor analyzers. *Tech. Report CRG-TR-96-1, Department of Computer Science, University of Toronto*, 1997.

[8] D. Grollman, O. Jenkins, and F. Wood. Discovering natural kinds of robot sensory experiences in unstructured environments. *Journal of field robotics*, 2006.

[9] M. Happold, M. Ollis, and N. Johnson. Enhancing supervised terrain classification with predictive unsupervised learning. *Robotics: Science and Systems Conference*, 2006.

[10] A. Julosky, S. Weiland, and W. Heemels. A Bayesian approach to identification of hybrid systems. *IEEE Trans. on Automatic Control*, 50(10):1520–1533, 2005.

[11] D. Kim, J. Sun, S. Oh, J. Rehg, and A. Bobick. Traversability classification using unsupervised on-line visual learning for outdoor robot navigation. *Int. Conference on Robotics and Automation*, 2006.

[12] S. Kumar, F. Ramos, B. Upcroft, and H. Durrant-Whyte. A statistical framework for natural feature representation. *International Conference on Intelligent Robots and Systems*, 2005.

[13] D. Lieb, A. Lookingbill, and S. Thrun. Adaptive road following using self-supervised learning and reverse optical flow. *Robotics: Science and Systems Conference*, 2005.

[14] R. Lindemann and C. Voorhees. Mars Exploration Rover mobility assembly design, test and performance. *IEEE International Conference on Systems, Man and Cybernetics*, 2005.

[15] L. Matthies and S. Schafer. Error modeling in stereo navigation. *IEEE Journal of Robotics and Automation*, RA-3(3), June 1987.

[16] L. Matthies, M. Turmon, A. Howard, A. Angelova, B. Tang, and E. Mjolsness. Learning for autonomous navigation: Extrapolating from underfoot to the far field. *NIPS, Workshop on Machine Learning Based Robotics in Unstructured Environments*, 2005.

[17] L. Saul and S. Roweis. Think globally, fit locally: Unsupervised learning of low dimensional manifolds. *Journal of Machine Learning Research*, 4:119–155, 2003.

[18] G. Seber and C. Wild. *Nonlinear Regression*. John Wiley & Sons, New York, 1989.

[19] B. Sofman, E. Lin, J. Bagnell, N. Vandapel, and A. Stentz. Improving robot navigation through self-supervised online learning. *Robotics: Science and Systems Conference*, 2006.

[20] M. Sugiyama. Local Fisher Discriminant Analysis for supervised dimensionality reduction. *Int. Conference on Machine learning*, 2006.

[21] J. Tenenbaum, V. de Silva, and J. Langford. A global geometric framework for nonlinear dimensionality reduction. *Science*, 290(5500):2319–2323, December 2000.

[22] B. Upcroft et al. Multi-level state estimation in an outdoor decentralised sensor network. *ISER*, 2006.

[23] M. Varma and A. Zisserman. Texture classification: Are filter banks necessary? *Conf. on Computer Vision and Pattern Recognition*, 2003.

[24] C. Wellington and A. Stentz. Online adaptive rough-terrain navigation in vegetation. *Int. Conference on Robotics and Automation*, 2004.

[25] X. Yang, H. Fu, H. Zha, and J. Barlow. Semi-supervised nonlinear dimensionality reduction. *Int. Conference on Machine learning*, 2006.

The Stochastic Motion Roadmap: A Sampling Framework for Planning with Markov Motion Uncertainty

Ron Alterovitz
LAAS-CNRS, University of Toulouse
Toulouse, France
ron@ieor.berkeley.edu

Thierry Siméon
LAAS-CNRS, University of Toulouse
Toulouse, France
nic@laas.fr

Ken Goldberg
Departments of IEOR and EECS
University of California, Berkeley
goldberg@ieor.berkeley.edu

Abstract— **We present a new motion planning framework that explicitly considers uncertainty in robot motion to maximize the probability of avoiding collisions and successfully reaching a goal. In many motion planning applications ranging from maneuvering vehicles over unfamiliar terrain to steering flexible medical needles through human tissue, the response of a robot to commanded actions cannot be precisely predicted. We propose to build a roadmap by sampling collision-free states in the configuration space and then locally sampling motions at each state to estimate state transition probabilities for each possible action. Given a query specifying initial and goal configurations, we use the roadmap to formulate a Markov Decision Process (MDP), which we solve using Infinite Horizon Dynamic Programming in polynomial time to compute stochastically optimal plans. The *Stochastic Motion Roadmap (SMR)* thus combines a sampling-based roadmap representation of the configuration space, as in PRM's, with the well-established theory of MDP's. Generating both states and transition probabilities by sampling is far more flexible than previous Markov motion planning approaches based on problem-specific or grid-based discretizations. We demonstrate the SMR framework by applying it to nonholonomic steerable needles, a new class of medical needles that follow curved paths through soft tissue, and confirm that SMR's generate motion plans with significantly higher probabilities of success compared to traditional shortest-path plans.**

I. Introduction

In many applications of motion planning, the motion of the robot in response to commanded actions cannot be precisely predicted. Whether maneuvering a vehicle over unfamiliar terrain, steering a flexible needle through human tissue to deliver medical treatment, guiding a micro-scale swimming robot through turbulent water, or displaying a folding pathway of a protein polypeptide chain, the underlying motions cannot be predicted with certainty. But in many of these cases, a probabilistic distribution of feasible outcomes in response to commanded actions can be experimentally measured. This stochastic information is fundamentally different from a deterministic motion model. Though planning shortest feasible paths to the goal may be appropriate for problems with deterministic motion, shortest paths may be highly sensitive to uncertainties: the robot may deviate from its expected trajectory when moving through narrow passageways in the configuration space, resulting in collisions.

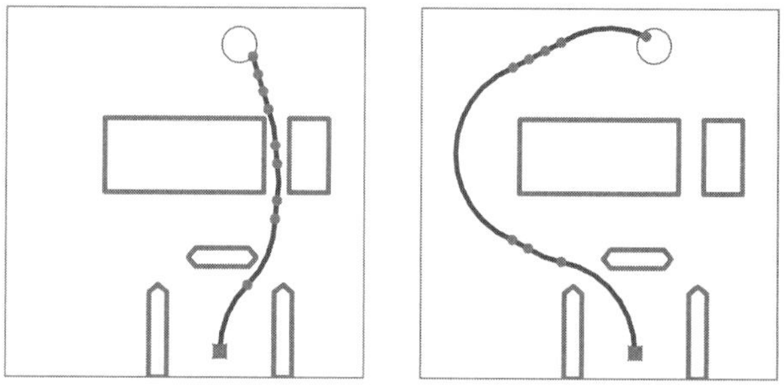

(a) Minimizing path length $p_s = 27\%$

(b) Maximize probability of success using SMR $p_s = 57\%$

Fig. 1. The expected results of two plans to steer a Dubins-car mobile robot with left-right bang-bang steering and normally distributed motion uncertainty from an initial configuration (solid square) to a goal (open circle). Dots indicate steering direction changes. The Stochastic Motion Roadmap (SMR) introduces sampling of the configuration space and motion uncertainty model to generate plans that maximize the probability p_s that the robot will successfully reach the goal without colliding with an obstacle. Evaluation of p_s using multiple randomized simulations demonstrates that following a minimum length path under motion uncertainty (a) is substantially less likely to succeed than executing actions from an SMR plan (b).

In this paper, we develop a new motion planning framework that explicitly considers uncertainty in robot motion at the planning stage. Because future configurations cannot be predicted with certainty, we define a plan by actions that are a function of the robot's current configuration. A plan execution is successful if the robot does not collide with any obstacles and reaches the goal. The idea is to compute plans that maximize the probability of success.

Our framework builds on the highly successful approach used in Probabilistic Roadmaps (PRM's): a learning phase followed by a query phase [20]. During the learning phase, a random (or quasi-random) sample of discrete states is selected in the configuration space, and a roadmap is built that represents their collision-free connectivity. During the query phase, the user specifies initial and goal states, and the roadmap is used to find a feasible path that connects the initial state to the goal, possibly optimizing some criteria such as minimum length. PRM's have successfully solved many path planning problems for applications such as robotic manipulators and mobile robots [12, 22]. The term "probabilistic" in PRM comes from the random sampling of states. An underlying assumption is that the collision-free connectivity of states is specified using boolean values rather than distributions.

In this paper, we relax this assumption and combine a roadmap representation of the configuration space with a stochastic model of robot motion. The input to our method is a geometric description of the workspace and a motion model for the robot capable of generating samples of the next configuration that the robot may attain given the current configuration and an action. We require that the motion model satisfy the Markovian property: the distribution of the next state depends only on the action and current state, which encodes all necessary past history. As in PRM's, the method first samples the configuration space, where the sampling can be random [20], pseudo-random [23], or utility-guided [11]. We then sample the robot's motion model to build a *Stochastic Motion Roadmap (SMR)*, a set of weighted directed graphs with vertices as sampled states and edges encoding feasible state transitions and their associated probability of occurrence for each action.

The focus of our method is not to find a *feasible* motion plan, but rather to find an *optimal* plan that maximizes the probability that the robot will successfully reach a goal. Given a query specifying initial and goal configurations, we use the SMR to formulate a Markov Decision Process (MDP) where the "decision" corresponds to the action to be selected at each state in the roadmap. We solve the MDP in polynomial time using Infinite Horizon Dynamic Programming. Because the roadmap is a discrete representation of the continuous configuration space and transition probabilities, the computed optimal actions are approximations of the optimal actions in continuous space that converge as the roadmap increases in size. Although the plan, defined by the computed actions, is fixed, the path followed by the robot may differ each time the plan is executed because different state transitions may occur due to motion uncertainty. As shown in Fig. 1, plans that explicitly consider uncertainty to maximize the probability of success can differ substantially from traditional shortest path plans.

In SMR, "stochastic" refers to the motion of the robot, not to the sampling of states. PRM's were previously extended to explore stochastic motion in molecular conformation spaces [4, 5], but without a planning component to optimize actions. Our SMR formulation is applicable to a variety of decision-based robotics problems. It is particularly suited for nonholonomic robots, for which a deflection in the path due to motion uncertainty can result in failure to reach the goal, even if the deflection does not result in an immediate collision. The expansion of obstacles using their Minkowski sum [22] with a circle corresponding to an uncertainty tolerance is often sufficient for holonomic robots for which deflections can be immediately corrected, but this does not address collisions resulting from a nonholonomic constraint as in Fig. 2. By explicitly considering motion uncertainty in the planning phase, we hope to minimize such failures.

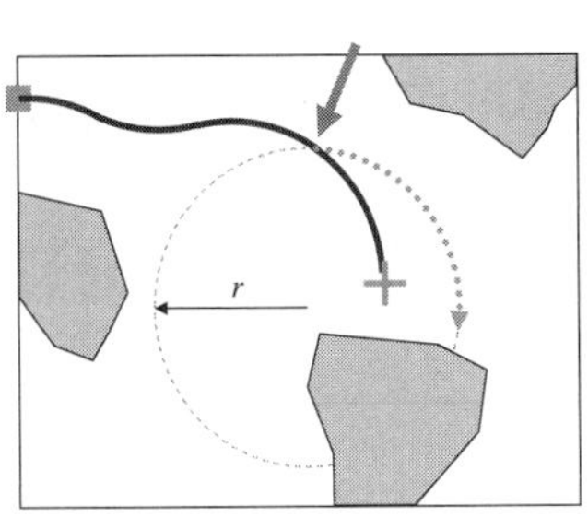

Fig. 2. The solid line path illustrates a motion plan from the start square to the goal (cross) for a nonholonomic mobile robot constrained to follow paths composed of continuously connected arcs of constant-magnitude curvature with radius of curvature r. If a deflection occurs at the location of the arrow, then the robot is unable to reach the goal due to the nonholonomic constraint, even if this deflection is immediately detected, since the robot cannot follow a path with smaller radius of curvature than the dotted line.

Although we use the terms robot and workspace, SMR's are applicable to any motion planning problem that can be modeled using a continuous configuration space and discrete action set with uncertain transitions between configurations. In this paper, we demonstrate a SMR planner using a variant of a Dubins car with bang-bang control, a nonholonomic mobile robot that can steer its wheels far left or far right while moving forward but cannot go straight. This model can generate motion plans for steerable needles, a new class of flexible bevel-tip medical needles that clinicians can steer through soft tissue around obstacles to reach targets inaccessible to traditional stiff needles [2, 36]. As in many medical applications, considering uncertainty is crucial to the success of medical needle insertion procedures: the needle tip may deflect from the expected path due to tissue inhomogeneities that cannot be detected prior to the procedure. Due to uncertainty in predicted needle/tissue interactions, needle steering is ill-suited to shortest-path plans that may guide the needle through narrow passageways between critical tissue such as blood vessels or nerves. By explicitly considering motion uncertainty using an SMR, we obtain solutions that result in possibly longer paths but that improve the probability of success.

A. Related Work

Motion planning can consider uncertainty in *sensing* (the current state of the robot and workspace is not known with certainty) and *predictability* (the future state of the robot and workspace cannot be deterministically predicted even when the current state and future actions are known) [22]. Extensive work has explored uncertainty associated with robot sensing, including SLAM and POMDP's to represent uncertainty in the current state [33, 12]. In this paper, we assume the current state is known (or can be precisely determined from sensors), and we focus on the latter type of uncertainty, predictability.

Predictability can be affected by uncertainty in the workspace and by uncertainty in the robot's motion. Previous and ongoing work addresses many aspects of uncertainty in the workspace, including uncertainty in the goal location, such as in pursuit-evasion games [22, 24], and in dynamic environments with moving obstacles [35, 34, 27]. A recently developed method for grasp planning uses POMDP's to consider uncertainty in the configuration of the robot and the state of the objects in the world [18]. In this paper we focus explicitly on the case of uncertainty in the robot's motion rather than in goal or obstacle locations.

Apaydin et al. previously explored the connection between probabilistic roadmaps and stochastic motions using Stochastic Roadmap Simulation (SRS), a method designed specifically for molecular conformation spaces [4, 5]. SRS, which formalizes random walks in a roadmap as a Markov Chain, has been successfully applied to predict average behavior of molecular motion pathways of proteins and requires orders of magnitude less computation time than traditional Monte-Carlo molecular simulations. However, SRS cannot be applied to more general robotics problems, including needle steering, because the probabilities associated with state transitions are specific to molecular scale motions and the method does not include a planning component to optimize actions.

Considering uncertainty in the robot's response to actions during planning results in a stochastic optimal control problem where feedback is generally required for success. Motion planners using grid-based numerical methods and geometric analysis have been applied to robots with motion uncertainty (sometimes combined with sensing uncertainty) using cost-based objectives and worst-case analysis [26, 10, 25]. MDP's, a general approach that requires explicitly defining transition probabilities between states, have also been applied to motion planning by subdividing the workspace using regular grids and defining transition probabilities for motions between the grid cells [14, 16, 22]. These methods differ from SMR's since they use grids or problem-specific discretization.

Many existing planners for deterministic motion specialize in finding feasible paths through narrow passageways in complex configuration spaces using specialized sampling [3, 9] or learning approaches [29]. Since a narrow passageway is unlikely to be robust to motion uncertainty, finding these passageways is not the ultimate goal of our method. Our method builds a roadmap that samples the configuration space with the intent of capturing the uncertain motion transition probabilities necessary to compute optimal actions.

We apply SMR's to needle steering, a type of nonholonomic control-based motion planning problem. Nonholonomic motion planning has a long history in robotics and related fields [12, 22]. Past work has addressed deterministic curvature-constrained path planning with obstacles where a mobile robot's path is, like a car, constrained by a minimum turning radius [19, 21, 8, 31, 28]. For steerable needles, Park et al. applied a numeric diffusion-based method but did not consider obstacles or motion uncertainty [30]. Alterovitz et al. proposed an MDP formulation to find a stochastic shortest path for a steerable needle to a goal configuration, subject to user-specified "cost" parameters for direction changes, insertion distance, and collisions [2]. Because these costs are difficult to quantify, Alterovitz et al. introduced the objective function of maximizing probability of success [1]. These methods use a regular grid of states and an ad-hoc, identical discretization of the motion uncertainty distribution at all states. The methods do not consider the advantages of sampling states nor the use of sampling to estimate motion models.

B. *SMR Contributions*

SMR planning is a general framework that combines a roadmap representation of configuration space with the theory of MDP's to explicitly consider motion uncertainty at the planning stage to maximize the probability of success. SMR's use sampling to both learn the configuration space (represented as states) and to learn the stochastic motion model (represented as state transition probabilities). Sampling reduces the need for problem-specific geometric analysis or discretization for planning. As demonstrated by the success of PRM's, sampling states is useful for modeling complex configuration spaces that cannot be easily represented geometrically and extends well to higher dimensions. Random or quasi-random sampling reduces problems associated with regular grids of states, including the high computational complexity in higher dimensions and the sensitivity of solutions and runtimes to the selection of axes [22]. Sampling the stochastic motion model enables the use of a wide variety of motion uncertainty representations, including directly sampling experimentally measured data or using parameterized distributions such as a Gaussian distribution. This greatly improves previous Markov motion planning approaches that impose an ad-hoc, identical discretization of the transition probability distributions at all states [1].

Although SMR is a general framework, it provides improvements for steerable needle planning compared to previously developed approaches specifically designed for this application. Previous planners do not consider uncertainty in needle motion [30], or apply simplified models that only consider deflections at decision points and assume that all other motion model parameters are constant [2, 1]. Because we use sampling to approximate the motion model rather than a problem-specific geometric approximation [2, 1], we eliminate the discretization error at the initial configuration and can easily include a more complex uncertainty model that considers arbitrary stochastic models for both insertion distance and radius of curvature. SMR's increase flexibility and decrease computation time; for problems with equal workspace size and expected values of motion model parameters, a query that requires over a minute to solve using a grid-based MDP due to the large number of states needed to bound discretization error [1] requires just 6 seconds using an SMR.

II. Algorithm

A. *Input*

To build an SMR, the user must first provide input parameters and function implementations to describe the configuration space and robot motion model. A configuration of the robot and workspace is defined by a vector $x \in C = \Re^d$, where d is the number of degrees of freedom in the configuration space C. At any configuration x, the robot can perform an action from a discrete action set U of size w. The bounds of the configuration space are defined by B_i^{min}

and B_i^{max} for $i = 1, \ldots, d$, which specify the minimum and maximum values, respectively, for each configuration degree of freedom i. The functions `isCollisionFree`(x) and `isCollisionFreePath`(x, y) implicitly define obstacles within the workspace; the former returns `true` if configuration x collides with an obstacle and `false` otherwise, and the latter returns `true` if the path (computed by a local planner [20]) from configuration x to y collides with an obstacle and `false` otherwise. (We consider exiting the workspace as equivalent to colliding with an obstacle.) The function `distance`(x, y) specifies the distance between two configurations x and y, which can equal the Euclidean distance in d-dimensional space or some other user-specified distance metric. The function `generateSampleTransition`(x, u) implicitly defines the motion model and its probabilistic nature; this function returns a sample from a known probability distribution for the next configuration given that the robot is currently in configuration x and will perform action u.

B. Building the Roadmap

We build the stochastic motion roadmap using the algorithm `buildSMR` defined in Procedure 1. The roadmap is defined by a set of vertices V and sets of edges E^u for each action $u \in U$. The algorithm first samples n collision-free states in the configuration space and stores them in V. In our implementation, we use a uniform random sampling of the configuration space inside the bounds defined by (B_i^{min}, B_i^{max}) for $i = 1, \ldots, d$, although other random distributions or quasi- random sampling methods could be used [12, 23]. For each state $s \in V$ and an action $u \in U$, `buildSMR` calls the function `getTransitions`, defined in Procedure 2, to obtain a set of possible next states in V and probabilities of entering those states when action u is performed. We use this set to add to E^u weighted directed edges (s, t, p), which specify the probability p that the robot will transition from state $s \in V$ to state $t \in V$ when currently in state s and executing action u.

The function `getTransitions`, defined in Procedure 2, estimates state transition probabilities. Given the current state s and an action u, it calls the problem-specific function `generateSampleTransition`(x, u) to generate a sample configuration q and then selects the state $t \in V$ closest to q using the problem-specific `distance` function. We repeat this motion sampling m times and then estimate the probability of transitioning from state s to t as the proportion of times that this transition occurred out of the m samples. If there is a collision in the transition from state s to t, then the transition is replaced with a transition from s to a dedicated "obstacle state," which is required to estimate the probability that the robot collides with an obstacle.

This algorithm has the useful property that the transition probability from state s to state t in the roadmap equals the fraction of transition samples that fall inside state t's Voronoi cell. This property is implied by the use of nearest neighbor checking in `getTransitions`. As $m \to \infty$, the probability p of transitioning from s to t will approach, with probability 1, the integral of the true transition distribution over the Voronoi

Procedure 1 `buildSMR`

Input:
n: number of nodes to place in the roadmap
U: set of discrete robot actions
m: number of sample points to generate for each transition

Output:
SMR containing states V and transition probabilities (weighted edges) E^u for each action $u \in U$

$V \leftarrow \emptyset$
for all $u \in U$ **do**
  $E^u \leftarrow \emptyset$
while $|V| < n$ **do**
  $q \leftarrow$ random state sampled from the configuration space
  if `isCollisionFree`(q) **then**
    $V \leftarrow V \cup \{q\}$
for all $s \in V$ **do**
  for all $u \in U$ **do**
    for all $(t, p) \in$ `getTransitions`(V, s, u, m) **do**
      $E^u \leftarrow E^u \cup \{(s, t, p)\}$
return weighted directed graphs $G^u = (V, E^u)\ \forall\ u \in U$

cell of t. As the number of states $n \to \infty$, the expected volume V_t of the Voronoi cell for state t equals $V/n \to 0$, where V is the volume of the configuration space. Hence, the error in the approximation of the probability p due to the use of a discrete roadmap will decrease as n and m increase.

C. Solving a Query

We define a query by an initial configuration s^* and a set of goal configurations T^*.

Using the SMR and the query input, we build an $n \times n$ transition probability matrix $P(u)$ for each $u \in U$. For each tuple $(s, t, p) \in E^u$, we set $P_{st}(u) = p$ so $P_{st}(u)$ equals the probability of transitioning from state s to state t given that action u is performed. We store each matrix $P(u)$ as a sparse matrix that only includes pointers to a list of non-zero elements in each row and assume all other entries are 0.

We define $p_s(i)$ to be the probability of success given that the robot is currently in state i. If the position of state i is inside the goal, $p_s(i) = 1$. If the position of state i is inside an obstacle, $p_s(i) = 0$. Given an action u_i for some other state i, the probability of success will depend on the response of the robot to the action and the probability of success from the next state. The goal of our motion planner is to compute an optimal action u_i to maximize the expected probability of success at every state i:

$$p_s(i) = \max_{u_i} \{E[p_s(j)|i, u_i]\}, \tag{1}$$

where the expectation is over j, a random variable for the next state. Since the roadmap is a discrete approximation of the continuous configuration space, we expand the expected value in Eq. 1 to a summation:

$$p_s(i) = \max_{u_i} \left\{ \sum_{j \in V} P_{ij}(u_i) p_s(j) \right\}. \tag{2}$$

Procedure 2 `getTransitions`

Input:
V: configuration space samples
s: current robot state, $s \in V$
u: action that the robot will execute, $u \in U$
m: number of sample points to generate for this transition
Output:
List of tuples (t, p) where p is the probability of transitioning from state $s \in V$ to state $t \in V$ after executing u.

```
R ← ∅
for i = 1 to m do
  q = generateSampleTransition(s,u)
  if isCollisionFreePath(s, q) then
    t ← arg min_{t∈V} distance(q, t)
  else
    t ← obstacle state
  if (t, p) ∈ R for some p then
    Remove (t, p) from R
    R ← R ∪ {(t, p + 1/m)}
  else
    R ← R ∪ {(t, 1/m)}
return R
```

We observe that Eq. 2 is an MDP and has the form of the Bellman equation for a stochastic shortest path problem [7]:

$$J^*(i) = \max_{u_i} \sum_{j \in V} P_{ij}(u_i)\left(g(i, u_i, j) + J^*(j)\right). \quad (3)$$

where $g(i, u_i, j)$ is a "reward" for transitioning from state i to j after action u_i. In our case, $g(i, u_i, j) = 0$ for all i, u_i, and j, and $J^*(i) = p_s(i)$.

Stochastic shortest path problems of the form in Eq. 3 can be optimally solved using infinite horizon dynamic programming. For stationary Markovian problems, the configuration space does not change over time, which implies that the optimal action at each state is purely a function of the state without explicit dependence on time. Infinite horizon dynamic programming is a type of dynamic programming (DP) in which there is no finite time horizon [7]. Specifically, we use the value iteration algorithm [7], which iteratively updates $p_s(i)$ for each state i by evaluating Eq. 3. This generates a DP look-up table containing the optimal action u_i and the probability of success $p_s(i)$ for each $i \in V$.

The algorithm is guaranteed to terminate in n (the number of states) iterations if the transition probability graph corresponding to some optimal stationary policy is acyclic [7]. Violation of this requirement can occur in rare cases in which a cycle is feasible and deviating from the cycle will result in imminent failure. To remove this possibility, we introduce a small penalty γ for each transition by setting $g(i, u_i, j) = -\gamma$ in Eq. 3. Increasing γ has the effect of giving preference to shorter paths at the expense of a less precise estimate of the probability of success, where the magnitude of the error is (weakly) bounded by γn.

D. Computational Complexity

Building an SMR requires $O(n)$ time to create the states V, not including collision detection. Generating the edges in E^u requires $O(wn)$ calls to `generateTransitions`, where $w = |U|$. For computational efficiency, it is not necessary to consolidate multiple tuples with the same next state t; the addition $p + 1/m$ can be computed automatically during value iteration. Hence, each call requires $O(mdn)$ time using brute-force nearest neighbor checking. For certain low-dimensional configuration spaces, this can be reduced to $O(m \exp(d) \log(n))$ using kd-trees [6]. Hence, the total time complexity of building an SMR is $O(wmdn^2)$ or $O(wm \exp(d)\, n \log(n))$. This does not include the cost of n state collision checks and nm checks of collision free paths, which are problem-specific and may increase the computational complexity depending on the workspace definition.

Solving a query requires building the transition probability matrices and executing value iteration. Although the matrices $P_{ij}(u)$ each have n^2 entries, we do not store the zero entries as described above. Since the robot will generally only transition to a state j in the spatial vicinity of state i, each row of $P_{ij}(u)$ has only k nonzero entries, where $k << n$. Building the sparse matrices requires $O(wkn)$ time. By only accessing the nonzero entries of $P_{ij}(u)$ during the value iteration algorithm, each iteration for solving a query requires only $O(wkn)$ rather than $O(wn^2)$ time. Thus, the value iteration algorithm's total time complexity is $O(wkn^2)$ since the number of iterations is bounded by n. To further improve performance, we terminate value iteration when the maximum change ϵ over all states is less than some user-specified threshold ϵ^*. In our test cases, we used $\epsilon^* = 10^{-7}$, which resulted in far fewer than n iterations.

III. SMR for Medical Needle Steering

Diagnostic and therapeutic medical procedures such as extracting tissue samples for biopsies, injecting drugs for anesthesia, or implanting radioactive seeds for brachytherapy cancer treatment require insertion of a needle to a specific location in soft tissue. A new class of needles, steerable needles, are composed of a flexible material and with a bevel-tip that can be steered to targets in soft tissue that are inaccessible to traditional stiff needles [36, 37, 2]. Steerable needles are controlled by two degrees of freedom actuated at the needle base: insertion distance and bevel direction. When the needle is inserted, the asymmetric force applied by the bevel causes the needle to bend and follow a path of constant curvature through the tissue in the direction of the bevel [36]. Webster et al. experimentally demonstrated that, under ideal conditions, the curve has a constant radius of curvature r.

We assume the workspace is extracted from a medical image, where obstacles represent tissues that cannot be cut by the needle, such as bone, or sensitive tissues that should not be damaged, such as nerves or arteries. In this paper we consider motion plans in an imaging plane since the speed/resolution trade-off of 3-D imaging modalities is generally poor for 3-D interventional applications. We assume the needle moves a distance δ between image acquisitions that are used to

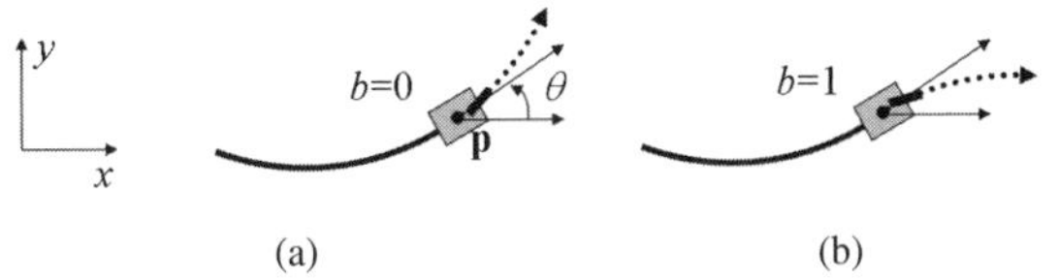

Fig. 3. The state of a bang-bang steering car is defined by point $\mathbf{p}$, orientation θ, and turning direction b (a). The car moves forward along an arc of constant curvature and can turn either left (a) or right (b).

determine the current needle position and orientation. We do not consider motion by the needle out of the imaging plane or needle retraction, which modifies the tissue and can influence future insertions. When restricted to motion in a plane, the bevel direction can be set to point left ($b = 0$) or right ($b = 1$) [36]. Due to the nonholonomic constraint imposed by the bevel, the motion of the needle tip can be modeled as a bang-bang steering car, a variant of a Dubins car that can only turn its wheels far left or far right while moving forward [36, 1].

Clinicians performing medical needle insertion procedures must consider uncertainty in the needle's motion through tissue due to patient differences and the difficulty in predicting needle/tissue interaction. Bevel direction changes further increase uncertainty due to stiffness along the needle shaft. Medical imaging in the operating room can be used to measure the needle's current position and orientation to provide feedback to the planner [13, 15], but this measurement by itself provides no information about the effect of future deflections during insertion due to motion uncertainty.

Stochastic motion roadmaps offer features particularly beneficial for medical needle steering. First, SMR's explicitly consider uncertainty in the motion of the needle. Second, intra-operative medical imaging can be combined with the fast SMR queries to permit control of the needle in the operating room without requiring time-consuming intra-operative re-planning.

A. SMR Implementation

We formulate the SMR for a bang-bang steering car, which can be applied to needle steering. The state of such a car is fully characterized by its position $\mathbf{p} = (x, y)$, orientation angle θ, and turning direction b, where b is either left ($b = 0$) or right ($b = 1$). Hence, the dimension of the state space is $d = 4$, and a state i is defined by $s_i = (x_i, y_i, \theta_i, b_i)$, as illustrated in Fig. 3. We encode b_i in the state since it is a type of history parameter that is required by the motion uncertainty model. Since an SMR assumes that each component of the state vector is a real number, we define the binary b_i as the floor of the fourth component in s_i, which we bound in the range $[0, 2)$.

Between sensor measurements of state, we assume the car moves a distance δ. The set U consists of two actions: move forward turning left ($u = 0$), or move forward turning right ($u = 1$). As the car moves forward, it traces an arc of length δ with radius of curvature r and direction based on u. We consider r and δ as random variables drawn from a given distribution. In this paper, we consider $\delta \sim N(\delta_0, \sigma_{\delta_a})$ and $r \sim N(r_0, \sigma_{r_a})$, where N is a normal distribution with given mean and standard deviation parameters and $a \in \{0, 1\}$ indicates direction change. We implement `generateSampleTransition` to draw random samples from these distributions. Although the uncertainty parameters can be difficult to measure precisely, even rough estimates may be more realistic than using deterministic transitions when uncertainty is high.

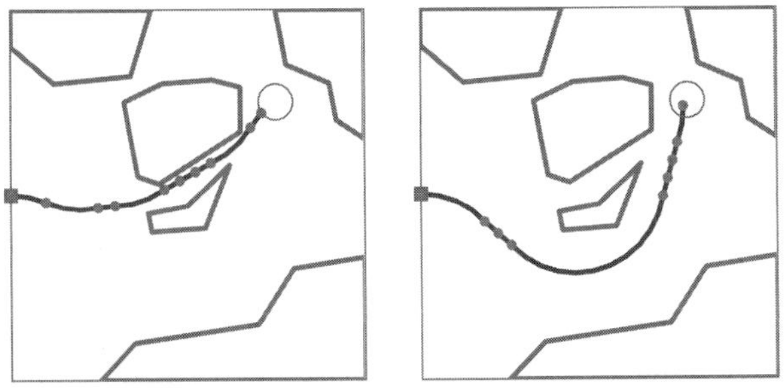

(a) Minimizing path length $p_s = 35\%$ (b) Maximizing p_s using SMR $p_s = 83\%$

Fig. 4. Explicitly considering motion uncertainty using an SMR planner improves the probability of success.

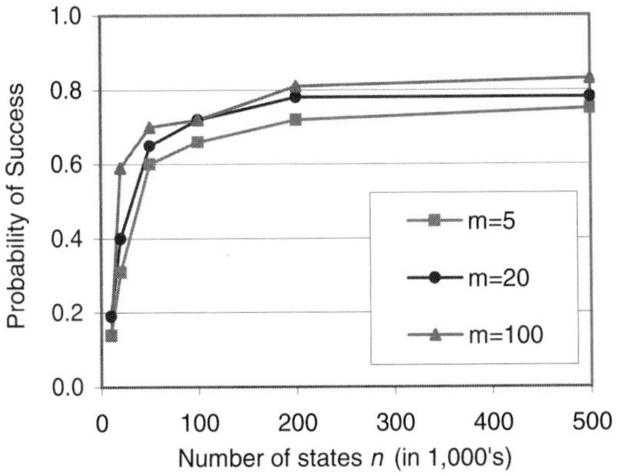

Fig. 5. Effect of the number of states n and number of motion samples m on the probability of success p_s.

We define the workspace as a rectangle of width x_{max} and height y_{max} and define obstacles as polygons in the plane. To detect obstacle collisions, we use the zero-winding rule [17]. We define the distance between two states s_1 and s_2 to be the weighted Euclidean distance between the poses plus an indicator variable to ensure the turning directions match: $\texttt{distance}(s_1, s_2) = \sqrt{(x_1 - x_2)^2 + (y_1 - y_2)^2 + \alpha(\theta_1 - \theta_2)^2 + M}$, where $M \to \infty$ if $b_1 \neq b_2$, and $M = 0$ otherwise. For fast nearest-neighbor computation, we use the CGAL implementation of kd-trees [32]. Since the `distance` function is non-Euclidean, we use the formulation developed by Atramentov and LaValle to build the kd-tree [6]. We define the goal T^* as all configuration states within a ball of radius t^r centered at a point $\mathbf{t}^*$.

B. Results

We implemented the SMR planner in C++ and tested the method on workspaces of size $x_{max} = y_{max} = 10$ with polygonal obstacles as shown in shown in Fig. 1 and Fig. 4. We set the robot parameters $r_0 = 2.5$ and $\delta_0 = 0.5$ with motion uncertainty parameters $\sigma_{\delta_0} = 0.1$, $\sigma_{\delta_1} = 0.2$, $\sigma_{r_0} = 0.5$, and $\sigma_{r_1} = 1.0$. We set parameters $\gamma = 0.00001$ and $\alpha = 2.0$. We tested the motion planner on a 2.2 GHz AMD Opteron PC. Building the SMR required approximately 1 minute for $n = 50{,}000$ states, executing a query required 6 seconds, and additional queries for the same goal required less than 1 second of computation time for both example problems.

We evaluate the plans generated by SMR with multiple randomized simulations. Given the current state of the robot, we query the SMR to obtain an optimal action u. We then execute this action and compute the expected next state. We repeat until the robot reaches the goal or hits an obstacle, and we illustrate the resulting expected path. Since the motion response of the robot to actions is not deterministic, success of the procedure can rarely be guaranteed. To estimate p_s, we run the simulation 100 times, sampling the next state from the transition probability distribution rather than selecting the expected value, and we compute the number of goal acquisitions divided by the number of obstacle collisions.

In Fig. 4(b), we illustrate the expected path using an SMR with $m = 100$ motion samples and $n = 500{,}000$ states. As in Fig. 1(b), the robot avoids passing through a narrow passageway near the goal and instead takes a longer route. The selection of the longer path is not due to insufficient states in the SMR; there exist paths in the SMR that pass through the narrow gaps between the obstacles. The plan resulting in a longer path is selected purely because it maximizes p_s.

The probability of success p_s improves as the sampling density of the configuration space and the motion uncertainty distribution increase, as shown in Fig. 5. As n and m increase, $p_s(s)$ is more accurately approximated over the configuration space, resulting in better action decisions. However, p_s effectively converges for $n \geq 100{,}000$ and $m \geq 20$, suggesting the inherent difficulty of the motion planning problem. Furthermore, the expected path does not substantially vary from the path shown in Fig. 4(b) for $n \geq 50{,}000$ and $m \geq 5$. The number of states required by the SMR planner is far smaller than the 800,000 states required for a similar problem using a grid-based approach with bounded error [1].

In Fig. 4(a), we computed the optimal shortest path assuming deterministic motion of the robot using a fine regular discrete grid with 816,080 states for which the error due to discretization is small and bounded [1]. We estimate p_s using the same simulation methodology as for an SMR plan, except that we compute the shortest path for each query. The expected shortest path passes through a narrow passage between obstacles and the resulting probability of success is substantially lower compared to the SMR plan. The result was similar for the example in Fig. 1; explicitly considering motion uncertainty improved the probability of success.

To further illustrate the importance of explicitly considering uncertainty during motion planning, we vary the standard deviation parameters σ_{δ_0}, σ_{δ_1}, σ_{r_0}, and σ_{r_1}. In Fig. 6, we compute a plan for a robot with each standard deviation parameter set to a quarter of its default value. For this low uncertainty case, the uncertainty is not sufficient to justify

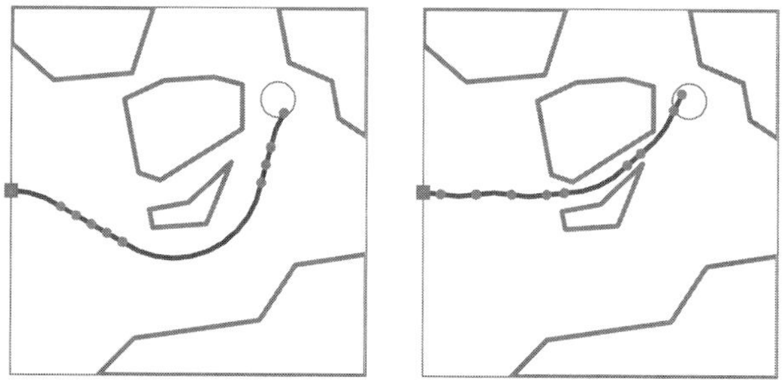

(a) High uncertainty $p_s = 78\%$ (b) Low uncertainty $p_s = 87\%$

Fig. 6. The level of uncertainty affects SMR planning results. In cases of low uncertainty (with 75% reduction in distribution standard deviations), the expected path resembles a deterministic shortest path due to the small influence of uncertainty on p_s and the effect of the penalty term γ. In both these examples, the same $n = 200{,}000$ states were used in the roadmap.

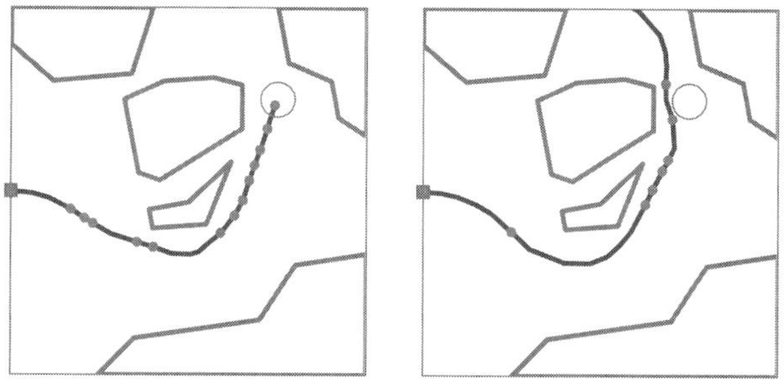

(a) Successful procedure (b) Unsuccessful procedure

Fig. 7. Two simulated procedures of needle steering, one successful (a) and one unsuccessful due to effects of uncertain motion (b), using an SMR with $n = 50{,}000$ states.

avoiding the narrow passageway; the penalty γ causes the plan to resemble the deterministic shortest plan in Fig. 4(a). Also, p_s is substantially higher because of the lower uncertainty.

In Fig. 7, we execute the planner in the context of an image-guided procedure. We assume the needle tip position and orientation is extracted from a medical image and then execute a query, simulate the needle motion by drawing a sample from the motion uncertainty distribution, and repeat. The effect of uncertainty can be seen as deflections in the path. In practice, clinicians could monitor $p_s(s)$ for the current state s as the procedure progresses.

IV. Conclusion

In many motion planning applications, the response of the robot to commanded actions cannot be precisely predicted. We introduce the Stochastic Motion Roadmap (SMR), a new motion planning framework that explicitly considers uncertainty in robot motion to maximize the probability that a robot will avoid obstacle collisions and successfully reach a goal. SMR planners combine the roadmap representation of configuration space used in PRM with the theory of MDP's to explicitly consider motion uncertainty at the planning stage.

To demonstrate SMR's, we considered a nonholonomic mobile robot with bang-bang control, a type of Dubins-car robot model that can be applied to steering medical needles through soft tissue. Needle steering, like many other medical procedures, is subject to substantial motion uncertainty and is therefore ill-suited to shortest-path plans that may guide medical tools through narrow passageways between critical tissues. Using randomized simulation, we demonstrated that SMR's generate motion plans with significantly higher probabilities of success compared to traditional shortest-path approaches.

In future work, we will extend the SMR framework and explore new applications with higher dimensional configuration spaces. We plan to extend the formulation to also consider actions in a continuous range rather than solely from a discrete set, to investigate more sophisticated sampling methods for generating configuration samples and for estimating transition probabilities, and to integrate the effects of sensing uncertainty. We also plan to explore new biomedical and industrial applications where uncertainty in motion should be considered to maximize the probability of success.

Acknowledgment

This work has been partially supported by the ITAV project ALMA.

References

[1] R. Alterovitz, M. Branicky, and K. Goldberg, "Constant-curvature motion planning under uncertainty with applications in image-guided medical needle steering," in *Proc. Int. Workshop on the Algorithmic Foundations of Robotics*, July 2006.

[2] R. Alterovitz, A. Lim, K. Goldberg, G. S. Chirikjian, and A. M. Okamura, "Steering flexible needles under Markov motion uncertainty," in *Proc. IEEE/RSJ Int. Conf. on Intelligent Robots and Systems (IROS)*, Aug. 2005, pp. 120–125.

[3] N. M. Amato, O. B. Bayazit, L. K. Dale, C. Jones, and D. Vallejo, "OBPRM: An obstacle-based PRM for 3D workspaces," in *Robotics: The Algorithmic Perspective*, P. Agarwal, L. E. Kavraki, and M. Mason, Eds. Natick, MA: AK Peters, Ltd., 1998, pp. 156–168.

[4] M. S. Apaydin, D. L. Brutlag, C. Guestrin, D. Hsu, and J.-C. Latombe, "Stochastic roadmap simulation: an efficient representation and algorithm for analyzing molecular motion," in *Proc. RECOMB*, 2002, pp. 12–21.

[5] M. Apaydin, D. Brutlag, C. Guestrin, D. Hsu, and J.-C. Latombe, "Stochastic conformational roadmaps for computing ensemble properties of molecular motion," in *Algorithmic Foundations of Robotics V*. Berlin, Germany: Springer-Verlag, 2003, pp. 131–148.

[6] A. Atramentov and S. M. LaValle, "Efficient nearest neighbor searching for motion planning," in *Proc. IEEE Int. Conf. on Robotics and Automation (ICRA)*, 2002.

[7] D. P. Bertsekas, *Dynamic Programming and Optimal Control*, 2nd ed. Belmont, MA: Athena Scientific, 2000.

[8] A. Bicchi, G. Casalino, and C. Santilli, "Planning shortest bounded-curvature paths for a class of nonholonomic vehicles among obstacles," *Journal of Intelligent and Robotic Systems*, vol. 16, pp. 387–405, 1996.

[9] V. Boor, N. H. Overmars, and A. F. van der Stappen, "The Gaussian sampling strategy for probabilisitic roadmap planners," in *Proc. IEEE Int. Conf. on Robotics and Automation (ICRA)*, 1999, pp. 1018–1023.

[10] B. Bouilly, T. Siméon, and R. Alami, "A numerical technique for planning motion strategies of a mobile robot in presence of uncertainty," in *Proc. IEEE Int. Conf. on Robotics and Automation (ICRA)*, Nagoya, Japan, May 1995.

[11] B. Burns and O. Brock, "Toward optimal configuration space sampling," in *Proc. Robotics: Science and Systems*, Cambridge, MA, June 2005.

[12] H. Choset, K. M. Lynch, S. Hutchinson, G. Kantor, W. Burgard, L. E. Kavraki, and S. Thrun, *Principles of Robot Motion: Theory, Algorithms, and Implementations*. MIT Press, 2005.

[13] K. Cleary, L. Ibanez, N. Navab, D. Stoianovici, A. Patriciu, and G. Corral, "Segmentation of surgical needles for fluoroscopy servoing using the Insight Software Toolkit (ITK)," in *Proc. Int. Conf. of the IEEE Engineering In Medicine and Biology Society*, Sept. 2003, pp. 698–701.

[14] T. Dean, L. P. Kaelbling, J. Kirman, and A. Nicholson, "Planning under time constraints in stochastic domains," *Artificial Intelligence*, vol. 76, no. 1-2, pp. 35–74, July 1995.

[15] S. P. DiMaio, D. F. Kacher, R. E. Ellis, G. Fichtinger, N. Hata, G. P. Zientara, L. P. Panych, R. Kikinis, and F. A. Jolesz, "Needle artifact localization in 3T MR images," in *Medicine Meets Virtual Reality 14*, J. D. Westwood *et al.*, Eds. IOS Press, Jan. 2006, pp. 120–125.

[16] D. Ferguson and A. Stentz, "Focussed dynamic programming: Extensive comparative results," Robotics Institute, Carnegie Mellon University, Pittsburgh, PA, Tech. Rep. CMU-RI-TR-04-13, Mar. 2004.

[17] D. Hearn and M. P. Baker, *Computer Graphics with OpenGL*, 3rd ed. Prentice Hall, 2003.

[18] K. Hsiao, L. P. Kaelbling, and T. Lozano-Perez, "Grasping POMDPs," in *Proc. IEEE Int. Conf. on Robotics and Automation (ICRA)*, Rome, Italy, Apr. 2007, pp. 4685–4692.

[19] P. Jacobs and J. Canny, "Planning smooth paths for mobile robots," in *Proc. IEEE Int. Conf. on Robotics and Automation (ICRA)*, May 1989, pp. 2–7.

[20] L. E. Kavraki, P. Svestka, J.-C. Latombe, and M. Overmars, "Probabilistic roadmaps for path planning in high dimensional configuration spaces," *IEEE Trans. on Robotics and Automation*, vol. 12, no. 4, pp. 566–580, 1996.

[21] J.-P. Laumond, P. E. Jacobs, M. Taïx, and R. M. Murray, "A motion planner for nonholonomic mobile robots," *IEEE Trans. on Robotics and Automation*, vol. 10, no. 5, pp. 577 – 593, Oct. 1994.

[22] S. M. LaValle, *Planning Algorithms*. Cambridge, U.K.: Cambridge University Press, 2006.

[23] S. M. LaValle, M. S. Branicky, and S. R. Lindemann, "On the relationship between classical grid search and probabilistic roadmaps," *Int. J. of Robotics Research*, vol. 23, no. 7/8, pp. 673–692, July 2004.

[24] S. M. LaValle, D. Lin, L. J. Guibas, J.-C. Latombe, and R. Motwani, "Finding an unpredictable target in a workspace with obstacles," in *Proc. IEEE Int. Conf. on Robotics and Automation (ICRA)*, 1997, pp. 737–742.

[25] S. M. LaValle and S. A. Hutchinson, "An objective-based framework for motion planning under sensing and control uncertainties," *Int. J. of Robotics Research*, vol. 17, no. 1, pp. 19–42, Jan. 1998.

[26] A. Lazanas and J. Latombe, "Motion planning with uncertainty: A landmark approach," *Artificial Intelligence*, vol. 76, no. 1-2, pp. 285–317, 1995.

[27] S. R. Lindemann, I. I. Hussein, and S. M. LaValle, "Real time feedback control for nonholonomic mobile robots with obstacles," in *Proc. IEEE Conf. Decision and Control*, 2006.

[28] R. Mason and J. Burdick, "Trajectory planning using reachable-state density functions," in *Proc. IEEE Int. Conf. on Robotics and Automation (ICRA)*, vol. 1, May 2002, pp. 273–280.

[29] A. W. Moore and C. G. Atkeson, "The parti-game algorithm for variable resolution reinforcement learning in multidimensional state spaces," *Machine Learning*, vol. 21, no. 3, pp. 199–233, 1995.

[30] W. Park, J. S. Kim, Y. Zhou, N. J. Cowan, A. M. Okamura, and G. S. Chirikjian, "Diffusion-based motion planning for a nonholonomic flexible needle model," in *Proc. IEEE Int. Conf. on Robotics and Automation (ICRA)*, Apr. 2005, pp. 4611–4616.

[31] J. Sellen, "Approximation and decision algorithms for curvature-constrained path planning: A state-space approach," in *Robotics: The Algorithmic Perspective*, P. K. Agarwal, L. E. Kavraki, and M. T. Mason, Eds. Natick, MA: AK Peters, Ltd., 1998, pp. 59–67.

[32] H. Tangelder and A. Fabri, "dD spatial searching," in *CGAL-3.2 User and Reference Manual*, CGAL Editorial Board, Ed., 2006. [Online]. Available: http://www.cgal.org/Manual/

[33] S. Thrun, W. Burgard, and D. Fox, *Probabilistic Robotics*. MIT Press, 2005.

[34] J. van den Berg and M. Overmars, "Planning the shortest safe path amidst unpredictably moving obstacles," in *Proc. Int. Workshop on the Algorithmic Foundations of Robotics*, July 2006.

[35] D. Vasquez, F. Large, T. Fraichard, and C. Laugier, "High-speed autonomous navigation with motion prediction for unknown moving obstacles," in *Proc. IEEE/RSJ Int. Conf. on Intelligent Robots and Systems (IROS)*, 2004, pp. 82–87.

[36] R. J. Webster III, J. S. Kim, N. J. Cowan, G. S. Chirikjian, and A. M. Okamura, "Nonholonomic modeling of needle steering," *Int. J. of Robotics Research*, vol. 25, pp. 509–525, May 2006.

[37] R. J. Webster III, J. Memisevic, and A. M. Okamura, "Design considerations for robotic needle steering," in *Proc. IEEE Int. Conf. on Robotics and Automation (ICRA)*, Apr. 2005, pp. 3599–3605.

A Fundamental Tradeoff between Performance and Sensitivity within Haptic Rendering

Paul G. Griffiths
Dept. of Mechanical Engineering
University of Michigan
Ann Arbor, MI 48109
Email: paulgrif@umich.edu

R. Brent Gillespie
Dept. of Mechanical Engineering
University of Michigan
Ann Arbor, MI 48109
Email: brentg@umich.edu

Jim S. Freudenberg
Dept. of Electrical Engineering
& Computer Science
University of Michigan
Ann Arbor, MI 48109
Email: jfr@eecs.umich.edu

***Abstract*—In this paper we show that for haptic rendering using position feedback, the structure of the feedback loop imposes a fundamental tradeoff between accurate rendering of virtual environments and sensitivity of closed-loop responses to hardware variations and uncertainty. Due to this tradeoff, any feedback design that achieves high-fidelity rendering incurs a quantifiable cost in terms of sensitivity. Analysis of the tradeoff reveals certain combinations of virtual environment and haptic device dynamics for which performance is achieved only by accepting very poor sensitivity. This analysis may be used to show that certain design specifications are feasible and may guide the choice of hardware to mitigate the tradeoff severity. We illustrate the predicted consequences of the tradeoff with an experimental study.**

I. Introduction

Design of a feedback controller involves compromises between various conflicting objectives. These tradeoffs are imposed by factors external to the controller design such as the location of sensors and actuators, and limitations of the hardware such as sample-rate, delay, bandwidth, and quantization. By quantifying these tradeoffs we reveal certain relationships between hardware and feedback properties which are satisfied for all controller designs. Such knowledge allows one, for instance, to identify infeasible specifications before any feedback design is attempted. If design specifications cannot be relaxed, interpretation of the underlying mathematical relationships can provide guidance in selecting different hardware to reduce the severity of tradeoffs. Furthermore, if feasible feedback designs do exist, knowledge of fundamental tradeoffs can confirm that a particular design strikes a favorable compromise between conflicting goals.

Feedback control in haptic rendering is used to shape the closed-loop dynamics of a haptic device to match the dynamics of a desired virtual environment. This use of feedback, however, must be weighed against the costs associated with feedback, notably the potential for instability. Previous work in haptic interface systems has addressed tradeoffs between performance and stability due to sampling [1]–[9], quantization [1], [5], [6], [8], [9], hardware damping [1]–[3], [5]–[9], and nonlinearities [1], [2], [4]–[9]. An important characteristic of feedback design which has not been analyzed for haptic rendering is the sensitivity of the closed-loop response to parameter variations in the hardware and model uncertainty. Predictions about stability or performance become less reliable as the sensitivity to hardware dynamics increases, and although feedback may be used attenuate sensitivity, feedback may also *amplify* sensitivity.

Tradeoffs between performance and sensitivity are well characterized for typical servo-control applications [12]; however feedback design for haptic interface cannot be treated as a typical servo-control problem. An important goal of feedback in servo-control applications is to reject disturbances that enter at the actuator input. In haptic rendering with back-drivable devices, the operator's input enters at the actuator input and rejecting this input is not the goal of feedback. Instead the virtual environment provides the desired response to the human operator. Feedback design for haptic rendering may then be treated as a *model-matching* problem which is not solved by standard loop-shaping techniques. A different control strategy is required for haptic rendering, and additional design tradeoffs exist which have no counterpart in servo-control systems.

The Bode sensitivity function characterizes multiple important properties of a feedback system including stability robustness and sensitivity of closed-loop transfer functions to variations in the hardware dynamics [12]. In typical servo-control problems, the Bode sensitivity function also describes the disturbance response of performance outputs; however, for a certain class of feedback systems, attenuation of this disturbance response is not achieved by attenuating the Bode sensitivity function. This class of systems is characterized by performance outputs that differ from the measured outputs *and* disturbance inputs that affect the plant through different dynamics than the control inputs. As detailed in Freudenberg et al. [13], an algebraic (frequency-by-frequency) tradeoff exists within these systems between performance goals and feedback properties described by the Bode sensitivity function.

In this paper we apply the results of [13] to reveal a tradeoff between performance and sensitivity not previously analyzed within haptic interface systems. Our analysis shows that, at a frequency, all feedback designs must compromise between rendering the virtual environment accurately and reducing sensitivity to haptic device dynamics. We capture the severity of this tradeoff in a single frequency-dependent parameter that depends on the virtual environment and haptic device dynamics. Based on the tradeoff, the cost in terms sensitivity to render a virtual environment accurately is independent of the feedback design, and is large at frequencies where inherent dynamics of the haptic device would mask the virtual environment dynamics. We introduce a controller design that cancels the haptic device dynamics as needed and is subject to the attendant poor sensitivity for certain virtual environments. Due to the tradeoff, the only way to reduce sensitivity without sacrificing performance is to re-design the haptic device hardware with reduced inherent dynamics such as smaller inertia and damping.

The tradeoff we show is fundamental to the hardware and does not depend on the controller implementation or complexity. Our analysis assumes a haptic device, equipped with position sensing, whose dynamic response to the control input is the same as the dynamic response to the human operator's applied force. For simplicity, we treat single-axis haptic devices. We work with continuous linear time-invariant models of the haptic interface system and do not capture sampled-data effects, quantization, and other nonlinearities. While these factors present their own limitations and tradeoffs, the tradeoff we discuss exists in addition to these, and its severity cannot be diminished by increasing sample rate, improving sensor quantization, or minimizing nonlinear dynamics.

II. Rendering Virtual Environments using Position Feedback Control

A. Hardware & Controller

In a standard configuration of haptic rendering, a human operator grasps and applies forces to a motorized, computer-controlled manipulator. Figure 1 shows this standard setup using a direct-drive, single-axis, rotary handwheel equipped with an encoder for position measurement. The controller reads the handwheel position X, as measured by the encoder, and computes a motor command. The amplified motor command drives a torque u that acts on the handwheel. The human operator also applies a torque F on the handwheel, which affects the position X through the same dynamics as u. The motor, amplifier, encoder, and handwheel together comprise the haptic device. The haptic device may be linear or rotary, and we henceforth refer to F and u as forces rather than torques without loss of generality.

We assume a linear systems framework, where variables are defined by the Laplace transform of their respective signals and transfer functions are defined by the Laplace transform of their respective operators. Let us capture the haptic device dynamics in the transfer function P. Then the position of the haptic device is

$$X = P(F+u). \tag{1}$$

The controller describes the transfer function from position measurement to motor command. We capture the controller dynamics in the transfer function C. Then the motor command is given by

$$u = -CX. \tag{2}$$

The human operator perceives the closed-loop response from F to X. Closing the loop between P and C, we find that this response is $P/(1+PC)$.

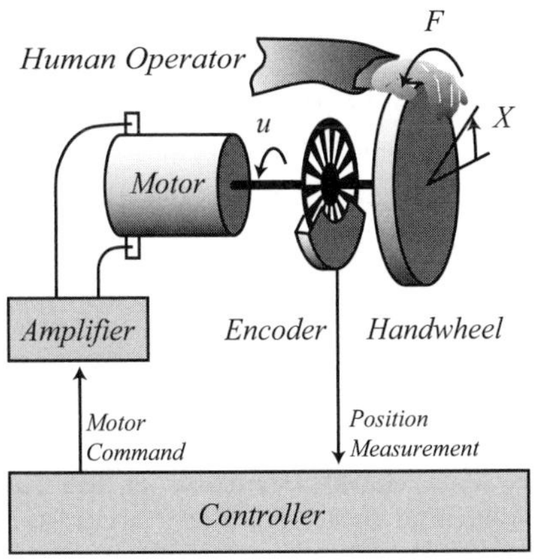

Fig. 1. Schematic of the haptic interface system hardware and controller.

B. Posing feedback design as a model-matching problem

The purpose of feedback control in haptic rendering is to shape the closed-loop response of the haptic device position X to the human operator's input F. The desired response is generated by the *virtual environment* dynamics, denoted by the transfer function R_d. Let us define X_d to be the desired closed-loop response of X to F. Then

$$X_d = R_d F \tag{3}$$

and accurate rendering of the environment dynamics is achieved by attenuating the response of the error signal $X - X_d$ to the human operator force F.

We use the standard form of the *general control configuration* [12] to capture the feedback design problem of haptic rendering. As shown in Fig. 2, the standard form consists of a generalized multivariable plant G in feedback with a generalized controller K. The generalized plant G describes the input/output responses from disturbance inputs w and control inputs u to performance outputs z and measured outputs y:

$$\begin{bmatrix} z \\ y \end{bmatrix} = \begin{bmatrix} G_{zw} & G_{zu} \\ G_{yw} & G_{yu} \end{bmatrix} \begin{bmatrix} w \\ u \end{bmatrix}. \tag{4}$$

We denote the closed-loop disturbance response from w to z by T_{zw}. If all signals are scalar, the disturbance response is given by

$$T_{zw} = G_{zw} + G_{zu}K(1 - G_{yu}K)^{-1} G_{yw}. \tag{5}$$

Performance goals are achieved by designing the generalized controller K to attenuate the disturbance response T_{zw}.

The block diagram shown in Fig. 3 depicts the feedback design of haptic interface posed in the general control configuration. The exogenous input w is generated by the human operator force F, and performance output z is defined by the normalized error signal

$$z \triangleq R_d^{-1}(X - X_d). \tag{6}$$

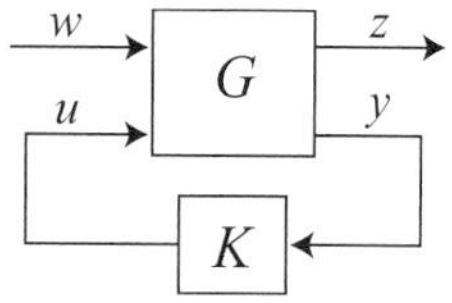

Fig. 2. Block diagram of the *general control configuration.*

The generalized control input u is synonymous with the motor torque u, and the measured output y is simply X. Then the elements of the generalized plant G (contained within the dashed box in Fig. 3) are

$$G = \begin{bmatrix} P/R_d - 1 & P/R_d \\ P & P \end{bmatrix}. \quad (7)$$

We may compute the disturbance response T_{zw} by substituting (7) into (5) and $-C$ for K:

$$T_{zw} = \frac{1}{R_d}\left(\frac{P}{1+PC} - R_d\right). \quad (8)$$

Note that the term $P/(1+PC)$ describes the actual closed-loop response of X to the human operator input F. We refer to this transfer function as the *rendered virtual environment*, which we denote by

$$R \triangleq \frac{P}{1+PC}. \quad (9)$$

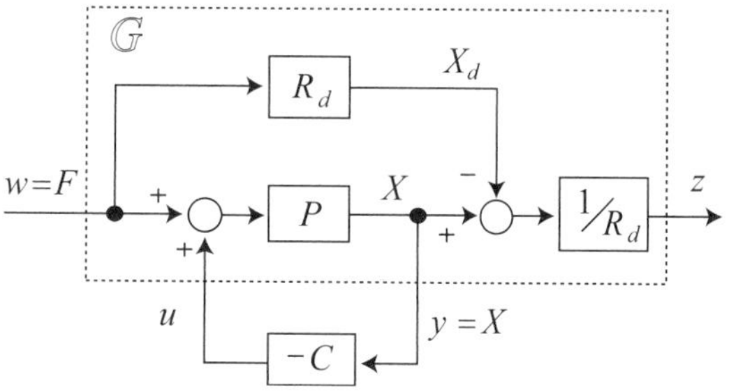

Fig. 3. A model-matching block diagram of haptic rendering. The human operator feels the *rendered virtual environment* R, the response X/F; whereas the desired response is given by virtual environment R_d. The performance variable z measures *distortion*, the normalized error between the actual and desired responses.

The disturbance attenuation problem captured by (8) may be treated as a model-matching problem. To attenuate the disturbance response T_{zw}, we must design C to reduce the mismatch between rendered virtual environment R and the desired virtual environment R_d. We denote the closed-loop disturbance response (8) for haptic rendering by Θ_c and call this relative error between R and R_d *distortion*. Exact model-matching $R \equiv R_d$, known as *perfect transparency* [14], is achieved when $\Theta_c \equiv 0$.

We note that the benefit of feedback in reducing closed-loop distortion should be quantified relative to the distortion of the open-loop system. We denote distortion of the open-loop system by $\Theta_o \triangleq (P - R_d)/R_d$. Distortion of the open-loop system may be smaller than closed-loop distortion at some frequencies implying a cost rather than benefit of feedback.

C. Sensitivity to parameter variations in the haptic device

In addition to nominal performance, any practical feedback design must provide a degree of robustness to variations in the haptic device. The Bode sensitivity function S describes several important feedback properties including the sensitivity of closed-loop transfer functions to variations in hardware dynamics [12]. In terms of the haptic device dynamics P and the controller C, the Bode sensitivity function is

$$S = \frac{1}{1+PC}. \quad (10)$$

Let us then consider the sensitivity of the rendered virtual environment R to the haptic device dynamics P. One may show that

$$\frac{P}{R}\frac{dR}{dP} = S. \quad (11)$$

Thus S describes the differential change in the rendered environment dR/R to a differential change in the haptic device model dP/P. Then, to a first-order approximation, relative error in the haptic device dynamics results in a relative error between the nominal and actual rendered virtual environment scaled by the Bode sensitivity function S. To reduce the sensitivity of R to variations in the haptic device dynamics and model uncertainty, the feedback design must attenuate S.

D. Structure of the haptic interface controller

The haptic interface controller is typically partitioned into two parts: a simulation of the virtual environment and a virtual coupler [15]. This latter element connects the hardware with the virtual environment. As shown in Fig. 4, the virtual coupler produces an input F_e to the virtual environment and receives the desired position of the haptic device X_d. Partitioned in this way, design of the controller is split into two problems: creating an accurate simulation of the virtual environment dynamics, and designing a generalized virtual coupler to render that virtual environment with low distortion. With this structure, various virtual environments can, in theory, be interchanged without redesigning the virtual coupler. Note that for haptic applications involving scaling between the haptic device and the virtual environment, we assume without loss of generality in our analysis that these scaling factors are internal to the virtual environment.

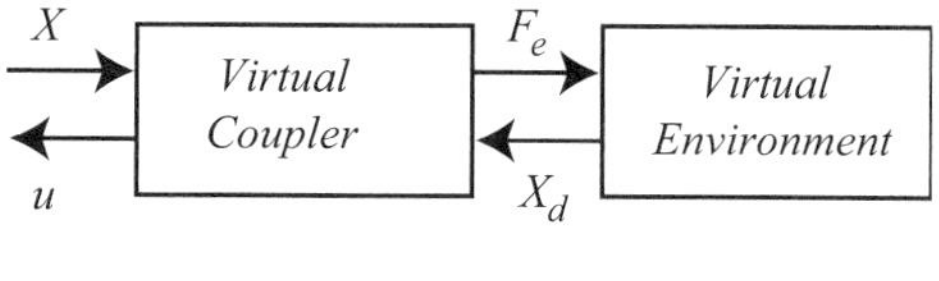

Fig. 4. Architecture of the controller C, which describes the transfer function from X to u.

We note that the virtual coupler may be modeled after mechanical elements such as springs and dampers, but can more generally be described by four transfer functions relating the two inputs and two outputs. In Section IV we find design directives for these four transfer functions which minimize distortion. For now we defer discussion about design of the virtual coupler, because the tradeoff between performance and sensitivity, which is the focus of the present work, is independent of the structure imposed on C.

III. A Tradeoff Between Performance and Sensitivity

A. Background

For many common feedback systems such as servo-control applications, the disturbance response T_{zw} is described by the Bode sensitivity function S. However, within multivariable control systems where (a) the output and performance variables differ, *and* (b) the control and exogenous input affect the plant through different dynamics, S does not describe T_{zw}. Then, at a frequency, the feedback design cannot attenuate both the disturbance response and sensitivity as dictated by an algebraic (frequency-by-frequency) identity presented in [13]. We now briefly reproduce this identity in terms of the general control configuration.

Recall that according to (5), the disturbance response T_{zw} in terms of the elements of the generalized plant G is $G_{zw}+G_{zu}K(1-G_{yu}K)^{-1}G_{yw}$. The Bode sensitivity function for the general control configuration is $S=1/(1-G_{yu}K)$. Combining the expressions for T_{zw} and S we can find an expression that does not depend explicitly on the controller K. Let $\Gamma \triangleq G_{zu}G_{yw}/(G_{zw}G_{yu})$. Then we have the algebraic identity

$$\frac{T_{zw}}{G_{zw}} = 1+\Gamma(S-1). \tag{12}$$

Only when $\Gamma = 1$ is disturbance attenuation described by the Bode sensitivity function. There are two special cases where $\Gamma \equiv 1$: (a) systems whose performance variable is also the measured output, that is $G_{zu}=G_{yu}$ and $G_{zw}=G_{yw}$, and (b) systems whose control and exogenous input enter the system through the same dynamics, that is $G_{zu}=G_{zw}$ and $G_{yu}=G_{yw}$.

For systems where $\Gamma \not\equiv 1$, achieving good performance and low sensitivity are competing goals. The severity of the tradeoff is determined by Γ and is generally frequency dependent. Recall that we wish to attenuate both T_{zw} and S. However, at frequencies where $|\Gamma(j\omega)| << 1$, the cost to attenuate the closed-loop disturbance response relative to the open-loop disturbance response is large amplification of $S(j\omega)$. On the other hand, at frequencies where $|\Gamma(j\omega)| >> 1$, the cost to attenuate the Bode sensitivity function is large amplification of the closed-loop disturbance response relative to the open-loop disturbance response. We note, furthermore, that in (12) both S and T_{zw} may be large. Indeed, at high-frequencies, both the Bode sensitivity function and the ratio of the closed-loop disturbance response to the open-loop disturbance must approach 1 for any proper feedback design.

B. Application to haptic rendering

We now interpret the tradeoff implied by (12) for haptic rendering with position feedback. We recall that in (12) the closed-loop disturbance response T_{zw} is Θ_c and the open-loop disturbance response G_{zw} is Θ_o. Then performance afforded by feedback control is gauged by attenuation of Θ_c/Θ_o. We note that without feedback control, the Bode sensitivity function S and the ratio Θ_c/Θ_o are both unity.

The tradeoff severity Γ is found by substituting (7) into $G_{zu}G_{yw}/(G_{zw}G_{yu})$. Further substituting open-loop distortion Θ_o for $(P-R_d)/R_d$, we find that

$$\Gamma = 1+\frac{1}{\Theta_o}. \tag{13}$$

Given a fixed device model P, the term $1/\Theta_o$ approaches 0 as R_d approaches 0. Then, for any $R_d \not\equiv 0$, the tradeoff severity $\Gamma \not\equiv 1$ and there exists a tradeoff between attenuating distortion and attenuating the Bode sensitivity function. The tradeoff is most severe at frequencies where $\Theta_o \to -1$ or frequencies where $\Theta_o \to 0$.

We first consider the situation where $\Theta_o \to -1$ and $\Gamma \to 0$. From the definition of Θ_o, we see that $\Theta_o \to -1$ at frequencies where $|R_d(j\omega)| >> |P(j\omega)|$. At these frequencies, the magnitude of the desired closed-loop response given by the virtual environment R_d is much greater than the response of the haptic device dynamics P, and partial cancellation of the device dynamics P is required. The cost of partially cancelling device dynamics is large amplification of the Bode sensitivity function.

Let us alternatively consider the situation where $\Theta_o \to 0$ and $\Gamma \to \infty$. This situation arises as the desired virtual environment R_d approaches the open-loop dynamics of the haptic device P. Little or no feedback control is required to achieve low distortion since open-loop distortion is already nearly 0. However, the Bode sensitivity function approaches 1 as the feedback gain approaches 0. We may use feedback to attenuate sensitivity, but only by accepting large amplification of the ratio Θ_c/Θ_o.

An important consequence of (12) is that, at frequencies where $\Gamma \neq 1$, any feedback design that attenuates the ratio Θ_c/Θ_o cannot also attenuate the Bode sensitivity function S. Furthermore, regardless of the controller synthesis technique or controller complexity, $S \to 1-1/\Gamma$ at frequencies where $\Theta_c/\Theta_o \to 0$. Substituting (13) for Γ, we reduce this limit to

$$S \to \frac{R_d}{P}. \tag{14}$$

At frequencies where $|R_d(j\omega)|/|P(j\omega)|$ is large, accurate rendering of the virtual environment can only be achieved by accepting very poor robustness to variations in the haptic device dynamics.

The virtual environment R_d and haptic device model P are transfer functions from force to motion; thus a large magnitude of either transfer function corresponds to a small mechanical impedance. It is then not surprising that, as given by (14), poor sensitivity results when we accurately render a virtual environment with a small mechanical impedance relative to the mechanical impedance of the haptic device. We note, however, that this sensitivity is not that typically recognized in feed-forward control—the controller C is indeed in feedback with the haptic device. Furthermore, the sensitivity to hardware dynamics induced by feedback control may be much greater than the unity sensitivity of feed-forward control. Perhaps less intuitive is the frequency dependent nature of the tradeoff. For instance, rendering a pure spring with low distortion using a haptic device with inertia induces sensitivity that increases with frequency.

IV. Virtual Coupler Design

The design tradeoff introduced above predicts the cost to achieve low distortion for any feedback design. With this analysis one may evaluate whether a feedback design efficiently trades off performance for sensitivity; however, the analysis does not provide a design technique. We now discuss design of the virtual coupler introduced in Section II-D. Our approach is to find a parameterization of the virtual coupler design which provides useful terms for tuning the closed-loop distortion. We then generate design directives for optimizing performance, and use such a design in Section V to demonstrate experimentally the sensitivity induced as the distortion is reduced. We note, however, that this design optimized for performance may not necessarily be appropriate. While it highlights one point in the tradeoff between performance and sensitivity, one may choose other points in the tradeoff which provide a different balance of feedback properties.

A. *Performance of the generalized virtual coupler*

As discussed in Section II-D, the haptic interface feedback controller C is typically partitioned into the virtual environment and virtual coupler. The virtual coupler is generally fixed and should accommodate a range of virtual environments. Referring to Fig. 4, the virtual coupler describes the response from the haptic device position X and desired position X_d to the motor command u and the virtual environment force F_e. This set of input/output responses is often modeled after a physical system such as a spring [15]. Let us simply express these input/output relationships by the matrix of transfer functions $B \triangleq \begin{bmatrix} B_{11} & B_{12} \\ B_{21} & B_{22} \end{bmatrix}$ such that

$$\begin{bmatrix} u \\ F_e \end{bmatrix} = B \begin{bmatrix} X \\ X_d \end{bmatrix}. \qquad (15)$$

Figure 5 shows the block diagram of the haptic interface system with B interposed between the virtual environment and haptic device. We note that B is more general than the virtual coupler described in [1] or extended in [15]. Each controller element of B may be any transfer function, and we impose no relationships between elements. We refer to B as a generalized virtual coupler.

To focus on the design of B, we find it useful to remove the virtual environment from the problem and just consider the feedback interconnection of the haptic device P with

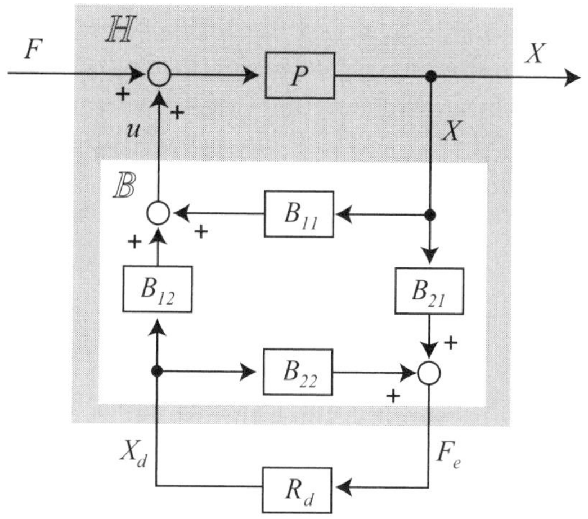

Fig. 5. Block diagram with the controller partitioned into a generalized virtual coupler B and virtual environment R_d. Design directives may be developed for B by considering the multivariable response of H.

the generalized virtual coupler B. Referring to Fig. 5, the input/output response of P in feedback with the B is

$$\begin{bmatrix} X \\ F_e \end{bmatrix} = \begin{bmatrix} \dfrac{P}{1-B_{11}P} & \dfrac{B_{12}P}{1-B_{11}P} \\ \dfrac{B_{21}P}{1-B_{11}P} & B_{22}+\dfrac{B_{21}B_{12}P}{1-B_{11}P} \end{bmatrix} \begin{bmatrix} F \\ X_d \end{bmatrix}. \qquad (16)$$

As we will show, the terms of the four input-output responses of (16) are easily related to closed-loop distortion when R_d is reconnected to the virtual coupler.

Let us denote the matrix of four transfer functions in (16) by H in the form of the hybrid matrix (where our mapping is between force and position rather than force and velocity):

$$\begin{bmatrix} X \\ F_e \end{bmatrix} = \begin{bmatrix} H_{11} & H_{12} \\ H_{21} & H_{22} \end{bmatrix} \begin{bmatrix} F_h \\ X_d \end{bmatrix}. \qquad (17)$$

The multivariable responses of H are indicated by the dark box in Fig. 5. We note that, if $H_{11} \neq 0$, then H uniquely determines the generalized virtual coupler B according to

$$\begin{bmatrix} B_{11} & B_{12} \\ B_{21} & B_{22} \end{bmatrix} = \begin{bmatrix} \dfrac{H_{11}-P}{PH_{11}} & \dfrac{H_{12}}{H_{11}} \\ \dfrac{H_{21}}{H_{11}} & H_{22}-\dfrac{H_{12}H_{21}}{H_{11}} \end{bmatrix}. \qquad (18)$$

We then use H as a re-parameterization of the generalized virtual coupler B.

Before we connect the virtual environment to H, let us remark on the role of the elements of H. Referring to (17), the response from the desired position of the haptic device X_d to the actual position X is described by H_{12}. Clearly, for small error between X and X_d, H_{12} must be close to 1. The virtual environment describes the desired response of X_d to the human operator force F; however, referring to Fig. 5, we see that the virtual environment generates X_d in response to F_e not F. Thus, to generate the correct desired position X_d, the term H_{21}, which describes the response from F to F_e, must be close to 1. The remaining terms H_{11}, which describes the feed-through from F to X, and H_{22}, which describes the feed-through from X_d to F_e, should be attenuated.

Let us now re-introduce the virtual environment in the feedback loop and compute distortion to verify the intuition just developed. Computing the response from F to X in Fig. 5, we find the rendered virtual environment in terms of R_d and the elements of H:

$$R = H_{11} + \frac{H_{12}H_{21}R_d}{1 - H_{22}R_d}. \tag{19}$$

Recall that closed-loop distortion Θ_c in terms of R is $(R - R_d)/R_d$. Substituting (19) into the expression for Θ_c, we find that

$$\Theta_c = \frac{H_{11}}{R_d} + \frac{H_{12}H_{21}}{1 - H_{22}R_d} - 1. \tag{20}$$

If H_{11} and H_{22} are 0, and if H_{12} and H_{21} are 1, then Θ_c is 0 regardless of the virtual environment.

B. Optimizing for performance

In practice, it is not possible to design the generalized virtual coupler to make distortion small for all environments and across all frequencies. Examining (20), we see that a general strategy to reduce distortion is to attenuate both H_{11} and $(H_{12}H_{21})/(1 - H_{22}R_d) - 1$. To attenuate the latter term, we can select H_{22} such that $|H_{22}R_d| << 1$ and $H_{12}H_{21} \approx 1$. Referring to (16), we see that H_{22} can in fact be made identically zero if $B_{22} = B_{21}B_{12}P/(1 - B_{11}P)$. The cancellation, accomplished by summation rather than inversion, is not in practice difficult using proper elements in B. We call generalized virtual couplers with $H_{22} \equiv 0$ *cancellation coupler*. The term H_{11} which describes the feed-through from F to X can be attenuated at low-frequencies, but due to bandwidth limitations, H_{11} must approach the open-loop dynamics P at high-frequencies. Similarly, H_{12} and H_{21} can be selected to have a response near 1 at low frequencies but must roll-off at high-frequencies.

The closed-loop distortion achieved by the cancellation coupler is tuned by the free parameters H_{11}, H_{12}, and H_{21}. Through our choice of parameters, we can guarantee that Θ_c satisfies an upper bound for a range of virtual environments. Examining (20), we can upper bound distortion by

$$|\Theta_c| \leq |H_{11}|/|R_d| + |H_{12}H_{21} - 1|. \tag{21}$$

We design the term $|H_{12}H_{21} - 1|$ and it is independent of the virtual environment. The term $|H_{11}|/|R_d|$ is upper bounded for any virtual environments satisfies a lower bound.

The design directives we have provided reduce closed-loop distortion but at the cost of other feedback goals such as sensitivity. In the next section, we use a cancellation coupler optimized for performance to highlight the costs associated with achieving low distortion; however one may reasonably choose a different virtual coupler design that accepts some performance penalty to improve sensitivity.

V. Experimental Results

We now use a cancellation coupler to render two sample virtual environments experimentally. We tune the feedback design to achieve good performance across as wide a frequency band as possible given our hardware. At frequencies where distortion is low, the Bode sensitivity function must approach the limit R_d/P. We provide experimental verification of the predicted sensitivity to hardware dynamics by varying the haptic device dynamics.

We render two mass-spring-damper virtual environments, both with a natural frequency of 2 Hz and a damping ratio of 0.2, but the gain of the second system is five times the gain of the first. Let R_{d1} be the first virtual environment, given by

$$R_{d1} = \frac{474}{s^2 + 5.03s + 158} \quad \text{(rad/N-m)}. \tag{22}$$

Then let R_{d2} be the second virtual environment, given by

$$R_{d2} = 5R_{d1} \quad \text{(rad/N-m)}. \tag{23}$$

Values for the mass, damping coefficient, and spring rate of R_{d2} are 1/5 of those for R_{d1}.

A. Hardware and Controller Design

We render R_{d1} and R_{d2} on a rotary, single-axis, impedance-type haptic device described in [16]. The haptic device dynamics, as determined experimentally, are

$$P = \frac{1550}{s^2 + 0.775s} \quad \text{(rad/N-m)}. \tag{24}$$

Figure 6a shows the frequency response of R_{d1} and R_{d2} plotted along with the haptic wheel dynamics P. For R_{d1} and R_{d2} rendered on the haptic wheel, we have the tradeoff severity Γ shown in Fig. 6b.

Design parameters of the cancellation coupler H_{11}, H_{12}, and H_{21} are designed according to the design directives given in Section IV. Frequency responses of the parameters are shown in Fig. 7. The magnitudes of H_{11} and $1 - H_{12}H_{21}$ are attenuated subject to practical limitations of our hardware such as sampling, quantization, and high-frequency resonances in the haptic device. Digital implementation of the controller is a Tustin discretization at a sampling frequency of 1 kHz. The resulting cancellation coupler design achieves the closed-loop distortion shown in Fig. 8, and the Bode sensitivity function for each virtual environment is shown in Fig. 9.

B. Results

Experimental step responses of the rendered virtual environments are shown in Fig. 10 with the simulated step responses of R_{d1} and R_{d2} for reference. We evaluate sensitivity of the step-response to a known change in the haptic device dynamics by removing part of the handwheel from our device. System identification shows that this modification reduces the rotational inertia by approximately 30%. Step responses of the rendered virtual environments using the modified hardware are the dashed traces shown in Fig. 10. We have scaled the applied step torque to produce 180° DC response of the virtual environments.

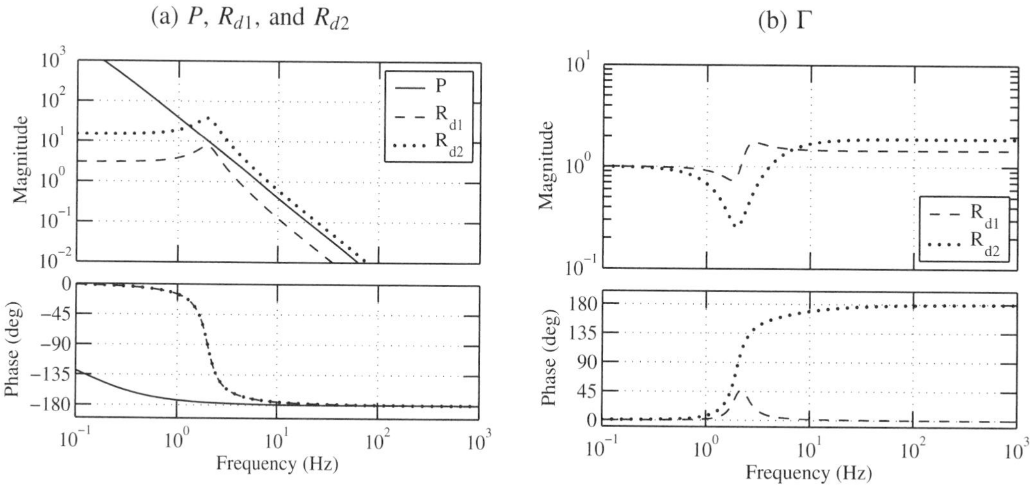

Fig. 6. Predicted tradeoff severity. The frequency responses of R_{d1} and R_{d2} relative to the haptic device P are shown in (a). The resulting tradeoff severities Γ are shown in (b). As indicated by $\Gamma \approx 1$, no tradeoff exists at low frequencies where $|P|$ is much larger than $|R_{d1}|$ or $|R_{d2}|$. The most severe tradeoff between performance and sensitivity occurs for R_{d2} near 2 Hz. At this frequency, Γ approaches 0 as $|R_{d2}|$ significantly exceeds $|P|$. Note that phase also provides important information; where $|\Gamma|$ crosses 1 near 4–5 Hz, the phase plot indicates that the complex value of Γ is not in fact close to 1, and thus a significant tradeoff still exists.

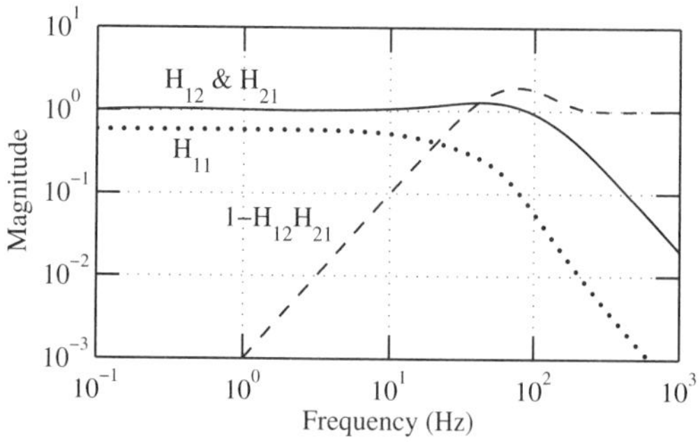

Fig. 7. Generalized virtual coupler design parameters H_{11}, H_{12} and H_{21}. As given by (21), distortion Θ_c is bounded above according to $|\Theta_c| \leq |H_{11}|/|R_d| + |H_{12}H_{21} - 1|$.

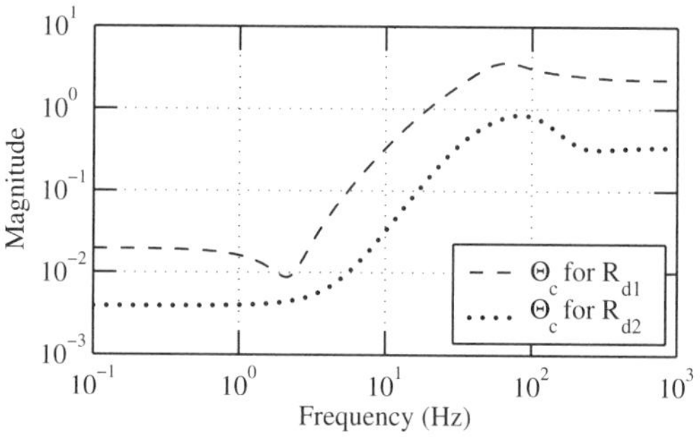

Fig. 8. Predicted closed-loop distortion Θ_c for the cancellation coupler rendering R_{d1} and R_{d2}.

As predicted by Γ shown in Fig. 6b, rendering R_{d2} with low distortion near 2 Hz involves a more severe tradeoff between transparency and sensitivity than does R_{d1}. For both R_{d1} and R_{d2}, the peak in S at 2.1 Hz is a direct consequence of the limit (14). For R_{d1}, the cost of low distortion near 2 Hz is a peak magnitude in S of 0.8; whereas for R_{d2}, the cost of low distortion near 2 Hz induces a peak in the Bode sensitivity function of 3.9.

Experimental step-responses of R_{d1} and R_{d2} as rendered by the cancellation coupler closely match the desired simulated step-responses of R_{d1} and R_{d2}; however, the step-response of R_{d2} on the modified haptic device reveals a much greater sensitivity to the hardware. Thus, using a well-tuned controller and the unperturbed device dynamics, we can achieve low distortion for both R_{d1} and R_{d2}; however larger sensitivity induced to render R_{d2} causes greater error between the actual response and the desired response given the same variation in the device dynamics.

VI. Discussion

In this paper we have revealed a fundamental design tradeoff between low distortion rendering of virtual environments and sensitivity of closed-loop responses to parameter variations in the haptic device and model uncertainty. The limit (14) quantifies the cost in terms of sensitivity to attenuate distortion and cannot be circumvent by feedback design. However, reduction of the inherent dynamics of the haptic device, such as reducing damping and inertia, will reduce the sensitivity required to render a virtual environment.

As a practical matter, the range virtual environments that can be rendered well with a particular haptic device is limited. As the magnitude of the virtual environment dynamics R_d increase relative to the haptic device dynamics P, error in the model of the haptic device P must be reduced to maintain low distortion in the face of high sensitivity. However, the accuracy of available models for a haptic device is typically limited and the hardware dynamics are themselves subject to some variation over time.

The algebraic tradeoff between performance and sensitivity

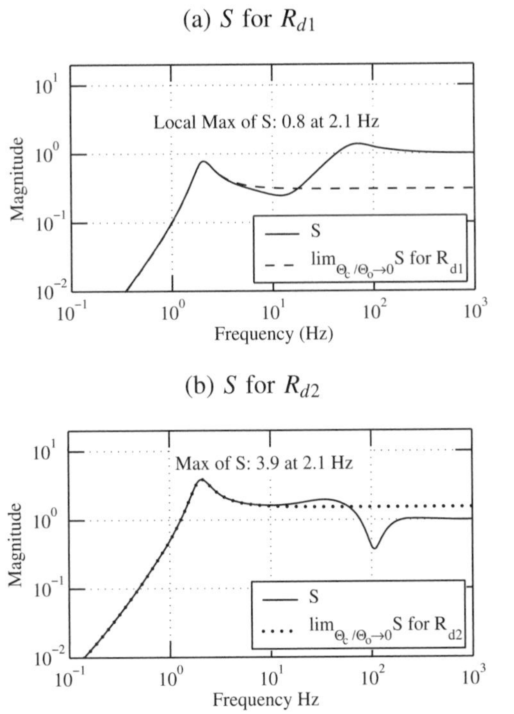

Fig. 9. Predicted Bode sensitivity function S for the cancellation coupler rendering R_{d1} and R_{d2}, and the limit of S as distortion Θ_c approaches 0 for each virtual environment.

EXPERIMENTAL RESULTS

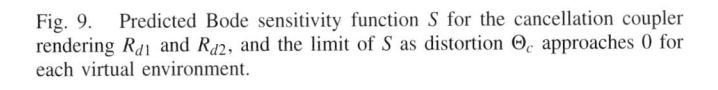

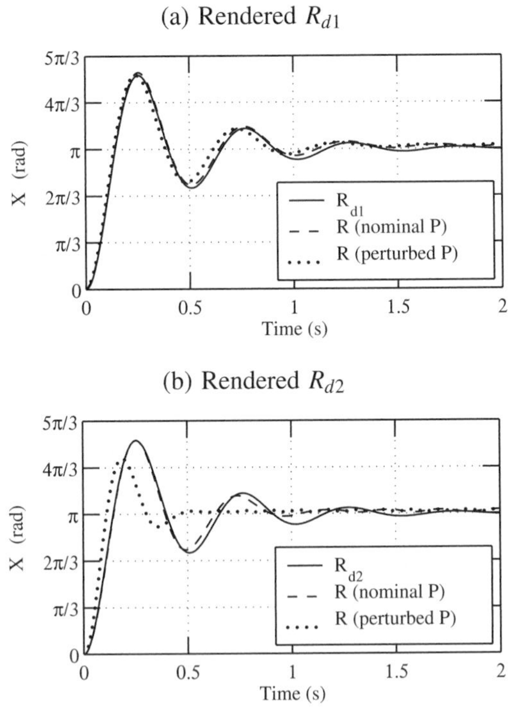

Fig. 10. Experimentally determined and desired step responses of virtual environments R_{d1} and R_{d2}. Dashed traces are step-responses rendered by the cancellation coupler on a haptic device with dynamics given by (24). Dotted traces are step-responses rendered on a modified device with 30% less rotational inertia.

to hardware dynamics assumes a control architecture with position feedback. Additional sensors can mitigate the tradeoffs inherent to the position feedback architecture. A logical choice is the addition of a force sensor measuring the human operator's force F on the haptic device. With this measurement we may compute the desired position X_d by R_dF. Then typical high-gain control techniques, not subject to the algebraic tradeoff between performance and sensitivity, could be applied to make the haptic device position X track X_d.

Acknowledgment

This research was supported by the National Science Foundation under Grant EHS-0410553.

References

[1] J. E. Colgate, M. C. Stanley, and J. M. Brown, "Issues in the haptic display of tool use," in *Proc. IEEE/RSJ Int'l Conf. on Intelligent Robots and Control*, Pittsburgh, PA, 1995, pp. 140–45.

[2] J. E. Colgate and G. G. Schenkel, "Passivity of a class of sampled-data systems: application to haptic interfaces," *J. of Rob. Sys.*, vol. 14, no. 1, pp. 37–47, Jan. 1997.

[3] M. Minsky, M. Ouh-Young, O. Steele, F. P. Brooks, and M. Behensky, "Feeling and seeing: Issues in forces in display," in *Proc. Symp. on Interact. 3D Graphics*, vol. 24, no. 2, Snowbird, UT, 1990, pp. 235–41.

[4] M. Mahvash and V. Hayward, "High-fidelity passive force-reflecting virtual environments," *IEEE Trans. Robot.*, vol. 21, no. 1, pp. 38–46, Feb. 2005.

[5] J. S. Mehling, J. E. Colgate, and M. A. Peshkin, "Increasing the impedance range of haptic display by adding electrical damping," in *1st Joint Eurohaptics Conf. & Symp. on Haptic Interfaces for Virtual Environments and Teleoperator Systems*, Pisa, Italy, 2005, pp. 257–62.

[6] N. Diolaiti, G. Niemeyer, F. Barbagli, and J. K. J. Salisbury, "Stability of haptic rendering: Discretization, quantization, time delay, and Coulomb effects," *IEEE Trans. Robot.*, vol. 22, no. 2, pp. 256–68, Apr. 2006.

[7] X. Shen and M. Goldfarb, "On the enhanced passivity of pneumatically actuated impedance-type haptic interfaces," *IEEE Trans. Robot.*, vol. 22, no. 3, pp. 470–80, June 2006.

[8] M. C. Çavuşoğlu, D. Feygin, and F. Tendick, "A critical study of the mechanical and electrical properties of the PHANToM haptic interface and improvements for high-performance control," *Presence*, vol. 11, no. 6, pp. 555–68, Dec. 2002.

[9] J. J. Abbott and A. M. Okamura, "Effects of position quantization and sampling rate on virtual-wall passivity," *IEEE Trans. Robot.*, vol. 21, no. 5, pp. 952–64, Oct. 2005.

[10] B. E. Miller, J. E. Colgate, and R. A. Freeman, "Guarenteed stability of haptic systems with nonlinear virtual environments," *IEEE Trans. on Robot. & Automat.*, vol. 16, no. 6, pp. 712–19, Dec. 2000.

[11] M. Fardad and B. Bamieh, "A frequency domain analysis and synthesis of the passivity of sampled-data systems," in *Proc. 43rd IEEE Conf. on Decision and Contr.*, vol. 3, Nassau, Bahamas, 2004, pp. 2358–63.

[12] S. Skogestad and I. Postlethwaite, *Multivariable Feedback Control: Analysis and Design*. New York: Wiley, 1997.

[13] J. S. Freudenberg, C. V. Hollot, R. H. Middleton, and V. Toochinda, "Fundamental design limitations of the general control configuration," *IEEE Tran. Autom. Contr.*, vol. 48, no. 8, pp. 1355–70, Aug. 2003.

[14] D. A. Lawrence, "Stability and transparency in bilateral teleoperation," *IEEE Trans. on Robot. & Automat.*, vol. 9, no. 5, pp. 624–37, Oct. 1993.

[15] R. J. Adams and B. Hannaford, "Stable haptic interaction with virtual environments," *IEEE Trans. on Robot. & Automat.*, vol. 15, no. 3, pp. 465–74, June 1999.

[16] R. B. Gillespie, M. B. Hoffman, and J. Freudenberg, "Haptic interface for hands-on instruction in system dynamics and embedded control," in *Proc. Symp. on Haptic Interfaces for Virtual Environment and Teleoperator Systems*, Los Angeles, CA, 2003, pp. 410–15.

Motion Strategies for Surveillance

Sourabh Bhattacharya Salvatore Candido Seth Hutchinson
Department of Electrical and Computer Engineering
University of Illinois at Urbana Champaign
Urbana, Illinois
Email: {sbhattac, candido, seth}@uiuc.edu

Abstract— **We address the problem of surveillance in an environment with obstacles. We show that the problem of tracking an evader with one pursuer around one corner is completely decidable. The pursuer and the evader have complete information about each other's instantaneous position. The pursuer has complete information about the instantaneous velocity of the evader. We present a partition of the visibility region of the pursuer where based on the region in which the evader lies, we provide strategies for the evader to escape the visibility region of the pursuer or for the pursuer to track the target for all future time. We also present the solution to the inverse problem: given the position of the evader, the positions of the pursuer for which the evader can escape the visibility region of the target. These results have been provided for varying speeds of the pursuer and the evader. Based on the results of the inverse problem we provide an $O(n^3 \log n)$ algorithm that can decide if the evader can escape from the visibility region of a pursuer for some initial pursuer and evader positions. Finally, we extend the result of the target tracking problem around a corner in two dimensions to an edge in three dimensions.**

I. Introduction

Surveillance is related to target tracking and the game of pursuit-evasion. The goal of the pursuer is to maintain a line of sight to the evader that is not occluded by any obstacle. The goal of the evader is to escape the visibility polygon of the pursuer (and break this line of sight) at any instant of time.

This problem has several interesting applications. It may be useful for a security robot to track a malicious evader that is trying to escape. The robot must maintain visibility to ensure the evader will not slip away while another party or the pursuer itself attempts to eventually trap or intercept the evader. Also, an "evader" may not be intentionally trying to slip out of view. A pursuer robot may simply be asked to continuously follow and monitor at a distance an evader performing a task not necessarily related to the target tracking game. The pursuer may somehow be supporting the evader or relaying signals to and from the evader. The pursuer may also be monitoring the evader for quality control, verifying the evader does not perform some undesired behavior, or ensuring that the evader is not in distress. Finally, the results are useful as an analysis of when escape is possible. If it is impossible to slip away, it may be desirable for the evader to immediately surrender or undertake a strategy not involving escape.

A great deal of research exists on pursuit-evasion. Pursuit-evasion games are analyzed in $\mathbb{R}^n$ [19], in non-convex domains of arbitrary dimension [1], and in graphs [34], [33]. A large volume of game theoretic formulations and analysis can be found in [2], [15], [35] and [21]. Also, [9] presents an approach that takes into account the pursuer's positioning uncertainty. While this analysis is pertinent, it often focuses on "capturing", moving within a certain distance, of the evader. We only seek to maintain a line of sight to the evader.

A related but different problem in the robotics community is to find an evader with one or more pursuers in various environments. Exact [12], [23], [24], [25], [38], [40], [11] and probabilistic [13], [41] strategies have been found to locate an evader while preventing it from slipping into a region that has already been searched. Randomized strategies have been employed to locate and capture an unpredictable evader in any simply connected polygon [17] and [16]. [39] deals with one pursuer searching one evader with k flashlights in a general environment. The paper presents necessary and sufficient conditions for various searchers. Our problem assumes the evader starts in a position visible to the pursuer. The pursuer's goal is to track the evader rather than search the environment for a hidden evader.

The problem of maintaining visibility of a moving evader has been traditionally addressed with a combination of vision and control techniques [8], [26], [27], and [14]. Pure control approaches are local by nature and do not take into account the global structure of the environment. Our interest is in deriving pursuer strategies that guarantee successful surveillance taking into account both constraints on motion due to obstacles and constraints on visibility due to occlusion.

The problem of target tracking has also been analyzed at a fixed distance between the pursuer and evader. In [28] and [29], optimal motion strategies are proposed for a pursuer and evader based on critical events and in [30] a target tracking problem is analyzed with delay in sensing. These papers are summarized in [32]. [3] deals with the problem of *stealth target tracking* where a robot equipped with visual sensors tries to track a moving target among obstacles and, at the same time, remain hidden from the target. Obstacles impede both the tracker's motion and visibility, and also provide hiding places for the tracker. A tracking algorithm is proposed that applies a local greedy strategy and uses only local information from the tracker's visual sensors and assumes no prior knowledge of target tracking motion or a global map of the environment.

[20] presents a method of tracking several evaders with multiple pursuers in an uncluttered environment. In [18] the problem of tracking multiple targets is addressed using a network of communicating robots and stationary sensors. A

region-based approach is introduced which controls robot deployment at two levels, namely, a coarse deployment controller and a target-following controller.

In [22] the problem of computing robot motion strategies that maintain visibility of a moving target is studied under deterministic and stochastic settings. For the deterministic target, an algorithm that computes optimal, numerical solutions is presented. For the stochastic case, two online algorithms are presented that each attempt to maintain future visibility with limited perception. In [31] one or more observers try to track one or more targets by maximizing the shortest distance the target needs to move in order to escape the observer's visibility region. In [10][4] a target tracking problem is analyzed for an unpredictable target and an observer lacking prior model of the environment. It computes a risk factor based on the current target position and generates a feedback control law to minimize it. [7] presents an algorithm that maximizes the evader's minimum time to escape for an evader moving along a known path.

There have been successful efforts in the past to deploy tracking and surveillance systems in real world. We only present a few of them. [37] and [36] developed a distributed heterogeneous robotic team that is based mainly on a miniature robotic system. Most of the robots are extremely small because some operations require covert action. They team the scouts with larger ranger robots that can transport the scouts over distances of several kilometers, deploy them rapidly over a large area, coordinate their behavior, and collect and present the resulting data. [5] presents a mobile robot called the *Intelligent Observer* which moves through an environment while autonomously observing moving targets selected by a human operator. The robot carries one or more cameras which allow it track objects while at the same time sensing its own location.

We address the problem of target tracking in an environment with one corner, one pursuer, and one evader. This is the first result, to our knowledge, that the surveillance problem is decidable around one corner and gives partitions of the workspace that demonstrate the outcome of the game. While the general problem of deciding whether the evader can escape or the pursuer can track the evader forever in any arbitrary polygonal environment is still, so far as we know, an open problem, we offer partial solutions to two important problems. First, we provide sufficient conditions for escape. These conditions could be used to solve the evader's problem (i.e., he would construct an escape strategy that exploited these conditions) when they are satisfied, thus providing a partial solution to this pursuit-evasion problem. Second, our analysis is in the direction of providing open loop policies for the pursuer to track the evader. Closed loop policies for the pursuer depend on the current state of the evader resulting in a delay in the reaction of the pursuer due to the time required to process sensor data. The model changes from a continuous time system to a discrete time system that in general lead to computationally intractable algorithms. Moreover due to the delay introduced in processing, the sufficient conditions weaken.

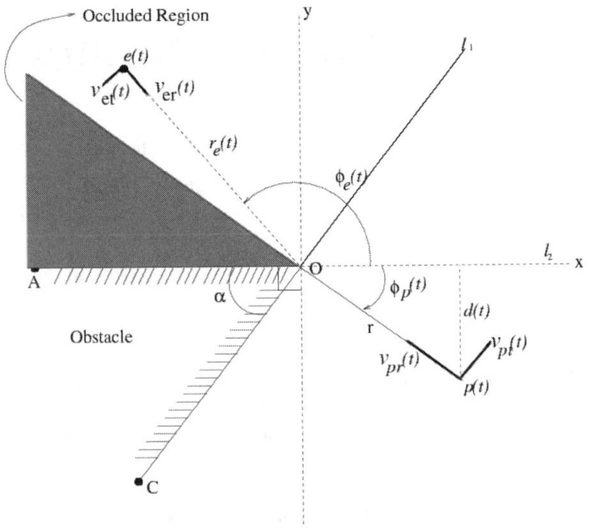

Fig. 1. The problem environment

In Section 2, based on the geometry of the corner, the ratio of maximum pursuer and evader velocities, and the initial position of the pursuer, we segment the free space into regions. The pursuer policy and the possibility of evader escape are determined by the region in which the initial evader position is contained. Pursuer policies are given that guarantee, for some regions of initial conditions, that no evader policy will lead to escape of the evader at any future time. It is then proved that outside these regions, there is an evader policy by which escape is inevitable irregardless of the pursuer policy. In Section 3, the same analysis is performed with respect to the initial position of the evader. In Section 4, we use the results of the previous section to construct a algorithm that can decide whether the evader can escape the pursuer for certain initial positions of the pursuer and the evader. In Section 5, we extend the above results to target tracking in $\mathbb{R}^3$ around an edge.

II. Pursuer-Based Partition

A mobile pursuer and evader exist on a plane at points $p(t)$ and $e(t)$, respectively. They are point robots and move with bounded speeds, $v_p(t)$ and $v_e(t)$. Therefore, $v_p(t) : [0, \infty) \rightarrow [0, \overline{v}_p]$ and $v_e(t) : [0, \infty) \rightarrow [0, \overline{v}_e]$. We assume that $v_p(t)$ and $v_e(t)$ can be discontinuous functions of time.

The workspace contains a semi-infinite obstacle with one corner that restricts pursuer and evader motions and may occlude the pursuer's line of sight to the evader. Without loss of generality, this corner is placed at the origin and one of the sides lies along the -x axis as shown in Figure 1. The unshaded region is the visibility region of the pursuer. $v_{pt}(t)$ and $v_{et}(t)$ describe the pursuer and evader tangential velocities. $v_{pr}(t)$ and $v_{er}(t)$ describe the radial velocities. $\phi_e(t)$ and $\phi_p(t)$ are the angles the evader and pursuer, respectively, make with the +x axis. Note that $\phi_e(t)$ is positive in the counterclockwise direction while $\phi_p(t)$ is positive in the clockwise direction. The minimum distance of the pursuer from line l_2 is denoted

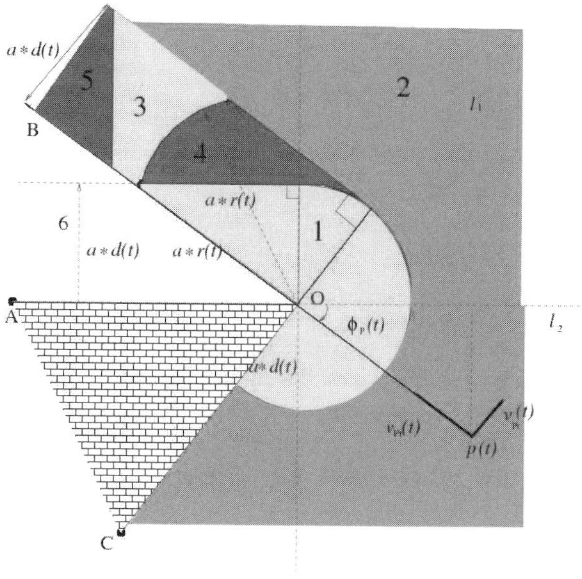

Fig. 2. Pursuer-based partition

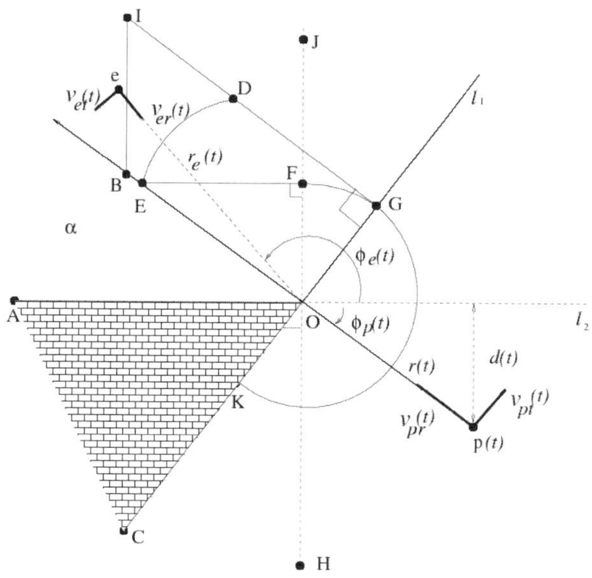

Fig. 3. The geometry of the partition

by $d(t)$. The distance of the pursuer from the corner is denoted by $r(t)$. The distance of the evader from the corner is denoted by $r_e(t)$.

The two edges meeting at this corner are considered to extend for an infinite length so there is no other geometry that the evader can hide behind in the workspace. The two sides of the obstacle form an angle α. If $\alpha \geq \pi$ then every point in the free workspace is visible to every other point and the pursuer will trivially be able to track the evader indefinitely. Thus, we only consider obstacles where $\pi > \alpha \geq 0$.

To prevent the evader from escaping, the pursuer must keep the evader in its visibility polygon, $V(p(t))$. The visibility polygon of the pursuer is the set of points from which a line segment from the pursuer to that point does not intersect the obstacle region. The evader escapes if at *any* instant of time it can break the line of sight to the pursuer. Visibility extends uniformly in all directions and is only terminated by workspace obstacles (omnidirectional, unbounded visibility).

We define the *star region* associated with a corner as the region bounded by the supporting lines of the two edges of a corner. We define the *star region* as the collection of all star points. As can be seen in Figure 1, the star-region extends outward from the corner of the obstacle. It is semi-infinite and bounded by rays l_1 and l_2. From any point in the star-region, the entire free space is visible. If the pursuer can enter the star region before losing sight of the evader, it will trivially be able to track the evader at all future times.

We want to address the following question. Given the initial position of the pursuer and the evader, the map of the environment, $\overline{v}_e$ and $\overline{v}_p$:

1) Does there exist a policy or an algorithm that takes finite steps to provide a policy for the pursuer to track the evader for all future times?
2) Does there exist a policy or an algorithm that takes finite steps to provide a policy for the evader to escape the visibility region in finite time?

We refer to the above questions by the term *decidability*. If the answer to one of the questions is affirmative at every configuration for an environment, we say the problem is *decidable* in that environment. Around one corner the surveillance problem is decidable.

Let $a = \overline{v}_e/\overline{v}_p$, the ratio of maximum evader and pursuer speeds, and define $d = d(t = 0)$. The outcome of the game can be decided based on the velocity ratio and initial positions of the evader and pursuer. This leads to a decomposition of the visibility region of the pursuer into regions in which the evader may lay. Region 1 is the set of all points closer than $a \cdot d(t)$ to segment AO, the far side of the obstacle. Region 2 is the set of points farther away than $a \cdot d(t)$ to segment OB, the edge of $V(p(t))$ laying in free space. Region 3 consists of points laying within distance $a \cdot d(t)$ to segment OB and farther than $a \cdot r(t)$ from point O, the corner. Region 4 is the set of points within distance $a \cdot d(t)$ from segment OB, closer than $a \cdot r(t)$ from point O and farther than $a \cdot d(t)$ from segment AO. Region 5 is the set of points within distance $a \cdot d(t)$ from segment OB and at a distance greater than $\frac{a * r(t)}{\cos(\theta)}$ from the origin. Region 6 is the portion of the free workspace

TABLE I

Evader Policies	Evader Region	Control Law
A	1 and $\phi_e \in [\alpha - \pi, \frac{\pi}{2}]$ 1 and $\phi_e \in [\frac{\pi}{2}, \pi - \phi_p]$	$\dot{r}_e(t) = \overline{v}_e$ $\dot{y}_e(t) = -\overline{v}_e$
Pursuer Policies	**Evader Region**	**Control Law**
B	2, 4	$\dot{y}_p(t) = \overline{v}_p$
C	3	$v_{pt}(t) = -\frac{r}{r_e}\,\lvert v_{et}(t)\rvert$ $v_{pr}(t) = -\frac{r}{r_e}\,\lvert v_{er}(t)\rvert$
D	5	$v_{pt}(t) = v_p$

not belonging to $V(p(t))$. The pursuer and evader policies necessary to decide the problem can be determined by this partition of $V(p(t))$ shown in Figure 2. These policies are summarized in Table I.

For the remainder of this section, refer to Figures 2 and 3 and Table I. Consider the case where $\alpha < \frac{\pi}{2}$ and $\phi_p \in \left(0, \frac{\pi}{2}\right]$.

Proposition 1 If the evader lies in Region 1 and follows Policy A, no pursuer policy exists that can prevent the escape of the evader.
Proof: If the evader lies in Region 1, the maximum time required by the evader to reach line AO by following Policy A is $t_e < \frac{a \cdot d}{\overline{v}_e} = \frac{d}{\overline{v}_p}$. The minimum time required by the pursuer to reach line l_2 with any policy is at least $t_p > \frac{d}{\overline{v}_p}$. Thus, $t_p > t_e$. Therefore the evader reaches the line AO before the pursuer can reach line l_2. If the evader lies on AO and the pursuer has not yet reached l_2 the evader will be outside the visibility region of the pursuer. Hence the evader escapes. ■

Proposition 2 If the evader lies in Region 2 and the pursuer follows Policy B, no evader policy exists that can escape the visibility region of the pursuer.
Proof: The time required by the pursuer to reach line l_2 by following Policy B is $t_p = \frac{d}{\overline{v}_p}$. If the evader lies in Region 2, the minimum time required by the evader to reach ray OB is $t_e > \frac{a \cdot d}{\overline{v}_e} = \frac{d}{\overline{v}_p}$. Thus, $t_e \geq t_p$. If the pursuer follows Policy B, $V(p(t = 0)) \subseteq V(p(t > 0))$. Since the evader cannot reach ray OB, the only free boundary of $V(p(t = 0))$, before the pursuer reaches the boundary of the star region, $e(t) \in V(p(t)) \forall t \in [0, t_p]$. Once the pursuer reaches the line l_2, the entire free workspace belongs to the $V(p(t_p))$. The pursuer stops hence the evader remains in sight of the pursuer for all future times. ■

Proposition 3 If the evader lies in Region 3 and the pursuer follows Policy C, for every evader policy the evader can either stay in Region 3 or move to region 2 of V$(p(t))$.
Proof: If the pursuer follows Policy C, then it follows both the radial and angular movements of the evader. The geometry of Region 3 is such that $r_e(t) \geq a \cdot r(t)$ so $r(t)/r_e(t) \leq 1/a$. Multiply that with the velocity bound of the evader, $v_e(t) \leq \overline{v}_e$. This quantity is equal to the pursuer velocity of the control law of Policy C.

$$v_p(t) \quad = \quad v_e(t)\frac{r(t)}{r_e(t)} \leq \frac{\overline{v}_e}{a} = \overline{v}_p$$

Thus, the pursuer velocities of Policy C are always attainable in Region 3. In order for the pursuer to maintain sight of the evader, the following equation must hold.

$$\phi_e(t) + \phi_p(t) \quad \leq \quad \pi$$

The tangential component of the control law implies

$$\begin{aligned} \dot{\phi}_e(t) &\leq -\dot{\phi}_p(t) \\ \Rightarrow \dot{\phi}_e(t) + \dot{\phi}_p(t) &\leq 0 \\ \Rightarrow \phi_e(t > 0) + \phi_p(t > 0) &\leq \phi_e(t = 0) + \phi_p(t = 0) \leq \pi \,. \end{aligned}$$

Thus, the evader cannot enter Region 5. The radial component of the control law implies

$$\begin{aligned} \frac{\dot{r}_e(t)}{r_e(t)} &= \frac{\dot{r}(t)}{r(t)} \\ \Rightarrow \frac{r_e(t)}{r(t)} &= \frac{r_e(0)}{r(0)} \geq a \end{aligned}$$

Thus, the evader cannot enter Region 4. Hence for any policy the evader can either stay in Region 3 or only enter Region 2. ■

Proposition 4 If the evader lies in Region 4 and the pursuer follows Policy B, for every evader policy the evader can either stay in Region 4 or move to regions 2 or 3 of V$(p(t))$.
Proof: If the pursuer follows Policy B, all points on segment EF move with velocity $a \cdot \overline{v}_p = \overline{v}_e$ toward the ray OA. Similarly, all points on the arc FG move with radial velocity $\overline{v}_e$ toward O. In order to enter Region 1 from Region 4, the evader must move toward the boundary of Region 1 with a velocity greater than the velocity at which the boundary is moving away from the evader. That is not possible since the boundary of Region 1 moves with velocity $\overline{v}_e$, the maximum possible evader velocity, away from the evader. Hence the evader cannot enter Region 1 from Region 4. Hence for all evader policies, the evader can only reach Region 3 or Region 2 from Region 4. ■

Proposition 5 For all initial positions of the evader in Regions 3 and 4, the pursuer can track of the evader by following a reactive motion and switching between policies B and C appropriately.
Proof: If the evader starts in Region 3 and remains in Region 3 then we have proved in Proposition 3 that Policy C for the pursuer can keep the evader in sight for all future time. While the pursuer is following policy C, if the evader enters Region 2, by Proposition 2, the pursuer can track the evader indefinitely by following Policy B. Hence the pursuer can keep sight of the evader for all future time.

If the evader starts in Region 4 and the evader remains in Region 4 then Proposition 4 proves that Policy B for the pursuer can keep the evader in sight for all future time. While the pursuer is following policy B, if the evader moves to Region 3, the strategy provided in the previous paragraph can keep the evader in sight for all future times. While the pursuer is following policy B, if the evader moves to Region 2, by Proposition 2, the pursuer can indefinitely track the evader by following Policy B. Thus, the pursuer will keep the evader in sight for all future time. ■

Proposition 6 For all initial positions of the evader in region 5, the pursuer can track the evader by following policy D.
Proof: Refer to Figure 4. After time t, the evader lies in the closure of a circle of radius $\overline{v}_e t$ centered at $e(0)$. A sufficient condition for the pursuer to keep the evader in sight for all future times is to keep the angular velocity of the line of the sight, OP, to be greater than the angular velocity of the line tangent to the growing circle, OL, for all future time until the

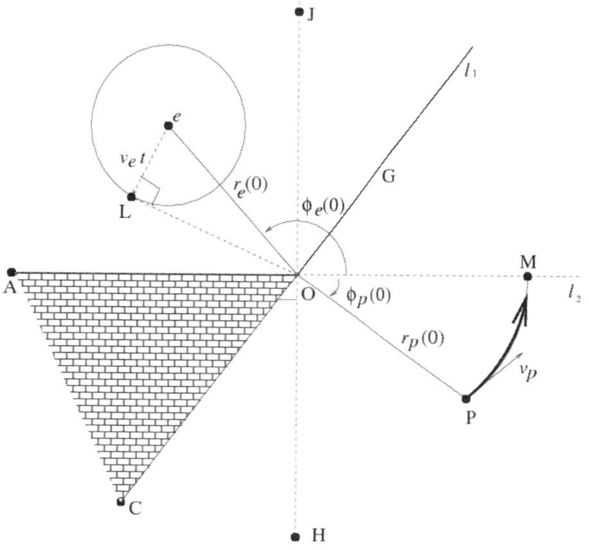

Fig. 4. Evader in region 6

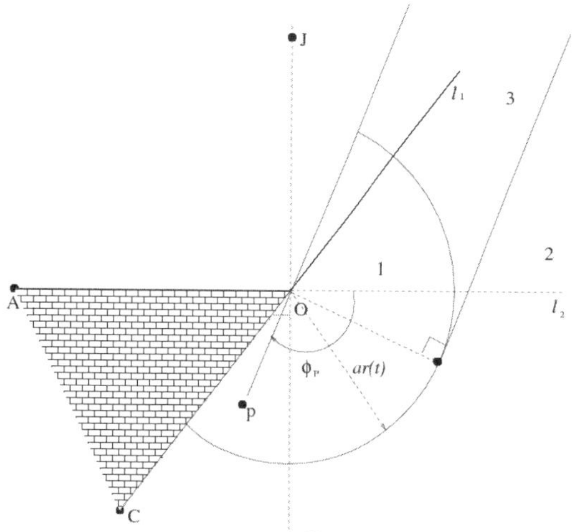

Fig. 5. Pursuer-based partition for the pursuer in region COH

pursuer reaches the *star region*. The angular velocity of the line OP is given by $\omega_p = \frac{v_p}{r_p}$. The maximum angular velocity of the evader is given by $\omega_{emax} = \frac{-v_e}{r_e(0)\cos(\phi_e(0))}$. Solving for $\omega_p \geq \omega_{emax}$ leads to the following condition

$$r_e(0) \geq -\frac{a * r_p}{\cos(\phi_e(0))}$$

which is satisfied for all points in Region 5. ∎

If $\phi_p > \frac{\pi}{2}$ the analysis still holds. The only changes are that Region 1 expands, the area of Region 4 is reduced to zero and Region 5 ceases to exist. Figure 5 shows the partition of the visibility region of the pursuer in this case. Note that if $\alpha \in \left[\frac{\pi}{2}, \pi\right]$, then ϕ_p must be less than $\frac{\pi}{2}$ and this case is not a consideration.

Corollary 1 There exists an evader policy that no pursuer policy can track if and only if the evader lies in Region 1.

Proof: The proof can be concluded from the proofs of Propositions 1, 2 and 5. ∎

All the above analysis was done for initial positions of the pursuer outside of the star region. If the initial position of the pursuer is in the star region the entire free space is visible to the pursuer. The policy of the pursuer will be to remain stationary and it will trivially be able to track the evader indefinitely.

III. Evader-Based Partition

In the previous section, a partition of the $V(p(t))$ has been given. From only one region a evader policy exists that allows the evader to escape for any pursuer policy. This section presents a partition of the visibility region of the evader, $V(e(t))$. The regions of the partition determine whether there will be an evader policy that guarantees escape. In short, we address the following question - Given the initial position of the evader, from what pursuer positions will the evader be able to escape as a function of the ratio of their speeds?

To find the set of initial pursuer placements from which the evader may escape we must consider two cases depending on whether the closest point to the evader on the obstacle lies on the corner or belongs uniquely to one of the sides.

Refer to Figure 6(a) where the partition for the case where the closest point is on one of the sides of the obstacle is considered. For this situation to occur the evader must lie outside the shaded region of Figure 6(a). Consider, without loss of generality, the case where the evader lies in quadrant II. It can be concluded from Corollary 1 that if the pursuer lies below l_2, the pursuer must be a distance of at least d_e/a from l_2 for the evader to escape. If the pursuer lies in the region between l_1 and OA, it is possible for the evader to escape to the side of the obstacle opposite the evader, OC. Thus, the distance between the pursuer and the closest point on l_1 must be greater than r_e/a for the evader to escape. If the pursuer lies in the region enclosed by l_1 and l_2, the evader cannot escape as the pursuer already lies in the star region. In summary, an evader policy that guarantees escape exists only if the pursuer lies outside of the shaded region of Figure 6(b).

For all points inside the shaded region in figure 6(a), the closest point is the corner. If the closest point to the evader on the obstacle is the corner, to escape to a hidden region on either side of the obstacle as quickly as possible the evader must reach the corner. The evader can escape by following this policy if the pursuer lies in the unshaded region in Figure 6(c), points laying farther than r_e/a from the star region. If the pursuer lies in the star region, no escape is possible.

Using the above idea we present an algorithm in the next section that can decide if the evader can escape for certain initial positions of the pursuer and the evader.

IV. Decidable Regions in Case of a Polygonal Environment

In the previous section, we provided a partition of $V(e(t))$ to decide the outcome of the target tracking game. We can conclude that if the pursuer lies outside the shaded region in Figures 6(b) and (c), a strategy exists where evader will

Fig. 6. Evader-based partitions

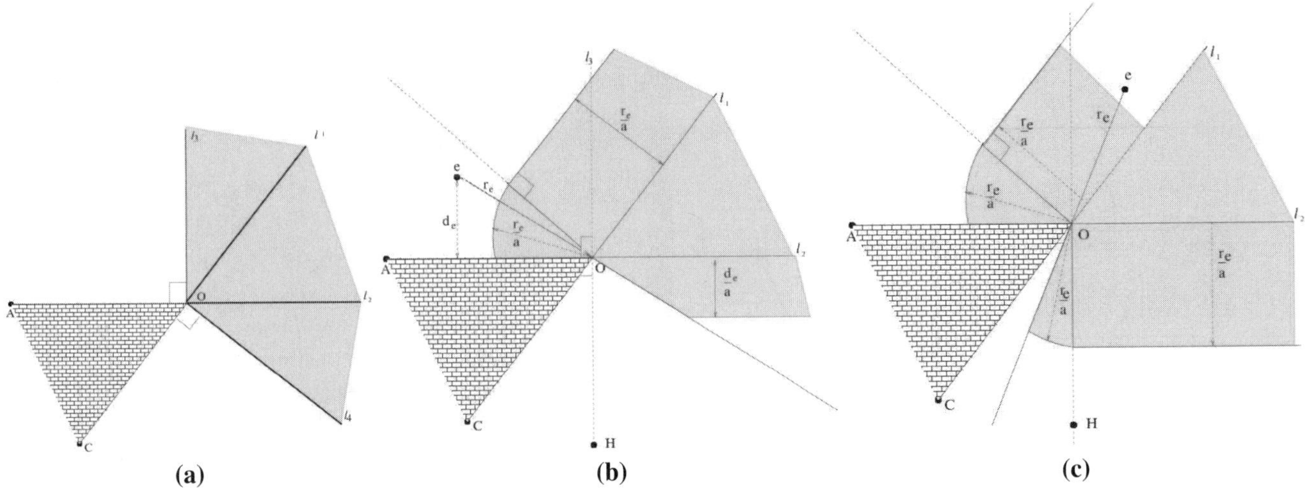

win irrespective of the pursuer policies. The presence of other obstacles does not affect this result.

If the pursuer lies in the shaded region, a strategy, guaranteed to track the evader for all future times and a strategy to do that has been proposed. At this moment, we do not have an extension of this strategy in the presence of multiple obstacles. Hence, we cannot decide the result of the target tracking game if the pursuer starts in the shaded region.

Using the ideas outlined so far in this paper, we propose an algorithm that can decide sufficient conditions for escape of an evader in a general polygonal environment.

Refer to Figures 6(b) and (c). From the previous section, we can conclude that, given a corner, if the pursuer lies outside the region enclosed by rays l_1 and l_2 and the minimum time required by the pursuer to reach rays l_1 or l_2 is greater than the minimum time required by the evader to reach the corner then the evader wins the game. This statement is true even in the presence of other obstacles. To check this condition, compute the shortest distance of the pursuer to the rays l_1 and l_2 and the shortest distance of the evader to the corner. Repeat the process for every corner in the environment. If the condition is satisfied for any corner, the Decidability Algorithm concludes that the evader can escape and the strategy for the evader to escape is to reach that corner along the shortest path. If the condition is not satisfied by any corner, the Decidability Algorithm does not know the result of the target tracking game. The psuedocode of this algorithm is provided in the appendix.

The shortest distance of the evader from the corner can be found by applying Dijkstra's Algorithm to the Visibility Graph of the environment between the initial position of the evader and the corner. To find the shortest distance of the pursuer to rays l_1 and l_2 we construct a *Modified Visibility Graph* (MVG) for a given vertex, v and apply Dijkstra's Algorithm. We present the main steps for the construction of the MVG for a given vertex v.

1) Construct the visibility graph of the environment with edge weights as the Euclidean distance between the two vertices.
2) For every vertex w, check if perpendicular line segments can be drawn from w to rays l_1 and l_2(corresponding to v), without intersecting other obstacles. If only one perpendicular line segment can be drawn to either l_1 or l_2, then compute the length of the perpendicular line segment. If a perpendicular line segment can be drawn to both l_1 and l_2, then compute the length of the smaller perpendicular line segment. Denote the length by d. If the edge wv already exists in the visibility graph, update the weight of the edge to d. Otherwise add an edge with weight d.
3) If for an vertex w, there is no perpendicular line segment to rays l_1 and l_2, do nothing.

The psuedocode for constructing the MVG for a vertex is given in the appendix. The algorithm DECIDABILITYTEST has time complexity of $O(n^3 \log n)$, where n is the number of vertices in the polygonal environment.

V. Extension to One Edge in $\mathbb{R}^3$

Consider an edge in $\mathbb{R}^3$ formed by the intersection of two half planes at an angle of α. Consider a plane π_p perpendicular to both the half planes and passing through the pursuer and a plane π_e perpendicular to both the half planes passing through the evader. Let e_p be the projection of the evader on π_p and p_e the projection of the pursuer on π_e.

Proposition 6 The line $\overline{pe}$ intersects the obstacle iff projected lines $\overline{ep_e}$ and $\overline{pe_p}$ intersect the obstacle.

Proof: Consider any point $(x*, y*, z*)$ on the line $\overline{pe}$. Its projection on π_p, $(x*, y*, z^p)$, lies on $\overline{pe_p}$ and its projection on π_e, $(x*, y*, z^e)$, lies on $\overline{ep_e}$. Since all three points are collinear on a line parallel to the z-axis and the obstacle's shape is invariant along the z-axis if any of the three points intersects the obstacle then the other two will as well. If the visibility line $\overline{pe}$ is broken, then the straight line connecting p to e intersects

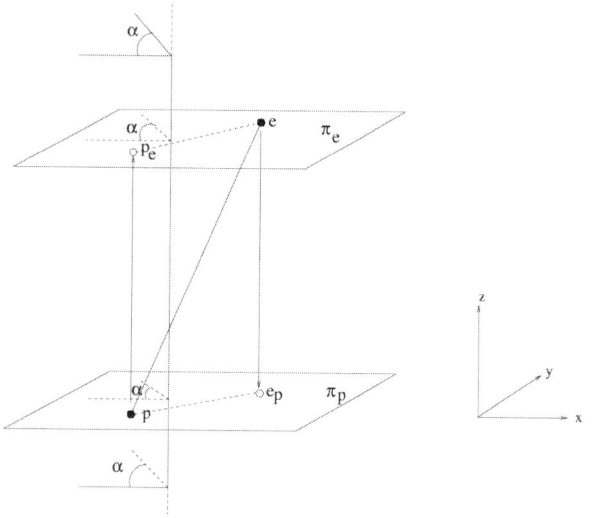

Fig. 7. An edge in three dimensions.

the obstacle at least at one point and the lines $\overline{ep_e}$ and $\overline{pe_p}$ will be broken as well. Otherwise, no point on $\overline{pe}$ is broken and so no point on $\overline{ep_e}$ and $\overline{pe_p}$ will be broken. ■

Proposition 5 shows that, in this particular geometry, the problem in $\mathbb{R}^3$ can be reduced to to a problem in $\mathbb{R}^2$. Given the maximum velocities of the pursuer and evader, the decomposition of the visible space is simply the extrusion of the planar decomposition of the visibility region along the z axis.

VI. Conclusions

This paper addresses the problem of surveillance in an environment with obstacles. We show that the problem of tracking an evader with one pursuer around one corner is completely decidable. We present a partition of the visibility region of the pursuer where based on the region in which the evader lies, we provide strategies for the evader to escape the visibility region of the pursuer or for the pursuer to track the target for all future time. We also present the solution to the inverse problem: given the position of the evader, the positions of the pursuer for which the evader can escape the visibility region of the target. These results have been provided for varying speeds of the pursuer and the evader. Based on the results of the inverse problem we provide an $O(n^3 \log n)$ algorithm that can decide if the evader can escape from the visibility region of a pursuer for some initial pursuer and evader positions. Finally, we extend the result of the target tracking problem around a corner in two dimensions to an edge in three dimensions. We have shown that the target tracking game in $\mathbb{R}^3$ can be reduced to a target tracking game in $\mathbb{R}^2$

In our future work, we plan to address the decidability of the target tracking problem in general polygonal environment. We are using game theory as a tool to increase the decidable regions for the problem. We also plan to use tools in topology and computational geometry to understand the problem to a greater depth. Finally, we would like to extend the problem to multiple pursuers and multiple evaders and use control theory to provide strategies for successful tracking.

References

[1] S. Alexander, R. Bishop, and R. Ghrist. Pursuit and evasion in non-convex domains of arbitrary dimensions. In *Proceedings of Robotics: Science and Systems*, Philadelphia, USA, August 2006.

[2] T. Başar and G. J. Olsder. *Dynamic Noncooperative Game Theory, 2nd Ed.* Academic, London, 1995.

[3] T. Bandyopadhyay, Y. Li, M.H. Ang Jr, and D. Hsu. Stealth Tracking of an Unpredictable Target among Obstacles. *Proceedings of the International Workshop on the Algorithmic Foundations of Robotics*, 2004.

[4] T. Bandyopadhyay, Y. Li, M.H. Ang Jr, and D. Hsu. A greedy strategy for tracking a locally predictable target among obstacles. *Proceedings of the International Workshop on the Algorithmic Foundations of Robotics*, 2006.

[5] C. Becker, H. Gonzalez-Banos, J.C. Latombe, and C. Tomasi. An intelligent observer. *Proceedings of International Symposium on Experimental Robotics*, pages 94–99, 1995.

[6] H. Choset, K.M. Lynch, S. Hutchinson, G. Kantor, W. Burgard, L. Kavraki, and S. Thrun. *Principles of robot motion: theory, algorithms, and implementations (Intelligent robotics and autonomous agents)*. The MIT Press, Cambridge, MA, 2005.

[7] A. Efrat, HH Gonzalez-Banos, SG Kobourov, and L. Palaniappan. Optimal strategies to track and capture a predictable target. *Robotics and Automation, 2003. Proceedings. ICRA'03. IEEE International Conference on*, 3, 2003.

[8] B. Espiau, F. Chaumette, and P. Rives. A new approach to visual servoing in robotics. *Robotics and Automation, IEEE Transactions on*, 8(3):313–326, 1992.

[9] P. Fabiani and J.C. Latombe. Tracking a partially predictable object with uncertainty and visibility constraints: a game-theoretic approach. Technical report, Technical report, Univeristy of Stanford, December 1998. http://underdog. stanford. edu/.(cited on page 76).

[10] HH Gonzalez-Banos, C.Y. Lee, and J.C. Latombe. Real-time combinatorial tracking of a target moving unpredictably among obstacles. *Robotics and Automation, 2002. Proceedings. ICRA'02. IEEE International Conference on*, 2, 2002.

[11] Leonidas J. Guibas, Jean-Claude Latombe, Steven M. LaValle, David Lin, and Rajeev Motwani. A visibility-based pursuit-evasion problem. *International Journal of Computational Geometry and Applications*, 9(4/5):471–, 1999.

[12] L. Guilamo, B. Tovar, and S. M. LaValle. Pursuit-evasion in an unknown environment using gap navigation trees. In *IEEE/RSJ International Conference on Intelligent Robots and Systems*, 2004.

[13] JP Hespanha, M. Prandini, and S. Sastry. Probabilistic pursuit-evasion games: a one-step Nash approach. *Decision and Control, 2000. Proceedings of the 39th IEEE Conference on*, 3, 2000.

[14] S. A. Hutchinson, G. D. Hager, and P. I. Corke. A tutorial on visual servo control. *IEEE Transactions on Robotics & Automation*, 12(5):651–670, October 1996.

[15] R. Isaacs. *Differential Games*. Wiley, New York, 1965.

[16] V. Isler, S. Kannan, and S. Khanna. Locating and capturing an evader in a polygonal environment. *Workshop on Algorithmic Foundations of Robotics (WAFR04)*, 2004.

[17] V. Isler, S. Kannan, and S. Khanna. Randomized pursuit-evasion in a polygonal environment. *Robotics, IEEE Transactions on [see also Robotics and Automation, IEEE Transactions on]*, 21(5):875–884, 2005.

[18] B. Jung and G.S. Sukhatme. Tracking Targets Using Multiple Robots: The Effect of Environment Occlusion. *Autonomous Robots*, 13(3):191–205, 2002.

[19] Swastik Kopparty and Chinya V. Ravishankar. A framework for pursuit evasion games in rn. *Inf. Process. Lett.*, 96(3):114–122, 2005.

[20] Parker L. Algorithms for Multi-Robot Observation of Multiple Targets. *Journal on Autonomous Robots*, 12:231–255, 2002.

[21] S. M. LaValle. *A Game-Theoretic Framework for Robot Motion Planning*. PhD thesis, University of Illinois, Urbana, IL, July 1995.

[22] S. M. LaValle, H. H. Gonzalez-Banos, C. Becker, and J. C. Latombe. Motion strategies for maintaining visibility of a moving target. In *Robotics and Automation, 1997. Proceedings., 1997 IEEE International Conference on*, volume 1, pages 731–736, Albuquerque, NM, USA, April 1997.

[23] S. M. LaValle and J. Hinrichsen. Visibility-based pursuit-evasion: The case of curved environments. *IEEE Transactions on Robotics and Automation*, 17(2):196–201, April 2001.

[24] S. M. LaValle, D. Lin, L. J. Guibas, J.-C. Latombe, and R. Motwani. Finding an unpredictable target in a workspace with obstacles. In *Proceedings IEEE International Conference on Robotics and Automation*, pages 737–742, 1997.

[25] S. M. LaValle, B. Simov, and G. Slutzki. An algorithm for searching a polygonal region with a flashlight. *International Journal of Computational Geometry and Applications*, 12(1-2):87–113, 2002.

[26] E. Malis, F. Chaumette, and S. Boudet. 2D 1/2 Visual Servoing. *IEEE Transactions on Robotics and Automation*, 15(2):238–250, 1999.

[27] E. Marchand, P. Bouthemy, F. Chaumette, and V. Moreau. Robust real-time visual tracking using a 2d-3d model-based approach. *IEEE Int. Conf. on Computer Vision, ICCV99*, 1:262–268.

[28] T. Muppirala, S. Hutchinson, and R. Murrieta-Cid. Optimal Motion Strategies Based on Critical Events to Maintain Visibility of a Moving Target. *Robotics and Automation, 2005. Proceedings of the 2005 IEEE International Conference on*, pages 3826–3831, 2005.

[29] R. Murrieta, A. Sarmiento, S. Bhattacharya, and SA Hutchinson. Maintaining visibility of a moving target at a fixed distance: the case of observer bounded speed. *Robotics and Automation, 2004. Proceedings. ICRA'04. 2004 IEEE International Conference on*, 1.

[30] R. Murrieta, A. Sarmiento, and S. Hutchinson. On the existence of a strategy to maintain a moving target within the sensing range of an observer reacting with delay. *Intelligent Robots and Systems, 2003.(IROS 2003). Proceedings. 2003 IEEE/RSJ International Conference on*, 2, 2003.

[31] R. Murrieta-Cid, H. H. Gonzalez-Banos, and B. Tovar. A reactive motion planner to maintain visibility of unpredictable targets. In *Robotics and Automation, 2002. Proceedings. ICRA '02. IEEE International Conference on*, volume 4, pages 4242–4248, 2002.

[32] R. Murrieta-Cid, T. Muppirala, A. Sarmiento, S. Bhattacharya, and S. Hutchinson. Surveillance strategies for a pursuer with finite sensor range. *Robotics Research, 2007. International Journal on*, 2007. To Appear.

[33] TD Parsons. The search number of a connected graph. *9th Southeastern Conference on Combinatorics, Graph Theory and Computing*, pages 549–554.

[34] TD Parsons. Pursuit-evasion in a graph. *Theor. Appl. Graphs, Proc. Kalamazoo*, 1976.

[35] L. A. Petrosjan. *Differential Games of Pursuit.* World Scientific, Singapore, 1993.

[36] PE Rybski, NP Papanikolopoulos, SA Stoeter, DG Krantz, KB Yesin, M. Gini, R. Voyles, DF Hougen, B. Nelson, and MD Erickson. Enlisting rangers and scouts for reconnaissance and surveillance. *Robotics & Automation Magazine, IEEE*, 7(4):14–24, 2000.

[37] P.E. Rybski, S.A. Stoeter, M.D. Erickson, M. Gini, D.F. Hougen, and N. Papanikolopoulos. A team of robotic agents for surveillance. *Proceedings of the fourth international conference on autonomous agents*, pages 9–16, 2000.

[38] S. Sachs, S. Rajko, and S. M. LaValle. Visibility-based pursuit-evasion in an unknown planar environment. *International Journal of Robotics Research*, 23(1):3–26, January 2004.

[39] I. Suzuki and M. Yamashita. Searching for a Mobile Intruder in a Polygonal Region. *SIAM Journal on Computing*, 21:863, 1992.

[40] B. Tovar and S. M. LaValle. Visibilty-based pursuit-evasion with bounded speed. In *Proceedings Workshop on Algorithmic Foundations of Robotics*, 2006.

[41] R. Vidal, O. Shakernia, HJ Kim, DH Shim, and S. Sastry. Probabilistic pursuit-evasion games: theory, implementation, and experimental evaluation. *Robotics and Automation, IEEE Transactions on*, 18(5):662–669, 2002.

APPENDIX

The subroutine VG(S), computes the visibility graph of a polygonal environment S.

Algorithm DECIDABILITYTEST(S,P,E,v_p,v_e)
Input: A set S of disjoint polygonal obstacles, The Pursuer position P, The evader position E, Pursuer maximum velocity v_p, Evader maximum velocity v_e
Output: result: If Evader wins YES, else DONOTKNOW

1) G_{visE} =VG(S ∪ E)
2) G_{mvisP} =MVG(S ∪ P)
3) i←1
4) Until $i = n + 1$ or $flag = false$
5) l_1 =DIJKSTRA(VG(S),E,V_i)
6) l_2 =DIJKSTRA(MVGCONSTRUCT(S,V_k),P,V_i)
7) if $\frac{l_2}{v_p} > \frac{l_1}{v_e}$
8) result=WINS
9) else $i = i + 1$
10) if $i = n + 1$, result=DONOT KNOW

In the psuedocode below, the subroutine VISIBLEVERTICES(v,S) is an algorithm which provides the vertices of a polygonal environment S, visible to a vertex v. It can be implemented using a rotational plane sweep algorithm [6].

Algorithm MVGCONSTRUCT(S,V_k)
Input: A set S of disjoint polygonal obstacles, A vertex V_k
Output: The Modified Visibility Graph $G_{MVIS}(S)$

1) Initialize a graph $G = (V, E)$ where V is the set of all vertices of the polygons in S and E= ∅
2) **for** all vertices $v \in V$
3) **do** W ← VISIBLEVERTICES(v,S)
4) **for** every vertex $w \in W$ and $w \neq V_k$
5) add edge (v, w) to E with weight as the
6) euclidean length of segment $\overline{vw}$.
7) **for** all vertices $v \in$ G
8) d←MINLSDIST(v,V_k)
9) If d$\neq \infty$
10) if edge vV_k exists, update weight of vV_k
11) to d
12) else add edge vV_k to E with weight d

In the following psuedocode the subroutine

1) ORTHOGONALITYCHECK(v,k) is to check if v lies outside rays l_1 and l_2 and if we can draw a perpendicular from v to rays l_1 and l_2 without intersecting other obstacles. It returns TRUE if either of the conditions is satisfied
2) CLOSER(v, l_1, l_2) is used to find which ray among l_1 and l_2 has a lesser distance from v.
3) MINDISTANCE(v,k) is to compute the minimum distance from v to a ray k.

Algorithm MINLSDISTANCE(v, V_k)
Input: A vertex v in G, a vertex V_k in G
Output: d:the length of the line perpendicular from v to l_1 and l_2 corresponding to V_k. If the perpendicular line intersects any obstacles then it gives an output of ∞.

1) If ORTHOGONALITYCHECK(v,l_1, l_2)=TRUE
2) k=CLOSER(v, l_1, l_2)
3) If COLLISIONCHECK(v,k)=TRUE
4) d← ∞
5) else d←MINDISTANCE(v,k)
6) else d← ∞

Learning Omnidirectional Path Following Using Dimensionality Reduction

J. Zico Kolter and Andrew Y. Ng
Computer Science Department
Stanford University
Stanford, CA 94305
Email: {kolter, ang}@cs.stanford.edu

Abstract—**We consider the task of omnidirectional path following for a quadruped robot: moving a four-legged robot along any arbitrary path while turning in any arbitrary manner. Learning a controller capable of such motion requires learning the parameters of a very high-dimensional policy class, which requires a prohibitively large amount of data to be collected on the real robot. Although learning such a policy can be much easier in a model (or "simulator") of the system, it can be extremely difficult to build a sufficiently accurate simulator. In this paper we propose a method that uses a (possibly inaccurate) simulator to identify a low-dimensional subspace of policies that is robust to variations in model dynamics. Because this policy class is low-dimensional, we can learn an instance from this class on the real system using much less data than would be required to learn a policy in the original class. In our approach, we sample several models from a distribution over the kinematic and dynamics parameters of the simulator, then use the Reduced Rank Regression (RRR) algorithm to identify a low-dimensional class of policies that spans the space of controllers across all sampled models. We present a successful application of this technique to the task of omnidirectional path following, and demonstrate improvement over a number of alternative methods, including a hand-tuned controller. We present, to the best of our knowledge, the first controller capable of omnidirectional path following with parameters optimized simultaneously for *all* directions of motion and turning rates.**

I. Introduction

In this paper we consider the task of omnidirectional path following: moving a four-legged robot along any arbitrary path while turning in any arbitrary manner. Examples of such motion include walking in a circle while facing the circle's center, following a straight line while spinning around, or any other such maneuver. In this paper we focus on "trot" gaits: gaits where the robot moves two feet at a time.

A key technical challenge in building such a robust gait is maintaining balance of the robot. Since the trot gait moves two feet at once, the polygon formed by the supporting feet reduces to a line, and it is very difficult to maintain any standard static or dynamic stability criterion. Fortunately, when executing a "constant" maneuver — that is, moving in a constant direction while possibly turning at a constant rate — it is possible to approximately balance the robot by offsetting the center of mass by some fixed distance. However, in omnidirectional path following we are frequently changing the direction of motion and turning rate, so a complete policy must determine the proper center offset for *any* given direction and turning rate. This results in a high-dimensional class of control policies[1], which therefore tends to require a prohibitively large amount of data from the real robot in order to learn.

It can be much easier to learn this high-dimensional policy in a model (or "simulator") of the system than on the real robot; there are many advantages to *model-based reinforcement learning* (RL): the simulator can be made deterministic, we can save and replay states, we can take gradients of the parameters, there is no risk of damage to the real robot, etc. However, it can be extremely difficult to build a sufficiently accurate simulator. Often times a policy learned in the simulator will perform very poorly on the real system.

In this paper, we propose a method that makes use of a (possibly inaccurate) simulator to identify a low-dimensional subspace of policies that is robust to variations in model dynamics. Our method is as follows: First, we sample randomly from a *distribution* over the kinematic and dynamic parameters of the simulator. We learn parameters for the high-dimensional policy in each of these simulation models. To identify a low-dimensional class of policies that can represent the learned policies across all the sampled models we formulate a dimensionality reduction optimization problem that can be solved via the Reduced Rank Regression (RRR) algorithm [26]. Finally, we learn a policy in this low-dimensional class on the real robot; this requires much less data than would be required to learn a policy in the original, high-dimensional class of policies. We present a successful application of this technique to the task of omnidirectional path following, and demonstrate improvement over a number of alternative control methods, including a hand-tuned controller. We also present, to the best of our knowledge, the first controller capable of omnidirectional motion with parameters optimized simultaneously for *all* directions of motion and turning rates — in the next section we clarify exactly how this work differs from previous work in quadruped locomotion.

The rest of this paper is organized as follows. In Section II we present background material and related work. In Section III we give an overview of our controller (the full parametrization of the controller is given in the Appendix) and present an online learning algorithm that is capable of learning balancing

[1] In this paper we focus on linearly-parametrized policies — that is, policies that are specified by a set of linear coefficients. In this case, the dimension of the policy is simply the number of free parameters.

offsets for any *fixed* direction of motion and turning rate. In Section IV we formally present our method for identifying a low-dimensional policy class by sampling from a distribution over simulator models. In Section V we present and analyze our results on the real robotic system. Finally, in Section VI we give concluding remarks.

II. Background and Related Work

A. Quadruped Locomotion

There has been a great deal of work in recent years, both in the robotics and the machine learning communities, on quadruped locomotion. Much of the research has focused on static walking (moving one leg at a time, thereby keeping the robot statically stable) including statically stable gaits capable of omnidirectional path following [19]. In addition, there has been much work on designing static gaits that can navigate irregular terrain [6, 5, 11, 18]. However, in this paper we focus on the dynamic trot gait over flat terrain, which requires very different approaches.

There has also been a great deal of work on developing omnidirectional trot gaits, in particular for the Sony AIBO robot. Much of this work is based on a trot gait, originally developed by Hengst et. al. [13], that is potentially capable of moving in any direction while turning at any rate. While the parameters of the gait presented in [13] are hand-tuned, there has been much subsequent work on gait optimization of this and similar gaits. Several gait optimization methods have been investigated, including evolutionary search methods [14, 8, 27], direction set minimization [16] and policy gradient techniques [17]. However, while the controllers in these papers are *capable* of omnidirectional motion in that they can walk in any direction while turning at any rate, all the papers listed above focus on optimizing parameters only for one *single* direction of motion at a time (usually forward walking). Additionally, as noted in [27], gaits optimized for one type of motion typically perform poorly on other maneuvers — for example, a gait optimized for forward walking typically performs poorly when attempting to walk backwards. In contrast, in this paper we focus on learning a policy that performs well for *all* directions of motion and turning rates.

Another vein of research in dynamic quadruped gaits follows work by Raibert [24, 25]. This and more recent work [22, 21] achieve dynamic gaits — both trot gaits and gallop gaits, which include a flight phase — by "hopping" on compliant legs, which typically employ some form of hydraulics. While this is a powerful technique that can allow for very fast locomotion, we discovered very quickly that our robot was not capable of generating sufficient force to jump off the ground, effectively disallowing such methods.

There is also a great deal of work on dynamic stability criteria for legged robots [12, 23]. One of the most well-known criteria is to ensure that the Zero Moment Point (or ZMP) [29] — which is similar to the center of mass projected on to the ground plane, except that it accounts for inertial forces acting on the robot — never leaves the supporting polygon. However, when the robot has only two rounded feet on the ground the supporting polygon reduces to a line, making it difficult to maintain the ZMP exactly on this line. In addition, we found it difficult to calculate the ZMP precisely on the real robot due to the inaccuracy of the on-board sensors. For these reasons, in this paper we focus on an approximate method for balancing the robot.

B. Learning and Control

The method we propose in this paper is related to the area of robust control theory. For a general overview of robust control, see [31] and [9]. However, our work differs from standard robust control theory in that the typical goal of robust control is to find a *single* policy that performs well in a *wide variety* of possible models. In contrast, we make no assumption that any such policy exists — indeed, in our application the optimal policy depends very much on the particular model dynamics — but rather we want to identify a *subspace* of policies that is robust to variation in the model dynamics. In the final step of our method, we then search this subspace to find a policy that is specific to the dynamics of the real system.

To identify the low-dimensional subspace of policies, we pose an optimization problem that can be solved via the Reduced Rank Regression (RRR) algorithm. The RRR setting was first discussed by Anderson [1]. Izenman [15] developed the solution that we apply in this paper, coined the term "Reduced Rank Regression," and discussed the relationship between this algorithm, Canonical Correlation Analysis (CCA) and Principle Component Analysis (PCA). RRR is discussed in great detail in [26], and there is a great deal of active research on this algorithm, both from theoretical [2] and numerical perspectives [10].

In the machine learning literature, the problem we formulate can be viewed as an instance of multi-task learning [7, 3]. However, our setting does differ slightly from the prototypical multi-task learning paradigm, since we do not ultimately care about performance on most of the tasks, except insofar as it helps us learn the one task we care about — i.e., we don't care how well the policies perform in the simulation models, just how well the final controller performs on the real system.

Finally, we note that there has been recent work on the application of dimensionality reduction techniques to control and reinforcement learning. Mahadevan [20] uses Graph Laplacian methods to learn a low-dimensional representation of value functions on a Markov Decision Process (MDP). Roy et. al. [28] use dimensionality reduction to compactly represent belief states in a Partially Observable MDP. Our work is similar in spirit, except that we apply dimensionality reduction to the space of controllers, to learn a low-dimensional representation of the control policies themselves.

III. A Controller for Omnidirectional Path Following

In this section we present a parametrized gait for the quadruped robot that is capable of omnidirectional path following. The design builds upon recent work on trot gaits for quadruped locomotion [13, 17]. The robot used in this work

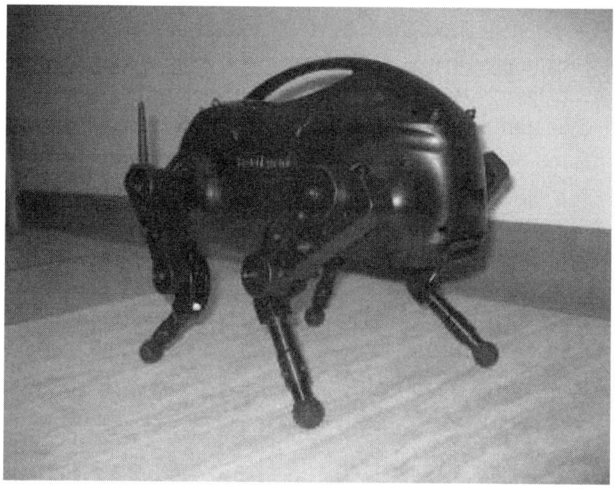

Fig. 1. The LittleDog robot, designed and built by Boston Dynamics.

is shown in Figure 1. The robot, known as "LittleDog," was designed and built by Boston Dynamics, Inc and is equipped with an internal IMU and foot force sensors. State estimation is performed via a motion capture system that tracks reflective markers on the robot.

Our controller uses inverse kinematics to specify locations for the four feet in Euclidean coordinates relative to the robot's body. While two feet move along the ground, the other two feet moving through the air in a box pattern; this moves the robot forward [13].[2] We achieve lateral movement by rotating the angle of all the four feet, and turn by skewing the angles of the front and back or left and right feet. We specify paths for the robot as linear splines, with each point specifying a desired position and angle for the robot. The controller is closed-loop: every time step (10 milliseconds) we use the current state estimate to find a direction and turning angle that forces the robot to follow the specified path. A more detailed description of the gait is given in the appendix.

A. Learning To Balance

We found that by far the most challenging aspect of designing a robust controller was balancing the robot as it moved. To balance the robot, our controller offsets the center of mass of the robot by some specified amount $(x_{\text{off}}, y_{\text{off}})$.[3] Without a proper center offset, the robot will "limp" visibly while walking. The challenge is to find a function that determines the proper center offset for any given direction of motion and turning rate. That is, given a direction angle ψ and turning rate ω, we want to find functions f_x and f_y such that

$$x_{\text{off}} = f_x(\psi, \omega), \quad y_{\text{off}} = f_y(\psi, \omega).$$

Since the direction angle and turning rates are inherently periodic — i.e., a direction angle of 2π is identical to a directon angle of 0 — the Fourier bases are a natural means of representing these functions. We therefore represent f_x as

$$f_x(\psi, \omega) = \theta_x^T \phi(\psi, \omega)$$

where $\theta_x \in \mathbb{R}^k$ is a vector of coefficients and

$$\begin{aligned}\phi(\psi, \omega) \quad = \quad & [\cos(i\psi)\cos(j\omega), \cos(i\psi)\sin(j\omega), \\ & \sin(i\psi)\cos(j\omega), \sin(i\psi)\sin(j\omega)], \\ & i, j = 1, 2, \ldots\end{aligned}$$

denotes first k principle Fourier basis functions of ψ and ω — here the range of i and j are chosen so that the dimension of ϕ is also k. The function f_y is represented in the same manner, and learning a parametrization of the controller requires learning the coefficients θ_x and θ_y of this approximation.

With this motivation, we first consider the problem of finding the center offset for a *fixed* direction angle ψ and turning rate ω. We designed a online learning algorithm that, for fixed ψ and ω, dynamically adjusts the center offsets during walking so as to balance the robot. The intuition behind this algorithm is that if the robot is balanced, then the two moving feet should hit the ground simultaneously. If the two feet do not hit the ground simultaneously, then the algorithm looks at which of the two feet hit the ground first, and adjusts the center offsets accordingly. If, for example, the back leg hits the ground before the front leg, then the algorithm will shift the center of mass forward, thereby tilting the robot forward, and encouraging the front leg to hit the ground sooner. The precise updates are given by

$$\begin{aligned}x_{\text{off}} &:= x_{\text{off}} + \alpha(g(t_{\text{FL}} - t_{\text{BR}}) + g(t_{\text{FR}} - t_{\text{BL}})) \\ y_{\text{off}} &:= y_{\text{off}} + \alpha g((t_{\text{FL}} - t_{\text{BR}}) - (t_{\text{FR}} - t_{\text{BL}}))\end{aligned} \tag{1}$$

where α is a learning rate, $t_{\text{FL}}, t_{\text{FR}}, t_{\text{BL}}, t_{\text{BR}}$ are the foot contact times for the four feet respectively, and $g(x) = x^2 \cdot \text{sgn}(x)$. So, for example, if the back left leg hits before the front right, $t_{\text{FR}} - t_{\text{BL}} > 0$, so x_{off} is increased, shifting the center of mass forward. The algorithm is similar in spirit to the previously mentioned gait optimization algorithms for the Sony AIBO gaits [14, 16, 8, 17, 27] in that it performs online optimization of the gait parameters for a fixed direction angle and turning rate. We implemented this algorithm both in simulation and on the actual robot; for the actual robot we used foot force sensors to determine when a foot hit the ground.[4] Convergence to a stable center position is generally quite fast, about a minute or two on the real robot (assuming no adverse situations arise, which we will discuss shortly).

Given a collection of direction angles, turning rates, and their corresponding center offsets, we can learn the coefficients θ_x and θ_y by least squares regression. Specifically, if we are given a set of n direction angles, turning rates, and resulting

[2]We also experimented with other locus shapes for moving the feet that are common in the literature, such as a half ellipse or a trapezoid [17], but found little difference on our robot in terms of performance.

[3]In the coordinate system we use, the positive x axis points in the direction that the robot is facing, and the positive y axis points to the robot's left.

[4]Although the foot sensors on the LittleDog robot are not particularly accurate in many situations, the trot gait hits the feet into the ground hard enough that the foot sensors can act as a simple Boolean switch indicating whether or not the feet have hit the ground.

x center offsets, $\{\psi_i, \omega_i, x_{\text{off},i}\}$, $i = 1 \dots n$, then we can learn the parameters $\theta_x \in \mathbb{R}^k$ by solving the optimization problem

$$\min_{\theta_x} \|y - X\theta_x\|_2^2 \tag{2}$$

where $X \in \mathbb{R}^{n \times k}$ and $y \in \mathbb{R}^n$ are design matrices defined as

$$X = [\phi(\psi_1, \omega_1) \dots \phi(\psi_n, \omega_n)]^T \quad y = [x_{\text{off},1} \dots x_{\text{off},n}]^T. \tag{3}$$

The solution to this problem is given by, $\theta_x = (X^T X)^{-1} X^T y$, and by a well known sample complexity result [4], we need $\Omega(k)$ data points to find such a solution.

Unfortunately, computing a sufficient number center offsets on the real robot is a time-consuming task. Although the algorithm described above can converge in about a minute under ideal circumstances, several situations arise that can slow convergence considerably. Communication with the robot is done over a wireless channel, and packet loss can make the robot slip unexpectedly, which causes incorrect adjustments to the robot. Additionally, an improper center offset (as would occur before the algorithm converged) can make the robot move in a way that degrades the joint angle calibration by bashing its feet into the ground. Although it is significantly easier to find proper joint offsets in simulation, it is difficult to create a simulator that accurately reflects the center offset positions in the real robot. Indeed, we invested a great deal of time trying to learn parameters for the simulator that reflected the real system as accurately as possible, but still could not build a simulator that behaved sufficiently similarly to the real robot. The method we present in the next section allows us to deal with this problem, and efficiently learn center offsets for the real robot by combining learning in both simulation and the real robot.

IV. Identifying a Low Dimensional Policy Class Using Dimensionality Reduction

In this section we present a method for identifying a low-dimensional policy class by making use of a potentially innaccurate simulator. Although we focus in this paper on the application of this method to the specific task of learning a policy for omnidirectional path following, the formulations we present in this section are general.

The intuition behind our algorithm is that even if it is very difficult to find the precise dynamic parameters that would make a simulator accurately reflect the real world, it may be much easier to specify a *distribution* over the potential variations in model dynamics.[5] Therefore, even if the simulator does not mimic the real system exactly, by considering a distribution over the model parameters, it can allow us identify a smaller subspace of policies in which to search. The method we propose is as follows:

1) Draw m random samples from a distribution over the dynamic and kinematic parameters of the simulator. Each set of parameters now defines a differently perturbed simulation model.
2) In each simulation model, collect n data points. For example, in our setting each of these data points corresponds to a particular direction angle ψ, turning rate ω, and the resulting center offset (x_{off} or y_{off}) found by the online balancing algorithm described previously.
3) Use the Reduced Rank Regression algorithm to learn a small set of basis vectors that span the major axes of variation in the space of controllers over all the sampled models.
4) For the real robot, learn the parameters of the policy in this low-dimensional class.

More formally, suppose we are given a matrix of feature vectors $X \in \mathbb{R}^{n \times k}$ and we collect a set of of output vectors $\{y^{(i)} \in \mathbb{R}^n\}, i = 1, \dots, m$, where each of the $y^{(i)}$'s corresponds to the data points collected from the ith simulator. In our setting these matrices are given by (3), i.e., the rows of X are the k-dimensional Fourier bases, and the entries of $y^{(i)}$ are the center offsets found by the online balancing algorithm (notice that we now require the i superscript on the y vectors, since the resulting center offsets will vary between the different simulator models). Rather than learn the coefficients for each simulator individually, as in (2), we consider all the m simulator models jointly, by forming the design matrices $Y \in \mathbb{R}^{n \times m}$ and $\Theta \in \mathbb{R}^{k \times m}$,

$$Y = \left[y^{(1)} \dots y^{(m)}\right], \quad \Theta = \left[\theta^{(i)} \dots \theta^{(m)}\right]$$

and considering the problem[6]

$$\min_{\Theta} \|Y - X\Theta\|_F^2. \tag{4}$$

However, solving this problem is identical to solving each of the least squares problems (2) individually for each i; if we want to learn the coefficients for a policy on the real robot, this approach offers us no advantage.

Instead, we consider matrices $A \in \mathbb{R}^{k \times \ell}, B \in \mathbb{R}^{\ell \times m}$, with $\ell \ll k$ and the problem

$$\min_{A,B} \|Y - XAB\|_F^2. \tag{5}$$

In this setting, the A matrix selects ℓ linear combinations of the columns of X, and the B matrix contains the coefficients for these linear combinations. In other words, this approximates the coefficients as $\theta^{(i)} \approx Ab^{(i)}$, were $b^{(i)} \in \mathbb{R}^\ell$ is the ith column of B. The matrix A forms a basis for representing the coefficients in *all* the m simulation models.

The key advantage of this approach comes when we consider learning the parameters $\theta \in \mathbb{R}^k$ of a policy on the real robot. We approximate θ as a linear combination of the columns of A, i.e., $\theta = Ab$. However, since A, as defined above, has only ℓ columns, we only need to learn the ℓ-dimensional coefficient vector b in order to approximate θ. By a standard sample complexity result [4], this requires only $O(\ell)$ examples, and since $\ell \ll k$, this greatly reduces the amount of data required to learn the policy on the real system.

[5] This claim is discussed further in Subsection IV-B.

[6] Here, $\|\bullet\|_F^2$ denotes the squared Frobenius norm, $\|A\|_F^2 = \sum_{i,j} A_{ij}^2$.

In order to motivate the exact optimization problem presented in (5), we discuss other possible approaches to the problem. First, we could solve (4) to find the least squares solution Θ, then run Principal Component Analysis (PCA) to find a set of basis vectors that could accurately represent all the columns of Θ. However, this approach ultimately minimizes the wrong quantity: we do not truly care about the error in approximating the coefficients $\theta^{(i)}$ themselves, but rather the error in approximating the actual data points $y^{(i)}$. Second, we could run PCA on the Y matrix, which would result in a set of basis vectors that could represent the data points across all the different simulation models. However, we would require some way of extending these bases to new data points not in the training set, and this can be difficult.[7] Instead, the minimization problem (5) truly represents the error quantity that we are interested in — how well our coefficients can approximate the data points Y.

Despite the fact that the optimization problem (5) is non-convex, it can be solved efficiently (and exactly) by the Reduced Rank Regression algorithm. We begin by noting that (5) is identical to the problem

$$\begin{array}{ll} \min_\Theta & \|Y - X\Theta\|_F^2 \\ \text{s.t.} & \operatorname{rank}(\Theta) = \ell. \end{array} \tag{6}$$

The solution to this Reduced Rank Regression problem, (6), is given by

$$\Theta = (X^T X)^{-1} X^T Y V V^T \tag{7}$$

where the columns of V are the ℓ principle eigenvectors of

$$Y^T X (X^T X)^{-1} X^T Y. \tag{8}$$

This result is proved in [26, Theorem 2.2]. Optimal values of A and B can be read directly from this solution,

$$A = (X^T X)^{-1} X^T Y V \quad B = V^T.$$

Notice that the Reduced Rank Regression solution can be interpreted as the least squares solution $(X^T X)^{-1} X^T Y$ projected into the subspace spanned by V. When V is full rank — i.e., there is no rank constraint — then $VV^T = I$, and the solution naturally coincides with the least squares solution.

After learning A and B, the final step in our algorithm is to learn a policy on the real robot. To do this we collect, from the real system, a small set of data points $y \in \mathbb{R}^p$ for some (new) set of feature vectors $X \in \mathbb{R}^{p \times k}$, and solve the least squares problem

$$\min_b \|y - XAb\|_2^2$$

For $b \in \mathbb{R}^\ell$. For the reasons mentioned above, b can be estimated using much less data that it would take to learn the full coefficient vector $\theta \in \mathbb{R}^k$. After learning b, we approximate the robot's policy parameters as $\theta = Ab$.

[7] There are certainly methods for doing this. For example, the Nyström approximation [30] could be used compute a non-parametric approximation to the output y' for previously unseen feature vector x'. However, this would require computing the outputs corresponding to x' for all (or at least many) of the simulators, making this technique less applicable for the real-time situations we are concerned with.

A. Non-uniform Features

Note that one restriction of the model as presented above is that the feature vectors X must be the same for all simulations. In our particular application of Reduced Rank Regression, this does not seem to be overly restrictive, since the data points are generated by a simulator. Therefore, we can typically choose the data points to be whatever we desire, and so can simply restrict them to be the same across all the simulation models. Alternatively, the framework can be extended to handle the case where the different simulation models have different feature vectors, though we no longer know of any method which is guarenteed to find the globally optimal solution.

Let $X^{(i)} \in \mathbb{R}^{n^{(i)} \times k}$ denote the feature vector for the ith simulation model. We now want to minimize

$$\min_{A,B} \sum_{i=1}^{m} \|y^{(i)} - X^{(i)} A b^{(i)}\|_2^2 \tag{9}$$

where $b^{(i)}$ is the ith column of B. This is referred to as the Seemingly Unrelated Regressions (SUR) model of Reduced Rank Regression [26, Chapter 7]. We know of no closed form for the solution in this case, but must instead resort to approximate iterative methods. Note that (9), while not convex in A and B jointly, is quadratic in either argument alone. Therefore, we can apply alternating minimization: we first hold A fixed and optimize over B, then hold B fixed and optimize over A; because each step involves unconstrained minimization of a quadratic form, it can be solved very efficiently by standard numerical codes. We repeat this process until convergence to a fixed point. More elaborate algorithms typically impose some form of normalization constraints on the two matrices [26].

We conducted extensive experiments with alternating minimization algorithms, when the feature vectors were chosen to be different across the different simulations, and we found performance to be nearly identical to the standard case for our particular data set. Since the general method we propose does allow for the feature vectors to be the same across all the simulation models, we focus on the previous setting, where the problem can be solved exactly.

B. Further Discussion

Since our approach requires specifying a distribution over kinematic and dynamic parameters of the simulator, the question naturally arises as to how we may come up with such a distribution, and whether specifying this distribution is truly easier than simply finding the "correct" set of parameters for the simulator. However, in reality it is likely that there does not exist *any* set of parameters for the simulator that reflects the real world. As mentioned in the previous section, we expended a great deal of energy trying to match the simulator to the real system as closely as possible, and still did not achieve a faithful representation. This is a common theme in robust control: stochasticity of the simulator is important not only because we believe the real world to be stochastic to a degree, but because the stochasticity acts as a surrogate

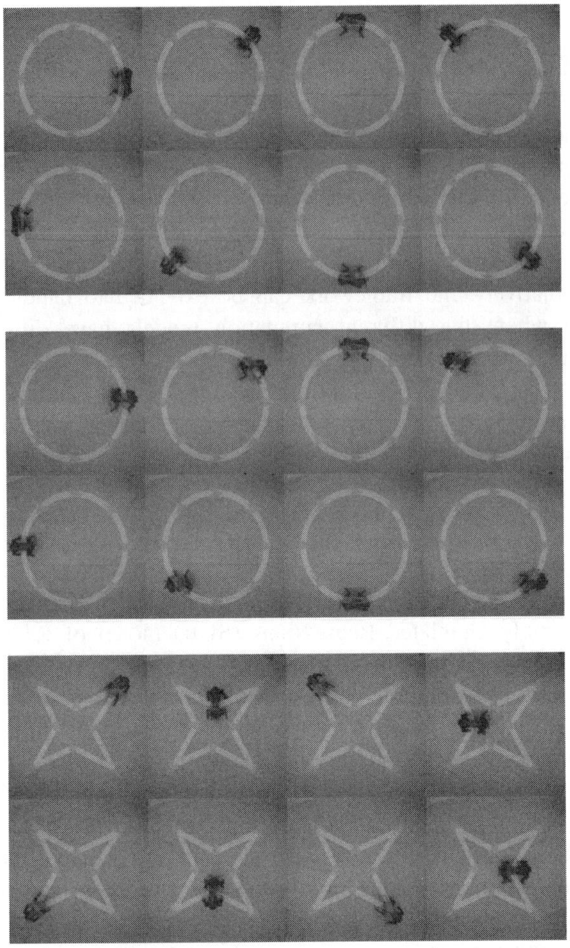

Fig. 2. Pictures of the quadruped robot following several paths.

for the unmodelled effects of the real world. Therefore, we use a straightforward approach to modeling the distribution over simulators: we use independent Gaussian distributions over several of the kinematic and dynamic parameters. This approach worked very well in practice, as we will show in the next section, and it was generally insensitive to the variance of the distributions.

Second, in the method presented here, learning the high-dimensional policy in each perturbed model is framed as a least-squares regression task. This is advantageous because it allows us to optimally solve the joint policy learning problem over all perturbed models with a rank constraint — squared error loss is the only loss we are aware of can be optimally solved with such a rank constraint. Were we to employ a different method for learning the high-dimensional policies in each perturbed model, such as policy gradient, then we would have to resort to heuristics for enforcing the rank constraint on the policy coefficient matrix. This may be necessary in some cases, if learning a policy cannot be framed as an ordinary least squares problem, but here we focus on the case where the problem can be solved optimally, as this was sufficient for our task of omnidirectional path following.

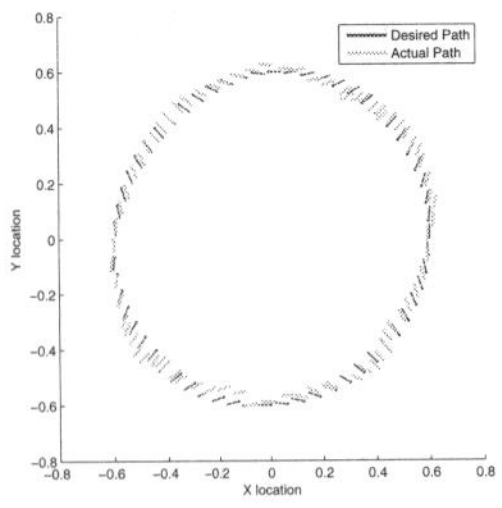

(a) Desired and actual trajectories for the learned controller on path 1.

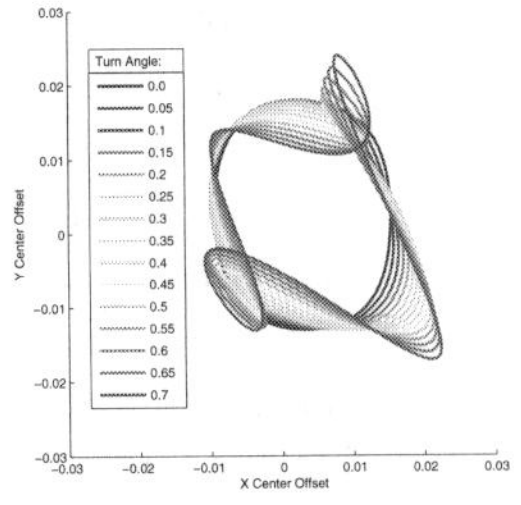

(b) Learned center offset curves for several different turning angles.

Fig. 3. Trajectory and center offsets for the learned controller.

V. Experimental Results

In this section we present experimental results on applying our method to learn a controller for omnidirectional path following. The simulator we built is based on the physical specifications of the robot and uses the Open Dynamics Engine (ODE) physics environment.[8]

Our experimental design was as follows: We first sampled 100 simulation models from a distribution over the simulator parameters.[9] In each of these simulation models, we used the online balancing algorithm described in Section III to find center offsets for a variety of fixed directions and turning rates.[10] In our experiments, we constrained the centering function (in both the x and y directions), to be a linear combination of the first $k = 49$ Fourier bases of the direction angle and turning rate. We then applied the Reduced Rank Regression algorithm to learn a low-dimensional representation of this function with only $\ell = 2$ parameters, effectively reducing the number of parameters by more than 95%.[11] Finally, to learn a policy on the actual robot, we used the online centering algorithm to compute proper center locations for 12 fixed maneuvers on the robot and used these data points to estimate the parameters of the low-dimensional policy.

To evaluate the performance of the omnidirectional gait and the learned centering function, we used three benchmark path splines: 1) moving in a circle while spinning in a direction opposite to the circle's curvature; 2) moving in a circle, aligned with the circle's tangent curve; and 3) moving in a circle keeping a fixed heading. To quantify performance of the robot

[8]ODE is available at http://www.ode.org.

[9]Experimental details: we varied the simulators primarily by adding a constant bias to each of the joint angles, where these bias terms were sampled from a Gaussian distribution. We also experimented with varying several other parameters, such as the centers of mass, weights, torques, and friction coefficients, but found that none of these had as great an effect on the resulting policies as the joint biases. This is somewhat unsurprising, since the real robot has constant joint biases. However, we reiterate the caveat mentioned in the previous section: it is not simply that we need to learn the correct joint biases in order to achieve a perfect simulator; rather, the results suggest that perturbing the joint biases results in a class of policies that is robust to the typical variations in model dynamics.

[10]For each model, we generated 100 data points, with turning angles spaced evenly between -1.0 and 1.0, and direction angles from 0 to 2π.

[11]Two bases was the smallest number that achieved a good controller with the data we collected: one basis vector was not enough, three basis vectors performed comparably to two, but had no visible advantage, and four basis vectors began to over-fit to the data we collected from the real robot, and started to perform worse.

Metric	Path	Learned Centering	No Centering	Hand-tuned Centering
Loop Time (sec)	1	**31.65 ± 2.43**	46.70 ± 5.94	34.33 ± 1.19
	2	**20.50 ± 0.18**	32.10 ± 1.79	31.69 ± 0.45
	3	**25.58 ± 1.46**	40.07 ± 0.62	28.57 ± 2.21
Foot Hit RMSE (sec)	1	**0.092 ± 0.009**	0.120 ± 0.013	0.098 ± 0.009
	2	**0.063 ± 0.007**	0.151 ± 0.016	0.106 ± 0.010
	3	**0.084 ± 0.006**	0.129 ± 0.007	0.097 ± 0.006
Distance RMSE (cm)	1	**1.79 ± 0.09**	2.42 ± 0.10	1.84 ± 0.07
	2	**1.03 ± 0.36**	2.80 ± 0.41	1.98 ± 0.21
	3	**1.58 ± 0.11**	2.03 ± 0.07	1.85 ± 0.16
Angle RMSE (rad)	1	0.079 ± 0.006	0.075 ± 0.009	**0.067 ± 0.013**
	2	**0.070 ± 0.011**	**0.070 ± 0.002**	0.077 ± 0.006
	3	**0.046 ± 0.007**	0.058 ± 0.012	0.071 ± 0.009

TABLE I
PERFORMANCE OF THE DIFFERENT CENTERING METHODS ON EACH OF THE THREE BENCHMARK PATHS, AVERAGED OVER 5 RUNS, WITH 95% CONFIDENCE INTERVALS.

on these different tasks, we used four metrics: 1) the amount of time it took for the robot to complete an entire loop around the circle; 2) the root mean squared difference of of the foot hits (i.e., the time difference between when the two moving feet hit the ground); 3) the root mean squared error of the robot's Euclidean distance from the desired path; and 4) the root mean squared difference between the robot's desired angle and its actual angle.

Note that these metrics obviously depend on more than just the balancing controller — speed, for example, will of course depend on the actual speed parameters of the trot gait. However, we found that good parameters for everything but the balancing controller were fairly easy to choose, and the same values were optimal, regardless of the balancing policy used. Therefore, the differences in speed/accuracy between the different controllers we present is entirely a function of how well the controller is capable of balancing — for example, if the robot is unable to balance it will slip frequently and its speed will be much slower than if it can balance well.

We also note that prior to beginning our work on learning basis functions, we spent a significant amount of time attempting to hand-code a centering controller for the robot. We present results for this hand-tuned controller, since we feel it represents an accurate estimate of the performance attainable by hand tuning parameters. We also evaluated the performance of the omnidirectional gait with no centering.

Figure 2 shows pictures of the robots following some of the benchmark paths, as well as an additional star-shaped path. Videos of these experiments are available at:

`http://www.stanford.edu/~kolter/omnivideos`

Table I shows the performance of each centering method, for each of the four metrics, on all three benchmark paths.[12] As can be seen, the learned controller outperforms the other methods in nearly all cases. As the distance and angle errors indicate, the learned controller was able to track the desired trajectory fairly accurately. Figure 3(a) shows the actual and desired position and orientation for the learned centering controller on the first path. Figure 3(b) shows the learned center offset predictor trained on data from the real robot. This figure partially explains why hand-tuning controller can be so difficult: at higher turning angles the proper centers form unintuitive looping patterns.

VI. CONCLUSIONS

In this paper, we presented a method for using a (possibly inaccurate) simulator to identify a low-dimensional subspace of policies that is robust to variations in model dynamics. We formulate this task as a optimization problem that can be solved efficiently via the Reduced Rank Regression algorithm. We demonstrate a successful application of this technique to the problem of omnidirectional path following for a quadruped robot, and show improvement over both a method with no balancing control, and a hand-tuned controller. This technique enables us to achieve, to the best of our knowledge, the first omnidirectional controller with parameters optimized simultaneously for all directions of motion and turning rates.

VII. ACKNOWLEDGMENTS

Thanks to Morgan Quigley for help filming the videos and to the anonymous reviewers for helpful suggestions. Z. Kolter was partially supported by an NSF Graduate Research Fellowship. This work was also supported by the DARPA Learning Locomotion program under contract number FA8650-05-C-7261.

REFERENCES

[1] T. W. Anderson. Estimating linear restrictions on regression coefficients for multivariate normal distributions. *Annals of Mathematical Statistics*, 22(3):327–351, 1951.
[2] T. W. Anderson. Asymptotic distribution of the reduced rank regression estimator under general conditions. *Annals of Statistics*, 27:1141–1154, 1999.
[3] Rie Kubota Ando and Tong Zhang. A framework for learning predictive structures from multiple tasks and unlabeled data. *Journal of Machine Learning Research*, 6:1817–1853, 2005.
[4] Martin Anthony and Peter L. Bartlett. *Neural Network Learning: Theoretical Foundations*. Cambridge University Press, 1999.
[5] Shaoping Bai and K.H. Low. Terrain evaluation and its appliaction to path planning for walking machines. *Advanced Robotics*, 15(7):729–748, 2001.
[6] Shaoping Bai, K.H. Low, Gerald Seet, and Teresa Zielinska. A new free gaint generation for quadrupeds based on primary/secondary gait. In *Proceedings of the 1999 IEEE ICRA*, 1999.

[12] The foot hit errors should not be interpreted too literally. Although they give a sense of the difference between the three controllers, the foot sensors are rather imprecise, and a few bad falls greatly affect the average. They therefore should not be viewed as an accurate reflection of the foot timings in a typical motion.

[7] Rich Caruana. Multitask learning. *Machine Learning*, 28(1):41–75, 1997.
[8] Sonia Chernova and Maneula Veloso. An evolutionary approach to gait learning for four-legged robots. In *Proceedings of the 2004 IEEE IROS*, 2004.
[9] Geir E. Dullerud and Fernando Paganini. *A Course in Robust Control Theory: A Convex Approach* vol 36 in *Texts in Applied Mathematics*. Springer, 2000.
[10] Lars Eldén and Berkant Savas. The maximum likelihood estimate in reduced-rank regression. *Numerical Linear Algebra with Applications*, 12:731–741, 2005.
[11] Joaquin Estremera and Pablo Gonzalex de Santos. Free gaits for quadruped robots over irregular terrain. *The International Journal of Robotics Research*, 21(2):115–130, 2005.
[12] Elena Garcia, Joaquin Estremera, and Pablo Gonzalez de Santos. A comparative study of stability margins for walking machines. *Robotica*, 20:595–606, 2002.
[13] Bernhard Hengst, Darren Ibbotson, Son Bao Pham, and Claude Sammut. Omnidirectional locomotion for quadruped robots. In *RoboCup 2001: Robot Soccer World Cup V*, pages 368–373, 2002.
[14] Gregory S. Hornby, S. Takamura, J. Yokono, T. Yamamoto O. Hanagata, and M. Fujita. Evolving robust gaits with aibo. In *Proceedings of the 2000 IEEE ICRA*, pages 3040–3045, 2000.
[15] Alan J. Izenman. Reduced-rank regression for the multivariate linear model. *Journal of Multivariate Analysis*, 5:248–264, 1975.
[16] Min Sub Kim and William Uther. Automatic gait optimization for quadruped robots. In *Proceedings of the 2003 Australian Conference on Robotics and Automation*, 2003.
[17] Nate Kohl and Peter Stone. Machine learning for fast quadrupedal locomotion. In *The Nineteenth AAAI*, pages 611–616, July 2004.
[18] Honglak Lee, Yirong Shen, Chih-Han Yu, Gurjeet Singh, and Andrew Y. Ng. Quadruped robot obstacle negociation via reinforcement learning. In *The 2006 IEEE ICRA*, 2006.
[19] Shugen Ma, Tomiyama Takashi, and Hideyuki Waka. Omnidirectional staic walking of a quadruped robot. *IEEE Transactions of Robotics*, 21(2):152–161, 2005.
[20] Sridhar Mahadevan. Proto-value functions: Developmental reinforcement learning. In *Proceedings of the 22nd ICML*, 2005.
[21] J. Gordon Nichol, Surya P.N. Singh, Kennetch J. Waldron, Luther R. Palmer III, and David E. Orin. System design of a quadrupedal galloping machine. *International Journal of Robotics Research*, 23(10–11):1013–1027, 2004.
[22] Ioannis Poulakakis, James A. Smith, and Martin Buehler. Experimentally validated boudning bodels for the scout II quadrupedal robot. In *Proceedings of the 2004 IEEE ICRA*, 2004.
[23] Jerry E. Pratt and Russ Tedrake. Velocity-based stability margins for fast bipedal walking. In *Proceedings of the First Ruperto Carola Symposium on Fast Motions in Biomechanics and Robotics: Optimization and Feedback Control*, 2005.
[24] Marc H. Raibert. *Legged Robots that Balance*. MIT Press, 1986.
[25] Marc H. Raibert. Trotting, pacing, and bounding by a quadruped robot. *Journal of Biomechanics*, 23:79–98, 1990.
[26] Gregory C. Reinsel and Raja P. Velu. *Multivariate Reduced-Rank Regression: Theory And Appplications*. Springer-Verlag, 1998.
[27] Thomas Rofer. Evolutionary gait-optimization using a fitness function based on proprioception. In *RobotCup 2004: Robot World Cup VIII*, pages 310–322, 2005.
[28] Nicholas Roy, Geoffrey Gordon, and Sebastian Thrun. Finding approximate pomdp solutions through belief compression. *Journal of Artificial Intelligence Research*, 23:1–40, 2005.
[29] Miomir Vukobratovic and Branislav Vorovac. Zero-moment point – thirty five years of its life. *International Journal of Humanoid Robotics*, 1(1):157–173, 2004.
[30] Christopher K. I. Williams and Matthias Seeger. Using the nystrom method to speed up kernel machines. In *NIPS 13*, pages 682–688, 2001.
[31] Kemin Zhou, John Doyle, and Keither Glover. *Robust and Optimal Controler*. Prentice Hall, 1996.

APPENDIX

The omnidirectional controller is parameterized by 15 parameters:

- (f_x, f_y, f_z): the x, y, z coordinates of the front left foot in its center position (other feet are properly reflected).
- (b_x, b_y, b_z): maximum size of the foot locus boxes.
- $(x_{\text{off}}, y_{\text{off}})$: x and y offfsets for the center of mass of the robot. As discussed in the paper, these are not specified to be constant values, but depend on the direction angle and turning rate.
- t: time it takes for one cycle through the gait (moving all four feet).
- $d_{\text{ang}}, d_{\text{dist}}$: the d parameters specify how closely the dog should follow its path spline (see below).
- $\omega_{\max}, c_\omega$: turning parameters. We restrict the turning angle of the robot ω to always be less than $\omega_{\max}$. c_ω is a factor that multiplies the turning angle (see below).
- ψ, ω: direction angle and turning angle for the robot respectively. Determined by the position of the robot relative to the path spline.

During the first $t/2$ seconds of the gait period, the robot moves the front left and back right legs in the in the box pattern, while moving the front right and back left legs on the ground (and vice versa during the next $t/2$ seconds) [13]. The legs are moved at a constant velocity along each side of the box, at whatever speed is necessary to move them through the box pattern in the allotted time.

The x and y positions of the ends of the box relative to the foot are determined by ψ and ω. Let α_{FL} be the desired tilt angle for the front left foot. The two endpoints of the foot's motion are given by

$$\begin{aligned}(x_0, y_0) &= (-(b_x/2)\cos(\alpha_{\text{FL}}), -(b_y/2)\sin(\alpha_{\text{FL}}))\\ (x_{t/2}, y_{t/2}) &= ((b_x/2)\cos(\alpha_{\text{FL}}), (b_y/2)\sin(\alpha_{\text{FL}})).\end{aligned}$$

More intuitively, we can think of rotating the box for the front foot by α_{FL} and setting its length so that it lies in the ellipse formed by the tangent points b_x and b_y. The desired tilt is given by the following formula (with the signs reversed appropriatedly for the other feet):

$$\alpha_{\text{FL}} = \psi + \cos(\psi)\omega + \sin(\psi)\omega.$$

Trajectories for the robot are described as linear splines, with each point representing a location in two dimensional space (paths do not have a z component), and an angle. The direction angle of the robot is set to be the direction to the point on the spline that is d_{dist} ahead of the robot. The turning rate is specified in much the same way: we look at the desired angle for the point on the spline d_{ang} ahead of the robot, and let ω be c_ω times the difference between this desired angle and the current angle of the robot.

A Fast and Practical Algorithm for Generalized Penetration Depth Computation

Liangjun Zhang [1] Young J. Kim [2] Dinesh Manocha [1]

[1] *Dept. of Computer Science, University of North Carolina at Chapel Hill, USA, {zlj,dm}@cs.unc.edu*

[2] *Dept. of Computer Science and Engineering, Ewha Womans University, Korea, kimy@ewha.ac.kr*

http://gamma.cs.unc.edu/PDG

Abstract— **We present an efficient algorithm to compute the generalized penetration depth (PD^g) between rigid models. Given two overlapping objects, our algorithm attempts to compute the minimal translational and rotational motion that separates the two objects. We formulate the PD^g computation based on model-dependent distance metrics using displacement vectors. As a result, our formulation is independent of the choice of inertial and body-fixed reference frames, as well as specific representation of the configuration space. Furthermore, we show that the optimum answer lies on the boundary of the contact space and pose the computation as a constrained optimization problem. We use global approaches to find an initial guess and present efficient techniques to compute a local approximation of the contact space for iterative refinement. We highlight the performance of our algorithm on many complex models.**

I. Introduction

Penetration depth (PD) is a distance measure that quantifies the amount of interpenetration between two overlapping objects. Along with collision detection and separation distance, PD is one of the proximity queries that is useful for many applications including dynamics simulation, haptics, motion planning, and CAD/CAM. Specifically, PD is important for computing collision response [1], estimating the time of contact in dynamics simulation [2], sampling for narrow passages in retraction-based motion planing [3], [4], and C-obstacle query in motion planning [5].

There has been considerable work on PD computation, and good algorithms are known for convex polytopes. As for non-convex models, prior approaches on PD computation can be classified into local or global algorithms. The local algorithms only take into account the translational motion, i.e. *translational PD* (PD^t), and the results may be overly conservative. In many applications, including torque computation for 6-DOF haptic rendering or motion planning for articulated models, it is important to compute a penetration measure that also takes into account the rotational motion, i.e. *generalized penetration depth* (PD^g). However, the computational complexity of global PD between non-convex models is high. For PD^t, it can be computed using *Minkowski sum* formulation with the combinatorial complexity $O(n^6)$, where n is the number of features in the models [6]. For PD^g, it can be formulated by computing the arrangement of contact surfaces, and the combinatorial complexity of the arrangement is $O(n^{12})$ [7]. As a result, prior algorithms for global PD only compute an approximate solution [5], [8]. Moreover, these algorithms perform convex decomposition on non-convex models and can be rather slow for interactive applications. Overall, there are no good and practical solutions to compute the PD between non-convex models, thereby limiting their applications [4], [9], [10].

A key issue in PD^g computation is the choice of an appropriate distance metric. It is non-trivial to define a distance metric that can naturally combine the translational and rotational motion for an undergoing model, such that the resulting distance metric is *bi-invariant* with the choice of inertial and body-fixed reference frames, as well as of specific representations of the configuration space [11]. Specifically, it is well-known that for the spatial rigid body motion group SE(3), it is impossible to define a bi-invariant distance metric unless the shape of the model is known a priori [12], [13]. Finally, the distance metric should be easy to evaluate in order to devise an efficient PD^g computation algorithm.

A. Main Results

We present an efficient algorithm for computing PD^g for rigid, non-convex models. We formulate PD^g computation as a constrained optimization problem that minimizes an objective function defined by any proper distance metric that combines both translational and rotation motions, such as DISP [14] and *object norm* [15]. We use global approaches, based on motion coherence and random sampling, to compute an initial guess and incrementally walk on the *contact space* along the maximally-decreasing direction of the objective function to refine the solution. The algorithm computes a local approximation of the contact space, and we present culling techniques to accelerate the computation. As compared to the prior approaches, our algorithm offers the following benefits:

- **Generality:** Our approach is general and applicable to both convex and non-convex rigid models. The algorithm can be also extended to articulated or deformable models.
- **Practicality:** Unlike the prior approaches, our algorithm is relatively simple to implement and useful for many applications requiring both translational and rotation measures for inter-penetration.
- **Efficiency:** We use a local optimization algorithm and reduce the problem of PD^g computation to multiple collision detection and contact queries. As a result, our algorithm is efficient and can be used for interactive applications with high motion coherence.

We have implemented our PD^g algorithm and applied it to many non-convex polyhedra. In practice, our algorithm takes about a few hundred milli-seconds on models composed of a few thousand triangles.

B. *Organization*

The rest of our paper is organized as follows. We provide a brief survey of related work on PD^g computations in Sec. 2. In Sec. 3, we present a formulation of PD^g and give an overview of distance metrics. In Sec. 4, we provide our optimization-based algorithm to compute PD^g. We present its implementation and highlight its performance in Sec 5.

II. PREVIOUS WORK

There has been considerable research work done on proximity queries including collision detection, separation distance, and PD computation [16], [17]. In this section, we briefly discuss prior approaches to PD computation and distance metrics.

A. *PD Computation*

Most of the work in PD computation has been restricted to PD^t, and these algorithms are based on Minkowski sums [6], [18]. A few good algorithms are known for convex polytopes [19], [20] and general polygonal models [8]. Due to the difficulty of computing a global PD^t between non-convex models, some local PD^t algorithms have been proposed [9], [10], [21].

A few authors have addressed the problem of PD^g computation. Ong's work [22], [23] can be considered as one of the earliest attempts. The optimization-based method using a quadratic objective function can be regarded as implicitly computing PD^g [24]. Ortega et al. [25] presented a method to locally minimize the *kinetic distance* between the configurations of a haptic probe and its proxy using constraint-based dynamics and continuous collision detection. Zhang et al. [5] proposed the first rigorous formulation of computing PD^g. They presented an efficient algorithm to compute PD^g for convex polytopes, and provide bounds on PD^g of non-convex polyhedra. The problem of PD^g computation is closely related to the containment problem [26]. The notion of *growth distance* has been introduced to unify separation and penetration distances [22]. Recently, Nawratil et al. [27] have also described a constrained optimization based algorithm for PD^g computation.

B. *Distance Metrics in Configuration Space*

The distance metric in configuration space is used to measure the distance between two configurations in the space. It is well-known that *model-independent* metrics are not bi-invariant, and thus most approaches use *model-dependent metrics* for proximity computations [11], [14], [28].

1) Distance Metrics in SE(3): The spatial rigid body displacements form a group of rigid body motion, SE(3). Throughout the rest of the paper, we will refer to a model-independent distance metric in SE(3) as a distance metric in SE(3). In theory, there is no natural choice for distance metrics in SE(3) [12], [13]. Loncaric [29] showed that there is no bi-invariant Riemannian metric in SE(3).

2) Model-dependent Distance Metrics: Using the notion of a *displacement vector* for each point in the model, the DISP distance metric is defined as the maximum length over all the displacement vectors [14], [28], [30]. The object norm, proposed by [15], is defined as an average squared length of all displacement vectors. Hofer and Pottmann [31] proposed a similar metric, but consider only a set of feature points in the model. All of these displacement vector-based metrics can be efficiently evaluated. The length of a trajectory travelled by a point on a moving model can be also used to define model-dependent metrics [5], [32]. However, it is difficult to compute the exact value of these metrics.

III. GENERALIZED PENETRATION DEPTH AND DISTANCE METRICS

In this section, we introduce our notation and highlight issues in choosing an appropriate distance metric for defining PD^g for polyhedral models. We then show that our metrics can naturally combine translational and rotational motions, have invariance properties, and can be rapidly calculated. We also show that the optimal solution for PD^g computation with respect to each metric exists on the contact space.

A. *Notation and Definitions*

We first introduce some terms and notation used throughout the rest of the paper. We define the *contact space*, $\mathscr{C}_{contact}$, as a subset of the configuration space, $\mathscr{C}$, that consists of the configurations at which a robot A only touches one or more obstacles without any penetration. The union of free space $\mathscr{F}$ and contact space constitutes the valid space, $\mathscr{C}_{valid}$, of the robot, and any configuration in $\mathscr{C}_{valid}$ is a *valid* configuration. The complement of $\mathscr{F}$ in $\mathscr{C}$ is the C-obstacle space or $\mathscr{O}$.

PD^g is a measure to quantify the amount of interpenetration between two overlapping models. Given a distance metric δ in configuration space, PD^g between two polyhedral models A and B can be defined as:

$$PD^g_\delta(A,B) = \{min\{\delta(\mathbf{q_o},\mathbf{q})\} \,\|\, \text{interior}(A(\mathbf{q})) \cap B = \emptyset, \mathbf{q} \in \mathscr{C}\}, \tag{1}$$

where $\mathbf{q_o}$ is the initial configuration of A, and $\mathbf{q}$ is any configuration in $\mathscr{C}$.

PD^g can be formulated as an optimization problem under non-penetration constraints (Fig. 1(a)), where the optimization objective is described by some distance metric to measure the extent of a model transformed from one configuration to another. Therefore, the computation of PD^g is directly governed by the underlying distance metric.

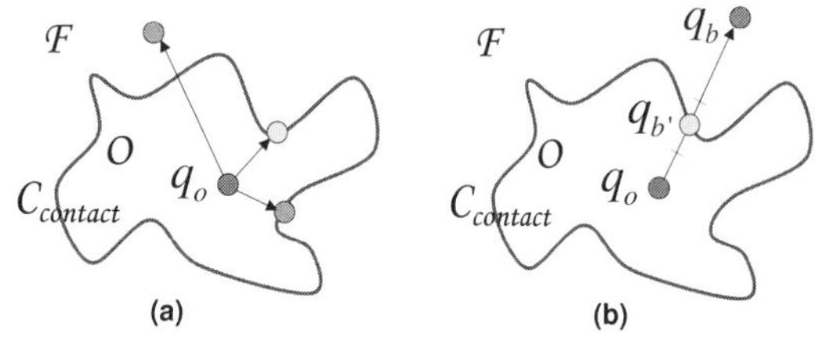

Fig. 1. PD^g **Definition and Contact Space Realization**: *(a)* PD^g *is defined as the minimal distance between the initial collision configuration* $\mathbf{q_o}$ *and any free or contact configuration, with respect to some distance metric. (b) The optimal configuration* $\mathbf{q_b}$, *which realizes* PD$^g{}_{\text{DISP}}$ *or* PD$^g{}_\sigma$, *must be on the contact space* $\mathscr{C}_{contact}$; *otherwise, one can compute another contact configuration* $\mathbf{q_b}'$, *which further reduces the objective function.* $\mathbf{q_b}'$ *is computed by applying the* **bisection** *method on the* screw motion *that interpolates* $\mathbf{q_o}$ *and* $\mathbf{q_b}$.

B. Distance Metric

We address the issue of choosing an appropriate distance metric to define PD^g. In principle, any distance metric in C-space can be used to define PD^g. We mainly use two distance metrics for rigid models, *displacement distance metric* DISP [28], [30] and *object norm* [15].

1) Displacement distance metric: Given a model A at two different configurations $\mathbf{q_a}$ and $\mathbf{q_b}$, the *displacement distance metric* is defined as the longest length of the displacement vectors of all the points on A [28], [30]:

$$\text{DISP}_A(\mathbf{q_a},\mathbf{q_b}) = \max_{\mathbf{x}\in A} ||\mathbf{x}(\mathbf{q_b}) - \mathbf{x}(\mathbf{q_a})||_2. \qquad (2)$$

2) Object norm: Also based on displacement vectors, Kazerounian and Rastegar [15] make use of an integral operator to define the *object norm*:

$$\sigma_A(\mathbf{q_a},\mathbf{q_b}) = \frac{1}{V}\int_A \rho(\mathbf{x})||\mathbf{x}(\mathbf{q_b}) - \mathbf{x}(\mathbf{q_a})||^2 dV, \qquad (3)$$

where V and $\rho(\mathbf{x})$ are the volume and mass distribution of A, respectively.

3) Properties of DISP and σ: Both metrics can combine the translational and rotational components of SE(3) without relying on the choice of any weighting factor to define PD^g. Since both metrics are defined by using displacement vectors, they have some *invariance properties*; they are independent of the choice of inertial reference frame and body-fixed reference frame [11], and also independent of the representation of $\mathscr{C}$.

Moreover, DISP and σ metrics can be computed efficiently. In [14], we show that for a rigid model, the DISP distance is realized by a vertex on its convex hull. This leads to an efficient algorithm, C-DIST, to compute DISP. For σ, by using a quaternion representation, we can further simplify the formula originally derived by Kazerounian and Rastegar [15] into:

$$\sigma_A(\mathbf{q_a},\mathbf{q_b}) = \frac{4}{V}(I_{xx}q_1^2 + I_{yy}q_2^2 + I_{zz}q_3^2) + q_4^2 + q_5^2 + q_6^2, \qquad (4)$$

where $diag(I_{xx}, I_{yy}, I_{zz})$ forms a diagonal matrix computed by diagonalizing the inertia matrix I of A. (q_0, q_1, q_2, q_3) is the quaternion for the **relative** orientation of A between $\mathbf{q_a}$ and $\mathbf{q_b}$, and (q_4, q_5, q_6) is the relative translation.

C. Properties of PD^g_{DISP} and PD^g_σ

Geometrically speaking, the generalized penetration depth under DISP, PD$^g{}_{\text{DISP}}$, can be interpreted as the minimum of the maximum lengths of the displacement vectors for all the points on A, when A is placed at any collision-free or contact configuration. Also, the generalized penetration depth under σ, PD$^g{}_\sigma$, can be interpreted as the minimum cost to separate A from B, where the cost is related to the kinetic energy of A.

Due to the underlying distance metric, both PD$^g{}_{\text{DISP}}$ and PD$^g{}_\sigma$ are independent of the choice of inertial and body-fixed reference frames. In practice, these invariance properties are quite useful since one can choose any arbitrary reference frame and representation of the configuration space to compute PD$^g{}_{\text{DISP}}$ and PD$^g{}_\sigma$.

D. Contact Space Realization

For rigid models, PD$^g{}_{\text{DISP}}$ (or PD$^g{}_\sigma$) has a contact space realization property. This property implies that any valid configuration $\mathbf{q_b}$ that minimizes the objective DISP (or σ) for PD^g must lie on the contact space of A and B, or equivalently, at this configuration $\mathbf{q_b}$, A and B just touch with each other.

Theorem 1 (Contact Space Realization) *For a rigid model A placed at* $\mathbf{q_o}$, *and a rigid model B, if* $\mathbf{q_b} \in \mathscr{C}_{valid}$ *and* $\text{DISP}_A(\mathbf{q_o},\mathbf{q_b}) = \text{PD}^g{}_{\text{DISP}}(A,B)$, *then* $\mathbf{q_b} \in \mathscr{C}_{contact}$. *A similar property holds for* PD$^g{}_\sigma$.

Proof: We prove it by contradiction. Suppose the configuration $\mathbf{q_b}$ realizing PD$^g{}_{\text{DISP}}$ does not lie on the contact space $\mathscr{C}_{contact}$. Then, $\mathbf{q_b}$ must lie in the free space $\mathscr{F}$ ((Fig. 1(b)).

We use Chasles' theorem in Screw theory [33], which states that a rigid body transformation between any two configurations can be realized by rotation about an axis followed by translation parallel to that axis, where the amount of rotation is within $[0,\pi]$. The screw motion is a curve in C-space, and we denote that curve between $\mathbf{q_o}$ to $\mathbf{q_b}$ as $s(t)$, where $s(0) = \mathbf{q_o}$ and $s(1) = \mathbf{q_b}$. Since $\mathbf{q_o}$ is in $\mathscr{O}$, and $\mathbf{q_b}$ is in $\mathscr{F}$, there is at least one intersection between the curve $\{s(t)|t \in [0,1]\}$ and the contact space (Fig. 1). We denote the intersection point as $\mathbf{q_b}'$.

Based on Chasles theorem, we can compute the length of the displacement vector for any point $\mathbf{x}$ on A between $\mathbf{q_o}$ and any configuration on the screw motion $s(t)$. Furthermore, we can show that this length strictly increases with the parameter t. Therefore, for each point on A, the length of the displacement vector between $\mathbf{q_o}$ and $\mathbf{q_b}$ is less than the one between $\mathbf{q_o}$ and $\mathbf{q_b}'$. Since DISP metric uses the maximum operator for the length of the displacement vector over all points on A, we can

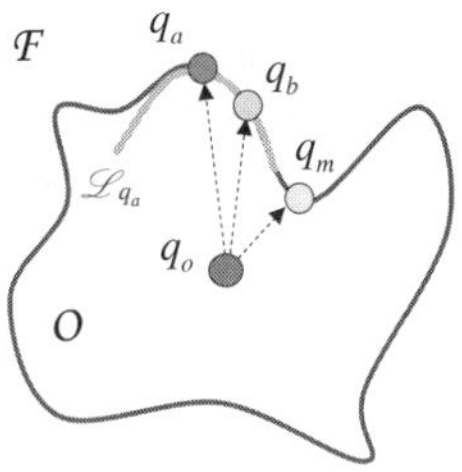

Fig. 2. **Optimization-based** PD^g **Algorithm**: *our algorithm walks on the contact space* $\mathscr{C}_{contact}$*, i.e. from* $\mathbf{q_a}$ *to* $\mathbf{q_b}$*, to find a local minimum* $\mathbf{q}_m$ *under any distance metric.*

infer that $DISP_A(\mathbf{q_o}, \mathbf{q_b}') < DISP_A(\mathbf{q_o}, \mathbf{q_b})$. This contradicts our assumption that $\mathbf{q_b}$ is the realization for $PD^g{}_{DISP}$.

Similarly, we can infer $\sigma_A(\mathbf{q_o}, \mathbf{q_b}') < \sigma_A(\mathbf{q_o}, \mathbf{q_b})$, and thus prove the property for $PD^g{}_\sigma$. ■

According to Thm. 1, in order to compute PD^g, it is sufficient to search only the contact space $\mathscr{C}_{contact}$, which is one dimension lower than that of $\mathscr{C}$. Our optimization-based algorithm for PD^g uses this property.

IV. *PD^g* Computation Algorithm

In this section, we present our PD^g computation algorithm. Our algorithm can optimize any distance metric (or objective) presented in Sec. 3 by performing incremental refinement on the contact space. As Fig. 2 illustrates, our iterative optimization algorithm consists of three major steps:

1) Given an initial contact configuration $\mathbf{q_a}$, the algorithm first computes a local approximation $\mathscr{L}_{\mathbf{q_a}}$ of the contact space around $\mathbf{q_a}$.
2) The algorithm searches over the local approximation to find a new configuration $\mathbf{q_b}$ that minimizes the objective function.
3) The algorithm assigns $\mathbf{q_b}$ as a starting point for the next iteration (i.e. *walk* from $\mathbf{q_a}$ to $\mathbf{q_b}$) if $\mathbf{q_b}$ is on the contact space with smaller value of the objective function as compared to $\mathbf{q_a}$'s. Otherwise, we compute a new contact configuration $\mathbf{q_b}'$ based on $\mathbf{q_b}$.

These steps are iterated until a local minimum configuration $\mathbf{q_m}$ is found or the maximum number of iterations is reached. Next, we discuss each of these steps in more detail. Finally, we address the issue of computing an initial guess.

A. Local Contact Space Approximation

Since it is computationally prohibitive to compute a global representation of the contact space $\mathscr{C}_{contact}$, our algorithm computes a local approximation. Given a configuration $\mathbf{q_a}$, where A is in contact with B, we enumerate all contact constraints according to the pairs of contact features [28], [34]. We further decompose each contact constraint into *primitive contact constraints*, i.e. vertex/face $(v-f)$, face/vertex$(f-v)$ or edge/edge $(e-e)$. Conceptually, each primitive contact constraint represents a halfspace, and the set of all primitive constraints are used to characterize the local non-penetration

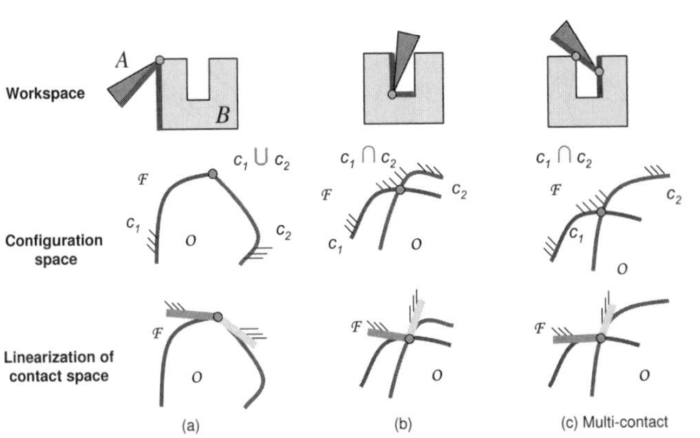

Fig. 3. **Local Contact Space Approximation**: *the local contact space is algebraically represented as a set of contact constraints concatenated with intersection or union operators (Eq. 5). Columns (a) and (b) explain how to obtain proper operators when decomposing a constraint into primitive contact constraints using 2D examples (cf. Sec IV.A). Column (c) shows a multiple contact situation. The last row illustrates the corresponding linearization for each local contact space.*

computation between A and B. Therefore, we obtain a local approximation $\mathscr{L}_{\mathbf{q_a}}$ of $\mathscr{C}_{contact}$ around the contact configuration $\mathbf{q_a}$ after concatenating all these primitive constraints $\{C_i\}$ using proper intersection or union operators $\{\circ_i\}$:

$$\mathscr{L}_{\mathbf{q_a}} = \{C_1 \circ_1 C_2 \cdots \circ_{n-1} C_n\}. \quad (5)$$

It should be noted that we do not explicitly compute a geometric representation of $\mathscr{L}_{\mathbf{q_a}}$. Instead, it is algebraically represented, and each primitive constraint is simply recorded as a pair of IDs, identifying the contact features from A and B, respectively.

When decomposing each constraint into primitive constraints, we need to choose proper Boolean operators to concatenate the resulting primitive constraints. This issue has been addressed in the area of dynamics simulation [35] and we address it in a similar manner for PD^g computation. Fig. 3 shows a 2D example with a triangle-shaped robot A touching a notch-shaped obstacle B. When decomposing a $v-v$ contact constraint into two $v-e$ constraints C_1 and C_2, if both of the contact vertices of A and B are convex (Fig. 3(a)), we use a union operator, because if either constraint C_1 or C_2 is enforced, there is no local penetration. Otherwise, if one contact vertex is non-convex (Fig. 3(b)), the intersection operation is used. For 3D models, a similar analysis is performed by identifying the convexity of edges based on their dihedral angles. In case of multiple contacts, one can first use intersection operations to concatenate all the constraints. Each individual constraint is then further decomposed into primitive constraints.

B. Searching over Local Contact Space

Given a local contact space approximation $\mathscr{L}$ of the contact configuration $\mathbf{q_a}$, we search over $\mathscr{L}$ to find $\mathbf{q_b}$ that minimizes the objective function. Since the contact space is a non-linear subspace of $\mathscr{C}$, we use two different search methods: random sampling in $\mathscr{L}$ and optimization over a first-order approximation of $\mathscr{L}$. Each of them can be performed independently.

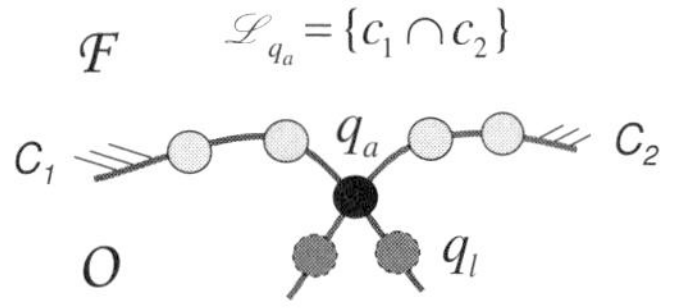

Fig. 4. **Sampling in Local Contact Space**: *$\mathscr{L}_{\mathbf{q_a}}$ is a local approximation of contact space around $\mathbf{q_a}$, represented by the intersection of its contact constraints C_1 and C_2. Our algorithm randomly generates samples on C_1 and C_2. Many potentially infeasible samples, such as $\mathbf{q}_l$, can be discarded since they are lying outside the halfspace of $\mathscr{L}_{\mathbf{q_a}}$.*

1) Sampling in Local Contact Space: Our algorithm randomly generates samples on the local contact approximation $\mathscr{L}_{\mathbf{q_a}}$ around $\mathbf{q_a}$ (Fig. 4), by placing samples on each primitive contact constraint C_i as well as on their intersections [36]. We discard any generated sample $\mathbf{q}$ if it lies outside of the halfspace formulated by $\mathscr{L}_{\mathbf{q_a}}$ by simply checking the sign of $\mathscr{L}_{\mathbf{q_a}}(\mathbf{q})$. Since $\mathscr{L}_{\mathbf{q_a}}$ is a local contact space approximation built from all contact constraints, this checking of $\mathscr{L}$ allows us to cull potentially many infeasible colliding configurations. For the rest of the configuration samples, we evaluate their distances δ to the initial configuration $\mathbf{q_o}$, and compute the minimum.

These samples are efficiently generated for each non-linear contact constraint C_i. First, we generate random values for the rotation parameters. By plugging these values into a non-linear contact constraint, we formulate a linear constraint for the additional translation parameters. Under the formulated linear constraint, random values are generated for these translation parameters.

In practice, an optimal solution for PD^g may correspond to multiple contacts, suggesting that one needs to generate more samples on the boundary formed by multiple contact constraints. As a result, we set up a system of non-linear equations for each combination of these constraints, generate random values for the rotation parameters in the system (thereby making the system linear), and sample the resulting linear system for the translation parameters.

2) Linearizing the Local Contact Space: We search for a configuration with smaller distance to the contact space by linearly approximating the contact space. For each basic contact constraint C_i, we compute its Jacobian, which is the normal of the corresponding parameterized configuration space. Using this normal, we obtain a half-plane, which is a linearization of the contact surface [21], [37]. By concatenating the half-planes using Boolean operators $\circ_i$, we generate a non-convex polyhedral cone, which serves as a local linear approximation of $\mathscr{C}_{contact}$.

3) Local Search: The sampling-based method is general for any distance metric. Moreover, we can generate samples on each non-linear contact constraint efficiently. Finally, using the local contact space approximation, our method can cull many potentially infeasible samples.

On the other hand, the method of linearizing the contact space

---

Algorithm 1 Optimization-based Local PD^g Algorithm
Input: two intersecting polyhedra: A - movable, B - static.
$\mathbf{q_o}$:= the initial collision configuration of A, $\mathbf{q_o} \in \mathscr{O}$.
$\mathbf{q_a}$:= a seed contact configuration of A, $\mathbf{q_a} \in \mathscr{C}_{contact}$.
Output: $PD^g(A, B)$

```
repeat
  i++;
  L_qa := Local contact space approximation at qa;
  qb := argmin{δ(qo, q), q ∈ L_qa};
  if δ(qo, qb) == δ(qo, qa) then
    return δ(qo, qa);
  else if qb ∈ C_contact then
    qa := qb;
  else if qb ∈ F then
    qa := CCD_Bisection(qo, qb);
  else
    qb' := CCD(qa, qb);
    L_qa := L_qa ∩ L'_qb;
    goto 3;
  end if
until i < MAX_ITERATION
```

is suitable for optimizing PD^g, if the underlying objective has a closed form. For example, for the *object norm*, we transform the coordinate in the quadratic function in Eq. (4), from an elliptic form to a circular one. Now, the problem of searching over $\mathscr{L}$ reduces to finding the closest point in the Euclidean space from $\mathbf{q_a}$ to the non-convex polyhedral cone, formulated using the linearization of $\mathscr{L}$. Since the polyhedral cone is formulated as a local approximation of $\mathscr{C}_{contact}$, it typically has a small size. Therefore, the closest point query can be performed by explicitly computing the non-convex polyhedral cone.

C. Refinement

Although searching over the local contact space $\mathscr{L}$ around $\mathbf{q_a}$ can yield a new configuration $\mathbf{q_b}$ that improves the optimization objective of $\mathbf{q_a}$, we still need to check whether $\mathbf{q_b}$ is a valid contact configuration before advancing to it because $\mathbf{q_b}$ is computed based upon a local approximation of contact space and $\mathbf{q_b}$ may not be on the contact space.

For instance, the new configuration $\mathbf{q_b}$ may be a collision-free configuration due to the first-order approximation. To handle this case, we project $\mathbf{q_b}$ back to $\mathscr{C}_{contact}$ by computing the intersection $\mathbf{q_b}'$ between the contact space and a curve interpolating from $\mathbf{q_o}$ to $\mathbf{q_b}$ using screw motion (Fig. 1). Since $\mathbf{q_o}$ is in $\mathscr{O}$ and $\mathbf{q_b}$ is free, the intersection $\mathbf{q_b}'$ can be efficiently computed by bisection (*CCD_Bisection* in Alg. 1). Also, according to the contact space realization theorem in Sec. III.D, $\delta(\mathbf{q_o}, \mathbf{q_b}') < \delta(\mathbf{q_o}, \mathbf{q_b})$. Therefore, we are guaranteed to obtain a new configuration $\mathbf{q_b}'$, which is closer to $\mathbf{q_o}$, and thus it can be used for successive iterations.

It is also possible the new configuration $\mathbf{q_b}$ may be a colliding configuration. As Fig. 5 on the left shows, when moving from

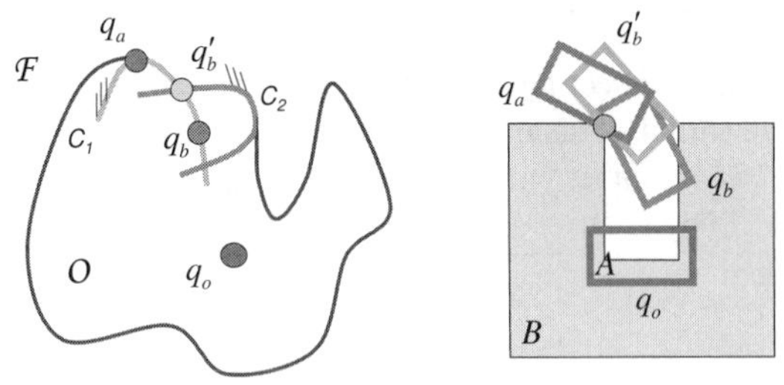

Fig. 5. **Refinement**. *Left: using the local contact space representation of* $\mathbf{q_a}$*, which includes only one constraint* C_1*, we obtain new configuration* $\mathbf{q_b}$*. Though* $\mathbf{q_b}$ *is still on* C_1*, it may not be on the contact space any more, since it will violate other constraint, such as* C_2 *here. The right figure shows a dual example happening in the workspace. When A slides on B, i.e. from* $\mathbf{q_a}$ *to* $\mathbf{q_b}$*, a collision can be created by other portions of the models. Our algorithm uses CCD to compute a correct, new contact configuration* $\mathbf{q_b}'$*.*

$\mathbf{q_a}$ to $\mathbf{q_b}$, the contact constraint C_1 is maintained. However, $\mathbf{q_b}$ is a colliding configuration as it does not satisfy the new constraint C_2. The figure on the right highlights this scenario in the workspace. When A moves from $\mathbf{q_a}$ to $\mathbf{q_b}$, the contact is still maintained. In order to handle this case, we use *continuous collision detection* (CCD) to detect the time of first collision when an object continuously moves from one configuration to another using a linearly interpolating motion in $\mathscr{C}$ [38]. In our case, when A moves from $\mathbf{q_a}$ to $\mathbf{q_b}$, we ignore the sliding contact of $\mathbf{q_a}$, and use CCD to report the first contact $\mathbf{q_b}'$ before the collision [39]. The new configuration $\mathbf{q_b}'$ can be used to update the local approximation of $\mathbf{q_a}$. This yields a more accurate contact space approximation and consequently improves the local search, e.g. culling away additional invalid samples.

D. Initial Guess

The performance of the PD^g algorithm depends on a good initial guess. For many applications, including dynamic simulation and haptic rendering, the motion coherence can be used to compute a good initial guess. Since no such motion coherence could be exploited in some other applications (e.g. sample-based motion planning), we propose a heuristic. Our method generates a set of samples on the contact space as a preprocess. At runtime, given a query configuration $\mathbf{q_o}$, our algorithm searches for the K nearest neighbors from the set of precomputed samples, and imposes the inter-distance between any pair of these K samples should be greater than some threshold. The distance metric used for nearest neighbor search is the same as the one to define PD^g. The resulting K samples serve as initial guesses for our PD^g algorithms. To generate samples on the contact space, we randomly sample the configuration space and enumerate all pairs of free and collision samples. For each pair, a contact configuration can be computed by a bisection method (Fig. 1(b)).

V. Implementation and Performance

We have implemented our PD^g algorithm using local contact space sampling for general non-convex polyhedra. In this section, we discuss some important implementation issues and highlight the performance of our algorithm on a set of complex polyhedral models. All the timeings reported here were taken on a Windows PC, with 2.8GHZ of CPU and 2GB of memory.

A. Implementation

Since our PD^g formulation is independent of the representation of the configuration space, we use a *quaternion* to represent the rotation because of its simplicity and efficiency. In our PD^g algorithm, any proximity query package supporting collision detection or contact determination can be employed. In our current implementation, we use the SWIFT++ collision detection library, because of its efficiency and it provides both these proximity queries [40]. Based on SWIFT++, our algorithm computes all the contacts between A at a contact configuration $\mathbf{q_a}$ with B. We sample the contact space locally around $\mathbf{q_a}$. For each primitive contact constraint C_i, we derive its implicit equation with respect to the parameters of a rotation component (a quaternion) and a translation component (a 3-vector). In order to sample on a constraint C_i, we first slightly perturb its rotational component by multiplying a random quaternion with a small rotational angle. The resulting rotational component is plugged back into the constraint C_i. This yields a linear constraint with only translational components, and therefore can be used to generate additional samples. To linearize C_i, we compute the Jacobian of its implicit equation for C_i . For other types of contacts, we decompose them into primitive contact constraints. Proper operators to concatenate them are identified by computing the dihedral angle of contacting edges, thereby determining whether the contact features are convex or not.

In the refinement step of the algorithm, we perform collision detection using SWIFT++ to check whether $\mathbf{q_b}$ from the local search step still lies on the contact space. When $\mathbf{q_b}$ is on contact space, our algorithm proceeds to the next iteration. Otherwise, when $\mathbf{q_b}$ is free, a new contact configuration $\mathbf{q_b}'$ is computed for the next iteration by performing recursive bisections (Fig. 1(b)) on the screw motion interpolating between $\mathbf{q_o}$ and $\mathbf{q_b}$. Finally, when $\mathbf{q_b}$ is in C-obstacle space, we compute a new contact configuration $\mathbf{q_b}'$ by using CCD. In our current implementation, we check for collision detection on a set of discrete samples on a linear motion between $\mathbf{q_a}$ and $\mathbf{q_b}$. In order to ignore the old contact during CCD query, the idea of *security distance* is used [39]. After computing a new contact configuration $\mathbf{q_b}'$ from the CCD query, our algorithm updates the local approximation around $\mathbf{q_a}$ and resumes a local search again.

B. Performance

We use different benchmarks to test the performance of our algorithm. Fig. 6(a) shows a typical setup of our experiment including two overlapping models, where A ('Pawn') is movable and B ('CAD Part') is stationary. In (b), our algorithm computes PD$^g{}_{\text{DISP}}$ or PD$^g{}_\sigma$ to separate the model A, initially placed at A_0, from the model B. The three images on the right highlight the intermediate configurations of A_1 and A_2 and a PD$^g{}_{\text{DISP}}$ solution A_3 with yellow color. The sequence of images (b,c,d,e) illustrates that our algorithm successfully finds

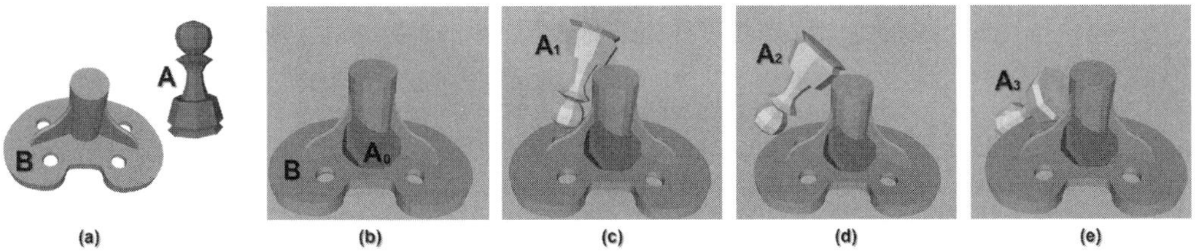

Fig. 6. **The 'CAD Part' example**: *(a) shows the models A - 'pawn' and B - 'CAD Part' used in this test. (b) illustrates a typical* PD^g *query scenario where the model A at* A_0 *overlaps with B.* A_1 *and* A_2 *are intermediate placements of A during the optimization for* $PD^g{}_{DISP}$. A_3 *is the solution for an upper bound of* $PD^g{}_{DISP}$. *The sequence of images (c,d,e) illustrates that our algorithm incrementally slides the model 'pawn' on the model 'CAD Part' to minimize* DISP *distance.*

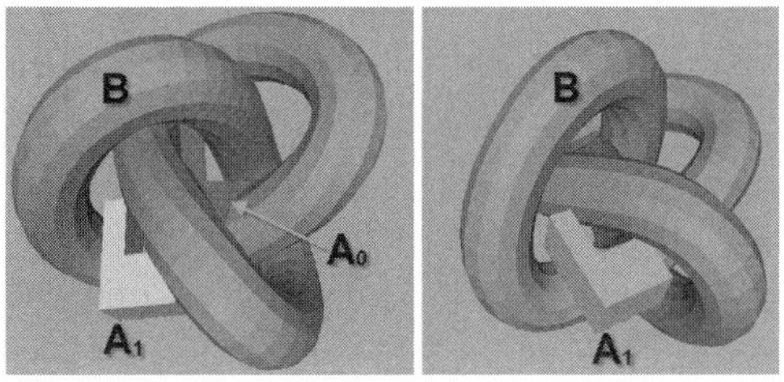

Fig. 7. **The 'torus-with-knot' example**: *the left image highlights a* PD^g *query between a model 'torus-with-knot' B intersecting with a model 'L-shaped box' at* A_0 *(red).* A_1 *is a collision-free placement of the 'L-shaped box' model as a result of* $PD^g{}_\sigma$*; the right image shows the same result but from another viewpoint.*

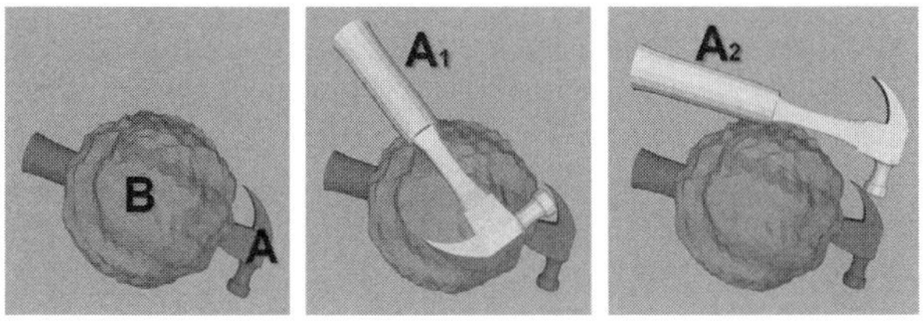

Fig. 8. **The 'hammer' example**: *from left to right:* $PD^g{}_{DISP}$ *query between A and B, an intermediate configuration* A_1*, and the solution* A_2.

an upper bound of $PD^g{}_{DISP}$ by gradually sliding the 'pawn' model on the 'CAD' model.

Figs. 7 and 8 show two more complex benchmarks that we have tested. In Fig. 7, the model 'torus-with-a-knot' has hyperbolic surfaces. This benchmark is difficult for the *containment optimization*-based algorithm [5], as that algorithm computes the convex decomposition of the complement of the model. On the other hand, our PD^g algorithm can easily handle this benchmark, and compute a tight upper bound on PD^g.

Table I summarizes the performance of our algorithm on different benchmarks. In our implementation, we set the maximum number of iterations as 30. For the most of the models we have tested, our algorithm can perform $PD^g{}_{DISP}$ query within 300*ms*, and $PD^g{}_\delta$ query with 450*ms*. Our current implementation is not optimized and the timings can be further improved.

C. Comparison and Analysis

Compared to prior method for PD^g computation in [5], our method can handle more complex non-convex models. This is because we reduce PD^g computation to proximity queries such as collision detection and contact determination. Since there are well known efficient algorithms for both these queries, our method can handle complex non-convex models. Instead, the method in [5] reduces PD^g computation to *containment optimization*, which suffers from enumerating convex containers using convex decomposition, and can result in overly conservative query results for non-convex models.

	1	2	3
A	L-Shape	Pawn	Hammer
tris #	20	304	1,692
B	Torus-with-a-Knot	CAD	Bumpy-Sphere
tris #	2,880	2,442	2,880
Avg $PD^g{}_{DISP}$(ms)	219	297	109
Avg $PD^g{}_\sigma$(ms)	156	445	138

TABLE I

Performance: *this table highlights the geometric complexity of different benchmarks we tested, as well as the performance of our algorithm.*

Our algorithm computes an upper bound on PD^g, since the resulting configuration is guaranteed to be on the contact space. Moreover, in general, the algorithm converges to a local minimum due to the constrained optimization formulation. The termination condition for the iterative optimization is to check whether the gradient of the distance metric is proportional to that of the contact constraint after each iteration. However, some issues arise in checking this condition in practice. For example, in the case of DISP metric, one can only compute an approximation of the gradient, since no closed form is available for DISP metric. Furthermore, a convergence analysis is difficult, due to the discontinuity in contact space caused by multiple contacts.

VI. Conclusion

We present a practical algorithm to compute PD^g for non-convex polyhedral models. Using model-dependent distance metrics, we reduce the PD^g computation to a constrained optimization problem. Our algorithm performs optimization on the contact space, and the experimental results show that we can efficiently compute a tight upper bound of PD^g.

The main limitation of our approach is that our algorithm can not guarantee a global solution for PD^g computation. Its performance depends on the choice of an initial guess. For future work, it is worthwhile to analyze the convergence properties of the algorithm, as well as an error bound on the approximation. We would also like to apply our algorithm to motion planning, dynamic simulation, and haptic rendering.

Acknowledgements: We would like to thank Ming C. Lin and

Stephane Redon for providing helpful discussions. This project was supported in part by ARO Contracts DAAD19-02-1-0390 and W911NF-04-1-0088, NSF awards 0400134 and 0118743, ONR Contract N00014-01-1-0496, DARPA/RDECOM Contract N61339-04-C-0043 and Intel. Young J. Kim was supported in part by the grant 2004-205-D00168 of KRF, the STAR program of MOST and the ITRC program.

References

[1] B. V. Mirtich, "Impulse-based dynamic simulation of rigid body systems," Ph.D. dissertation, University of California, Berkeley, 1996.

[2] D. E. Stewart and J. C. Trinkle, "An implicit time-stepping scheme for rigid body dynamics with inelastic collisions and coulomb friction," *International Journal of Numerical Methods in Engineering*, vol. 39, pp. 2673–2691, 1996.

[3] D. Hsu, L. Kavraki, J. Latombe, R. Motwani, and S. Sorkin, "On finding narrow passages with probabilistic roadmap planners," *Proc. of 3rd Workshop on Algorithmic Foundations of Robotics*, pp. 25–32, 1998.

[4] M. Saha, J. Latombe, Y. Chang, Lin, and F. Prinz, "Finding narrow passages with probabilistic roadmaps: the small step retraction method," *Intelligent Robots and Systems*, vol. 19, no. 3, pp. 301–319, Dec 2005.

[5] L. Zhang, Y. Kim, G. Varadhan, and D.Manocha, "Generalized penetration depth computation," in *ACM Solid and Physical Modeling Symposium (SPM06)*, 2006, pp. 173–184.

[6] D. Dobkin, J. Hershberger, D. Kirkpatrick, and S. Suri, "Computing the intersection-depth of polyhedra," *Algorithmica*, vol. 9, pp. 518–533, 1993.

[7] D. Halperin, "Arrangements," in *Handbook of Discrete and Computational Geometry*, J. E. Goodman and J. O'Rourke, Eds. Boca Raton, FL: CRC Press LLC, 2004, ch. 24, pp. 529–562.

[8] Y. J. Kim, M. C. Lin, and D. Manocha, "Fast penetration depth computation using rasterization hardware and hierarchical refinement," *Proc. of Workshop on Algorithmic Foundations of Robotics*, 2002.

[9] B. Heidelberger, M. Teschner, R. Keiser, M. Mueller, and M. Gross, "Consistent penetration depth estimation for deformable collision response," in *Proceedings of Vision, Modeling, Visualization VMV'04*, November 2004, pp. 339–346.

[10] A. Sud, N. Govindaraju, R. Gayle, I. Kabul, and D. Manocha, "Fast proximity computation among deformable models using discrete voronoi diagrams," *Proc. of ACM SIGGRAPH*, pp. 1144–1153, 2006.

[11] Q. Lin and J. Burdick, "Objective and frame-invariant kinematic metric functions for rigid bodies," *The International Journal of Robotics Research*, vol. 19, no. 6, pp. 612–625, Jun 2000.

[12] J. Loncaric, "Normal forms of stiffness and compliance matrices," *IEEE Journal of Robotics and Automation*, vol. 3, no. 6, pp. 567–572, December 1987.

[13] F. Park, "Distance metrics on the rigid-body motions with applications to mechanism design," *ASME J. Mechanical Design*, vol. 117, no. 1, pp. 48–54, March 1995.

[14] L. Zhang, Y. Kim, and D. Manocha, "C-DIST: Efficient distance computation for rigid and articulated models in configuration space," in *ACM Solid and Physical Modeling Symposium (SPM07)*, 2007, pp. 159–169.

[15] K. Kazerounian and J. Rastegar, "Object norms: A class of coordinate and metric independent norms for displacement," in *Flexible Mechanism, Dynamics and Analysis: ASME Design Technical Conference, 22nd Biennial Mechanisms Conference*, G. K. et al, Ed., vol. 47, 1992, pp. 271–275.

[16] M. Lin and D. Manocha, "Collision and proximity queries," in *Handbook of Discrete and Computational Geometry*, 2003.

[17] C. Ericson, *Real-Time Collision Detection*. Morgan Kaufmann, 2004.

[18] P. Agarwal, L. Guibas, S. Har-Peled, A. Rabinovitch, and M. Sharir, "Penetration depth of two convex polytopes in 3d," *Nordic J. Computing*, vol. 7, pp. 227–240, 2000.

[19] G. van den Bergen, "Proximity queries and penetration depth computation on 3d game objects," *Game Developers Conference*, 2001.

[20] Y. Kim, M. Lin, and D. Manocha, "Deep: Dual-space expansion for estimating penetration depth between convex polytopes," in *Proc. IEEE International Conference on Robotics and Automation*, May 2002.

[21] S. Redon and M. Lin, "Practical local planning in the contact space," *ICRA*, 2005.

[22] C. Ong, "Penetration distances and their applications to path planning," Ph.D. dissertation, Michigan Univ., Ann Arbor., 1993.

[23] ——, "On the quantification of penetration between general objects," *International Journal of Robotics Research*, vol. 16, no. 3, pp. 400–409, 1997.

[24] V. Milenkovic and H. Schmidl, "Optimization based animation," in *ACM SIGGRAPH 2001*, 2001.

[25] M. Ortega, S. Redon, and S. Coquillart, "A six degree-of-freedom god-object method for haptic display of rigid bodies," in *IEEE Virtual Reality*, 2006.

[26] V. Milenkovic, "Rotational polygon containment and minimum enclosure using only robust 2d constructions," *Computational Geometry*, vol. 13, no. 1, pp. 3–19, 1999.

[27] G. Nawratil, H. Pottmann, and B. Ravani, "Generalized penetration depth computation based on kinematical geometry," Geometry Preprint Series, Vienna Univ. of Technology, Tech. Rep. 172, March 2007.

[28] J. Latombe, *Robot Motion Planning*. Kluwer Academic Publishers, 1991.

[29] J. Loncaric, "Geometrical analysis of compliant mechanisms in robotics," Ph.D. dissertation, Harvard University, 1985.

[30] S. M. LaValle, *Planning Algorithms*. Cambridge University Press (also available at http://msl.cs.uiuc.edu/planning/), 2006.

[31] M. Hofer and H. Pottmann, "Energy-minimizing splines in manifolds," in *SIGGRAPH 2004 Conference Proceedings*, 2004, pp. 284–293.

[32] D. Hsu, J.-C. Latombe, and R. Motwani, "Path planning in expansive configuration spaces," *International Journal of Computational Geometry and Applications*, vol. 9, no. (4 & 5), pp. 495–512, 1999.

[33] R. Murray, Z. Li, and S. Sastry, *A Mathematical Introduction to Robotic Manipulation*. CRC Press, 1994.

[34] J. Xiao and X. Ji, "On automatic generation of high-level contact state space," *International Journal of Robotics Research*, vol. 20, no. 7, pp. 584–606, July 2001.

[35] K. Egan, S. Berard, and J. Trinkle, "Modeling nonconvex constraints using linear complementarity," Department of Computer Science, Rensselaer Polytechnic Institute (RPI), Tech. Rep. 03-13, 2003.

[36] X. Ji and J. Xiao, "On random sampling in contact configuration space," *Proc. of Workshop on Algorithmic Foundation of Robotics*, 2000.

[37] D. Ruspini and O. Khatib, "Collision/contact models for the dynamic simulation of complex environments," *Proc. IEEE/RSJ Int. Conf. on Intelligent Robots and Systems*, 1997.

[38] X. Zhang, M. Lee, and Y. Kim, "Interactive continuous collision detection for non-convex polyhedra," in *Pacific Graphics 2006 (Visual Computer)*, 2006.

[39] S. Redon, "Fast continuous collision detection and handling for desktop virtual prototyping," *Virtual Reality*, vol. 8, no. 1, pp. 63–70, 2004.

[40] S. Ehmann and M. C. Lin, "Accurate and fast proximity queries between polyhedra using convex surface decomposition," *Computer Graphics Forum (Proc. of Eurographics'2001)*, vol. 20, no. 3, pp. 500–510, 2001.

Planning and Control of Meso-scale Manipulation Tasks with Uncertainties

Peng Cheng David J. Cappelleri Bogdan Gavrea Vijay Kumar
GRASP Laboratory
University of Pennsylvania
Philadelphia, PA 19104 USA
{chpeng, dcappell, gavrea, kumar}@grasp.upenn.edu

***Abstract*— We develop a systematic approach to incorporating uncertainty into planning manipulation tasks with frictional contacts. We consider the canonical problem of assembling a peg into a hole at the meso scale using probes with minimal actuation but with visual feedback from an optical microscope. We consider three sources of uncertainty. Because of errors in sensing position and orientation of the parts to be assembled, we must consider uncertainty in the sensed configuration of the system. Second, there is uncertainty because of errors in actuation. Third, there are geometric and physical parameters characterizing the environment that are unknown. We discuss the synthesis of robust planning primitives using a single degree-of-freedom probe and the automated generation of plans for meso-scale manipulation. We show simulation and experimental results in support of our work.**

I. Introduction

Manipulation and assembly tasks are typically characterized by many nominally rigid bodies coming into frictional contacts, possibly involving impacts. Manipulation tasks are difficult to model because uncertainties associated with friction and assembly tasks are particularly hard to analyze because of the interplay between process tolerance and geometric uncertainties due to manufacturing errors. Manipulation at the meso (hundred microns to millimeters) and micro (several microns to tens of microns) scale is even harder because of several reasons. It is difficult to measure forces at the micro-netwon level reliably using off-the-shelf force sensors and good force-feedback control schemes have not proved successful. It is hard to manufacture general-purpose end effectors at this scale and it is even more difficult to grasp and manipulate parts at the micro and meso level than it is at the macro level. Finally, the lack of good models of the mechanics of contact interactions at this scale means that model-based approaches to planning and control are difficult.

The mechanics of pushing operations and sliding objects have been extensively studied in a quasi-static setting in [19, 22]. There is also extensive work addressing the analysis and simulation of mechanical systems with frictional contacts [3, 14, 5]. In particular, semi-implicit and instantaneous-time models for predicting motion and contact forces for quasi-static multi-rigid-body systems have recently been developed [25, 27]. We build on these models and time-stepping algorithms discussed in these papers.

Modeling dry friction is a notoriously difficult problem area. Estimations of friction parameters for pushed objects to

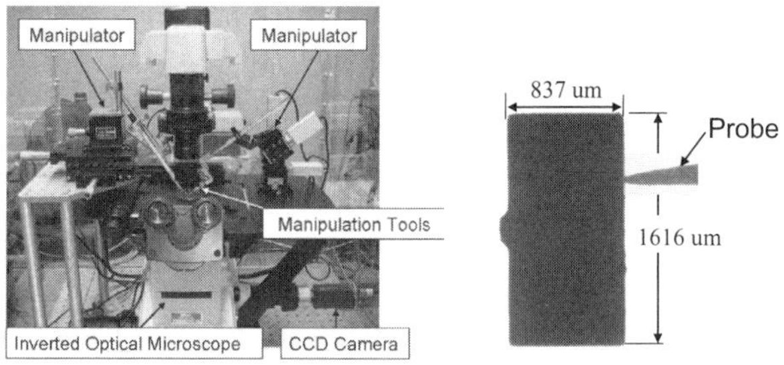

Fig. 1. Our experimental setup (left) and an image from the optical microscope showing the peg and a probe (right).

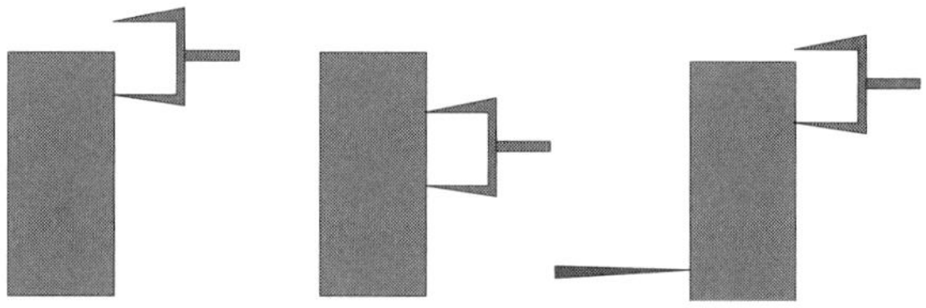

Fig. 2. Planar manipulation with a single degree-of-freedom, dual-tip probe and a passive single-tip probe. There are three sets of operations that can be performed.

improve the control of pushing have been investigated previously on larger objects and with different strategies than the ones presented here. In [17], test pushes on different objects with known support surfaces are used to estimate support surfaces experimentally. It leaves the open question of how the hypothesized support points for an unknown object should be chosen. Similarly in [28], a method for estimating the friction distribution of an object and the center of friction from pushing the object several times is presented. In both of these papers, a grid system of N possible support points is applied to the base of the object being pushed. The respective algorithms determine the distribution of the normal force of the object at these support locations. Similarly, estimates of surface friction for meso-scale manipulation are experimentally determined in [7]. In our experiments the support surface is coated with a thin film of oil which circumvents the difficulties of modeling dry friction.

A good survey of motion planning under uncertainty is available in [10, 15]. Pushing operations and the instantaneous motions of a sliding object during multiple contact pushing is examined and the manipulation primitive of stable rotational pushing is defined in [16]. In [2], the bounds of the possible motions of a pushed object are investigated. [23] presents a comparison between the dynamic and quasistatic motions of a push object. It is well-known that open-loop motion strategies, without the use of sensors, can be used to eliminate uncertainty and to orient polygonal parts [13, 11, 1]. In many cases, the problem of positioning and orienting a planar object with a random initial condition can be reduced to a planning problem which can be solved using a complete, polynomial-time algorithm.

In particular, the problem of finding motion primitives that rely on pushing and are robust to errors has received significant attention. A pushing control system with visual feedback for open-loop pushing is described in [24] as a way to mitigate the instability of pushing with point contacts. To remove the uncertainty associated with robot pushing tasks, [4] establishes stable orientation and positions by pushing objects with two-point fingers. The problem of planning pushing paths using stable pushes with line contact is discussed in [18], and conditions on the pushing directions are derived that ensure that line sticking contact will always be maintained.

Sensorless orientation of parts is applied to micro-scale parts in [20]. At the micro scale, sticking effects due to Van der Walls forces and static electricity make the manipulator motions and part release more complicated [12, 6]. Micro-manipulators also have limited degrees of freedom when compared to manipulators at the macro-scale. These problems are addressed in [20] with a parallel-jaw gripper and squeeze and roll primitives to orient a randomly oriented polygonal part up to 180° symmetry.

In this paper, we develop a formulation of the motion planning problem for manipulation with friction contacts incorporating uncertainty at three levels: (a) Errors in estimates of states (positions and orientations) of the manipulated object; (b) Errors in actuation or input; and (c) Errors in geometric and physical parameters characterizing the assembly task. We consider the canonical problem of assembling a planar, rectangular part into a planar, rectangular slot, but with a single degree-of-freedom probe with two rigid finger tips and a second passive probe (Fig. 2), and with visual feedback from an optical microscope (see Fig. 1). We address the automated generation of motion plans for this assembly task. We argue that it is appropriate to consider a quasi-static model of the manipulation task with Coulomb friction at the contact(s) between the probe and the object. The interaction between the manipulated part and the oil-coated surface is modeled using a viscous friction model. We explicitly model all three sources of uncertainty through experiments and show how motion primitives that are robust to these uncertainties can be identified. We describe a motion planner that can find motion plans composed of motion primitives for this task and illustrate the application through simulation and experiments.

II. Modeling and Definitions

The manipulation problem considered in this paper can be studied in the framework of motion planning for systems that are subject to both differential equations and uncertainties. In this section, we will briefly describe the framework for the general problem and a general planning methodology based on robust motions. The application of the general method in the manipulation problem is described in Section IV.

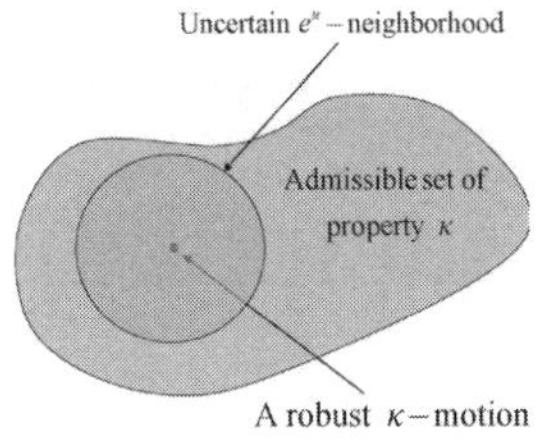

Fig. 3. Robust motion primitives

A. Problem description

Assume that the motion of the control system in the given environment is characterized by $\dot{x} = f(x, u, p)$, in which $x \in X \subset \Re^n$ is the state, $u \in U \subset \Re^m$ is the input, and $p \in P \subset \Re^l$ is the parameters for the system and environment. Given a control $\tilde{u} : [0, t_{\tilde{u}}] \to U$, a parameter history $\tilde{p} : [0, t_{\tilde{p}}] \to P$, and a state $x_0 \in X$ for some $t_{\tilde{u}} > 0$ (varies with $\tilde{u}$), the trajectory (a.k.a. motion) under $\tilde{u}$ and $\tilde{p}$ from x_0 is $\tilde{x}(\tilde{u}, \tilde{p}, x_0, t) = x_0 + \int_0^t f(\tilde{x}(\eta), \tilde{u}(\eta), \tilde{p}(\eta))\, d\eta$.

We consider three bounded uncertainties stemming from sensing, control (actuation), and the environment.

1. Sensing uncertainty We assume that sensors can estimate the global state of the system with bounded error s^u_x. Let x and x^s respectively represent the actual and sensed states of the system. We have $x \in B_{s^u_x}(x^s)$, in which $B_r(x') = \{x \mid \|x, x'\| \le r\}$ is the r-neighborhood of state x with respect to a metric $\|\cdot,\cdot\|$ on X.

2. Control uncertainty We assume that actuators will realize the commanded control with a bounded error $c^u_{\tilde{u}}$. Let $\tilde{u}$ and $\tilde{u}^i$ respectively represent the actual and intended controls for the system. We have $\tilde{u} \in B_{c^u_{\tilde{u}}}(\tilde{u}^i)$.

3. Modeling uncertainty We assume that the geometry and the physics of the underlying model are parameterized by $\tilde{p}$ with bounded error $e^u_{\tilde{p}}$. Let $\tilde{p}$ and $\tilde{p}^n$ respectively represent the actual and nominal parameter history. We have $\tilde{p} \in B_{e^u_{\tilde{p}}}(\tilde{p}^n)$.

Given a sensed initial state x_{init} and a goal set $X_{\text{goal}} = B_\tau(x_{\text{goal}})$ for a specified τ and x_{goal}, the objective is to compute a control $\tilde{u}$ (that may depend on feedback information) which will drive the system from x_{init} to X_{goal} under uncertainties.

B. Planning with robust motion primitive

To solve the above problem is quite difficult. Because complete algorithms are difficult to find except for the simplest of problems, we pursue the synthesis of plans that are obtained by composing *robust motion primitives*. Robust motion primitives are used to define controls whose resulting trajectories will preserve a specified property of interest in the presence of uncertainties. We model a *property of interest* by a characteristic function, κ, which maps a trajectory into 0 or 1. If $\kappa(\tilde{x}) = 1$, then we say that the trajectory $\tilde{x}$ satisfies the given property and is called a *κ-motion*. The *admissible set* for a property κ (see Fig. 3) is $\mathcal{A}_\kappa = \{\tilde{x} \mid \kappa(\tilde{x}) = 1\}$. If the system has uncertainty bound $e^u = (s^u_x, c^u_{\tilde{u}}, e^u_{\tilde{p}})$, the *uncertainty neighborhood* of trajectory $\tilde{x} = (x_0, \tilde{u}, \tilde{p})$ is $\{\tilde{x}' \mid \|x'_0, x_0\| \le s^u_x, \|\tilde{u}', \tilde{u}\| \le c^u_{\tilde{u}}, \|\tilde{p}', \tilde{p}\| \le e^u_{\tilde{p}}\}$. A κ-motion is a robust motion primitive only if its uncertainty neighborhood is contained within the admissible set.

We can now consider the *composition of robust motion primitives*. Let κ_1 and κ_2 be two properties. If there exists a robust κ_1-motion and a robust κ_2-motion such that the κ_1-motion can be reliably appended to the κ_2-motion under uncertainties, then we say that it is possible to sequentially compose the motion primitives.

Thus our approach to planning will involve the construction of a set of robust motion primitives followed by their sequential composition. At this point, a graph search based motion planning algorithm in [15] can be used to synthesize the complete motion plan. It is worth mentioning that such algorithms are not complete because they restrict the search space from the original control space to a smaller one consisting only of robust motion primitives.

In the next section we will describe our experimental testbed and the specifics of the manipulation task before developing models of the manipulation task and robust motion primitives for the task.

III. The Experimental Testbed

The mico-manipulation system (Fig. 1 left) consists of an inverted optical microscope and CCD camera (for sensing the configuration), 4 axis micro-manipulator, controller, 5 μm tip tungsten probes, and control computer. There is a 4X objective on the microscope along with a 0.6X optical coupler producing a field of view (FOV) of 3.37 mm x 2.52 mm. The CCD camera records the images in the FOV and sends them to the control computer at 30 Hz (lower frequency with image processing). The micro-manipulator with controller has a minimum incremental motion of 0.1 μm along four axes, with a maximum travel of 20 mm and with speeds ranging from 1.6 μm/sec to 1.7 mm/sec. We consider two types of probes, a passive Single-Tip Probe (STP) and an active Dual-Tip Probe (DTP). The STP is passive and although it can be positioned, its motion is not controlled during manipulation. The DTP is actuated along one direction (the x-axis) and can be used either for single or two point contact (see Fig. 2).

The control of the DTP is fully characterized by $u = (d_2, v_p, p^t)$ (see Fig. 3), denoting a push in x direction with relative distance d_2 with duration p^t and constant speed v_p. In the experiments in this paper, we choose from one of three discrete values of speeds: v_p = 140.0, 75.0 or 7.4 μm/sec. The other two inputs are continuous.

As mentioned before, there are three sources of uncertainty. The sensing uncertainty arises because of the limitation on the magnification and resolution of the camera. Because with our objective, each pixel subtends only 5.26 μm, our errors in positions are approximately $\pm$ 5 μm and the error in estimating the orientation of our 1616 $\mu\text{m} \times 837\ \mu\text{m}$ part is $\pm$ 0.3 degrees. The control uncertainty exists only in the probe position. The errors in probe position relative to the part are also of the order of $\pm$ 5 μm. Errors in geometric parameters stem from manufacturing imperfections. The part is not a perfect rectangle as shown in Fig. 1. The tips in the DTP are of different length, in which one tip is longer than the other, reflected in the angle β in Fig. 4 (right). However, we assume the exact dimensions are known. The principal source of modeling error stems from surface friction and the coefficient of friction between the probe(s) and the part. We will discuss the dynamic model and the parameters governing this model in greater detail in the next section.

IV. Motion Planning with Uncertainty

A. System dynamics

We use a quasi-static model for the system (inertial forces are of the order of nano-newtons for the accelerations involved, while the frictional forces are on the order of micro-newtons). We assume the support plane to be uniform, and all pushing motions of the probes to be parallel to this plane. The most important assumption is about the support friction. Because we coat the support surface with oil (Extra Heavy Mineral Oil, LSA, Inc.), it is reasonable to assume viscous damping at the interface. Based on experimental data we chose the model $f = Ev$ in which $v = [v_x, v_y, v_\theta]^T$ is the velocity of the part (peg) configuration x, y, θ; f is the corresponding vector of forces and moments; and E is the damping diagonal matrix with diagonal elements e_x, $e_y = e_x$, and e_θ. The coefficient of friction between the probe and the part is μ. These parameters were computed by parameter fitting with experimental results (see Section V-A). Finally, we assume the only contacts that occur are between the probe and the part. Although we consider the assembly task as our goal, we only consider the problem of guiding the part into the designated slot without any collisions with the environment.

From quasi-static analysis, we have $Ev = \sum_i \left(w_n^i \lambda_n^i + w_t^i \lambda_t^i\right)$, where w denotes the wrench vector and λ the magnitude of the contact force, with subscripts n and t indicating normal and tangential directions, and the superscript i denoting the ith contact. Because of space constraints, we do not write the entire model which includes complementarity constraints for sticking, sliding and separation, but instead refer the reader to [21, 27].

We note that the existence of a trajectory for the rigid-body quasi-static model described above may be guaranteed under the assumption that the generalized friction cone is pointed (by pointed cone, we mean a cone that doesn't contain any proper linear subspace). The proof of this result follows the lines of [26] but is omitted because of space constraints. Therefore, for the one-point contact case in Fig. 2 (left), existence is immediately obtained from the linear independence of the normal and tangential wrenches. When the probe pushes the same side of the part with a two-point contact (Fig. 2 (center)), it is also easy to see that the friction cone is again pointed, and thus a solution will always exist. The remaining two-point contact case corresponds to the two point contact in Fig. 2 (right), for which it can be shown that the pointedness of the friction cone holds if the distance between the points of contact is large enough. This motion, which can be used to rotate the part is discussed later in the next subsection. Finally, if we derive robust motion primitives that guarantee sticking is maintained at one or two contact points, we automatically obtain uniqueness of the trajectories by using traditional arguments with the underlying differential algebraic equation. We note that the uniqueness of contact forces does not hold in general, even though part's trajectory is guaranteed to be unique.

B. Properties of motions and admissible sets

There are many properties of interest for pushing primitives for our meso-scale manipulation task, *e.g.*, inputs that guarantee motions with one (or two) sticking contacts or input that guarantee desired clockwise (or counter clockwise

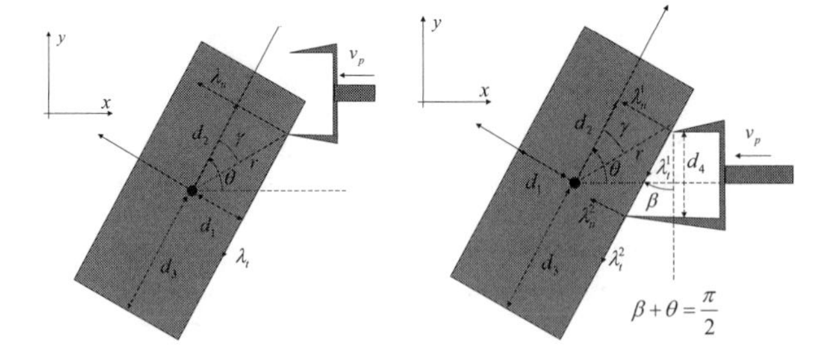

Fig. 4. Pushing with one-point (left) and two-point contact (right). In the right picture, the DTP is shown with the exaggerated misalignment between its two tips for better visualization.

rotation) of the part. In the following, we will specially discuss three types of properties for which robust motions can be systematically constructed. The first property is to maintain the one-point sticking contact with counter clockwise (or clockwise) rotation. The second property is to maintain the two-point sticking contact for the DTP. The third property is that the orientation of the final state of the motion is close to 0 or π radians (because the slot is horizontally oriented). Sufficient conditions for motion primitives that guarantee each of these properties are presented below.

1) One-point sticking contact with counter clockwise rotation: We only consider the case in which $\theta \in (0, \pi)$ and the probe pushes on the long edge of the part (see Fig. 4 left). However, other cases, such as pushing on the short edge or the right side of the part, can be analyzed similarly.

The following provides the conditions for a static point:

$$\begin{aligned} \left|\frac{\lambda_n}{\lambda_t}\right| &= \left|\frac{d_1 d_2 e_x \cos\theta + (e_\theta + d_1^2 e_x)\sin\theta}{(e_\theta + d_2^2 e_x)\cos\theta + d_1 d_2 e_x \sin\theta}\right| > \frac{1}{\mu} \\ \lambda_n &= -\frac{e_x v_p (d_1 d_2 e_x \cos\theta + (e_\theta + d_1^2 e_x)\sin\theta)}{f_d(x,u,E)} > 0 \\ v_\theta &= \frac{e_x v_p (d_1\cos\theta - d_2\sin\theta)}{f_d(x,u,E)} > 0 \end{aligned} \quad (1)$$

in which $f_d(x, u, E) = e_\theta + (d_2^2 + d_1^2)e_x > 0$ and $v_p < 0$.

From (1), we can infer the property of the whole motion just from its initial point, which is stated in the following lemma:

Lemma 1: If the part starts a counter clockwise rotation with sticking contact at the initial point with orientation $\theta \in (0, \pi)$ (satisfying (1)) as shown in Fig. 4 (left), then the part will keep counter clockwise rotation with sticking contact until its orientation θ reaches

$$\pi - \max\{\tan^{-1}\frac{d_1}{d_2}, \tan^{-1}\frac{d_1 d_2 e_x}{e_\theta + d_1^2 e_x}\}. \quad (2)$$

Proof: The derivatives of $\frac{\lambda_n}{\lambda_t}$ and v_θ with respect to θ are as follow:

$$\frac{\partial(\lambda_n/\lambda_t)}{\partial\theta} = \frac{e_\theta(e_\theta + (d_1^2 + d_2^2)e_x)}{((e_\theta + d_2^2 e_x)\cos\theta + d_1 d_2 e_x \sin\theta)^2}. \quad (3)$$

$$\frac{\partial v_\theta}{\partial\theta} = -\frac{e_x v_p (d_2\cos\theta + d_1\sin\theta)}{e_\theta + (d_1^2 + d_2^2)e_x}. \quad (4)$$

It can be observed that both derivatives are strictly positive before θ reaches (2). Therefore, if the part rotates counter clockwise ($v_\theta > 0$) in the sticking mode ($\left|\frac{\lambda_n}{\lambda_t}\right| > \frac{1}{\mu}$) at the initial point, then the part will keep staying in the sticking mode because $\frac{\lambda_n}{\lambda_t}$ will keep increasing and v_θ will keep strictly positive as θ increases. ■

2) The two-point sticking contact: We only describe the case in which $\theta \in (0, \pi)$ and the DTP pushes on the long edge of the part and the contact is sticking (see Fig. 4 right).

The following equations ensure that two point contact will be sticking at a static point:

$$\begin{aligned} \left|\frac{\lambda_n}{\lambda_t}\right| &= |\tan\theta| = |1/\tan\beta| > \frac{1}{\mu} \\ \lambda_n^1 &= \frac{e_x v_p \cos\beta(d_1\sin\beta + d_2\cos\beta - d_4)}{d_4} > 0 \\ \lambda_n^2 &= -\frac{e_x v_p \cos\beta(d_1\sin\beta + d_2\cos\beta)}{d_4} > 0 \end{aligned} \quad (5)$$

The following lemma shows whether the whole motion has a two-point sticking contact can be determined from the initial point.

Lemma 2: If the part starts with two-point sticking contact as shown in Fig. 4 (right), then the pushing will stay in the two-point sticking contact mode.

Proof: It is because (5) depends on the orientation and the orientation is invariant when the initial point has the two-point sticking contact. ■

3) The orientation of the final state is close to 0 or π radians: This property will be achieved in a motion by pushing the part with the active DTP with a separation

$$d_v \geq d_w + 2s_x^u + 2c_{\tilde{u}}^u. \quad (6)$$

to the passive STP to guarantee the intended rotation under sensing and control uncertainties (see Fig. 5 left). Such pushing will ensure that the final orientation will be in θ_t-neighborhood of π, in which

$$\theta_t = \sin^{-1}\frac{d_w + 2s_x^u + 2c_{\tilde{u}}^u}{\sqrt{d_w^2 + d_l^2}} - \alpha. \quad (7)$$

Remark: In order to guarantee existence, the pointed cone assumption requires $d_v \geq d_w \tan\sigma$ where $\sigma = tan^{-1}\mu$ is the angle of the friction cone. This is clearly satisfied by (6). However, for this motion we cannot guarantee uniqueness of the resulting trajectory. In this case, the property of interest (the desired change in orientation) does not depend on the specifics of the trajectory and thus the lack of a guarantee on uniqueness is not a problem.

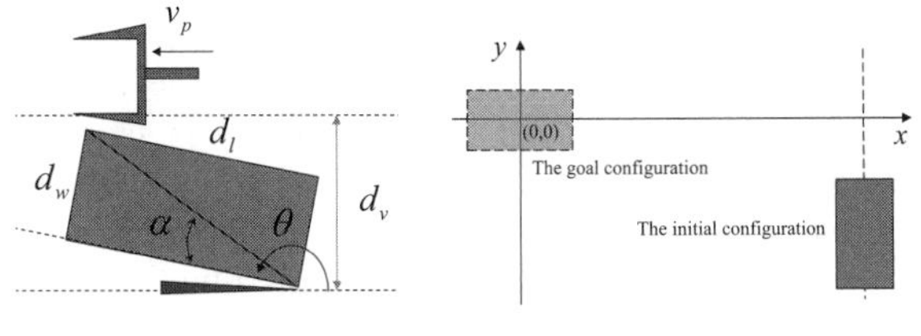

Fig. 5. The robust rotational motion and planning problem setup

C. Computing robust motions from the admissible sets

We use a simple sampling-based algorithm to find a robust motion with respect to a given property at a given state. We incrementally decrease the sampling dispersion along each dimension of the input space until the dispersion reaches the respective control uncertainty bounds. Initially, the sampling dispersion in each dimension of the input space is chosen to be the half of the maximal distance. In each iteration, the sample points in each dimension with respect to its current sampling dispersion are combined to generate all possible inputs. Each input is tested for membership in the admissible set under the nominal state and parameters. If no input is in the admissible set, then the sampling dispersion is decreased by half and the algorithm goes into the next iteration. If an input is in the admissible set under the nominal state and parameters, then this input is tested to see whether it is still in the admissible set under the maximal uncertainties in sensing, control, and parameters. If yes, then the motion from this input is returned as a robust motion. If no robust motion is found when the algorithm stops, then there exists no robust motion with respect to the given property under such uncertainties.

D. Comparison of robust motions

As we show in Section IV-B, there might exist many types of robust motions with respect to different properties. In this section, we will provide a measure, the Lipschitz constant of the motion equation, to compare robustness of these motions with respect to uncertainties.

The Lipschitz constants have been used before to provide an upper bound on the variation of the trajectory with respect to changes in the state, control, and parameters [8, 9]. The magnitude of Lipschitz constants characterizes the worst case trajectory variation of the system under uncertainties. If the Lipschitz constant is smaller, then the upper bound on the trajectory variation with respect to uncertainties is smaller, i.e., the corresponding motion will be more robust.

We compute the Lipschitz constants with respect to the fixed initial part configuration (x, y, θ), d_2, E (damping matrix), and μ (friction coefficient) for motion equations of the part under the push of the STP and DTP with the same probe initial position (the top tip for the DTP and the tip of the STP have the same position), constant fixed velocity, and time duration. It is shown that those constants for the STP are greater, and therefore the DTP has less uncertainty than the STP with respect to this measure. This result is supported by the experimental results in Section V.

E. Planning with robust motion primitives

The assembly task considered in the paper has the initial configuration with orientation $\frac{\pi}{2}$ and a goal configuration of $(0, 0, 0)$ with position tolerance of $\epsilon_p = 76\mu m$ and orientation tolerance $\epsilon_\theta = 5°$ (see Fig. 5 right). Note that we currently ignore obstacles in the environment. For such a task, our planning algorithm relies on composing the simple robust motions defined above. We first construct the following three higher level robust motion primitives using these simple ones.

1. Robust translation in the x direction

This robust translation is achieved by using DTP to push the part in the x direction while maintaining two-point sticking contact. However, because the two tips of a DTP may not be aligned (see β in Fig. 4 right) or sensing errors exist, two point contact might not be established, or can only be established after the one-point contact is established first at either the top or the bottom tip. To increase robustness, we define a *to-two-contact* property, denoted as t^2, by a sequential composition of a one-point sticking contact motion with a counter clockwise rotation followed by a two-point sticking contact motion (see Fig. 6). Lemmas 1 and 2 respectively provide conditions for one-point and two-point sticking contact motions. The following lemma will ensure that two sticking motions can be combined to ensure a t^2 motion.

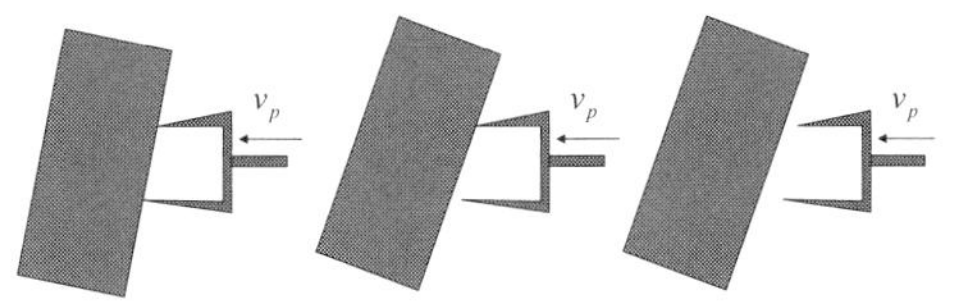

Fig. 6. The t^2 motion starting from the right figure to the left.

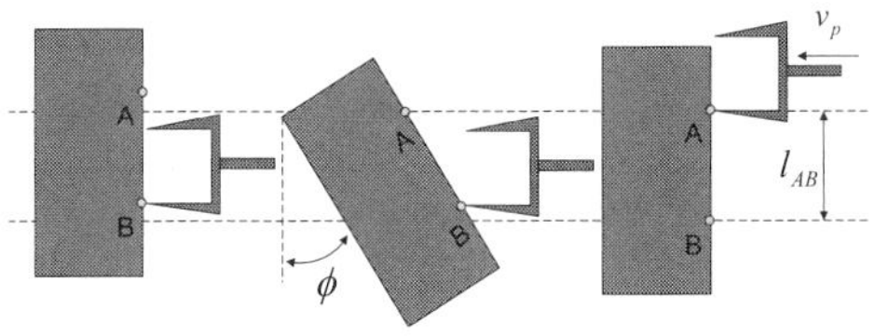

Fig. 7. Vertical translational motion starting from the right figure to the left.

Lemma 3: Assume that the top tip first establishes the contact. When the misalignment parameter, β, of the DTP satisfies

$$|\tan\beta| < \min\{\mu, \frac{d_2}{d_1}, \frac{d_1 d_2 e_x}{e_\theta + d_2^2 e_x}\}; \beta + \theta < \frac{\pi}{2}, \quad (8)$$

the counter clockwise rotation with one-point sticking contact can be followed by a two-point sticking motion.

Proof: The first inequality in (8) ensures that two-point sticking contact is admissible and can be established before the one-point sticking contact motion stops. The second inequality ensures that a counter clockwise rotation with one-point sticking contact will precede the two-point sticking contact motion. ■

2. Robust translation in the y direction

This translation is achieved by composing a robust motion with one point sticking contact and intended rotation followed by a robust t^2 motion (see Fig. 7). The amount of the net vertical translation is $l_{AB}(1 - \cos\phi)$ under nominal conditions (no uncertainty).

3. Robust rotation

This motion is achieved with the pushing described in Section IV-B.3.

Planning algorithm: With the above three higher level robust motion primitives, the planning algorithm consists of the following steps:

Step 1: Move in the y direction by pushing along the long edge of the part such that $y \in [-\frac{\epsilon_p}{2}, \frac{\epsilon_p}{2}]$. We use a sequence of y-direction motions in Fig. 7, guaranteeing that the net y translation of $l_{AB}(1 - \cos\phi)$ in Fig. 7 will have the following error bound $d_v^u = \max\{d_1^e, d_2^e\}$, in which $d_1^e = |l_{AB}(1-\cos\phi) - (l_{AB} - 2s_p^u - 2c_p^u)(1 - \cos(\phi - 2s_\theta^u))|$, $d_2^e = |l_{AB}(1-\cos\phi) - (l_{AB} + 2s_p^u + 2c_p^u)(1 - \cos(\phi + 2s_\theta^u))|$, s_p^u and c_p^u are respectively the sensing and control error bounds in the position, and s_θ^u is the sensing error bound in the orientation. To ensure that $y \in [-\frac{\epsilon_p}{2}, \frac{\epsilon_p}{2}]$ can be achieved using the vertical primitive under sensing and control uncertainties, the following conditions on the uncertainty bounds must be satisfied: $s_p^u + d_v^u \le \frac{\epsilon_p}{2}$, $\phi > 2s_\theta^u$, $l_{AB} > 2s_p^u + 2c_p^u$.

Step 2: Rotate to $\theta = \pi$. As shown in (7) and Fig. 5 (left), the distance of the orientation of the part to the horizontal line

TABLE I
Y-TRANSLATION PRIMITIVE: NET DISPLACEMENT OF THE PART

Test No.	X (μm)	Y (μm)	θ
1	1996	19	0.6°
2	1975	18	0.5°
3	1559	20	1.1°
Average	1843	19	0.7°
Simulation	1443	11	0.0°
Theory	*NA*	11	0.0°

will be bounded. To ensure that the final t^2 pushing can be robustly applied, we require that uncertainty bounds satisfy: $\theta_t = \sin^{-1}\frac{d_w+2s_p^u+2c_p^u}{\sqrt{d_w^2+d_l^2}} - \alpha < \theta_{t^2}^{\max}$, in which $\theta_{t^2}^{\max}$ is the maximal orientation of the part allowing a robust t^2 pushing and can be computed using the algorithm in Section IV-D.

Step 3: If necessary, move in the y direction by pushing along the short edge of the part such that $y \in [-\frac{\epsilon_p}{2}, \frac{\epsilon_p}{2}]$.

Step 4: Translate the part in x direction to the goal $(0,0,0)$. With the robust t^2 motion primitives, the final configuration of part will be $x \in [p^x + r\cos(\gamma+\beta) - c_p^u, p^x + r\cos(\gamma+\beta) + c_p^u]$, $y \in [p_y - r\sin(\gamma+\beta) - c_p^u, p_y - r\sin(\gamma+\beta) + c_p^u]$, and $\theta = \beta$ in which p^x, p^y is the position of the top tip of the DTP, d_2, r and γ are as shown in Fig. 4 (right). These equations also impose restrictions on the uncertainty bounds to ensure the intended tolerance: $r_{\max}(\cos(\gamma_{\max} - \beta_{\max}) - \cos\gamma_{\max}) + 2c_p^u < \epsilon_p$ and $r_{\max}\sin\beta_{\max} + 2c_p^u < \epsilon_p, \beta_{\max} < \epsilon_\theta$, in which $\gamma_{max} = \tan^{-1}\frac{d_1}{d_3}$, $r_{\max} = \sqrt{d_1^2 + d_3^2}$, and $\beta_{\max}$ is the maximal magnitude for β (see Fig. 4 right).

V. SIMULATION AND EXPERIMENTAL RESULTS

We did a series of experiments to estimate the parameters (including the damping matrix and friction coefficient) for the system and to compare robust and non-robust motions using both the DTP and STP. In the next two subsections, we show representative results for the system identification and for the different motion primitives. In Section V-D, we used the designed planner to compute probe controls to complete a given task in both the simulation and experiment.

A. Estimating of system parameters

The parameter fitting was done with the experimental data obtained using the STP. Figure 8 shows experimental trajectories versus predicted trajectories for one trial that was used in the parameter estimation (top) and one trial that was not (bottom). To estimate the parameters, a root-mean-square metric is used. The optimization algorithm is derived from the Nelder-Mead method. The diagonal elements of damping matrix E are estimated to be $e_x = e_y = 160.89N \cdot \text{sec.}/\text{m}$ and $e_\theta = 60.64N \cdot m \cdot \text{sec}$. The coefficient of friction between the part and the probe is estimated to be $\mu = 0.3 \sim 0.36$. These figures show 30-40 μm position errors across a translation of about 600μm and about 3° orientation errors for a 45° rotation.

B. Comparison between robust and non-robust motions

Trajectories from robust motion primitives show less variation (and are therefore more predictable) than trajectories from other motion primitives. Figure 9 shows the experiment setup (top) and experimental trajectory plots for comparison of the robust and non-robust motions using the DTP and STP. Tests 1 and 2 are for robust and non-robust t^2 motions with the DTP. Test 1 was verified to satisfy the robust t^2 motion conditions in Sections IV-B.1 and IV-B.2. The experiments showed that the two-point contact is well maintained because the orientation θ is almost constant after the two point contact is established. Test 2 did not satisfy the two-point sticking contact conditions, and therefore the two point contact was broken once it was established. We also observed that Test 1 has maximal trajectory differences of 20μm in x, 15μm in y, and 0.023 radians in θ, which are smaller than the corresponding numbers for Test 2 (maximal trajectory differences at 15μm in x, 25μm in y, and 0.1 radians in θ).

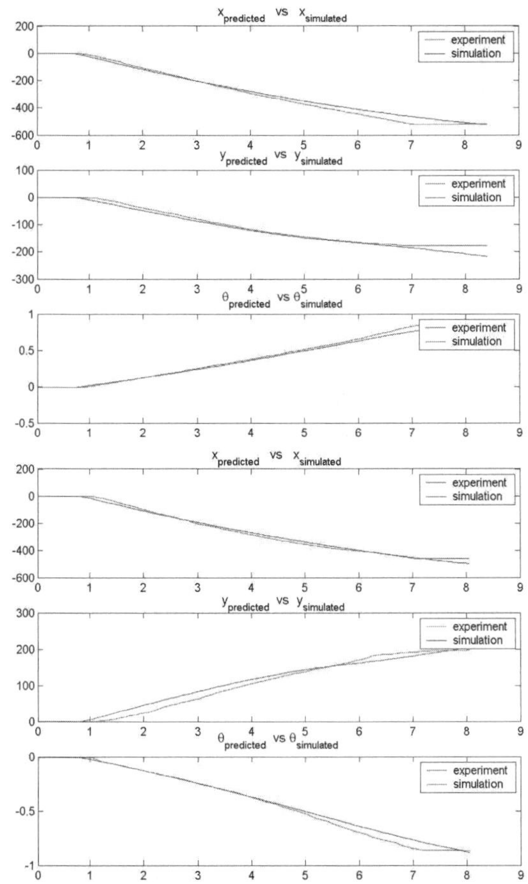

Fig. 8. System identification: The top three plots show a representative trial used for system identification and the bottom three plots show a representative trial used to verify the model.

C. Comparison between the DTP and STP

Trajectories using the DTP show less variation than those obtained from the STP. Tests 1 and 3 in Fig. 9 are results from robust motion primitives for the DTP and STP respectively. The top tip of the DTP had the same y position as the STP. Trajectories from Test 1 have less variation than those from Test 3, whose maximal trajectory differences are 75μm in x, 75μm in y, and 0.2 radians in θ.

D. Planning in both the simulation and experiment

Table I shows the comparison between theoretical, simulated and experimental results for robust translation in the y direction, for the motion primitive described in Section IV-E. Tables II and III compare the experimental and simulated results

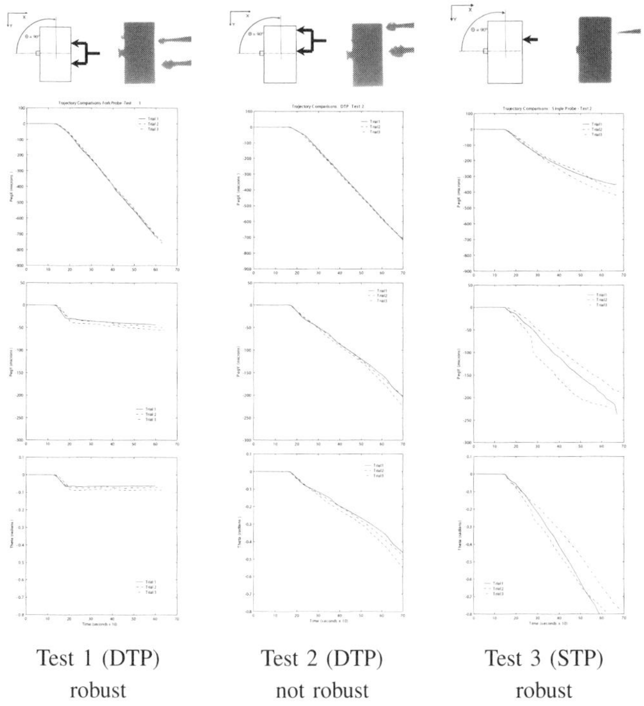

Fig. 9. Experimental results for robust and non-robust motions with the DTP and the STP.

TABLE II

ROTATIONAL MOTION: NET DISPLACEMENT OF THE PART

Test No.	X (μm)	Y (μm)	θ
1	381	34	88°
2	434	32	90°
3	370	14	90°
Average	382	27	89°
Simulation	295	0.05	90°

for executing robust rotational motion and robust translation in the x direction motion, respectively. At least 3 experimental tests were done for each motion type and the average values of the tests are shown in the tables. For the robust y translation tests, the initial robust one-point sticking contact is maintained until a desired part rotation angle, ϕ, is achieved. This is then followed by a robust t^2 push to restore the part to the upright position. Simulation and theoretical results match very well for the ϕ tested. Experiments show a somewhat higher (7-9 μm) net displacement than the predicted y translation, but it is likely due to measurement errors — errors in estimating position are $\pm$ 5 μm. We did not observe sliding in the pushing from the image analysis.

In the robust rotational motion experiments, the separation

TABLE III

X-TRANSLATION PRIMITIVE: NET DISPLACEMENT OF THE PART

Test No.	X (μm)	Y (μm)	θ
1	954	5	0.1°
2	944	11	0.7°
3	965	11	0.7°
4	959	5	0.5°
5	954	0	0.0°
Average	955	6	0.4°
Simulation	949	0.2	0.0°

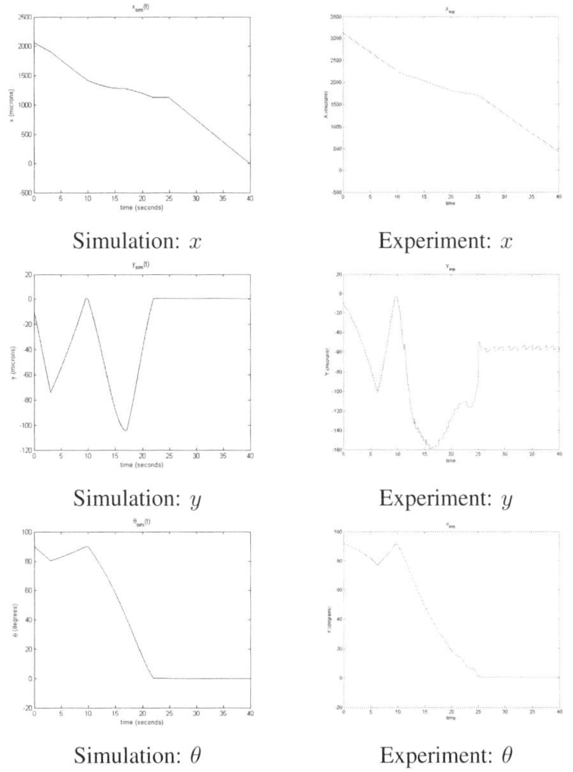

Fig. 10. Simulation (left) and experimental (right) results for a planning task

distance is determined from (6) and the two probe tips are centered about the center of the part that has orientation $\pi/2$. The STP probe is to the left of the part and is held stationary. The DTP, on the right side, pushes the part with its bottom probe tip for a distance of about 1100μm. From these experiments, we can see that the orientation of the peg is robustly rotated closed to π even though uncertainties cause significant mismatches in x and y displacements.

A push of approximately 950μm was used for the robust x translation experiments. The predicted results are within the error margins of the experimental observations.

Combining these three types of robust motions together allows us to execute the planned algorithm described in Section IV-E. Because of the limited controllability of x position of the peg in the current experimental platform and planning algorithm, the initial configuration is set to be ($x_{\text{init}} = 2060.4\mu\text{m}, y_{init} = -9.2\mu\text{m}, \theta_{\text{init}} = \frac{\pi}{2}$) such that one robust y motion followed by one robust rotation followed by one robust x motion is able to push the peg into the hole. A simulation of the planned motion is shown in Fig. 10 (left). In the experiments, the peg is successfully pushed into the hole three times over three trials. The experimental data from one trial is shown in Fig. 10 (right). The snapshots of the experiment are shown in Fig. 11 with the associated robust controls.

VI. CONCLUSION

In this paper, we established a framework for motion planning under differential constraints and uncertainties in sensing, control (actuation), and geometric/dynamic parameters. We show how we can characterize robust motion primitives with applications to quasi-static manipulation and assembly

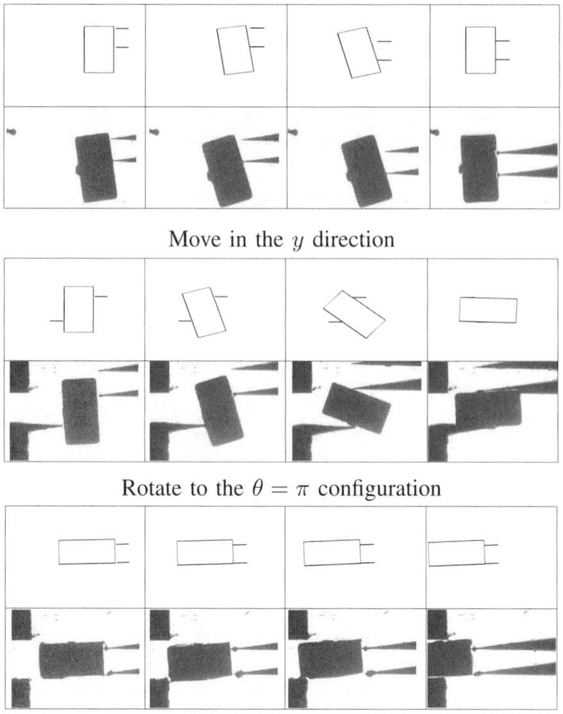
Move in the y direction
Rotate to the $\theta = \pi$ configuration
Translate the part in x direction to the goal

Fig. 11. Snapshots illustrating the assembly of the part into the slot.

tasks and propose measures to quantify robustness of motion primitives. Further, we describe an algorithm to automatically synthesize motion plans which sequentially compose robust motion primitives to move parts to goal positions with minimal actuation.

The main contribution in this paper is the quantitative treatment of uncertainty and the incorporation of models of uncertainty into the synthesis of motion primitives and the motion plans. It is clear that this paper is only a starting point and does not address the problems associated with multi-point contact which characterize assembly tasks. Further, we simplified the modeling of the contact friction by considering lubricated surfaces which appear to be well modeled by viscous damping. Nevertheless, the ability to plan and reliably execute the plan for positioning and orienting parts using visual feedback with only a single degree-of-freedom actuator represents a significant accomplishment over previous studies on quasi-static manipulation.

Acknowledgments

We gratefully acknowledge the support of NSF grants DMS01-39747, IIS02-22927, and IIS-0413138, and ARO grant W911NF-04-1-0148.

References

[1] S. Akella and M. T. Mason. Posing polygonal objects in the plane by pushing. *International Journal of Robotics Research*, 17(1):70–88, January 1998.

[2] J.C. Alexander and J.H. Maddocks. Bounds on the friction-dominated motion of a pushed object. *The International Journal of Robotics Research*, 12(3):231–248, June 1993.

[3] M. Anitescu and F.A. Potra. Formulating dynamic multi-rigid-body contact problems with friction as solvable linear complementarity problems. *Nonlinear Dynamics*, 14:231–247, 1997.

[4] Z. Balorda. Reducing uncertainty of objects by robot pushing. *IEEE Int. Conf. on Robotics and Automation*, pages 1051–1056, 1990.

[5] S. Berard, B. Nguyen, B. Roghani, J. Trinkle, J. Fink, and V. Kumar. Davincicode: A multi-model simulation and analysis tool for multi-body systems. In *Proceedings of the IEEE International Conference on Robotics and Automation*, 2007.

[6] K. Boehringer, R.Fearing, and K. Goldberg. *Handbook of Industrial Robotics, 2nd Ed.*, chapter Microassembly, pages 1045–1066. John Wiley and Sons, 1999.

[7] D.J. Cappelleri, J. Fink, and V. Kumar. Modeling uncertainty for planar meso-scale manipulation and assembly. *ASME 2006 International Design Engineering Technical Conference (IDETC), Philadelphia, PA*, 2006.

[8] P. Cheng. *Sampling-Based Motion Planning with Differential Constraints*. PhD thesis, University of Illinois, Urbana, IL, August 2005.

[9] P. Cheng and V. Kumar. Sampling-based falsification and verification of controllers for continuous dynamic systems. In S. Akella, N. Amato, W. Huang, and B. Misha, editors, *Workshop on Algorithmic Foundations of Robotics VII*, 2006.

[10] H. Choset, K. M. Lynch, S. Hutchinson, G. Kantor, W. Burgard, L. E. Kavraki, and S. Thrun. *Principles of Robot Motion: Theory, Algorithms, and Implementations*. MIT Press, Cambridge, MA, 2005.

[11] M.A. Erdmann and M.T. Mason. An exploration of sensorless manipulation. *IEEE Journal of Robotics and Automation*, 4(4), August 1998.

[12] R.S. Fearing. Survey of sticking effects for micro parts handling. *IEEE/RSJ Int. Conf. on Intelligent Robotics and Sys.(IROS), Pittsburgh, PA*, 2:212–217, August 5-9 1995.

[13] K.Y. Goldberg. Orientating polygonal parts without sensing. *Algorithmica*, 10(2/3/4):210–225, August/September/October 1993.

[14] Suresh Goyal, Elliott N. Pinson, and Frank W. Sinden. Simulation of dynamics of interacting rigid bodies including friction i: General problem and contact model. *Engineering with Computers*, 10:162–174, 1994.

[15] S. M. LaValle. *Planning Algorithms*. Cambridge University Press, Cambridge, U.K., 2006. Available at http://planning.cs.uiuc.edu/.

[16] K.M. Lynch. The mechanics of fine manipulation by pushing. *IEEE Int. Conf. on Robotics and Automation, Nice, France*, pages 2269–2276, May 1992.

[17] K.M. Lynch. Estimating the friction parameters of pushed objects. *Proc. 1993 IEEE/RSJ Int. Conf.*, July 1993.

[18] K.M. Lynch and M.T. Mason. Stable pushing: Mechanics, controllability, and planning. *International Journal of Robotics Research*, 15(6):553–556, December 1996.

[19] M.T. Mason. Mechanics and planning of manipulator pushing operations. *International Journal of Robotics Research*, 5(3):53–71, 1986.

[20] M. Moll, K. Goldberg, M.A. Erdmann, and R. Fearing. Orienting micro-scale parts with squeeze and roll primitives. *IEEE Int. Conf. on Robotics and Automation, Washington, DC*, May 11-15 2002.

[21] J.S. Pang and J.C. Trinkle. Complementarity formulations and existence of solutions of dynamic multi-rigid-body contact problems with coulomb friction. *Mathematical Programming*, 73:199–226, 1996.

[22] M.A. Peshkin and A.C. Sanderson. The motion of a pushed, sliding object, part1: Sliding friction. Technical Report CMU-RI-TR-85-18, Robotics Institute, Carnegie Mellon University, Pittsburgh, PA, September 1985.

[23] D.T. Pham, K.C. Cheung, and S.H. Yeo. Initial motion of a rectangular object being pushed or pulled. *IEEE Int. Conf. on Robotics and Automation*, 1046-1050, 1990.

[24] M. Salganicoff, G. Metta, A. Oddera, and G. Sandini. A vision-based learning method for pushing manipulation. *AAAI Fall Symp. on Machine Learning in Computer Vision.*, 1993b.

[25] P. Song, J.S. Pang, and V. Kumar. A semi-implicit time-stepping model for frictional compliant contact problems. *International Journal for Numerical Methods in Engineering*, Accepted for publication 2004.

[26] D. E. Stewart and J.C. Trinkle. An implicit time-stepping scheme for rigid body dynamics with inelastic collisions and coulomb friction. *International J. Numer. Methods Engineering*, 39(15):281–287, 1996.

[27] J.C. Trinkle, S. Berard, and J.S. Pang. A time-stepping scheme for quasistatic multibody systems. *International Symposium of Assembly and Task Planning*, July 2005.

[28] T. Yoshikawa and M. Kurisu. Identification of the center of friction from pushing an object by a mobile robot. *IEEE/RSJ International Workshop on Intelligent Robots and Systems IROS*, November 1991.

Data Association in $O(n)$ for Divide and Conquer SLAM

Lina M. Paz
Instituto de Investigación
en Ingeniería en Aragón
Universidad de Zaragoza, Spain
Email: linapaz@unizar.es

José Guivant
Australian Centre for Field Robotics
The University of Sydney, Australia
Email: jguivant@acfr.usyd.edu.au

Juan D. Tardós and José Neira
Instituto de Investigación
en Ingeniería en Aragón
Universidad de Zaragoza, Spain
Email: tardos,jneira@unizar.es

***Abstract*—In this paper we show that *all* processes associated to the move-sense-update cycle of EKF SLAM can be carried out in time *linear* in the number of map features. We describe Divide and Conquer SLAM, an EKF SLAM algorithm where the computational complexity per step is reduced from $O(n^2)$ to $O(n)$ (the total cost of SLAM is reduced from $O(n^3)$ to $O(n^2)$). In addition, the resulting vehicle and map estimates have better consistency properties than standard EKF SLAM in the sense that the computed state covariance more adequately represents the real error in the estimation. Both simulated experiments and the Victoria Park Dataset are used to provide evidence of the advantages of this algorithm.**

***Index Terms*—SLAM, Computational Complexity, Consistency, Linear Time.**

I. Introduction

The Simultaneous Localization and Mapping (SLAM) problem deals with the construction of a model of the environment being traversed with an onboard sensor, while at the same time maintaining an estimation of the sensor location within the model [1], [2]. Solving SLAM is central to the effort of conferring real autonomy to robots and vehicles, but also opens possibilities in applications where the sensor moves with six degrees of freedom, such as egomotion and augmented reality. SLAM has been the subject of much attention since the seminal work in the late 80s [3], [4], [5], [6].

The most popular solution to SLAM considers it a stochastic process in which the Extended Kalman Filter (EKF) is used to compute an estimation of a state vector $\mathbf{x}$ representing the sensor and environment feature locations, together with the covariance matrix $\mathbf{P}$ representing the error in the estimation. Currently, most of the processes associated to the move-sense-update cycle of EKF SLAM are linear in the number of map features n: vehicle prediction and inclusion of new features [7], [8], continuous data association [9], global localization [10]. The exception is the update of the covariance matrix of the stochastic state vector that represents the vehicle and map states, which is $O(n^2)$. The EKF solution to SLAM has been used successfully in small scale environments, however the $O(n^2)$ computational complexity limits the use EKF-SLAM in large environments. This has been a subject of much interest in research. Postponement [11], the Compressed EKF filter [8], and Local Map Sequencing [12] are alternatives that work on local areas of the stochastic map and are essentially constant time most of the time, although they require periodical $O(n^2)$ updates (given a certain environment and sensor characteristics, an optimal local map size can be derived to minimize the total computational cost [13]). More recently, researchers have pointed out the approximate sparseness of the Information matrix $\mathbf{Y}$, the inverse of the full covariance matrix $\mathbf{P}$. This suggests using the Extended Information Filter, the dual of the Extended Kalman Filter, for SLAM updates. The Sparse Extended Information Filter (SEIF) algorithm [14] approximates the Information matrix by a sparse form that allows $O(1)$ updates on the information vector. Nonetheless, data association becomes more difficult when the state and covariance matrix are not available, and the approximation can yield overconfident estimations of the state [15]. This overconfidence is overcome by the Exactly Sparse Extended Information Filter (ESEIF) [16] with a strategy that produces an exactly sparse Information matrix with no introduction of inaccuracies through sparsification.

The Thin Junction Tree Filter algorithm [17] works on the Gaussian graphical model represented by the Information matrix, and achieves high scalability by working on an *approximation*, where weak links are broken. The Treemap algorithm [18] is a closely related technique, which also uses a weak link breakage policy. Recently insight was provided that the full SLAM problem, the complete vehicle trajectory plus the map, is sparse in information form (although ever increasing) [19], [20]. Sparse linear algebra techniques allow to compute the state, without the covariance, in time linear with the whole trajectory and map size. The T-SAM algorithm [21] provides a local mapping version to reduce the computational cost. However, the method remains a batch algorithm and covariance is not available to solve data association.

A second important limitation of standard EKF SLAM is the effect that linearizations have in the consistency of the final vehicle and feature estimates. Linearizations introduce errors in the estimation process that can render the result inconsistent, in the sense that the computed state covariance does not represent the real error in the estimation [22], [23], [24]. Among other things, this shuts down data association, which is based on contrasting predicted feature locations with observations made by the sensor. Thus, important processes in SLAM like loop closing are crippled. The Unscented Kalman

Filter [25] avoids linearization via a parametrization of means and covariances through selected points to which the nonlinear transformation is applied. Unscented SLAM has been shown to have improved consistency properties [26]. These solutions however ignore the computational complexity problem. All algorithms for EKF SLAM based on efficiently computing an approximation of the EKF solution [17], [18] will inevitably suffer from this problem.

In this paper we describe Divide and Conquer SLAM (D&C SLAM), an EKF SLAM algorithm that overcomes these two fundamental limitations:

1) The computational cost per step is reduced from $O(n^2)$ to $O(n)$; the total cost of SLAM is reduced from $O(n^3)$ to $O(n^2)$;
2) the resulting vehicle and map estimates have better consistency properties than standard EKF SLAM in the sense that the computed state covariance adequately represents the real error in the estimation.

Unlike many current large scale EKF SLAM techniques, this algorithm computes an exact solution, without relying on approximations or simplifications to reduce computational complexity. Also, estimates and covariances are available when needed by data association without any further computation. Empirical results show that, as a by-product of reduced computations, and without losing precision because of approximations, D&C SLAM has better consistency properties than standard EKF SLAM.

This paper is organized as follows: in section II we briefly review the standard EKF SLAM algorithm and its computational properties. Section III contains a description of the proposed algorithm. We study of its computational cost in comparison with EKF SLAM, as well as its consistency properties. In section IV we describe **RJC**, an algorithm for carrying out data association in D&C SLAM also in linear time. In section V we use the Victoria Park dataset to carry out an experimental comparison between EKF SLAM and D&C SLAM. Finally in section VI we draw the main conclusions of this work.

II. The EKF SLAM algorithm

The EKF SLAM algorithm (see alg. 1) has been widely used for mapping. Several authors have described the computational complexity of this algorithm [7], [8]. With the purpose of comparing EKF SLAM with the proposed D&C SLAM algorithm, in this section we briefly analyze its computational complexity.

A. Computational complexity of EKF SLAM per step

For simplicity, assume that in the environment being mapped features are distributed more or less uniformly. If the vehicle is equipped with a sensor of limited range and bearing, the amount of measurements obtained at any location will be more or less constant. Assume that at some step k the map contains n features, and the sensor provides m measurements, r of which correspond to re-observed features, and $s = m - r$ which correspond to new features.

Algorithm 1 : `ekf_slam`

$$\mathbf{z}_0, \mathbf{R}_0 = \mathit{get\_measurements}$$
$$\hat{\mathbf{x}}_0, \mathbf{P}_0 = \mathit{new\_map}(\mathbf{z}_0, \mathbf{R}_0)$$

for $k = 1$ to steps **do**

$$\hat{\mathbf{x}}_{R_k}^{R_{k-1}}, \mathbf{Q}_k = \mathit{get\_odometry}$$
$$\hat{\mathbf{x}}_{k|k-1}, \mathbf{F}_k, \mathbf{G}_k = \mathit{prediction}(\hat{\mathbf{x}}_{k-1}, \hat{\mathbf{x}}_{R_k}^{R_{k-1}})$$
$$\mathbf{P}_{k|k-1} = \mathbf{F}_k \mathbf{P}_{k-1} \mathbf{F}_k^T + \mathbf{G}_k \mathbf{Q}_k \mathbf{G}_k^T \quad (1)$$

$$\mathbf{z}_k, \mathbf{R}_k = \mathit{get\_measurements}$$
$$\mathcal{H}_k, \mathbf{H}_{\mathcal{H}_k} = \mathit{data\_assoc}(\hat{\mathbf{x}}_{k|k-1}, \mathbf{P}_{k|k-1}, \mathbf{z}_k, \mathbf{R}_k)$$

$$\mathbf{S}_{\mathcal{H}_k} = \mathbf{H}_{\mathcal{H}_k} \mathbf{P}_{k|k-1} \mathbf{H}_{\mathcal{H}_k}^T + \mathbf{R}_{\mathcal{H}_k} \quad (2)$$
$$\mathbf{K}_{\mathcal{H}_k} = \mathbf{P}_{k|k-1} \mathbf{H}_{\mathcal{H}_k}^T / \mathbf{S}_{\mathcal{H}_k} \quad (3)$$
$$\mathbf{P}_k = (\mathbf{I} - \mathbf{K}_{\mathcal{H}_k} \mathbf{H}_{\mathcal{H}_k}) \mathbf{P}_{k|k-1} \quad (4)$$
$$\nu_{\mathcal{H}_k} = \mathbf{z}_k - \mathbf{h}_{\mathcal{H}_k}(\hat{\mathbf{x}}_{k|k-1}) \quad (5)$$
$$\hat{\mathbf{x}}_k = \hat{\mathbf{x}}_{k|k-1} + \mathbf{K}_{\mathcal{H}_k} \nu_{\mathcal{H}_k} \quad (6)$$
$$\hat{\mathbf{x}}_k, \mathbf{P}_k = \mathit{add\_feat}(\hat{\mathbf{x}}, \mathbf{P}_k, \mathbf{z}_k, \mathbf{R}_k, \mathcal{H}_k)$$

end for

`return` $\mathbf{m} = (\mathbf{x}_k, \mathbf{P}_k)$

The computational complexity of carrying out the move-sense-update cycle of algorithm 1 at step k involves the computation of the *predicted map* $\hat{\mathbf{x}}_{k|k-1}, \mathbf{P}_{k|k-1}$, which requires obtaining also the computation of the corresponding jacobians $\mathbf{F}_k, \mathbf{G}_k$, and the *updated map* $\mathbf{x}_k, \mathbf{P}_k$, which requires the computation of the corresponding jacobian $\mathbf{H}_{\mathcal{H}_k}$, the Kalman gain matrix $\mathbf{K}_{\mathcal{H}_k}$, as well as the innovation $\nu_{\mathcal{H}_k}$, and its covariance $\mathbf{S}_k$ (the complexity of data association is analyzed in section IV).

The fundamental issue regarding computational complexity is that all jacobians are *sparse* matrices [7], [8], [20]. Thus, their computation is $O(1)$, but more importantly, since they take part in the computation of both the predicted and updated map, the computational cost of eqs. (1) to (6) can also be reduced. Consider as an example the innovation covariance matrix $\mathbf{S}_k$ in eq. (2). Normally, the computation of this $r \times r$ matrix would require $rn^2 + r^2 n$ multiplications and $rn^2 + r^2 n + r^2$ sums, that is, $O(n^2)$ operations (see fig. 1). But given that matrix $\mathbf{H}_k$ is sparse, with an effective size of $r \times c$, the computation requires $rcn + r^2 c$ multiplications and $rcn + r^2 c + r^2$ sums, that is, $O(n)$ operations. Similar analysis leads to the conclusion that the cost of computing both the predicted covariance $\mathbf{P}_{k|k-1}$ and the Kalman gain matrix $\mathbf{K}_{\mathcal{H}_k}$ is $O(n)$, and that the greatest cost in an EKF SLAM update is the computation of the covariance matrix $\mathbf{P}_k$, which is $O(n^2)$. Thus, the computational cost per step of EKF SLAM is quadratic on the size of the map:

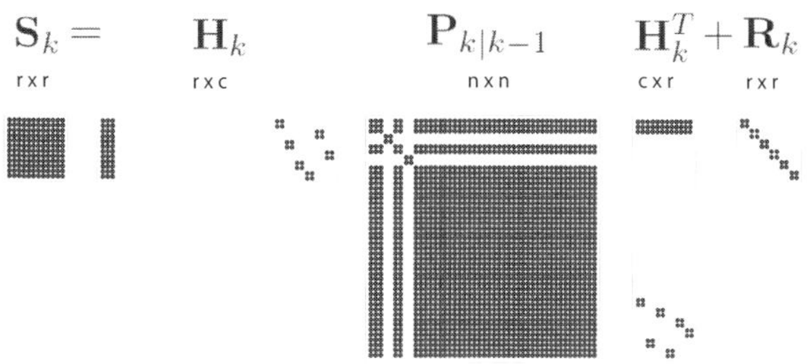

Fig. 1. Computation of the innovation covariance $\mathbf{S}_k$ matrix: the computation requires $O(n)$ operations ($rcn + r^2c$ multiplications and $rcn + r^c2 + r^2$ sums).

$$C_{EFK,k} = O(n^2) \quad (7)$$

Figure 3 shows the results of carrying out EKF SLAM in four simulated scenarios. In an environment with uniform distribution of point features, the vehicle performs a $1m$ motion at every step. The odometry of the vehicle has standard deviation error of $10cm$ in the x direction (the direction of motion), $5cm$ in y direction, and $(0.5deg)$ for orientation. We simulate an onboard range and bearing sensor with a range of $3m$, so that 16 features are normally seen at every step. The standard deviation error is 5% of the distance in range, and $1deg$ in bearing. Four different trajectories are carried out: straight forward exploration (first column); loop closing (second column), lawn mowing (third column), and snail path (fourth column). The execution time of EKF SLAM per step for each of these trajectories is shown in fig. 3, second row.

B. Total computational complexity of EKF SLAM

Assume that the process of building a map of size n features is carried out with an exploratory trajectory, in which the sensor obtains m measurements per step as said before, s of which are new (all four examples in fig. 3, straight forward, loop closing, lawn mowing and spiral path, are exploratory trajectories). Given that s new features are added to the map per step, n/s steps are required to obtain the final map of size n, and thus the total computational complexity will be:

$$\begin{aligned} C_{EKF} &= O\left(\sum_{k=1}^{n/s} (ks)^2\right) \\ &= O\left(s^2 \sum_{k=1}^{n/s} k^2\right) \\ &= O\left(s^2 \frac{(n/s)(n/s+1)(2n/s+1)}{6}\right) \\ &= O\left(\frac{1}{6} 2n^3/s + 3n^2 + ns\right) \\ &= O(n^3) \end{aligned} \quad (8)$$

The total cost of computing a map is cubic with the final size of the map. The total execution time of EKF SLAM for each of these trajectories is shown in fig. 3, third row.

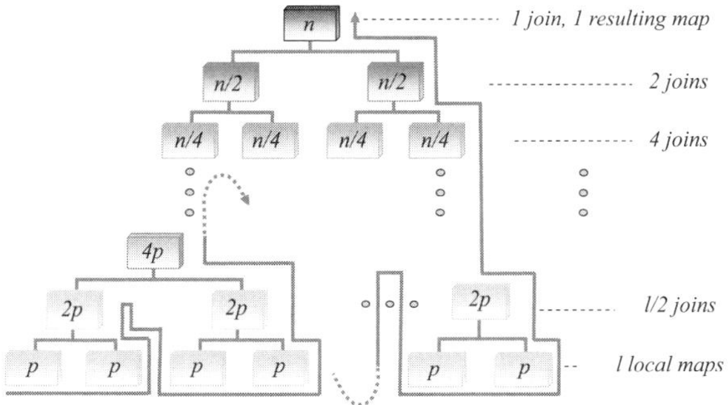

Fig. 2. Binary tree representing the hierarchy of maps that are created and joined in D&C SLAM. The red line shows the sequence in which maps are created and joined.

III. The Divide and Conquer Algorithm

The Divide and Conquer algorithm for SLAM (D&C SLAM) is an EKF-based algorithm in which a sequence of local maps of minimum size p is produced using the standard EKF SLAM algorithm [27]. These maps are then joined using the map joining procedure of [12], [28] (or the improved version 2.0 detailed in [27]) to produce a single final stochastic map.

Instead of joining each new local map to a global map sequentially, as Local Map Sequencing does [12], D&C SLAM carries out map joining in a binary hierarchical fashion, as depicted in fig. 2. Although algorithms like Treemap [18] use a similar structure, the tree is not used here to sort features, it represents the hierarchy of local maps that are computed. The leaves of the tree are the sequence of local maps of minimal size p that the algorithm produces with standard EKF-SLAM. The intermediate nodes represent the maps resulting form the intermediate map joining steps that are carried out, and the root of the tree represents the final map that is computed. D&C follows algorithm 2, which performs a *postorder* traversal of the tree using a stack to save intermediate maps. This allows a sequential execution of D&C SLAM.

A. Total computational complexity of D&C SLAM

In D&C SLAM, the process of building a map of size n produces $l = n/p$ maps of size p, at cost $O(p^3)$ each (see eq. (8)), which are joined into $l/2$ maps of size $2p$, at cost $O((2p)^2)$ each. These in turn are joined into $l/4$ maps of size $4p$, at cost $O((4p)^2)$ each. This process continues until two local maps of size $n/2$ are joined into 1 local map of size n, at a cost of $O(n^2)$. Thus, the total computational complexity of D&C SLAM is (note that the sum represents all costs associated to map joining, which is $O(n^2)$ [12]):

$$\begin{aligned} C_{DC} &= O\left(p^3 l + \sum_{i=1}^{\log_2 l} \frac{l}{2^i} (2^i\, p)^2\right) \\ &= O\left(p^3 n/p + \sum_{i=1}^{\log_2 n/p} \frac{n/p}{2^i} (2^i\, p)^2\right) \end{aligned}$$

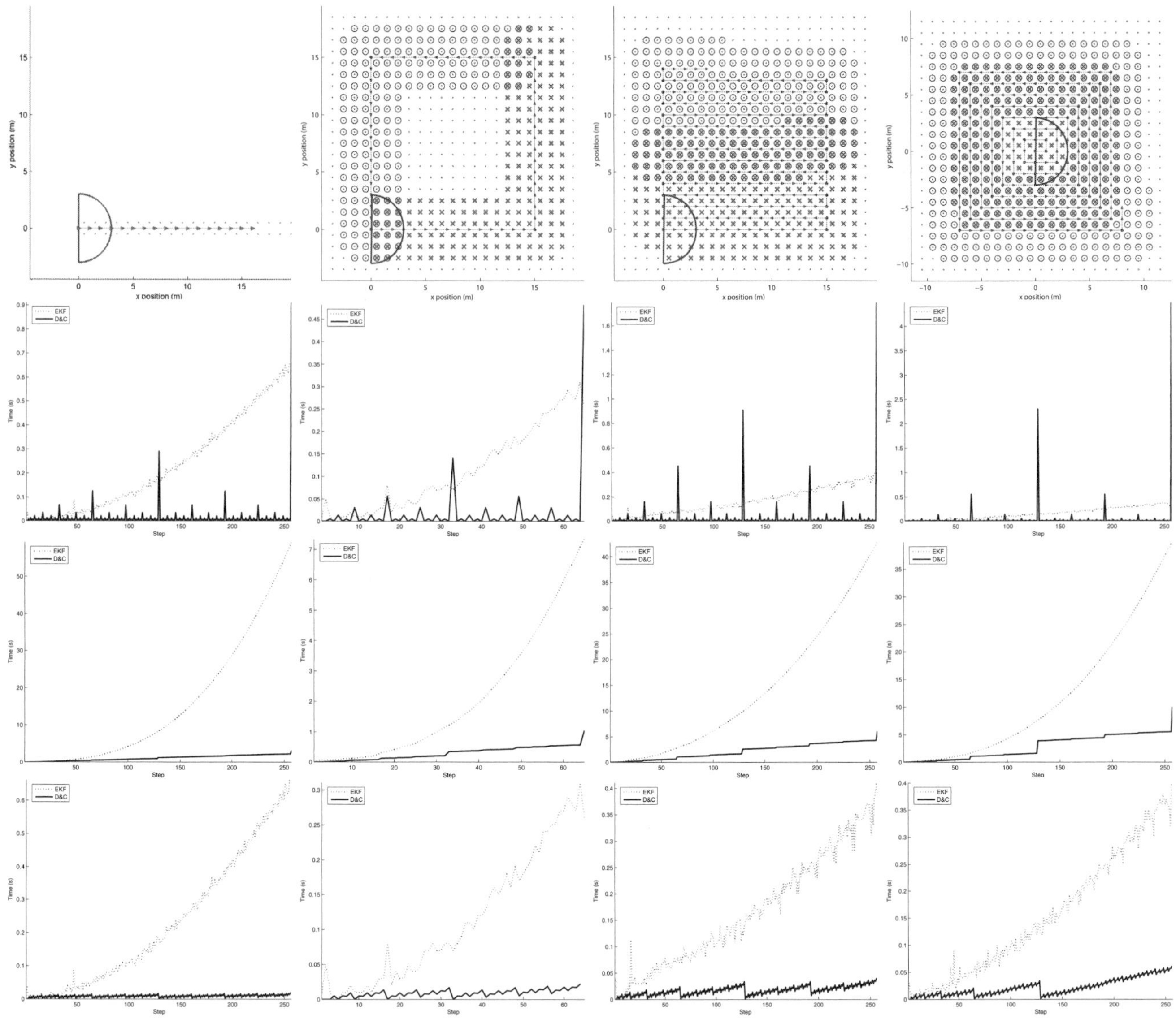

Fig. 3. Four simulated experiments for comparing the EKF and D&C SLAM algorithms: detail of a straight forward trajectory (first colum); loop closing (second column); lawn mowing (third column); snail path (fourth column). Ground truth environment, trajectory and first and second halves $n/2$ of maps features for data association analysis (top row); execution time per step of EKF .vs. D&C SLAM (second row); total execution time of EKF .vs. D&C SLAM (third row); execution time per step of EKF .vs. amortized execution time per step of D&C SLAM (bottom row).

$$= O\left(p^2 n + \sum_{i=1}^{\log_2 n/p} p \frac{n}{2^i}(2^i)^2\right)$$
$$= O\left(p^2 n + p\,n \sum_{i=1}^{\log_2 n/p} 2^i\right)$$

The sum in the equation above is a geometric progression of the type:

$$\sum_{i=1}^{k} r^i = \frac{r - r^{k+1}}{1-r}$$

Thus, in this case:

$$\begin{aligned} C_{DC} &= O\left(p^2 n + p\,n\frac{2^{\log_2 n/p+1} - 2}{2-1}\right) \\ &= O\left(p^2 n + p\,n(2\,n/p - 2)\right) \\ &= O\left(p^2 n + 2n^2 - 2pn\right) \\ &= O(n^2) \end{aligned} \quad (9)$$

This means that D&C SLAM performs SLAM with a total cost quadratic with the size of the environment, as compared with the cubic cost of standard EKF-SLAM. The difference between this approach and other approaches that also use map joining, such as Local Map Sequencing, is that in D&C SLAM the number of map joining operations carried out is proportional to $\log(n)$, instead of n. This allows the total cost

Algorithm 2 : `dc_slam`
`sequential implementation using a stack.`

```
stack = new()
m0 = ekf_slam()
stack = push(m0, stack)
{
Main loop: postorder traversing of the map tree.
}
repeat
  mk = ekf_slam()
  while ¬ empty(stack) and then
  size(mk) ≥ size(top(stack)) do
    m = top(stack)
    stack = pop(stack)
    mk = join(m, mk)
  end while
  stack = push(mk, stack)
until end_of_map
{
Wrap up: join all maps in stack for full map recovery.
}
while ¬ empty(stack) do
  m = top(stack)
  stack = pop(stack)
  mk = join(m, mk)
end while
return (mk)
```

to remain quadratic with n.

Figure 3, second and third rows, show the execution time per step and total execution time, respectively, for D&C SLAM .vs. EKF SLAM for the four simulations of straight forward, loop closing, lawn mowing and spiral path. It can be seen that the total cost of D&C SLAM very quickly separates from the total cost of EKF SLAM. The reason is that the computational cost per step of D&C SLAM is lower than that of EKF SLAM most of the time. EKF SLAM works with a map of non-decreasing size, while D&C SLAM works on local maps of small size most of the time. In some steps though (in the simulation those which are a multiple of 2), the computational cost of D&C is higher than EKF. In those steps, one or more map joining operations take place (in those that are a power of 2, 2^l, l map joining operations take place).

B. Computational complexity of D&C SLAM per step

In D&C SLAM, the map to be generated at step k will not be required for joining until step $2\,k$. We can therefore amortize the cost $O(k^2)$ at this step by dividing it up between steps k to $2\,k-1$ in equal $O(k)$ computations for each step. We must however take into account all joins to be computed at each step. If k is a power of 2 ($k = 2^l$), $i = 1 \cdots l$ joins will take place at step k, with a cost $O(2^2) \dots O((2^l)^2)$. To carry out join i we need join $i-1$ to be complete. Thus if we wish to amortize all joins, we must wait until step $k+k/2$ for join $i-1$ to be complete, and then start join i. For this reason, the amortized version of this algorithm divides up the largest join at step k into steps $k + k/2$ to $2\,k-1$ in equal $O(2\,k)$ computations for each step. Amortization is very simple, the computation of the elements of $P_{k|k}$ is divided in $k/2$ steps. If $P_{k|k}$ is of size $n \times n$, $2n^2/k$ elements have to be computed per step.

Fig. 3 (bottom row) shows the resulting amortized cost per step for the four simulated experiments. Note that at steps $i = 2^l$, the cost falls steeply. As said before, in these steps l joins should be computed, but since join i required the map resulting from join $i-1$, all l joins are postponed. We can see that the amortized cost of D&C SLAM is $O(n)$ always lower than that of EKF SLAM. D&C SLAM is an anytime algorithm, if at any moment during the map building process the full map is required for another task, it can be computed in a single $O(n^2)$ step.

C. Consistency in Divide and Conquer SLAM

Apart from computational complexity, another important aspect of the solution computed by the EKF has gained attention recently: map consistency. When the ground truth solution $\mathbf{x}$ for the state variables is available, a statistical test for filter consistency can be carried out on the estimation $(\hat{\mathbf{x}}, \mathbf{P})$, using the Normalized Estimation Error Squared (NEES), defined as:

$$D^2 = (\mathbf{x} - \hat{\mathbf{x}})^T \, \mathbf{P}^{-1} \, (\mathbf{x} - \hat{\mathbf{x}}) \tag{10}$$

Consistency is checked using a chi-squared test:

$$D^2 \leq \chi^2_{r,1-\alpha} \tag{11}$$

where $r = dim(\mathbf{x})$ and α is the desired significance level (usually 0.05). If we define the consistency index of a given estimation $(\hat{\mathbf{x}}, \mathbf{P})$ with respect to its true value $\mathbf{x}$ as:

$$CI = \frac{D^2}{\chi^2_{r,1-\alpha}} \tag{12}$$

when $CI < 1$, the estimation is consistent with ground truth, and when $CI > 1$, the estimation is inconsistent (optimistic) with respect to ground truth.

We tested consistency of both standard EKF and D&C SLAM algorithms by carrying 20 Monte Carlo runs on the simulated experiments. We have used simulated experiments to test consistency because this allows to have ground truth easily available. Additionally, Monte Carlo runs allow to gather statistically significant evidence about the consistency properties of the algorithms being compared, while a single experiment allows to carry out only one run of the algorithms.

Figure 4 (top) shows the evolution of the mean consistency index of the vehicle orientation during all steps of the straight forward trajectory simulation. We can see that the D&C estimate on vehicle location is always more consistent than the standard EKF estimate, EKF falls out of consistency while D&C remains consistent. In order to obtain a value for consistency in all steps, we emptied the stack and carried out all joins at every step to obtain the full map, but this is not done normally.

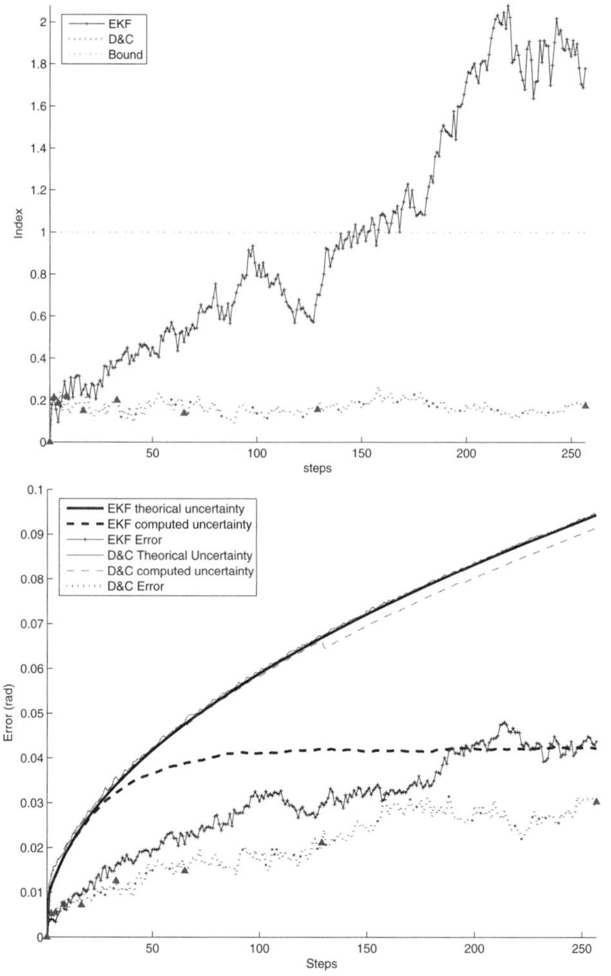

Fig. 4. Mean consistency index for the robot orientation(top); Mean absolute angular robot error (bottom).

Figure 4 (bottom) shows the evolution of the mean absolute angular error of the vehicle. The 2σ bounds for the theoretical (without noise) and computed (with noise) uncertainty of both standard EKF and *Divide and Conquer* SLAM algorithms are also drawn. We can see how the error increases more slowly in the case of D&C SLAM, but we can also see that the main cause of inconsistency in the standard EKF SLAM is the fast rate at which the computed uncertainty falls below its theoretical value.

IV. Data Association for Divide and Conquer SLAM

The data association problem in continuous SLAM consists in establishing a correspondence between each of the m sensor measurements and one (on none) of the n map features. The availability of a stochastic model for both the map and the measurements allows to check each measurement-feature correspondence for *individual compatibility* using a hypothesis test on the innovation of the pairing and its covariance [9]. In standard EKF SLAM, and for a sensor of limited range and bearing, m is constant and thus individual compatibility is $O(nm) = O(n)$, linear on the size of the map. This cost can be easily reduced to $O(m)$, constant, by a simple tessellation or grid of the map computed during map building, which allows to determine individual candidates for a measurement in constant time, simply by checking the grid element in which the predicted feature falls.

In cases where clutter or vehicle error are high, there may be many more than one possible correspondence for each measurement. More elaborate algorithms are required to disambiguate in these cases. Nevertheless, the overlap between the measurements and the map is the size of the sensor range plus the vehicle uncertainty, and thus more or less constant. After individual compatibility is sorted out, any disambiguation algorithm, such as **JCBB** [9], will then disambiguate between the m measurements and a region of the map of constant size, regardless of map size, and thus will execute in constant time.

We use **JCBB** in the case of building the local maps of size p, given that it is a standard EKF-SLAM process. However, data association for D&C SLAM is a critical issue because map joining involves finding correspondences between two local maps of similar size, in accordance with their level in the tree. For instance, before of obtaining the final map, the data association problem has to be solved between two maps of size $n/2$ maps and so computing individual compatibility becomes $O(n^2)$. Fortunately, this can be easily reduced to linear again using a simple tessellation or grid for the maps.

The size of the region of overlap between two maps in D&C SLAM depends on the environment and type of trajectory. Consider the simulated examples of fig. 3 where two $n/2$ maps are shown (features in the first map are red crosses, features in the second are blue circles). In the second case, the square loop, the region of overlap between two maps will be of constant size, basically dependent on the sensor range. In the case of the lawn mowers trajectory, the overlap will be proportional to the length of the trajectory before the vehicle turns back, still independent of map size, and thus constant. In the fourth case, the snail path, the region of overlap between the inner map and the encircling map is proportional to the final map size. In these cases, data association algorithms like **JCBB** will not execute in constant time.

In order to limit the computational cost of data association between local maps in D&C SLAM, we use a *randomized joint compatibility* algorithm. Our **RJC** approach (see algorithm 3) is a variant of the linear **RS** algorithm [10]) used for global localization.

Consider two consecutive maps $\mathbf{m}_1$ and $\mathbf{m}_2$, of size n_1 and n_2 respectively, to be joined. First, the overlap between the two maps is identified using individual compatibility. Second, instead of performing branch and bound interpretation tree search in the whole overlap as in **JCBB**, we randomly select b features in the overlapped area of the second map and use **JCBB***: a version of **JCBB** *without exploring the star node*, i.e., considering all b measurements good. This produces a hypothesis $\mathcal{H}$ of b jointly compatible features in the first map. Associations for the remaining features in the overlap are obtained using the simple nearest neighbor rule given hypothesis $\mathcal{H}$, which amounts to finding pairings that are compatible with the first b features. In the spirit of adaptative **RANSAC** [29], we repeat this process t times, so that the probability of missing a correct association is limited to P_{fail}.

Algorithm 3 : RJC

$\mathbf{P}_{fail} = 0.01$, $\mathbf{P}_{good} = 0.8$, $b = 4$
$i = 1$, $Best$ = []
while $(i \leq t)$ **do**
  $\mathbf{m}_2^*$ = random_select($\mathbf{m}_2$, b)
  $\mathcal{H}$ = JCBB*($\mathbf{m}_1$, $\mathbf{m}_2^*$)
  $\mathcal{H}$ = NN($\mathcal{H}$, $\mathbf{m}_1$, $\mathbf{m}_2^*$)
  if pairings($\mathcal{H}$) > pairings($Best$) **then**
    $Best = \mathcal{H}$
  end if
  $\mathbf{P}_{good} = max(\mathbf{P}_{good}, pairings(Best)\backslash m)$
  $t = \log \mathbf{P}_{fail} / \log(1 - \mathbf{P}_{good}^b)$
  $i = i + 1$
end while

Since **JCBB*** is executed using a fixed number of features, its cost remains constant. Finding the nearest neighbor for each remaining feature among the ones that are individually compatible with it, a constant number, will be constant. The cost of each try is thus $O(n)$. The number of tries depends on the number of features randomly selected (b), on the probability that a selected feature in the overlap can be actually found in the first map (P_{good}), and on the acceptable probability of failure in this probabilistic algorithm (P_{fail}). It does not depend on the size of either map. In this way, we can maintain data association in D&C SLAM linear with the size of the joined map.

V. Experiments

We have used the well known Victoria Park data set to validate the algorithms D&C SLAM and **RJC**. This experiment is particulary adequate for testing SLAM due its large scale, and the significant level of spurious measurements. The experiment also provides critical loops in absence of reliable features.

For **RJC**, we chose $b = 4$ as the number of map features to be randomly selected as seed for hypothesis generation. Two features are sufficient in theory to fix the relative location between the maps, but we have found 4 to adequately disambiguate. The probability that a selected feature in the overlap is not spurious, $\mathbf{P}_{good}$ is set to 0.8, and the probability of not finding a good solution when one exists, $\mathbf{P}_{fail}$ is set to 0.01. These parameters make the data association algorithm carry out 9 random tries.

Figure 5 shows the resulting maps from standard EKF SLAM .vs. D&C SLAM, which are essentially equivalent; there are some minor differences due to missed associations in the case of EKF. Figure 6, top, shows the amortized cost of D&C SLAM. We can see that in this experiment an EKF step can take 0.5 seconds, while the amortized D&C SLAM step will take at most 0.05 seconds. In this experiment, the total cost of D& C SLAM is one tenth of the total cost of standard EKF (fig. 6, bottom).

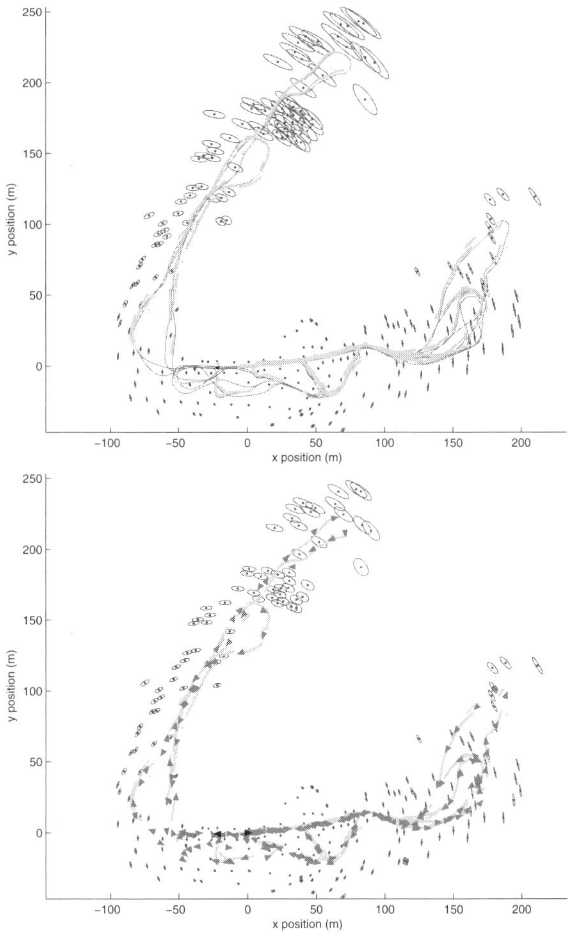

Fig. 5. Map for Victoria Park dataset: according to the standard EKF SLAM algorithm (top); according to the D &C SLAM algorithm. The results are essentially equivalent; some missed associations may result in minor differences. The estimated position along the whole trajectory is shown as a red line for EKF SLAM, and the vehicle locations are drawn as red triangles when available in D&C SLAM. Green points are GPS readings in both cases.

VI. Conclusions

In this paper we have shown that EKF SLAM can be carried out in time *linear* with map size. We describe and EKF SLAM variant: *Divide and Conquer* SLAM, a simple algorithm to implement. In contrast with many current efficient SLAM algorithms, all information required for data association is available when needed with no further processing. D&C SLAM computes the exact EKF SLAM solution, the state *and* its covariance, with no approximations, and with the additional advantage of providing always a more precise and consistent vehicle and map estimate. Data association can also be carried out in linear time per step. We hope to have shown that D&C SLAM is the algorithm to use in all applications in which the Extended Kalman Filter solution is to be used.

Despite of the differences with other methods presented in section I, a very important fact to be emphasized is that the

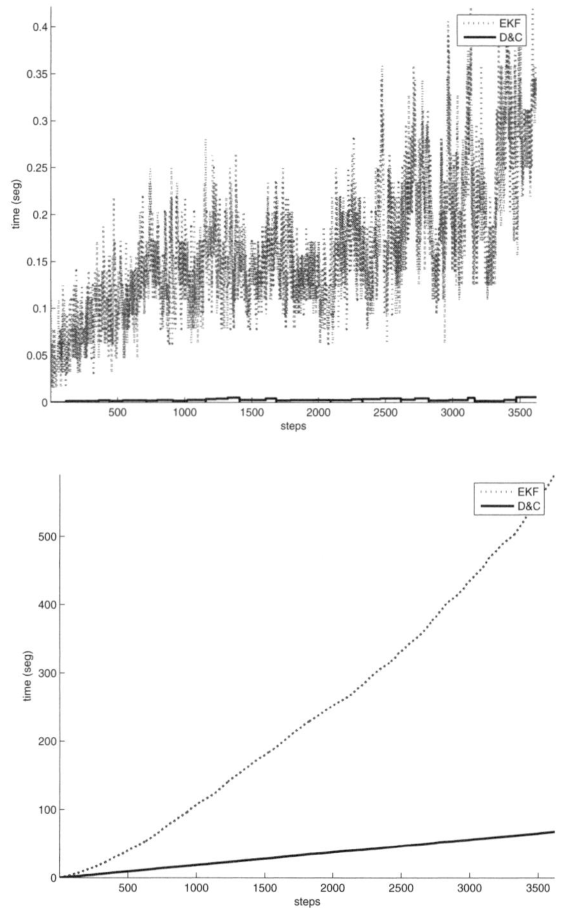

Fig. 6. Time per step of EKF and D&C SLAM for Victoria experiment (top); time per step of EKF SLAM .vs. amortized time per step of D& SLAM (middle); accumulated time of EKF SLAM .vs. D& SLAM.

D&C map splitting strategy can also be incorporated in those recent algorithms. This idea is part of our future work.

ACKNOWLEDGMENT

This research has been funded in part by the Dirección General de Investigación of Spain under project DPI2006-13578.

REFERENCES

[1] H. Durrant-Whyte and T. Bailey, "Simultaneous localization and mapping: part I," *IEEE Robotics & Automation Magazine*, vol. 13, no. 2, pp. 99–110, 2006.

[2] T. Bailey and H. Durrant-Whyte, "Simultaneous localization and mapping (SLAM): part II," *Robotics & Automation Magazine, IEEE*, vol. 13, no. 3, pp. 108–117, 2006.

[3] R. Chatila and J. Laumond, "Position referencing and consistent world modeling for mobile robots," *Robotics and Automation. Proceedings. 1985 IEEE International Conference on*, vol. 2, 1985.

[4] R. C. Smith and P. Cheeseman, "On the representation and estimation of spatial uncertainty," *Int. J. of Robotics Research*, vol. 5, no. 4, pp. 56–68, 1986.

[5] R. Smith, M. Self, and P. Cheeseman, "A stochastic map for uncertain spatial relationships," in *Robotics Research, The Fourth Int. Symposium*, O. Faugeras and G. Giralt, Eds. The MIT Press, 1988, pp. 467–474.

[6] J. Leonard and H. Durrant-Whyte, "Simultaneous Map Building and Localization for an Autonomous Mobile Robot," in *1991 IEEE/RSJ Int. Conf. on Intelligent Robots and Systems*, Osaka, Japan, 1991, pp. 1442–1447.

[7] J. A. Castellanos and J. D. Tardós, *Mobile Robot Localization and Map Building: A Multisensor Fusion Approach.* Boston, Mass.: Kluwer Academic Publishers, 1999.

[8] J. E. Guivant and E. M. Nebot, "Optimization of the Simultaneous Localization and Map-Building Algorithm for Real-Time Implementation," *IEEE Trans. on Robotics and Automation*, vol. 17, no. 3, pp. 242–257, 2001.

[9] J. Neira and J. D. Tardós, "Data Association in Stochastic Mapping Using the Joint Compatibility Test," *IEEE Trans. Robot. Automat.*, vol. 17, no. 6, pp. 890–897, 2001.

[10] J. Neira, J. D. Tardós, and J. A. Castellanos, "Linear time vehicle relocation in SLAM," in *IEEE Int. Conf. on Robotics and Automation*, Taipei, Taiwan, September 2003, pp. 427–433.

[11] J. Knight, A. Davison, and I. Reid, "Towards constant time SLAM using postponement," in *IEEE/RSJ Int'l Conf on Intelligent Robots and Systems*, Maui, Hawaii, 2001, pp. 406–412.

[12] J. Tardós, J. Neira, P. Newman, and J. Leonard, "Robust mapping and localization in indoor environments using sonar data," *Int. J. Robotics Research*, vol. 21, no. 4, pp. 311–330, 2002.

[13] L. Paz and J. Neira, "Optimal local map size for ekf-based slam," in *2006 IEEE/RSJ International Conference on Intelligent Robots and Systems*, Beijing, China., October 2006.

[14] S. Thrun, Y. Liu, D. Koller, A. Y. Ng, Z. Ghahramani, and H. Durrant-Whyte, "Simultaneous Localization and Mapping with Sparse Extended Information Filters," *The International Journal of Robotics Research*, vol. 23, no. 7-8, pp. 693–716, 2004.

[15] R. Eustice, M. Walter, and J. Leonard, "Sparse extended information filters: Insights into sparsification," in *Proceedings of the IEEE/RSJ International Conference on Intelligent Robots and Systems*, Edmonton, Alberta, Canada, August 2005.

[16] M. Walter, R. Eustice, and J. Leonard, "A provably consistent method for imposing sparsity in feature-based slam information filters," in *Proc. of the Int. Symposium of Robotics Research (ISRR)*, 2004.

[17] M. A. Paskin, "Thin Junction Tree Filters for Simultaneous Localization and Mapping," in *Proc. of the 18th Joint Conference on Artificial Intelligence (IJCAI-03)*, San Francisco, CA., 2003, pp. 1157–1164.

[18] U. Frese, *Treemap: An o(logn) algorithm for simultaneous localization and mapping.* Springer Verlag, 2005, ch. Spatial Cognition IV, p. 455476.

[19] R. M. Eustice, H. Singh, and J. J. Leonard, "Exactly Sparse Delayed-State Filters for View-Based SLAM," *Robotics, IEEE Transactions on*, vol. 22, no. 6, pp. 1100–1114, Dec 2006.

[20] F. Dellaert and M. Kaess, "Square Root SAM: Simultaneous Localization and Mapping via Square Root Information Smoothing," *Intl. Journal of Robotics Research*, vol. 25, no. 12, December 2006.

[21] K. Ni, D. Steedly, and F. Dellaert, "Tectonic SAM: Exact, Out-of-Core, Submap-Based SLAM," in *2007 IEEE Int. Conf. on Robotics and Automation*, Rome, Italy, April 2007.

[22] S. J. Julier and J. K. Uhlmann, "A Counter Example to the Theory of Simultaneous Localization and Map Building," in *2001 IEEE Int. Conf. on Robotics and Automation*, Seoul, Korea, 2001, pp. 4238–4243.

[23] J. Castellanos, J. Neira, and J. Tardós, "Limits to the consistency of EKF-based SLAM," in *5th IFAC Symposium on Intelligent Autonomous Vehicles*, Lisbon, Portugal, 2004.

[24] T. Bailey, J. Nieto, J. Guivant, M. Stevens, and E. Nebot, "Consistency of the ekf-slam algorithm," in *IEEE/RSJ International Conference on Intelligent Robots and Systems*, 2006.

[25] S. Julier and J. Uhlmann, "A new extension of the Kalman Filter to nonlinear systems," in *International Symposium on Aerospace/Defense Sensing, Simulate and Controls*, Orlando, FL, 1997.

[26] R. Martinez-Cantin and J. A. Castellanos, "Unscented SLAM for large-scale outdoor environments," in *2005 IEEE/RSJ Int. Conference on Intelligent Robots and Systems*, Edmonton, Alberta, Canada, 2005, pp. pp. 328–333.

[27] L. Paz, P. Jensfelt, J. D. Tards, and J. Neira, "EKF SLAM Updates in O(n) with Divide and Conquer," in *2007 IEEE Int. Conf. on Robotics and Automation*, Rome, Italy., April 2007.

[28] S. B. Williams, "Efficient solutions to autonomous mapping and navigation problems," Ph.D. dissertation, Australian Centre for Field Robotics, University of Sydney, September 2001, available at http://www.acfr.usyd.edu.au/.

[29] R. Hartley and A. Zisserman, *Multiple View Geometry in Computer Vision.* Cambridge, U. K.: Cambridge University Press, 2000.

An Experimental Study of Exploiting Multipath Fading for Robot Communications

Magnus Lindhé and Karl Henrik Johansson
School of Electrical Engineering
Royal Institute of Technology
Stockholm, Sweden
{lindhe | kallej}@ee.kth.se

Antonio Bicchi
Interdepartmental Research
Center "Enrico Piaggio"
University of Pisa, Italy
bicchi@ing.unipi.it

Abstract—**A simple approach for mobile robots to exploit multipath fading in order to improve received radio signal strength (RSS), is presented. The strategy is to sample the RSS at discrete points, without deviating too far from the desired position. We first solve the problem of how many samples are needed for given communications performance and how they should be spaced. Second, we propose a circular and a grid trajectory for sampling and give lower bounds on how many samples they will yield. Third, we estimate the parameters of our strategy from measurements. Finally we demonstrate the validity of our analysis through experiments.**

I. Introduction

In many applications for multi-agent robotics, such as surveillance, mapping of unknown environments and searching, the result is of no value unless it can be communicated to a base station or operator. A need arises to ensure that robots perform their tasks in a "communications aware" manner, to avoid losing contact with their team-members or the outside world. There are several approaches to doing this, for example by restricting the movement to preserve connectivity [8], preserving a clear line of sight [1] or deploying relaying agents if the link is getting too weak [15, 14, 10]. A similar idea is to use mobile agents to bridge disconnected groups of networked sensors [2]. Most of the proposed coordination schemes have been developed based on assumptions of very simplified channel models. A popular approach is to use a binary link model where agents can communicate perfectly within a certain radius, but not at all outside it [3]. Slightly more sophisticated approaches assume that the signal strength decays according only to the path loss [11, 9, 6]. In indoor environments, this overlooks the very significant multipath fading effect that has been extensively studied in the field of electromagnetic wave propagation. In this paper, we show that awareness of this enables us to make significant improvements in the achieved communication performance for multi-robot systems. There are several other ways of counteracting the multipath fading, such as antenna diversity, adaptive antenna arrays or frequency spreading. But antenna diversity or arrays require a certain physical size of the robot and so may not be feasible for very small platforms. Frequency spreading, on the other hand, is not always very effective, as shown in our measurements and by Puccinelli *et al.* [12].

The contribution of this paper is to formulate the problem of positioning the robot to alleviate the effects of multipath fading, and to validate the proposed approach against real data. We consider static environments, where only the robot is moving so that the fading does not vary over time. Such scenarios are relevant in contexts of, e.g., robotic search and surveillance. The problem is divided into two parts: first we provide an estimate of the number of points that the robot needs to sample to find a position that has a given signal strength. Then we suggest two possible sampling strategies to collect the required number of independent samples, without deviating too far from the desired path of the robot. It is shown that the approach is robust and works in several kinds of propagation environments, and almost certainly avoids the deep fades where the signal can be attenuated as much as 20 dB. In most cases, the strategy gives a gain of 5 dB or more compared to the local mean signal power. A demonstration in a simple experiment with a single robot is provided.

This paper is organized as follows: In Section II, we develop our channel model and explain the phenomenon of multipath fading. We then formally state the problem in Section III and solve it in the case of perfect Rayleigh fading in Section IV. The assumptions for Rayleigh fading are not, in general, true in actual environments, so in Section V we present measurements to estimate the statistical properties of fading in representative environments. Finally we present experimental validation of our approach on a robot platform in Section VI, and end with some conclusions in Section VII.

II. Communications Preliminaries

In this section we review some results on what determines the received signal strength (RSS) in a radio receiver and how this is affected by the surrounding environment. Unless stated separately, this section follows the book by Stüber [13].

Given a transmission power of P_t, and a distance d m to the receiver, the *nominal* signal power P_r in the receiver is

$$P_r(d) = P_t + G_t - PL_0 - 10\ n\ \log_{10} d + G_r\ \text{[dB]}.$$

Here PL_0 is the path loss at 1 m from the antenna and n is the path loss exponent. G_t and G_r are antenna gains in the transmitter and receiver, respectively. The path loss exponent n is 2 for free-space propagation and can reach as high as 4 in

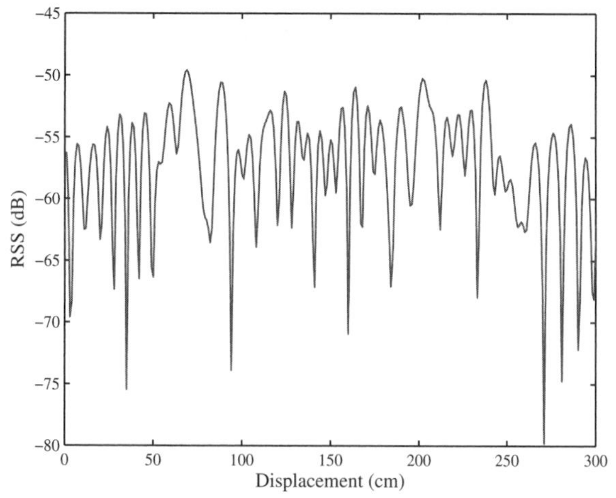

Fig. 1. Simulated Rayleigh fading at 2.4 GHz as the receiver moves along a 3 m line, taking samples of the RSS at every cm.

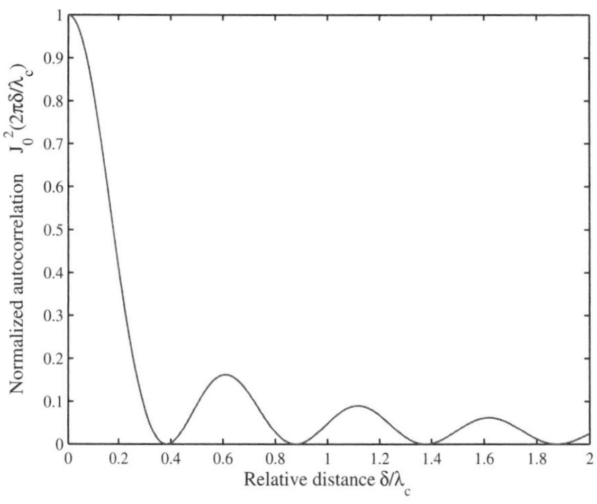

Fig. 2. Normalized spatial autocorrelation of the RSS in a Rayleigh fading environment.

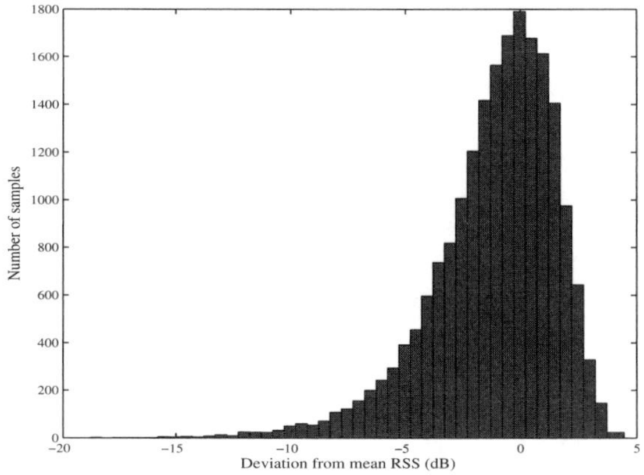

Fig. 3. A histogram based on 20 000 simulated measurements, showing the typical Rayleigh distribution.

some environments. At 2.4 GHz and in an office environment, values around 2.5 have been reported [5].

In an urban or indoor environment, there are however large fluctuations around the nominal level due to shadowing and multipath fading. Shadowing is caused by objects obstructing the signal path and varies over distances the same order as the obstructions.

Multipath fading, on the other hand, is caused by destructive or constructive interference between the signal and its reflections and it varies over very short distances, in the order of a wavelength. Even a stationary receiver will experience multipath fading if the environment is changing, for example due to cars and people moving or doors opening and closing. If all signal components that reach the receiver are of equal strength, the multipath fading is called Rayleigh fading, while if there is a line-of-sight (LoS) component that is significantly stronger, we have Ricean fading.

It should be pointed out that *for given antenna positions*, the fading is reciprocal and thus affects the signal path equally in both directions. But its spatial properties in general are not. Specifically, if a base station with very open surroundings is communicating with a robot in a cluttered environment, the multipath fading varies over much longer distances at the base station than at the robot.

Due to the difficulty of predicting the Rayleigh fading, it is usually modeled as a stochastic effect. The probability density function (pdf) of the RSS in a Rayleigh fading environment is

$$f_P(x) = \frac{1}{P_r} \exp\left(\frac{-x}{P_r}\right). \qquad (1)$$

The expected value is P_r.

A simulated Rayleigh fading signal power plot is depicted in Figure 1. The spatial autocorrelation of the fading envelope as a function of the distance δ between two samples is

$$R(\delta) = k\, J_0^2(2\pi\delta/\lambda_c), \qquad (2)$$

where k is a constant, J_0 is the zero-order Bessel function of the first kind and λ_c is the carrier wavelength. This is illustrated in Figure 2. It shows that two samples taken $0.38\lambda_c$ apart (4.75 cm at 2.4 GHz) should have zero correlation, and samples taken at a greater distance should always have small correlation. In practice, samples taken more than about half a wavelength apart are considered to have independent Rayleigh fading. For comparison with measurements in later sections, we also include a histogram of 20 000 simulated samples of Rayleigh fading in Figure 3. The histogram is normalized by subtracting the mean (computed in dB) to allow comparison with the measurement results in later sections.

III. Problem Formulation

In this section we formulate the problem of coordinating a robot to improve its communication capability by counteracting multipath fading. We consider situations when the application allows the robot to deviate slightly from the desired

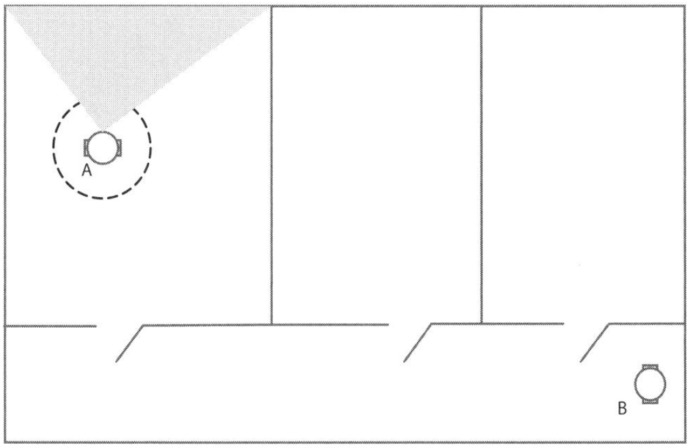

Fig. 4. A scenario where a robot (A) enters a room to monitor part of it, and send the data back to another robot (B). The task can be fulfilled from anywhere inside the dashed circle, so the robot can move inside it to improve the communications with B.

position, in order to improve the communication. A typical scenario is demonstrated in Figure 4, where robot A monitors a room and needs to send data to robot B in the corridor outside. The task allows A to deviate within a given region in search of higher signal strength.

Since the RSS is practically impossible to predict and has multiple local maxima, finding the optimal position would require visiting all of the feasible region. We therefore suggest sampling the RSS in a finite number of points, and then going back to the best. This requires high-accuracy navigation, which is not always available. An alternative is to sample a few points to estimate the nominal RSS and then continue the sampling, stopping at a point that offers a given improvement over the nominal level. We can then statistically express what improvement of the RSS this will give. Our first problem is thus formulated as:

Problem A: Find the number N of independent samples that we need to get an improvement of G dB over the nominal RSS, with probability P.

The trajectory should allow for simple control laws and not deviate outside the feasible region. We have concentrated on what we believe to be the most common robot types today – car-like robots or robots with differential drive. The kinematics of a differential drive robot are

$$\begin{aligned} \dot{x} &= v \cos\theta \\ \dot{y} &= v \sin\theta \\ \dot{\theta} &= u, \end{aligned} \tag{3}$$

where the robot has position (x,y) and bearing θ, and the control inputs are the angular velocity u and linear velocity v. For car-like robots,

$$u = \frac{v}{L}\tan\alpha, \tag{4}$$

with L being the distance between rear and front axels, and α the steering angle, limited by $|\alpha| \leq \alpha_{max}$.

Problem B: Find a trajectory that is simple to follow with a car-like or differential drive robot, and offers N sampling points, spaced at least Δ, without deviating more than R from the original position.

As a remark, applications such as transportation or patrolling may require the robot to follow a trajectory. In this case the robot may sample the RSS along its trajectory, and stop to communicate when it finds a good position. Problem A is then of interest to give an upper bound on the number of samples (thus also the time or distance) between successful communication attempts.

The general problem has thus been divided into two parts: First finding the number of samples N required to achieve the desired performance and, second, finding a suitable trajectory for the robot to visit that many sampling points. In the following, we provide solutions first in the theoretical case of Rayleigh fading, and then based on properties of real environments.

IV. Solution in Perfect Rayleigh Fading Environment

In this section we give a solution with provable properties in the case of perfect Rayleigh fading. We first give a proposition on the number of samples required to achieve a certain gain and then suggest two alternative strategies of fitting the required number of independent samples within the region where the robot is allowed to move.

Proposition 4.1 (Number of samples): For a Rayleigh fading environment, the number of independent samples N needed to achieve a power gain of G dB with probability P compared to the nominal RSS is given by

$$N = \frac{\ln(1-P)}{\ln(1-\exp(-10^{G/10}))}.$$

Proof: From Equation 1, we have the pdf of the signal power, which gives the cumulative distribution function (cdf)

$$C(P_n) := \text{Prob}(X < P_n) = 1 - e^{-P_n/P_r}$$

i.e., the probability that the power in a single sample is lower than the threshold P_n. Taking N independent samples, the probability that all of them are lower than P_n is $C(P_n)^N$. We note that at least one sample being greater than P_n is the complementary event to the above, and since $P_n/P_r = 10^{G/10}$, the probability of this is

$$\text{Prob}(G) = 1 - \left[1 - \exp(-10^{G/10})\right]^N.$$

Solving for N gives the proposition. □

As described in Section II, samples taken at a distance of $0.38\lambda_c$ can be considered independent. This can be viewed as each sample being surrounded by a disc of radius $0.19\lambda_c$ where no more samples should be taken. So taking N independent samples inside the feasible region with radius R is essentially a two-dimensional sphere-packing problem.

We propose two possible sampling trajectories; driving in a circle and sweeping a hexagonal lattice. They represent different trade-offs between ease of navigation and maximizing the number of samples.

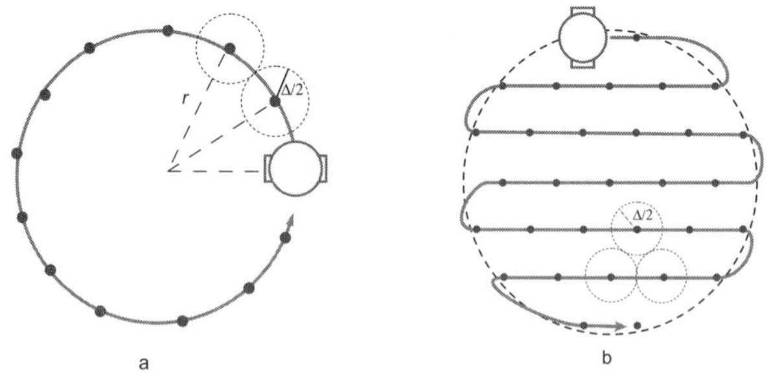

Fig. 5. Two possible sampling trajectories: a circle (a) and a hexagonal lattice (b). In both cases, the distance between sampling points is at least Δ.

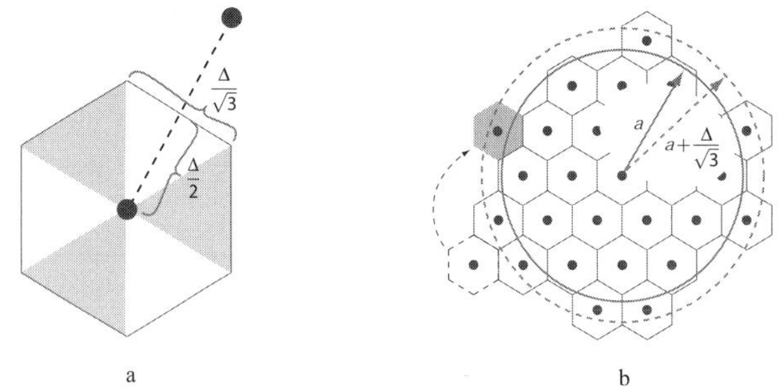

Fig. 6. (a) The sampling pattern follows a hexagonal lattice. (b) If one of the N hexagons (dashed) is completely outside the circle of radius a, there must exist a free space partially inside the circle, where it can be moved (in gray).

Proposition 4.2 (Circular sampling trajectory): If N samples are taken on a circle, and the samples are at a distance not less than Δ, the radius of the circle must be

$$r \geq \frac{\Delta}{\sqrt{2}\sqrt{1-\cos(2\pi/N)}}.$$

This is illustrated in Figure 5a. Another possible sampling pattern is the intersection of a hexagonal lattice and a circle with radius r. A hexagonal lattice can be defined as

$$\left\{(x,y) = \Delta(k+a, \ell\sqrt{3}/2) : a = \frac{1}{2} \bmod (\ell, 2), \; k, \ell \in \mathbb{Z}\right\}$$

which was proven by Thue [7] to be optimal for two dimensional sphere-packing. The distance Δ is the vertex distance. This arrangement of sampling points is also suitable for being covered by differential drive or car-like robots and with collision sensors also in the back, one could reverse along every second line to simplify maneuvering. Sensors such as a camera can be pointed in the interesting direction during the whole sampling procedure. If the robot detects an obstacle, it can simply turn earlier and start following the next line back. A hexagonal lattice with sampling trajectory is depicted in Figure 5b. The required size of the sampling region is stated by the following proposition:

Proposition 4.3 (Hexagonal lattice of samples): A hexagonal lattice with vertex distance Δ has at least N vertices within the distance

$$r = \left[\sqrt{\frac{\sqrt{3}(N+1)}{2\pi}} + \frac{1}{\sqrt{3}}\right]\Delta \quad (5)$$

from the origin.

Proof: Each vertex can be regarded as the center of a hexagon with area $\sqrt{3}\Delta^2/2$, as shown in Figure 6a. A circle of radius a has an area equal to or greater than the area covered by

$$N = \left\lfloor \frac{2\pi a^2}{\sqrt{3}\Delta^2} \right\rfloor \quad (6)$$

such hexagons. The hexagons can be tiled so that their centers all fit within a circle of radius $a + \Delta\sqrt{3}$, see Figure 6b. This can be proved as follows.

Assume that any hexagon is completely outside the circle. Since the remaining hexagons cannot fill the circle, there must be some free space partially inside, and since hexagons can be tiled with no gaps, this space must be on the perimeter. So the hexagon can be moved there instead. To complete the proof, we also note that no part of a hexagon is more than $\Delta/\sqrt{3}$ from its center, so since all hexagons have some part inside the circle of radius a, their centers must then fit inside a concentric circle of radius $a + \Delta/\sqrt{3}$.

Solving Equation 6 for a, using that $N+1 \geq \lfloor N \rfloor$ and adding the margin $\Delta/\sqrt{3}$, gives the proposition. □

Other trajectories than the two described here are of course also possible; the fewer samples needed, the greater the flexibility to choose a trajectory.

V. Measurements in Real Environments

To apply the proposed strategy in a real environment, we need to estimate the spatial correlation (to determine the sample spacing) and the cdf of the signal strength. We have chosen a basement corridor and a cluttered lab room as representative environments for the measurements.

To automate the measurements, we have mounted a radio based on the CC2420 chip on a robot. It communicates with the same chip on a TMote Sky wireless sensor node, connected to a PC, see Figure 7. The CC2420 operates at 2.4 GHz with a maximal output power of 0 dBm and has a software-accessible received signal strength indicator (RSSI) [4]. It is worth noting that the CC2420 uses direct sequence spread spectrum modulation. This is supposed to alleviate the effects of multipath fading, but as shown below and in [12], the problem of deep fades remains.

The TMote has an antenna integrated on its circuit board, while the robot has a quarter-wave antenna on top. The integrated antenna is by design not omnidirectional, and measurements show that the antenna mounted on the robot also has some directional dependence, due to the influence from the rest of the robot. This makes it important that, when the robot has found the best position and returns there, it also returns to the same orientation.

To estimate the spatial correlation in the different environments, we have driven the robot along straight 200 cm lines, stopping and sampling the RSSI each centimeter. Each sample

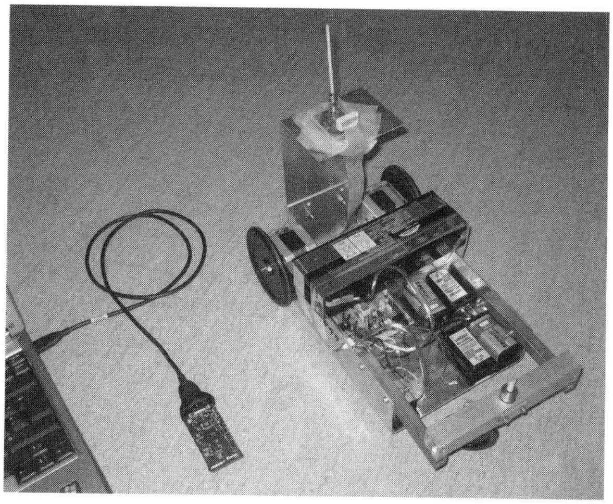

Fig. 7. The measurement system, with the robot and the TMote connected to a PC. The robot has two driving wheels and a third caster wheel, and its antenna is positioned about 25 cm above ground.

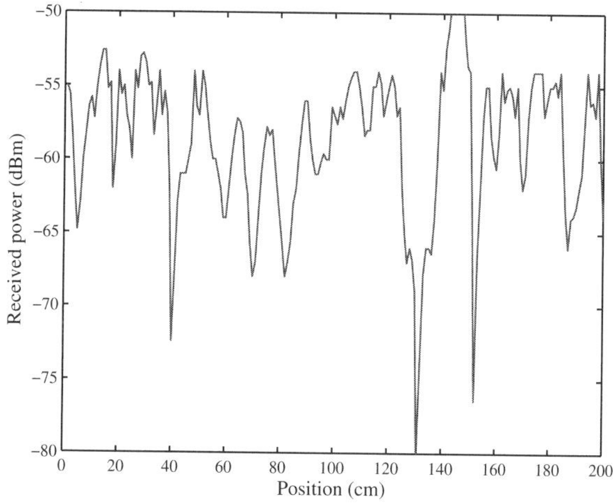

Fig. 8. Measurement results from the lab room, where the RSS varies over 30 dB. Note the deep fade at 130 cm, where the connection was temporarily lost.

is formed from averaging the RSSI readings from four radio packets. For each sequence of $N = 200$ samples, we computed the unbiased estimate of the autocorrelation

$$\hat{R}(k) = \frac{1}{N-|k|} \sum_{m=0}^{N-k-1} [z(m) - \bar{z}][z(m+k) - \bar{z}],$$

where $z(n)$ is the signal envelope in sample n and $\bar{z}$ is the mean value.

The cdf of the RSS samples was estimated using the same measurement setup, but driving a distance greater than Δ between samples. We then assumed that the samples could be regarded as independent. Since the nominal RSS is difficult to calculate, we estimate it by the local average. The result is plotted in a histogram. Summing over this histogram gives an estimate of the cdf.

The lab room contains lots of computers, metal cabinets and other effective scatterers, so it is our belief that this environment produces near-Rayleigh fading. This is also confirmed by the measurements. One representative measurement series is depicted in Figure 8, and the estimated autocorrelations for five measurements are superimposed in Figure 9. The autocorrelation decays rapidly, and reaches the noise floor at $\Delta = 6$ cm in all measurements in this environment. (This matches the predicted $\lambda_c/2 = 6.25$ cm.) Finally, the histogram in Figure 10 indicates that the distribution is very similar to the Rayleigh distribution. The samples were taken with a distance of 10 cm to ensure independence.

The corridor has metal grates and cable ducts along its sides, and large metal cabinets at the entrance. This means that the radio signal may be efficiently reflected from a few large surfaces, so the robot does not receive signals of equal strength from all directions as required for Rayleigh fading. As shown by the measurements, this gives somewhat different spatial properties to the fading. The RSS fluctuates as in the lab room, but also over longer distances, much like shadowing.

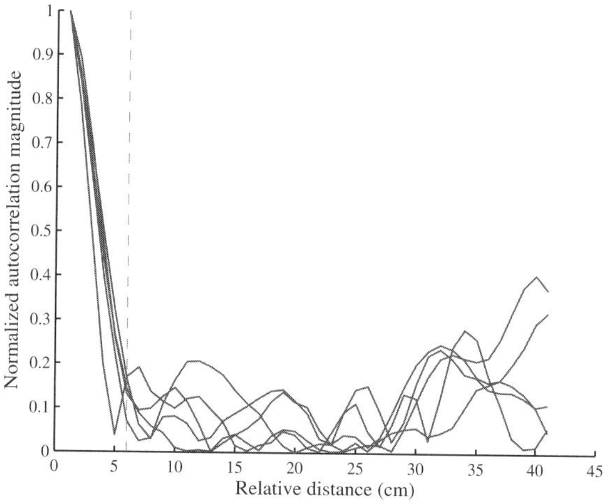

Fig. 9. Autocorrelation estimates for five measurement series in the lab room. The dashed line is the estimated decorrelation distance Δ. The autocorrelation drops rapidly and the spuriouses at 40 cm are probably due to the estimate being noise sensitive at high lags.

A representative measurement result is illustrated in Figure 11, and autocorrelation estimates for eight measurements are superimposed in Figure 13. The measurement in Figure 11 corresponds to the slowest decaying autocorrelation estimate. At $\Delta = 15$ cm, all autocorrelation estimates seem to have reached the noise floor for this environment.

To estimate the cdf for the corridor, we took samples 15 cm apart and collected them in the histogram in Figure 13. Despite the difference in spatial properties, this distribution also resembles that of Rayleigh fading.

The CC2420 data sheet states an RSSI accuracy of ± 6 dB and linearity within ± 3 dB [4]. Since we are not interested in the absolute signal power, we therefore consider the measurements to have an uncertainty of 3 dB. During our measurements in static environments, the typical standard

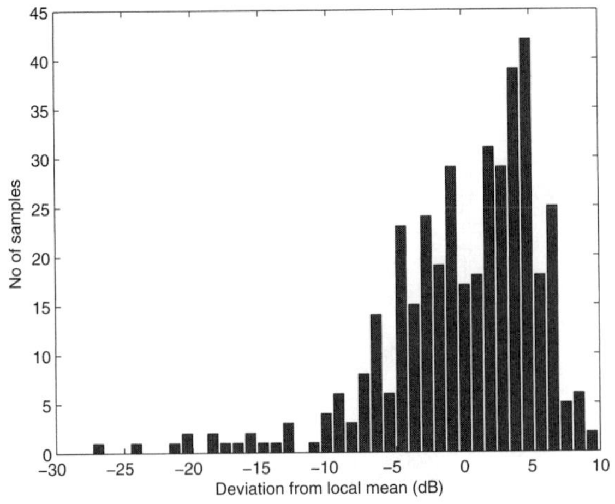

Fig. 10. Histogram of 400 RSS samples, taken in the lab room with sample spacing 10 cm. The distribution resembles the Rayleigh distribution.

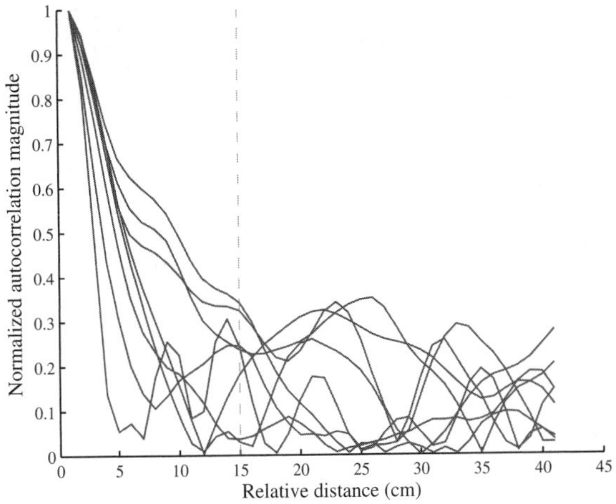

Fig. 12. Autocorrelation estimates for eight measurement series in the corridor. The dashed line is the estimated decorrelation distance Δ. The autocorrelation decays slowly for some series (cf. Figure 9), probably due to shadowing effects.

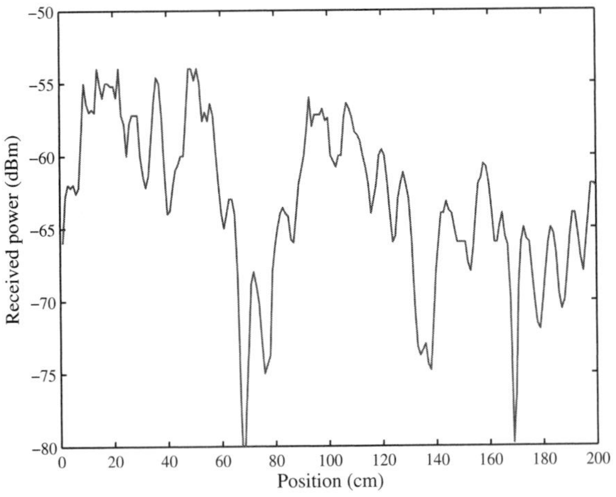

Fig. 11. Measurement results from the corridor. The multipath fading is probably superimposed on a shadowing effect.

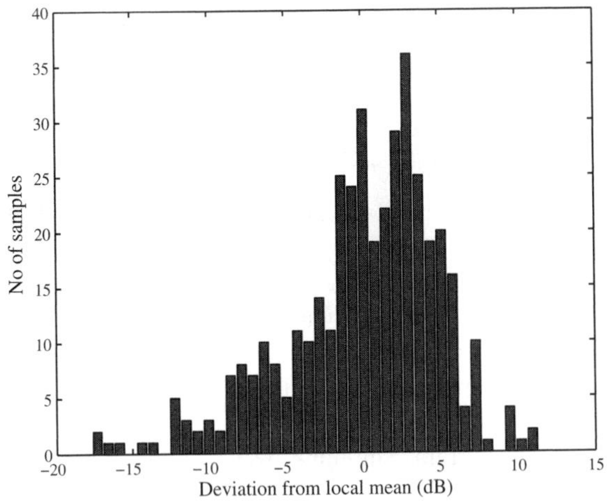

Fig. 13. Histogram of 400 RSS samples, taken in the corridor with sample spacing 15 cm. The distribution resembles the Rayleigh distribution.

deviation within 20 packets was less than 1 dB.

Since our motivating application is autonomous exploration or surveillance of indoor environments, we expect those environments to be static, i.e. with no humans present and little or no movement except for that of the robots themselves. Therefore, the fast fading should not change over time, but only as a function of the position of the transmitter and receiver. To verify this, we made two measurements, first driving the robot forward 100 cm and then back again along the same line. As illustrated in Figure 14, the RSS as a function of position is very similar between the measurements. The RMS deviation between the measurements is 1.2 dB, i.e., well within the expected uncertainty.

Using the above measurements, we can compute the estimated cdf $\hat{C}(P_n)$. This yields a result similar to Proposition 4.1, but where the signal gain is expressed in relation to the local average: The probability of achieving gain G when taking N samples can be estimated as

$$\text{Prob}(G,N) = 1 - \hat{C}(G)^N.$$

Several curves of $\text{Prob}(G,N)$, for some values of G, are plotted for the lab environment as well as the corridor, in Figures 15 and 16, respectively. These figures summarize the answer to Problem A, showing how many samples are needed to reach a given gain with a specified probability.

In practice this means that if the robot can take approximately 9 samples in the lab room (or 14 in the corridor), it has a 95% chance of finding a position where the signal strength is 5 dB better than the local average. Under the same conditions, the probability of finding a point where the signal strength is

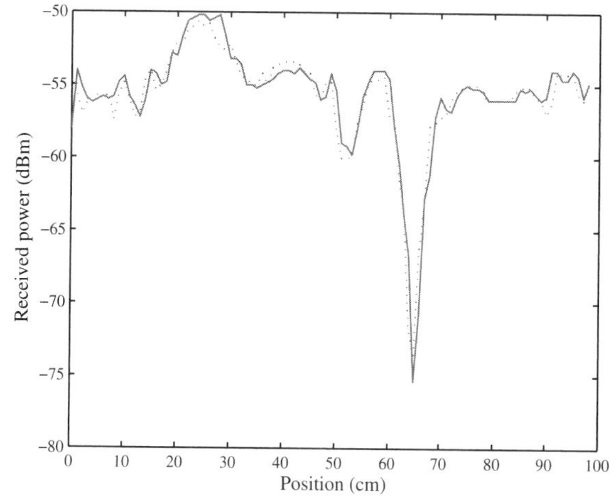

Fig. 14. Two measurement series along the same trajectory. The RSS is clearly a function of the position, and does not vary over time. The RMS deviation between the measurements is 1.2 dB, i.e., within the measurement accuracy.

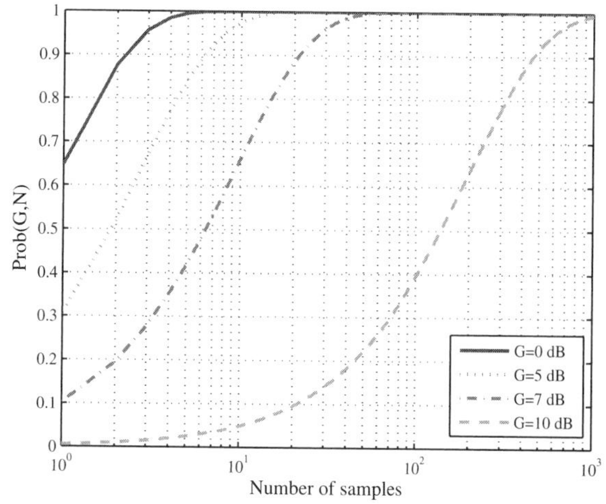

Fig. 15. Results for the lab room: The required number of independent samples of the RSS to achieve a given gain (compared to the local average) with a certain confidence level. We have plotted curves for several gains in the interval 0-10 dB.

at least equal to the local average (and thus avoiding any of the deep fades) is greater than 99.99%. Taking 9 samples in the lab room can be done by going in a circle of radius 8 cm. Conversely, the curves can be used as a guideline for an application designer, choosing the allowed deviation based on what signal gain is needed.

VI. Experimental verification

As a simple demonstration of our suggested approach, we have made experiments positioning the robot at random positions and orientations within a 1-by-1 m square in the lab room, as if a task such as patrolling or mapping had made it drive there. We then measured the signal strength between the

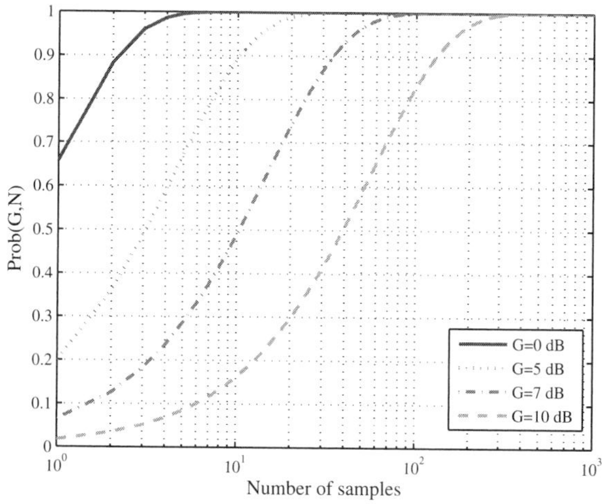

Fig. 16. Results for the corridor: The required number of independent samples of the RSS to achieve a given gain (compared to the local average) with a certain confidence level. We have plotted curves for several gains in the interval 0-10 dB.

robot and its base station (the TMote), in the other end of the room.

First we performed 20 such trials, *series 1*, allowing the robot to deviate slightly from the initial position, sampling the RSS in 9 points, separated by 5 cm. It then moved to the point with the best RSS before the measurement was made. Then we performed 20 more trials, *series 2*, where the robot was not allowed to deviate. The result of the two experiment series is plotted as two histograms in Figure 17. When just staying at the initial position, seven out of twenty trials yielded signal strengths worse than the local average, in one case by as much as 15 dB. The theoretical analysis predicted a gain of at least 5 dB compared to the local average in 95% of the cases, but in practice this happened in 80% of the trials. It is worth noticing, however, that all trials avoided negative gains as predicted.

To illustrate the benefit of gaining 5 dB, we quote an expression for the packet reception rate (PRR) as function of the signal-to-noise ratio (SNR), derived by Zuniga *et al.* [16]. They use MICA2 motes (which work at lower frequencies but have similar performance as the TMotes) with a data rate of 19.2 kbit/s and a PRR of

$$p(SNR) = \left(1 - 1/2e^{-\frac{SNR}{1.28}}\right)^{8f},$$

where f is the frame size, i.e., the number of bytes in each packet. With a frame size of $f = 50$, a MICA2 mote on the limit of losing contact, with SNR=5 dB, would receive 1.8% of the packets and thus have an effective bandwidth of 340 bits/s. Sending a 10 kbyte camera image would then take 3 min 55 s. Gaining 5 dB would raise the bandwidth to 17.7 kbit/s, reducing the sending time to 4.5 seconds.

VII. Conclusion and future work

We have formulated the problem of positioning an autonomous robot to avoid the negative effect of multipath fading

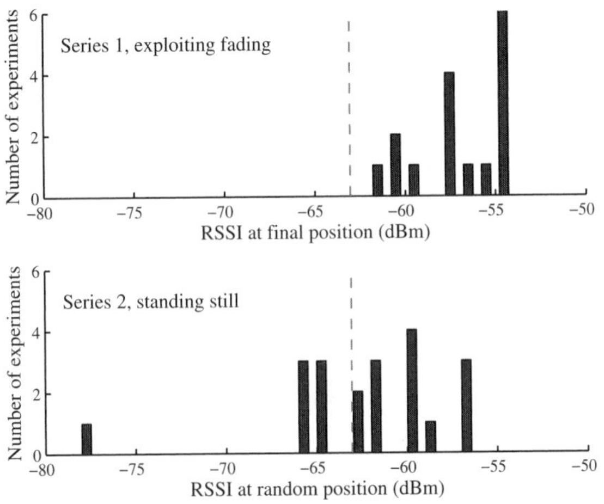

Fig. 17. Experiment results, measuring the RSS at the final position of the robot. The upper histogram shows the result when using the proposed approach, and the lower histogram shows what happened when the robot did not move from its (random) original position. The dashed line shows the local average RSS.

in radio communication. The problem is divided into two parts: first we provide an estimate of the number of points that it needs to sample to find a position that has a given signal strength. Then we suggest two possible sampling strategies to collect the required number of independent samples, without deviating too far from the original position.

This is a robust approach that works in several kinds of propagation environments and almost certainly avoids the deep fades where the signal can be attenuated as much as 20 dB. In most cases it also gives a gain of 5 dB or more compared to the local mean signal power. This can be interpreted as using the unstructured environment as a directional antenna. The performance of this strategy was demonstrated in a simple experiment.

As mentioned in the introduction, the problem of multipath fading can also be countered in other ways, for example by antenna diversity as used in WLAN base stations. Our approach can be seen as a sort of antenna diversity over time, in case the robot is too small to host two antennas separated by at least half a wavelength. (Due to the asymmetry pointed out earlier, in some cases antenna diversity at the base station does not give the same advantages.) This also motivates moving groups of robots in hexagonal lattice formations with distance Δ between agents. Such a lattice represents the tightest possible formation offering antenna diversity gain for the group as a whole: ensuring that at least some of the robots have good signal strength.

In the future we would like to find criteria for how the variance of the RSS over time can be used to detect if the environment is no longer static. Then the robot could adapt its strategy to increase the search space, exploit shadow fading instead or simply stop making adjustments to its position. Finally, we would like to measure the fading properties of other environments to further validate our approach.

Acknowledgments

The authors would like to thank Carlo Fischione and Lucia Pallottino for useful discussions, and Simon Berg for building the measurement robot. This work was partially supported by the Swedish Defence Materiel Administration, the European Commission through the RUNES and HYCON projects, the Swedish Research Council, and the Swedish Foundation for Strategic Research.

References

[1] R. Arkin and T. Balch. Line-of-sight constrained exploration for reactive multiagent robotic teams. 7th International Workshop on Advanced Motion Control, July 2002.

[2] K. E. Årzén, A. Bicchi, G. Dini, S. Hailes, K. H. Johansson, J. Lygeros, and A. Tzes. A component-based approach to the design of networked control systems. European Journal of Control, 2007. To appear.

[3] Z. Butler and D. Rus. Event-based motion control for mobile sensor networks. *IEEE Pervasive Computing*, 2(4), 2003.

[4] Chipcon AS. *CC2420 Datasheet 1.4*. www.chipcon.com, 2006.

[5] F. Darbari, I. McGregor, G. Whyte, R. W. Stewart, and I. Thayne. Channel estimation for short range wireless sensor network. In *Proceedings of the 2nd IEE/Eurasip Conference on DSP Enabled Radio*, 2005.

[6] A. Drenner, I. Burtz, B. Kratochvil, B. J. Nelson, N. Papanikolopoulos, and K. B. Yesin. Communication and mobility enhancements to the scout robot. In *Proceeding of the IEEE/RSJ International Conference on Intelligent Robots and System*, 2002.

[7] T. C. Hales. An overview of the Kepler conjecture. http://www.math.pitt.edu/ thales/kepler98/sphere0.ps, 1998.

[8] M. A. Hsieh, A. Cowley, V. Kumar, and C. J. Taylor. Towards the deployment of a mobile robot network with end-to-end performance guarantees. In *Proceedings of the IEEE International Conference on Robotics and Automation*, 2006.

[9] M. F. Mysorewala, D.O. Popa, V. Giordano, and F.L. Lewis. Deployment algorithms and in-door experimental vehicles for studying mobile wireless sensor networks. In *Proceedings of the7th ACIS International Conference on Software Engineering, Artificial Intelligence, Networking, and Parallel/Distributed Computing*, 2006.

[10] H. G. Nguyen, N. Farrington, and N. Pezeshkian. Maintaining Communication Link for Tactical Ground Robots. In *AUVSI Unmanned Systems North America*, 2004.

[11] D. O. Popa and C. Helm. Robotic deployment of sensor networks using potential fields. In *Proceedings of the IEEE International Conference on Robotics & Automation*, 2004.

[12] D. Puccinelli and M. Haenggi. Multipath fading in wireless sensor networks: Measurements and interpretation. In *Proceedings of the International Wireless Communications and Mobile Computing Conference*, 2006.

[13] Gordon L. Stüber. *Principles of mobile communication*. Kluwer academic publishers, 1996.

[14] J.D. Sweeney, R.A. Grupen, and P. Shenoy. Active QoS flow maintenance in controlled mobile networks. In *Proceedings of the Fourth International Symposium on Robotics and Automation*, 2004.

[15] A. Wagner and R. Arkin. Multi-robot communication-sensitive reconnaissance. In *Proceedings of the IEEE International Conference on Robotics and Automation*, 2004.

[16] M. Zuniga and B. Krishnamachari. Analyzing the transitional region in low power wireless links. In *Proceedings of the First IEEE International Conference on Sensor and Ad hoc Communications and Networks*, 2004.

Mapping Large Loops with a Single Hand-Held Camera

Laura A. Clemente
Instituto de Investigación
en Ingeniería de Aragón
Universidad de Zaragoza, Spain
laura.clemente@unizar.es

Andrew J. Davison
Dept. of Computing
Imperial College London
ajd@doc.ic.ac.uk

Ian D. Reid
Dept. of Eng. Science
University of Oxford
ian@robots.ox.ac.uk

José Neira and Juan D. Tardós
Instituto de Investigación
en Ingeniería de Aragón
Universidad de Zaragoza, Spain
jneira,tardos@unizar.es

***Abstract*— This paper presents a method for Simultaneous Localization and Mapping (SLAM), relying on a monocular camera as the only sensor, which is able to build outdoor, closed-loop maps much larger than previously achieved with such input. Our system, based on the Hierarchical Map approach [1], builds independent local maps in real-time using the EKF-SLAM technique and the inverse depth representation proposed in [2]. The main novelty in the local mapping process is the use of a data association technique that greatly improves its robustness in dynamic and complex environments. A new visual map matching algorithm stitches these maps together and is able to detect large loops automatically, taking into account the unobservability of scale intrinsic to pure monocular SLAM. The loop closing constraint is applied at the upper level of the Hierarchical Map in near real-time.**

We present experimental results demonstrating monocular SLAM as a human carries a camera over long walked trajectories in outdoor areas with people and other clutter, even in the more difficult case of forward-looking camera, and show the closing of loops of several hundred meters.

I. Introduction

Simultaneous Localization And Mapping (SLAM) is one of the most active research fields in robotics, with excellent results obtained during recent years, but until recently mainly restricted to the use of laser range-finder sensors and predominantly building 2D maps (see [3] [4] for a recent review). Under these conditions, robust large-scale indoor and outdoor mapping has now been demonstrated by several groups around the world.

It is more challenging to attempt SLAM with standard cameras as the main sensory input, since the essential geometry of the world does not 'pop-out' of images in the same way as it does from laser data. Nevertheless, the combination of detailed 3D geometric and photometric information available from cameras means that they have great promise for SLAM applications of all types, and recent progress has been very encouraging. In particular, recent robotic SLAM systems which use odometry and single cameras [5], [6], stereo rigs [7], omnidirectional cameras [8] or inertial sensors [9] have all demonstrated reliable and accurate vision-based localisation and mapping, often in real-time and on increasingly large scales. Also impressive have been stereo vision-based 'visual odometry' approaches [10], [11] which match large numbers of visual features in real-time over sequences and obtain highly accurate local motion estimates, but do not necessarily aim to build globally consistent maps.

In this paper, we consider the extreme case where the only sensory input to SLAM is a single low-cost 'webcam', with no odometry, inertial sensing or stereo capability for direct depth perception – a camera carried by a walking person, for example. Under these conditions, successful real-time SLAM approaches have been limited to indoor systems [12]–[14] which can build maps on the scale of a room. Such work on estimating motion and maps from a single moving camera must also be compared with the wealth of work in visual structure from motion. (e.g. [15]) where high quality reconstructions from image sequences are now routinely obtained, but requiring significant off-line optimisation processing.

Now we show that an approach which builds and joins local SLAM maps, previously proven in laser-based SLAM, can be used to obtain much larger outdoor maps than previously built with single camera only visual input and works in near real-time. The keys are the efficient and accurate building of local submaps, and robust matching of these maps despite high localisation uncertainty. Other approaches to vision-based closing of large loops in SLAM have used appearance-based methods separated from the main mapping representation [8], [16]. While these methods are certainly valuable, here we show that under the conditions of the experiments in our paper we are able to directly match up local maps by photometric and geometric correspondences of their member features.

One of the main difficulties of monocular visual SLAM is landmark initialization, because feature depths cannot be initialized from a single observation. In this work we have adopted the inverse-depth representation proposed by Montiel et al. [2], which performs undelayed initialization of point features in EKF-SLAM from the first instant they are detected. In that work, data association was performed by predicting the feature locations in the next image and matching them by correlation. In this paper we demonstrate that adding a Joint Compatibility test [17] makes the method robust enough to perform for the first time real-time monocular SLAM walking with a hand-held camera in urban areas. In our experiments, the inverse depth representation allows SLAM to benefit from features which are far away from the camera, which are revealed to be essential to maintaining good angular accuracy

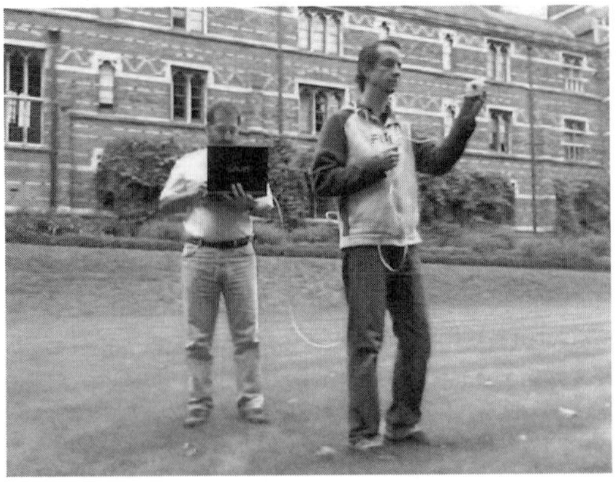

Fig. 1. Experimental setup: a hand-held camera, a firewire cable and a laptop.

in open areas. The joint compatibility technique is able to successfully reject incorrect data associations which jeopardize the operation of SLAM in repetitive or dynamic environments.

To attack the problem of mapping large areas, the technique is applied to build several independent local maps that are integrated into the Hierarchical Map approach proposed by Estrada et al. [1]. Two of the main factors that fundamentally limit EKF-based SLAM algorithms are (i) the processing time associated with the EKF update which is $O(n^2)$ in the number of map features; and (ii) cumulative linearisation errors in the EKF that ultimately contribute to biased and overconfident state estimates which eventually break the filter, usually via poor data association. Hierachical SLAM addresses both of these issues. First, by segmenting the problem into smaller chunks of bounded size, the computational time of the filter is bounded (i.e. $O(1)$). Second, since each local map effectively resets the base frame, linearisation errors only accumulate *within* a local map and not between maps. The main difficulty appearing here is that the scale in pure monocular vision is not observable, so the scale of the different local maps is not consistent. We propose a novel scale invariant map matching technique in the spirit of [18], able to detect loop closures, that are imposed in the upper level of the Hierarchical Map, obtaining a sub-optimal SLAM solution in near real time.

The rest of the paper is structured as follows. Section II describes in detail the local map building technique proposed and presents some experiments showing its robustness in real environments. Section III presents the map matching algorithm and the loop optimization method used at the global level of the Hierarchical Map. Section IV demonstrates the technique by mapping a courtyard by walking with the camera in hand (see fig. 1) along a loop of several hundred meters. The conclusions and future lines of research are drawn in section V.

II. Building Monocular Local Maps

A. EKF SLAM with inverse depth representation

To achieve scalability to large environments we have adopted the Hierarchical Map method proposed in [1]. This technique builds a sequence of local maps of limited size using the EKF-SLAM approach [19]. We have observed that, in the case of monocular SLAM, the use of the Iterated Extended Kalman Filter (IEKF) [20] improves the accuracy of the map and the camera trajectory, at the price of a small increase in the computational cost. In any case, by limiting the maximum size of the local maps, the computation time required per step during the local map building is bounded by a constant.

The state vector of each local map $\mathbf{M}_i$ comprises the final camera location $\mathbf{x}_v^i$ and the 3D location of all features $(\mathbf{y}_1^i \dots \mathbf{y}_n^i)$, using as base reference B, the camera location at the beginning of the local map. We also store the complete camera trajectory inside each local map, that is used only for displaying results. For the state representation inside each local map, we use the inverse-depth model proposed by Montiel *et al.* [2]:

$$\mathbf{x}^T = (\mathbf{x}_v^T, \mathbf{y}_1^T, \mathbf{y}_2^T, \dots, \mathbf{y}_n^T) \quad (1)$$

where:

$$\mathbf{x}_v = \begin{pmatrix} \mathbf{r}^{BC} \\ \mathbf{q}^{BC} \\ \mathbf{v}^B \\ \mathbf{w}^C \end{pmatrix} \quad (2)$$

$$\mathbf{y}_i = (x_i \, y_i \, z_i \, \theta_i \, \phi_i \, \rho_i)^T \quad (3)$$

This feature representation codes the feature state as the camera optical center location $(x_i \, y_i \, z_i)$ when the feature point was first observed, and the azimuth and elevation $(\theta_i \, \phi_i)$ of the ray from the camera to the feature point. Finally, the depth d_i along this ray is represented by its inverse $\rho_i = 1/d_i$. The main advantage of the inverse-depth parametrization is that it allows consistent undelayed initialization of the 3D point features, regardless of their distance to the camera. In fact, distant points, or even points at *infinity* are modelled and processed in the same way. This is in contrast with most current techniques that delay the use of a feature until the baseline is big enough to compute its depth [8], [12].

The camera state $\mathbf{x}_v$ is composed of the camera position $\mathbf{r}^{BC}$ and orientation quaternion $\mathbf{q}^{BC}$ and its linear and angular velocities $\mathbf{v}^B$ and $\mathbf{w}^C$. The process model used for the camera motion is a constant velocity model with white Gaussian noise in the linear and angular accelerations. Using pure monocular vision, without any kind of odometry, the scale of the map is not observable. However, by choosing appropriate values for the initial velocities and the covariance of the process noise, the EKF-SLAM is able to "guess" an approximate scale for each local map, as will be shown in the experimental results.

B. Feature extraction and matching

Now we focus on the features selection which make up the local maps. Our goal is to be able to recognize the same features repeatedly during local map building and also for loop closing detection and optimization. So what we need are persistent and realiable features that ensure us with high probability a quality tracking process. For this very purpose we have followed the approach of Davison *et al.* [12], [21], who showed that selecting salient image patches (11 x 11 pixels) is useful for performing long-term tracking.

To detect salient image regions we use the Shi and Tomasi operator [22] with some modifications which result in more salient and better trackable features. The first modification is the application of a gaussian weighted window to the Hessian matrix (4) which makes the response of the detector isotropic and results in patches better centered around the corner or salient point.

$$H = \begin{pmatrix} G_\sigma * (I_x I_x) & G_\sigma * (I_x I_y) \\ G_\sigma * (I_x I_y) & G_\sigma * (I_y I_y) \end{pmatrix} \quad (4)$$

Apart from using the Shi and Tomasi response:

$$\lambda_{min} > \lambda_{threshold} \quad (5)$$

where λ_{max} and λ_{min} are the maximum and minimum eigenvalues of the Hessian image matrix (4) respectively we only accept as good feature points those whose two eigenvalues have similar magnitude:

$$\lambda_{max}/\lambda_{min} < ratio_{threshold} \quad (6)$$

This avoids selecting regions with unidirectional patterns that cannot be tracked reliably. Instead of using all the features that passed both tests, we have implemented a simple selection algorithm that forces a good distribution of features on the image. The features that pass the tests are stored in a 2D-spatial structure. When the number of tracked features falls bellow a threshold, the spatial structure is used to find the best feature (with higher Shi-Tomasi response) from the image region with less visible features.

For tracking the features on the sequence of images we use the active search approach of Davison and Murray [21]. The stochastic map is used to predict the location of each feature in the next image and compute its uncertainty ellipse. The features are searched by correlation inside the uncertainty ellipse using normalised sum-of-squared-differences. This gives enough robustness with respect to light condition changes and also to small viewpoint changes.

C. Joint compatibility

It is well known that data association is one of the most critical parts in EKF-SLAM, since a few association errors may ruin the quality of an otherwise good map. The active search strategy presented gives good feature matchings *most* of the time. However, since we are building maps of large outdoor dynamic environments we have to deal with two well differentiated problems. The first problem is that moving objects produce valid matches – in that they correspond to the same point on the object – which nevertheless violate the basic assumption of static features made by most SLAM techniques. The second problem arises in the presence of ambiguous matches caused, for example, by repeated texture in the environment. Such ambiguous matches are more likely with "young" features whose 3D locations (especially depth) are still uncertain, leading to large search regions.

One approach to tackling the problems of repeated patterns is to try to avoid it altogether by selecting features that are highly salient [16]. In this work we take the complementary approach of explicitly detecting and rejecting these matches using the notion of Joint Compatibility, as proposed in [17]. The idea is that the whole set of matchings accepted in one image must be *jointly* consistent.

As the active search gives only one candidate for each matched feature, and the matchings are usually good, we have implemented a simplified version of the Joint Compatibility algorithm that reduces the computational cost of including a backtracking algorithm at each update of the EKF (see Alg. 1). The algorithm tries first the optimistic hypothesis $\mathcal{H}$ that each image feature E_i pairs correctly with its corresponding map feature F_i verifying the *joint compatibility* using the Mahalanobis distance and the Chi-squared distribution:

$$D^2_{\mathcal{H}} = \nu^T_{\mathcal{H}} C^{-1}_{\mathcal{H}} \nu_{\mathcal{H}} < \chi^2_{d,\alpha} \quad (7)$$

where $d = 2\cdot$ num_pairings($\mathcal{H}$), α is the desired confidence level (0.95 by default) and $\nu_{\mathcal{H}}$ and $C_{\mathcal{H}}$ are the joint innovation and its covariance:

Algorithm 1 Simplified Joint Compatibility:
$\mathcal{H}$ = simplified_JCBB ()

```
H ⇐ [true]^m
if not joint_compatibility(H) then
   Best ⇐ []
   JCBB([], 1)
   H ⇐ Best
end if
```

Algorithm 2 Recursive Joint Compatibility:
JCBB ($\mathcal{H}$, i) : *find pairings for observation* E_i

```
if i = m then {Leaf node}
   if num_pairings(H) > num_pairings(Best) then
      Best ⇐ H
   else if num_pairings(H) = num_pairings(Best) then
      if D^2(H) < D^2(Best) then
         Best ⇐ H
      end if
   end if
else {Not leaf node}
   if joint_compatibility([H true]) then
      JCBB([H true], i + 1) {pairing (E_i, F_j) accepted}
   end if
   if num_pairings(H) + m - i ≥ num_pairings(Best) then
      {Can do better}
      JCBB([H false], i + 1) {Star node: E_i not paired}
   end if
end if
```

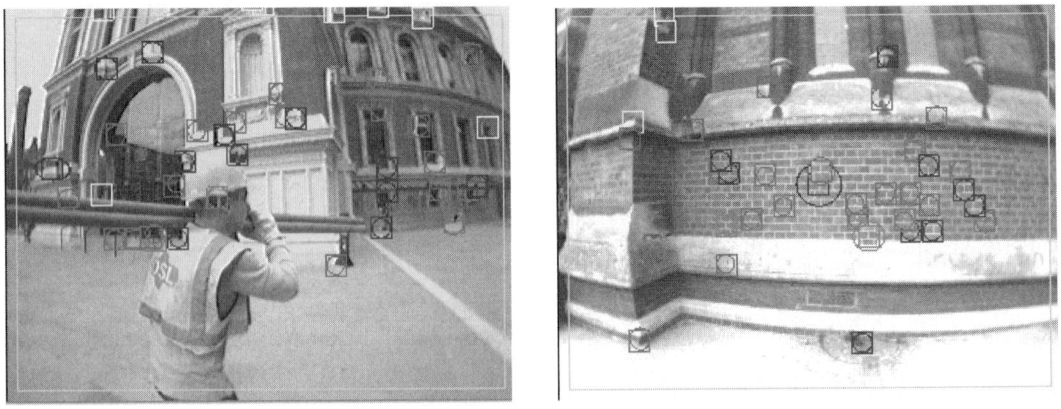

Fig. 2. Incorrect matches successfully rejected by the joint compatibility algorithm (marked in magenta).

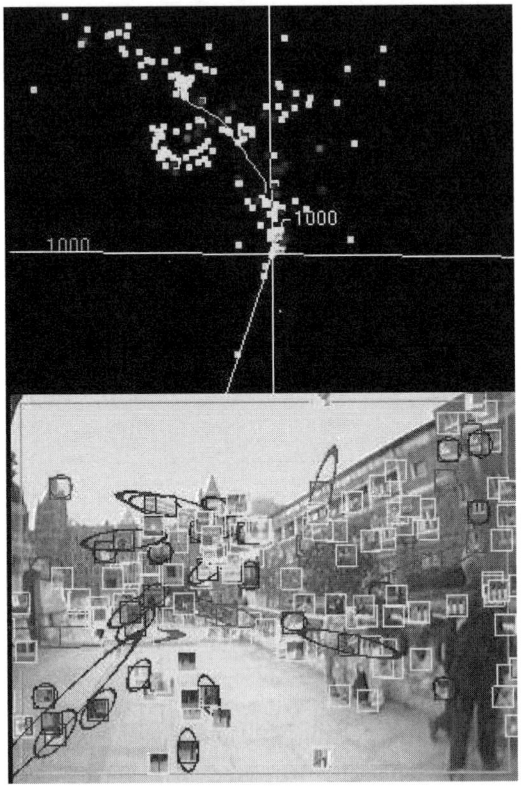

(a) Without Joint Compatibility

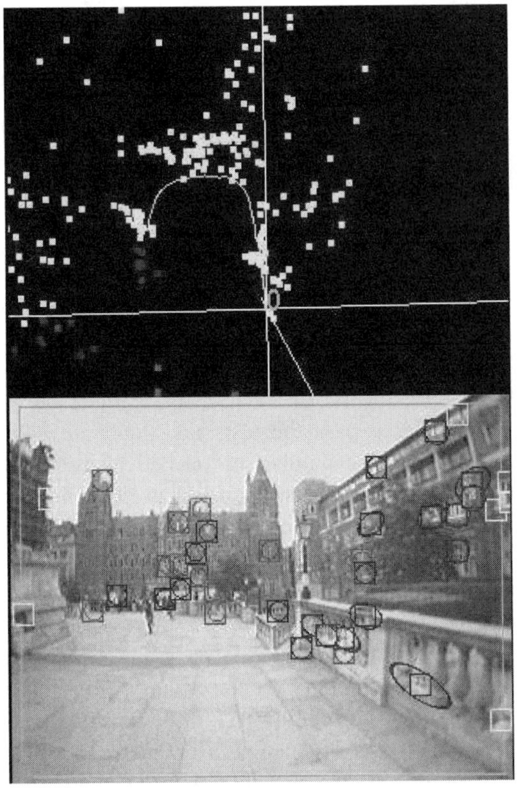

(b) With Joint Compatibility

Fig. 3. Local map obtained in a dynamic outdoor environment along a U-shaped trajectory. The colored ellipses represent the uncertainty of feature localization and the yellow line on the map corresponds to the computed camera trajectory. Red: features predicted and matched, Blue: features predicted but not found, Yellow: features not predicted.

$$\nu_{\mathcal{H}} = \mathbf{z}_{\mathcal{H}} - \mathbf{h}_{\mathcal{H}}(\hat{\mathbf{x}}) \quad (8)$$

$$C_{\mathcal{H}} = H_{\mathcal{H}} P H_{\mathcal{H}}^T + R_{\mathcal{H}} \quad (9)$$

This compatibility test only adds the computation of equation (7), because we already need the innovation and its covariance for the update phase of the Kalman filter. Only when the innovation test of the complete hypothesis is not satisfied, our simplified algorithm performs the branch and bound search of the JCBB algorithm to find the largest subset of jointly compatible matchings, as shown in Alg. 2. As the active search has only found one possible pair F_i for each measurement E_i, the solution space consists of a binary tree (accepting or rejecting each possible pair) whose depth is the number of measurements m. It should be noted that the results of JCBB are order-independent. Even when a matching has been accepted, the "star node" part of the algorithm also analyzes all the hypothesis that do not contain that matching.

An alternative technique that that could be used to detect false matchings is RANSAC (see for example [10]). Our

JCBB technique has two advantages: it allows matchings with features that have been occluded for a while and it is able to reject outliers using all points in the same way, even points with high depth uncertainty or points at infinity.

To verify the robustness of this technique we conducted a mapping experiment in an outdoor populated area. Most of the time, all matchings found by correlation were correct, and the branch and bound algorithm was not executed. Figure 2 shows two typical cases where the Joint Compatibility successfully rejected wrong matchings on dynamic and repetitive parts of the environment. Even in this cases, the computational cost added is negligible. The key question is: if the number of bad matchings is so small, how bad can they be for the SLAM process? The answer is given in figure 3. We followed a long U-shaped trajectory walking with a hand-held camera looking forward. Figure 3(a) shows the dramatic effect of the moving people and the repetitive environment patterns. The estimated trajectory is completely wrong and the monocular SLAM algorithm is trying to find in the image features that are actually well behind the camera (drawn in yellow). Running on the same dataset, the inverse-depth SLAM algorithm with the Joint Compatibility test gives the excellent map of figure 3(b). To our knowledge this is the first demonstration of a real-time SLAM algorithm walking with a camera in hand in a large outdoor environment.

III. Hierarchical SLAM

A. Building sequences of local maps

To achieve scalability to large environments, the technique described in the previous section is used to build a sequence of independent local maps of limited size that are later combined using the Hierarchical Map technique [1]. In this way, the computational cost of the EKF-SLAM iteration is constrained to real-time operation. Once the current map reaches the maximum number of features, it is freezed and a new local map is initialized, using as base reference the current camera location. To maintain the statistical independence between local maps, no information is transferred from the previous map to the new one.

When a new map is started, the set of features currently visible in the old map are inserted into the new map as *new* inverse-depth features, using their image locations as the *only* knowledge of their 3D locations. This is important since, though it may seem to be throwing away the prior knowledge of their locations from the previous map, it is only through doing so that the local maps remain independent, yielding the desired O(1) update. It is, however important that there is a group of features which are represented in adjacent maps, since only through these common features can loop-closing and trajectory refinement be effected. These common features are used to estimate the change on the scale factor that may exists between consecutive maps.

At this point, we have a series of local maps $(M_1, \ldots, M_n)$ containing the state vector defined in eq. (1) and its covariance matrix. The final camera location $\mathbf{x}_C^i$ in map i corresponds to the base reference of map $i+1$. The transformations between the successive local maps and their covariances constitute the global level of the Hierarchical Map:

$$hmap.\hat{\mathbf{x}} = \begin{pmatrix} \mathcal{T}_1^W \\ \mathcal{T}_2^1 \\ \vdots \\ \mathcal{T}_n^{n-1} \end{pmatrix} = \begin{pmatrix} \hat{\mathbf{x}}_C^1 \\ \hat{\mathbf{x}}_C^2 \\ \vdots \\ \hat{\mathbf{x}}_C^n \end{pmatrix} \tag{10}$$

$$hmap.P = \begin{pmatrix} P_1 & 0 & \ldots & 0 \\ 0 & P_2 & & \vdots \\ \vdots & & \ddots & 0 \\ 0 & \ldots & 0 & P_n \end{pmatrix} \tag{11}$$

B. Scale-invariant map matching

By composing the camera state locations $\mathcal{T}_i^{i-1}$ we are able to compute the current camera location and hypothesize loop closures. To verify the loop we have developed a map matching algorithm (see Algorithms 3 and 4) able to deal with the presence of an unknown scale factor between the overlapping maps. First, the method uses normalized correlation to find features in both maps that are compatible (unary constraints). Then, a specialized version of the Geometric Constraints Branch and Bound (GCBB) algorithm [23] is used to find the maximal subset of geometrically compatible matchings, by comparing the relative distances between feature points is space (binary constraints). Although the results of the algorithm is also order independent, its running time may benefit from a good ordering (most promising matchings first). Then, the homogeneous transformation $\mathcal{T}_j^{i-1}$ and the scale change between both maps is estimated from the subset of matched features. An example of loop detection using this technique is shown in figure 4. This technique is able to find loops when the view points in both maps are similar. To detect loops with arbitrary viewpoints, invariant features such as SIFT would be needed.

C. Loop optimization

We have local independent maps scaled to the same reference and we also know the relation between two overlapping maps that close a loop. Then, the Iterated Extended Kalman Filter [20] is used for re-estimating the transformations between the maps that from the loop, as proposed in [1]. The measurement $\mathbf{z}$ corresponds to the transformation estimated by the map matching algorithm, and the measurement function is given by the compositions of all local map states that form the hierarchical map:

$$\mathbf{z} = \mathbf{h}(\mathbf{x}) = \mathcal{T}_i^{i-1} \oplus \mathcal{T}_{i+1}^{i} \oplus \ldots \oplus \mathcal{T}_j^{j-1} \tag{12}$$

IV. Experimental results

To validate the proposed SLAM method we have conducted an experiment in a large and dynamic outdoor environment. The experimental setup consists of a low cost Unibrain IEEE1394 camera with a 90 degree field of view, acquiring monochrome image sequences of 320x240 resolution at 30

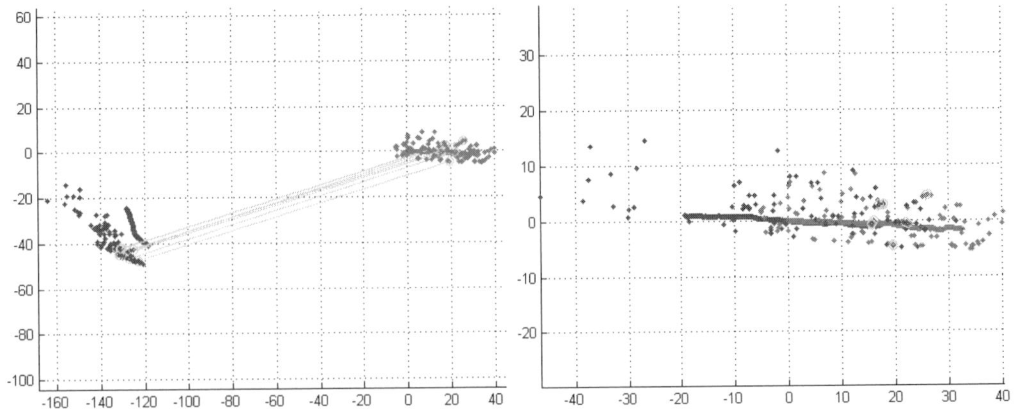

Fig. 4. Loop closure detection. Matchings found between two maps (left) and aligned maps (right).

Algorithm 3 Map Matching GCBB with variable scale:
$\mathcal{H}$ = map_matching_GCBB (observations, features)

```
unary ⇐ compute_unary_constraints(features, observations)
binary.O.distances ⇐ estimate_distances(observations)
binary.F.distances ⇐ estimate_distances(features)
Best.H ⇐ []
Best.scale ⇐ -1.0
variable_scale_GCBB([], 0)
H ⇐ Best
```

fps, a firewire cable and a laptop (see Fig. 1). We acquired a real image sequence of 6300 frames walking in a courtyard along a loop trajectory of around 250 meters, with the camera in hand, looking to one side. The sequence was processed with the proposed algorithms on a desktop computer with an Intel Core 2 processor at 2,4GHz. Figure 5(a) shows the sequence of independent local maps obtained with the inverse-depth EKF-SLAM using joint compatibility. As it can be seen in the figure, the algorithm "guesses" an approximate scale that is different for each local map. When a map is finished, it is matched with the previous map and the relative change on the scale is corrected, as shown in figure 5(b). When a loop is hypothesized, the map matching algorithm is executed to find the loop closure, and the loop constraint is applied at the upper level of the Hierarchical Map, giving the result of figure 5(c).

The local map building process has been tested to run in real-time (at 30Hz) with maps up to 60 point features. During the experiments, the joint compatibility algorithm consumed $200\mu s$ at every step and, when occasionally the complete search is executed, the computation cost increases only up to $2ms$, which is an acceptable cost for the great increase in robustness and precision obtained.

The map matching, scale adjustment and loop optimization phases have been implemented in Matlab. The scale factor estimation between two adjacent maps takes about $120ms$ and the loop optimization using the IEKF takes $800ms$ when performing 6 iterations. The most expensive part is the

Algorithm 4 Recursive Modified Geometric Constraints:
variable_scale_GCBB ($\mathcal{H}$, i) : *find pairings for observation E_i*

```
if i > m then {Leaf node}
  if num_pairings(H) > num_pairings( Best.H) then
    Best.H ⇐ H
    Best.scale ⇐ binary.scale
  end if
else {Not leaf node}
  if num_pairings(H) == 0 then {This is the first pair}
    for ∀j | (unary(i,j) == true) do
      variable_scale_GCBB([H j], i+1)
    end for
  else if num_pairings(H) == 1 then {This is the 2nd pair}
    k ⇐ { K | H(K) ≠ 0 }
    distance_obs ⇐ binary.O.distances(i,k)
    for ∀j | (unary(i,j) == true) do
      distance_feat ⇐ binary.F.distances(j,H(k))
      if distance_feat ≠ 0 then
        binary.scale ⇐ distance_obs ÷ distance_feat
        binary.satisfies ⇐ binary_constraints(binary.scale)
        variable_scale_GCBB([H j], i+1)
      end if
    end for
  else {Normal recursion with binary constraints calculated}
    for ∀j | ( (unary(i,j) == true) AND ({∀k | H(k) ≠ 0 }
            AND binary.satisfies(i, k, j, H(k))}) ) do
      variable_scale_GCBB([H j], i+1)
    end for
  end if
end if
{Checking if can do better}
if num_pairings(H) + m - i > num_pairings(Best.H) then
  variable_scale_GCBB([H 0], i+1) {Star node: E_i no paired}
end if
```

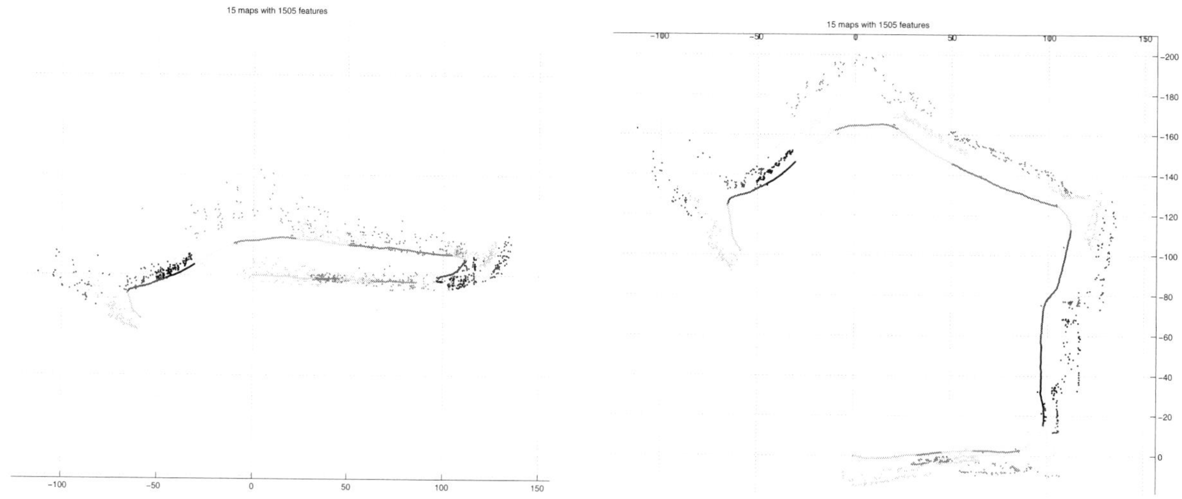

(a) Local maps obtained with pure monocular SLAM

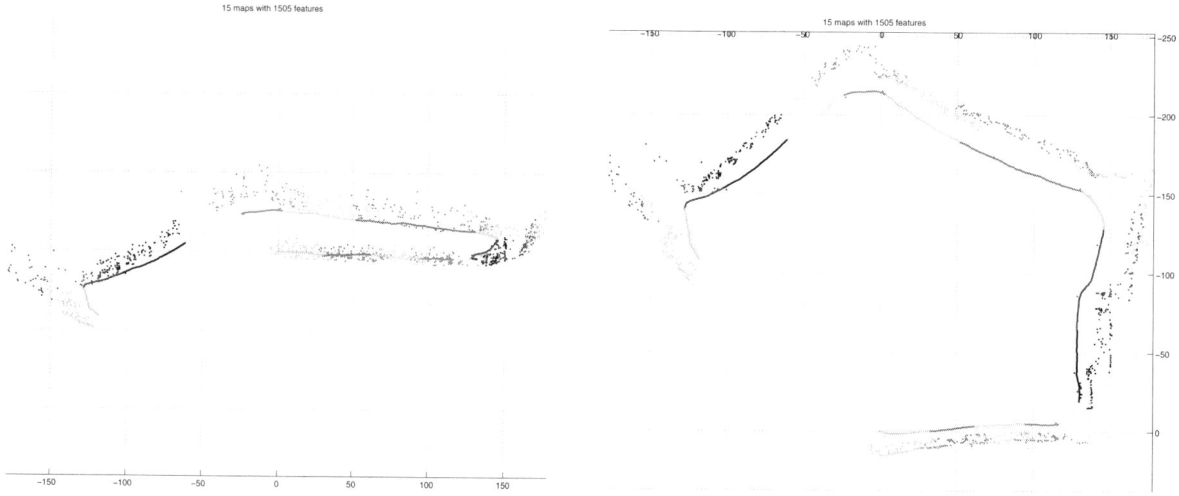

(b) Hierarchical map auto-scaled, before loop detection

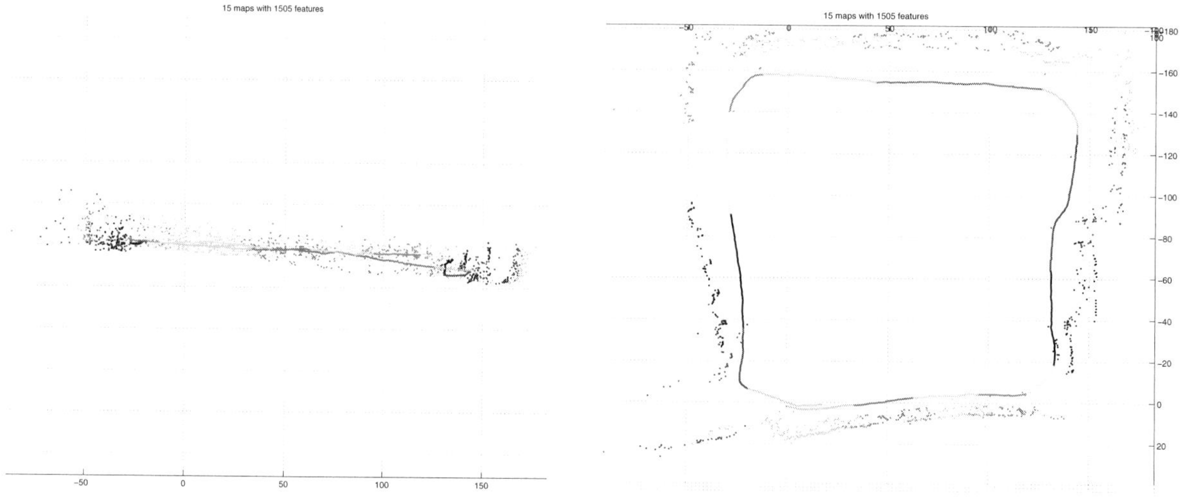

(c) Hierarchical map after loop closing

Fig. 5. Results obtained mapping a loop of several hundred meters with a camera in hand: side view (left) and top view (right).

scale-invariant map matching algorithm that takes around one minute in our current Matlab implementation. We expect that an optimized C++ implementation running on background will provide close to real time loop detection.

V. Conclusion

In this paper we have demonstrated for the first time that monocular vision-only SLAM – with a single hand-held camera providing the *only* data input – can achieve large-scale outdoor closed-loop mapping in near real-time. Achieving these results which such basic hardware opens up new application areas for vision-based SLAM, both in flexible (possibly low-cost) robotic systems and related areas such as wearable computing. The success of our system lies in the careful combination of the following elements:

- An inverse depth representation of 3D points. It allows the use of partial information, inherent to monocular vision, in a simple and stable way. All features, even those far from the camera, immediately contribute valuable information.
- A branch and bound joint compatibility algorithm that allows the rejection of measurements coming from moving objects that otherwise plague and corrupt the map. Although texture gives a powerful signature for matching points in images, the spatial consistency that this algorithm enforces is essential here.
- A Hierarchical SLAM paradigm in which sequences of local maps of limited size are managed, allowing the system to work with bounded complexity on local maps during normal operation. By running the map matching algorithm in the background, the system can attain real time execution.
- A new map matching algorithm to detect large loops which takes into account the unobservability of the scale intrinsic to pure monocular SLAM. This algorithm allows us to detect loop closures even when the maps involved have been computed with different scales.

Future work includes improving the map matching algorithm to reach real time performance, possibly using invariant feature descriptors. A current limitation of Hierarchical SLAM is the fact that it does not make use of matchings between neighboring maps. We plan to investigate new large mapping techniques that can overcome this limitation, obtaining maps closer to the optimal solution.

Acknowledgments

This research has been funded in part by the Royal Society International Joint Project 2005/R2, DGA(CONSI+D)-CAI under Programa Europa, Dirección General de Investigación of Spain grant DPI2006-13578, the Engineering and Physical Sciences Research Council (EPSRC) grant GR/T24685/01 and an EPSRC Advanced Research Fellowship to AJD.

Thanks to José M. M. Montiel and Javier Civera for providing their code for inverse-depth visual SLAM and the fruitful discussions maintained and to Paul Smith for his help with the initial experiments.

References

[1] C. Estrada, J. Neira, and J. D. Tardós, "Hierarchical SLAM: real-time accurate mapping of large environments," *IEEE Transactions on Robotics*, vol. 21, no. 4, pp. 588–596, August 2005.

[2] J. M. M. Montiel, J. Civera, and A. J. Davison, "Unified inverse depth parametrization for monocular SLAM," in *Proceedings of Robotics: Science and Systems*, Philadelphia, USA, August 2006.

[3] H. Durrant-Whyte and T. Bailey, "Simultaneous localization and mapping: part I," *IEEE Robotics & Automation Magazine*, vol. 13, no. 2, pp. 99–110, 2006.

[4] T. Bailey and H. Durrant-Whyte, "Simultaneous localization and mapping (SLAM): part II," *IEEE Robotics & Automation Magazine*, vol. 13, no. 3, pp. 108–117, 2006.

[5] N. Karlsson, E. D. Bernardo, J. Ostrowski, L. Goncalves, P. Pirjanian, and M. E. Munich, "The vSLAM algorithm for robust localization and mapping," in *IEEE Int. Conf. on Robotics and Automation*, 2005, pp. 24–29.

[6] J. Folkesson, P. Jensfelt, and H. Christensen, "Vision SLAM in the Measurement Subspace," *IEEE Int. Conf. Robotics and Automation*, pp. 30–35, 2005.

[7] R. Sim and J. J. Little, "Autonomous vision-based exploration and mapping using hybrid maps and rao-blackwellised particle filters," in *IEEE/RSJ Conf. on Intelligent Robots and Systems*, 2006.

[8] T. Lemaire and S. Lacroix, "SLAM with panoramic vision," *Journal of Field Robotics*, vol. 24, no. 1-2, pp. 91–111, 2007.

[9] R. M. Eustice, H. Singh, J. J. Leonard, M. Walter, and R. Ballard, "Visually navigating the RMS titanic with SLAM information filters," in *Proceedings of Robotics: Science and Systems*, 2005.

[10] D. Nistér, O. Naroditsky, and J. Bergen, "Visual odometry," in *IEEE Conference on Computer Vision and Pattern Recognition*, 2004.

[11] K. Konolige, M. Agrawal, R. C. Bolles, C. Cowan, M. Fischler, and B. P. Gerkey, "Outdoor mapping and navigation using stereo vision," in *Proc. of the Intl. Symp. on Experimental Robotics (ISER)*, July 2006.

[12] A. J. Davison, "Real-time simultaneous localisation and mapping with a single camera," in *9th International Conference on Computer Vision, Nice*, 2003.

[13] E. Eade and T. Drummond, "Scalable monocular SLAM," in *IEEE Conf. on Computer Vision and Pattern Recognition, New York*, 2006.

[14] D. Chekhlov, M. Pupilli, W. W. Mayol, and A. Calway, "Real-time and robust monocular SLAM using predictive multi-resolution descriptors," in *2nd International Symposium on Visual Computing*, 2006.

[15] "2d3 web based literature," URL http://www.2d3.com/, 2005.

[16] P. Newman and K. Ho, "SLAM-Loop Closing with Visually Salient Features," *IEEE Int. Conf. on Robotics and Automation*, pp. 635–642, 2005.

[17] J. Neira and J. D. Tardós, "Data association in stochastic mapping using the joint compatibility test," *IEEE Transactions on Robotics and Automation*, vol. 17, no. 6, pp. 890–897, 2001.

[18] W. Grimson, "Recognition of object families using parameterized models," *Proceedings of the First International Conference on Computer Vision*, pp. 93–100, 1987.

[19] R. Smith, M. Self, and P. Cheeseman, "A stochastic map for uncertain spatial relationships," in *Robotics Research, The Fourth Int. Symposium*, O. Faugeras and G. Giralt, Eds. The MIT Press, 1988, pp. 467–474.

[20] Y. Bar-Shalom, X. Li, and T. Kirubarajan, *Estimation with Applications to Tracking and Navigation*. New York: John Willey and Sons, 2001.

[21] A. J. Davison and D. W. Murray, "Simultaneous localization and map-building using active vision," *IEEE Transactions on Pattern Analysis and Machine Intelligence*, vol. 24, no. 7, pp. 865–880, 2002.

[22] J. Shi and C. Tomasi, "Good features to track," in *IEEE Conf. on Computer Vision and Pattern Recognition*, Seattle, Jun 1994, pp. 593–600.

[23] J. Neira, J. D. Tardós, and J. A. Castellanos, "Linear time vehicle relocation in SLAM," in *IEEE Int. Conf. on Robotics and Automation*, Taipei, Taiwan, September 2003, pp. 427–433.

Dynamic Coverage Verification in Mobile Sensor Networks via Switched Higher Order Laplacians

Abubakr Muhammad and Ali Jadbabaie
Department of Electrical and Systems Engineering
University of Pennsylvania, Philadelphia, PA 19104, USA
{abubakr, jadbabai}@seas.upenn.edu

***Abstract*— In this paper, we study the problem of verifying dynamic coverage in mobile sensor networks using certain switched linear systems. These switched systems describe the flow of discrete differential forms on time-evolving simplicial complexes. The simplicial complexes model the connectivity of agents in the network, and the homology groups of the simplicial complexes provides information about the coverage properties of the network. Our main result states that the asymptotic stability the switched linear system implies that every point of the domain covered by the mobile sensor nodes is visited infinitely often, hence verifying dynamic coverage. The enabling mathematical technique for this result is the theory of higher order Laplacian operators, which is a generalization of the graph Laplacian used in spectral graph theory and continuous-time consensus problems.**

I. Introduction and Motivation

Recent years have witnessed a surge of research interest in science and engineering of networked dynamic systems. Due to the advances in computing, communication, sensing and actuation technologies, networks composed of hundreds or even thousands of inexpensive mobile sensing platforms have become closer to reality. The field of *sensor networks* is undergoing a revolutionary transformation from a subject of academic curiosity to a mature enabling technology in many industrial and engineering solutions. Although, producing low cost and tiny sensor nodes is still an active area of research, the miniaturization and lowering of cost are understood to follow from recent and future progress in the fields of MEMS and NEMS. Thus, the main challenge in this area has now shifted from the manufacturing of cheap hardware and deployment tools to the development of analytical tools for predicting and controlling the complexities arising in such large-scale networks.

An interdisciplinary research area of *network science* is emerging which combines disciplines such as communications and networking, graph theory, statistical physics, probability and statistics, social sciences, with robotics, optimization, and control [1]. A major theme in the broad research topic of network science is understanding of the emergence of *global behavior* from *local interactions* via adaptation and cooperation. Examples of such phenomena span domains from biological sciences to systems and controls, and from statistical physics to computer graphics, where researchers have been trying to develop an understanding of how a group of moving objects (natural or man-made) can collectively reach a global behavior such as flocking, synchronization or consensus with local interaction rules [2].

A useful abstraction for modeling and analysis of such complex systems has been developed using graph theory. Typically interaction among agents is modeled with graphs in which nodes represent agents and edges represent some form of proximity or other binary relationships [3]. The term graph *topology* is frequently used to denote the interconnection structure among agents. The use of the term topology (which is typically associated with a branch of mathematics involved with studying sets, the nature of space, its fine structures, and its global properties) is by no means an accident. In fact both disciplines of graph theory and topology where born at the same time, by the famous 1736 paper of Leonard Euler on Seven Bridges of Knigsberg [4], even though algebraic or combinatorial topology became a formal discipline as recent as 1930s [5]. In fact, the very central question of inferring global properties from local information (which is the main goal of research in network science and networked dynamic systems) is the realm of algebraic topology which deals with topics like homology, homotopy, and in general topology of discrete and combinatorial sets.

The philosophy underlying this paper is that meeting the challenges of network science requires a *minimalist* thinking, i.e. the development of abstractions that retain the essential features of the system using minimum information to manage the explosion of redundancies with scale. We believe that many of these challenges are *geometrical* and a minimalist approach to solving geometrical problems in networks is essentially *topological*. In this paper, we use this philosophy to obtain low-cost, provably correct algorithms for the verification of distributed coverage problems in mobile sensing networks.

II. Problem Formulation

The primary task of a sensor is to detect, discern or locate certain physical changes within its proximity. In many situations, the environment where such physical changes are expected to take place has such large physical dimensions that it is impossible to monitor it with a single sensor. To accomplish the task of monitoring such an environments, a network of sensors can be deployed in which the individual

† This work is supported by DARPA DSO # HR0011-07-1-0002 via the project SToMP: Sensor Topology & Minimal Planning, and by ARO MURI grant # W911NF-05-1-0381.

sensor nodes can process and communicate individual sensory data collected in their proximities.

Thus, the first requirement for the deployment of a sensor network is to ensure that the sensors reliably *cover* the environment in such a way that there are no gaps left in the coverage. Coverage problems for sensor networks have been extensively studied in the literature [6]. When feasible, it is preferred to provide a continual coverage to the environment by deploying a network of static sensing nodes. This is known as the *blanket coverage* problem. In many cases, it is not feasible to deploy a large number of static sensors for continual coverage. If ubiquitous coverage is not necessary, then one solution is to deploy a *mobile sensor network* made up of robotic sensing platforms. The mobile agents patrol the environment in such a manner that each point of the environment is visited infinitely often, possibly with a certain frequency. As a result, no point remains undetected for 'too long', as depicted in Figure 1. The problem of verifying such coverage is called a *dynamic coverage* problem. The dynamic coverage problems have been categorized by Gage in his survey [6] as the *sweep* and *barrier* coverage problems. In barrier coverage, the objective is to achieve a static arrangement of elements that minimizes the probability of undetected penetration through the barrier.

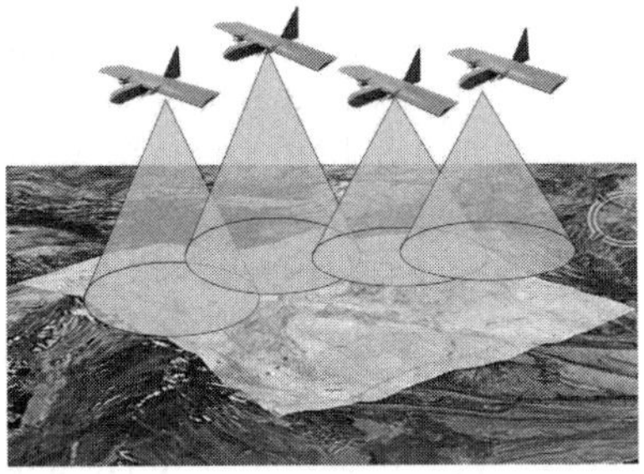

Fig. 1. A dynamic coverage scenario : An area of interest (white) under surveillance by a group of UAV's.

Coverage verification, whether static or dynamic is inherently a geometrical problem and needs some geometrical information about the nodes for computing a solution. Therefore, it is no surprise that many network coverage problems have been studied in the computational geometry literature in various other contexts. For example, the *Art Gallery Problem* is to determine the number of observers necessary to cover an art gallery (or an area of interest) such that every point in the art gallery is monitored by at least one observer. Inspired by these computational geometry results, many researchers in sensor networks have studied the blanket coverage for static nodes in various settings [7, 8, 9, 10]. In [7], the authors have studied k-*coverage*: whether every point in the service area of the sensor network is covered by at least k number of sensors. In [11, 12], the authors consider network coverage using wireless sensors that conserve power by alternating between active and sleep states with an average sleep period (much) longer than the active period. Other researchers have also examined the area coverage of random sensor networks in bounded domains [13].

The problem of dynamically covering an area of interest has been another subject of interest for various researchers in control theory, networks and robotics. One aspect of bringing mobility into the picture is to design trajectories for network reconfiguration in changing environments or in situations where some areas of the environment are of more interest than others [14, 15, 16, 17]. The coverage provided by the network is still *blanket*, and the mobility of the nodes only cater to a changing surveillance domain. The dynamic coverage problems, as described above, are the *sweep* and *barrier* coverage problems. A variant of sweep coverage has been studied under the title of *effective* coverage for mobile agents in [18]. In [19], sweep coverage schemes have been proposed and studied using a frequency coverage metric that measures the frequency at which every-point in the environment is visited. Similarly, barrier coverage has been studied [20], although for static sensors only, but the results can be extended to studying a dynamic barrier [6].

Such geometrical approaches often suffer from the drawback that they are too expensive to compute in real-time. Moreover, they typically require exact knowledge of the locations of the sensing nodes. Although, this information can be made available in real-time by a localization algorithm or by fitting the sensors with localization sensors (such as GPS), it can only be used most effectively in an off-line pre-deployment analysis for large networks or when there are strong assumptions about the geometrical structure of the network and the environment. Due to the challenges in designing effective distributed algorithms, there may be a massive accumulation of information at a small subset of nodes. Furthermore, most computational geometry algorithms suffer from a lack of robustness due to the required precision in exact geometrical information. In addition, if the network topology changes due to node mobility or node failure, a continuous monitoring of the network coverage becomes prohibitive if the algorithm is too expensive to run or is sensitive to location uncertainty. Finally, localization equipment adds to the cost of the network.

To address these issues, we use a fresh approach for distributed coverage verification using *minimal geometry*. Our goal here is to show that surprisingly simple, and *minimal* topological information, such as the ID's of nearest neighbors can be used to this end. The geometrical information that is needed for solving such problems is remarkably minimal: no means of measuring distance, orientation, or location of the nodes in an environment are required. As described below, the basic idea is to infer geometric proximity of other nodes by the mere existence of radio communication (without any aid of specialized localization equipment or refinement in proximity). From this connectivity data alone, coverage for the entire network is verifiable using the tools of algebraic topology.

Our approach is quite interdisciplinary in nature and combines recent advances in multiagent systems and control on agreement and consensus problems [2, 21], with recent advances in coverage maintenance in sensor networks using computational algebraic topology methods [22, 23, 24, 25]. to develop distributed algorithms that are computationally inexpensive and robust to failures.

III. Beyond Graphs: Simplicial Models of Networks

A. From graphs to simplicial complexes

Graphs can be generalized to more expressive combinatorial objects known as a simplicial complexes. For a thorough treatment of simplicial complexes and their topological invariants, see for example [26]. While graphs model binary relations, simplicial complexes can be used to model higher order relationships. Given a set of points V, a *k-simplex* is an unordered set $\{v_0, v_1, \cdots, v_k\} \subset V$ such that $v_i \neq v_j$ for all $i \neq j$. The *faces* of the k-simplex $\{v_0, v_1, \cdots, v_k\}$ are defined as all the $(k-1)$-simplices of the form $\{v_0, \cdots, v_{j-1}, v_{j+1}, \cdots, v_k\}$ with $0 \leq j \leq k$. A *simplicial complex* is a collection of simplices which is closed with respect to the inclusion of faces. One can define an *orientation* for a simplicial complex by defining an ordering on all of its k-simplices. We denote the k-simplex $\{v_0, \cdots, v_k\}$ with an ordering by $[v_0, \cdots, v_k]$, where a change in the orientation corresponds to a change in the sign of the coefficient as $[v_0, \cdots, v_i, \cdots, v_j \cdots, v_k] = -[v_0, \cdots, v_j, \cdots, v_i \cdots, v_k]$.

With these definitions, it is clear that a graph is a simplicial complex which its vertices and edges are in fact the 0- and 1-simplices respectively, and each vertex is a face of all its incident edges. An oriented simplicial complex is merely a generalization of a directed graph[1]

One can also generalize the concepts of adjacency and degree to simplicial complexes. Two k-simplices σ_i and σ_j are *upper adjacent* (denoted by $\sigma_i \frown \sigma_j$) if both are faces of a $(k+1)$-simplex in X. The two k-simplices are said to be *lower adjacent* (denoted by $\sigma_i \smile \sigma_j$) if both have a common face (see Figure 2 below).

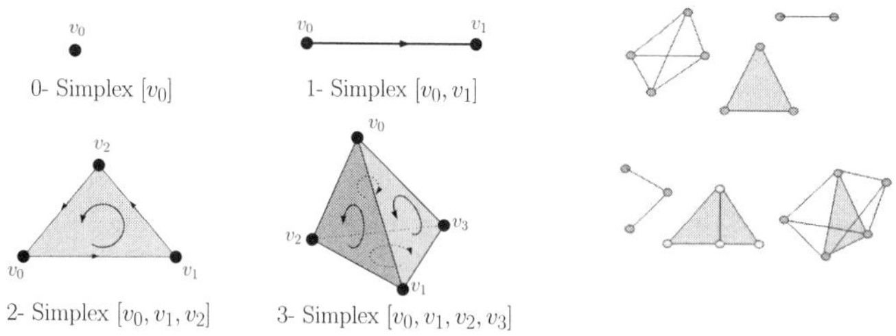

Fig. 2. (Left:) Examples of 1,2, and 3 simplices. (Top Right): Two nodes that are connected by an edge; two edges that have a common face are upper adjacent. (Bottom Right) Two faces that share an edge and two tetrahedra that share a face are lower adjacent.

Having defined the adjacency, one can define the upper and lower adjacency matrices, A_u and A_l respectively, in a similar fashion to a graph's adjacency matrix. The upper and lower degree matrices D_u and D_l are also defined similarly.

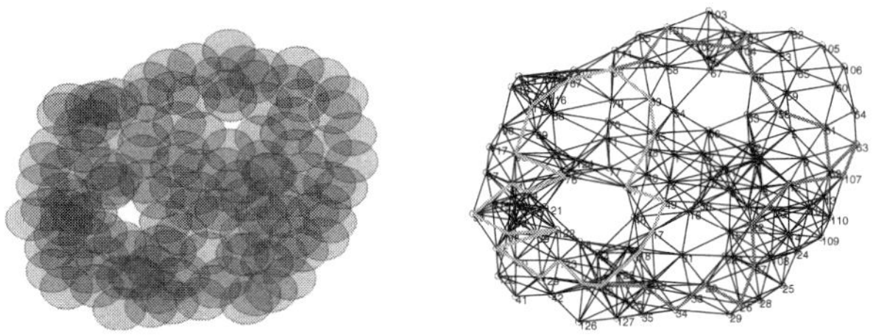

Fig. 3. (Left) Sensor coverage with disk footprints.(Right) The corresponding Čech complex. Coverage holes are represented by holes in the Čech complex.

If X is a finite simplicial complex, for each $k \geq 0$, define $C_k(X)$ to be the vector space whose basis is the set of oriented k-simplices of X. We let $C_k(X) = 0$, if k is larger than the dimension of X. Each element of these vector spaces is a linear combination of the basis elements. Over these vector spaces the *k-th boundary map* is defined to be the linear transformation $\partial_k : C_k(X) \to C_{k-1}(X)$, which acts on the basis elements of its domain via

$$\partial_k[v_0, \cdots, v_k] = \sum_{j=0}^{k}(-1)^j[v_0, \cdots, v_{j-1}, v_{j+1}, \cdots, v_k]. \quad (1)$$

Intuitively, the boundary map ∂_k operated on a k-simplex, returns a linear combination of its $k+1$ faces. In fact the zeroth boundary map ∂_1 is nothing but the edge-vertex incidence matrix of a graph which maps edges (1 simplices) to nodes (0 simplices), mapping a simplex to its boundary (hence the name boundary map).

Using (1), it is easy to show that the composition $\partial_k \circ \partial_{k-1}$ is uniformly zero for all k and as a result, we have *Im* $\partial_k \subset$ *Ker* ∂_{k-1}. Then, the *k-th homology group* of X is defined as

$$H_k(X) = \textit{Ker}\ \partial_{k-1} / \textit{Im}\ \partial_k.$$

Since the boundary operators ∂_i map a simplex to a linear combination of its faces (i.e. its boundary), the homology groups can be used to distinguish topological spaces from one another. More precisely, the dimension of the k-th homology group (known as its *Betti number*) identifies the number of k-dimensional "holes" in the given topological space. For example, the dimension of $H_0(X)$ is the number of connected components of the 1-skeleton (or collection of 0 and 1 simplices, i.e., nodes and edges) of X. On the other hand, the dimension of $H_1(X)$ is the number of its non-contractible[2] cycles or *holes*.

Homology groups are used to distinguish topological spaces from one another by identifying the number of 'holes' of various dimension, contained in these spaces. Each non-trivial homology class in a certain dimension helps identify a corresponding hole in that dimension. Crudely speaking, the dimension of $H_0(X)$ is the number of connected components (0-dimensional holes) of X. the dimension of $H_1(X)$ is the

[1]Note that this should not be confused with the notion of a hyper graph in which any subset of the power set of vertices could correspond to a hyper edge.

[2]A space is contractible if it is homotopy equivalent to a single point.

number of non-contractable cycles in X. For a surface, it identifies the 'punctures' in that surface. For a graph it is the number of loops or circuits. $H_2(X)$ identifies the number of 3-dimensional voids in a space and so on.

B. Čech and Rips complexes and coverage problems

Since we aim to apply the topological properties of the simplicial complexes to the coverage problem and therefore dealing with union of disks, it would be useful to define what is known as the Čech or Nerve complex. Given a collection of sets $\mathcal{U} = \{\mathcal{U}_i\}$, the Čech complex of $\mathcal{U}$ denoted by $\mathcal{C}(\mathcal{U})$ is the abstract simplicial complex whose k-simplices correspond to non-empty intersections of $k+1$ distinct elements of $\mathcal{U}$. This complex is simply formed by associating convex sets (e.g. disks) to each node, and then representing edges or higher order simplices from the overlap of such sets. For example, given a network of sensors with disk footprints, we can construct the Čech complex(also known as the nerve complex) by book-keeping over lap of sensor footprints. In the complex, nodes or zero simplices represent the sensors. When two footprints overlap, we draw an edge or a 1-simplex between two nodes (see Figure 3). However, we also keep track of further set overlaps by representing three disk overlaps with faces or filled triangles and 4 set overlaps with tetrahedra and so forth. A closely related complex, known as the Rips-Vietoris complex (hereafter called the Rips complex), is the natural analog of proximity graphs (also known as connectivity graphs, r-disk graphs or geometric graphs) which have become ubiquitous in networking, control theory and network science research. In such graphs, nodes represent the agents or sensors and edges represent distance proximity. There is an edge between two nodes if they are within a certain distance of each other. A Rips complex is just a generalization of this concept to higher order relations (see Figure 4). If 2 nodes are within the distance of each other, there is an edge between them. When 3 nodes are within a distance of each other then the three nodes will form a face. If four nodes are within a certain distance, they form a tetrahedron. While the Rips complex is the suitable abstraction for modeling nearest neighbor communications, the Čech complex is suitable for coverage problems.

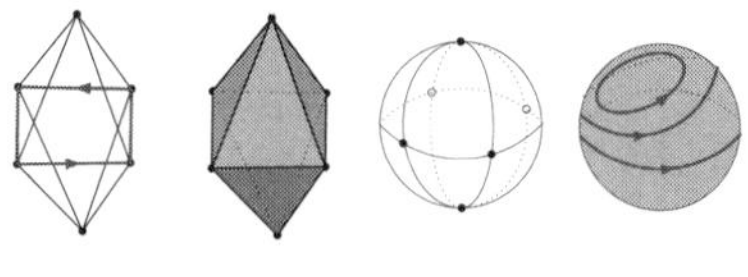

Fig. 4. A connectivity graph [left], its Rips complex [center left] and a depiction of its topology as a sphere [right].

A well-known result in algebraic topology known as the Čech theorem [27] implies that if collection of sets $\mathcal{U} = \{\mathcal{U}_i\}$ and their non-empty finite intersections are contractible (i.e., the sets themselves have no holes and their intersections have no holes) then the Čech complex $\mathcal{C}$ has the same homotopy type as the union of $\mathcal{U}_i$s. The above result indicates that in order to verify coverage in a given domain by a set of sensors, one only needs to look at the homology groups (which represents holes) of the underlying Čech complex. If the Čech complex has no holes, neither does the sensor cover. However, computation of the Čech complex is not an easy task, as it requires localization of each sensor as well as measurement of distances in order to verify that sensor footprints overlap. Furthermore, the Čech complex is very fragile with respect to uncertainties in distance and location measurements [25]. On the other hand, the Rips complex can be easily formed by merely communication with nearest neighbors, solely based on local connectivity information. However, the Rips complex is unfortunately not rich enough to contain all the topological and geometric information of the Čech complex and in general does not tell us information about coverage holes. Recently, the authors in [22] have shown that in certain cases, the Rips complex *does* carry the necessary information. Namely, it is shown that any Čech complex $\mathcal{C}_{r_c}$ made from sensor footprints of disks of radius r_c, can be "sandwiched" between two Rips complexes formed from nearest neighbor communication with broadcast disk radius $r_{b1} = \sqrt{3}r_c$ from left and $r_{b2} = 2r_c$ from right. In other words,

$$\mathcal{R}_{\sqrt{3}r_c} \subseteq \mathcal{C}_{r_c} \subseteq \mathcal{R}_{2r_c},$$

where $\mathcal{R}_\epsilon$ is a Rips complex formed from broadcast disks of radius ϵ, and $\subseteq$ represents containment of one simplicial complex in another. Therefore, if the Rips complex with broadcast disks of radius r_{b1} is hole free, then so is the sensor coverage, and if the Rips complex with broadcast radius of r_{b2} has holes, so does the coverage. In other words, this result gives us a necessary and a a sufficient homological criterion for coverage verification

In principle, one can detect holes in a simplicial complex by looking at the homology groups associated with that complex. However such linear algebra computation are generally centralized [22]. In what follows, we resent our preliminary results on detection of coverage holes in a decentralized fashion. Before doing so, however, we need to introduce the machinery of combinatorial Laplacians for simplicial complexes (which are again generalizations of the same notion for graphs).

C. Combinatorial Laplacians

The graph Laplacian [28] has various applications in image segmentation, graph embedding, dimensionality reduction for large data sets, machine learning , and more recently in consensus and agreement problems in distributed control of multi agent systems [2, 21]. We now define the Laplacian matrix of a graph and its generalizations to simplicial complexes.

If the vertex-by-edge-dimensional matrix B is the incidence matrix of a graph $\mathcal{G}$, then its Laplacian matrix is defined as $L = BB^T$. As it is evident from the definition, L is a positive semi-definite matrix. Also it is well-known that the Laplacian matrix can be written in terms of the adjacency and degree matrixes of $\mathcal{G}$ as well:

$$L = D - A,$$

which implies that the i-th row of the Laplacian matrix only depends on the local interactions between vertex i and its neighbors.

A similar operator can be defined for simplicial complexes using the boundary maps [29]. The operator $\mathcal{L}_k : C_k(X) \rightarrow C_k(X)$ defined as

$$\mathcal{L}_k = \partial_{k+1}\partial_{k+1}^* + \partial_k^*\partial_k \qquad (2)$$

is called the *k-th combinatorial Laplacian* of the simplicial complex X, where the operators have simple matrix representations and ∂_k^* is the adjoint operator of ∂_k. Note that the expression for $\mathcal{L}_0$ reduces to the definition of the graph Laplacian matrix. Also similar to the case of graph Laplacian, the combinatorial Laplacians can be represented in terms of the adjacency and degree matrices as follows [30]:

$$\mathcal{L}_k = D_u - A_u + (k+1)I + A_l \qquad k > 0, \qquad (3)$$

where I is the identity matrix of the proper size, A_l is the lower adjacency matrix between k-simplices and A_u is the corresponding upper adjacency matrix. This equation indicates that $\mathcal{L}_k$ is a positive semi-definite matrix, whose kernel represents the cohomology groups[3] [30]. This is merely extension of the fact that in the case of graphs, the kernel of the Laplacian matrix represents the connected components (0-dimensional holes). Moreover, (3) implies that as in the case of the graph Laplacian, the i-th row of $\mathcal{L}_k$ only depends on the local interactions between the i-th k-simplex and its upper and lower adjacent simplices. As a result, to verify a hole-free coverage in the plane, one needs to verify that the kernel of the k-Laplacian of the Rips complex with radius r_{b1} above is zero. This property makes the combinatorial Laplacian a suitable tool for distributed coverage verification. In the next section we study dynamic coverage verification based on these and other observations.

IV. Mobile Networks and Dynamic Coverage

A. Graph Laplacian, Consensus and Homology

The dimension of the null space of this graph Laplacian is well known to be equal to the the number of connected components of the graph [28]. Therefore, it is no coincidence that $\ker \mathcal{L}_0 \cong H_0(X)$ also counts the number of connected components of X. Moreover, when the graph is connected, the vector $\mathbf{1} = [1\ 1 \ldots 1]^T$ is always an eigenvector of $\mathcal{L}_0$ corresponding to the eigenvalue 0. Again, this can be explained in terms of the dual space to homology, i.e. the cohomology. Note that the zero dimensional cohomology is an equivalence class of functions on the vertices [27]. The vector $\mathbf{1}$ corresponds to the zero cohomology class of constant functions on the vertices. Any non-zero vector in the span of $\mathbf{1}$ represents a constant function in this cohomology class. If there are more than one connected components, there is a cohomology class of locally constant functions for each component [27]. The locally constant functions are precisely the *harmonic functions* predicted by Hodge theory.

Let us turn our attention to consensus algorithms for multi-agent systems that are based on the graph Laplacian. If $x_i : v_i \rightarrow \mathbb{R}$ is a scalar function on vertex v_i, then the state vector $\mathbf{x}(t) = [x_1(t)\ x_2(t) \ldots x_n(t)]^T$ can be used to model the evolution of the system,

$$\dot{\mathbf{x}}(t) = -\mathcal{L}_0\mathbf{x}(t), \qquad (4)$$

in which $\mathcal{L}_0$ captures the connectivity between the nodes in the graph. If the underlying graph is connected, then $\lim_{t\rightarrow\infty} \mathbf{x}(t) = c\mathbf{1}$, where c is a constant that depends on $\mathbf{x}(0)$. Thus connectedness is a sufficient condition for consensus.

In [2], the authors have proven consensus under a weaker condition of *joint connectedness*. A collection of finite graphs is said to be jointly connected if their union is a connected graph. The basic idea is to model the dynamics under changes in the graph topology as a switched linear system by

$$\dot{\mathbf{x}}(t) = -\mathcal{L}_0^{\sigma(t)}\mathbf{x}(t),$$

where $\sigma : [0, \infty) \rightarrow \mathcal{P}$ is a switching signal that indexes the appropriate graph topology at each switch. The main result in [2] says that consensus is possible even under the (weaker) condition of joint connectedness of the graphs encountered in an infinite sequence of contiguous bounded time-intervals.

Based on these results, one can ask if the consensus dynamics can be generalized to some switched linear systems that models the flow of the k-Laplacian under switching topologies. Moreover, if the underlying simplicial complexes are Rips complexes, the topologies change due to the motion of the nodes and one can replace the generalize the concept of joint connectedness to the absence of holes (homology) in certain finite unions of Rips complexes, one can hope to study the sweep coverage problem using the stability properties of that switched linear system. We make these concepts concrete in the following section.

B. k-Laplacian Flows on Simplicial Complexes

C. Fixed Network Topology

The dynamical system (4) runs at the node level in a network. Assume that the nodes are fixed, so that the connectivity graph G of the network is also fixed. Let us now run a dynamical system on the k-simplices of a network, as a k-Laplacian flow on the corresponding Rips complex $\mathcal{R}$. We described in [30], how to interpret this flow correctly using the language of discrete differential forms. We denote the k-forms on a simplicial complex as $C^k(\mathcal{R})$, to indicate that they are dual to the chains $C_k(\mathcal{R})$. We study the dynamical system

$$\frac{\partial\omega(t)}{\partial t} = -\mathcal{L}_k\omega(t), \qquad \omega(0) = \omega_0 \in C^k(\mathcal{R}). \qquad (5)$$

The equilibrium points of this dynamical system are the set $\ker(\mathcal{L}_k)$. The stability of this dynamical system is described by the following results, proven in [30].

Proposition IV-D: The dynamical system of Equation 5 is semi-stable.

[3] Strictly speaking, cohomology groups are dual spaces of homology groups. For the sake of our discussion however, this distinction is immaterial as either one represent k dimensional holes in the simplicial complex.

This implies that the system always converges to to the set $\ker(\mathcal{L}_k)$ in the limit. Note that the condition that $\dim\ker(\mathcal{L}_k) = 0$ is equivalent to saying that the k-th co-homology group (or the respective k-th homology group) is zero. Therefore, we have the following corollary.

Corollary IV-E: [30] The system in Equation 5 is asymptotically stable if and only if $H_k(\mathcal{R}) = 0$.
Thus the asymptotic stability of the system is an indicator of an underlying trivial topology of the complex. These results also prove that for any initial $\omega(0)$, the trajectory $\omega(t), t \geq 0$ always converges to some point in $\ker \mathcal{L}_k$. In other words, the dynamical system is a mechanism for producing discrete harmonic k-forms on simplicial complexes from any arbitrary k-forms [30].

F. Dynamic Network Topology

Let us now consider the case when the nodes are mobile, giving rise to a switching connectivity graph structure. This also gives rise to a switching structure for the induced Rips complex. Following the presentation in [2], let us write the flow on this changing simplicial complex using a piece-wise constant switching signal $\sigma : [0, \infty) \to \mathcal{P}$, where $\mathcal{P}$ indexes the set of connectivity graphs on n vertices, emerged by the motion of the nodes. We assume that under reasonable assumptions on the smoothness of the trajectories, there are a finite number of switches in a bounded interval of time. We will further assume that switching periods cannot be smaller than a finite dwell time τ_D.

Let the switching times be $t_1, t_2, \ldots$. If $G_{\sigma(t_0)}, G_{\sigma(t_1)}, \ldots$ are the graphs encountered along a certain evolution, then the corresponding Rips complexes are denoted by $\mathcal{R}(G_{\sigma(t_0)}), \mathcal{R}(G_{\sigma(t_1)}), \ldots$, respectively. Let

$$\mathcal{R} = \bigcup_i \mathcal{R}(G_{\sigma(t_i)}).$$

Since the Rips complex of a complete graph on n vertices is an $n-1$-simplex, denoted by $\boldsymbol{\Delta}^{n-1}$, it follows that $\mathcal{R} \subseteq \boldsymbol{\Delta}^{n-1}$. We next order the k-simplices of $\mathcal{R}$ in the order of their appearance in time and thus produce the spaces $C^k(\mathcal{R})$. Thus $C^k(\mathcal{R}(G_{\sigma(t_i)})) \subseteq C^k(\mathcal{R})$, so the dynamical system

$$\frac{\partial \omega(t)}{\partial t} = -\mathcal{L}_k^i \omega(t), \qquad \omega(t_i) \in C^k(\mathcal{R}(G_{\sigma(t_i)})),\ t \in [t_i, t_{i+1}),$$

can also be written as a zero-padded system

$$\frac{\partial \omega(t)}{\partial t} = -\tilde{\mathcal{L}}_k^i \omega(t), \qquad \omega(t_i) \in C^k(\mathcal{R}),\ \ t \in [t_i, t_{i+1}).$$

Here $\tilde{\mathcal{L}}_k^i$ is a zero-padded matrix representation of $\mathcal{L}_k^i$. This lets us write the flow for all $t \geq 0$ as a switched linear system

$$\frac{\partial \omega(t)}{\partial t} = -\tilde{\mathcal{L}}_k^\sigma \omega(t), \qquad \omega(0) \in C^k(\mathcal{R}). \tag{6}$$

We want to study the conditions under which this switched linear system is asymptotically stable. From the observation that

$$\ker \mathcal{L}_k^i \cong H_k(\mathcal{R}(G_{\sigma(t_i)}))$$

and Corollary IV-E, one can hope that if we are ensured that each simplicial complex encountered during the evolution is hole-free, i.e. $H_k(\mathcal{R}(G_{\sigma(t_i)})) \cong 0$, then the switched linear system may show asymptotic stability. We will show below that we can prove the asymptotic stability of the switched system under an even weaker condition, whereby the simplicial complexes encountered in bounded, non-overlapping time intervals are *jointly hole-free*. The proof of this result closely follows the presentation in [2]. As explained later, this will be a key result in generalizing the previous blanket coverage criteria of [22, 23] to a sweeping coverage criterion. Let us give the following definition.

Definition IV-G: Let $\{X^1, X^2, \ldots, X^m\}$ be a finite collection of simplicial complexes. Then, they are said to be jointly hole-free in dimension k if

$$H_k\left(\bigcup_{i=1}^m X^i\right) \cong 0.$$

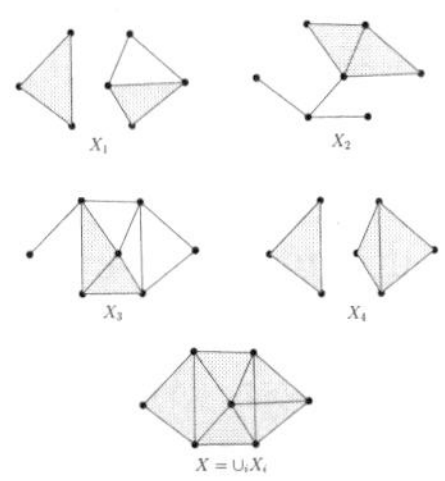

Fig. 5. Jointly hole-free simplicial complexes in dimension 0 and 1. Note that the complexes are *not* jointly hole-free in dimension 2 due to the hollow tetrahedron in the right part of X.

For an example, see Figure 5. We now state the following result:

Proposition IV-H: Let $\mathcal{X} = \{X^1, X^2, \ldots, X^m\}$ be a finite collection of simplicial complexes, whose union is X. Let the k-Laplacian operator of X^i be given by $\mathcal{L}_k^i : C^k(X^i) \to C^k(X^i)$, which can be zero-padded appropriately to get the operator $\tilde{\mathcal{L}}_k^i : C^k(X) \to C^k(X)$. If $\mathcal{X}$ is jointly hole-free in dimension k, then

$$\bigcap_{i=1}^m \ker \tilde{\mathcal{L}}_k^i = \{0\}.$$

Proof: First note that $\ker \tilde{\mathcal{L}}_k^i \cong \ker \mathcal{L}_k^i \oplus F^i \cong H_k(X^i) \oplus F^i$, where F^i is the space spanned by the eigenvectors corresponding to the 'fake' zero eigenvalues, due to zero-padding. Therefore, the zero-padding does not present an obstruction in analyzing the intersection of the kernels and is only there for proper comparison. For notational convenience, let us also write $\tilde{\mathcal{L}}_k^i$ as $\tilde{\mathcal{L}}_k(X^i)$. It is easy to see that

$$\tilde{\mathcal{L}}_k(X^p) + \tilde{\mathcal{L}}_k(X^q) = \tilde{\mathcal{L}}_k(X^p \cap X^q) + \tilde{\mathcal{L}}_k(X^p \cup X^q),$$

which can be generalized to the relation

$$\sum_{i=1}^m \tilde{\mathcal{L}}_k(X^i) = \tilde{\mathcal{L}}_k\left(\bigcup_{i=1}^m X^i\right) + \sum_{i=1}^{m-1} \tilde{\mathcal{L}}_k\left(X^{i+1} \cap \left(\bigcup_{j=1}^i X^j\right)\right).$$

Since the zero-padded k-Laplacians are all positive semi-definite, a k-form $w \in C^k(X)$ which is harmonic for each $\tilde{\mathcal{L}}_k^i$, also satisfies

$$\tilde{\mathcal{L}}_k \left(\bigcup_{i=1}^{m} X^i \right) \omega = 0.$$

Therefore,

$$\bigcap_{i=1}^{m} \ker \tilde{\mathcal{L}}_k(X^i) \subset \ker \tilde{\mathcal{L}}_k \left(\bigcup_{i=1}^{m} X^i \right) = H_k \left(\bigcup_{i=1}^{m} X^i \right).$$

Since by the given condition $\mathcal{X}$ is jointly hole-free, it immediately follows that

$$\bigcap_{i=1}^{m} \ker \tilde{\mathcal{L}}_k^i = \{0\}.$$

■

Given the trajectories $\{x_i(t)\}$ of the nodes, the switched linear system (6) for a fixed k, captures the evolution of the Rips complexes as well as the k-Laplacian flow on them. We can now state the following result, whose proof has been included in the appendix.

Theorem IV-I: The following are equivalent

1) The switched linear system (6) is globally asymptotically stable.
2) There is an infinite sequence of switching times $t_0, t_1, \ldots$, such that across each interval $[t_j, t_{j+1})$, the encountered collection of Rips complexes are jointly hole-free in dimension k.

Proof:

Suppose that the switched dynamical system (6) is asymptotically stable. Now, suppose to the contrary that there does not exist a sequence of switching times, such that the complexes encountered during the intervals are jointly hole-free. Then there exists a finite T, such that for $t \geq T$ there exists a non-zero k-form θ in the intersection of the kernels of all complexes encountered after T. In other words, $\tilde{\mathcal{L}}_k^{\sigma}\theta = 0$ for $t \geq T$. Without loss of generality pick $T = 0$. The evolution of (6) can be written as

$$\omega(t) = \left(\prod_{i=0}^{\infty} \exp\left(-\tilde{\mathcal{L}}_k^i (t_{i+1} - t_i) \right) \right) \omega(0).$$

Now let $\omega(0) = \theta$, then by virtue of being in the intersection of the kernels, $\omega(t) = \theta$ for all $t \geq 0$. Therefore, we have an initial condition that does not converge to zero. But it is given that the system is asymptotically stable. Therefore, we have a contradiction and the implication is true.

For the converse, Consider the time-interval $[t_j, t_{j+1})$. By assumption on the switching sequence, there are a finite number of switches in this interval. Let those switching times be given by $t_j, t_{j_1}, \ldots, t_{j_N}$. The Rips complexes encountered during the interval are $\mathcal{R}(G_{\sigma(t_j)}), \mathcal{R}(G_{\sigma(t_{j_1})}), \ldots, \mathcal{R}(G_{\sigma(t_{j_N})})$. By assumption, they are jointly hole-free so that

$$\ker \tilde{\mathcal{L}}_k^{\sigma(t_j)} \cap \left(\bigcap_{i=1}^{N} \ker \tilde{\mathcal{L}}_k^{\sigma(t_{j_i})} \right) = \{0\}.$$

Now let $\Phi(t_j, t_{j+1})$ be the product of the individual flows $\exp(-\tilde{\mathcal{L}}_k^{\sigma(t_{j_m})}(t_{j_m+1} - t_{j_m}))$ between switches, during this interval. By the semi-stability of each individual flow, $\| \exp(-\tilde{\mathcal{L}}_k^{\sigma(t_{j_m})}(t_{j_m+1} - t_{j_m}))\| \leq 1$ for each m. Therefore, $\|\Phi(t_j, t_{j+1})\| \leq 1$.

Let T be an upper bound on the lengths of the intervals $[t_{i+1}, t_i)$, and M be the smallest positive integer such that $M \geq T/\tau_D$. Now, by considering all sequences $\mathcal{S}^M$ of Rips complexes of length at most M that can occur within an interval of length bounded by T, and are jointly-hole free one can define,

$$\mu = \max_{\tau_1 \in [\tau_D, T]} .. \max_{\tau_N \in [\tau_D, T]} \max_{\mathcal{S}^M} \| \exp(-\tilde{\mathcal{L}}_k^{s_1}\tau_1)..\exp(-\tilde{\mathcal{L}}_k^{s_N}\tau_N)\|.$$

where $N \leq M$. It can be shown that $\|\Phi(t_j, t_{j+1})\| \leq \mu < 1$. Furthermore, the repeated application of this inequality across a sequence of intervals yields, $\|\Phi(t_p, t_{p+1}) \ldots \Phi(t_1, t_2)\| \leq \mu^p < 1$. From these contractions, it is can be inferred that the system is asymptotically stable ■

J. Dynamic Coverage Verification

The consequences of this theorem for verifying dynamic coverage are now easy to explain. The property of being jointly hole-free guarantees that in contiguous bounded time-intervals, the union of the Rips complexes has no hole. As mentioned in Section III-B, the topology of the Rips complex can be used to infer blanket coverage (or coverage loss) from network holes. In the same way, the topology of the Rips complexes encountered during the dynamic evolution of a network may be used to infer dynamic coverage of the sensors.

One example is to verify network coverage in which sensors alternate between active and sleep states to conserve power, as depicted in Figure 6. Several constructive schemes have been proposed by various researchers [11, 12, 31, 32, 33, 34]. In order to verify these schemes in real time, we run the dynamical system (6) as a cheap distributed verification method. Since the node are stationary (even though the network topology is changing), the Rips complexes generated during each change in the topology (via wake-ups or sleeping of the nodes) can be compared for the joint hole-free condition. If the Rips complexes are indeed jointly hole-free, then the dynamical system will converge to zero, thus indicating dynamical coverage. Such a scenario has been depicted in Figure 6.

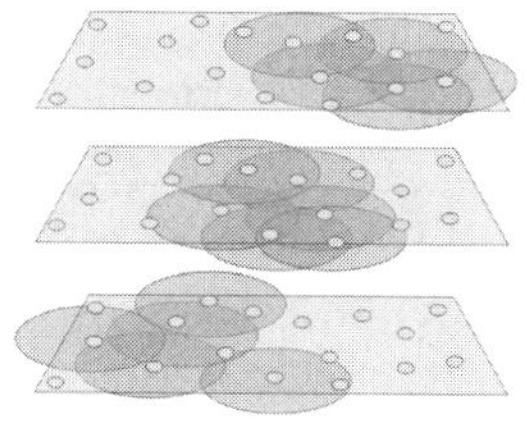

Fig. 6. Sensor wake-up/sleep scheme in a sensor network for providing dynamic coverage of the area under surveillance.

In the case of dynamic nodes, the above mentioned results are still applicable but the implementation need further work. Due to the changes in node locations, the Rips complexes generated during the evolution cannot be compared directly for a joint hole-free condition. In fact, such a comparison can lead to misleading interpretations of coverage. To fix this, one can use fixed landmarks in the surveillance domain and then compare the complexes in which the landmarks serve as the nodes and the simplices are generated by the visitation of the agents to each landmark. The details of this method will be given in a future work.

Finally, we should mention that our results on dynamic coverage are similar in spirit but quite different in technique to the results on pursuit-evasion problem in [22]. In [22], the authors present a homological criterion for detecting wandering holes in time-varying Rips complexes. The notion of a wandering hole is different from the loss of sweep coverage (one can have sweep coverage with wandering holes). Moreover, our emphasis on decentralized methods results in a technique which is quite different as compared to the method of prism complexes provided in [22].

V. Conclusions

In this paper, we have used the flows of k-Laplacian operators on time-varying simplicial complexes for verifying dynamic coverage in mobile sensor networks. The verification has been shown to be equivalent to the asymptotic stability of the switched linear systems that describe those flows. The k-Laplacians are a natural generalization of the graph Laplacian. The techniques used for proving consensus under the condition of joint connectedness of switching graphs have been utilized for this coverage problem. This approach gives new insights into the working of consensus algorithms. It also allows us to view algebraic graph theory as a special case of the spectral theory of simplicial complexes. This viewpoint has proven useful in the context of networked sensing and control.

References

[1] M. E. J. Newman, "The structure and function of complex networks," *SIAM Review*, vol. 45, no. 2, pp. 167–256, 2003.

[2] A. Jadbabaie, J. Lin, and A. S. Morse, "Coordination of groups of mobile autonomous agents using nearest neighbor rules," *IEEE Transactions on Automatic Control*, vol. 48, no. 6, pp. 988–1001, 2003.

[3] M. Penrose, *Random Geometric Graphs*. Oxford University Press, 2003.

[4] L. Euler, "Solutio pproblematis ad geometrium situs pertinentis," *Commentarii academiae scientiarum petropolitanae*, vol. 8, pp. 128–140, 1741.

[5] J. Dieudonné and J. Dieudonné, *A History of Algebraic and Differential Topology, 1900-1960*. Birkhäuser, 1989.

[6] D. Gage, "Command control for many-robot systems," in *Nineteenth Annual AUVS Technical Symposium*, pp. 22–24, 1992.

[7] C. F. Huang and Y.-C. Tseng, "The coverage problem in a wireless sensor network," in *ACM Intl. Workshop on Wireless Sensor Networks and Applications*, 2003.

[8] P. J. W. X. Y. Li and O. Frieder, "Coverage in wireless ad-hoc sensor networks," *IEEE Transaction on Computers*, vol. 52, no. 6, pp. 753–763, 2003.

[9] B. Liu and D. Towsley, "A study of the coverage of large-scale sensor networks," 2004.

[10] M. P. S. Meguerdichian, F. Koushanfar and M. Srivastava, "Coverage problems in wireless ad-hoc sensor network," in *IEEE INFOCOM*, pp. 1380–1387, 2001.

[11] C. F. Hsin and M. Liu, "Network coverage using low duty-cycled sensors: random and coordinated sleep algorithms," in *IPSN*, 2004.

[12] F. Ye, G. Zhong, S. Lu, and L. Zhang, "PEAS: a robust energy conserving protocol for long-lived sensor networks," *Proceedings of the 10th IEEE International Conference on Network Protocols*, pp. 200–201, 2002.

[13] H. Koskinen, "On the coverage of a random sensor network in a bounded domain," in *16th ITC Specialist Seminar*, pp. 11–18, 2004.

[14] M. Batalin and G. Sukhatme, "Spreading out: A local approach to multi-robot coverage," *Distributed Autonomous Robotic Systems*, vol. 5, pp. 373–382, 2002.

[15] M. Batalin and G. Sukhatme, "Sensor coverage using mobile robots and stationary nodes," *Procedings of the SPIE*, vol. 4868, pp. 269–276.

[16] J. Cortes, S. Martinez, T. Karatas, and F. Bullo, "Coverage control for mobile sensing networks," *Robotics and Automation, 2002. Proceedings. ICRA'02. IEEE International Conference on*, vol. 2, 2002.

[17] S. Poduri and G. Sukhatme, "Constrained coverage for mobile sensor networks," *Robotics and Automation, 2004. Proceedings. ICRA'04. 2004 IEEE International Conference on*, vol. 1.

[18] I. Hussein and A. Bloch, "Dynamic coverage optimal control of interferometric imaging spacecraft formations," *Decision and Control, 2004. CDC. 43rd IEEE Conference on*, vol. 2, 2004.

[19] M. Batalin and G. Sukhatme, "Multi-robot dynamic coverage of a planar bounded environment," *IEEE/RSJ International Conference on Intelligent Robots and Systems (Submitted)*, 2002.

[20] S. Kumar, T. Lai, and A. Arora, "Barrier coverage with wireless sensors," *Proceedings of the 11th annual international conference on Mobile computing and networking*, pp. 284–298, 2005.

[21] R. Olfati-Saber, J. Fax, and R. Murray, "Consensus and cooperation in networked multi-agent systems," *Proceedings of the of IEEE*, 2006.

[22] V. de Silva and R. Ghrist, "Coordinate-free coverage in sensor networks with controlled boundaries via homology." To appear in Intl. J. Robotics Research.

[23] V. de Silva and R. Ghrist, "Coverage in sensor networks via persistent homology." To appear in Algebraic and Geometric Topology.

[24] R. G. V. de Silva and A. Muhammad, "Blind swarms for coverage in 2-d," in *Robotics: Science and Systems*, June 2005.

[25] R. Ghrist and A. Muhammad, "Coverage and hole-detection in sensor networks via homology," in *The Fourth International Conference on Information Processing in Sensor Networks*, April 2005.

[26] A. Hatcher, *Algebraic Topology*. Cambridge University Press, 2002.

[27] R. Bott and L. Tu, *Differential Forms in Algebraic Topology*. Springer-Verlag, Berlin, 1982.

[28] R. Merris, "Laplacian matrices of graphs: A survey," *Linear Algebra and its Applications*, vol. 197, pp. 143–176, 1994.

[29] B. Eckmann, "Harmonische funktionen und randwertaufgaben einem komplex," *Commentarii Math. Helvetici*, vol. 17, pp. 240–245, 1945.

[30] A. Muhammad and M. Egerstedt, "Control using higher order laplacians in network topologies," in *17th International Symposium on Mathematical Theory of Networks and Systems*, 2006.

[31] T. H. L. S. Kumar and J. Balogh, "On k-coverage in a mostly sleeping sensor network," in *10th Intl. Conf. on Mobile Computing and Networking*, 2004.

[32] T. He, S. Krishnamurthy, J. Stankovic, T. Abdelzaher, L. Luo, R. Stoleru, T. Yan, L. Gu, J. Hui, and B. Krogh, "Energy-efficient surveillance system using wireless sensor networks," 2004.

[33] M. Nosovic and T. Todd, "Low power rendezvous and RFID wakeup for embedded wireless networks," *Annual IEEE Computer Communications Workshop*, pp. 3325–3329, 2000.

[34] M. Cardei and J. Wu, "Energy-efficient coverage problems in wireless ad-hoc sensor networks," *Computer Communications*, vol. 29, no. 4, pp. 413–420, 2006.

Discrete Search Leading Continuous Exploration for Kinodynamic Motion Planning

Erion Plaku, Lydia E. Kavraki, and Moshe Y. Vardi
Department of Computer Science, Rice University, Houston, Texas 77005
Email: {plakue, kavraki, vardi}@cs.rice.edu

***Abstract*— This paper presents the Discrete Search Leading continuous eXploration (DSLX) planner, a multi-resolution approach to motion planning that is suitable for challenging problems involving robots with kinodynamic constraints. Initially the method decomposes the workspace to build a graph that encodes the physical adjacency of the decomposed regions. This graph is searched to obtain *leads*, that is, sequences of regions that can be explored with sampling-based tree methods to generate solution trajectories. Instead of treating the discrete search of the adjacency graph and the exploration of the continuous state space as separate components, DSLX passes information from one to the other in innovative ways. Each lead suggests what regions to explore and the exploration feeds back information to the discrete search to improve the quality of future leads. Information is encoded in edge weights, which indicate the importance of including the regions associated with an edge in the next exploration step. Computation of weights, leads, and the actual exploration make the core loop of the algorithm.**

Extensive experimentation shows that DSLX is very versatile. The discrete search can drastically change the lead to reflect new information allowing DSLX to find solutions even when sampling-based tree planners get stuck. Experimental results on a variety of challenging kinodynamic motion planning problems show computational speedups of two orders of magnitude over other widely used motion planning methods.

I. Introduction

Robot motion planning with complex kinodynamic constraints is a topic that has attracted a lot of recent attention [1]–[5]. It is greatly motivated by the availability of new robotic systems and the need to produce paths that respect the physical constraints in the motion of the robotic systems and hence can be translated into trajectories executed by the real platforms with minimum effort. Sampling-based tree planners, such as Rapidly-exploring Random Tree (RRT) [6], Expansive Space Tree (EST) [7], and others [1]–[5], [8]–[10] have in recent years been widely successful in kinodynamic motion planning. Such planners typically explore the state space using a single or a bidirectional tree [1]–[5], [7], [9], or multiple trees, as in the case of Sampling-based Roadmap of Trees (SRT) [10]. Recent books [1], [2] contain additional references and descriptions of many successful sampling-based tree planners.

Given the large amount of work on kinodynamic motion planning and the dominant place of sampling-based tree planners, a new planner for robots with kinodynamic constraints can be justified only if it offers significant advantages over previous work. This paper describes DSLX, a multi-resolution approach that as other existing and highly successful sampling-based tree planners (e.g., RRT, EST, SRT) uses tree exploration of the state space, but displays a superior performance when compared to them.

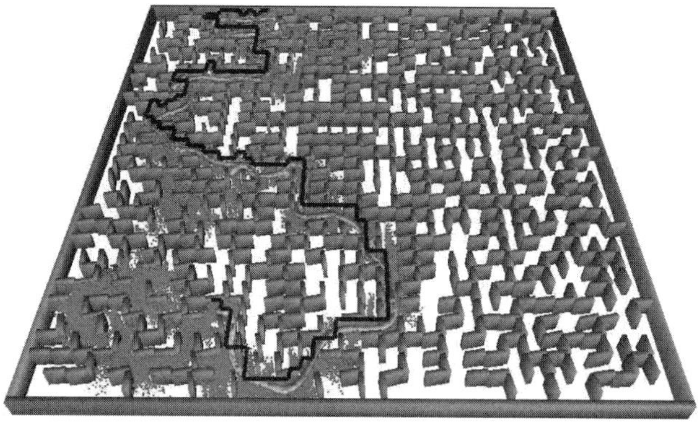

Fig. 1. Example of benchmark "RandomSlantedWalls," a kinodynamic motion planning problem solved by DSLX two orders of magnitude faster than other tree-based motion planning methods. The robot model is that of a smooth car (see Section III-A). Red dots indicate projections of the states of the exploration tree onto the workspace. The black line indicates the current lead that is used to guide the tree exploration toward the goal.

DSLX utilizes information provided by the problem specification and information gathered during previous exploration steps to guide future explorations closer to obtaining a solution to the given motion planning problem. This is a concept that has been studied before mainly in the context of sampling-based planners that construct roadmaps. For example, the original Probabilistic Roadmap Method (PRM) [11] uses the information of the connectivity of the samples to create more samples in parts of the space where connectivity is low. The work in [12] uses nearest-neighbors information in the context of PRM to define the utility of each sample in an information-theoretic sense and only add to the roadmap those samples that increase the overall entropy. The planners in [13] and [14] also utilize information in the context of PRM to find appropriate sampling strategies for different parts of the configuration space. In contrast to roadmap methods, traditional tree-based methods such as RRT [6], ADDRRT [4], EST [7] rely on limited information, such as distance metrics or simple heuristics to guide the exploration. Although the tree may advance quickly towards its goal, if it gets stuck it becomes more and more difficult to find promising directions for the exploration.

DSLX is a tree-based planner that systematically uses the information gathered during previous explorations steps to lead future explorations toward increasingly promising directions. Initially, DSLX constructs a coarse-resolution representation of the motion planning problem by obtaining a decomposition of the workspace. The decomposition is represented in terms of a graph whose vertices are regions in the decomposition and whose edges denote physical adjacency between different regions. DSLX exploits the simple observation that any solution trajectory that connects the initial state to the goal state corresponds to some coarse-grained sequence of neighboring regions that starts and ends at regions associated with the initial and goal states, respectively. Although the converse does not hold, a sequence of coarse-grained neighboring regions can however serve as a guide to the exploration.

The idea of using decompositions of the workspace, configuration space and state space appear early in the motion planning literature. Key theoretical results in motion planning were obtained using decomposition based methods [15]. Some of the first planners obtained decompositions of the workspace into regions that were linked in a graph which was subsequently used to find a path. An extensive discussion of early exact and approximate cell decomposition methods can be found in [15] (Chapters 5 and 6) and in [16]–[18]. Initially only geometric planning was considered. More recently approaches that deal with kinodynamic planning have appeared in the context of sampling-based methods [19]–[22].

The decomposition graph is weighted and the initial weights are all set to a fixed value. The core part of DSLX proceeds by repeating the following three steps until a solution is found or a maximum amount of time has elapsed:

1) Obtain a guide sequence, called a *lead*, by some discrete search method on the decomposition graph.
2) Explore the continuous state space for a short period of time. DSLX attempts to extend the branches of an exploring tree from one region to its neighbor, as specified by the lead.
3) Update the weights of the decomposition graph. The weight of an edge represents an estimation of how important the exploration of the regions it connects is for the construction of solution trajectories. The weight depends on the total time spent so far exploring, the progress of the exploration in these regions and other quantities. The weights are updated after each exploration of a region to reflect the new information gathered during the most recent exploration.

A critical difference and advantage of DSLX over earlier workspace decomposition methods is the close interaction of the discrete search and the continuous exploration and the flexibility this interaction provides. The coarse-grained representation can provide DSLX with many alternative leads. A central issue is which lead to choose from the possibly combinatorially large number of possibilities. Since the weight estimates that bias the computation of leads are based on partial information, it is important not to ignore leads associated with lower weights, especially during the early stages of the exploration. DSLX aims to strike a balance between greedy and methodical search by selecting more frequently sequences of coarse-grained regions that are associated with higher weights and less frequently sequences of coarse-grained regions that are associated with lower weights.

Through extensive experimentation on a variety of kinodynamic motion planning problems it has been observed that DSLX can focus the exploration on promising search directions while able to radically change these directions if the information gathered during exploration suggests other promising leads. This flexibility tends to prevent the method from getting stuck in the way that other sampling-based tree planners do. Extensive comparisons have been done with RRT [6], a more recent version of RRT called Adaptive Dynamic Domain RRT (ADDRRT) [4], EST [7], and SRT [10] showing that DSLX can be up to two orders of magnitude more efficient. Fig. 1 shows one such case for kinodynamic motion planning where DSLX finds solutions more than 170 times faster than the single-tree based methods and 90 times faster than SRT.

This paper is organized as follows. Section II describes DSLX in detail. Experiments and results are described in Section III. Section IV discusses the experimental results and provides insights into the reasons for the observed computational efficiency of DSLX. The paper concludes in Section V with a summary and directions for future work.

II. DSLX

Pseudocode for the overall approach is given in Algorithm 1. The construction of the coarse-grained decomposition into neighboring regions is given in line 2 and described in Section II-A. The lead computation occurs in line 6 and is described in Section II-B. The other lines refer to the state space exploration, which is described in Section II-C.

Algorithm 1 Pseudocode for DSLX

Input:
$\mathcal{W}$, geometric description of the workspace
$\mathcal{R}$, motion model and geometric description of the robot
s, g, initial and goal specifications
$t_{\max} \in \mathbb{R}^{>0}$, upper bound on computation time
$t_e \in \mathbb{R}^{>0}$, short time allocated to each exploration step

Output: A solution trajectory or NIL if no solution is found

```
1: StartClock1
2: G = (V, E) ← CoarseGrainedDecomposition(W)
3: InitExplorationEstimates(G)
4: T ← exploration tree rooted at s
5: while ElapsedTime1 < t_max do
6:    [R_i1, ..., R_in] ← ComputeLead(G, s, g)
7:    StartClock2
8:    while ElapsedTime2 < t_e do
9:       R_ij ← SelectRegion([R_i1, ..., R_in])
10:      for several times do
11:         x ← SelectStateFromRegion(R_ij)
12:         PropagateForward(T, x, R_ij, R_ij+1)
13:   if a solution is found then
14:      return solution trajectory
15:   UpdateExplorationEstimates(G, [R_i1, ..., R_in])
16: return NIL
```

A. *Coarse-Grained Decomposition*

This paper uses a simple grid-based decomposition of the workspace. Even with this simple decomposition, DSLX is computationally efficient in solving challenging kinodynamic motion planning problems, as indicated by the experimental results in Section III-D.

The coarse-grained decomposition of the workspace provides a simplified layer to the motion planning problem. As discussed in Section II-B, DSLX uses the coarse-grained decomposition to compute general leads from the initial to the goal region. It is important to note that DSLX allows for a great degree of flexibility on the decomposition of the workspace. In particular, since DSLX relies on information collected during exploration to determine which lead to select, it does not even require that the decomposition be collision-free. When regions that are occupied by obstacles are selected as part of the lead, the exploration estimates will indicate that no progress can be made. Consequently, such regions will be selected less and less frequently. The use of better workspace decompositions (see [1], [2] for details on different decomposition methods) may certainly further improve the computational efficiency of DSLX, since it will in general provide better search directions.

B. Coarse-Grained Leads

The coarse-grained decomposition is used to obtain sequences of neighboring regions that provide promising leads for the state space exploration. A transition graph $G = (V, E)$ is constructed based on the coarse-grained decomposition as described in Section II-A. Each vertex $v_i \in V$ is associated with a region R_i of the coarse-grained decomposition and an edge $(v_i, v_j) \in E$ indicates that R_i and R_j are neighboring regions in the workspace decomposition. Let $v(s) \in V$ and $v(g) \in V$ be the two vertices whose corresponding regions are associated with the initial and goal states, s and g, respectively.

A lead is computed by searching G for sequences of edges from $v(s)$ to $v(g)$. A weight w_{ij} associated with each edge $(v_i, v_j) \in E$ is an estimate of DSLX on the importance of including R_i and R_j as an edge in the lead. w_{ij} is updated after each exploration of R_i and R_j. Sequences of edges associated with higher weights are selected more frequently since, according to current estimates such edge sequences provide promising leads.

1) Computation of w_{ij}: The coverage $c(\mathcal{T}, R_k)$ of a region R_k by the states of the tree $\mathcal{T}$ is estimated by imposing an implicit uniform grid on R_k and measuring the fraction of cells that contain at least the projection of one state from $\mathcal{T}$. Let $c(\mathcal{T}, R_k)$ denote the coverage estimate of R_k by $\mathcal{T}$ at the beginning of the current exploration step and let $c'(\mathcal{T}, R_k)$ denote the new coverage estimate at the end of the exploration step. Thus $\alpha(\mathcal{T}, R_k) = c'(\mathcal{T}, R_k) - c(\mathcal{T}, R_k)$ measures the change in the coverage of R_k by $\mathcal{T}$ as a result of the current exploration step. Then, the weight w_{ij} is defined as

$$w_{ij} = 0.5(\alpha(\mathcal{T}, R_i) + \alpha(\mathcal{T}, R_j))/t + \epsilon/t_{\mathrm{acc}}(i, j),$$

where t is the computational time devoted to the exploration of R_i, R_j during the current exploration step; $t_{\mathrm{acc}}(i, j)$ is the accumulated time over all explorations of R_i, R_j; and ϵ is a small normalization constant.

Large values of w_{ij} are obtained when branches of $\mathcal{T}$ quickly reach previously unexplored parts of R_i and R_j and are thus indicative of promising leads. The accumulated time $t_{\mathrm{acc}}(i, j)$ is used to give a higher weight to those regions that have been explored less frequently. Initially, since there is no exploration information available, each weight w_{ij} is set to a fixed value.

2) Computation of leads: Many possible strategies can be used to compute search directions. The computation of a lead is essentially a graph searching algorithm and the literature on this subject is abundant (see [23] for extensive references).

The combination of search strategies in this work aims to provide leads that are more frequently biased toward promising directions. However, random leads are also used, although less frequently, as a way to correct for errors inherent with the estimates. The use of random leads is motivated by observations made in [10], [24], where random restarts and random neighbors have been suggested as effective ways to unblock the exploration when tree planners or PRM get stuck.

COMPUTELEAD (line 6 of Algorithm 1) frequently returns the most probable sequence of edges in G from $v(s)$ to $v(g)$. The probability p_{ij} associated with $(v_i, v_j) \in E$ is computed by normalizing the weight w_{ij}, i.e., $p_{ij} = w_{ij}/\sum_{(v_k, v_\ell) \in E} w_{k\ell}$. The probability of a sequence of edges is then defined as the product of the probabilities associated with its edges. The most probable edge sequence from $v(s)$ to $v(g)$ is the one with the highest probability. In order to compute the most probable edge sequence as defined above using a shortest path algorithm, such as Dijkstra's, the weight function used in the graph search is set to $-\log(p_{ij})$ for $(v_i, v_j) \in E$.

Almost as frequently, COMPUTELEAD returns the acyclic sequence of edges from $v(s)$ to $v(g)$ with the highest sum of edge weights. In order to compute such sequence using Dijkstra's shortest path algorithm, the weight function used in the graph search is set to $w_{\max} - w_{ij}$ for $(v_i, v_j) \in E$, where $w_{\max}$ denotes the maximum value for the weights.

Less frequently, COMPUTELEAD computes a random sequence of edges from $v(s)$ to $v(g)$. This computation is carried out by using depth-first search, where the unvisited children are visited in a random order.

C. Exploration

The exploration starts by rooting a tree $\mathcal{T}$ at the specified initial state s (line 3). The objective is to quickly grow $\mathcal{T}$ toward the goal by using the coarse-grained leads as guides for the exploration. The exploration is an iterative process (lines 8–12). At each iteration a region R_{i_j} is selected from the coarse-grained lead $[R_{i_1}, \ldots, R_{i_n}]$ and explored for a short period of time. The exploration aims to extend branches of $\mathcal{T}$ from R_{i_j} to $R_{i_{j+1}}$. For this reason, several states are selected from the states associated with R_{i_j} and are propagated forward toward $R_{i_{j+1}}$.

SELECTREGION*:* Note that some regions in $[R_{i_1}, \ldots, R_{i_n}]$ may not contain any states from $\mathcal{T}$, since the branches of $\mathcal{T}$ have yet to reach such regions. Such regions are not considered for selection, since they do not contain any states from which to propagate forward, as required by line 12 of Algorithm 1.

The objective is to select a nonempty region $R_{i_j} \in [R_{i_1}, \ldots, R_{i_n}]$ whose exploration causes $\mathcal{T}$ to grow closer to the goal. Since $[R_{i_1}, \ldots, R_{i_n}]$ specifies a sequence of neighboring regions that end at the region associated with the goal state, the order in which the regions appear in the lead provides an indication of how close $\mathcal{T}$ is to the goal. For this reason, `DSLX` prefers to select regions that appear toward the end of $[R_{i_1}, \ldots, R_{i_n}]$ more frequently than regions that appear at the beginning. Preference is also given to regions that have been selected less frequently for exploration. The exploration of such regions is vital in order to provide a balance between greedy and methodical search. This objective is achieved by selecting a region R_{i_j} from $[R_{i_1}, \ldots, R_{i_n}]$ based on the weight $\alpha j/n + (1-\alpha)/\text{nsel}(R_{i_j})$, where $0 < \alpha < 1$ is selected uniformly at random and $\text{nsel}(R_{i_j})$ is the number of times R_{i_j} has been selected for exploration.

SELECTSTATEFROMREGION*:* Each state x associated with R_{i_j} is selected based on the weight $1/\text{nsel}(x)$, where $\text{nsel}(x)$ is the number of times x has been selected. Preference is thus given to the states associated with R_{i_j} that have been selected less frequently. A state $x \in \mathcal{T}$ is associated with region R_{i_j} iff $\text{proj}(x) \in R_{i_j}$, where $\text{proj}(x)$ denotes the projection of the state $x \in \mathcal{T}$ onto the workspace $\mathcal{W}$.

PROPAGATEFORWARD*:* A state x is propagated forward to a new state x_{new} by selecting a control u and applying u to x for several time steps or until a collision is found. If x_{new} is more than a minimum number of propagation steps away from x, then x_{new} and the edge connecting x to x_{new} is added to $\mathcal{T}$. The state x_{new} is also added to the appropriate region R_k. Since the objective is to guide the propagation from R_{i_j} toward the neighboring region $R_{i_{j+1}}$, the control u is selected as the control that brings $\text{proj}(x_{\text{new}})$ closer to $R_{i_{j+1}}$ out of several controls sampled uniformly at random. The propagation is computed by integrating the motion equations of the robot. This work uses a fourth-order Runge-Kutta integrator [1], [2].

III. EXPERIMENTS AND RESULTS

The design of `DSLX` was motivated by challenging kinodynamic motion planning problems with vehicles and the experiments in this paper were chosen to test the efficiency of `DSLX` in solving such problems. The performance of `DSLX` is compared against several existing state-of-the-art methods. Results presented in Section III-D show the competitiveness of the proposed method, `DSLX`, and highlight the benefits of incorporating discrete-search and coarse-grained decomposition into sampling-based approaches, as proposed in this work.

A. Robot Models

The motions of the robot are defined by a set of ordinary differential equations. The robot models used in this work consist of a kinematic car (`KCar`), a smooth (second-order) car (`SCar`), smooth unicycle (`SUni`), and a smooth differential drive (`SDDrive`). Detailed descriptions of these models can be found in [1], [2]. The range of controls and bounds on the state variables are empirically determined based on the workspaces used for the experiments.

1) Kinematic Car (`KCar`): The motion equations are $\dot{x} = u_0 \cos(\theta); \dot{y} = u_0 \sin(\theta); \dot{\theta} = u_0 \tan(u_1)/L$, where (x, y, θ) is the car configuration; u_0 and u_1 are the speed and steering wheel controls; L is the distance between the front and rear axles. The speed and steering control are restricted to $|u_0| \leq \text{v}_{\text{max}} = 1$ and $|u_1| \leq \psi_{\text{max}} = \pi/4$.

2) Smooth Car (`SCar`): The kinematic car model can be extended to a second-order model by expressing the velocity v and steering angle ϕ as differential equations of the acceleration u_0 and the rotational velocity of the steering wheel u_1 controls, as follows: $\dot{x} = v\cos(\theta)$; $\dot{y} = v\sin(\theta)$; $\dot{\theta} = v\tan(\phi)/L; \dot{v} = u_0$; $\dot{\phi} = u_1$. The acceleration and rotational velocity of the steering wheel controls are restricted to $|u_0| \leq 0.0015\text{v}_{\text{max}}$ and $|u_1| \leq 0.0015\psi_{\text{max}}$.

3) Smooth Unicycle (`SUni`): The motion equations are $\dot{x} = v\cos(\theta)$; $\dot{y} = v\sin(\theta)$; $\dot{\theta} = \omega; \dot{v} = u_0$; $\dot{\omega} = u_1$, where (x, y, θ) is the configuration; ω and v are the rotational and translational velocities, respectively. The translation u_0 and rotational u_1 accelerations are restricted to $|u_0| \leq 0.0015r$ and $|u_1| \leq 0.0015r$, where r is the radius of the unicycle.

4) Smooth Differential Drive (`SDDrive`): The motion equations are $\dot{x} = 0.5r(\omega_\ell + \omega_r)\cos(\theta)$; $\dot{y} = 0.5r(\omega_\ell + \omega_r)\sin(\theta)$; $\dot{\theta} = r(\omega_r - \omega_\ell)/L$; $\dot{\omega_\ell} = u_0$; $\dot{\omega_r} = u_1$, where (x, y, θ) is the configuration; ω_ℓ and ω_r are the rotational velocities of the left and right wheels, respectively; r is the wheel radius; and L is the length of the axis connecting the centers of the two wheels. In this work, the controls u_0 and u_1 are restricted to $|u_0| \leq 0.15$ and $|u_1| \leq 0.15$.

B. Benchmarks

The benchmarks used in the experiments are designed to vary in type and difficulty and to test different aspects of motion planning methods. Illustrations of benchmarks and robot geometries can be found in Fig. 1 and Fig. 2.

Benchmark “Misc” consists of several miscellaneous obstacles arranged as in Fig. 2(a). Random queries are created that place the robot in opposite corners of the workspace. In this way, the robot must wiggle its way through the various obstacles and the narrow passages in the workspace.

Benchmark “WindingCorridors” consists of long and winding corridors, as shown in Fig. 2(b). Random queries are created by placing the robot in two different corridors, either 4 and 5 or 5 and 4 (counting from left to right), respectively. This benchmark is chosen to illustrate the efficacy of motion planning methods in solving problems where even though the initial and goal specification place the robot in neighboring places in the workspace, the solution trajectory is rather long and the robot travels through a large portion of the workspace.

Benchmark “RandomObstacles” consists of a large number of obstacles (278 obstacles) of varying sizes placed at random throughout the workspace, as shown in Fig. 2(c). The random placement of the obstacles creates many narrow passages, posing a challenging problem for motion planning methods, since research [1], [2] has shown that many motion planners

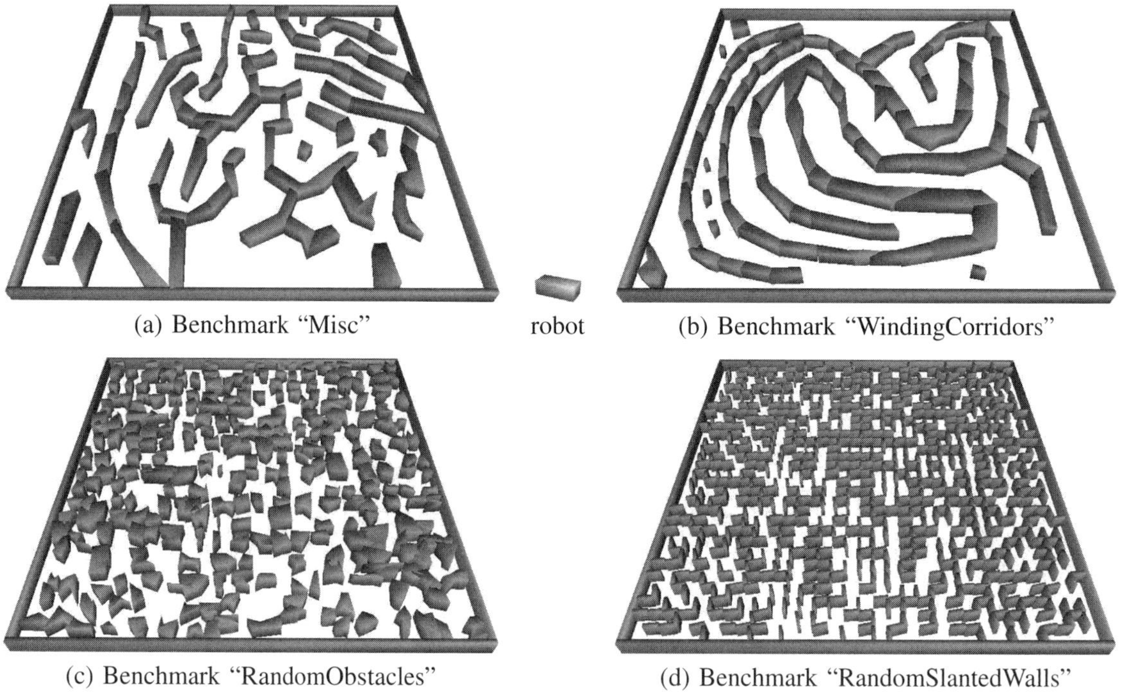

Fig. 2. Several benchmarks used for the experimental comparisons of DSLX. In each case, the robot geometry is a box and the workspace is a unit box. The body length and width of the robot in benchmarks "RandomObstacles" and "RandomSlantedWalls" are 1/40 and 1/60, respectively. In the case of benchmarks "Misc" and "WindingCorridors" the robot is twice the size of the one used in benchmarks "RandomObstacles" and "RandomSlantedWalls," since "Misc" and "WindingCorridors" have in general more open areas and wider passages.

have a tendency of getting stuck in such random environments with narrow passages. Random queries place the robot in opposite sides of the workspace.

Benchmark "RandomSlantedWalls" consists of 890 obstacles resembling slanted walls, as illustrated in Fig. 1. Initially, a random maze is created using the disjoint set strategy and then only 97% of the maze walls are kept. Knocking down of the maze walls creates multiple passages in the workspace for connecting any two points. The dimensions of the remaining walls are set uniformly at random from the interval $[1/60, 1/90]$ in order to create obstacles of different sizes. Each of the remaining walls is rotated by some angle chosen at random from $[2°, 15°]$, so that the walls are aligned at different angles. This benchmark tests the efficiency of motion planning methods in finding solutions for problems with multiple passages. Random queries place the robot in opposite sides of the workspace.

C. Other Motion Planning Methods used in Comparisons

This work presents comparisons with RRT [2], [6], ADDRRT [4], EST [7], and SRT [10]. Standard implementations were followed as suggested in the respective research papers and motion planning books [2], [6]. These implementations utilize the OOPS-MP (Online Open-source Programming System for Motion Planning) framework [25] and are well-tested, robust, and efficient, as they have been widely used by our research group. Every effort was made to fine-tune the performance of these motion planners for the experimental comparisons presented in this paper.

In addition to single-tree versions, bidirectional versions of RRT, ADDRRT, and EST also exist. It has also been shown that SRT [10] takes the bidirectional search a step further and uses single-tree based methods such as RRT, EST, etc., to grow multiple trees in different regions of the state space and then connects the neighboring trees to obtain a solution trajectory. Note that for nonholonomic problems tree connections however may contain gaps [1], [2]. Trajectories obtained by DSLX, RRT, ADDRRT, and EST do not contain any gaps, while trajectories obtained by SRT contain gaps. Such gaps could be closed using steering or numerical methods [26] at the expense of incurring additional computational costs.

D. Results

For each benchmark, experiments are run using each of the robot models described in Section III-A. For each combination of benchmark and robot model, 30 random queries are generated as described in Section III-B. Each motion planning method is then used to solve all the input random queries. In each instance, the computational time required to solve the query is measured. Rice PBC and Cray XD1 ADA clusters were used for code development. Experiments were run on ADA, where each of the processors runs at 2.2GHz and has up to 8GB of RAM.

Table I contains a summary of the experimental results. For each benchmark and robot model combination, the table indicates the average computational time required by each motion planning method to solve 30 random queries. In addition, Table I indicates the computational speedup obtained by DSLX in comparison to the other motion planning methods used in the experiments of this work. The experimental comparisons of DSLX with the single-tree methods are summarized in Table I (columns 1–3), while the comparisons with the multi-tree method SRT are summarized in Table I (column 4).

1) Comparison with the single-tree methods: Table I shows that DSLX is consistently more efficient than RRT, ADDRRT, and EST. For each benchmark and robot model combination, the average time required to solve a query is considerably lower for DSLX.

When the simple kinematic car model (KCar) is used, DSLX is between 3–12 times faster on "Misc."; 9–13 on "WindingTunnels"; 3–10 on "RandomObstacles"; and 23–29 times faster on "RandomSlantedWalls."

When the other robot models are used, DSLX is between 7–32 times faster on "Misc.,"; 9–29 on "WindingTunnels"; 36–69 on "RandomObstacles"; and 102–255 times faster on "RandomSlantedWalls."

2) Comparison with the multi-tree method SRT: The best computational times for SRT are obtained when several trees are also grown in other parts of the state space in addition to the trees grown at the initial and goal states. The computational times obtained by SRT tend to be lower than the computational times required by the bidirectional versions of RRT, ADDRRT, and EST. Recall that, as discussed in Section III-C, the trajectories computed by SRT contain gaps, while trajectories computed by DSLX do not contain any gaps. Results indicated in Table I are obtained when SRT grows 75 trees using EST as its building block. Similar results were obtained when RRT is used as a building block of SRT.

As indicated in Table I, DSLX is in each case computationally faster than SRT. The computational speedup of DSLX varies from a factor of 3–10 on the easier problems to a factor of 48–90 times on the more challenging problems.

IV. A Closer Look at the State Space Exploration

Experimental results presented in Table I indicate that DSLX offers considerable computational advantages over the other motion planning methods used in this work across a variety of challenging benchmarks and robot models. The experimental results show that DSLX is capable of solving challenging motion planning methods in a matter of one to three minutes as opposed to several hours required by other methods. DSLX computationally outperforms powerful motion planning methods, such as RRT, ADDRRT, EST, and SRT, by an order of magnitude on easy problems and as much as two orders of magnitude on more challenging problems.

The understanding of the main reasons for the success of a motion planning method is in general a challenging issue and subject of much research. This section takes a closer look at the exploration done by RRT, ADDRRT, EST, and SRT and compares it to the exploration done by DSLX in order to provide some insights behind the computational efficiency of DSLX.

By using nearest neighbors to random states as exploration points, RRT [2], [6] is frequently led toward obstacles where it may remain stuck for some time [1], [2], [4], [10]. Adjusting the exploration step size of RRT, as ADDRRT does, has been shown to alleviate the problem to a certain extent but not in all situations [4]. The use of ADDRRT incurs additional computational costs, which in some cases, as those observed in this work, outweigh the benefits offered by ADDRRT. However, both in the case of RRT and ADDRRT, as the tree grows large, it becomes more frequent for the nearest neighbors to random states not to be at the frontier of the tree but instead at "inner" nodes of the tree. Consequently, especially in challenging problems where propagation is difficult, these methods end up exploring the same region many times, thus wasting computational time.

EST [7] on the other hand suffers from a different kind of problem. EST directs the search toward less explored regions of the state space. As the tree grows large, the growth of the tree slows down as there are many regions with similar low density distributions. Consequently, EST ends up slowly expanding the tree in all possible directions, which do not necessarily bring the exploration closer to the goal region.

SRT [10] approaches the above problems by using multiple trees in randomly selected regions of the state space and only growing each tree for a shorter period of time to avoid the slow down on the growth of the trees. SRT is however not particularly well suited for nonholonomic motion planning problems, since tree connections create gaps that may require considerable additional computational time to be closed [26].

Although these methods have been shown to work well in a variety of settings, the main drawback common to all these methods is that they only use a limited amount of information to guide the exploration. There is generally much more information available to motion planning methods that, if properly used, can significantly speed up the computations.

The main strength of DSLX is the systematic use of information gathered during previous explorations steps to guide future explorations. As detailed in Section II, DSLX takes into account all the available workspace information, the initial and goal specifications, and carefully and closely integrates the information gathered during exploration into a discrete search and exploration of the state space. The discrete search provides DSLX with leads that guide the exploration closer to the goal specification. The exploration of the state space provides valuable feedback information that is used by DSLX to refine the lead for the next exploration step. As the exploration progresses, the leads produced by DSLX become more accurate and thus cause the tree to reach the goal specification. Fig. 3 provides a snapshot of the exploration done by DSLX at different time intervals. The tree quickly grows and reaches the goal in a short amount of time.

V. Discussion

We have presented DSLX, a tree-based motion planning method that relies on a decomposition of the workspace to obtain a coarse-grained representation of the motion planning problem and discrete search to find promising leads that bring the tree exploration closer to the goal. Information gathered

TABLE I

SUMMARY OF EXPERIMENTAL COMPARISONS WITH SINGLE- AND MULTIPLE-TREE METHODS

	Average time in seconds to solve one query					Speedup Obtained by DSLX			
	RRT	ADDRRT	EST	SRT	DSLX	RRT	ADDRRT	EST	SRT
"Misc"									
KCar	3.51	5.87	13.5	3.12	1.02	3.45	5.76	12.91	3.06
SCar	248.72	279.05	95.84	70.52	13.27	18.74	21.02	7.22	5.31
SUni	417.06	461.42	151.63	144.23	14.14	29.50	32.64	10.73	10.20
SDDrive	73.82	94.36	47.82	34.04	4.52	16.35	20.89	10.59	7.54
"WindingTunnels"									
KCar	15.33	19.29	22.94	12.21	1.70	9.03	11.37	13.52	7.18
SCar	282.49	231.11	90.76	92.32	9.46	29.85	24.42	9.59	9.75
SUni	161.16	175.92	83.14	106.73	8.57	18.21	20.53	9.70	12.45
SDDrive	178.10	213.35	142.59	108.75	7.60	23.43	28.06	18.76	14.30
"RandomObstacles"									
KCar	3.46	5.87	13.15	1.21	0.97	3.55	7.12	10.72	1.24
SCar	440.52	528.62	831.36	221.64	11.94	36.90	44.28	69.65	18.56
SUni	374.34	413.43	562.70	232.80	9.26	40.43	44.66	60.78	25.14
SDDrive	224.60	276.55	269.02	125.63	4.22	53.27	65.60	63.81	29.77
"RandomSlantedWalls"									
KCar	29.22	33.21	36.86	27.31	1.23	23.69	26.92	29.89	22.20
SCar	6330.95	6637.62	3716.27	1772.34	36.43	173.79	182.21	102.01	48.60
SUni	7207.27	6571.33	4819.99	2536.24	28.17	255.83	233.26	171.09	90.03
SDDrive	615.56	579.35	478.36	240.01	3.64	169.26	159.31	131.54	65.93

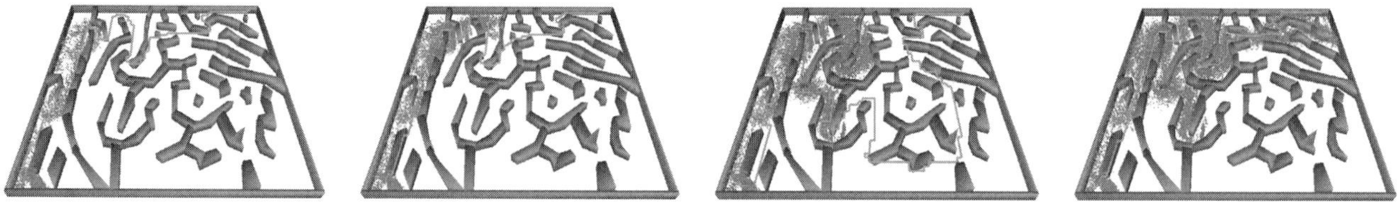

(a) Exploration of benchmark "Misc." after 2s, 4s, 6s, 8s of running time

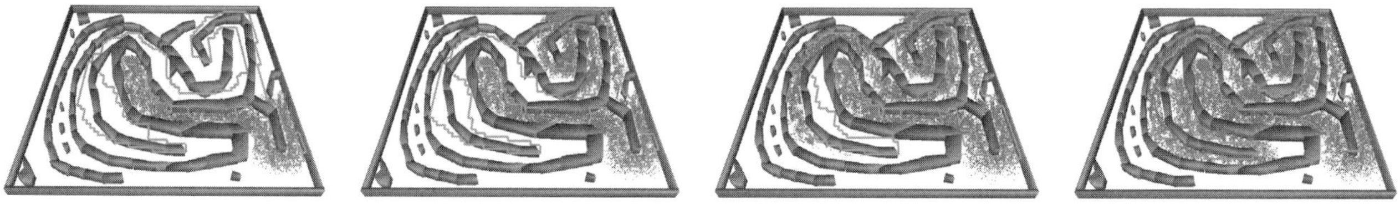

(b) Exploration of benchmark "WindingTunnels" after 2s, 4s, 6s, 8s of running time

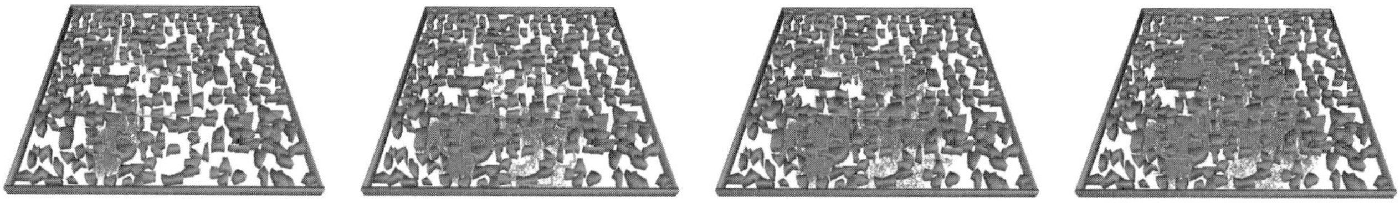

(c) Exploration of benchmark "RandomObstacles" after 2s, 4s, 6s, 8s of running time

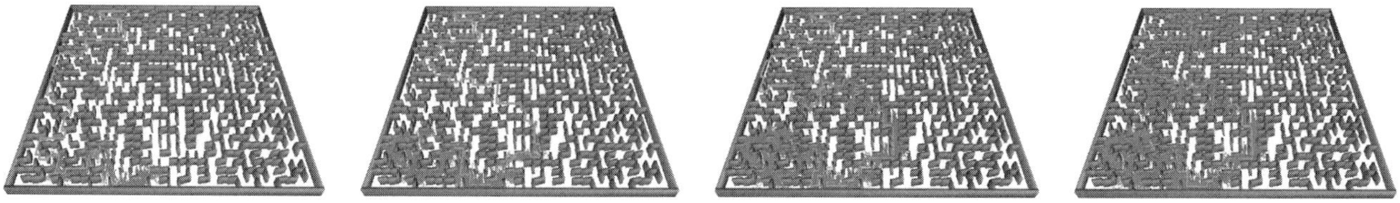

(d) Exploration of benchmark "RandomSlantedWalls" after 6s, 12s, 18s, 24s of running time

Fig. 3. Snapshots of the tree exploration by DSLX of different benchmarks with the smooth car (SCar) as the robot model. Red dots indicate projections of the states of the exploration tree onto the workspace. The green line in each figure indicates the current lead.

during exploration is used to further refine the discrete search and improve the quality of future explorations. `DSLX` was shown to offer considerable computational advantages over other methods across a variety of kinodynamic motion planning problems. Experimental results indicated that `DSLX` is capable of solving challenging motion planning problems two orders of magnitude faster than other widely used motion planning methods.

The combination of coarse-grained representation, discrete search, and continuous exploration in the framework of `DSLX` results in an effective motion planner that allocates most of the available computational time to the exploration of the parts of the state space that lead to a solution for a given motion planning problem. This combination raises many interesting research issues, such as finding the best workspace decomposition for a motion planning problem, improving the discrete search, continuous exploration, and interaction between the different components, that we intend to investigate in future research. We are currently investigating extensions to the theoretical framework developed in [27] to analyze `DSLX`. Furthermore, we would like to apply and extend the framework of `DSLX` to increasingly challenging and high-dimensional problems in motion planning and other settings, such as hybrid-system testing [28]. As we address these challenging problems, it becomes important to extend the framework [29], [30] to obtain an effective distribution of `DSLX` that makes use of all the available computational resources.

ACKNOWLEDGMENT

This work has been supported in part by NSF CNS 0615328 (EP, LEK, MYV), NSF 0308237 (EP, LEK), a Sloan Fellowship (LEK), and NSF CCF 0613889 (MYV). Experiments were carried out on equipment supported by NSF CNS 0454333 and NSF CNS 0421109 in partnership with Rice University, AMD, and Cray.

REFERENCES

[1] H. Choset, K. M. Lynch, S. Hutchinson, G. Kantor, W. Burgard, L. E. Kavraki, and S. Thrun, *Principles of Robot Motion: Theory, Algorithms, and Implementations*. Cambridge, MA: MIT Press, 2005.

[2] S. M. LaValle, *Planning Algorithms*. Cambridge, MA: Cambridge University Press, 2006.

[3] J. Bruce and M. Veloso, "Real-time randomized path planning for robot navigation," in Robocup 2002: Robot Soccer World Cup VI, *Lecture Notes in Computer Science*, G. A. Kaminka, P. U. Lima, and R. Rojas, Eds. Berlin, Heiderlberg: Springer, 2003, vol. 2752, pp. 288–295.

[4] L. Jaillet, A. Yershova, S. M. LaValle, and T. Simeon, "Adaptive tuning of the sampling domain for dynamic-domain RRTs," in *IEEE/RSJ International Conference on Intelligent Robots and Systems*, Edmonton, Canada, 2005, pp. 4086–4091.

[5] A. M. Ladd and L. E. Kavraki, "Motion planning in the presence of drift, underactuation and discrete system changes," in *Robotics: Science and Systems I*. Boston, MA: MIT Press, 2005, pp. 233–241.

[6] S. M. LaValle and J. J. Kuffner, "Randomized kinodynamic planning," in *IEEE International Conference on Robotics and Automation*, 1999, pp. 473–479.

[7] D. Hsu, R. Kindel, J.-C. Latombe, and S. Rock, "Randomized kinodynamic motion planning with moving obstacles," *International Journal of Robotics Research*, vol. 21, no. 3, pp. 233–255, 2002.

[8] B. Burns and O. Brock, "Single-query motion planning with utility-guided random trees," in *IEEE Interantional Conference on Robotics and Automation*, Rome, Italy, Apr. 2007.

[9] G. Sánchez and J.-C. Latombe, "On delaying collision checking in PRM planning: Application to multi-robot coordination," *International Journal of Robotics Research*, vol. 21, no. 1, pp. 5–26, 2002.

[10] E. Plaku, K. E. Bekris, B. Y. Chen, A. M. Ladd, and L. E. Kavraki, "Sampling-based roadmap of trees for parallel motion planning," *IEEE Transactions on Robotics*, vol. 21, no. 4, pp. 597–608, 2005.

[11] L. E. Kavraki, P. Švestka, J.-C. Latombe, and M. H. Overmars, "Probabilistic roadmaps for path planning in high-dimensional configuration spaces," *IEEE Transactions on Robotics and Automation*, vol. 12, no. 4, pp. 566–580, June 1996.

[12] B. Burns and O. Brock, "Toward optimal configuration space sampling," in *Robotics: Science and Systems*, Cambridge, MA, June 2005.

[13] M. Morales, L. Tapia, R. Pearce, S. Rodriguez, and N. M. Amato, "A machine learning approach for feature-sensitive motion planning," in *International Workshop on Algorithmic Foundations of Robotics*, M. Erdmann, D. Hsu, M. Overmars, and A. F. van der Stappen, Eds. Utrecht/Zeist, Netherlands: Springer-Verlag, July 2004, pp. 361–376.

[14] D. Hsu, G. Sánchez-Ante, and Z. Sun, "Hybrid PRM sampling with a cost-sensitive adaptive strategy," in *IEEE Interantional Conference on Robotics and Automation*. Barcelona, Spain: IEEE Press, May 2005, pp. 3885–3891.

[15] J.-C. Latombe, *Robot Motion Planning*. Boston, MA: Kluwer, 1991.

[16] K. Kondo, "Motion planning with six degrees of freedom by multistrategic bidirectional heuristic free-space enumeration," *IEEE Transactions on Robotics and Automation*, vol. 7, no. 3, pp. 267–277, 1991.

[17] J. Lengyel, M. Reichert, B. Donald, and P. Greenberg, "Real-time robot motion planning using rasterizing computer graphics hardware," *Computer Graphics (SIGGRAPH'90)*, pp. 327–335, 1990.

[18] R. Bohlin, "Path planning in practice: Lazy evaluation on a multi-resolution grid," in *IEEE/RSJ International Conference on Intelligent Robots and Systems*, 2001, pp. 49–54.

[19] C. Holleman and L. E. Kavraki, "A framework for using the workspace medial axis in PRM planners," in *IEEE Interantional Conference on Robotics and Automation*, 2000, pp. 1408–1413.

[20] M. Foskey, M. Garber, M. C. Lin, and D. Manocha, "A voronoi-based hybrid motion planner," in *IEEE/RSJ International Conference on Intelligent Robots and Systems*, vol. 1, Maui, HI, 2001, pp. 55–60.

[21] Y. Yang and O. Brock, "Efficient motion planning based on disassembly," in *Robotics: Science and Systems*, S. Thrun, G. S. Sukhatme, and S. Schaal, Eds. Cambridge, USA: MIT Press, June 2005.

[22] H. Kurniawati and D. Hsu, "Workspace-based connectivity oracle: An adaptive sampling strategy for PRM planning," in *International Workshop on Algorithmic Foundations of Robotics*, S. Akella *et al.*, Eds. New York, NY: Springer-Verlag, 2006, in press.

[23] W. Zhang, *State-space Search: Algorithms, Complexity, Extensions, and Applications*. New York, NY: Springer Verlag, 2006.

[24] R. Geraerts and M. Overmars, "A comparitive study of probabilistic roadmap planners," in *International Workshop on Algorithmic Foundations of Robotics*, J.-D. Boissonnat, J. Burdick, K. Goldberg, and S. Hutchinson, Eds. Springer-Verlag, 2002, pp. 43–58.

[25] E. Plaku, K. E. Bekris, and L. E. Kavraki, "OOPS for Motion Planning: An Online Open-source Programming System," in *IEEE International Conference on Robotics and Automation (ICRA)*, Rome, Italy, 2007, pp. 3711–3716.

[26] P. Cheng, E. Frazzoli, and S. M. LaValle, "Improving the performance of sampling-based planners by using a symmetry-exploiting gap reduction algorithm," in *IEEE Interantional Conference on Robotics and Automation*, 2004, pp. 4362–4368.

[27] A. M. Ladd and L. E. Kavraki, "Measure theoretic analysis of probabilistic path planning," *IEEE Transactions on Robotics and Automation*, vol. 20, no. 2, pp. 229–242, 2004.

[28] E. Plaku, L. E. Kavraki, and M. Y. Vardi, "Hybrid systems: From verification to falsification," in *International Conference on Computer Aided Verification*, W. Damm and H. Hermanns, Eds. Berlin, Germany: Lecture Notes in Computer Science, Springer-Verlag, 2007, vol. 4590, pp. 468–481.

[29] E. Plaku and L. E. Kavraki, "Distributed sampling-based roadmap of trees for large-scale motion planning," in *IEEE Inter. Conf. on Robotics and Automation*, Barcelona, Spain, 2005, pp. 3879–3884.

[30] ——, "Distributed computation of the knn graph for large high-dimensional point sets," *Journal of Parallel and Distributed Computing*, vol. 67, no. 3, pp. 346–359, 2007.

Active Policy Learning for Robot Planning and Exploration under Uncertainty

Ruben Martinez-Cantin*, Nando de Freitas†, Arnaud Doucet†, José A. Castellanos*
* Department of Computer Science and Systems Engineering, University of Zaragoza
Email: {rmcantin,jacaste}@unizar.es
† Department of Computer Science, University of British Columbia
Email: {nando,arnaud}@cs.ubc.ca

Abstract— **This paper proposes a simulation-based active policy learning algorithm for finite-horizon, partially-observed sequential decision processes. The algorithm is tested in the domain of robot navigation and exploration under uncertainty, where the expected cost is a function of the belief state (filtering distribution). This filtering distribution is in turn nonlinear and subject to discontinuities, which arise because constraints in the robot motion and control models. As a result, the expected cost is non-differentiable and very expensive to simulate. The new algorithm overcomes the first difficulty and reduces the number of simulations as follows. First, it assumes that we have carried out previous evaluations of the expected cost for different corresponding policy parameters. Second, it fits a Gaussian process (GP) regression model to these values, so as to approximate the expected cost as a function of the policy parameters. Third, it uses the GP predicted mean and variance to construct a statistical measure that determines which policy parameters should be used in the next simulation. The process is iterated using the new parameters and the newly gathered expected cost observation. Since the objective is to find the policy parameters that minimize the expected cost, this active learning approach effectively trades-off between exploration (where the GP variance is large) and exploitation (where the GP mean is low). In our experiments, a robot uses the proposed method to plan an optimal path for accomplishing a set of tasks, while maximizing the information about its pose and map estimates. These estimates are obtained with a standard filter for SLAM. Upon gathering new observations, the robot updates the state estimates and is able to replan a new path in the spirit of open-loop feedback control.**

I. Introduction

The direct policy search method for reinforcement learning has led to significant achievements in control and robotics [1, 2, 3, 4]. The success of the method does often, however, hinge on our ability to formulate expressions for the gradient of the expected cost [5, 4, 6]. In some important applications in robotics, such as exploration, constraints in the robot motion and control models make it hard, and often impossible, to compute derivatives of the cost function with respect to the robot actions. In this paper, we present a direct policy search method for continuous policy spaces that relies on active learning to side-step the need for gradients.

The proposed active policy learning approach also seems to be more appropriate in situations where the cost function has many local minima that cause the gradient methods to get stuck. Moreover, in situations where the cost function is very expensive to evaluate by simulation, an active learning approach that is designed to minimize the number of evaluations

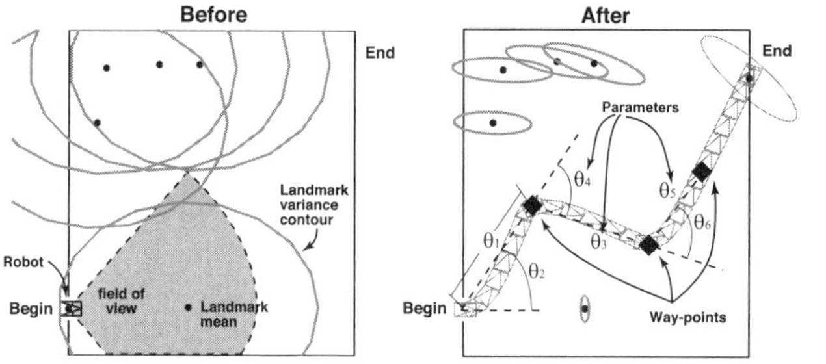

Fig. 1. The robot plans a path that allows it to accomplish the task of going from "Begin" to "End" while simultaneously reducing the uncertainty in the map and pose (robot location and heading) estimates. The robot has a prior over its pose and the landmark locations, but it can only see the landmarks within its field of view (left). In the planning stage, the robot must compute the best policy vector consisting of the three way-points describing the path. After running the stochastic planning algorithm proposed in this paper, the robot has reached the "End" while following a planned trajectory that minimizes the posterior uncertainty about its pose and map (right). The uncertainty ellipses are scaled for clarification.

might be more suitable than gradient methods, which often require small step sizes for stable convergence (and hence many cost evaluations).

We demonstrate the new approach on a hard robotics problem: planning and exploration under uncertainty. This problem plays a key role in simultaneous localization and mapping (SLAM), see for example [7, 8]. Mobile robots must maximize the size of the explored terrain, but, at the same time, they must ensure that localization errors are minimized. While exploration is needed to find new features, the robot must return to places were known landmarks are visible to maintain reasonable map and pose (robot location and heading) estimates.

In our setting, the robot is assumed to have a rough *a priori* estimate of the map features and its own pose. The robot must accomplish a series of tasks while simultaneously maximizing its information about the map and pose. This is illustrated in Figure 1, where a robot has to move from "Begin" to "End" by planning a path that satisfies logistic and physical constraints. The planned path must also result in improved map and pose estimates. *As soon as the robot accomplishes a task, it has a new a posteriori map that enables it to carry out future tasks in the same environment more efficiently.* This sequential decision making problem is exceptionally difficult because the actions and states are continuous and high-dimensional. Moreover, the cost function is not differentiable and depends on the posterior belief (filtering distribution). Even a toy problem requires enormous computational effort. As a result, it is not surprising that most existing approaches relax the constrains. For instance, full observability is assumed in [9, 7], known robot location is assumed in [10], myopic planning is adopted in [8], and discretization of the state and/or actions spaces appears in [11, 12, 7]. *The method proposed in this paper does not rely on any of these assumptions.*

Our direct policy solution uses an any-time probabilistic active learning algorithm to predict what policies are likely to result in higher expected returns. The method effectively balances the goals of exploration and exploitation in policy search. It is motivated by work on experimental design [13, 14, 15]. Simpler variations of our ideas appeared early in the reinforcement literature. In [16], the problem is treated in the framework of exploration/exploitation with bandits. An extension to continuous spaces (infinite number of bandits) using locally weighted regression was proposed in [17]. Our paper presents richer criteria for active learning as well suitable optimization objectives.

This paper also presents posterior Cramér-Rao bounds to approximate the cost function in robot exploration. The appeal of these bounds is that they are much cheaper to simulate than the actual cost function.

Although the discussion is focused on robot exploration and planning, our policy search framework extends naturally to other domains. Related problems appear the fields of terrain-aided navigation [18, 9] and dynamic sensor nets [19, 6].

II. Application to Robot Exploration and Planning

Although the algorithm proposed in this paper applies to many sequential decision making settings, we will restrict attention to the robot exploration and planning domain. In this domain, the robot has to plan a path that will improve its knowledge of its pose (location and heading) and the location of navigation landmarks. In doing so, the robot might be subject to other constraints such as low energy consumption, limited time, safety measures and obstacle avoidance. However, for the time being, let us first focus on the problem of minimizing posterior errors in localization and mapping as this problem already captures a high degree of complexity.

There are many variations of this problem, but let us consider the one of Figure 1 for illustration purposes. Here, the robot has to navigate from "Begin" to "End" while improving its estimates of the map and pose. For the time being, let us assume that the robot has no problem in reaching the target. Instead, let us focus on how the robot should plan its path so as to improve its map and pose posterior estimates. Initially, as illustrated by the ellipses on the left plot, the robot has vague priors about its pose and the location of landmarks. We want the robot to plan a path (parameterized policy $\pi(\boldsymbol{\theta})$) so that by the time it reaches the target, it has learned the most about its pose and the map. This way, *if the robot has to repeat the task, it will have a better estimate of the map and hence it will be able to accomplish the task more efficiently.*

In this paper, the policy is simply a path parameterized as a set of ordered way-points $\boldsymbol{\theta}_i$, although different representations can be used depending on the robot capabilities. A trajectory with 3 way-points, whose location was obtained using our algorithm, is shown on the right plot of Figure 1. We use a standard proportional-integral-derivative (PID) controller to generate the motion commands $\mathbf{a} = \{\mathbf{a}_{1:T}\}$ to follow the path for T steps. The controller moves the robot toward each way-point in turn while taking into account the kinematic and dynamic constrains of the problem.

It should be noticed that the robot has a limited field of view. It can only see the landmarks that "appear" within an *observation gate*.

Having restricted the problem to one of improving posterior pose and map estimates, a natural cost function is the average mean square error (AMSE) of the state:

$$C^{\pi}_{AMSE} = \mathbb{E}_{p(\mathbf{x}_{0:T},\mathbf{y}_{1:T}|\boldsymbol{\pi})}\left[\sum_{t=1}^{T} \lambda^{T-t}(\widehat{\mathbf{x}}_t - \mathbf{x}_t)(\widehat{\mathbf{x}}_t - \mathbf{x}_t)'\right],$$

where $\widehat{\mathbf{x}}_t = \mathbb{E}_{p(\mathbf{x}_t|\mathbf{y}_{1:t},\boldsymbol{\pi})}[\mathbf{x}_t]$. The expectation is with respect to $p(\mathbf{x}_{0:T},\mathbf{y}_{1:T}|\boldsymbol{\pi}) = p(\mathbf{x}_0)\prod_{t=1}^{T} p(\mathbf{x}_t|\mathbf{a}_t,\mathbf{x}_{t-1})p(\mathbf{y}_t|\mathbf{x}_t,\mathbf{a}_t)$, $\lambda \in [0,1]$ is a discount factor, $\pi(\boldsymbol{\theta})$ denotes the policy parameterized by the way-points $\boldsymbol{\theta}_i \in \mathbb{R}^{n_\theta}$, $\mathbf{x}_t \in \mathbb{R}^{n_x}$ is the hidden state (robot pose and location of map features) at time t, $\mathbf{y}_{1:T} = \{\mathbf{y}_1,\mathbf{y}_2,\ldots,\mathbf{y}_T\} \in \mathbb{R}^{n_y T}$ is the history of observations along the planned trajectory for T steps, $\mathbf{a}_{1:T} \in \mathbb{R}^{n_a T}$ is the history of actions determined by the policy $\pi(\boldsymbol{\theta})$ and $\widehat{\mathbf{x}}_t$ is the posterior estimate of the state at time t.

In our application to robotics, we focus on the uncertainty of the posterior estimates at the end of the planning horizon. That is, we set λ so that the cost function reduces to:

$$C^{\pi}_{AMSE} = \mathbb{E}_{p(\mathbf{x}_T,\mathbf{y}_{1:T}|\boldsymbol{\pi})}\left[(\widehat{\mathbf{x}}_T - \mathbf{x}_T)(\widehat{\mathbf{x}}_T - \mathbf{x}_T)'\right], \quad (1)$$

Note that the true state $\mathbf{x}_T$ and observations are unknown in advance and so one has to marginalize over them.

The cost function hides an enormous degree of complexity. It is a matrix function of an *intractable filtering distribution* $p(\mathbf{x}_T|\mathbf{y}_{1:T},\boldsymbol{\pi})$ (also known as the belief or information state). This belief can be described in terms of the observation and odometry (robot dynamics) models using marginalization and Bayes rule. The computation of this belief is known as the simultaneous localization and mapping problem (SLAM) and it is known to be notoriously hard because of nonlinearity and non-Gaussianity. Moreover, in our domain, the robot only sees the landmarks within and observation gate.

Since the models are not linear-Gaussian, one cannot use standard linear-quadratic-Gaussian (LQG) controllers [20] to solve our problem. Moreover, since the action and state spaces are large-dimensional and continuous, one cannot discretize the problem and use closed-loop control as suggested in [21]. That is, the discretized partially observed Markov decision process is too large for stochastic dynamic programming [22].

As a result of these considerations, we adopt the direct policy search method [23, 24]. In particular, the initial policy is set either randomly or using prior knowledge. Given this policy, we conduct simulations to estimate the AMSE. These simulations involve sampling states and observations using the prior, dynamic and observation models. They also involve estimating the posterior mean of the state with suboptimal filtering. After evaluating the AMSE using the simulated

1) Choose an initial policy π_0.
2) For $j = 1 : MaxNumberOfPolicySearchIterations$:
   a) For $i = 1 : N$:
      i) Sample the prior states $\mathbf{x}_0^{(i)} \sim p(\mathbf{x}_0)$.
      ii) For $t = 1 : T$:
         A) Use a PID controller regulated about the path π_j to determine the current action $\mathbf{a}_t^{(i)}$.
         B) Sample the state $\mathbf{x}_t^{(i)} \sim p(\mathbf{x}_t|\mathbf{a}_t^{(i)}, \mathbf{x}_{t-1}^{(i)})$.
         C) Generate observations $\mathbf{y}_t^{(i)} \sim p(\mathbf{y}_t|\mathbf{a}_t^{(i)}, \mathbf{x}_t^{(i)})$ as described in Section II-A. There can be missing observations.
         D) Compute the filtering distribution $p(\mathbf{x}_t|\mathbf{y}_{1:t}^{(i)}, \mathbf{a}_{1:t}^{(i)})$ using a SLAM filter.
   b) Evaluate the approximate AMSE cost function of equation (2) using the simulated trajectories.
   c) Use the active learning algorithm with Gaussian processes, described in Section III, to generate the new policy π_{j+1}. The choice of the new policy is governed by our desire to exploit and our need to explore the space of policies (navigation paths). In particular, we give preference to policies for which we expect the cost to be minimized and to policies where we have high uncertainty about what the cost might be.

Fig. 2. The overall solution approach in the open-loop control (OLC) setting. Here, N denotes the number of Monte Carlo samples and T is the planning horizon. In replanning with open-loop feedback control (OLFC), one simply uses the present position and the estimated posterior distribution (instead of the prior) as the starting point for the simulations. One can apply this strategy with either approaching or receding control horizons. It is implicit in the pseudo-code that we freeze the random seed generator so as to reduce variance.

trajectories, we update the policy parameters and iterate with the goal of minimizing the AMSE. Note that in order to reduce Monte Carlo variance, the random seed should be frozen as described in [24]. The pseudo-code for this open-loop simulation-based controller (OLC) is shown in Figure 2.

Note that as the robot moves along the planned path, it is possible to use the newly gathered observations to update the posterior distribution of the state. This distribution can then be used as the prior for subsequent simulations. This process of replanning is known as open-loop feedback control (OLFC) [20]. We can also allow for the planning horizon to recede. That is, as the robot moves, it keeps planning T steps ahead of its current position. This control framework is also known as receding-horizon model-predictive control [25].

In the following two subsections, we will describe a way of conducting the simulations to estimate the AMSE. The active policy update algorithm will be described in Section III.

A. *Simulation of the cost function*

We can approximate the AMSE cost by simulating N state and observation trajectories $\{\mathbf{x}_{1:T}^{(i)}, \mathbf{y}_{1:T}^{(i)}\}_{i=1}^N$ and adopting the Monte Carlo estimator:

$$C_{AMSE}^{\pi} \approx \frac{1}{N}\sum_{i=1}^{N}(\widehat{\mathbf{x}}_T^{(i)} - \mathbf{x}_T^{(i)})(\widehat{\mathbf{x}}_T^{(i)} - \mathbf{x}_T^{(i)})'. \quad (2)$$

Assuming that π is given (we discuss the active learning algorithm to learn π in Section III), one uses a PID controller to obtain the next action $\mathbf{a}_t$. The new state $\mathbf{x}_t$ is easily simulated using the odometry model. The process of generating observations is more involved. As shown in Figure 3, for each landmark, one draws a sample from its posterior. If the sample falls within the observation gate, it is treated as an observation. As in most realistic settings, most landmarks will remain unobserved.

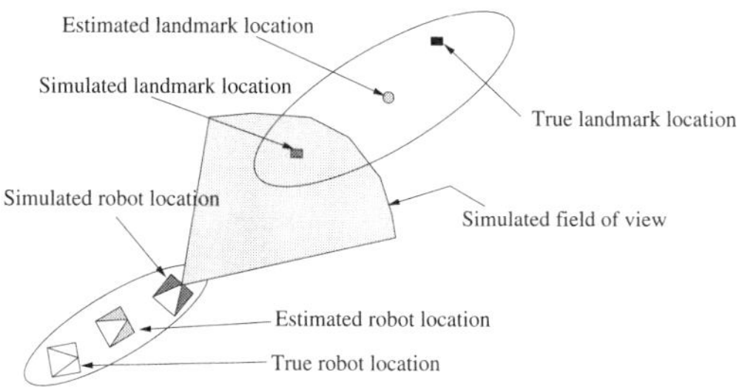

Fig. 3. An observation is generated using the current map and robot pose estimates. Gating information is used to validate the observation. In this picture, the simulation validates the observation despite the fact that the true robot and feature locations are too distant for the given field of view. New information is essential to reduce the uncertainty and improve the simulations.

After the trajectories $\{\mathbf{x}_{1:T}^{(i)}, \mathbf{y}_{1:T}^{(i)}\}_{i=1}^N$ are obtained, one uses a SLAM filter (EKF, UKF or particle filter) to compute the posterior mean state $\widehat{\mathbf{x}}_{1:T}^{(i)}$. (In this paper, we adopt the EKF-SLAM algorithm to estimate the mean and covariance of this distribution. We refer the reader to [26] for implementation details.) *The evaluation of the cost function is therefore extremely expensive.* Moreover, since the model is nonlinear, it is hard to quantify the uncertainty introduced by the suboptimal filter. Later, in Section IV, we will discuss an alternative cost function, which consists of a lower bound on the AMSE. Yet, in both cases, it is imperative to minimize the number of evaluations of the cost functions. This calls for an active learning approach.

III. Active Policy Learning

This section presents an active learning algorithm to update the policy parameters after each simulation. In particular, we adopt the expected cost simulation strategy presented in [24]. In this approach, a scenario consists of an initial choice of the state and a sequence of random numbers. Given a policy parameter vector and a set of fixed scenarios, the simulation is deterministic and yields an empirical estimate of the expected cost [24].

The simulations are typically very expensive and consequently cannot be undertaken for many values of the policy parameters. Discretization of the potentially high-dimensional and continuous policy space is out of the question. The standard solution to this problem is to optimize the policy using gradients. However, the local nature of gradient-based optimization often leads to the common criticism that direct policy search methods "get stuck" in local minima. Even more pertinent to our setting, is the fact that the cost function is discontinuous and hence policy gradient methods do not apply. We present an alternative approach to gradient-based optimization for continuous policy spaces. This approach, which we refer to as active policy learning, is based on experimental design ideas [27, 13, 28, 29]. Active policy learning is an any-time, "black-box" statistical optimization

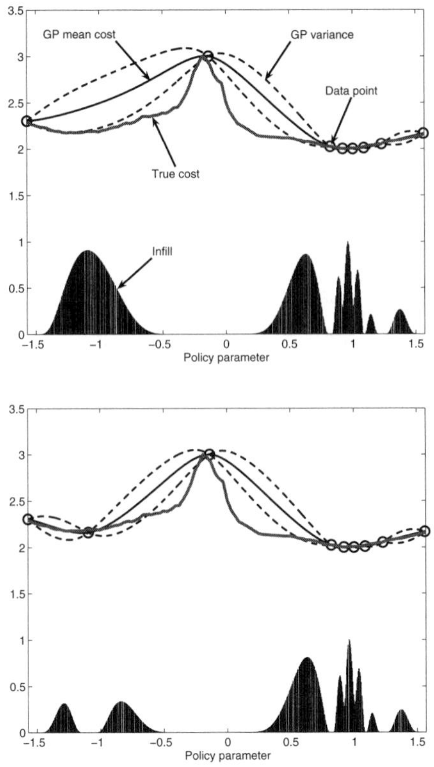

Fig. 4. An example of active policy learning with a univariate policy using data generated by our simulator. The figure on top shows a GP approximation of the cost function using 8 simulated values. The variance of the GP is low, but not zero, at these simulated values. In reality, the true expected cost function is unknown. The figure also shows the expected improvement (infill) of each potential next sampling location in the lower shaded plot. The infill is high where the GP predicts a low expected cost (exploitation) and where the prediction uncertainty is high (exploration). Selecting and labelling the point suggested by the highest infill in the top plot produces the GP fit in the plot shown below. The new infill function, in the plot below, suggests that we should query a point where the cost is expected to be low (exploitation).

approach. Figure 4 illustrates it for a simple one-dimensional example. The approach is iterative and involves three steps.

In the first step, a Bayesian regression model is learned to map the policy parameters to the estimates of the expected cost function obtained from previous simulations. In this work, the regression function is obtained using Gaussian processes (GPs). Though in Figure 4 the GPs provide a good approximation to the expected cost, it should be emphasized that the objective is not to predict the value of the regression surface over the entire feasible domain, but rather to predict it well near the minima. The details of the GP fit are presented in Section III-A.

The second step involves active learning. Because the simulations are expensive, we must ensure that the selected samples (policy parameter candidates) will generate the maximum possible improvement. Roughly speaking, it is reasonable to sample where the GP predicts a low expected cost (exploitation) or where the GP variance is large (exploration). These intuitions can be incorporated in the design of a statistical measure indicating where to sample. This measure is known as the infill function, borrowing the term from the geostatistics literature. Figure 4 depicts a simple infill function that captures our intuitions. More details on how to choose the infill are presented in Section III-B.

Having defined an infill function, still leaves us with the problem of optimizing it. This is the third and final step in the approach. Our thesis is that the infill optimization problem is more amenable than the original problem because in this case the cost function is known and easy to evaluate. Furthermore, for the purposes of our application, it is not necessary to guarantee that we find the global minimum, merely that we can quickly locate a point that is likely to be as good as possible.

To deal with this nonlinear constrained optimization problem, we adopted the DIvided RECTangles (DIRECT) algorithm [30, 31]. DIRECT is a deterministic, derivative-free sampling algorithm. It uses the existing samples of the objective function to decide how to proceed to divide the feasible space into finer rectangles. For low-dimensional parameter spaces, say up to 10D, DIRECT provides a better solution than gradient approaches because the infill function tends to have many local optima. Another motivating factor is that DIRECT's implementation is easily available [32]. However, we conjecture that for large dimensional spaces, sequential quadratic programming or concave-convex programming [33] might be better algorithm choices for infill optimization.

A. *Gaussian processes*

A *Gaussian process*, $\mathbf{z}(\cdot) \sim GP(m(\cdot), K(\cdot,\cdot))$, is an infinite random process indexed by the vector $\boldsymbol{\theta}$, such that any realization $\mathbf{z}(\boldsymbol{\theta})$ is Gaussian [34]. We can parameterize the GP hierarchically

$$\begin{aligned} C^{\pi}(\boldsymbol{\theta}) &= \mathbf{1}\mu + \mathbf{z}(\boldsymbol{\theta}) \\ \mathbf{z}(\cdot) &\sim GP(0, \sigma^2 K(\cdot,\cdot)) \end{aligned}$$

and subsequently estimate the posterior distributions of the mean μ and scale σ^2 using standard Bayesian conjugate analysis, see for example [14]. The symbol $\mathbf{1}$ denotes a column vector of ones. Assuming that n simulations have been conducted, the simulated costs $\{C^{\pi}_{1:n}\}$ and the predicted cost C^{π}_{n+1} for a new test point $\boldsymbol{\theta}_{n+1}$ are jointly Gaussian:

$$\begin{bmatrix} C^{\pi}_{n+1} \\ C^{\pi}_{1:n} \end{bmatrix} \sim \mathcal{N}\left(\begin{bmatrix} 1 \\ \mathbf{1} \end{bmatrix} \mu, \sigma^2 \begin{bmatrix} k & \mathbf{k}^T \\ \mathbf{k} & \mathbf{K} \end{bmatrix} \right),$$

where $\mathbf{k}^T = [k(\boldsymbol{\theta}_{n+1}, \boldsymbol{\theta}_1) \cdots k(\boldsymbol{\theta}_{n+1}, \boldsymbol{\theta}_n)]$, $k = k(\boldsymbol{\theta}_{n+1}, \boldsymbol{\theta}_{n+1})$ and $\mathbf{K}$ is the training data kernel matrix with entries $k(\boldsymbol{\theta}_i, \boldsymbol{\theta}_j)$ for $i = 1, \ldots, n$ and $j = 1, \ldots, n$. Since we are interested in regression, the Matern kernel is a suitable choice for $k(\cdot|\cdot)$ [14].

We assign a normal-inverse-Gamma conjugate prior to the parameters: $\mu \sim \mathcal{N}(0, \sigma^2\delta^2)$ and $\sigma^2 \sim \mathcal{IG}(a/2, b/2)$. The priors play an essential role at the beginning of the design process, when there are only a few data. Classical Bayesian analysis allow us to obtain analytical expressions for the posterior modes of these quantities:

$$\begin{aligned} \widehat{\mu} &= (\mathbf{1}^T\mathbf{K}^{-1}\mathbf{1} + \delta^{-2})^{-1}\mathbf{1}^T\mathbf{K}^{-1}C^{\pi} \\ \widehat{\sigma}^2 &= \frac{b + C^{\pi T}\mathbf{K}^{-1}C^{\pi} - (\mathbf{1}^T\mathbf{K}^{-1}\mathbf{1} + \delta^{-2})\widehat{\mu}^2}{n + a + 2} \end{aligned}$$

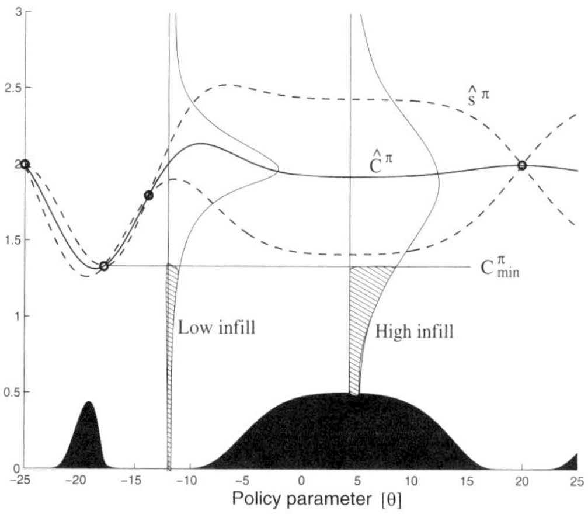

Fig. 5. Infill function.

Using the previous estimates, the GP predictive mean and variance are given by

$$\begin{aligned}\widehat{C}^{\pi}(\boldsymbol{\theta}) &= \widehat{\mu} + \mathbf{k}^T\mathbf{K}^{-1}(C^{\pi}_{1:n} - \mathbf{1}\widehat{\mu})\\ \widehat{s}^2(\boldsymbol{\theta}) &= \widehat{\sigma}^2\left\{k - \mathbf{k}^T\mathbf{K}^{-1}\mathbf{k} + \frac{(1-\mathbf{1}^T\mathbf{K}^{-1}\mathbf{k})^2}{(\mathbf{1}^T\mathbf{K}^{-1}\mathbf{1}+\delta^{-2})}\right\}.\end{aligned}$$

Since the number of query points is small, the GP predictions are very easy to compute.

B. Infill Function

Let $C^{\pi}_{\min}$ denote the current lowest (best) estimate of the cost function. As shown in Figure 5, we can define the probability of improvement at a point $\boldsymbol{\theta}$ to be

$$p(C^{\pi}(\boldsymbol{\theta}) \le C^{\pi}_{\min}) = \Phi\left(\frac{C^{\pi}_{\min} - \widehat{C}^{\pi}(\boldsymbol{\theta})}{\widehat{s}(\boldsymbol{\theta})}\right),$$

where $C^{\pi}(\boldsymbol{\theta}) \sim \mathcal{N}(\widehat{C}^{\pi}(\boldsymbol{\theta}), \widehat{s}(\boldsymbol{\theta})^2)$ and Φ denotes CDF of the standard Normal distribution. This measure was proposed several decades ago by [27], who used univariate Wiener process. However, as argued by [13], it is sensitive to the value of $C^{\pi}_{\min}$. To overcome this problem, Jones defined the improvement over the current best point as $I(\boldsymbol{\theta}) = \max\{0, C^{\pi}_{\min} - C^{\pi}(\boldsymbol{\theta})\}$. This resulted in the following expected improvement (infill function):

$$EI(\boldsymbol{\theta}) = \begin{cases}(C^{\pi}_{\min} - \widehat{C}^{\pi}(\boldsymbol{\theta}))\Phi(d) + \widehat{s}(\boldsymbol{\theta})\phi(d) & \text{if } \widehat{s} > 0\\ 0 & \text{if } \widehat{s} = 0\end{cases}$$

where ϕ is the PDF of the standard Normal distribution and

$$d = \frac{C^{\pi}_{\min} - \widehat{C}^{\pi}(\boldsymbol{\theta})}{\widehat{s}(\boldsymbol{\theta})}.$$

IV. A Cheaper Cost: The Posterior Cramér-Rao Bound

As mentioned in Section II-A, it is not possible to compute the AMSE cost function exactly. In that section, we proposed a simulation approach that required that we run an SLAM filter for each simulated scenario. This approximate filtering step is not only expensive, but also a possible source of errors when approximating the AMSE with Monte Carlo simulations.

The posterior Cramér-Rao bound (PCRB) for nonlinear systems leads to an alternative objective function that is cheaper to evaluate and *does not require that we run a SLAM filter*. That is, the criterion presented next does not require the adoption of an EKF, UKF, particle filter or any other suboptimal filter in order to evaluate it. The PCRB is a "measure" of the maximum information that can be extracted from the dynamic system when both the measurements and states are assumed random. It is defined as the inverse of the Fisher information matrix $\mathbf{J}$ and provides the following lower bound on the AMSE:

$$C^{\pi}_{AMSE} \ge C^{\pi}_{PCRB} = \mathbf{J}^{-1}$$

Tichavský [35], derived the following Riccati-like recursion to compute the PCRB for any unbiased estimator:

$$\mathbf{J}_{t+1} = \mathbf{D}_t - \mathbf{C}'_t(\mathbf{J}_t + \mathbf{B}_t)^{-1}\mathbf{C}_t + \mathbf{A}_{t+1}, \qquad (3)$$

where

$$\begin{aligned}\mathbf{A}_{t+1} &= \mathbb{E}[-\Delta_{x_{t+1},x_{t+1}} \log p(\mathbf{y}_{t+1}|\mathbf{x}_{t+1})]\\ \mathbf{B}_t &= \mathbb{E}[-\Delta_{x_t,x_t} \log p(\mathbf{x}_{t+1}|\mathbf{x}_t,\mathbf{a}_t)]\\ \mathbf{C}_t &= \mathbb{E}[-\Delta_{x_t,x_{t+1}} \log p(\mathbf{x}_{t+1}|\mathbf{x}_t,\mathbf{a}_t)]\\ \mathbf{D}_t &= \mathbb{E}[-\Delta_{x_{t+1},x_{t+1}} \log p(\mathbf{x}_{t+1}|\mathbf{x}_t,\mathbf{a}_t)],\end{aligned}$$

where the expectations are with respect to the simulated trajectories and Δ denotes the Laplacian operator. By simulating (sampling) trajectories, using our observation and transition models, one can easily approximate these expectations with Monte Carlo averages. *These averages can be computed offline and hence the expensive recursion of equation (3) only needs to be done once for all scenarios.*

The PCRB approximation method of [35] applies to nonlinear (NL) models with additive noise only. This is not the case in our setting and hence a potential source of error. An alternative PCRB approximation method that overcomes this shortcoming, in the context of jump Markov linear (JML) models, was proposed by [36]. We try both approximations in our experiments and refer to them as NL-PCRB and JML-PCRB respectively.

The AMSE simulation approach of Section II-A using the EKF requires that we perform an expensive Ricatti update (EKF covariance update) for each simulated trajectory. In contrast, the simulation approach using the PCRB only requires one Ricatti update (equation (3)). Thus, the latter approach is considerably cheaper. Yet, the PCRB is only a lower bound and hence it is not guaranteed to be necessarily tight. In the following section, we will provide empirical comparisons between these simulation approaches.

V. Experiments

We present two sets of experiments. The first experiment is very simple as it is aimed at illustrating the approach. It involves a fixed-horizon stochastic planning domain. The second set of experiments is concerned with exploration with receding horizon policies in more realistic settings. In all cases, the aim is to find the optimal path in terms of posterior information about the map and robot pose. For clarification, other terms contributing to the cost, such as time and obstacles are not considered, but the implementation should be straightforward.

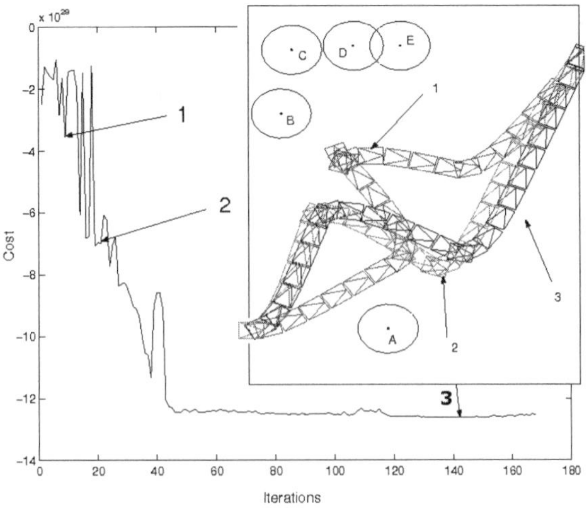

Fig. 6. Empirical AMSE cost as a function of policy improvement iterations. For this 6D parameter space, the solution converges to the minimum in few iterations, while allowing for several exploration steps in the first 40 iterations. The figure also shows the actual computed trajectories at three different iteration steps.

A. *Fixed-horizon planning*

The first experiment is the one described in Figure 1. Here, the start and end positions of the path are fixed. The robot has to compute the coordinates of three intermediate way-points and, hence, the policy has six parameters. For illustration purposes we chose a simple environment consisting of 5 landmarks (with vague priors). We placed an informative prior on the initial robot pose. Figure 6 shows three different robot trajectories computed during policy optimization. The trajectories are also indicated in the Monte Carlo AMSE cost evolution plot. The 6D optimization requires less than 50 iterations. We found that the optimal trajectory allowed the robot to observe the maximum number of features. However, since the prior on the robot's initial pose is informative (narrow Gaussian), feature A is originally detected with very low uncertainty. Consequently, the robot tries to maintain that feature in the field of view to improve the localization. A greedy strategy would have focused only on feature A, improving the estimation of that feature and the robot, but dropping the global posterior estimate.

B. *Receding-horizon planning*

In this experiment, the a priori map has high uncertainty (1 meter standard deviation – see Figure 7). The robot is a differential drive vehicle equipped with odometers and a stereo camera that provides the location of features. The field of view is limited to 7 meters and 90°, which are typical values for reliable stereo matching. We assume that the camera and a detection system that provides a set of observations every 0.5 seconds. The sensor noise is Gaussian for both range and bearing, with standard deviations $\sigma_{range} = 0.2 \cdot range$ and $\sigma_{bearing} = 0.5^o$. The policy is given by a set of ordered way-points. Each way-point is defined in terms of heading and distance with respect to the robot pose at the preceding way-point. The distance between way-points is limited to 10 meters and the heading should be in the interval $[-3\pi/4, 3\pi/4]$

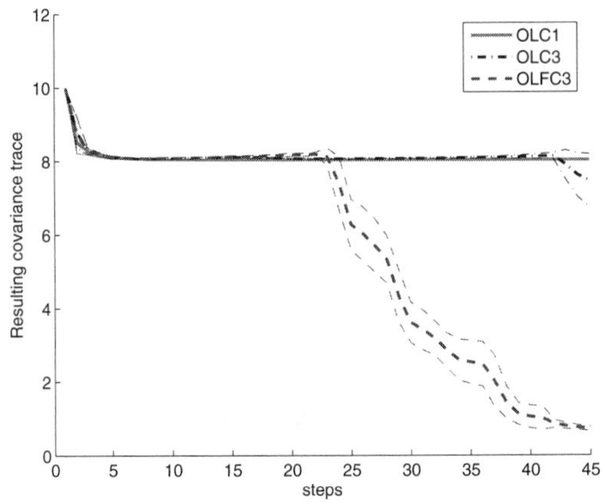

Fig. 8. Evolution of the trace of the state covariance matrix for 15 runs using OLC1, OLC3 and OLFC3, with 99% confidence intervals, for the map in Figure 7.

to avoid backwards trajectories. The motion commands are computed by a PID controller to guarantee that the goal is reached in 10 seconds.

First, we compare the behavior of the robot using different planning and acting horizons. The three methods that we implemented are:

OLC1 : This is an open loop greedy algorithm that plans with only 1 way-point ahead. Once the way-point is reached, the robot replans a path. Strictly speaking this is a hybrid of OLC and OLFC as replanning using actual observations takes place. For simplicity, however, we refer to it as OLC.

OLC3 : This is an open loop algorithm that plans with 3 way-points ahead. Once the third way-point is reached, the robot replans using actual observations.

OLFC3 This is an open loop feedback controller with a receding horizon. The planning horizon is 3 way-points, but the execution horizon is only 1 step. Thus, the last 2 way-points plus a new way-point are recomputed after a way-point is reached.

It is obvious that the OLC algorithms have a lower computational cost. Using the AMSE cost and the map of Figure 7, the times for OLC1, OLC3 and OLFC3 are approximately 6, 30 and 75 minutes (using an un-optimized Matlab implementation). On the other hand, the OLC methods can get trapped in local minima, as shown in Figure 7. Due to the limited planning horizon of OLC1, it barely explores new areas. OLC3 tends to overshoot as it only replans at the third way-point. OLFC3, on the other hand, replans at each step and as a result is able to steer to the unexplored part of the map. Figure 8 plots the evolution of the uncertainty in this trap situation for 15 experimental runs. The controller with feedback is clearly the winner because it avoids the trap. This behavior is stable across different runs.

We repeated this experiment with different initial random maps (5 landmarks). Figure 9 shows the methods perform similarly as worst-case situations are rare.

The next set of simulations is used to experimentally

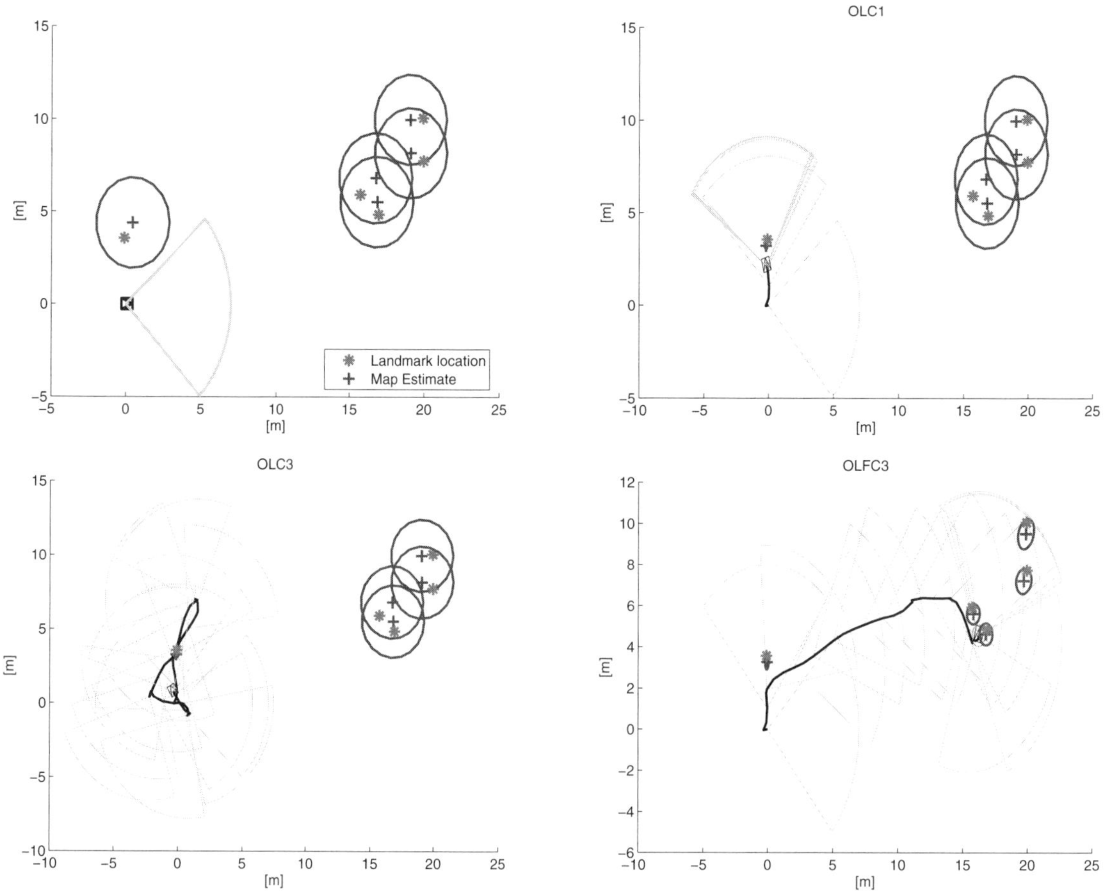

Fig. 7. Trajectories generated using OLC1, OLC3 and OLFC3. The blue and red ellipses represent the landmark and robot location 95% confidence intervals. The robot field of view is shown in green. OLC3 is more exploratory than OLC1, which "gets stuck" repeatedly updating the first landmark it encounters. Yet, only OLFC3, because of being able to replan at each step, is able to fully explore the map and reduce the uncertainty.

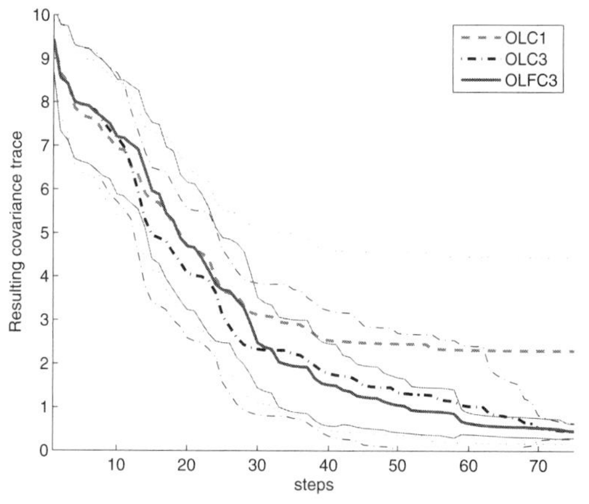

Fig. 9. Evolution of the trace of the state covariance matrix for 15 random maps using OLC1, OLC3 and OLFC3, with 99% confidence intervals.

validate the PCRB approximation of the AMSE cost. We increase the size and the complexity of the environment to 30 landmarks in a 25 by 25 meters squared area. Figure 10 shows the trace of the covariance matrix of the map and robot location, estimated using OLFC3, for the three approximations discussed in this paper. The JML-PCRB remains close to the simulated AMSE. This indicates that this bound is tight and a good choice in this case. On the other hand, NL-PCRB seems to be too loose. In this larger map, the computational times for the approximate AMSE and JML-PCRB were approximately 180 and 150 minutes respectively.

VI. Discussion and Future Work

We have presented an approach for stochastic exploration and planning rooted in strong statistical and decision-theoretic foundations. The most important next step is to test the proposed simulator on a real robotic domain. One needed step in this implementation is enabling the robot to add new landmarks to the existing map, while still within our decision-theoretic framework. We also note that our method is directly applicable to the problem of planning the architecture of a dynamic sensor network. In terms of modelling, we need to

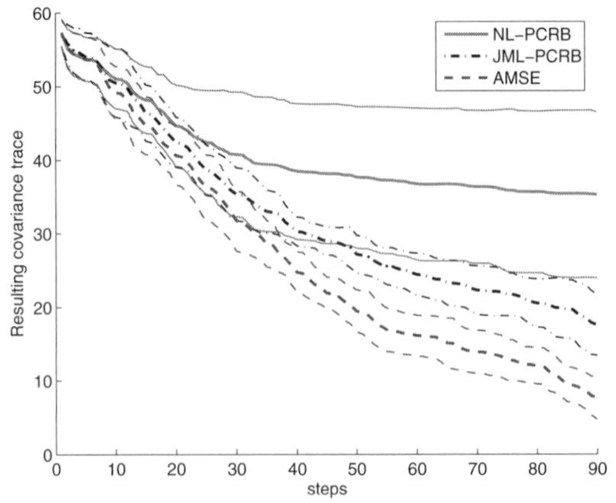

Fig. 10. Evolution of the trace of the state covariance matrix for 10 random maps with the AMSE, NL-PCRB and JML-PCRB cost functions, while using OLFC3.

introduce richer cost functions and constraints. In terms of algorithm improvement, we must design infill optimization strategies for high-dimensional policies. Whenever gradients are available, the approach presented here could be improved by ensuring that the regression function matches the gradients at the query points. Finally, on the theoretical front, we plan to build upon early work on correlated bandits to obtain theoretical performance bounds and, hence, confidence intervals that could potentially do better than the current infill criterion.

ACKNOWLEDGMENTS

We would like to thank Eric Brochu, Nick Roy, Hendrik Kueck and Andrew Ng for valuable discussions. This project was supported by NSERC, MITACS and the Dirección General de Investigación of Spain, project DPI2006-13578.

REFERENCES

[1] N. Kohl and P. Stone, "Policy gradient reinforcement learning for fast quadrupedal locomotion," in *IEEE International Conference on Robotics and Automation*, 2004.

[2] G. Lawrence, N. Cowan, and S. Russell, "Efficient gradient estimation for motor control learning," in *Uncertainty in Artificial Intelligence*. San Francisco, CA: Morgan Kaufmann, 2003, pp. 354–36.

[3] A. Ng, A. Coates, M. Diel, V. Ganapathi, J. Schulte, B. Tse, E. Berger, and E. Liang, "Inverted autonomous helicopter flight via reinforcement learning," in *International Symposium on Experimental Robotics*, 2004.

[4] J. Peters and S. Schaal, "Policy gradient methods for robotics," in *IEEE International Conference on Intelligent Robotics Systems*, 2006.

[5] J. Baxter and P. L. Bartlett, "Infinite-horizon policy-gradient estimation," *Journal of Artificial Intelligence Research*, vol. 15, pp. 319–350, 2001.

[6] S. Singh, N. Kantas, A. Doucet, B. N. Vo, and R. J. Evans, "Simulation-based optimal sensor scheduling with application to observer trajectory planning," in *IEEE Conference on Decision and Control and European Control Conference*, 2005, pp. 7296– 7301.

[7] R. Sim and N. Roy, "Global A-optimal robot exploration in SLAM," in *Proceedings of the IEEE International Conference on Robotics and Automation (ICRA)*, Barcelona, Spain, 2005.

[8] C. Stachniss, G. Grisetti, and W. Burgard, "Information gain-based exploration using Rao-Blackwellized particle filters," in *Proceedings of Robotics: Science and Systems*, Cambridge, USA, June 2005.

[9] S. Paris and J. P. Le Cadre, "Planification for terrain-aided navigation," in *Fusion 2002*, Annapolis, Maryland, July 2002, pp. 1007–1014.

[10] C. Leung, S. Huang, G. Dissanayake, and T. Forukawa, "Trajectory planning for multiple robots in bearing-only target localisation," in *IEEE International Conference on Intelligent Robotics Systems*, 2005.

[11] M. L. Hernandez, "Optimal sensor trajectories in bearings-only tracking," in *Proceedings of the Seventh International Conference on Information Fusion*, P. Svensson and J. Schubert, Eds., vol. II. Mountain View, CA: International Society of Information Fusion, Jun 2004, pp. 893–900.

[12] T. Kollar and N. Roy, "Using reinforcement learning to improve exploration trajectories for error minimization," in *Proceedings of the IEEE International Conference on Robotics and Automation (ICRA)*, 2006.

[13] D. R. Jones, M. Schonlau, and W. J. Welch, "Efficient global optimization of expensive black-box functions," *Journal of Global Optimization*, vol. 13, no. 4, pp. 455–492, 1998.

[14] T. J. Santner, B. Williams, and W. Notz, *The Design and Analysis of Computer Experiments*. Springer-Verlag, 2003.

[15] E. S. Siah, M. Sasena, and J. L. Volakis, "Fast parameter optimization of large-scale electromagnetic objects using DIRECT with Kriging metamodeling," *IEEE Transactions on Microwave Theory and Techniques*, vol. 52, no. 1, pp. 276–285, January 2004.

[16] L. P. Kaelbling, "Learning in embedded systems," Ph.D. dissertation, Stanford University, 1990.

[17] A. W. Moore and J. Schneider, "Memory-based stochastic optimization," in *Advances in Neural Information Processing Systems*, D. S. Touretzky, M. C. Mozer, and M. E. Hasselmo, Eds., vol. 8. The MIT Press, 1996, pp. 1066–1072.

[18] N. Bergman and P. Tichavský, "Two Cramér-Rao bounds for terrain-aided navigation," *IEEE Transactions on Aerospace and Electronic Systems*, 1999, in review.

[19] M. L. Hernandez, T. Kirubarajan, and Y. Bar-Shalom, "Multisensor resource deployment using posterior Cramèr-Rao bounds," *IEEE Transactions on Aerospace and Electronic Systems Aerospace*, vol. 40, no. 2, pp. 399– 416, April 2004.

[20] D. P. Bertsekas, *Dynamic Programming and Optimal Control*. Athena Scientific, 1995.

[21] O. Tremois and J. P. LeCadre, "Optimal observer trajectory in bearings-only tracking for manoeuvring sources," *IEE Proceedings on Radar, Sonar and Navigation*, vol. 146, no. 1, pp. 31–39, 1999.

[22] R. D. Smallwood and E. J. Sondik, "The optimal control of partially observable Markov processes over a finite horizon," *Operations Research*, vol. 21, pp. 1071–1088, 1973.

[23] R. J. Williams, "Simple statistical gradient-following algorithms for connectionist reinforcement learning," *Machine Learning*, vol. 8, no. 3, pp. 229–256, 1992.

[24] A. Y. Ng and M. I. Jordan, "Pegasus: A policy search method for large MDPs and POMDPs." in *Uncertainty in Artificial Intelligence (UAI2000)*, 2000.

[25] J. M. Maciejowski, *Predictive control: with constraints*. Prentice-Hall, 2002.

[26] J. A. Castellanos and J. D. Tardós, *Mobile Robot Localization and Map Building. A Multisensor Fusion Approach*. Kluwer Academic Publishers, 1999.

[27] H. J. Kushner, "A new method of locating the maximum of an arbitrary multipeak curve in the presence of noise," *Journal of Basic Engineering*, vol. 86, pp. 97–106, 1964.

[28] D. R. Jones, "A taxonomy of global optimization methods based on response surfaces," *Journal of Global Optimization*, vol. 21, pp. 345–383, 2001.

[29] M. J. Sasena, "Flexibility and efficiency enhancement for constrained global design optimization with Kriging approximations," Ph.D. dissertation, University of Michigan, 2002.

[30] D. R. Jones, C. D. Perttunen, and B. E. Stuckman, "Lipschitzian optimization without the Lipschitz constant," *Journal of Optimization Theory and Applications*, vol. 79, no. 1, pp. 157–181, October 1993.

[31] J. M. Gablonsky, "Modification of the DIRECT algorithm," Ph.D. dissertation, Department of Mathematics, North Carolina State University, Raleigh, North Carolina, 2001.

[32] D. E. Finkel, *DIRECT Optimization Algorithm User Guide*. Center for Research in Scientific Computation, North Carolina State University, 2003.

[33] A. J. Smola, S. V. N. Vishwanathan, and T. Hofmann, "Kernel methods for missing variables," in *AI Stats*, 2005, pp. 325–332.

[34] C. E. Rasmussen and C. K. I. Williams, *Gaussian Processes for Machine Learning*. Cambridge, Massachusetts: MIT Press, 2006.

[35] P. Tichavský, C. Muravchik, and A. Nehorai, "Posterior Cramér-Rao bounds for discrete-time nonlinear filtering," *IEEE Transactions on Signal Processing*, vol. 46, no. 5, pp. 1386–1396, 1998.

[36] N. Bergman, A. Doucet, and N. J. Gordon, "Optimal estimation and Cramér-Rao bounds for partial non-Gaussian state space models," *Annals of the Institute of Mathematical Statistics*, vol. 52, no. 1, pp. 1–17, 2001.

Robot Manipulation: Sensing and Adapting to the Real World

Charles C. Kemp
Georgia Tech

Aaron Edsinger
MIT

Robert Platt
NASA JSC

Neo Ee Sian
AIST

***Abstract*— There is a resurgence of interest in robot manipulation as researchers seek to push autonomous manipulation out of controlled laboratory settings and into real-world applications such as health care, generalized factory automation, domestic assistance, and space exploration. This workshop facilitated progress towards these goals by focusing on autonomous robot manipulation and the requisite sensing and adaptation. The workshop was a full-day event consisting of nine talks, four discussions, two poster sessions, and demos which are summarized below.**

A. Tactile Sensing

Several different approaches to tactile manipulation were presented in the morning session through talks and posters, including statistical matching of geometric models to tactile sensor readings, POMDPs for grasping of planar block-shaped objects, and behavior-based grasping of books from a shelf. Most of the work focused on grasping objects. The accompanying discussion heavily emphasized the need for off-the-shelf tactile sensing technology. It was suggested that tactile sensing for robots might be the next significant sensor to drive manipulation research, similar to the way that SICK laser range finders influenced navigation research.

B. Low-dimensional Control

Prior to lunch, two presentations on low-dimensional control for robotic grasping were given, followed by discussion. Both talks proposed a low-dimensional representation of the full manipulator configuration space. One talk used the low-dimensional representation to simplify manipulation teleoperation. The other used it to simplify search in a manipulation planning task. In the discussion that followed, it was noted that neither approach integrated realistic sensor feedback. Nevertheless, participants were interested in the potential of this approach. It was conjectured that if the majority of grasping interactions can be described in terms of a low-dimensional embedding in the manipulator configuration space, then some important autonomous manipulation problems may not be as difficult as generally believed. Participants also noted that underactuated manipulators can take advantage of this low-dimensional representation mechanically, instead of computationally.

C. Keynote Talk

Andrew Ng from Stanford University gave the keynote talk after lunch. He discussed a diverse set of research projects in progress at Standford including the STAIR Project (Stanford AI Robot) that he leads. This research addressed the themes of the workshop with a strong emphasis on visual sensing. It culminated in several examples of integrated real-world robotic systems performing manipulation tasks using visual, laser-based, and tactile perception. For example, the STAIR robot fetched a stapler in another room for a person sitting at a table in a conference room. One interesting challenge that Andrew Ng mentioned in his talk is the lack of robust and effective visual object recognition in real-world settings. He asserted that because visual recognition techniques are currently not as robust as necessary, it may be worthwhile for manipulation research to circumvent this problem by using RFIDs or other technologies.

D. Poster and Demo Session II

As part of a second poster and demo session after the keynote talk, posters focused on a variety of different topics ranging from learning to human safety. The demonstrations showed off-the-shelf robot arms and hands from Barrett Technologies, Neuronics, and Shadow.

E. Visual Sensing

The session on visual sensing looked at methods for visually locating door handles, detecting rotary joints in planar objects through manipulation, and using human-in-the-loop control to train visual-motor controllers. The discussion touched on a number of issues, including the desire for perceptual systems that attend to what is important in a given task. Attendees also expressed interest in the potential for sensors that provide reliable, dense depth information over an entire plane. The general consensus was that current sensors do not do this sufficiently well, but that high-quality off-the-shelf sensors of this nature could potentially have a dramatic impact on robot manipulation.

F. The Future

The workshop concluded with three very short impromptu talks and a moderated discussion of potential future applications for autonomous robot manipulation. In the discussion, participants identified some promising applications, technical challenges impeding these applications, and milestones for research. The notes from this discussion, a schedule, accepted submissions, and more can be found at the workshop website: http://manipulation.csail.mit.edu/rss07/ and in the workshop proceedings: http://www.archive.org/details/sensing_and_adapting_to_the_real_world_2007.

Robotic Sensor Networks: Principles and Practice

Gaurav S. Sukhatme
Department of Computer Science
University of Southern California
941 W. 37th Place
Los Angeles, CA 90089-0781
Email: gaurav@usc.edu

Wolfram Burgard
Institut für Informatik
Albert-Ludwigs-Universität Freiburg
Georges-Köhler-Allee, Geb. 079
D-79110 Freiburg, Germany
Email: burgard@informatik.uni-freiburg.de

Abstract—**We summarize the activities and organization of the workshop entitled *Robotic Sensor Networks: Principles and Practice* which took place on Saturday, June 30, 2007 at the Georgia Institute of Technology as part of *Robotics Science and Systems 2007*.**

Sensor network research has risen to prominence in recent years. *RSS 2007* featured a one-day focused workshop on *robotic* sensor networks, namely sensor networks which incorporate robotic mobility or articulation. Such systems include, e.g., a networked multi-robot group, a network of immobile computing and sensing nodes and mobile robot(s), a network of immobile nodes each with computing and actuated sensing (allowing e.g., each node to change the direction in which a sensor is pointed). The design of such systems raises algorithmic and theoretical challenges as well as challenges associated with the practicalities of of conducting real deployments. This workshop brought together people interested in the algorithmic aspects, mathematical and statistical foundations, and experimentalists who have fielded robotic sensor networks in the context of specific applications.

The workshop was structured as a mix of short technical (though relatively informal) presentations and discussion panels. There were three technical sessions, each followed by a panel discussion. The first technical session (5 talks) was about algorithmic issues in sensor selection, placement, deployment, and belief propagation. The second and third sessions (5 talks each) were on applications of robotic sensor networks. The average attendance at the workshop was approximately 35, though we estimate that approximately 50 people attended overall over the course of the day.

In Session I, the speakers concentrated on *algorithmic problems* associated with accomplishing certain sensing tasks. The talk by Isler outlined an approach to networked motion planning, sensor planning and topology management in the context of a network composed of robots equipped with cameras. Bullo's talk focused on routing problems for autonomous vehicles, boundary estimation for robotic sensor networks, and an algorithm for sensor placement and control in the context of placing asynchronous guards in art galleries. Poduri's talk gave a theoretical bound on the rate of convergence of a coalescence algorithm in robotic sensor networks, while Jenkins discussed inference problems in robotic sensor networks as instances of multi-robot belief propagation. Berman concluded the session with a discussion of her work on the analysis and synthesis of bio-inspired controllers for swarm robotic systems.

In sessions II and III, the speakers focused on *implementations and applications* of robotic sensor networks. Three of the five talks in Session II dealt with applications to the aquatic environment, while three of the five talks in Session III were focused on multi-robot systems. Three of the five talks in Session III were also about environmental monitoring.

Bayazit opened Session II with an interesting talk on roadmap-based navigation strategies with an application to route finding in dynamic environments such as those affected by a wildfire. Kansal followed this with a distributed optimization approach to the coordination of networked cameras, thus exploiting low-complexity motion for large sensing gains. Sukhatme's talk was on mapping and exploration using an aquatic sensor network posed as a sampling design problem. Deshpande followed this with a discussion of how to exploit natural (passive) mobility in water bodies. Paley concluded the session with an inspirational talk showing results in cooperative control of an AUV network for ocean sampling with experiments performed at-sea.

In Session III, the first three speakers focused on multi-robot applications: Stachniss on cooperative multi-robot exploration, Hsieh on multi-robot behaviors for the perimeter surveillance problem, and Singh on human-assisted robotic teams for environmental monitoring. Song and Stealey continued the environmental monitoring theme: Song with a discussion of his work on creating a collaborative observatory for natural environments and Stealey with a discussion of infrastructure-supported mobile systems.

The workshop led to an interesting exchange of ideas, and an informal survey of the participants indicates that they enjoyed the experience. Further, judging by the number of speakers interested in participating, and the significant number of non-speaker attendees, we believe the workshop was timely and useful. For a complete speaker list, and titles of the various presentations, please visit the *RSS 2007* website.

We are grateful to the the workshop speakers and the attendees. We thank Udo Frese for his work on workshop organization, Cyrill Stachniss for his work on publicity, and Frank Dellaert, Magnus Egerstedt, and Henrik Christensen for their excellent local arrangements.

Workshop on Algorithmic Equivalences Between Biological and Robotic Swarms

Paulina Varshavskaya and James McLurkin
Computer Science and Artificial Intelligence Lab
Massachusetts Institute of Technology
Cambridge, MA 02139
Email: {paulina, jamesm}@csail.mit.edu

I. Introduction: Workshop Goals

This workshop brought together researchers in biology and robotics who study distributed computational systems: insect colonies and multi-robot teams. The main goal was to enable rigorous discussion of the common algorithms employed by natural and artificial swarms.

The premise is that similarities in our respective high-level models of sensing, communication, processing, and mobility constrain the distributed algorithms that can be used. We set out to discuss many open questions: What fundamental mathematical theories underpin both biological and artificial systems? What analytical tools are commonly used in one field, which may be of benefit to the other? What specific common constraints apply to the solutions found by nature and robot engineers? When do roboticists wish they knew more biology? When do biologists wish they knew more computer science or control theory? What tasks and applications for swarm technology are most like those of their natural counterparts? What is the most productive way to use "natural algorithms" in distributed robotics?

II. Presentations

The workshop consisted of two invited talks, four research presentations, and a series of discussions. The workshop program and abstracts can be found at *http://projects.csail.mit.edu/rss2007workshop/program.html*.

1) Invited talk: *Collective decision-making by ant colonies: Linking group and individual behavior* Stephen Pratt School of Life Sciences, Arizona State University
2) Invited talk: *Cooperative Control of Multi-Agent Systems* Sonia Martinez Diaz Mechanical and Aerospace Engineering (MAE), University of California at San Diego
3) *Rational swarms for distributed on-line bayesian search* Alfredo Garcia, Chenyang Li and Fernan Pedraza
4) *Macroscopic information processing in natural systems as an insight for swarm robotics* Dylan Shell and Maja Mataric
5) *Genetic structure and cohesion of insect societies* Michael Goodisman
6) *Energy-efficient multi-robot rendezvous: Parallel solutions by embodied approximation* Yaroslav Litus, Pawel Zebrowski and Richard T. Vaughan
7) *Roadmap-based group behaviors* S. Rodriguez, R. Salazar, N.M. Amato, O.B. Bayazit, and J.-M. Lien

III. Discussion

The afternoon discussions identified broad categories of commonality between our two fields: 1) the assumptions and constraints underlying physical distributed systems, 2) the models we use in researching these systems, and 3) the algorithms and mechanisms used by biological and robotic swarms. In addition, we have discussed experiments and applications that are relevant to our respective communities.

A computational model of many insect societies shares several key components with a model of distributed computation on teams of robots: short-range, local communications, limited processing, and constraints on information gathering and algorithm execution. For example, the physical speed of an insect (or robot) affects how much territory it can survey in a given amount of time. The general theme is that the similarities in physical hardware should forces the theory underlying these two systems to be similar.

Our discussions highlighted many differences as well. In many insect societies, physical polymorphism is used to produce workers in many different sizes for different jobs. Large-scale heterogeneous systems are currently the exception in multi-robot teams. The scale of natural systems can be very large, with some global behaviors only evident above certain population thresholds, but there is no coherent view of how colony size affects algorithms in general. Individual variation seems to be important in natural systems and may help smooth community responses to external stimulus and prevent global oscillations.

Artificial distributed systems sometimes break the common assumptions for the sake of elegant proofs, which can limit the applicability of these results to biological systems.

The applications of natural and artificial systems differ fundamentally. Artificial systems are designed to perform a desired task, natural systems are selected by their ability to reproduce.

The final discussion topic constructed a bibliography for for a graduate student wishing to study this problem. A next step would be to develop an interdisciplinary curriculum for a course. The references can be found at *http://projects.csail.mit.edu/rss2007workshop/bibliography.html*.

Workshop on Research in Robots for Education

Keith O'Hara
College of Computing
Georgia Institute of Technology
keith.ohara@gatech.edu

Doug Blank
Computer Science Department
Bryn Mawr College
dblank@brynmawr.edu

Maria Hybinette
Computer Science Department
University of Georgia
maria@cs.uga.edu

Daniela Rus
EECS Department
MIT
rus@csail.mit.edu

Introduction

The landscape of robots in education has continued to change since the 2005 RSS Robotics Education Workshop[1]. Over the last two years, there has been a noticeable spike of interest in the use of robots in education. The focus of this workshop was to provide a venue for presentation of the research exploring the effectiveness of robots in education, and help shape future research in this area.

The workshop explored how robots are used as educational tools, in terms of platforms (hardware and software) and pedagogy in various disciplines and why certain types of robots may be more effective for different purposes. The workshop also explored how robots can be used as a platform to broaden participation, hoping to attract and retain students to science, technology, engineering, and mathematics (STEM) disciplines.

The workshop was organized into three sessions. Each session was composed of a series of 5-6 ten minute talks, followed by a twenty minute discussion period. The sessions were organized by the following themes: Pedagogy, Participation, and Platforms. In addition, other cross-cutting themes considered were the educational discipline being taught (e,g. Electrical Engineering, Computer Science, Mathematics, Art), the age of the students (e.g. K-5, 6-8, 9-12, Undergraduate, Graduate, General Education), and the educational environment (e.g. Contest, Classroom, Club, Gallery).

Participation

The participation session investigated how robots can be used to attract and retain new students STEM disciplines. We had a wide variety of presenters. For instance, we had middle and high school teachers (Fred Stillwell and Susan Crowe), who use robots in the classroom and in extracurricular clubs, discuss how they use robots to engage students. Michael Dumont discussed the work with his colleagues on using robots and video games to attract and retain computer science students. Fred Martin presented his iCode System which uses online community technology to help students program robots. Jerry Weinberg presented his work on assessing how robots (in a competition setting) impact females' attitudes toward science and engineering. Finally, Joel Weingarten discussed his work on using a legged robot to teach electrical engineering.

Platforms

The platforms session explored the hardware and software platforms available for using as robots as educational tools. Monica Anderson presented her work on using the Player/Stage robot software package for both simulation and real robot assignments. Kristie Brown, from LEGO education, spoke about the new LEGO NXT robot and its use by educators. Marco Morales presented a graphical tool for exploring motion planning. Eric Schweikardt discussed his work on using their reconfigurable robot platform, roBlocks, to teach students about emergent complexity. Finally, Stewart Tansley spoke about using Microsoft Robotics Studio as a robot education platform.

Pedagogy

The last session centered on the pedagogical needs in robot education and using robots as a tool for other educational goals. Doug Blank spoke about the Institute for Personal Robots, and their project concerning introductory computer science with robots. Michael Gennert spoke about creating the first undergraduate degree in robotics at the Worcester Polytechnic Institute. Aaron Dollar led a discussion on the Robotics OpenCourseWare Project. Wen-Jung Hsin presented her work on teaching assembly programming with simple robots. Fred Martin spoke about a course combining art and robotics. Finally, Marco Morales (in place of Jesus Savage Carmona) presented a system for teaching robot programming using both virtual and real robots.

Discussion

Many topics were discussed, but a few topics kept reemerging. First, service learning and community outreach, and the need for more cooperation between universities and middle and high schools. Second, the need for clear scientific evidence of the utility of using robots for education. In particular, to ease and accelerate adoption, many administrators will need convincing evidence that introducing robots into the curriculum adds academic value. Third, the need for a repository of educational robotics materials, and a stronger community. The papers, presentations, and pointers to additional material can be found on the workshop's website[2].

[1] http://projects.csail.mit.edu/rss/RobotEd/

[2] http://www.roboteducation.org/rss-2007/